C0-ATL-609

SUPRAMOLECULAR PHOTOCHEMISTRY

SUPRAMOLECULAR PHOTOCHEMISTRY

Controlling Photochemical Processes

EDITED BY

V. RAMAMURTHY
YOSHIHISA INOUE

A JOHN WILEY & SONS, INC., PUBLICATION

Published by John Wiley & Sons, Inc., Hoboken, New Jersey
Published simultaneously in Canada

For general information on our other products and services or for technical support, please contact our Customer Care Department within the United States at (800) 762-2974, outside the United States at (317) 572-3993 or fax (317) 572-4002.

Wiley also publishes its books in a variety of electronic formats. Some content that appears in print may not be available in electronic formats. For more information about Wiley products, visit our web site at www.wiley.com.

Library of Congress Cataloging-in-Publication Data

Supramolecular photochemistry : controlling photochemical processes / edited by V. Ramamurthy, Yoshihisa Inoue.
p. cm.
Includes bibliographical references and index.
ISBN 978-0-470-23053-4 (hardback)
1. Photochemistry. 2. Supramolecular chemistry. I. Ramamurthy, V. II. Inoue, Yoshihisa, 1949-
QD708.2.S87 2011
541'.226–dc22

2011010605

Printed in Singapore

obook ISBN: 978-1-118-09530-0
ePDF ISBN: 978-1-118-09527-0
ePub ISBN: 978-1-118-09529-4

10 9 8 7 6 5 4 3 2 1

CONTENTS

PREFACE

Ancient civilizations recognized the importance of light for the maintenance and sustenance of life on earth. As pointed out in ancient Indian scriptures, "The rising sun is the giver of energy, heat, all powers, happiness and prosperity." The sun, the ultimate source of light that provides the energy that drives life, has been an object of worship by various civilizations since antiquity. However, orderly investigations of the connection between the absorption of light by matter and its chemical and physical consequences were not reported in the scientific literature until about the turn of the last century, when systematic efforts revealed that exposure of matter to sunlight led to a rich range of transformations that are now termed *photochemical reactions*.

During the 1930s, gas-phase photochemistry was a popular area of study. In the early 1940s, the theoretical concept of the "triplet state" of organic molecules was confirmed with striking experimental evidence. The "rules" of the photophysics of organic molecules were established by 1950, and correlations were made between the spectroscopic properties of molecules and the orbital configurations of electronically excited states. The 1960s and 1970s marked a period when many new photoreactions were discovered, and the mechanisms of these reactions were investigated with a cluster of powerful new spectroscopic techniques and theories. During this period, the power of mechanistic investigations was demonstrated and the field of molecular photochemistry gained the status of a truly new and emerging discipline. It was only in the 1980, that photochemistry became an integral part of organic, inorganic and biological chemistry. This development led photochemists to import concepts from other disciplines and take the field to a next level that is now called "Supramolecular Photochemistry."

Inspiration from and aspiration to mimic biological systems have been a driving force for the advancement of supramolecular photochemistry. Tailoring and tuning molecules have been one of the main goals of researchers exploring this topic. "Control" has been a key element in most of the discoveries in this field. Twenty years ago one of us (VR) edited a monograph titled *Photochemistry in Organized and Constrained Media* that set the trend that was to come. In the Foreword, George S. Hammond wrote, "Study and exploitation of photochemistry in organized media is a field, if not in its infancy, barely past adolescence." Since then, the field of photochemistry has attracted a large number of new researchers and has become more relevant in solving pressing current problems related to sustainability, energy, and the environment. Its future role in harnessing solar energy and developing environmentally friendly synthetic methodologies cannot be overemphasized. In addition, during the last three decades, photochemistry has found its way into a number of industrial and medicinal applications; lithography and photodynamic therapy are just two examples. There is no doubt that "photochemistry with control" will lead to even more important useful outcomes in the future.

This book, with 14 chapters contributed by 28 chemists, covers both solution- and solid-state supramolecular assemblies as well as photophysics and photochemistry of molecules that are integrated into the supramolecular assembly. Reaction media discussed include crystals, zeolites, polymers, organic glass, and organic host-guest assemblies in the solid state and in solution, as well as within biological matrices. A comprehensive survey of "supramolecular photochemistry" (photoreactions with control) contributed by experts in the field is expected to be of great value to students, teachers, and researchers. This book can serve as a source of reference and as supplementary book in photochemistry courses.

The book begins with a chapter by C. Bohne in which the author summarizes various techniques that are used in probing the dynamics of supramolecular assemblies. The author elegantly illustrates the power of these techniques with examples chosen from micelles, cyclodextrins, and DNA. In the second chapter D. Bassani critically evaluates the various strategies available to obtain selectivity in photocycloaddition reactions in solution. This chapter also includes a summary of other photoreactions in which templation has played a key role. Recently, cucurbiturils have been extensively explored as a host in photochemical and photophysical studies. These studies have also brought to forefront the potential of other hosts such as cyclodextrins, calixarenes, and other cavitands. In Chapter 3, W. Nau, R. N. Dsouza, and U. Pischel have highlighted the value of these hosts in photophysical studies and in sensor applications. Chiral chemistry has been one of the passions of the editors of this volume. In Chapter 4, Y. Inoue and C. Yang review the advances in the field of supramolecular chiral photochemistry. The chapter includes studies in zeolites, cyclodextrins, and proteins, and with organic chiral templates.

Chapters 5 to 8 deal with photochemical studies in crystals. Each one emphasizes different aspects of solid-state photochemistry. Chapter 5 by

P. Coppens and S. L. Zheng emphasizes the need to bring in a new technique (time-resolved crystallography) to gain an in-depth understanding of photoreactions in the crystalline state. Challenges are many, yet this may be one of the few options we currently have to get an insight into what happens in pico-, and nano-second time scales when molecules are excited in the crystalline state. Photodimerization in solid state continues to draw the attention of chemists with both crystallographic and photochemical inclinations. Photodimerization provides valuable examples that could be subject to deeper investigation using more sophisticated modern and yet-to-be discovered techniques. With that in mind A. Natarajan and B. R. Bhogala have summarized what is known in the field of photodimerization for over five decades in Chapter 6. Since very early studies on urea-inclusion complexes, photochemical studies of host-guest chemistry in the solid state (also known as *inclusion chemistry* or *clathrate chemistry*) have made steady progress. The diverse literature on host-guest solid-state photochemistry has been beautifully summarized by M. Kaftory, who himself has made pioneering contributions to this topic (Chapter 7). In the next chapter (Chapter 8), M. Irie and M. Morimoto have summarized their work on the photochromism of diarylethylene crystals. This is one of the few photoreactions that has shown potential commercial value. If any device based on this reaction ever reaches the commercial market, the public may begin appreciating the value of solid-state photochemistry.

The next two chapters (Chapters 9 and 10) deal with the photochemistry and photophysics of organic molecules included within zeolites. G. Calzaferri has been a pioneer in exploring the highly ordered structure of zeolites to channel light and electrons in an organized manner between molecules. In Chapter 9, he and A. Devaux have provided a detailed review of photophysics of molecules included within zeolite L. This chapter is a must-read for anyone wishing to explore zeolites in the context of solar energy capture. Chapter 10 by Ramamurthy and Sivaguru summarizes their contributions to the use of zeolites to carry out selective photochemistry and photophysics. The authors emphasize the value of zeolite cations in controlling excited-state chemistry including chiral induction.

In addition to their widespread use in various commercial applications, polymers are also used as a supramolecular matrix to carry out photochemical and photophysical studies of doped organic molecules. R. G. Weiss and S. Abraham provide an exhaustive and critical survey of this field in Chapter 11. All aspects of the excited-state chemistry of molecules included in polymers have been covered. This chapter is not about photochemistry of polymers (a topic covered by a number of books), but rather, it describes how the photochemistry within polymers is unique and how concepts presented here are likely to lead to useful applications. In Chapter 12, T. Majima and M. Fujitsuka highlight the value of biopolymers as matrices for photoreactions. This chapter focuses on the use of ultra-fast time-resolved techniques to probe charge and energy migration in organized assemblies.

In Chapter 13, R. S. H. Liu and co-authors illustrate how the mechanism of a simple photoprocess such as *cis-trans* isomerization could change when molecules are confined in matrices such as organic glass. While most investigators have focused their studies on discovering new reactions and controlling excited-state chemistry, R. S. H. Liu, the proponent of the "hula-twist," demonstrates that the mechanism of known photoreactions could change within supramolecular assemblies. In his elegant presentation, he argues how the "hula-twist" can be important in crystals, proteins, and so on. The final chapter (Chapter 14) by H. Kandori is an important one for those of us who are inspired to undertake supramolecular photochemistry in the laboratory. He demonstrates that the best, most efficient, rapid, and highly selective supramolecular photoprocess occurs in the human eye. A detailed presentation of what is known on ultrafast photoisomerization of retinal that resides in a protein pocket is an ideal example of the value of photochemistry research. We hope that it motivates young investigators to undertake research on "supramolecular photochemistry" in general.

We are grateful to authors for their patience and cooperation during the entire process of production of this monograph. We are proud to have convinced pioneers in the field to contribute chapters to this volume. While we recognize that color figures would have enhanced the quality of the topics, due to cost control needs only black and white figures are included in the text. However, readers are encouraged to view several color figures free of charge at the Wiley ftp site: ftp://ftp.wiley.com/public/sci_tech_med/supramolecular_photochemistry.

VR is grateful to the U.S. National Science Foundation for supporting his research that has allowed him to keep his interest focused on this topic. YI thanks the Japan Science and Technology Agency and Japan Society for the Promotion of Science for their generous support of his research projects on "photochirogenesis" in molecular, supramolecular, and biomolecular regimes. VR and YI thank their former and present students and colleagues for their devotion and collaboration. They also thank Ms. Anita Lekhwani, Senior Commissioning Editor, Wiley-Blackwell for encouraging us to undertake this project and for her perseverance despite the slow progress of the project. VR, especially, has enjoyed the collaboration with Anita on various book projects for over a period of 15 years. And most importantly, the editing of this book would not have been possible without the understanding, patience, support, and encouragement of our wives Rajee (VR) and Masako (YI).

V. Ramamurthy
Yoshihisa Inoue

CONTRIBUTORS

Shibu Abraham, Department of Chemistry, Georgetown University, Washington, DC 20057-1227

Dario M. Bassani, Institut des Sciences Moléculaires, CNRS UMR 5255, Université Bordeaux 1, 351 Cours de la Libération, F-33405 Talence, France

Balakrishna R. Bhogala, Department of Chemistry, University of Miami, Coral Gables, FL 33124

Cornelia Bohne, Department of Chemistry, University of Victoria, PO Box 3065, Victoria, BC, Canada V8W

Gion Calzaferri, Department of Chemistry and Biochemistry, University of Bern, Freiestrasse 3, CH-3012 Bern, Switzerland

Philip Coppens, Chemistry Department, University at Buffalo, State University of New York, Buffalo, NY 14221

André Devaux, Physikalisches Institut, Westfälische Wilhelms-Universität Münster, Mendelstr. 7, D-48149 Münster, Germany

Roy N. Dsouza, School of Engineering and Science, Jacobs University Bremen, Campus Ring 1, D-28759 Bremen, Germany

Mamoru Fujitsuka, The Institute of Scientific and Industrial Research (SANKEN), Osaka University, Mihogaoka 8-1, Ibaraki, Osaka, 567-0047, Japan

Yoshihisa Inoue, Department of Applied Chemistry, Osaka University, 2-1 Yamada-oka, Suita 565-0871, Japan

Masahiro Irie, Department of Chemistry, Rikkyo University, 3-34-1 Nishi-Ikebukuro, Toshima-ku, Tokyo 171-8501, Japan

Menahem Kaftory, The Schulich Faculty of Chemistry, Technion-Israel Institute of Technology, Haifa 32000, Israel

Hideki Kandori, Department of Frontier Materials, Nagoya Institute of Technology, Showa-ku, Nagoya 466-855, Japan

Akira Kawanabe, Department of Material Sciences and Engineering, Nagoya Institute of Technology, Showa-ku, Nagoya 466-855, Japan

Robert S.H. Liu, Department of Chemistry, University of Hawaii, 2545 The Mall, Honolulu, HI 96822

Tetsuro Majima, The Institute of Scientific and Industrial Research (SANKEN), Osaka University, Mihogaoka 8-1, Ibaraki, Osaka, 567-0047, Japan

Masakazu Morimoto, Department of Chemistry, Rikkyo University, 3-34-1 Nishi-Ikebukuro, Toshima-ku, Tokyo 171-8501, Japan

Arunkumar Natarajan, Department of Chemistry, University of Miami, Coral Gables, FL 33124

Werner M. Nau, School of Engineering and Science, Jacobs University Bremen, Campus Ring 1, D-28759 Bremen, Germany

Uwe Pischel, Department of Chemical Engineering, Physical Chemistry, and Organic Chemistry, Faculty of Experimental Sciences, University of Huelva, Campus de El Carmen s/n, E-21071 Huelva, Spain

V. Ramamurthy, Department of Chemistry, University of Miami, Coral Gables, FL 33124

Jayaraman Sivaguru, Department of Chemistry and Biochemistry, North Dakota State University, Fargo, ND 58108-6050

Richard G. Weiss, Department of Chemistry, Georgetown University, Washington, DC 20057-1227

Cheng Yang, PRESTO and Department of Applied Chemistry, Osaka University, 2-1 Yamada-oka, Suita 565-0871, Japan

Lan-Ying Yang, Department of Chemistry, University of Hawaii, 2545 The Mall, Honolulu, HI 96822

Yao-Peng Zhao, Department of Chemistry, University of Hawaii, 2545 The Mall, Honolulu, HI 96822

Shao-Liang Zheng, Chemistry Department, University at Buffalo, State University of New York, Buffalo, NY 14221

1

DYNAMICS OF GUEST BINDING TO SUPRAMOLECULAR ASSEMBLIES

CORNELIA BOHNE

INTRODUCTION

Supramolecular systems are formed from molecular building blocks held together by intermolecular interactions such as electrostatic interactions, hydrogen bonds, π-stacking, or the hydrophobic effect.[1–9] These interactions are, in general, weaker than covalent bonds found in molecules, and consequently, supramolecular systems are inherently reversible. This reversibility is essential for some of the functions expressed by supramolecular systems, such as chemical sensing, catalysis, or transport. Reversibility and weak interactions are also responsible for the sensitivity of supramolecular systems to experimental conditions such as temperature or the solvent's nature. The structure of a molecule, that is, the connectivity between atoms and their spatial relationship, requires large changes in temperature and solvent properties to be altered. In contrast, the structure of supramolecular systems, defined by the stoichiometry of the building blocks and their spatial relationship, can be altered with small changes to the system's environment. For example, many molecules that are soluble and stable in polar solvents are also soluble and stable in polar hydrogen-bonding solvents. However, a supramolecular system formed primarily through hydrogen bonds can be inexistent in hydrogen-bonding solvents while being formed in polar or polarizable solvents with no hydrogen-bonding capability.

Supramolecular Photochemistry: Controlling Photochemical Processes, First Edition. Edited by V. Ramamurthy and Yoshihisa Inoue.

The differentiation between a molecular and a supramolecular system is not always clear-cut because a gradation can exist for the strength of the forces involved in holding atoms or molecules together. In addition, the temporal division into nonreversible and reversible systems is dictated by human experience, which establishes the timescales for "stable" and "unstable" systems. This chapter will focus on processes that are faster than a few seconds and, for this reason, require the use of fast kinetic techniques.

Each system is characterized by its structure, energetics, and dynamics. In the molecular world, primary characterization is based on structural assignments and determination of the system's energetics, that is, thermodynamics. Dynamics are not important in the characterization of individual molecules because the vast majority of molecules are stable. Dynamics play a role in the characterization of the reactivity of molecules. Therefore, for stable molecules, structural and thermodynamic characterization can be uncoupled from kinetic aspects. The reversibility of supramolecular systems dictates that knowledge on its dynamics is as essential as structural and energetic characterization. The structure of new supramolecular systems is frequently characterized by X-ray crystallography, nuclear magnetic resonance (NMR), or imaging experiments. Thermodynamic studies reveal information on the stability of the system (e.g., equilibrium constants) and on the distribution of species with different building block stoichiometries. Rate constants for the assembly and disassembly of supramolecular systems, as well as rate constants for supramolecular function, such as transport and catalysis, are obtained from kinetics studies.

The dynamics of a system can be investigated using real-time kinetic techniques or by the measurement of relative rates. In real-time kinetic measurements, the concentration of the species of interest is followed as a function of time, and rate constants are obtained from the fit of kinetic measurements to rate laws derived for each system. Rate constants can be derived from relative rate measurements where ratios of products are related to ratios of rate constants and a rate constant for a standard reaction is known. The latter method assumes the same mechanism for the standard reaction and the reaction under study. Relative rate measurements are limited in scope when mechanistic information is not reasonably well developed, because it is difficult to assess the validity of the assumption of equal mechanisms for the reactions being compared. This is still the case for supramolecular systems where the interplay of the various intermolecular forces is not understood well enough to make predictions on their relative importance. As a comparison, in the molecular world we have a good understanding on which bonds are likely to be involved in a particular reaction, based on bond strengths and polarizabilities.

The timescale for supramolecular dynamics studies is determined by the size of the system, leading to a time–length scale relationship for the fastest reactions that can be expected. Bimolecular reactions in a solution are limited by how fast the diffusion of two reactants occurs. The dimensions of supramolecular systems span 1–100 nm, and the diffusion of a small molecule, such as glucose in water at room temperature, is 3 ns for 1 nm and 30 μs for 100 nm.

This analysis shows that the dynamics for small host–guest systems is expected to occur in microseconds or faster.[10,11]

The diffusion limit for bimolecular reactions in solution establishes the fastest timescale for which one expects association processes to occur for reagents that are completely separated in solution. In an aqueous solution at 25°C ($k_{dif} = 7.4 \times 10^9\,M^{-1}\,s^{-1}$),[12] the association process will be slower than 1 ns for reactant concentrations equal or less than 0.1 M, suggesting that kinetic studies in the nanosecond time domain and slower are required. Dissociation of a molecule from a supramolecular system or relocation of the molecule within the system is a unimolecular process, which have rate constants of $10^{12}\,s^{-1}$ and lower. Complete dissociation normally occurs in timescales of nanoseconds or slower, whereas femto- to picosecond processes are frequently observed for internal relocation and changes in solvation.[13–16]

The requirement in supramolecular dynamics to use fast kinetic techniques in the nanosecond to microsecond timescale is ideal for the application of photophysical methods. Excited states or reactive intermediates are used to measure association and dissociation kinetics in supramolecular systems.[10] This chapter covers the concepts on how photophysical techniques, that is, fluorescence, transient absorption, fluorescence correlation spectroscopy, and laser temperature jump, are employed in supramolecular dynamics studies. The advantages and disadvantages of photophysical techniques compared with other fast kinetic methods will be summarized as a comprehensive review on the applications of all these techniques has been recently published.[17] Studies for small guest binding to cyclodextrins (CDs), micelles, bile salt aggregates, and DNA will be employed to exemplify on how photophysical techniques are used to gain kinetic information. This chapter focuses on the kinetic determination of the association and dissociation processes involved in supramolecular systems and does not focus on the internal mobility of guests within a confined system or changes in solvation.

PHOTOPHYSICS IN SUPRAMOLECULAR DYNAMICS STUDIES

Excited states and reactive intermediates in general are species that have finite lifetimes and can be formed in a fast manner by the use of lasers. The fast perturbation of a chemical system is a requirement for real-time kinetic studies. The finite lifetime of excited states compared with the time for the kinetic process of interest can be explored in different ways, that is, if the lifetimes are shorter, longer, or of the same order of magnitude as the time for the kinetic process being studied.[10,18] Each of these cases will be considered below for the simplest host–guest system, where the binding stoichiometry is 1:1 (Scheme 1.1). It will be assumed that the guest is the species being excited; however, the conceptual framework is the same if the host is excited. The kinetic processes of interest are the association (k_+) and dissociation (k_-) rate constants of the guest with the supramolecular system.

$$\mathrm{H} + \mathrm{G} \underset{k_-}{\overset{k_+}{\rightleftharpoons}} \mathrm{HG}$$

$$K_{11} = \frac{k_+}{k_-}$$

Scheme 1.1. Formation of a host–guest complex (HG) with 1:1 stoichiometry.

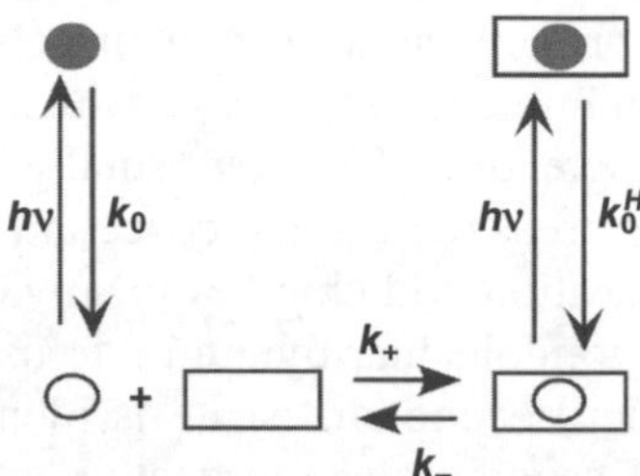

Figure 1.1. Schematic representation for the formation of the excited state of the guest (filled circles). The guest (open circles) in the homogeneous phase is in equilibrium with the host (squares) in the ground state, but no relocation occurs for the excited state.

Short-Lived Excited States

A short-lived excited state will not have the time to relocate between the homogeneous phase and the supramolecular host; that is, the association/dissociation kinetics is too slow to compete with the deactivation of the excited state (Fig. 1.1). Therefore, the excited state is a probe for the environment in which it is located. Differences in excited-state lifetimes or in their emission or absorption spectra are employed to identify the excited guest in the homogeneous phase or bound to the host. Spectral or lifetime differences are explored to determine the relative concentrations of the guest bound to the host (HG) and in the homogeneous solvent (G). Absorption and emission spectra are composite spectra for all species in solution, for example, G and HG, taking into account their molar absorptivities and emission quantum yields and their relative concentrations. In contrast, in the case of excited-state lifetime measurements, each species can be identified separately provided the lifetimes for the excited states of G and HG are different.

In general, for organic molecules, the short-lived excited state is the singlet excited state, which frequently leads to fluorescence emission. From the analytical point of view, fluorescence is a much more sensitive technique than absorption. Light intensity is detected as an absolute measurement and is dependent only on the sensitivity of the detector, that is, the discrimination between noise and signal. Absorption is a relative measurement where changes in light intensities are measured, leading to lower sensitivity. Therefore, for short-lived excited states, most studies are performed using emission and not absorption measurements.

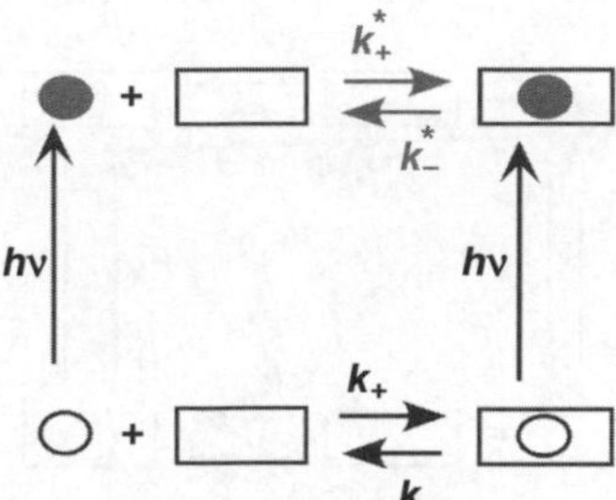

Figure 1.2. Schematic representation for the formation of a long-lived excited state of the guest (filled circles) for which re-equilibration between the excited state guest in the homogeneous phase and the host (squares) occurs before the excited states decay back to the ground state.

Long-Lived Excited States

The formation of an excited state with a lifetime much longer than the association/dissociation kinetics can be viewed as a permanent perturbation of the system (Fig. 1.2). The formation of the excited state is equivalent to the formation of a new chemical, because excited states have properties that differ from those of their ground states, such as different geometries, longer bond lengths, and higher or lower dipole moments. These different properties will lead to a re-equilibration of the system between the excited-state guest and the host. The kinetics will only reflect this re-equilibration, that is, k_+^* and k_-^*, because the excited state will not have time to decay back to its ground state while the re-equilibration occurs. Measurement of the kinetics for the re-equilibration using laser flash photolysis requires that the absorption for G* and HG* is different, so that changes in their relative concentrations can be measured. Fluorescence is only rarely suitable for these measurements because the singlet excited-state lifetimes are frequently shorter than the time required for the association/dissociation processes.

Excited-State Lifetimes Comparable to the Kinetics of Interest

Coupled kinetics are observed when the lifetime of the excited state is of the same order of magnitude as the kinetic process of interest (k_0, $k_0^H \approx k_+^*$, k_-^*) (Fig. 1.3). In this case, the observed rate constant (k_{obs}) for the kinetics will include both the re-equilibration of the excited-state guest with the host system and the decay of the excited state of the guest in the homogeneous solution (k_0) and inside the host (k_0^H). The values for the association (k_+^*) and dissociation (k_-^*) rate constants for the excited guest are determined when the excitation to G* and HG* leads to a nonequilibrium situation. These measurements require that a property of G*, for example, its absorption, be different when bound to the guest (HG*).

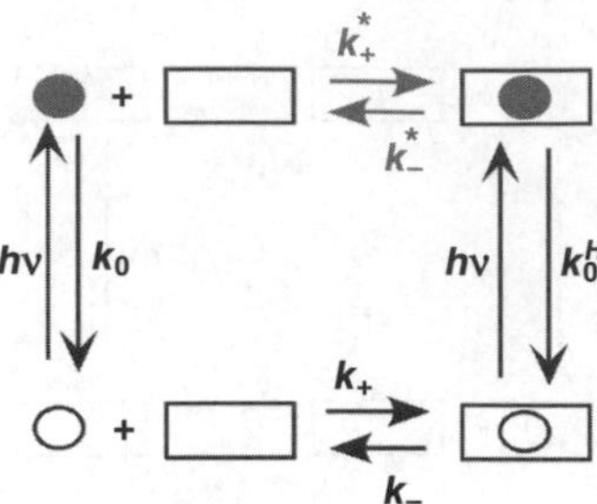

Figure 1.3. Schematic representation for the formation of an excited state of the guest (filled circles) for which re-equilibration between the excited state guest in the homogeneous phase and the host (squares) is competitive with the decay of the excited state to the ground state.

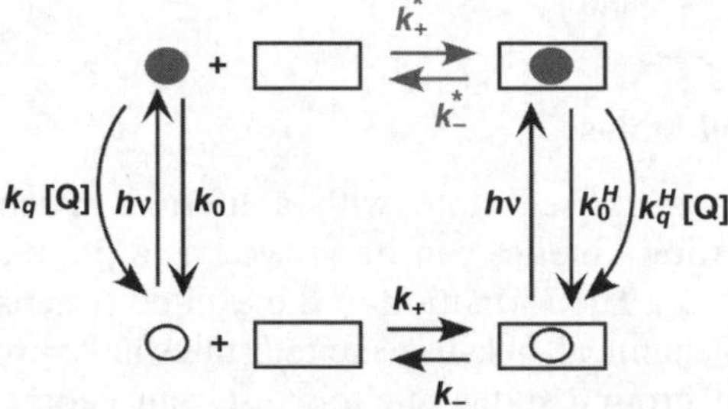

Figure 1.4. Schematic representation for the formation of a long-lived excited state of the guest (filled circles) for which re-equilibration between the excited state guest in the homogeneous phase and the host (squares) is competitive with the decays of the excited states. The lifetime of the excited state in the homogeneous phase and inside the host is shortened with the addition of a quencher (processes shown by curved arrows).

Quenching Studies

The reaction of excited states with molecules (quenchers) that lead to their deactivation can be exploited in two different ways to obtain information on the dynamics of supramolecular systems. In the case of coupled kinetics, the use of quenchers lifts the restriction that G* and HG* need to have different properties that can be followed in kinetic measurements. The introduction of a quencher leads to a competitive deactivation pathway for the excited states G* and HG* (Fig. 1.4), and information on the association and dissociation kinetics can be obtained if the quenching rate constants for the excited state in the homogeneous solution (k_q) and for the excited-state guest in the host (k_q^H) are different. For example, when $k_q \gg k_q^H$, the quenching of the guest in the homogeneous phase at high quencher concentrations leads to such a short lifetime for the free guest (G*) that the rate-limiting step for the deactivation of the excited state of the guest (G* and HG*) will be the guest's exit from

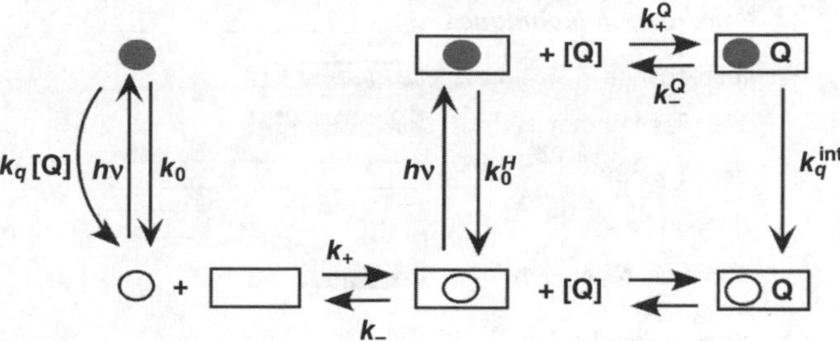

Figure 1.5. Schematic representation for the formation of the excited state of the guest (gray circles) in the singlet state. The guest (open circles) in the homogeneous phase is in equilibrium with the host (squares) in the ground state, but no relocation occurs for the excited state. Quenching (process shown by curved arrow) can occur for the excited state in the homogeneous phase and for the guest inside the host.

the host or its decay within the host. This differential quenching makes it possible to measure the association and dissociation rate constants even when the absorption of G* and HG* are the same. This quenching methodology broadens the scope of supramolecular dynamics studies because only a limited number of molecules have sufficiently different absorption spectra for G* and HG* to study their relocation kinetics without the use of quenchers.

Quenching studies can also be employed to measure the association rate constants of the quencher molecules with the host system. Short-lived excited-state guests located in the host act as probes for the host's interior. The quenching of these excited states provides a measure of the accessibility of the quencher to the interior of the host (Fig. 1.5). Quenching corresponds to a bimolecular process, which can be viewed as a process leading to an encounter complex, followed by the intrinsic quenching rate constant (k_q^{int}) (Fig. 1.5). A value close to the diffusion controlled limit for the quenching rate constant indicates that the intrinsic quenching rate constant (k_q^{int}) is very high and the overall quenching process is directly related to the diffusional rate constant to form the encounter complex. In such a scenario, the quenching rate constant for the excited-state guest inside the host (k_q^H) will be equal to the association rate constant of the quencher with the host (k_+^Q), assuming that the intrinsic quenching rate constant (k_q^{int}) does not change because of the different environments in the host and homogeneous solution. The latter assumption can be tested by measuring the quenching of the excited guests in solvents with different polarities and viscosities. For quenching rate constants that are not diffusion controlled, the analysis for the changes in k_q^H needs to take into account possible changes in the intrinsic quenching rate constant for the reaction in homogeneous solution and inside the host system.

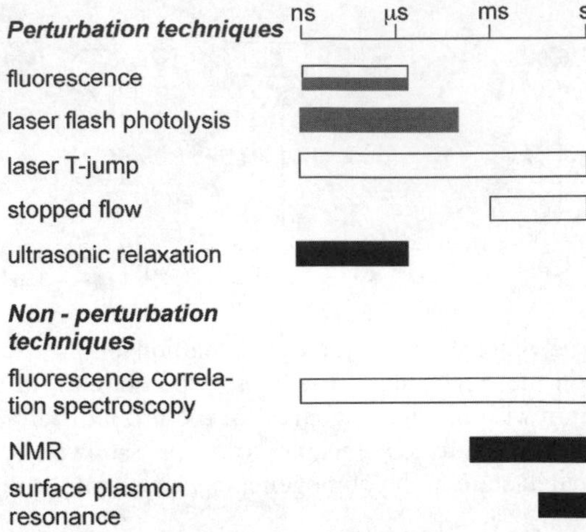

Figure 1.6. Time ranges for the techniques commonly used for supramolecular dynamics studies. The role of the photophysics of the guest is indicated by different colors. Gray: excited-state mobility; white: excited state as an analytical tool to differentiate between free and bound guest; black: no role for photophysics.

TECHNIQUES FOR SUPRAMOLECULAR DYNAMICS STUDIES

The techniques used to study the dynamics of supramolecular systems can be classified as perturbation and nonperturbation experiments (Fig. 1.6). The principle of perturbation measurements is to take a system away from equilibrium by inducing a perturbation that is faster than the kinetics being measured. Perturbations can be rapid changes in concentration, temperature, the simultaneous change in temperature and pressure, or changes in a chemical species. Each technique has a specific time domain for which kinetic measurements can be performed. The role of photophysics is either as an analytical tool, where absorption or emission spectroscopy is used to follow concentrations of reactants or products, or as a tool where the dynamics of excited states is employed to obtain kinetic information. The principles of each technique will be described with a focus on the time resolution between nanoseconds and seconds and the role of photophysics. Some of the techniques are suitable for kinetic measurements faster than nanoseconds or slower than seconds, but these aspects fall outside the scope of this chapter. The use of several different techniques frequently provides essential complementary information on a system, and for this reason, a description of the relevant nonphotophysical techniques is included in this chapter. More details on their applications can be found in a recent review.[17]

Fluorescence (Emission) Measurements

In the case of organic molecules, the most common emission from excited states is fluorescence. The principles here described can also be applied for

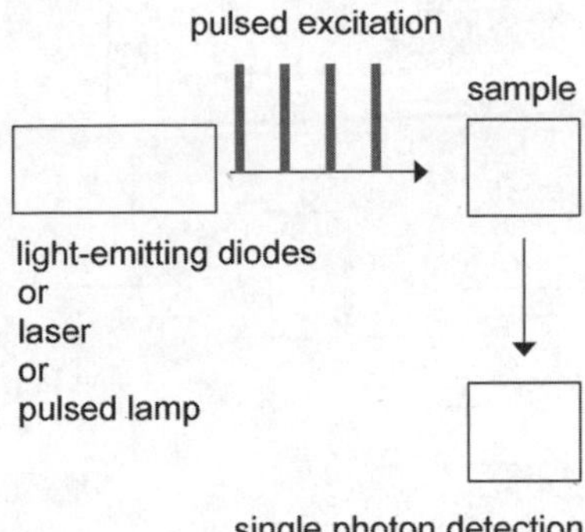

Figure 1.7. Schematic representation for the single photon counting experiment. A train of pulses excites the sample, and the first photon emitted is detected. The rate for the excitation pulses is much higher than the rate of detected photons.

other emissive species, such as phosphorescence, but their use to study supramolecular dynamics has been limited. Kinetic data are obtained from time-resolved experiments. Several time-resolved fluorescence methods exist,[19–23] but single photon counting is the method most commonly employed to study the dynamics of supramolecular systems (Fig. 1.7).[10,18] An excitation source (flash lamps, light-emitting diodes, or lasers) with high repetition rate and pico- or nanosecond resolution is employed to create the excited state of the guest. At a suitably low detection rate (≤2.5% of excitation rate), the first emitted photon from the sample is detected and a histogram of the number of photons as a function of time is built. The low detection rate ensures that the time distribution for the detected photons follows a Poisson distribution. Therefore, the change in the intensity with time recorded in single photon counting experiments is equivalent to the time profile that would have been obtained if many molecules were excited in one pulse and all emitted photons were detected. The advantage of single photon counting is that a large dynamic range for the intensities is accumulated with high signal-to-noise ratios. Such a large dynamic range is essential when differentiating between emitting species that have similar lifetimes.

The perturbation of the system in time-resolved fluorescence experiments is due to the formation of the excited state. For studies involving supramolecular systems, fluorescence can be used in an analytical mode when the lifetime of the guest is short and no relocation occurs between the guest in the host and the homogeneous phase. Time-resolved fluorescence studies can also be used to obtain dynamic information when the lifetime of the excited state and the timescale for the dynamics of interest are similar. The time domain covered by time-resolved fluorescence studies is from nanoseconds to a few microseconds and is limited by the lifetime of the excited states being studied. Use of this technique requires a guest or host that has a reasonably high fluorescence quantum yield and an unreactive excited state.

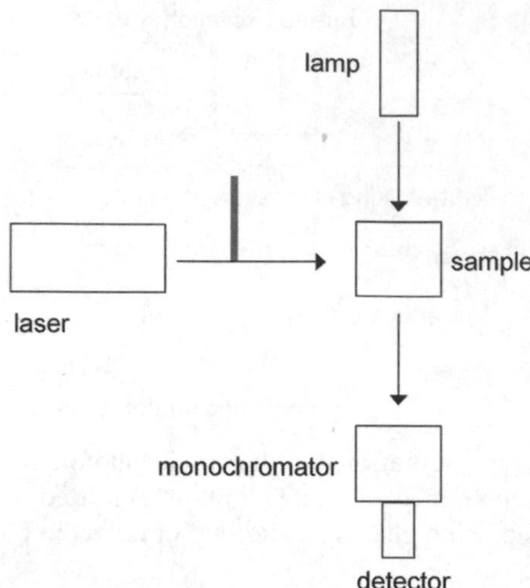

Figure 1.8. Schematic representation for the laser flash photolysis experiment. One laser pulse excites the sample, and the change in the lamp intensity, which is related to the transient absorbance of the sample, is detected.

Laser Flash Photolysis

A pulsed laser is used to form the excited state of the guest, and the absorption of the excited state is monitored over time (Fig. 1.8).[18,24,25] The laser excitation pulse and the monitoring beam are usually in an orthogonal arrangement. The changes in light intensities before and after each laser pulse are measured. This change in light intensity is related to the difference in absorbance of the sample before the laser pulse and the absorbance by the transient formed during the laser pulse.

The perturbation induced in this method is the formation of a different chemical species, that is, the excited state. The time domain covered by most laser flash photolysis systems is from nanoseconds to hundreds of microseconds, because these systems were originally optimized for studies in the nanosecond time domain. The technique can in principle be extended to the millisecond time domain if the instability of common monitoring lamps is minimized.[24,26] Use of laser flash photolysis studies for supramolecular dynamics requires efficient formation of an unreactive long-lived excited state.

Laser Temperature Jump

Temperature jump experiments can be employed to study any system for which the equilibrium is changed when the temperature is raised. In this experiment, increasing the temperature of the solvent causes the perturbation for which the relaxation kinetics is measured. The three methods available to change temperature in a pulsed fashion are Joule heating through an electric discharge, microwave heating, and heating through the absorption of laser light.[27] The time resolution for Joule heating is of microseconds, and this technique requires the use of solutions with high ionic strength, which, in some cases, can lead to artifacts unless special care is taken.[27,28] Microwave heating was not used frequently because the signals obtained were small. Laser temperature jump is the technique of choice when studying fast kinetics.[27,29–32] The type of laser depends on the absorption coefficient of the solvent being heated, which was water in most cases. The typical 1064-nm laser band of Nd:YAG (neodynium-doped yttrium aluminum garnet) lasers is not suitable because the water absorption coefficient is too low at this wavelength. Excitation of water in the 1.3–2.0 μm range was implemented by either using a chemical iodine laser or Raman-shifted Nd:YAG lasers with different gases. The wavelength of the laser determines the largest volume that can be heated without leading to inhomogeneous heating. In the case of Raman-shifted lasers, the volumes heated were typically 20 μL, while for the iodine laser volumes, up to 500 μL was heated. The time resolution for the experiment is determined by how fast the laser light is transferred into thermal energy and is mainly limited by the width of the laser pulse. The longest times that can be measured are determined by the cooling rate of the volumes heated, which is ca. 10 ms for a 20-μL volume and in excess of 1 s in the case of a 500-μL volume.[29,33] Absorption, fluorescence, or light scattering is employed to follow the relaxation kinetics after heating. The monitoring beam and detection are placed either at a 90° geometry or in a collinear fashion with respect to the laser beam. Detection by absorption or fluorescence requires that the absorption or fluorescence of the guest be different in water and when bound to the supramolecular system, and in the case of fluorescence, the excited state should be unreactive toward the components of the supramolecular system. Photophysics plays an analytical role in this technique.

Stopped Flow

Stopped flow is a technique in which solutions are quickly mixed by driving solutions in two (or more) syringes at high pressure through a mixing chamber and having the flow stop abruptly (Fig. 1.9). Changes in the system's chemical composition after the flow is stopped are followed over time by detecting changes in either absorption or fluorescence.[34,35] Photophysics plays an analytical role in stopped flow experiments, and the use of this technique requires that either the guest or the host have different absorption or emission properties when free in solution or as part of the HG complex. The time resolution is determined by the time required to mix the solutions and stop the flow. For most systems, the time resolution is of 1–2 ms, although shorter resolutions can be achieved with the use of cells with smaller volumes. The tradeoff in using smaller cells is the decrease in the signal-to-noise ratios for the detected signal.

Stopped flow experiments lead to a concentration jump that is employed to study the formation or disassembly of supramolecular systems. In the former

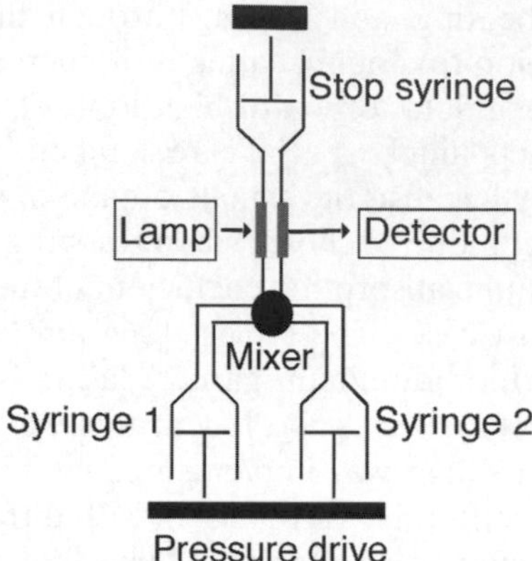

Figure 1.9. Schematic representation for a stopped flow experiment. Two solutions in syringes 1 and 2 are rapidly mixed in the mixer (black). The kinetics is observed in the cell (gray) after the flow is stopped. The lamp and detector are in line for absorption measurements while they are at a 90° arrangement for fluorescence experiments.

case, the host and guest are placed in different syringes, while in the latter case, the solution with the host–guest complex is placed in one syringe and the second syringe contains the solvent.

Ultrasonic Relaxation

Photophysics plays no role in the use of this technique, and therefore, it does not require the use of molecules that contain chromophores. A sound wave is passed through the cell containing the sample, which leads to periodic variations of pressure and temperature.[36,37] Chemical equilibria, which have a nonzero ΔH^0 or ΔV^0, respond to the sound wave when its frequency is similar to the relaxation rate of the chemical equilibrium. The frequency of the sound waves is varied and a maximum absorption is observed for the frequency that corresponds to the relaxation rate of the chemical equilibrium being studied. When more than one equilibrium is present, an absorption maximum will be observed for each equilibrium provided the relaxation rates differ by more than a factor of 5.

Fluorescence Correlation Spectroscopy

This technique is a nonperturbation method where fluorescence is required as an analytical tool because of its sensitivity. The technique is sensitive to two different properties: (1) the different diffusion coefficients for the guest and the host–guest complex, and (2) different emission quantum yields for G and HG.[38] A guest with a high fluorescence quantum yield and minimal intersystem crossing efficiency is required for these measurements.

Fluorescence correlation spectroscopy is based on the principle that intensity fluctuations for the emission of one molecule are related to events that lead to changes in the fluorescence efficiency of that one molecule.[39–41] The sample volume is continuously irradiated, and fluctuations occur because of events such as diffusion of the molecule out of the detection volume, formation of a dark triplet excited state, or association/dissociation from a supramolecular system. This technique requires the detection of fluorescence from a small volume containing a small number of molecules. The concentration of fluorophores has to be low so that fluctuations in the intensity from one molecule can be discriminated from the background emission.

NMR

NMR experiments are used to measure the exchange kinetics of a system in equilibrium, and no perturbation is required for thesc experiments.[42–44] There is no role for photophysics in NMR experiments. Relaxation processes with a time constant slower than 0.1 ms can be investigated. The biggest advantage of using NMR is the correlation between kinetic data and structural assignments.

Kinetic measurements are possible when the NMR signal for a guest or a host is different when located in the homogeneous solution or in the host–guest complex. Two limits exist for such a situation, that is, when the exchange between the host–guest complex and the homogeneous solution is slow or fast compared with the NMR relaxation processes. In the slow exchange limit, separate peaks are observed for the guest or host in the homogeneous phase and in the complex. The integration of each peak is proportional to the concentration of each species and is used for the determination of equilibrium constants. In the fast exchange limit, only one peak is observed where the position of the peak corresponds to the average values for the frequencies of the free and bound guest or host weighted by the fraction of each species present in the solution. Line broadening of the NMR signals is observed in the intermediate regime when the association/dissociation processes for the supramolecular system occur on the same timescale as the relaxation processes for the NMR experiment. The values for the association and dissociation rate constants are obtained from line-shape analysis. Line broadening is frequently induced by changes in temperature so that the system being studied changes between the slow and fast relaxation regimes.

Surface Plasmon Resonance (SPR)

SPR is used for the determination of equilibrium constants and the dynamics for complex formation.[45–47] This technique has found wide use for studies with biological samples. There is no role for photophysics in the use of this technique. The principle of SPR measurements is that the surface plasmon waves traveling parallel to the interface between a metal and a dielectric are affected when the refractive index at the interface changes. The evanescent wave produced by total reflection of an incoming beam can couple with the plasmons in the metal film at the interface. Changes in the refractive index alter this coupling, leading to changes in the intensity of the reflected light detected. One of the components of a bimolecular reaction leading to complexation (G or H) is bound to the metal surface. A solution with the other component flows over the surface, and formation of a complex changes the refractive index at the interface. The kinetics of the association process is observed until equilibration is achieved. The dissociation kinetics is measured by flowing a solution without the second reagent over the metal surface leading to the dissociation of the complex.

The advantage of SPR experiments is that small amounts of sample are required. This technique is sensitive to mass changes, and therefore, it is advantageous to immobilize the reagent with the smallest mass on the surface. Specific immobilization chemistries for each reagent have to be developed, and the surface coverage has to be kept low enough to ensure that the access of the second reagent from the solution phase is not restricted. The upper limit for the association rate constants that can be measured is 10^5–$10^6\,M^{-1}\,s^{-1}$, while the dissociation has to be slower than hundreds of milliseconds.

COMPARISON OF TECHNIQUES

The choice of a technique to study the dynamics of supramolecular systems depends on several factors, such as the timescale of the relaxation process, number of relaxation processes present, magnitude of the concentration change, and the amount of sample available.

For the simplest equilibrium possible, that is, formation of a 1:1 complex (Scheme 1.1), the relaxation rate constant (k_{obs}) is equal to the sum of the association and dissociation processes. Assuming that the host concentration is in excess over the concentration of the guest, the relaxation process leads to first-order kinetics where k_{obs} is expressed by Equation 1.1:

$$k_{obs} = k_{+}[H] + k_{-} \tag{1.1}$$

The timescale for a particular study is determined by the value of k_{obs}. A relaxation process can be slowed down by decreasing the concentration of host, which also may lead to a decrease in the concentration of guest so as to keep the host in excess over guest. The limitation of such a strategy is that at low concentrations of host and guest, a more sensitive method, for example, fluorescence, needs to be used to measure concentration changes. The minimum value for k_{obs} is determined by the dissociation rate constant; that is, the relaxation process cannot be slower than $1/k_{-}$.

Fast processes can be studied using time-resolved fluorescence, laser flash photolysis, laser temperature jump, fluorescence correlation spectroscopy, and ultrasonic relaxation. With the exception of ultrasonic relaxation, all these techniques rely on photophysical measurements (Fig. 1.6). The widest time range is achieved using laser temperature jump or fluorescence correlation spectroscopy. These two techniques are suitable when a system has relaxation processes that occur on different timescales. It is advantageous to measure all kinetic processes with one technique since similar experimental conditions such as concentration ratios can be used for all measurements. Fluorescence and laser flash photolysis are not suitable for kinetics measurements slower than a few microseconds and hundreds of microseconds, respectively. Stopped flow and NMR can be employed to investigate relaxation processes slower than milliseconds.

Fluorescence is the most sensitive of all the methods described and is used when the concentrations involved are low. Absorption is a relative measurement, and for this reason, it is less sensitive than fluorescence, making laser flash photolysis experiments less suitable when small concentration changes are involved. The least sensitive technique for concentration changes is NMR.

A further consideration when choosing a kinetic technique is the necessity for differentiating between relaxation processes with similar relaxation times. Time-resolved fluorescence using single photon counting is the best technique for the differentiation of similar relaxation times because of the ability to collect a large number of counts at a high signal-to-noise ratio. Laser flash photolysis, laser temperature jump, and stopped flow provide an intermediate ability to differentiate between kinetic processes with similar relaxation rate constants. Ultrasonic relaxation and fluorescence correlation spectroscopy have the lowest resolution to differentiate between kinetic processes.

Ultrasonic relaxation and NMR do not require the formation of an excited state, and systems can be studied that do not contain a chromophore. One requirement in using photophysical methods is to ensure that the excited states formed are unreactive toward all components of the system. This requirement is easier to meet when using fluorescence measurements because a molecule with high fluorescence quantum yield can be chosen, diminishing the likelihood of reaction from the excited state. In the case of triplet excited states, control studies are required to determine their photoreactivity because triplet excited states are, in general, more reactive than their corresponding singlet excited states.

Finally, when the amount of sample is limited, time-resolved fluorescence, laser temperature jump, and fluorescence correlation spectroscopy experiments are used because the sample volumes are small and low concentration of reactants can be employed. NMR is not a suitable technique because of the high concentrations required. In the case of stopped flow experiments, large volumes have to be employed because a number of kinetic traces are averaged.

METHODS

The various photophysical methods used to study supramolecular dynamics will be discussed in this section. Thermodynamic information, that is, equilibrium constants and host–guest stoichiometries, is required for modeling of the kinetics. These thermodynamic measurements frequently rely on photophysical experiments, such as fluorescence, absorption, and circular dichroism.[18,48] Use of these techniques for thermodynamic measurements will not be covered in this chapter, and it will be assumed that the species present at equilibrium are known before kinetic studies are performed.

Steady-State Excited-State Detection

Kinetic measurements using continuous irradiation of samples are employed for stopped flow, laser temperature jump, and fluorescence correlation spectroscopy measurements. For the first two techniques, the changes in concentration of chemicals can be followed by changes in absorption or fluorescence, while for fluorescence correlation spectroscopy, only fluorescence can be used as the detection mode. For all three techniques, the sample is continuously irradiated, and the lifetime of the excited state is not relevant to the experiment. In the case of stopped flow and laser temperature jump experiments, the kinetics is analyzed within the framework of relaxation processes.[49,50] Each one of the exponential functions can be related a particular relaxation process. More complex equations are employed for bimolecular reactions when the concentrations of reactants are similar.

Fluorescence correlation spectroscopy measurements are based on the intensity fluctuations of single molecules in the detection volume because of events that change the molecule's fluorescence efficiency or because of the diffusion of the molecule out of the detection volume. An autocorrelation function ($G(\tau)$) is measured, which defines the probability of detecting a photon emitted (I) from the same molecule at time zero and time τ. Loss of correlation at a particular time signifies that the molecule is not available for excitation.[38,51–54] A change in the guest's emission quantum yield because of a change in environment, such as host complexation, is detected as a change in I in the autocorrelation function. Two uncorrelated photons appear as an offset for $G(\tau)$:

$$G(\tau) = \langle I(t)I(t+\tau)\rangle \tag{1.2}$$

A typical correlation curve (Fig. 1.10) shows an increase in $G(\tau)$ at short times, called anti-bunching, which is related to the excitation of the fluorophore. The inflection points after the maximum for $G(\tau)$ are related to events that decreases the probability of forming the singlet excited state of the molecule.

The correlation curve is analyzed by assuming a profile, normally a Gaussian one, for the excitation pulse. For a molecule in a homogeneous solution, which has an appreciable formation of its triplet excited state, the correlation function is the product of the correlation function for fluorescence (G_F), the triplet state (G_T), diffusion of the fluorophore (G_D), and any reaction term (G_R) (Eq. 1.3):

$$G(\tau) = G_F \times G_T \times G_D \times G_R \tag{1.3}$$

The diffusion time of the fluorophore in a solution containing a host–guest system increases as more host–guest complexes are formed at higher concentrations of host, leading to a shift to longer times for the inflection point related to the diffusion of the fluorophore.[38,52] In addition, a relaxation process due to the association/dissociation of the guest becomes apparent in the correlation curve. The new inflection point that appears in the correlation curve in the presence of the host has a time constant τ_R. For a 1:1 complex between a guest and a host, the time constant for the correlation time associated with the host–guest dynamics is defined by:

$$\tau_R = \frac{1}{k_{obs}} = \frac{1}{k_+[H] + k_-} \tag{1.4}$$

The correlation time shifts to shorter times as the host concentrations is raised. The amplitude for the host–guest relaxation process is largest at intermediate host concentrations when both the free guest and host–guest complexes are present is appreciable amounts.

Mechanistic assumptions are required to analyze the correlation curve for a host–guest complex. The mechanism is simplified if one works with a guest for which intersystem crossing is negligible because then one can assume that the binding dynamics is only due to reaction between the ground state of the guest and the host. The correlation function including the host–guest dynamics includes the terms N_G and N_{HG}, which are the mean numbers of free guest and

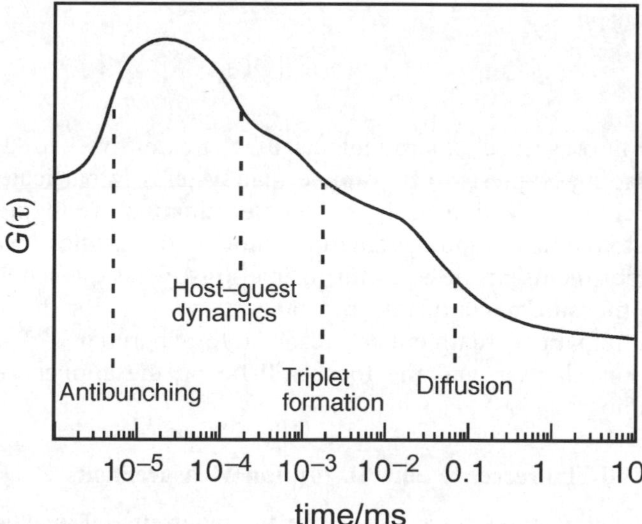

Figure 1.10. Correlation curve showing the various processes that occur on different timescales.

complexed guest in the sample volume; the term A_R, which is the amplitude for the reaction term of the correlation function; and the ratio w_{xy}/w_z, which is the aspect ratio of the sampling volume (Eq. 1.5).[38,52] Fitting the data of the correlation curve to Equation 1.5 leads to values of τ_R at different host concentrations, and the values of k_+ and k_- are determined from the dependence of $1/\tau_R$ with the concentration of host (Eq. 1.4). Other mechanistic assumptions will lead to different equations for $G_R(\tau)$ from which the relaxation times τ_R are determined:

$$G_R(\tau) = \frac{1}{N_G + N_{HG}} \left(1 + \frac{\tau}{\tau_D}\right)^{-1} \left(1 + \left(\frac{w_{xy}}{w_z}\right)^2 \frac{\tau}{\tau_D}\right)^{-1/2} (1 + A_R e^{-\tau/\tau_R}) \tag{1.5}$$

Steady-State Fluorescence Measurements

The singlet excited-state lifetimes of fluorophores are usually shorter than the association/dissociation binding dynamics of guests with hosts.[10,11] For this reason, the excited guest does not leave the host, and it can be used to determine the accessibility of quencher molecules to the interior of the host. Steady-state quenching is observed as a decrease of the emission intensity (I) compared with the intensity in the absence of quencher (I_0). The Stern–Volmer constant (K_{sv}) is related to the quenching efficiency and is equal to the product of the quenching rate constant and the lifetime of the fluorophore in the absence of quencher (τ_0) (Eq. 1.6). Nonlinear quenching plots are observed when more than one fluorophore with different quenching efficiencies are present (downward curvature) or when static, that is, immediate, quenching is present in addition to a bimolecular dynamic quenching reaction (upward curvature):[23]

$$\frac{I_0}{I} = 1 + K_{sv}[Q] \tag{1.6}$$

Steady-state fluorescence experiments have been extensively used to characterize the binding of guests to supramolecular systems, in particular, for thermodynamic characterization, such as the determination of binding constants of host–guest complexes and aggregation numbers of micelles.[18] Steady-state emission experiments are less useful to investigate the quencher dynamics because the measurements of rate constants is indirect. For this reason, the detailed mathematical treatment of steady-state experiments[23,55] will not be covered in this chapter, and the focus will be on the application of time-resolved studies.

Time-Resolved Fluorescence and Absorption Measurements

Short-lived excited states act as probes for the environment in which they are located, that is, the homogeneous phase or within the host system. The bimolecular reaction of the excited states with the quencher provides information on the mobility of the quencher between the homogeneous phase and the host system[18,56–63] and two separate cases will be considered: (1) The quencher deactivates the excited state in the host for every encounter between excited state and quencher ($k_q^{int} \gg k_-^Q$, Fig. 1.5), and (2) the quencher has a finite probability of exiting the host without quenching the excited state bound to the host ($k_q^{int} \sim k_-^Q$ or $k_q^{int} \ll k_-^Q$). Scenario (1) is more realistic for host–guest complexes with confined spaces, such as CD complexes, where the quencher "entering" the host can only lead to a close contact between the guest and quencher. Scenario (2) is likely for larger hosts, such as micelles, where the quencher can be incorporated in the micelle at a distance from the excited state and an intramicellar unimolecular quenching process (k_q^{int}) competes with the exit of the guest (k_-^Q).

Deactivation for Every Quencher–Excited State Encounter

The lifetimes for the excited states of the guest in the homogeneous phase and bound to the host can frequently be different, with a longer lifetime being generally observed for the guest in the host. The fluorescence decay will not follow a mono-exponential function when an appreciable amount of excited guest is present in the homogeneous solution and bound to the host, that is, when more than one emitting species with different lifetimes are present in the solution. In this case, the decay is fit to a sum of exponentials (Eq. 1.7), where A_i corresponds to the pre-exponential factor of each species "i." The sum of all pre-exponential factors is normalized to unity. The parameter k_i is the decay rate constant for each species "i" and corresponds to the inverse of the lifetime for "i":

$$I(t) = I_o \sum_1^i A_i\, e^{-k_i t} \quad \text{and} \quad \sum_1^i A_i = 1 \tag{1.7}$$

Quenching leads to a decrease of the observed decay rate constant of the excited fluorophore, which follows a linear relationship with the quencher concentration (Eq. 1.8). The values for the rate constant in the absence of quencher (k_0) and the quenching rate constant for the excited guest in the homogeneous phase (k_q) are obtained in an independent experiment. In the presence of the host system, the decays are fit to the sum of two exponentials where k_{obs}^w corresponds to the decay of the excited guest in the homogeneous phase and k_{obs}^H corresponds to the observed lifetime of the excited guest in the host:

$$k_{obs}^i = k_0^i + k_q^i[\mathrm{Q}] \tag{1.8}$$

The values of the pre-exponential factors are related to the concentrations of excited states and the excitation efficiencies for each species. A_i values can only be related to absolute concentrations if the system can be excited at an isosbestic point or if the molar absorptivity for the host–guest complex is known. In addition, the bandwidth for the excitation light has to be taken into account when calculating the excitation efficiencies when broad bandwidths are employed. The values for the pre-exponential factors are constant as the quencher concentration is raised because the excited-state guest does not move between the host and the homogeneous phase. The assumption of non-mobile fluorophore, that is, the excited guest, is not valid if the pre-exponential factors vary and a different mechanistic model needs to be employed to analyze the data.

The values for the guest's excited-state lifetime in the homogeneous solvent (k_0) and in the host (k_0^H) may not be very different and therefore a precise differentiation between k_{obs}^w and k_{obs}^H may be difficult. The values for the lifetime of the excited guest ($1/k_0^w$) and quenching rate constant (k_q) in the homogeneous solvent can be determined in an independent experiment. Therefore, the value of k_{obs}^w can be fixed for the analysis of the fluorescence decay kinetics in the presence of host. This procedure increases the precision of the k_{obs}^H values.

The reaction of the excited guest with the quencher corresponds to a bimolecular reaction in which an initial encounter complex is formed followed by the quenching reaction in the complex. In the case of the quenching for the guest bound to the host, the formation of the encounter complex is related to the association rate constant of the quencher with the host (k_+^Q), the dissociation of the quencher from the host (k_-^Q) and the intrahost quenching rate constant (k_q^{int}) (Fig. 1.5, Eq. 1.9). The quenching rate constant (k_q^H) is equal to the association rate constant of the quencher (k_+^Q) when each encounter of the quencher with the guest in the host leads to quenching; that is, k_-^Q is much smaller than k_q^{int}. In general, this condition is met for quenchers that have a diffusion controlled rate constant in solution, but this condition can also be met for lower values of k_q as long as no exit of the quencher from the host occurs before the quenching reaction:

$$k_q^H = \frac{k_+^Q\, k_q^{int}}{k_-^Q + k_q^{int}} \tag{1.9}$$

It is important to note that time-resolved measurements are not sensitive to the situation where the guest and quencher are solubilized, prior to excitation, in the same binding site within the host and the intrahost quenching rate constant is higher than the exit rate constant of the quencher from the host. In such case, the excited guest is quenched within the excitation pulse, and no lifetime is measured for this component. However, in steady-state experiments, the guest–quencher–host complex will lead to a decrease in the emission intensity due to static quenching.[64,65]

Competition between Intrahost Quenching and Quencher Exit from the Host

An intrahost quenching efficiency lower than 100% is observed when exit of the quencher from the host system competes with quenching. This situation is rare for host–guest complexes with defined binding sites, such as macrocycles, because of the limited volume within the host. However, for large self-assemblies, such as micelles, the quencher can bind in a region far from the excited guest and intrahost migration is required for quenching to occur. The simplest model developed to analyze the fluorescence decay of excited guests in the host assumes that all guest molecules are bound to the host and that the probability of the quencher to enter and exit the host is independent of the host containing a guest molecule; that is, k_+^Q and k_-^Q are the same for an empty host and a host containing a guest. The decay in the absence of the quencher follows an exponential function with a rate constant k_0, while in the presence of the quencher, the nonexponential decay is fit to Equation 1.10, where parameters A to D are defined in Equations 1.11–1.14.[57,60–62] Measurements are performed at different quencher concentrations because parameters B and C are dependent on the quencher concentration, leading to the recovery of all relevant rate constants:

$$I(t) = A\, e^{\left(-Bt - C\left[1 - e^{-Dt}\right]\right)} \tag{1.10}$$

$$A = I_0 \tag{1.11}$$

$$B = k_0 + \left[\frac{k_q^{\text{int}}\, k_+^Q}{\left(k_q^{\text{int}} + k_-^Q\right)\,(1 + K_Q[\text{H}])}\right][\text{Q}] \tag{1.12}$$

$$C = \left[\frac{\left(k_q^{\text{int}}\right)^2 k_+^Q}{\left(k_q^{\text{int}} + k_-^Q\right)^2 k_-^Q\,(1 + K_Q[\text{H}])}\right][\text{Q}] \tag{1.13}$$

$$D = k_q^{\text{int}} + k_-^Q \tag{1.14}$$

The model described above was expanded to more complex mechanisms, such as the inclusion of the intermicellar exchange of the quencher or guest during the lifetime of the excited guest.[62]

Equation 1.10 provides a general solution for quenching studies of guests in hosts, where parameters A, B, C, and D are related differently to the rate constants relevant to different mechanisms. A theoretical analysis was performed on the identifiability of several mechanisms for guest and quencher binding to micelles.[66] The mechanisms considered for the analysis of the fluorescence decays were (1) the quencher and guest are immobile, (2) mobile quencher and immobile guest, (3) immobile quencher and mobile guests, and (4) mobile guest and quencher. The models can be uniquely identified, and the minimum set of experiments required to identify each model was determined theoretically.

Mobility of Excited-State Guests

The association and dissociation rate constants of guests can be determined when the lifetime for the excited state guest is of the same order of magnitude as the dynamics being investigated. The lifetimes of triplet excited states of organic molecules are frequently in the tenth to hundreds of microsecond range, which is the timescale for the host–guest dynamics involving macrocyclic host systems. The kinetics of triplet states is normally measured by laser flash photolysis experiments, where changes in the transient absorption are followed. The conceptual framework developed below is also applicable to long-lived singlet excited states, where fluorescence is the experimental observable.

Formation of the triplet excited state is a perturbation where the new chemical formed (Fig. 1.3), that is, the excited state, relaxes to a new equilibrium. If the molar absorptivities of the triplet guest in the solvent and bound to the host are different, the relocation of the guest follows a kinetic behavior that can be fit to a sum of two exponentials (Eq. 1.15), where the two exponential factors γ_1 and γ_2 are related to the deactivation rate constants of the excited guest in the solvent (k_0) and in the host (k_0^H) and the association (k_+^*) and dissociation (k_-^*) rate constants of the triplet guest with the host (Eqs. 1.16–1.19). The mathematical treatment is equivalent to that developed for the dynamics of excimer emission:[67]

$$\Delta A = A_1\, e^{-\gamma_1 t} + A_2\, e^{-\gamma_2 t} \tag{1.15}$$

$$\gamma_{1,2} = -\frac{1}{2}\left[(A+B) \pm \sqrt{(A-B)^2 + 4C}\right] \tag{1.16}$$

$$A = k_0 + k_+^*\,[\mathrm{H}] \tag{1.17}$$

$$B = k_0^H + k_-^* \tag{1.18}$$

$$C = k_-\, k_+^*\,[\mathrm{H}] \tag{1.19}$$

Parameters A and C in Equation 1.16 are dependent on the concentration of host and the values of γ_1 and γ_2 are determined at various host concentrations. The sum and product of the exponential factors γ_1 and γ_2 are related to the individual rate constants by Equations 1.20 and 1.21:[68]

$$\gamma_1 + \gamma_2 = k_0 + k_0^H + k_-^* + k_+^*[\mathrm{H}] \tag{1.20}$$

$$\gamma_1\gamma_2 = k_0\left(k_0^H + k_-^*\right) + k_0^H k_+^*[\mathrm{H}] \tag{1.21}$$

The analysis is simplified when the lifetime of the excited state is much longer than the association and dissociation dynamics ($k_0 \ll k_+^*[\mathrm{H}]$ and $k_0^H \ll k_-^*$). When the k_0 and k_0^H values are smaller by a factor of 1000 compared with the host–guest dynamics, the relaxation process levels off at a value of A_2 and γ_2 is equal to zero. Conceptually, the lifetime of the excited guest can be considered constant compared with the host–guest dynamics (Fig. 1.2). The value of the exponential factor γ_1 is equal to the relaxation process of the host–guest complex (Eq. 1.22):[10]

$$\gamma_1 = k_{\mathrm{obs}} = k_-^* + k_+^*\,[\mathrm{H}] \tag{1.22}$$

Quenching experiments can also be employed to study the dynamics of the triplet excited-state guest with the host.[10,11,69,70] This methodology is suitable when the molar absorptivities of the guest in the homogeneous phase and bound to the host are the same. The quenching methodology can be applied when the quenching rate constant for the triplet guest in the homogeneous solution (k_q) is different from the quenching rate constant of the guest in the host (k_q^H). In general, quenchers are employed for which the quenching efficiency in the homogeneous solution is higher than the quenching efficiency for the excited guest in the host. Sufficiently high host concentrations are employed to assure that the concentration of guest in the homogeneous phase is small compared with the host–guest concentration. Under this condition, the steady-state condition is applied to the rate laws and the kinetics follows a mono-exponential decay, where the observed rate constant is given by the expression in Equation 1.23:[69,70]

$$k_{\text{obs}} = k_0^H + k_-^* + k_q^H[\text{Q}] - \frac{k_-^* \, k_+^*[\text{H}]}{k_0 + k_q[\text{Q}] + k_+^*[\text{H}]} \qquad (1.23)$$

The dependence of k_{obs} with the quencher concentration shows a downward curvature (Fig. 1.11) because at high quencher concentrations, the lifetime of the triplet guest in the homogeneous solution becomes shorter than the exit of the excited guest from the host. Qualitatively, the larger the difference between the linear relationship in homogeneous solution with the curved quenching plot, the slower is the dynamics for the host–guest complex. The slope of the linear relationship observed for the curved quenching plot at high quencher concentrations corresponds to the quenching rate constant of the excited guest in the host. The effect of changes to the various rate constants and the host concentration on the curvature of the quenching plot have been discussed previously.[10]

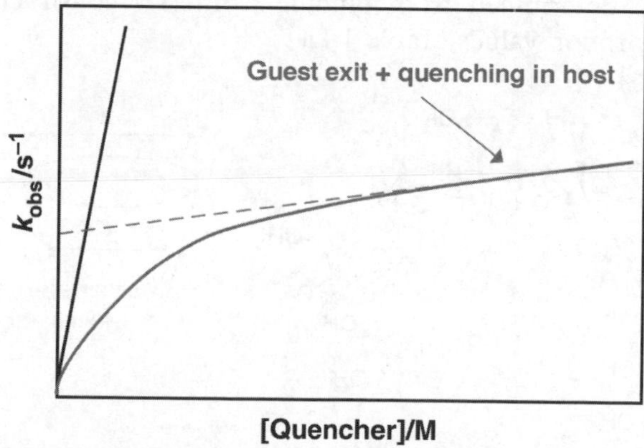

Figure 1.11. Linear quenching plot for the excited guest in homogeneous solution (black, Eq. 1.8) and curved quenching plot for the excited guest in equilibrium between the host and the homogeneous solution (gray, Eq. 1.23). At high quencher concentration, the linear portion of the graph corresponds to the exit rate constant of the excited guest from the host (intercept of the dashed line) and the quenching rate constant for the guest inside the host (slope).

APPLICATIONS

The objective of this section is to present selected examples on the application of the photophysical methods described above. Four classes of host systems will be covered, that is, CDs, DNA, micelles, and bile salt aggregates. A complete survey of all the studies on the dynamics of guests with these four host systems is beyond the scope of this chapter, and a selection to cover different types of mechanisms was chosen. CDs and DNA are hosts that form host–guest complexes with defined stoichiometries, while micelles and bile salt aggregates provide larger binding regions and progressive amounts of guest can be bound to these latter hosts.

CDs

CDs are cyclic oligosaccharides (Fig. 1.12) where the internal cavity size increases with the number of glucose units (6, 7, and 8 for α-, β-, and γ-CD).[71,72] Guest molecules are complexed inside the CD cavity, and the efficiency for complex formation is determined by the hydrophobicity of the guest and the steric constraints on how well the guest fits inside the relatively rigid CD cavity. The thermodynamics of guest complexation to CDs has been extensively characterized,[73,74] and enthalpic and entropic factors are relevant for complex formation.

Temperature jump experiments (Joule heating) were employed in the early studies on the host–guest characterization with CDs to determine the guest binding dynamics between azo dyes **1** (Scheme 1.2) with α-CD.[75] Binding of derivatives of **1** with α-CD led to changes in the absorption spectra of the azo dyes. Complexes with 1:1 stoichiometry were formed, and the equilibrium constants were determined from binding isotherms constructed from the changes in absorption values (Table 1.1).

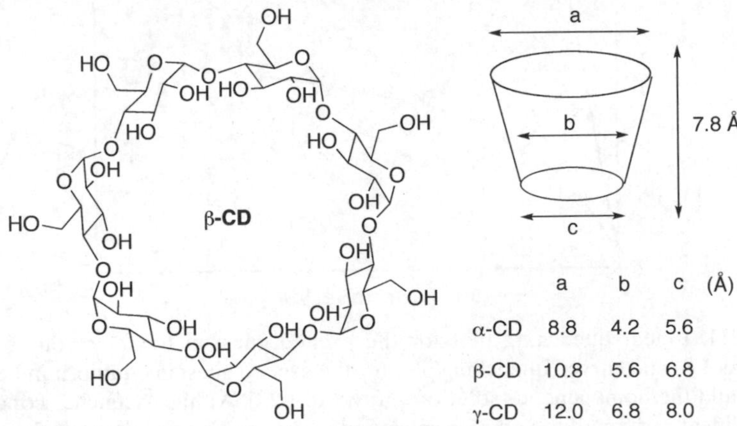

β-CD
a
b
c
7.8 Å
a b c (Å)
α-CD 8.8 4.2 5.6
β-CD 10.8 5.6 6.8
γ-CD 12.0 6.8 8.0

Figure 1.12. Structure of β-CD and dimensions of the various CDs.

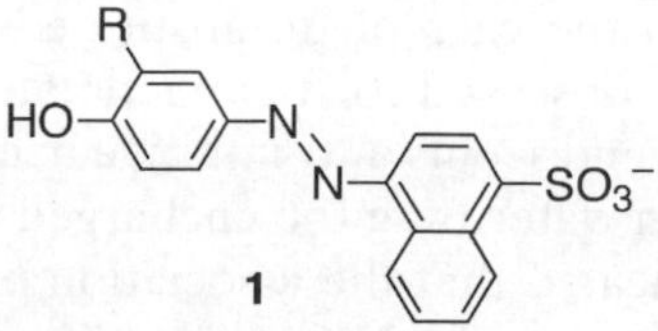

Scheme 1.2. 3′-alkyl-4′-hydroxyphenylazo-1-naphthalene-4-sulfonate guests.

TABLE 1.1. Equilibrium Constants Determined from UV-Vis Binding Isotherms (K_{11}), Association (k_+) and Dissociation (k_-) Rate Constants of Guests 1 with α-CD,[75] and the Calculated Value for K_{11} (K_{11}^{calc}) from the Kinetic Data (Scheme 1.1)

R	K_{11}/M^{-1}	k_+/10^4 M^{-1}s^{-1}	k_-/10^2 s^{-1}	K_{11}^{calc} /M^{-1}
H, phenol	270	1300	550	240
H, phenolate	650	17	2.6	650
CH_3, phenol	420	12	3.5	340
CH_3, phenolate	480	0.012	0.0028	430
CH_2CH_3, phenol	460	0.6	0.19	320
CH_2CH_3, phenolate	290	0.00028	0.0001	280

Data were collected at $T = 14°C$, $\mu = 0.5$ M, pH 3.5 for phenol, and pH 11 for phenolate.

A linear relationship was observed between the increase in the observed relaxation rate constant and CD concentration, and the association and dissociation rate constants were obtained using Equation 1.1 (Table 1.1). The cavity of α-CD is too small to accommodate the naphthyl moiety of **1** and only the phenolate moiety can be included in the CD. The first important general observation is that, as long as the phenolate moiety fits inside the CD cavity, increases in the size of this moiety has a small effect on the equilibrium constant. However, the dynamics is significantly affected by these structural changes with changes of 10^3 for the values of k_+ and k_- for the phenol derivatives and 10^4 for the phenolate derivatives (Table 1.1). These results are a good example that dynamic information cannot be obtained from thermodynamic studies. The second general feature of importance is that the values for the equilibrium constants calculated from the association and dissociation rate constants are in good agreement with the K_{11} values determined from thermodynamic binding experiments. This agreement indicates that the kinetics investigated corresponds to the process leading to equilibrium, and no further slow kinetics occur beyond the time window investigated.

The increase in the bulk of the phenyl ring by substitution with methyl and ethyl groups decreased the rate constants for the association process, indicating that a tighter fit into the cavity occurred, which may require the distortion of the CD framework. However, once the complex is formed, the exit of the bulkier guests was also slowed down, leading to a small increase in the values of K_{11}. *Di-ortho* alkyl substitution for derivatives of **1** did not lead to the formation of complexes with α-CD, showing that these guests are too bulky to enter the CD cavity.[75] In the case of unsubstituted **1**, a significantly higher equilibrium constant was observed for the phenolate ion compared with the phenol compound. This trend seems anti-intuitive because the charged species should be more soluble in water than the uncharged guest. However, analysis of the rate constants indicated that the association process was slowed down for the phenolate derivative as would be expected for the more highly solvated species. However, the larger decrease for the dissociation rate constant for the phenolate derivative is responsible for the increase in K_{11}, probably because the moiety with the negative charge has to pass through the hydrophobic cavity of the CD for the guest to exit into the aqueous phase. For the bulkier methyl and ethyl derivatives of **1**, the values for k_+ and k_- are lower for the phenolate derivative than the phenol. However, the equilibrium constant for the protonated and unprotonated species is the same in the case of the methyl derivative and smaller for the ethyl derivative, showing that the bulkiness of the incorporated moiety and the presence of a charge do not act independently on the binding dynamics. This example shows that generalizations for the host–guest thermodynamics and dynamics are not straightforward.

CDs are known to form complexes with more than one guest and host. Methyl orange (**2**),[76] pyronine Y (**3**),[77] and pyronine B (**4**)[78] (Scheme 1.3) were shown to form complexes with 1:1, 2:1 (guest:CD), and, in the case of **2** and **3**, 2:2 stoichiometries (Scheme 1.4). Temperature jump studies showed that the formation of the 1:1 and 2:2 complexes was fast and occurred within the time

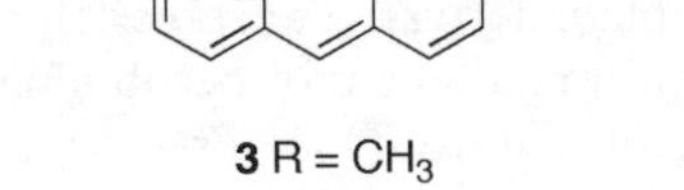

2

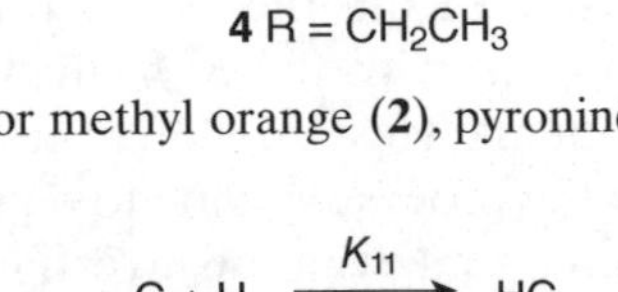

3 R = CH_3

4 R = CH_2CH_3

Scheme 1.3. Structures for methyl orange (**2**), pyronine Y (**3**), and pyronine B (**4**).

$$G + H \underset{}{\overset{K_{11}}{\rightleftharpoons}} HG$$

$$G + HG \underset{k_-^{21}}{\overset{k_+^{21}}{\rightleftharpoons}} HG_2$$

$$H + HG_2 \underset{}{\overset{K_{22}}{\rightleftharpoons}} H_2G_2$$

Scheme 1.4. Mechanism for the formation of host–guest complexes with multiple stoichiometries.

resolution (2–5 μs) of the equipment. The fast equilibria are not observable in the kinetic studies, but they are incorporated, in the form of equilibrium constants, into the dependence of k_{obs} with the CD concentration (Eq. 1.24):

$$k_{obs} = k_{+}^{21} K_{11}\left[\frac{[G]([G]+[HG]+4[H])}{1+K_{11}[G]+K_{11}[H]}\right] - k_{-}^{21}\left[\frac{1+K_{22}[HG_2]}{1+K_{22}[H]+K_{22}[HG_2]}\right] \quad (1.24)$$

The values for the equilibrium constants and the association and dissociation rate constants for the 2:1 complex were obtained from the nonlinear fit of the data to Equation 1.24 (Table 1.2). Guests **2–4** dimerize at high concentrations in aqueous solution. The equilibrium constants for the 2:1 complex with γ-CD are much higher than for the dimer formation in water. In addition, the equilibrium constants for the 2:1 complex are much higher than for the 1:1 complex, showing that the CD cavity stabilizes the dimer of these guests. The association rate constant for the second guest to the 1:1 host–guest complex is close to the diffusion controlled limit ($7.5 \times 10^9\ M^{-1}\,s^{-1}$ at 25°C),[12] showing that the γ-CD cavity with one guest is not fully occupied and provides enough space for the entry of the second guest. The residence time of guests **2–4** is between 70 and 200 μs. These slow dissociation rate constants are an expression of the stabilization of the dimer within the γ-CD cavity.

A recent study on the binding of **3** and **4** with γ-CD[79] agreed qualitatively with the previous investigations[77,78] that γ-CD stabilizes the dimers of **3** and **4** in the 2:1 complex when compared with dimer formation in the aqueous phase. However, the new K_{11} (40–70 M^{-1} for **3** and 190 M^{-1} for **4**) and K_{21} values ($2 \times 10^6\ M^{-1}$ for **3** and $2.2 \times 10^5\ M^{-1}$ for **4**) determined in the more detailed study are very different from the previously reported ones. This discrepancy was mainly attributed to the fact that the temperature jump experiments were performed in the presence of 1 M NaCl required for Joule heating. Xathene dye aggregation, such as **3** and **4**, is known to be sensitive to the ionic strength,[80] and the degree of aqueous dimer formation in the two sets of experiments is probably different. In addition, the assumption made in the previous work about the equality of the absorption spectra of the guest in water and in the 1:1 complex were shown to be incorrect.[79] This example shows that rigorous thermodynamic characterization is required to establish the possible species present when performing kinetic studies. In the absence of such a characterization, the kinetics may be consistent with a mechanism proposed, but may need

TABLE 1.2. Equilibrium Constants and Association and Dissociation Rate Constants for the 2:1 Complex of Guests 2–4 with γ-CD[a]

Guest	K_{11}/M^{-1}	$K_{21}/10^5\,M^{-1}$	$k_{+}^{21}/10^9\ M^{-1}\,s^{-1}$	$k_{-}^{21}/10^3\ s^{-1}$	K_{22}/M^{-1}
2[76]	45	20	9	4.8	6000
3[77]	1000	1	2	14	50
4[78]	430	1.3	0.82	6.4	—

[a] 25°C for **2–4**; for **2**: 0.1 M Na_2PO_4, pH = 9.0; for **3**: 1 M NaCl, pH = 6.1; for **4**: 1 M NaCl, pH = 5.7.

reinterpretation if further experiments show that a host–guest species is missing or one has to be added.

Stopped flow experiments were employed to study the dynamics of CD host–guest complexes on a timescale slower than milliseconds. These studies include the binding of metal ions, such as Cu^{2+} to α-CD,[81] and the competition between organic guests and Ni^{2+} or Zn^{2+}.[82,83] Examples for complex kinetics studied by stopped flow are the two-step formation of α-CD complexes (Scheme 1.5) with azo compounds **5** (Scheme 1.6)[84,85] and the kinetics for the formation of 2:2 complexes between pyrene (**6**, Scheme 1.6) and γ-CD.[86]

The binding kinetics for compound **5** with hydrogen and methyl substituents[84,85] showed one relaxation process because these guests are sufficiently small to slip into the α-CD cavity. For larger substituents with increasing chain lengths (ethyl, propyl) or branching chains (*iso*-propyl, *iso*-butyl, *tert*-butyl), two relaxation processes were observed, which were assigned to a sequential mechanism where an encounter complex "*a*" was formed followed by the internalization of the guest into the cavity in process "*b*" (Scheme 1.5). The rate constants for both processes were much less dependent on the size of the substituent (factors of 2–3) than observed for compound **1** containing the larger naphthyl moiety (Table 1.1). The exception is the significant slowdown observed for the *tert*-butyl derivative of **5**, where the substituent is too large for the formation of species "*a*." This latter result suggests that species "*a*" is not just an encounter complex but has a defined structural relationship between the CD and the guest.

Pyrene forms 1:1 and 2:2 complexes with γ-CD.[86–88] The fluorescence of the pyrene monomer was observed for pyrene in water and in the 1:1 complex. An excimer-like emission was observed for the 2:2 complex, due to the pres-

$$\mathrm{G + H} \underset{k_-^a}{\overset{k_+^a}{\rightleftharpoons}} (\mathrm{G-H})_a \underset{k_-^b}{\overset{k_+^b}{\rightleftharpoons}} (\mathrm{G-H})_b$$

Scheme 1.5. Sequential formation of a host–guest complex.

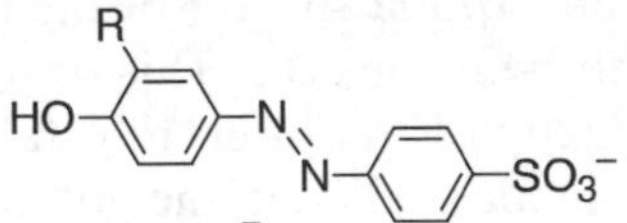

R
HO
N
N
SO_3^-
5

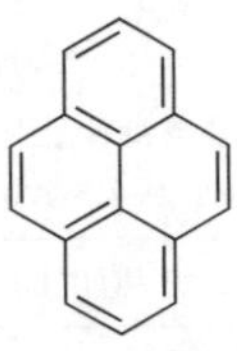

6

Scheme 1.6. Structures for 3′-alkyl-4′-hydroxyphenylazo-1-phenyl-4-sulfonate (**5**) and pyrene (**6**).

ence of the pyrene dimer. The excimer emission is red shifted in comparison to the monomer emission, and changes in the excimer emission intensity were used to follow the kinetics for the formation of the pyrene–γ-CD 2:2 complex from the association of two 1:1 complexes. The kinetics occurs within 0.2 s and the value for k_{+}^{22} was determined to be $6 \times 10^{7}\,M^{-1}s^{-1}$, while the value for k_{-}^{22} was $73\,s^{-1}$. The value for the association rate constant is between 5 and 40 times smaller than the rate constants observed for the formation of 1:1 complex with guests that fit into the CD cavity (see Table 1.4), while the value for the dissociation rate constant of the 2:2 complex is at least 200 times slower than the dissociation rate constants for 1:1 complexes.

In the original study of the pyrene–γ-CD system, a very slow kinetics over several seconds was assigned to the dynamics of the 1:2 (pyrene : CD) complex.[86] A temperature annealing effect observed for the pyrene–γ-CD system[88] showed that aggregates of γ-CD are present in aqueous solution, leading to a slowed down re-equilibration process. Preliminary studies suggested that the dynamics described for the 2:2 complex was not affected by the presence of γ-CD aggregates, but the slow dynamics assigned to the 1:2 complex is probably due to the dynamics with the small amount of aggregates present in the solution and not due to the formation of a distinct pyrene–γ-CD species.

Fluorescence correlation spectroscopy was employed to study the binding dynamics of **3** and **4** with β-CD.[52] Two features are apparent on the correlation curves when CD was added to an aqueous solution of **3** (Fig. 1.13) or **4**. The

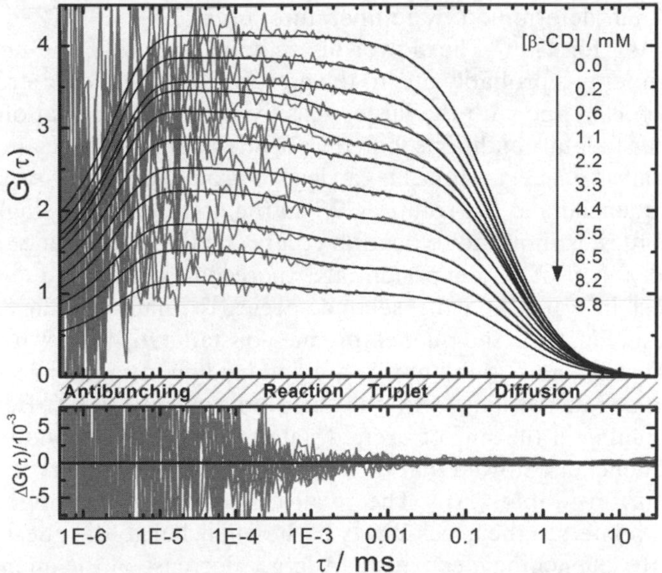

Figure 1.13. Scaled ($G(6\,\mu s) = 1.0$) and offset experimental correlation curves for **3** in water measured with increasing concentrations of β-CD (gray). The black lines correspond to the global fit of the data to Equation 1.3. Reprinted with permission from Reference 52. Copyright 2005 American Chemical Society.

diffusion term, which is characterized by the decrease of $G(\tau)$ at ca. 1 ms, shifted to longer times with the addition of CD. This trend is a consequence of the higher molar mass (ca. five times) for the host–guest complex when compared with the molar mass for the guest, which leads to a lengthening of the diffusion times of the fluorophore from 0.25 to 0.4 ms for **3** and 0.30 to 0.45 ms for **4**. The second feature was the appearance of a new correlation term between 1 and 10 μs due to the formation of the host–guest complex. This correlation time shifted to shorter times when the β-CD concentration was raised.

The experimental data were fit to Equation 1.3 using the global analysis method,[89,90] where all the experimental curves were fit simultaneously. The correlation time for the host–guest dynamics was then fit to Equations 1.4 and 1.5, and the values for k_- were $5.0 \times 10^5\,s^{-1}$ for **3** and $7.6 \times 10^4\,s^{-1}$ for **4**. The values for k_+ were calculated from the equilibrium constants and were $2 \times 10^8\,M^{-1}\,s^{-1}$ for **3** and $1.5 \times 10^8\,M^{-1}\,s^{-1}$ for **4**. The association rate constants are the same for both guests and are one order of magnitude lower than the diffusions controlled limit for a bimolecular reaction in water. The difference in the binding efficiency is dictated by the dissociation rate constant, which is one order of magnitude higher for the smaller guest **3** than for **4**. This result suggests that **4** had the optimum size complementarity with the cavity of β-CD.

The values for k_+ for the binding of **3** or **4** with β-CD obtained from fluorescence correlation spectroscopy are the same as the k_+ values measured in temperature jump experiments ($1.1 \times 10^8\,M^{-1}\,s^{-1}$ for **3** and **4**),[77,78] while the k_- values were much higher for the fluorescence correlation spectroscopy measurements than determined by temperature jump ($2.6 \times 10^4\,M^{-1}\,s^{-1}$ for **3** and $1.5 \times 10^4\,M^{-1}\,s^{-1}$ for **4**).[77,78] The slower dissociation process in the temperature jump experiments is probably due to the presence of a high ionic strength (1 M NaCl), which is in line with the larger sensitivity of the dissociation rate constant to experimental conditions than the sensitivity for the association process.

Incorporation of guests into CDs can lead to protection of the guest's excited state from quenchers in the solution. When the lifetime of the singlet excited state of the guest is short, no relocation occurs and the fluorescence decay corresponds to a sum of two exponentials, where one species is related to the excited guest in water and the second species is related to the CD-bound excited guest (Fig. 1.5). The quenching rate constants for the excited guest in the aqueous phase and bound to CD are obtained by fitting the dependencies of the observed rate constants with the quencher concentration to Equation 1.8.

The quenching of the singlet excited state of pyrene (**6**) bound to β-CD by a series of quenchers showed that the decrease in the quenching rate constants was not uniform (Table 1.3).[91] The relative decrease in the quenching rate constants is related to the accessibility of the quencher to the guest inside the CD and to the quenching mechanism. A larger decrease of the quenching rate constant is expected for quenchers that require close contact and are relatively large, such as Tl^+.[91]

The excited state of 2-naphthol (**7**, Scheme 1.7) bound to β-CD is protected from the quenching by iodide anions. Analysis using Equation 1.8 showed that the quenching rate constant decreased by a factor between 3 and 4.[92] The

TABLE 1.3. Rate Constants for the Quenching of Singlet Excited State of 6 Bound to β-CD[91]

	$k_q/10^9\,M^{-1}s^{-1}$	
Quencher	Water	β-CD
O_2	11	1.2
CH_3NO_2	8.1	0.56
Tl^+	6.3	0.02
Cu^{2+}	4.5	0.17

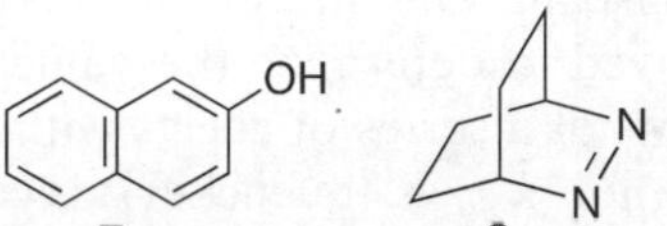

Scheme 1.7. Structures for 2-naphthol (**7**) and 2,3-diazabicyclo(2.2.2)oct-2-ene (**8**).

fluorescence decay of this system was also analyzed using global compartmental analysis where, in addition to the quenching rate constants, the association and dissociation rate constants of the excited state of **7** with the CD were recovered (Fig. 1.4). This analysis was possible despite the short lifetime of the excited state of **7** because the lifetimes changed with the CD concentration. The quenching rate constant for the singlet excited state of **7** inside β-CD using global analysis was the same as by analyzing each lifetime separately (Eq. 1.8). The association and dissociation rate constants of the singlet excited state of **7** with β-CD were determined to be $2.5 \times 10^9\,M^{-1}s^{-1}$ and $520\,s^{-1}$, respectively.

The singlet excited-state lifetime of **8** is sufficiently long in water (420 ns) and D_2O (730 ns) for it to relocate between the aqueous phase and the host. The lifetime of **8** is shortened by a quenching mechanism involving an "aborted" hydrogen transfer,[93] where the lifetime of the excited state is dependent on the presence of abstractable hydrogen atoms on the solvent molecules. An important feature of the quenching is that no net products are formed in contrast to the typical hydrogen abstraction reaction for excited ketones, and therefore, **8** can be used as an inert probe.

In the presence of α- and β-CD, the fluorescence decay for **8** showed two exponential terms, where the lifetime for the short-lived component was independent of the CD concentration, while the long-lived component was shortened with an increase in the CD concentration.[94] The singlet excited state of **8** inside the CD reacts with available hydrogen atoms from the glucose units, shortening the lifetimes for **8** (33 ns for α-CD and 95 ns for β-CD). These lifetimes are too short for the excited state to exit the cavity before it is deactivated. The long-lived component is related to the bimolecular reaction between the excited state of **8** and the CDs, and corresponds to the association rate constant between **8** and the CDs. The values for k_+^* were determined to be $1.9 \times 10^8 M^{-1}s^{-1}$, $4.0 \times 10^8 M^{-1}s^{-1}$, and $0.8 \times 10^8 M^{-1}s^{-1}$ for α-, β- and γ-CD, respectively. These values follow qualitatively the same order as was observed for the equilibrium constants of the ground state (K_{11}: $50 M^{-1}$ for α-CD, $1100 M^{-1}$ for β-CD, and $6 M^{-1}$ for γ-CD). The development of **8** for supramolecular binding studies is important because the modulation of its lifetime by aborted hydrogen transfer makes this probe uniquely suited to determine the values for the association rate constants. The requirement of using **8** as a probe to measure k_+^* is that the host has abstractable hydrogens that lead to a decrease in the excited-state lifetime of **8** once the complex is formed.

The triplet excited states of guests, due to their long lifetimes, can move between the aqueous phase and the interior of the CD cavity. The quenching methodology was employed to determine the values for the association and dissociation rate constants of a series of guests with CDs.[11,70] In the presence of a host, the quenching plot (k_{obs} vs. [quencher]) is curved because exit of the triplet guest from the CD becomes rate limiting; that is, exit is slower than the excited-state decay in water (Fig. 1.14). The larger the difference between the curved quenching plot and the linear relationship observed in water, the slower is the binding dynamics. In the example shown in Fig. 1.15, the binding dynamics of xanthone (**9**, Scheme 1.8) is markedly faster than the binding dynamics for flavone (**10**).

Fits of the experimental data in the presence of a host to Equation 1.23 led to the determination of the k_+^*, k_-^*, and k_q^H values (Table 1.4).[95,96] The rate constants for the association of guests **9–11** with β-CD are lower than the limit for a diffusion-controlled reaction,[12] and the k_+^* values showed a small dependence on the structure of the guest. Compound **9** was shown to bind to the

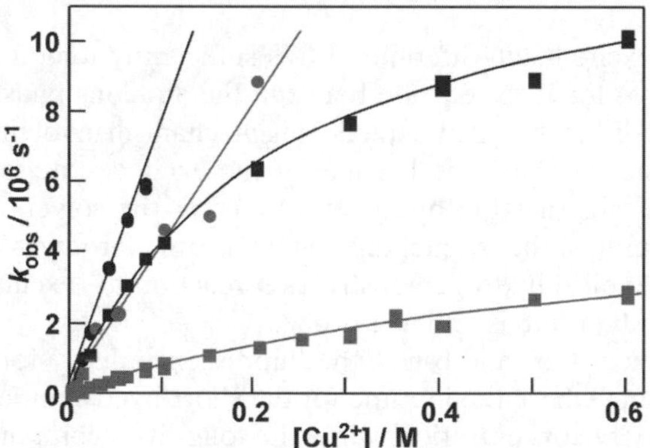

Figure 1.14. Quenching plot for the reaction between triplet excited states of **9** (black) or **10** (gray) with Cu^{2+} in the absence (circles, straight lines) and presence of 10 mM β-CD (squares, curved lines). The data for the quenching in water were fit to Equation 1.8, while the data in the presence of CD were fit to Equation 1.23. Reprinted with permission from Reference 11. Copyright 2006 American Chemical Society.

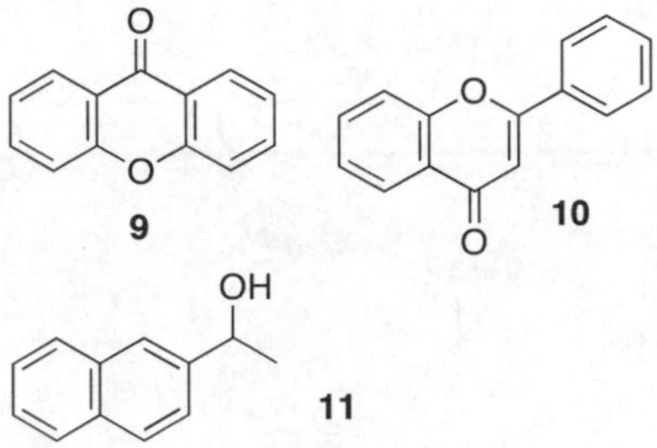

Scheme 1.8. Structures for xanthone (**9**), flavone (**10**), and 2-naphthyl-l-ethanol (**11**).

TABLE 1.4. Values for k_+^*, k_-^*, and k_q^H Determined from the Triplet-State Quenching of 9 and 10 by Cu^{2+} and 11 by Mn^{2+} in the Presence of β-CD[95, 96]

Guest	$k_+^*/10^9\ M^{-1}\ s^{-1}$	$k_-^*/10^6\ s^{-1}$	K_{11}/M^{-1}	K_{11}^{T*}/M^{-1}
9	1.1	12	1100	90
10	2.4	4.4	1090	550
11	0.29	0.18	1800	1600

The ground-state equilibrium constants (K_{11}) were determined from changes in the fluorescence or absorption intensities of the guest in the presence of CD, and the triplet state equilibrium constants (K_{11}^{T*}) were determined from the values of k_+^* and k_-^*.

rim of β-CD,[97] while **11** is deeply included in the CD cavity,[98] leading to a lower association rate constant for the latter guest. The dissociation of guests from CD is much more sensitive to the structure of the guest than observed for the association process. The value for the equilibrium constant of the triplet state of **9** is much lower than for its ground state, showing that the binding behavior for the excited state cannot be extrapolated from that determined for the ground state. The lower value of K_{11}^{T*} for **9** was attributed to the higher basicity[99] and dipole moment for the excited π-π* state.[100] However, this difference in equilibrium constants is not universal since for **11**, which also has a π-π* triplet state, no significant change was observed for the ground- and excited-state equilibrium constants. Guests with the same ground-state equilibrium constants can have very different excited-state dynamics as shown for **9** and **10**. In the case of **10**, the phenyl ring was proposed to act as an anchor for the guest, slowing down the exit from the cavity leading to an overall slower dynamics for **10** when compared with **9**.[95]

Xanthone is, to date, the only guest for which relocation of its triplet state from the CD to the aqueous phase can be measured directly,[101,102] because the triplet–triplet absorption spectrum for xanthone shifts significantly with solvent polarity.[100,103] The lifetime of triplet xanthone in water is sufficiently long compared with the binding dynamics of the excited state, and Equation 1.22 can be employed for the analysis of the kinetics.[102,104,105] The values for k_+^* and k_-^* recovered from the direct method ($6 \times 10^8\ M^{-1}s^{-1}$ and $8.1 \times 10^6 s^{-1}$) are

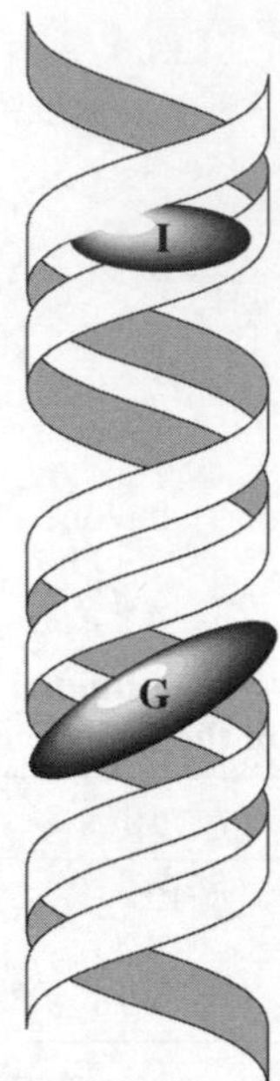

Figure 1.15. Cartoon representation of a DNA double helix with an intercalated guest (I) and groove bound guest (G).

somewhat lower than determined from the quenching experiments. Laser temperature jump experiments, where the kinetics for the ground state was measured, showed that the association rate constant was the same for ground and excited states, while k_-^* increased significantly for the triplet excited state when compared with the ground state.[102]

DNA

Small molecules bind to DNA either by intercalation between base pairs of the double helix or by binding to the minor or major groves of the helix (Fig. 1.15). In addition, nonspecific electrostastic interactions between charged guests and the phosphate backbone can also occur.[106] Good intercalators have planar aromatic backbones containing heteroatoms and positively charged flexible side chains, while guests for groove binding have a curved shape complementary to the DNA helix with repeating aromatic units and cationic end groups.[107] The dynamics of small guests with DNA was studied by a variety of techniques.[17] The examples below highlight the type of mechanisms that were proposed for DNA binding.

Ethidium bromide (**12**) is a widely used staining dye for DNA because it does not fluoresce in water but has a very high fluorescence quantum yield when intercalated between the base pairs of DNA. The binding kinetics of **12**

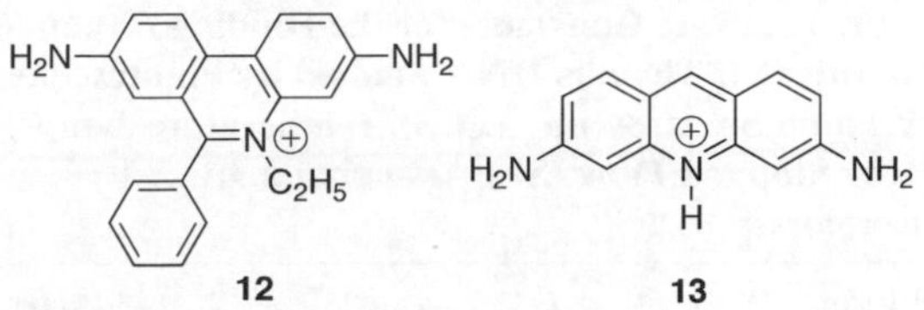

Scheme 1.9. Structures for the ethidium cation (**12**) and for the proflavine cation (**13**).

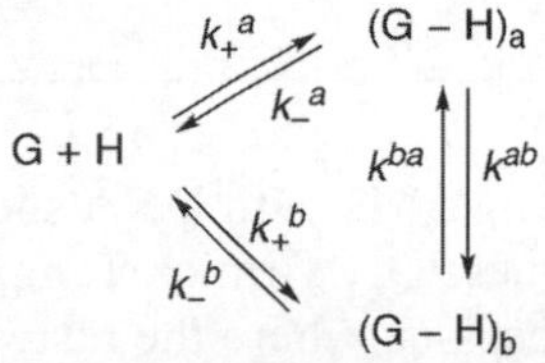

Scheme 1.10. Parallel mechanism for the binding of a guest to two different sites on the host and interconversion of the guest between these two sites. In the case of **12**, the interconversion between sites "*a*" and "*b*" is a bimolecular reaction involving a second DNA molecule.

with DNA was extensively studied using fluorescence correlation spectroscopy,[108] stopped flow,[109] and temperature jump.[109–111] The two mechanisms proposed for the binding of **12** with DNA are the formation of a 1:1 complex (Scheme 1.1) or the binding of **12** to two distinct binding sites in DNA ("*a*" and "*b*") with the possibility of interconversion between these two sites mediated by a second molecule of DNA (Scheme 1.10).

The binding of **12** with calf thymus DNA was studied in the original development of fluorescence correlation spectroscopy. The kinetics was analyzed as a first-order decay, despite the acknowledgement by the authors that binding would be expected to be more complex.[108] The association process (Table 1.5) was shown to be significantly slower than a diffusion controlled process, indicating that the association had a significant activation barrier. This result contrasts with the binding to CDs (see above) and suggests that DNA needs to significantly rearrange before intercalation can occur. The residence time of **12** in DNA (37 ms) is much longer than for guests in CDs, again suggesting that once intercalation occurs, exit of the guest is hindered. Temperature jump experiments were performed parallel to stopped flow experiments in order to compare the data from both techniques.[109] Experimental conditions were employed for which the kinetics followed a mono-exponential decay consistent with a 1:1 complexation model. In addition, the temperature jump experiments were performed at the magic angle for the detection in order to avoid artifacts that overlay the relaxation signal due to the high electric fields employed for the Joule heating of the samples.[28,109] The values for the rate constants obtained from temperature jump and stopped flow measurements were the same and were comparable to those obtained from fluorescence correlation spectroscopy.

TABLE 1.5. Rate Constants for the Binding Dynamics of 12 with Calf Thymus DNA Studied by Fluorescence Correlation Spectroscopy (FCS), Temperature Jump (TJ), or Stopped Flow (SF) Assuming a 1:1 Stoichiometry

Technique	$k_+/10^6\,M^{-1}\,s^{-1}$	$k_-/10^2\,s^{-1}$
FCS[108]	15	0.27
TJ[109]	6.4	0.16
SF[109]	5.4	0.39

The kinetics for the binding of **12** with DNA showed more than one relaxation process at different guest/DNA ratios. Temperature jump experiments showed three relaxation processes, where the fastest one was uncoupled from the other two processes.[110] This fast process is probably related to an artifact because the detection was not performed at the magic angle. The other two relaxation processes were coupled, and the rate constants corresponding to each relaxation process varied linearly with the DNA concentration. The kinetics was consistent with parallel reactions where two different complexes were formed and the interconversion between these two complexes was mediated by a second DNA molecule (Scheme 1.10). The binding dynamics to site "*a*" is faster than to site "*b*," and the values for the association and dissociation rate constants are within one order of magnitude of those measured for conditions where only one binding site was observed (Table 1.6). The interconversion between the two sites mediated by DNA is comparable or faster than the association process of the free guest with DNA, suggesting that the interconversion process is important when a significant amount of guest is bound to DNA compared with the fraction of free guest in the aqueous solution.

Two relaxation processes were also observed using temperature jump for the binding of proflavine (**13**) with calf thymus DNA.[112] The slow relaxation process increased as the concentration of DNA was raised but leveled off at higher DNA concentrations, while a linear relationship was observed for the observed rate constant of the fast process with the DNA concentration. The kinetics is consistent with a sequential mechanism (Scheme 1.5) or with a parallel mechanism (Scheme 1.10). The latter was discarded because it led to an unreasonably high association rate constant for the intercalation process. The fast relaxation process was analyzed using Equation 1.1 ($k_+^a = 1.4 \times 10^7\ M^{-1}\ s^{-1}$, $k_-^a = 1.3 \times 10^4\ s^{-1}$, $K_{11} = 1.1 \times 10^3\,M^{-1}$). The dependence of k_{obs} with the DNA concentration was fit to Equation 1.25 yielding k_+^b and k_-^b values of $6.9 \times 10^3\,s^{-1}$ and $4.2 \times 10^2\,s^{-1}$, respectively. It is important to note that the dynamics for step "*a*" is much faster than the dynamics observed for **12**, consistent with the interpretation that step "*a*" corresponds to the formation of a weak complex followed by the intercalation process. The kinetics for the binding of **13** with calf thymus DNA was studied using laser temperature jump experiments at several ionic strengths (0.05–0.5 M).[113] A mono-exponential decay was observed where the value for k_{obs} leveled off at high DNA concentrations. The values for K_{11}

TABLE 1.6. Rate Constants for the Binding Dynamics of 12 with Calf Thymus DNA Studied by Temperature Jump (TJ) or Stopped Flow (SF) Assuming the Mechanism in Scheme 1.10

Technique	k_+^a / $10^6 M^{-1}s^{-1}$	k_-^a / $10^2 s^{-1}$	k_+^b / $10^6 M^{-1}s^{-1}$	k_-^b / $10^2 s^{-1}$	k^{ab}/ $10^5 M^{-1} s^{-1}$	k^{ba}/ $10^5 M^{-1} s^{-1}$
TJ[110]	—	1.6	0.48	0.35	21	
TJ[a]	1.4	1.7	0.26	0.59	1.3	2.5
SF[109]	7.3	0.39	1.1	0.14	6.0	15

[a]Personal communication by Ryan and Crothers in Reference 111.

($[3.5-7.5]\times 10^3 M^{-1}$), k_+^b ($[3.3-4.3]\times 10^3 s^{-1}$) and k_-^b ($[2.5-3.0]\times 10^2 s^{-1}$) are similar to those obtained in temperature jump experiments:

$$k_{obs} = \frac{K_{11}\, k_+^b[H]}{1+K_{11}[H]} + k_-^b \tag{1.25}$$

Flash photolysis experiments were performed to compare the binding dynamics of ground and triplet excited states of **13** with DNA.[114] This qualitative study showed that the residence time for the triplet state of **13** was shorter than for its ground state. A quantitative laser flash photolysis study for the binding of triplet **13** with poly[d(A-T)] showed that binding followed a sequential mechanism as observed for the ground state.[115] The same value was observed for the pre-equilibrium step for the ground and excited states, but the intercalation step was favored for the ground state when compared with the excited state.

Micelles

Micelles are self-assembled structures formed from surfactants that create a pseudophase within aqueous solutions into which hydrophobic guests can be solubilized. These surfactants usually have a hydrophobic chain with a charged or polar head group (Fig. 1.16). Micelles are formed when the surfactant concentration surpasses its critical micellar concentration (CMC). The exchange of individual monomers between the micelle and the aqueous phase occurs in microseconds, while micelles as a whole have lifetimes of milliseconds.[116] Therefore, on the timescale of photophysical experiments, a micelle can be viewed as an integral host to which guest molecules are bound.

Quenching studies for guests solely solubilized in the micelle provide information on the accessibility of the quencher to the micelle interior, provided quenching inside the micelle is faster than the exit of the quencher. In this case, the fluorescence decay follows a mono-exponential function and the quenching rate constant is obtained from Equation 1.8. Micelles are formed from charged (negatively or positively) or nonionic surfactants. In the case of charged micelles and charged quenchers, electrostatic attraction or repulsion

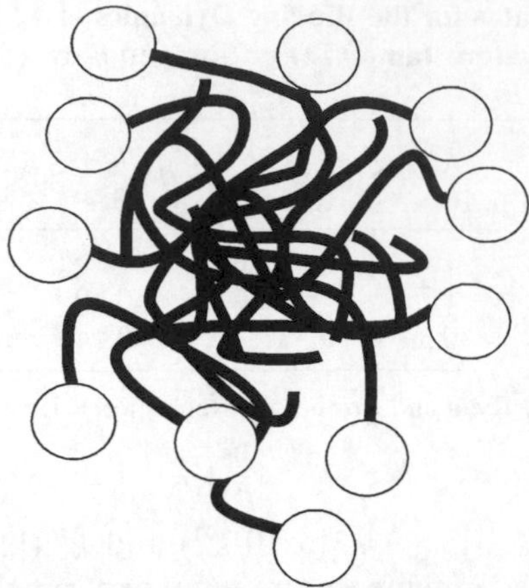

Figure 1.16. Cross section of a micelle where the charged (polar) groups (white) are located at the interface with water and the hydrophobic tails are located in the interior of the micelle.

plays a dominant role in determining the quencher's accessibility to the micelle. The quenching of the singlet excited state of pyrene (**6**, Scheme 1.6) in the negatively charged micelles of sodium dodecyl sulfate (SDS) or in the positively charged micelle of cetyltrimethylammonium bromide (CTAB) is only modestly diminished for oxygen (factors of 2.2–2.4) or nitromethane (factors of 1.2–2.0),[117] showing that these neutral quenchers diffuse readily through the micelles. On the other hand, when iodide anions were used as quenchers for **6** in SDS micelles, the quenching rate constant decreased by ca. 1200,[117] because of the electrostatic repulsion between the micelle and the quencher.

Equation 1.10 is employed when quenching inside the micelle is competitive with the exit of the quencher into the aqueous phase. This analysis has been extensively used to study the binding dynamics of quenchers with micelles.[10,62] Studies with different quenchers showed that the association process is controlled by diffusion when electrostatic repulsion or attraction is not present. In these cases, the partition of the quencher between the micelles and the aqueous phase is determined by the dissociation rate constant of the quencher from the micelle. The more hydrophobic the quencher, the lower the dissociation rate constants and the longer the residence time in the micelle as exemplified in Table 1.7 for the quenching of the singlet excited state of **6** in SDS micelles by alkyl iodides.[118]

At high concentrations of micelles, parameter D (Eq. 1.14) was shown to depend on the micelle concentration.[119,120] Such a dependence is inconsistent with the mechanism used to derive Equation 1.10. One proposal for the observed dependence was that at high micelle concentrations, the movement of quencher between micelles can occur due to the migration between two micelles without exit of the quencher into the aqueous phase.[120] Subsequent studies showed that the dissociation rate constant for the quencher depends

TABLE 1.7. Association (k_+^Q), Dissociation (k_-^Q), and Intramicellar Unimolecular Quenching Rate Constants (k_q^{int}) for the Quenching of the Singlet Excited State of 6 by Alkyl Iodides in SDS Micelles (0.04 M)[118]

Quencher	$k_+^Q / 10^9\ M^{-1}\ s^{-1}$	$k_-^Q / 10^6\ s^{-1}$	$k_q^{int} / 10^9\ s^{-1}$
Ethyl iodide	9.7	8.3	2.9
Butyl iodide	8.8	1.4	4.8
Hexyl iodide	7.8	0.75	5.4
Octyl iodide	6.6	0.4	5.9

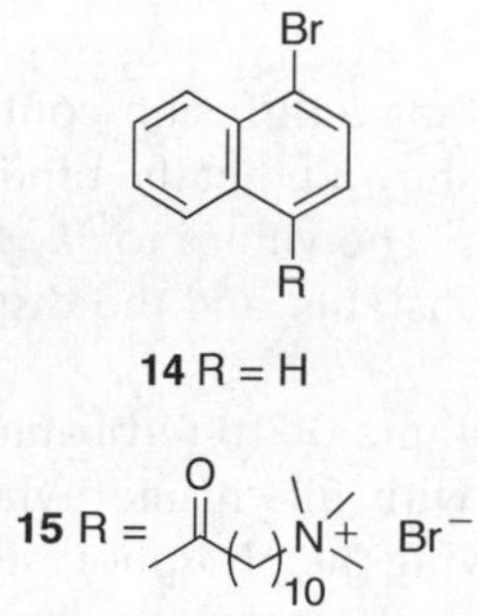

14 R = H

15 R =

Scheme 1.11. Structures for 1-bromonaphthalene (**14**) and 10-(4-bromo-1-naphthoyl) decyltrimethylammonium bromide (**15**).

strongly on the salt concentration in solution and only moderately on the micelle concentration. This result led the authors to propose that the dissociation rate constant of ionic quenchers from ionic micelles depends on the micellar surface potential and that micellar exchange is not important. This conclusion was reached for studies with SDS and cetyltrimethylammonium chloride (CTAC) micelles.[121,122] This example shows that complementary experiments are required to validate the models employed for the fitting of fluorescence decays to complex models such as those underlying Equation 1.10.

The migration of guests between micelles was studied using a quencher that was immobile in CTAC micelles.[123] The fluorescence decays were fit to competing mechanistic models, and global analysis was employed for the analysis of the kinetics. No migration was observed for the more hydrophobic 1-methylpyrene, while migration was observed for pyrene sulfonate.

The mobility of guests was studied for several triplet excited states using the quenching methodology (Eq. 1.23), which was originally developed to study the binding dynamics of polyaromatic hydrocarbons with SDS micelles.[69] In the case of 1-bromonaphthalene (**14**, Scheme 1.11), the value for the dissociation rate constant of the triplet state (k_-^*) was determined directly from the analysis of the quenching plot ($2.5 \times 10^4\ s^{-1}$). A very similar value was recovered when k_-^* ($3.3 \times 10^4\ s^{-1}$) was calculated from the equilibrium constant

TABLE 1.8. Dissociation Rate Constants for the Triplet State of 15 from SC_nS Micelles Obtained from the Quenching Plot Where $Fe(CN)_3$ Was Used as Quencher[124]

	$k_-^*/10^3\ s^{-1}$	
SC_nS	20°C	30°C
$SC_{10}S$	5.7	9.8
$SC_{12}S$ (SDS)	1.4	2.5
$SC_{14}S$	0.7	0.8

of the ground state and assuming a diffusion-controlled value for the association rate constant. This result showed that the binding dynamics for the ground and excited states was similar. The values for k_-^* for other guests were calculated from the equilibrium constants, and the dissociation was slower for the more hydrophobic guests.

The dissociation rate constants of 10-(4-bromo-1-naphthoyl)decyltrimethylammonium bromide (**15**) with alkyltrimethylammonium chloride (SC_nS) micelles was studied by following the phosphorescence decay of triplet **15** with the addition of ferric cyanide as quencher and by fitting the quenching plot to Equation 1.23.[124] The values for k_-^* decreased as the length of the alkyl chains increased because the volume of the micelles increased. The same trend was observed at both temperatures studied (Table 1.8).

The binding dynamics of several triplet excited-state ketones with SDS micelles was investigated by using a quencher that either was immobile inside the micelle (γ-methylvalerophenone) or resided exclusively in the aqueous phase (nitrite anions).[125,126] The k_-^* values for the triplet state of acetophenone ($7.7 \times 10^6\ s^{-1}$), propiophenone ($3.0 \times 10^6\ s^{-1}$), and isobutyrophenone ($1.6 \times 10^6\ s^{-1}$) were determined when the quencher was inside the micelle. The values for k_+^* for the triplet state of acetophenone ($1.6 \times 10^{10}\ M^{-1}\ s^{-1}$), propiophenone ($1.4 \times 10^{10}\ M^{-1}\ s^{-1}$), and isobutyrophenone ($1.2 \times 10^{10}\ M^{-1}\ s^{-1}$) were determined from the quenching studies with nitrite anions. The studies on the dynamics of the ketones parallel the ones with polyaromatic hydrocarbons as guests, where the association rate constants were diffusion controlled while the dissociation rate constants decreased when the hydrophobicity of the guest increased. However, it is important to note that the k_-^* values for the ketones are significantly higher than observed for the polyaromatic hydrocarbons, in line with the higher hydrophilicity of the ketones when compared with polyaromatic hydrocarbons.

The binding dynamics of the Rhodamine 123 cation (**16**) (Scheme 1.12) to micelles of Triton X-100 and Brij-35, two nonionic surfactants, was studied using fluorescence correlation spectroscopy.[38,127] A cationic guest was chosen to ensure its partition between the micelle and water, while nonionic surfactants were used to eliminate the role of electrostatic interactions on the guest

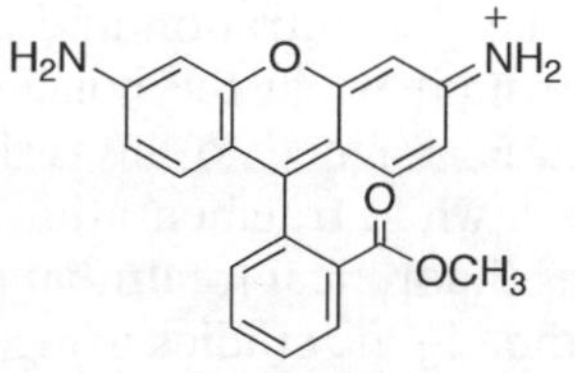

Scheme 1.12. Structure for the Rhodamine 123 cation (**16**).

binding dynamics. In the case of **16**, a significant amount of triplet states is formed and the correlation time for the decay of the triplet appears in the correlation curve. Two different sets of experiments were performed to identify the correlation time for the triplet state and differentiate it from the correlation time for the host–guest dynamics. The excitation power was varied because an increase in power leads to a higher population of triplets. When the micelle concentration was varied, the correlation time for the dynamics of the host–guest complex was mainly affected; that is, at higher micelle concentrations, the correlation time for the host–guest dynamics became shorter. The correlation time for the triplet state was shown to be at longer times than for the host–guest complex, and the amplitude for the term corresponding to the triplet state was smaller than for the host–guest dynamics. The values for k_- were determined from the fluorescence correlation experiment ($2.2 \times 10^5\,s^{-1}$ for Triton X-100 and $4 \times 10^5\,s^{-1}$ for Brij-35), while the values for k_+ ($1.4 \times 10^{10}\,M^{-1}s^{-1}$ for Triton X-100 and $8 \times 10^9\,M^{-1}s^{-1}$ for Brij-35) were calculated from the equilibrium constants and the k_- values. The value for the association rate constant of **16** with Triton X-100 was the same as calculated for a diffusional process, while the k_+ value for Brij-35 was slightly lower.

The redistribution of guests through micelle fusion or fragmentation was studied using the pyrene derivative **17** (Scheme 1.13) because this guest does not exit from the micelle over a period of several seconds.[128,129] Micelle fusion is the process where two micelles collide, and the content, that is, the guests, are redistributed between the two micelles followed by the separation of the fused complex. This process is bimolecular because it requires the collision of two micelles. The fragmentation mechanism is related to the formation of two smaller micelles where these micelles then grow to the average micellar size. This reaction is unimolecular because the rate-limiting step is the fragmentation of one micelle. The concentration of **17** used was sufficiently high that a significant number of micelles contained two guest molecules, which led to the observation of the excimer emission of pyrene. Stopped flow experiments were performed in which a solution containing Triton X-100 micelles and **17** was mixed with a solution containing only the micelles. The redistribution of **17** was observed as a decrease in the excimer emission or an increase in the monomer emission because more micelles contained only one guest. Two relaxation processes were observed, one of which showed a dependence of the relaxation time on the micelle concentration and a slower process for which the relaxation time was constant when the concentration of surfactant was changed. The first process was assigned to the fusion of micelles with a rate constant of ca. $1 \times 10^6 M^{-1} s^{-1}$, while fragmentation occurred with a rate constant of ca. $12 s^{-1}$. A rigorous theoretical treatment suggested that the qualitative conclusions based on the kinetic studies using **17**[128,129] were correct but a stochastic model of solubilization should be taken into account when determining the values for the rate constants.[130]

Bile Salt Aggregates

Bile salts (Scheme 1.14) are molecules that have planar amphiphilicity, where the concave face containing hydroxyl groups is more hydrophilic than the

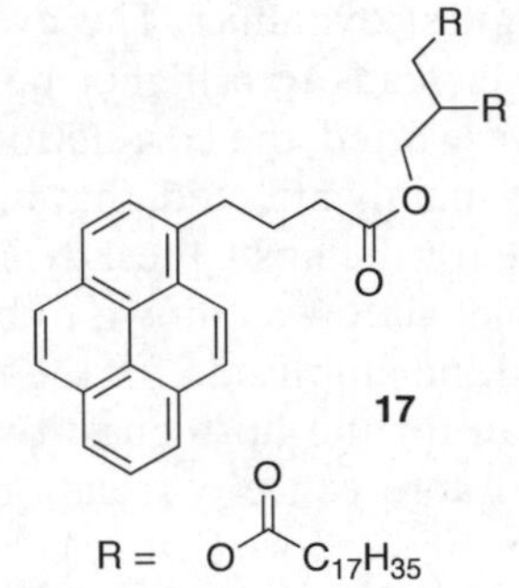

17

R = $O-C(=O)-C_{17}H_{35}$

Scheme 1.13. Structure for a pyrene-substituted triglyceride (**17**).

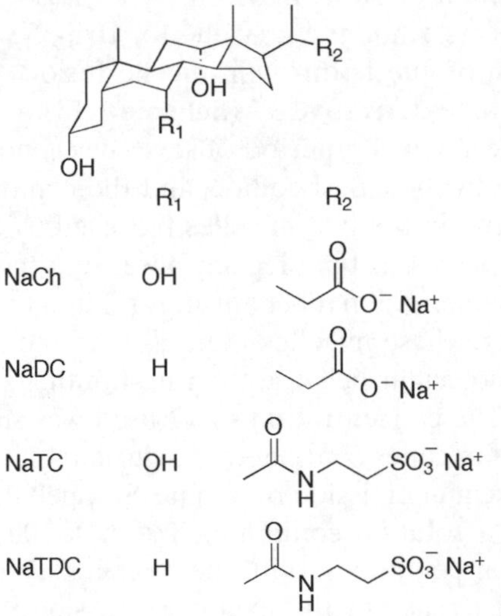

	R_1	R_2
NaCh	OH	$CH_2CH_2CO_2^- Na^+$
NaDC	H	$CH_2CH_2CO_2^- Na^+$
NaTC	OH	$C(=O)NHCH_2CH_2SO_3^- Na^+$
NaTDC	H	$C(=O)NHCH_2CH_2SO_3^- Na^+$

Scheme 1.14. Structures for sodium cholate (NaCh), sodium deoxycholate (NaDC), sodium taurocholate (NaTC), and sodium deoxytaurocholate (NaTDC).

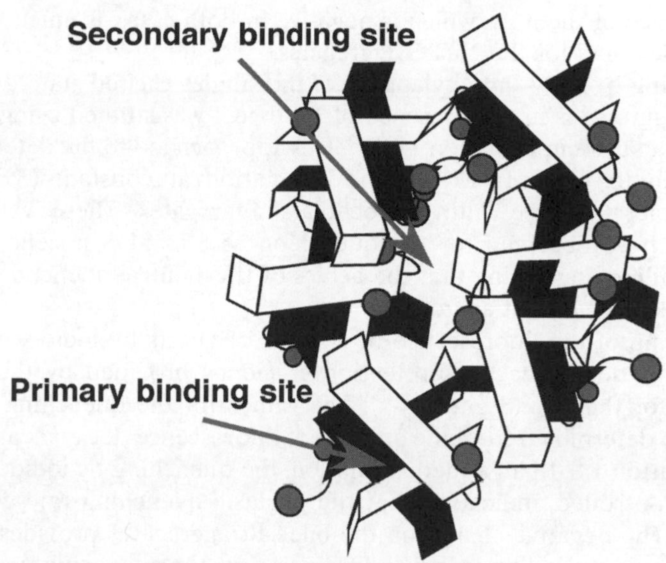

Figure 1.17. Cartoon representation for the aggregation of bile salts where the black areas correspond to the hydrophobic face of the monomer, the white areas correspond to the face of the monomer containing the hydroxyl groups, and the gray circles are the charged head groups.

convex face containing the methyl substituents. This structural framework is very different from surfactants, such as those discussed in the previous section, where the hydrophilic group is at one end of the molecule and the rest of the molecule is hydrophobic. Bile salts continuously aggregate as the monomer concentration is increased,[131–133] and the aggregate does not have a defined size as is the case for conventional micelles. The most widely adopted model for the structure of the aggregates is the formation of small primary aggregates at low monomer concentration, which then aggregate into larger structures at higher concentrations called secondary aggregates (Fig. 1.17).[134,135] Alternate models proposed that hydrogen bonds are required for the formation of primary aggregates,[136,137] or that the aggregation is stepwise.[138]

The quenching of the singlet excited state of pyrene (**6**) by oxygen or nitromethane was compared with the quenching for **6** in SDS micelles.[117] The aggregates of sodium taurocholate (NaTC) provide more protection than micelles of SDS. The quenching rate constant decreased by a factor of 2 for the quenching by oxygen, while it decreased by 25 times for nitromethane. This result shows that oxygen can have a reasonably good access to the aggregates of NaTC, while the access is more restricted for the larger nitromethane. However, the quenching rate constant for iodide anions is 14 times larger for the NaTC aggregates than for SDS micelles, showing that the charge density on the surface of the host, which is negative in both cases, is much higher for SDS micelles than for the NaTC aggregates.

Quenching by *N,N*-dimethylaniline of the singlet excited state of **6** bound to NaTC aggregates in the presence of 1 M NaCl was studied using the four-parameter analysis in Equation 1.10.[139] This approach led to the determination of the association ($1.4 \times 10^9\,M^{-1}\,s^{-1}$) and dissociation rate constants ($3.8 \times 10^6\,s^{-1}$) of *N,N*-dimethylaniline with taurocholate aggregates. These values were similar to those observed for the quenching of **6** in SDS micelles by *N,N*-dimethylaniline, suggesting that the access of this neutral quencher is similar for micelles and bile salt aggregates.

Quenching of the fluorescence of a series of guests by iodide anions was employed to determine the protection efficiency provided by the bile salt aggregate for the bound guest.[140–143] The values for the quenching rate constants were determined from the analysis of fluorescence decay measurements using Equation 1.8. In an aqueous solution, the quenching by iodide anions is diffusion controlled, indicating that the intrinsic quenching rate constant is very high. The negative charge on the bile salt aggregates provides a barrier for the entry of the iodide anion, and the value for the quenching rate constant (k_q^H) was equated to k_+^Q since the intrinsic quenching rate constant is not expected to change with a change of the nature of the environment around the guest.[142,143] Polyaromatic hydrocarbons, such as the naphthalenes **18–20** (Scheme 1.15), are bound to the primary aggregates of the bile salts, while introduction of a hydroxyl group (**21**) led to the binding of the guest to the secondary aggregates. Iodide anions quench the fluorescence of **18–21**, and the quenching rate constants in the presence of different bile salts (Table 1.9) showed that the association of iodide anions to primary aggregates is much slower than observed for the secondary aggregates, since for all bile salts, the quenching rate constants for **18–20** are much lower than for **21**. The quenching rate constants for **18–20** are significantly different, suggesting that the structure of the primary aggregates of sodium cholate (NaCh) change with the concentration of the guest, most likely because of the small number of monomers

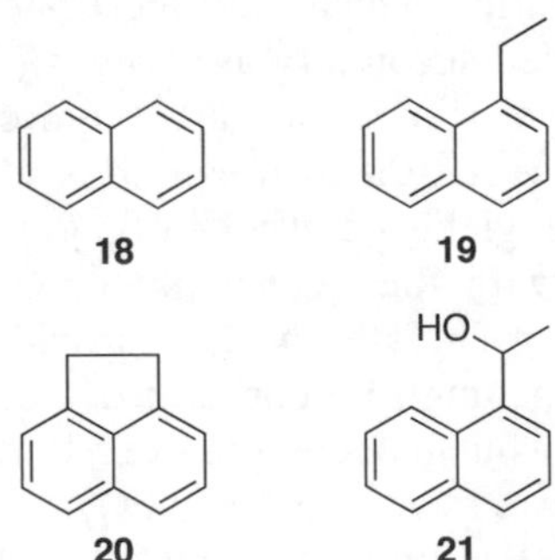

Scheme 1.15. Structures for naphthalene (**18**), 1-ethylnaphthalene (**19**), acenaphthene (**20**), and 1-naphthyl-1-ethanol (**21**).

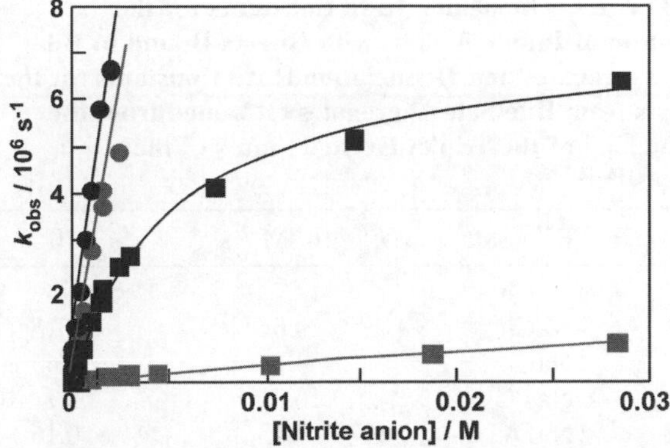

Figure 1.18. Quenching plot for the reaction of nitrite anions with the triplet excited states of **19** (gray symbols) and **21** (black symbols) in water (circles) and in the presence of 40 mM NaCh (squares). Adapted with permission from Reference 11. Copyright 2006 American Chemical Society.

involved in the formation of the primary aggregates. Bile salts with a smaller number of hydroxyl groups (sodium deoxycholate [NaDC] and sodium deoxytaurocholate [NaTDC]) form more compact primary aggregates leading to a decrease of the access for the iodide anions, while the number of hydroxyl groups has a much smaller effect on the structure of the secondary aggregates, since the quenching rate constants did not vary significantly for **21**.

Studies on the kinetics of the triplet excited state of guests led to information on their dissociation from bile salt aggregates. Estimates for the k_-^* values for guests in NaTC aggregates were obtained for Rose Bengal ($>1 \times 10^5\,s^{-1}$)[144] in flash photolysis experiments and from the analysis of the triplet–triplet annihilation process of anthracene ($2.8 \times 10^5\,s^{-1}$).[117] The quenching methodology was employed to measure the binding dynamics of a series of guests,[141–143,145] where the triplet states of the guests were quenched by nitrite anions and the values for k_-^* and k_q^H were obtained from the analysis of the data using Equation 1.23. Qualitatively, the dynamics is slower the farther away the curved quenching plot in the presence of bile salt aggregates is from the linear quenching plot in water. The dynamics for **19** is markedly slower than for **21** (Fig. 1.18), and this result in conjunction with the singlet excited-state quenching experiments led to the proposal that **19** was bound to primary aggregates, while **21** was bound to secondary aggregates.[141] In addition, in the case of **21**, no curvature was observe for the quenching of its triplet excited state in the presence of 10 mM NaCh, a bile salt concentration at which primary aggregates are formed but no secondary aggregates are present. The dissociation rate constant for **19** was 30 times lower than for **21**, showing that the residence time for the guests in the primary aggregates is much longer than in the secondary aggregates (Table 1.9).

TABLE 1.9. Quenching Rate Constants for the Reaction of Iodide Anions with Guests Bound to Bile Salt Aggregates and Dissociation Rate Constants for the Guests from Bile Salt Aggregates Obtained from the Quenching of the Triplet Excited States of the Guests[142,143]

Guest	Bile Salt	k_q^H /10^7 M^{-1} s^{-1}	k_-^* /10^6 s^{-1}
18	NaCh	17.7	1.4
19	NaCh	10.6	0.18
20	NaCh	3.9	0.4
	NaTC	10.8	1.7
	NaDC	0.43	0.16
	NaTDC	1.22	0.4
21	NaCh	70	5.5
	NaTC	80	8
	NaDC	34	4
	NaTDC	25	1.9

CONCLUSION

This chapter provides the conceptual framework for using photophysical methods to measure the dynamics of surpramolecular systems. The principles for the techniques and the analysis methodologies used in these kinetic studies were described and their advantages and disadvantages discussed. Examples were provided for the guest binding dynamics with CDs, DNA, micelles, and bile salt aggregates.

ACKNOWLEDGMENTS

I would like to thank the continuous support to my research program from the Natural Sciences and Engineering Research Council of Canada (NSERC) in the form of operating and equipment grants. I would also like to thank my coworkers and collaborators for their contributions in the development of the various research projects in supramolecular dynamics.

REFERENCES

1. Lehn, J.M. *Supramolecular Chemistry: Concepts and Perspectives*. Weinheim: VCH, **1995**.
2. Lehn, J.M. *Chem. Soc. Rev.* **2007**, *36*, 151–160.
3. Atwood, J.L.; Davies, J.E.D.; MacNicol, D.D.; Vögtle, F.; Lehn, J.M. *Comprehensive Supramolecular Chemistry*, Vol. 1–11. New York: Pergamon, **1996**.

4. Badjic, J.D.; Nelson, A.; Cantrill, S.J.; Turnbull, W.B.; Stoddart, J.F. *Acc. Chem. Res.* **2005**, *38*, 723–732.
5. Holliday, B.J.; Mirkin, C.A. *Angew. Chem. Int. Ed. Engl.* **2001**, *40*, 2022–2043.
6. Sauvage, J.-P. *Acc. Chem. Res.* **1998**, *31*, 611–619.
7. Seidel, S.R.; Stang, P.J. *Acc. Chem. Res.* **2002**, *35*, 972–983.
8. Whitesides, G.M.; Mathias, J.P.; Seto, C.T. *Science* **1991**, *254*, 1312–1319.
9. Whitesides, G.M.; Simanek, E.E.; Mathias, J.P.; Seto, C.T.; Chin, D.; Hammen, M.; Gordon, D.M. *Acc. Chem. Res.* **1995**, *28*, 37–44.
10. Kleinman, M.H.; Bohne, C. In: Ramamurthy, V.; Schanze, K.S., editors. *Molecular and Supramolecular Photochemistry*, Vol. 1. New York: Marcel Dekker, **1997**, pp. 391–466.
11. Bohne, C. *Langmuir* **2006**, *22*, 9100–9111.
12. Montalti, M.; Credi, A.; Prodi, L.; Gandolfi, M.T. *Handbook of Photochemistry*, 3rd edition. Boca Raton, FL: CRC Press, **2006**.
13. Vajda, S.; Jimenez, R.; Rosenthal, S.J.; Fidler, V.; Fleming, G.R.; Castner, E.W. Jr. *J. Chem. Soc. Faraday Trans.* **1995**, *91*, 867–873.
14. Nandi, N.; Bagchi, B. *J. Phys. Chem.* **1996**, *100*, 13914–13919.
15. Bhattacharyya, K. *Acc. Chem. Res.* **2003**, *36*, 95–101.
16. Douhal, A. *Chem. Rev.* **2004**, *104*, 1955–1976.
17. Pace, T.C.S.; Bohne, C. *Adv. Phys. Org. Chem.* **2008**, *42*, 167–223.
18. Bohne, C.; Redmond, R.W.; Scaiano, J.C. In: Ramamurthy, V., editor. *Photochemistry in Organized and Constrained Media*. New York: VCH Publishers, **1991**, pp. 79–132.
19. Eaton, D.F. *Pure Appl. Chem.* **1990**, *62*, 1631–1648.
20. Holden, D.A. In: Scaiano, J.C., editor. *Handbook of Organic Photochemistry*, Vol. I. Boca Raton, FL: CRC Press, **1989**, pp. 261–277.
21. O'Connor, D.V.; Phillips, D. *Time-Correlated Single Photon Counting*. Orlando, FL: Academic Press, **1984**.
22. Birch, D.J.S.; Imhof, R.E. In: Lakowisz, J., editor. *Topics in Fluorescence Spectroscopy*, Vol. 1. New York: Plenum Press, **1991**, pp. 1–95.
23. Lakowicz, J.R. *Principles of Fluorescence Spectroscopy*. New York: Plenum Press, **1983**.
24. Cosa, G.; Scaiano, J.C. *Photochem. Photobiol.* **2004**, *80*, 159–174.
25. Scaiano, J.C. In: Moss, R.A.; Platz, M.S.; Jones, M. Jr., editors. *Reactive Intermediate Chemistry*. New York: John Wiley & Sons, **2004**, pp. 847–871.
26. Mitchell, R.H.; Bohne, C.; Wang, Y.; Bandyopadhyay, S.; Wozniak, C.B. *J. Org. Chem.* **2006**, *71*, 327–336.
27. Crooks, J.E. *J. Phys. E* **1983**, *16*, 1142–1147.
28. Porschke, D. *Ber. Bunsenges. Phys. Chem.* **1996**, *100*, 715–720.
29. Holzwarth, J.F.; Schmidt, A.; Wolff, H.; Volk, R. *J. Phys. Chem.* **1977**, *81*, 2300–2301.
30. Callender, R.; Dyer, R.B. *Curr. Opin. Struct. Biol.* **2002**, *12*, 628–633.
31. Gruebele, M.; Sabelko, J.; Ballew, M.; Ervin, J. *Acc. Chem. Res.* **1998**, *31*, 699–707.
32. Ma, H.; Wan, C.; Zewail, A.H. *J. Am. Chem. Soc.* **2006**, *128*, 6338–6340.

33. Williams, A.P.; Longfellow, C.E.; Freier, S.M.; Kierzek, R.; Turner, D.H. *Biochemistry* **1989**, *28*, 4283–4291.
34. Berger, R.L.; Balko, B.; Borcherdt, W.; Friauf, W. *Rev. Sci. Instrum.* **1968**, *39*, 486–493.
35. Robinson, B.H. In: Wyn-Jones, E., editor. *Chemical and Biological Applications of Relaxation Spectrometry*. Boston: D. Reidel Publishing Co., **1975**, pp. 41–48.
36. Thurn, T.; Bloor, D.M.; Wyn-Jones, E. In: Abe, M.; Scamehorn, J.F., editors. *Mixed Surfactant Systems*, 2nd edition. New York: CRC Press, **2005**, pp. 709–768.
37. Verrall, R.E. *Chem. Soc. Rev.* **1995**, *24*, 135–142.
38. Al-Soufi, W.; Reija, B.; Felekyan, S.; Seidel, C.A.M.; Novo, M. *Chemphyschem* **2008**, *9*, 1819–1827.
39. Thompson, N.L. In: Lakowicz, J.R., editor. *Topics in Fluorescence Spectroscopy Volume 1 Techniques*. New York: Plenum Press, **1991**, pp. 337–378.
40. Webb, W.W. In: Rigler, R.; Elson, E.S., editors. *Fluorescence Correlation Spectroscopy*. Berlin: Springer, **2001**, pp. 305–330.
41. Widengren, J. In: Rigler, R.; Elson, E.S., editors. *Fluorescence Correlation Spectroscopy*. Berlin: Springer, **2001**, pp. 277–301.
42. Freeman, R. *A Handbook of Nuclear Magnetic Resonance*. New York: John Wiley & Sons, **1988**.
43. Perrin, C.L.; Dwyer, T.J. *Chem. Rev.* **1990**, *90*, 935–967.
44. Strehlow, H. *Rapid Reactions in Solution*. Weinheim: VCH, **1992**.
45. Englebienne, P.; Van Hoonacker, A.; Verhas, M. *Spectroscopy* **2003**, *17*, 255–273.
46. Green, R.J.; Frazier, R.A.; Shakesheff, K.M.; Davies, M.C.; Roberts, C.J.; Tendler, S.J.B. *Biomaterials* **2000**, *21*, 1823–1835.
47. Schuck, P. *Annu. Rev. Biophys. Biomol. Struct.* **1997**, *26*, 541–566.
48. Connors, K.A. *Binding Constants—The Measurement of Molecular Complex Stability*. New York: John Wiley & Sons, **1987**.
49. Bernasconi, C.F. *Relaxation Kinetics*. New York: Academic Press, Inc., **1976**.
50. Czerlinski, G.H. *Chemical Relaxation: An Introduction to Theory and Application of Stepwise Perturbation*. New York: Marcel Dekker, **1966**.
51. Enderlein, J.; Gregor, I.; Patra, D.; Fitter, J. *Curr. Pharm. Biotechnol.* **2004**, *5*, 155–161.
52. Al-Soufi, W.; Reija, B.; Novo, M.; Felekyan, S.; Kühnemuth, R.; Seidel, C.A.M. *J. Am. Chem. Soc.* **2005**, *127*, 8775–8784.
53. Krichevsky, O.; Bonnet, G. *Rep. Prog. Phys.* **2002**, *65*, 251–297.
54. Felekyan, S.; Kühnemuth, R.; Kudryavtsev, V.; Sandhagen, C.; Becker, W.; Seidel, C.A.M. *Rev. Sci. Instrum.* **2005**, *76*, 083104-1–083104-14.
55. Quina, F.H.; Lissi, E.A. *Acc. Chem Res.* **2004**, *37*, 703–710.
56. Yekta, A.; Aikawa, M.; Turro, N.J. *Chem. Phys. Lett.* **1979**, *63*, 543–548.
57. Tachiya, M. *Chem. Phys. Lett.* **1975**, *33*, 289–292.
58. Infelta, P.P.; Grätzel, M.; Thomas, J.K. *J. Phys. Chem.* **1974**, *78*, 190–195.
59. Atik, S.S.; Nam, M.; Singer, L.A. *Chem. Phys. Lett.* **1979**, *67*, 75–80.
60. Infelta, P.P.; Grätzel, M. *J. Chem. Phys.* **1979**, *70*, 179–186.

61. Van der Auweraer, M.; Dederen, C.; Palmans-Windels, C.; De Schryver, F.C. *J. Am. Chem. Soc.* **2002**, *104*, 1800–1804.

62. Gehlen, M.H.; De Schryver, F.C. *Chem. Rev.* **1993**, *93*, 199–221.

63. Reekmans, S.; De Schryver, F.C. In: Schneider, H.-J.; Dürr, H., editors. *Frontiers in Supramolecular Organic Chemistry and Photochemistry*. Weinheim: VCH Verlagsgesellschaft, **1992**, pp. 287–310.

64. Turro, N.J.; Yekta, A. *J. Am. Chem. Soc.* **1978**, *100*, 5951–5952.

65. Infelta, P.P. *Chem. Phys. Lett.* **1979**, *61*, 88–91.

66. Boens, N.; Van der Auweraer, M. *Chem. Phys. Chem.* **2005**, *6*, 2352–2358.

67. Birks, J.B. *Photophysics of Aromatic Molecules*. London: Wiley-Interscience, **1970**.

68. Cheung, S.T.; Ware, W.R. *J. Phys. Chem.* **1983**, *87*, 466–473.

69. Almgren, M.; Grieser, F.; Thomas, J.K. *J. Am. Chem. Soc.* **1979**, *101*, 279–291.

70. Turro, N.J.; Okubo, T.; Chung, C.-J. *J. Am. Chem. Soc.* **1982**, *104*, 1789–1794.

71. Szejtli, J.; Osa, T. In: Atwood, J.L.; Davies, J.E.; MacNicol, D.D.; Vögtle, F.; Lehn, J.-M., editors. *Comprehensive Supramolecular Chemistry*, Vol. 3. New York: Elsevier Science Ltd., **1996**.

72. Szejtli, J. *Chem. Rev.* **1998**, *98*, 1743–1753.

73. Connors, K.A. *Chem. Rev.* **1997**, *97*, 1325–1357.

74. Rekharsky, M.V.; Inoue, Y. *Chem. Rev.* **1998**, *98*, 1875–1917.

75. Cramer, F.; Saenger, W.; Spatz, H.-C. *J. Am. Chem. Soc.* **1967**, *89*, 14–20.

76. Clarke, R.J.; Coates, J.H.; Lincoln, S.F. *Carbohydr. Res.* **1984**, *127*, 181–191.

77. Schiller, R.L.; Lincoln, S.F.; Coates, J.H. *J. Chem. Soc. Faraday Trans. 1* **1987**, *83*, 3237–3248.

78. Schiller, R.L.; Lincoln, S.F.; Coates, J.H. *J. Chem. Soc. Faraday Trans. 1* **1986**, *82*, 2123–2132.

79. Bordello, J.; Reija, B.; Al-Soufi, W.; Novo, M. *Chemphyschem* **2009**, *10*, 931–939.

80. Valdes-Aguilera, O.; Neckers, D.C. *Acc. Chem. Res.* **1989**, *22*, 171–177.

81. Mochida, K.; Matsui, Y. *Chem. Lett.* **1976**, 963–966.

82. Hersey, A.; Robinson, B.H.; Kelly, H.C. *J. Chem. Soc. Faraday Trans. 1* **1986**, *82*, 1271–1287.

83. Demont, P.M.; Reinsborough, V.C. *Aust. J. Chem.* **1991**, *44*, 759–763.

84. Seiyama, A.; Yoshida, N.; Fujimoto, M. *Chem. Lett.* **1985**, 1013–1016.

85. Yoshida, N.; Seiyama, A.; Fujimoto, M. *J. Phys. Chem.* **1990**, *94*, 4246–4253.

86. Dyck, A.S.M.; Kisiel, U.; Bohne, C. *J. Phys. Chem. B* **2003**, *107*, 11652–11659.

87. Hamai, S. *J. Phys. Chem.* **1989**, *93*, 6527–6529.

88. Wright, P.J.; Bohne, C. *Can. J. Chem.* **2005**, *83*, 1440–1447.

89. Knutson, J.R.; Beecham, J.M.; Brand, L. *Chem. Phys. Lett.* **1983**, *102*, 501–507.

90. Beecham, J.M.; Ameloot, M.; Brand, L. *Chem. Phys. Lett.* **1985**, *120*, 466–472.

91. Hashimoto, S.; Thomas, J.K. *J. Am. Chem. Soc.* **1985**, *107*, 4655–4662.

92. van Stam, J.; De Feyter, S.; De Schryver, F.C.; Evans, C.H. *J. Phys. Chem.* **1996**, *100*, 19959–19966.

93. Nau, W.M.; Greiner, G.; Rau, H.; Wall, J.; Olivucci, M.; Scaiano, J.C. *J. Phys. Chem.* **1999**, *103A*, 1579–1584.

94. Nau, W.M.; Zhang, X. *J. Am. Chem. Soc.* **1999**, *121*, 8022–8032.
95. Christoff, M.; Okano, L.T.; Bohne, C. *J. Photochem. Photobiol. A* **2000**, *134*, 169–176.
96. Barros, T.C.; Stefaniak, K.; Holzwarth, J.F.; Bohne, C. *J. Phys. Chem. A* **1998**, *102*, 5639–5651.
97. Murphy, R.S.; Barros, T.C.; Barnes, J.; Mayer, B.; Marconi, G.; Bohne, C. *J. Phys. Chem. A* **1999**, *103*, 137–146.
98. Murphy, R.S.; Barros, T.C.; Mayer, B.; Marconi, G.; Bohne, C. *Langmuir* **2000**, *16*, 8780–8788.
99. Ireland, J.F.; Wyatt, P.A.H. *J. Chem. Soc. Faraday Trans. 1.* **1972**, *68*, 1053–1058.
100. Scaiano, J.C. *J. Am. Chem. Soc.* **1980**, *102*, 7747–7753.
101. Barra, M.; Bohne, C.; Scaiano, J.C. *J. Am. Chem. Soc.* **1990**, *112*, 8075–8079.
102. Liao, Y.; Frank, J.; Holzwarth, J.F.; Bohne, C. *J. Chem. Soc. Chem. Commun.* **1995**, 199–200.
103. Evans, C.H.; Prud'homme, N.; King, M.; Scaiano, J.C. *J. Photochem. Photobiol. A* **1999**, *121*, 105–110.
104. Liao, Y.; Frank, J.; Holzwarth, J.F.; Bohne, C. *J. Chem. Soc. Chem. Commun.* **1995**, 2435–2436.
105. Okano, L.T.; Barros, T.C.; Chou, D.T.H.; Bennet, A.J.; Bohne, C. *J. Phys. Chem.* **2001**, *105B*, 2122–2128.
106. Kumar, C.V. In: Ramamurthy, V., editor. *Photochemistry in Organized and Constrained Media*. New York: VCH Publishers, **1991**, pp. 783–816.
107. Denny, W.A. In: Demeunynck, M.; Bailly, C.; Wilson, W.D., editors. *DNA and RNA Binders: From Small Molecules to Drugs*. Weinheim: VCH, **2003**, pp. 482–502.
108. Magde, D.; Elson, E.S.; Webb, W.W. *Phys. Rev. Lett.* **1972**, *29*, 705–708.
109. Meyer-Almes, F.J.; Porschke, D. *Biochemistry* **1993**, *32*, 4246–4253.
110. Bresloff, J.L.; Crothers, D.M. *J. Mol. Biol.* **1975**, *95*, 103–123.
111. Wakelin, L.P.G.; Waring, M.J. *J. Mol. Biol.* **1980**, *144*, 183–214.
112. Li, H.J.; Crothers, D.M. *J. Mol. Biol.* **1969**, *39*, 461–477.
113. Marcandalli, B.; Winzek, C.; Holzwarth, J.F. *Ber. Bunsenges. Phys. Chem.* **1984**, *88*, 368–374.
114. Geacintov, N.E.; Waldmeyer, J.; Kuzmin, V.A.; Kolubayev, T. *J. Phys. Chem.* **1981**, *85*, 3608–3613.
115. Corin, A.F.; Jovin, T.M. *Biochemistry* **1986**, *25*, 3995–4007.
116. Aniansson, E.A.G.; Wall, S.N.; Almgren, M.; Hoffmann, H.; Kielmann, I.; Ulbricht, W.; Zana, R.; Lang, J.; Tondre, C. *J. Phys. Chem.* **1976**, *80*, 905–922.
117. Chen, M.; Grätzel, M.; Thomas, J.K. *J. Am. Chem. Soc.* **1975**, *97*, 2052–2057.
118. Löfroth, J.E.; Almgren, M. *J. Phys. Chem.* **1982**, *86*, 1636–1641.
119. Dederen, J.C.; Auweraer, M.V.D.; De Schryver, F.C. *Chem. Phys. Lett.* **1979**, *68*, 451–454.
120. Dederen, J.C.; Auweraer, M.V.D.; De Schryver, F.C. *J. Phys. Chem.* **1981**, *85*, 1198–1202.
121. Alonso, E.O.; Quina, F.H. *Langmuir* **1995**, *11*, 2459–2463.
122. Alonso, E.O.; Quina, F.H. *J. Braz. Chem. Soc.* **1995**, *6*, 155–159.

123. Gehlen, M.H.; Boens, N.; De Schryver, F.C.; Van der Auweraer, M.; Reekmans, S. *J. Phys. Chem.* **1992**, *96*, 5592–5601.

124. Bolt, J.D.; Turro, N.J. *J. Phys. Chem.* **1981**, *85*, 4029–4033.

125. Scaiano, J.C.; Selwyn, J.C. *Can. J. Chem.* **1981**, *59*, 2368–2372.

126. Scaiano, J.C.; Selwyn, J.C. *Photochem. Photobiol.* **1981**, *34*, 29–32.

127. Novo, M.; Felekyan, S.; Seidel, C.A.M.; Al-Soufi, W. *J. Phys. Chem. B* **2007**, *111*, 3614–3624.

128. Rharbi, Y.; Winnik, M.A. *Langmuir* **1999**, *15*, 4697–4700.

129. Rharbi, Y.; Li, N.; Winnik, M.A.; Hahn, K.G. Jr. *J. Am. Chem. Soc.* **2000**, *122*, 6242–6251.

130. Hilczer, M.; Barzykin, A.V.; Tachiya, M. *Langmuir* **2001**, *17*, 4196–4201.

131. Hinze, W.L.; Hu, W.; Quina, F.H.; Mohammadzai, I.U. In: Hinze, W.L., editor. *Organized Assemblies in Chemical Analysis*, Vol. 2: Bile Acid/Salt Surfactant Systems. Stamford, CT: JAI Press, **2000**, pp. 1–70.

132. Mazer, N.A.; Carey, M.C.; Kwasnick, R.F.; Benedek, G.B. *Biochemistry* **1979**, *18*, 3064–3075.

133. O'Connor, C.J.; Wallace, R.G. *Adv. Colloid Interface Sci.* **1985**, *22*, 1–111.

134. Small, D.M. In: Nair, P.P.; Kritchevsky, D., editors. *The Bile Salts*, Vol. 1. New York: Plenum Press, **1971**, pp. 249–256.

135. Small, D.M.; Penkett, S.A.; Chapman, D. *Biochim. Biophys. Acta* **1969**, *176*, 178–189.

136. Campanelli, A.R.; De Sanctis, S.C.; Giglio, E.; Pavel, N.V.; Quagliata, C. *J. Incl. Phenom. Mol. Recognit. Chem.* **1989**, *7*, 391–400.

137. Esposito, G.; Giglio, E.; Pavel, N.V.; Zanobi, A. *J. Phys. Chem.* **1987**, *91*, 356–362.

138. Sugioka, H.; Moroi, Y. *Biochim. Biophys. Acta* **1998**, *1394*, 99–110.

139. Hashimoto, S.; Thomas, J.K. *J. Colloid Interface Sci.* **1984**, *102*, 152–163.

140. Ju, C.; Bohne, C. *Photochem. Photobiol.* **1996**, *63*, 60–67.

141. Rinco, O.; Nolet, M.-C.; Ovans, R.; Bohne, C. *Photochem. Photobiol. Sci.* **2003**, *2*, 1140–1151.

142. Amundson, L.L.; Li, R.; Bohne, C. *Langmuir* **2008**, *24*, 8491–8500.

143. Li, R.; Carpentier, E.; Newell, E.D.; Olague, L.M.; Heafey, E.; Yihwa, C.; Bohne, C. *Langmuir* **2009**, *25*, 13800–13808.

144. Seret, A.; Van de Vorst, A. *J. Photochem. Photobiol. B* **1993**, *17*, 47–56.

145. Ju, C.; Bohne, C. *J. Phys. Chem.* **1996**, *100*, 3847–3854.

2

TEMPLATING PHOTOREACTIONS IN SOLUTION

DARIO M. BASSANI

INTRODUCTION

Many chemical transformations are exquisitely sensitive to the geometry of the reagents, though this is not always obvious in solution, where Brownian motion ensures the exploration of all possible thermally accessible reaction trajectories. Where even minimal external constraints, either conformational, as in intramolecular processes, or more generally intermolecular interactions, limit the available reaction space, one frequently observes selectivity toward specific products. This observation, a cornerstone of chemical synthesis, is of course valid for photochemical processes. Perhaps because of the significant energies involved in populating electronically excited states, the concept of using geometrical constraints to influence photochemical processes did not take hold until the seminal work of Schmidt in the early 1970s.[1] The striking regiocontrol obtained in the solid-state photodimerization of *E*-cinnamic acid was attributed to differences in its environment between polymorphs and dubbed topochemical control. Since then, a significant number of photochemical transformations in the solid state have been found to be under topochemical control, generating considerable interest in their use toward synthetic organic chemistry. Obvious limitations originate from the difficulty in directing (or even predicting) crystal packing motifs, coupled to technical challenges inherent to the irradiation of solid samples.

Supramolecular Photochemistry: *Controlling Photochemical Processes*, First Edition. Edited by V. Ramamurthy and Yoshihisa Inoue.

With the advent of supramolecular chemistry, and the renewed interest it generated in exploiting intermolecular interactions, attention once again turned to applying topochemical principles to photoreactions in solution. These now rely on the use of labile covalent or nonbonded interactions (e.g., hydrogen bonding [H-B], metal ion coordination, π-stacking interactions) to promote the assembly of supramolecular architectures in which the reagents are constrained in geometries suitable for selected reaction pathways. While this approach is freed from the constraints of the crystal lattice, it is not devoid of pitfalls. The principal difficulties include (1) ensuring a high concentration of the correct supramolecular assembly (large association constants), (2) maintaining good solubility of the individual subcomponents and assemblies, (3) guaranteeing the availability of a suitable reaction trajectory for the reagents within the supramolecular architecture, and (4) allowing selective generation of the intended excited state upon irradiation. The first recorded example of the use of a regenerable covalent template to promote a photochemical transformation is attributed to Damen and Neckers,[2] who used an imprinted polymer as a support for directing the preferential formation of specific cinnamate ester dimers (see the section on "Templated [2π + 2π] Photocyclization Reactions in Solution"). Breslow and Scholl had long since reported a supramolecular approach to achieve functionalization of a remote site using a noncovalent tether to position an aromatic ketone during the key photoinitiated hydrogen-atom abstraction step.[3]

GENERAL CONSIDERATIONS

First and foremost, the photoreactivity of the systems as a whole must be considered. In cases where the template absorbs at longer wavelengths than the chromophore of the intended reaction center, competitive absorption of light (internal filter effect) or energy transfer to the template will lower the efficiency or, if the template is not photostable, lead to its decomposition. For this reason, assemblies based on the coordination of transition metals, though widely used in supramolecular chemistry, have found little application due to the presence of strongly absorbing metal-to-ligand charge transfer transitions. Instead, complexation of alkali metal ions by crown ether derivatives gives rise to complexes that do not absorb in the near UV-Vis region that have been used in templating photochemical processes. As for all photochemical reactions, transformations leading to products that do not absorb the incident radiation are preferred.

Templated reactions in solution suffer from competition from nontemplated (background) reactions, and the concentration of the substrate–template assembly should be as high as possible with respect to the concentration of free substrate in solution. This ensures that the background reactions are minimized and that the observed reactivity originates from the intended templated process. When this is not possible, as in the case of noncovalent assemblies, corrections must be made for the proportion (variable over the course of the reaction) of bound versus free substrate. In such cases, it is convenient if the products formed from the templated reaction differ from those of the free substrates in solution, as this allows direct quantification of the templated versus background reactions. Understandably, association constants for the assembly of supramolecular substrate–template complexes should be large while taking into consideration other factors such as solubility and nonself-complementarity of the subunits. To illustrate the importance of this, Figure 2.1 shows the fraction of bound species (X_{AB}) versus [A] for a monotopic template for different values of K_a. It is readily seen that, unless relatively large association constants are obtained, one must resort to high concentrations to attain appreciable proportions of substrate–template complex.

Templated homogeneous photochemical transformations involving bond formation between two or more substrates generally lead to the formation of photoproducts that bind the template better than the reagents (Fig. 2.2). This observation can be tied to the fact that photochemical processes are inherently fast, and intermolecular processes requiring high activation energies are not competitive with unimolecular deactivation of the excited state. It follows that photochemical reactions generally involve least-atom motion,

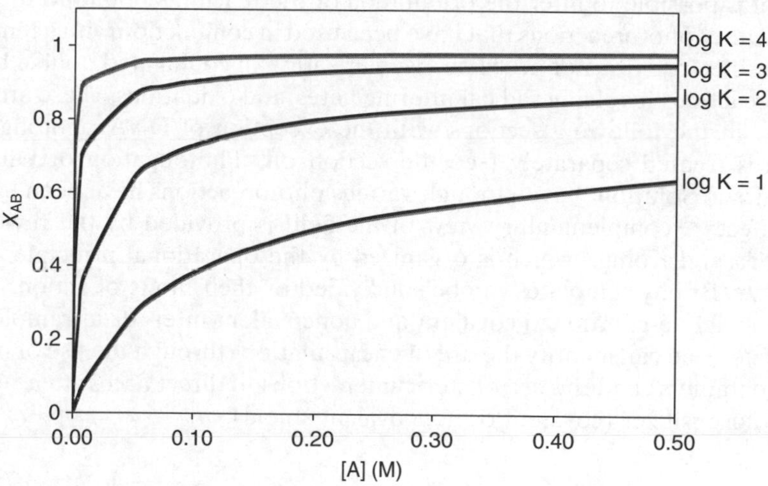

Figure 2.1. Fraction of bound substrate (X_{AB}) versus substrate concentration [A] for different values of K_{ass} ([A] = [B]) in the case of 1:1 reversible binding.

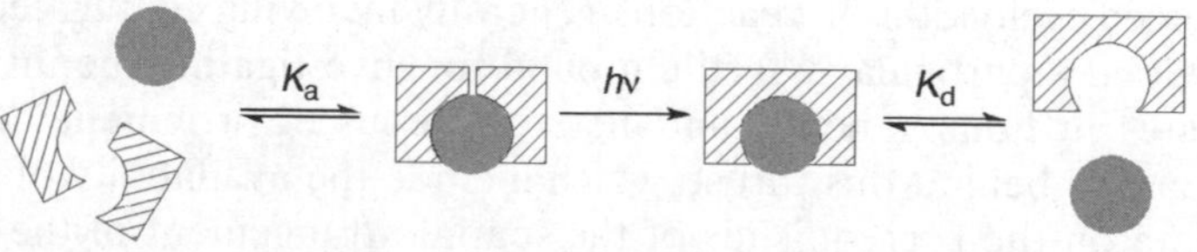

Figure 2.2. A template (gray circle) organizes one or more photoactive subunits during a self-assembly step. Irradiation of the supramolecular assembly induces a covalent transformation leading to a product that may bind the template.

and the products do not possess markedly different geometry (i.e., topology) from the reactants. Poisoning of the template by the product effectively limits the turnover to unity and is related to the process of product inhibition in enzymatic catalysis. Interesting exceptions to this have been reported by König and coworkers,[4,5] using a supramolecular photoinduced electron transfer (PET) sensitizer for the oxidation of alcohols, and Bach and coworkers,[6] who constructed an electron transfer sensitizer connected to a shield that promoted a photosensitized stereoselective cyclization reaction with turnover greater than unity (see the sections on "Templated Electron-Transfer Reactions in Solution" and "Templated Photooxydation Reactions in Solution").

The organization of this chapter is by type of reaction, as this highlights the weighted contributions of the different photoinduced processes that have been reported and may guide the reader toward as yet untested approaches that, by analogy, may prove successful. The most used reactions to investigate template effects in photochemistry are by far photocycloaddition reactions. These are further subdivided between photoinduced [2 + 2] and [4 + 4] cycloadditions. The reason for this is that these reactions are the archetype topochemical transformations *par excellence* and are particularly sensitive to geometrical constraints. Also, because the products reflect the initial geometry of the reactants, it is possible to infer the orientation of the reactants bound to the template. Other photoreactions that have been used in conjunction with a template include electron transfer processes, isomerization of conjugated double bonds, photoreactions involving radical intermediates, and oxidations. These are each treated in the following sections, with the exception of DNA photoligation, which is treated separately (see the section on "Photoligation of Oligonucleotides in Solution") even though various photoreactions have been used to this effect. A complementary view of the field is provided by the review by Svoboda and König,[7] which is organized by the operational principle of the template. Briefly, templates can be subdivided by their mode of action. Along with the above-mentioned covalent and noncovalent inter- or intramolecular templates, one can identify the use of encapsulation (through the use of molecular containers) and the use of steric interactions to direct access to a reagent or photosensitizer (covalent or noncovalent shield).

TEMPLATED (2Π + 2Π) PHOTOCYCLIZATION REACTIONS IN SOLUTION

Photoinduced cycloaddition reactions, generally involving conjugated alkenes, have provided a particularly fertile ground for investigating the intervention of templates in homogeneous and heterogeneous environments. There are several reasons behind this thrust, which include the availability of relatively precise data on the prerequisites of the spatial arrangement of the reactants thanks to Schmidt and Cohen's pioneering work. In general, it is assumed that the participating π-orbitals should be in near contact; that is, the carbon atoms should not be separated by more than 4 Å. Also, because the photoreaction generates two covalent bonds between the reagents, conformational movement in the products is limited, and it is frequently possible to directly correlate the relative orientation of the reagents prior to the photoreaction to the geometry of the products (in cases where the cycloaddition occurs in a concerted fashion). Lastly, the products, particularly the strained cyclobutanes, are of interest for synthetic organic chemistry in view of their presence in some natural products and their latent strain-induced reactivity. This section covers cycloadditions between C-centered π-orbitals, such as between simple arenes (naphthalene, acenaphthalene, fullerene), stilbenes and derivatives, cinnamates and related coumarins, and enones. Cross-cycloadditions between ketones and C–C double bonds (Paternò–Büchi) are covered in the section on "Templated Paternò–Büchi Reactions in Solution." Various strategies have proven successful to assist cyclodimerization reactions in solution. In particular, the use of molecular capsules to bring together and orient two photodimerizable molecules is well documented. Other approaches include H-B interactions, metal ion coordination, and labile covalent tethers.

The facile interconversion between carboxylic acids and their esters make these derivatives interesting as temporary covalent linkers to hold photoreactive molecules to a template. In principle, one might harness the variety of functional group protection/deprotection strategies available to conceive other means to temporarily tether template and reaction center(s) while preserving the possibility of regenerating the template. An advantage of this approach is that the strength of covalent bonds is such that the template–substrate assembly may be isolated and characterized prior to irradiation, which facilitates in-depth kinetic and mechanistic investigation. Demmen and Neckers first used styrene–divinylbenzene copolymers imprinted with cinnamate dimers as templates for cinnamate photodimerization.[2] The copolymers thus obtained were found to promote the formation of the photodimers with which they had been imprinted, along with ca. 50% of α-truxinic acid (Fig. 2.3). Though not strictly homogeneous (chemical and photochemical transformations of the functionalized cross-linked polymers took place as dispersed solids in solution), the underlying principle is akin to the more recent use by Hopf and coworkers of a cyclophane-diamine to tether two cinnamate moieties in a geometry that is specific for the formation of the *cis* head-to-head (HH) dimer.[8] Hydrolysis of the amide linker allows nearly quantitative recovery of the β-truxinic acid photoproduct and the template.

The nonspecificity of inclusion complexes, mostly driven by hydrophobic interactions in the case of water-soluble capsules, makes this approach particularly general to accelerate bimolecular photoreactions. Intramolecular cycloaddition of unconjugated olefins can also take place in cyclodextrins (CDs), provided sufficient room is available for the rearrangement. Inoue and coworkers pursued this strategy in view of inducing regioselectivity in the [2 + 2] cycloaddition of chiral tetronates. Under optimum conditions (aqueous solutions, γ-CD), regioselectivities >80% could be attained.[9] For intermolecular

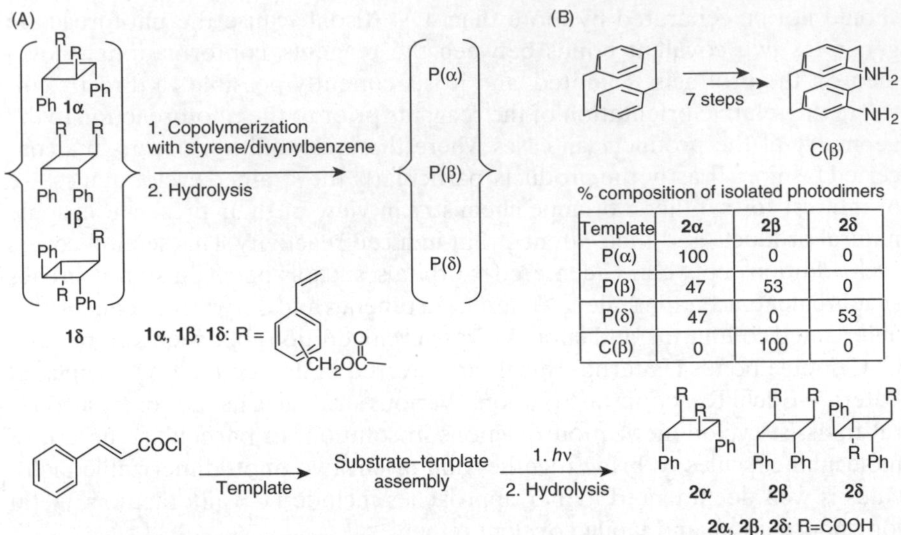

Template	2α	2β	2δ
P(α)	100	0	0
P(β)	47	53	0
P(δ)	47	0	53
C(β)	0	100	0

Figure 2.3. Two alternative strategies for preparing labile covalent templates. (A) Imprinted polymer approach (Damen and Neckers[2]); (B) directed organic synthesis (Hopf and coworkers[8]). The templates are loaded with cinnamoyl chloride and irradiated to afford the photodimers in the indicated proportions after hydrolysis.

reactions, the capsules should be large enough to comfortably enfold two chromophores and additional free volume associated with the reaction coordinate.[10] Solutions of the chromophores and capsules are allowed to stand for a period of time sufficient to attain equilibrium and are then directly irradiated, or evaporated to afford solid samples prior to irradiation. The products thus obtained show regioselectivities that differ from those formed in solution in the absence of capsules, indicative of a modification of the reaction course due to inclusion. Acceleration of the reaction due to the localized confinement of the reactants can also be expected, but this can be more difficult to establish due to the inherent difficulties in determining even relative rate constants for heterogeneous samples or samples in which the reactive species is distributed in a diversity of environments of unequal reactivity. Figure 2.4 shows the structures of selected molecular capsules that have been successfully employed to template photochemical cycloaddition reactions.

CDs are cyclic oligosaccharides known to encapsulate small molecules in their hydrophobic interior, whose propensity to accelerate certain organic reactions has been paralleled to the enzymatic activity of proteins.[11–13] Early work on the effect of CDs on photodimerizations involved substituted anthracenes and are covered in the section on "Templated [4π + 4π] Photocyclization Reactions in Solution." Concerning photoinduced [2 + 2] cycloadditions, initial conflicting reports focused on the propensity of coumarins to photodimerize when included in CD hosts and irradiated as solid powders. Whereas Tanaka et al.[14] found that inclusion in β-CD inhibited the photodimerization of coumarin, Venkatesan and coworkers[15] reported that irradiation of the same inclu-

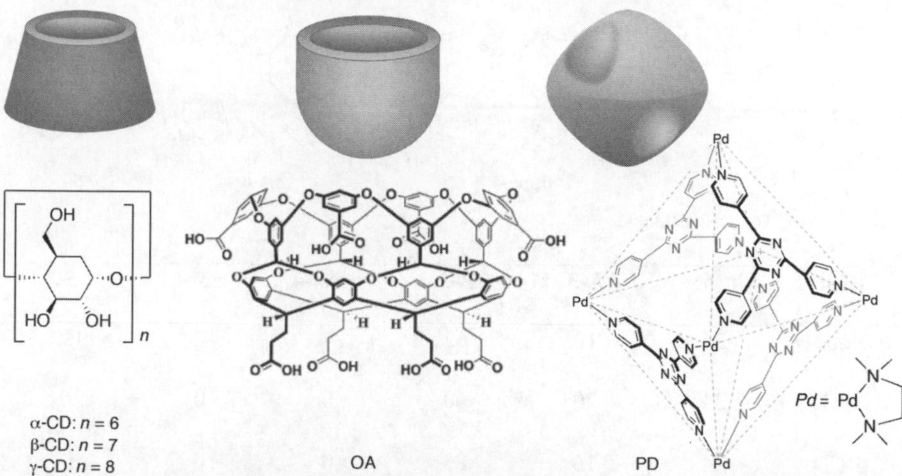

Figure 2.4. Chemical structures and graphical representation of various molecular capsules. Octa acid (OA) forms a dimeric capsule.

sion complexes favors the efficient formation of *syn* HH photodimers. Difficulties in precisely identifying the stoichiometry of the complexes in the solid state may explain these conflicting reports, but it is now widely recognized that CDs can host two chromophores and accelerate their photodimerization.

Unlike coumarins, stilbenes undergo competing unimolecular *E*,*Z* isomerization upon irradiation and, when symmetrical, give rise to only four cycloadducts. Herrmann et al.[16] examined the photochemistry of a water-soluble stilbene derivative in the presence of α-, β-, and γ-CDs in aqueous solutions. Their results show that while the smaller α-CD strongly binds the stilbene derivative, it has no effect on the photochemistry. This lack of effect is attributed to partial encapsulation of a single chromophore. In contrast, β-CD fully encapsulates a single stilbene moiety, favoring unimolecular isomerization while preventing electrocyclization of the *Z* isomer to phenanthrene (confirming a similar result previously obtained by Syamala and Ramamurthy[17]). The larger γ-CD afforded smooth conversion to the *trans* cyclodimer by encapsulating two molecules of stilbene (Fig. 2.5). Interestingly, the association constants (determined by microcalorimetry) show that γ-CD has a strong preference for the *trans* versus *cis* cycloadduct. Better fit of the *trans* photoadduct in the γ-CD would explain the observed switch in selectivity compared with the solution, where the *cis* cycloadduct is preferentially formed.

For CD-bound stilbenes, as for other photodimerizable chromophores (see below), the selectivity for one of several photocycloadducts is apparently linked to the substitution pattern. Hubig and coworkers examined a series of unsymmetrical substituted stilbenes containing electron-withdrawing and electron-donating groups, and found that the *cis* photodimers were preferred in all cases, with very modest preference for either HH or head-to-tail (HT) photodimers.[18] Related styrylpyridines (Fig. 2.6) were investigated by the

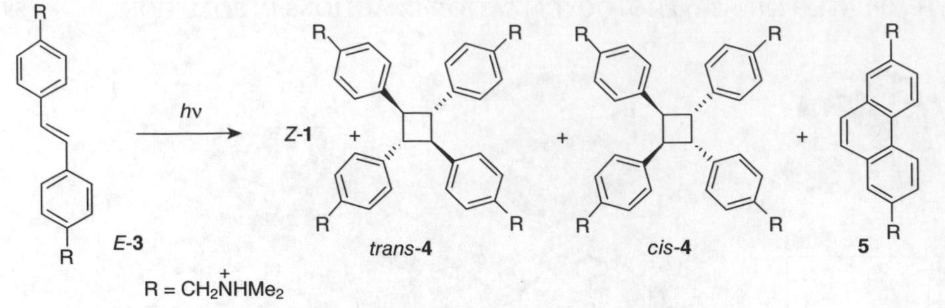

Host	t (h)	*E*-**3**	*Z*-**3**	*trans*-**4**	*cis*-**4**	**5**
None	24	10	62	7	2	19
α-CD	24	20	60	0	0	20
β-CD	24	16	83	0	0	1
γ-CD	72	0	0	79	19	2

Figure 2.5. Product distribution upon irradiation of aqueous solutions of **3** in the presence of CDs.[16]

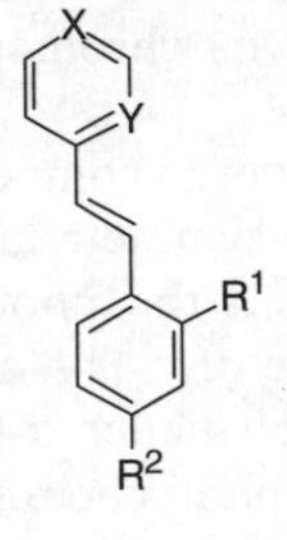

	X	Y	R^1, R^2	α-CD	β-CD	γ-CD
6	CH	N	H, H	*Z*-**6**	*Z*-**6**	*cis*-HT
7a	NH^+Cl^-	CH	H, H	*Z*-**7a**	*Z*-**7a**	*trans*-HT
7b	NH^+Cl^-	CH	H, Me	*Z*-**7b**	*Z*-**7b**	*trans*-HT
7c	NH^+Cl^-	CH	H, OMe	*Z*-**7c**	*Z*-**7c**	*cis*-HH
7d	NH^+Cl^-	CH	H, Cl	*Z*-**7d**	*Z*-**7d**	*cis*-HH
7e	NH^+Cl^-	CH	Cl, Cl	*Z*-**7e**	*Z*-**7e**	*cis*-HH

Figure 2.6. Product distribution upon irradiation of powders of **6** or **7** as inclusion complexes in CDs.[19,20]

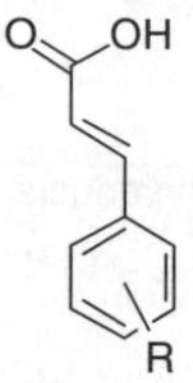

8a: R = H

8b: R = 4-OH

8c: R = 4-Me

8d: R = 4-$NH_3^+Cl^-$

8e: R = 3-OMe

8f: R = 3-Me

8g: R = 4-OMe

8h: R = 4-NO_2

8i: R = 4-NO_2

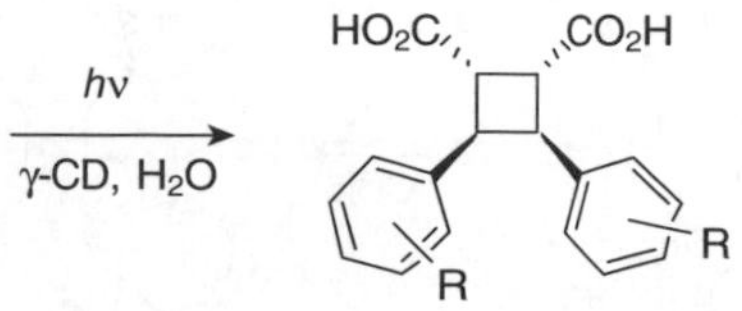

$h\nu$, γ-CD, H_2O

Figure 2.7. Irradiation of various cinnamic acids as aqueous solutions in the presence of γ-CD leads to the preferential formation of the *cis* head-to-head dimer.[22]

groups of Ramamurthy and Srinivasan as solid inclusion complexes in α-, β-, or γ-CDs. In the case of substituted 4-styrylpyridiniums in γ-CD hosts, the *cis* HH products are favored, in contrast to the parent 4-styrylpyridinium in γ-CD, which gives a mixture of *cis* HH and *trans* HT cycloadducts.[19] Free base 2-styrylpyridine was reported to yield the *cis* HT photoproduct when irradiated in γ-CD host.[20] Once again, irradiation of inclusion complexes in smaller α- or β-CD afforded predominantly isomerization and lower yields of dimers.

The effect of CD binding on the photochemistry of cinnamic acids was not investigated until later by Ramamurthy and coworkers. High yields of photodimers were obtained upon irradiation of various cinnamic acid derivatives in the presence of γ-CD. In all cases, the formation of β-truxinic acid (*cis* HH dimer) was observed (Fig. 2.7).[21,22] As noted in the introduction, cinnamic acids give solid-state photoproducts that depend on the crystal lattice and generally lead to the inefficient formation of mixtures of photocycloadducts when irradiated in fluid solutions (predominantly truxinic acids, i.e., HH dimers).[23]

Cucurbit[*n*]urils (CB[n]) and calix[*n*]arenes (CA[*n*]) are obtained from the condensation of glycouril and phenols with formaldehyde, respectively, and are in many ways related to the CD hosts described above, presenting hydrophobic pockets in which to encapsulate guests. Notable differences include the rigidity and low water solubility of cucurbiturils, whereas calixarenes are geometrically less well defined and can be made both lipophillic or hydrophilic depending on their substitution pattern. The obtention of larger CBs (in particular, $n = 8$ present a cavity of similar size to γ-CD) represents a significant achievement as it allows the inclusion of two average-sized aromatic molecules.[24] Many of the photoinduced dimerization reactions described above for CDs have also been investigated using CBs and CAs as hosts, which can form the basis for useful comparisons. For example, 4,4′-diaminostilbene forms 2:1 inclusion compounds in CB[8] hosts when deprotonated. Irradiation of this assembly leads to the exclusive formation (>95:5) of the *cis* photodimer,[25] analogously to what is observed for related *E*-**3** in γ-CD. Irradiation of cinnamates **8a**, **8b**, and **8d**–**8i** in the presence of CB[8] in water affords mostly the *cis* HH photodimers, and, also paralleling the cavity size effects of CDs, smaller CB[7] exclusively leads to the isomerization of the exocyclic double bond.[21]

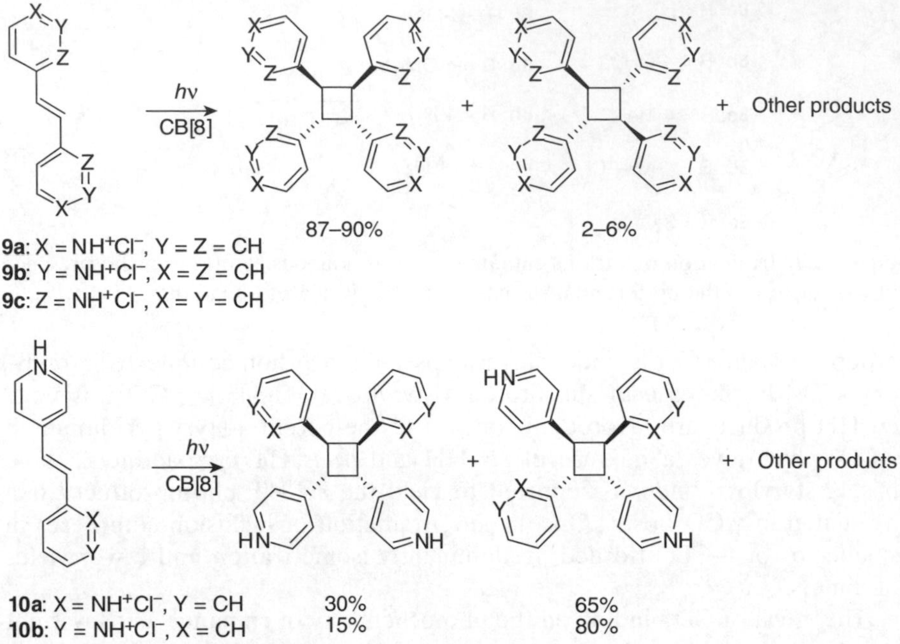

hν, CB[8] + Other products

87–90% 2–6%

9a: X = NH+Cl−, Y = Z = CH
9b: Y = NH+Cl−, X = Z = CH
9c: Z = NH+Cl−, X = Y = CH

hν, CB[8] + Other products

10a: X = NH+Cl−, Y = CH 30% 65%
10b: Y = NH+Cl−, X = CH 15% 80%

Figure 2.8. Inclusion of azastilbenes in CB[8] can steer the formation of the resulting photodimers.[19,26,27]

When protonated, azastilbenes are particularly prone to form inclusion compounds with CB[8]. Interestingly, irradiation of aqueous solutions of the inclusion complexes selectively leads to the formation of the *syn* photodimers in the case of symmetrical 1,2-dipyridiniumethylenes, whereas unsymmetrical analogs show preference for the *anti* photodimers (Fig. 2.8).[26] Protonated stilbazoles (styrylpyridinium salts) also form inclusion compounds in CB[8], which behave similarly to the aforementioned unsymmetrical 1,2-dipy ridiniumethylenes.[19,27] Anachenko et al.[28] and Kaliappan et al.[29] investigated the effect of calixarene hosts on the photochemistry of stilbenes and stilbazoles, respectively. Their findings are in agreement with the intervention of the calixarene host as moderately efficient template for the formation of photodimers, consistent with the more flexible nature of CAs compared with CDs and CBs.

Octa acid (OA) is a more recent addition to the library of cavitands and was first prepared by Gibb and Gibb[30] via an eightfold ullmann ether coupling. Ramamurthy and coworkers showed that it can be advantageously used to direct photoinduced events (*vide infra*), including [2 + 2] photodimerization of acenaphthylene and styrene. In the case of acenaphthylene, the exclusive formation of the *syn* photodimer is not altogether unexpected.[31] However, during the irradiation of styrene in the presence of an OA host, the formation of the 1,3-dimer is accompanied by the formation of an unusual styrene pho-

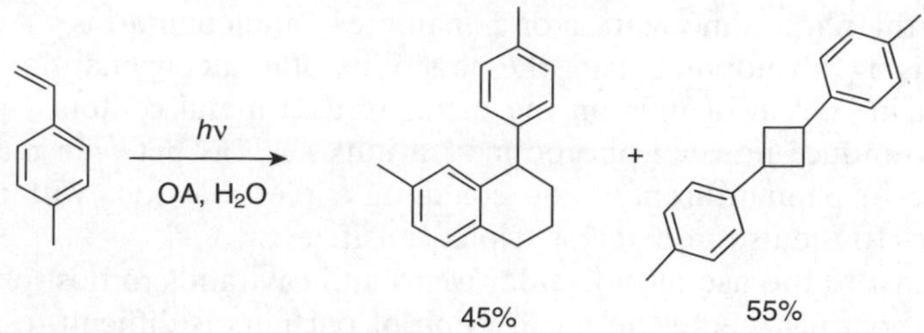

hν
OA, H_2O
45% 55%

Figure 2.9. Irradiation of 4-methylstyrene in the presence of an OA host yields the unexpected formation of a fused photodimer in addition of the expected cyclobutane.[32]

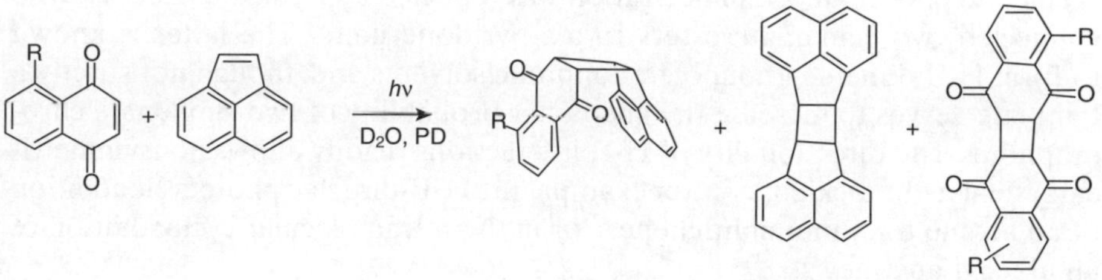

hν
D_2O, PD

R = H	35%	21%	14%
R = OMe	44%	6%	22%
R = OEt	92%	0%	0%

Figure 2.10. Product distribution in olefin cross-photodimerization templated by the PD cage (see Fig. 2.4 for structure).[35]

todimer (Fig. 2.9).[32] Its formation is ascribed to rearrangement of the intermediate 1,4-biradical intermediate within the OA capsule.

A supramolecular cage formed by the self-assembly of four tridentate ligands and eight palladium(II) metal ions (PD cage, Fig. 2.4) was reported by Fujita and coworkers to possess an interior cavity suitable for encapsulating small- to medium-sized aromatic molecules.[33] Initial experiments rapidly asserted the usefulness of this capsule to direct the dimerization of 1-methylacenaphthylene to afford exclusively the *syn* HT dimer (D_2O, >98% yield).[34]* Interestingly, the PD cage also permits the cross-photodimerization of olefins. By judiciously selecting photodimerizable chromophores of different sizes, such that two molecules of the larger chromophore (e.g., 5-ethoxynaphthoquinone) cannot be accommodated within the capsule, and that two molecules of the smaller chromophores fit "loosely," it is possible to favor the ternary complex in which one small and one large chromophore are included in the PD host (Fig. 2.10).[35] The PD cage has also been shown to

*The authors reported that no photodimerization takes place in benzene solutions in the absence of PD.

accelerate the photodimerization of cinnamates[36] and coumarins.[37] Additional very promising developments are the possibility of inducing enantioselectivity by introducing chiral ligands on the palladium(II) metal centers;[38] using the PD cage to induce unusual photodimerizations, such as between maleimides and pyrene or phenanthrene;[39] and confining triplet reactions within the PD cage thanks to the use of exterior triplet sensitizers.[40]

In contrast to the use of molecular cages and cavitands to host photoreactive chromophores, where the localization of partners is difficult to predict a priori, the use of H-B interactions requires precise planning of the geometrical requirements. In favorable cases, it is possible to steer bimolecular reactions through the use of an exterior template, or through the use of a molecular shield to restrict access to one of the reactants. Initial work on using H-B to accelerate photoinduced dimerization was reported by Beak and Zeigler, who connected two cinnamate esters to a 2-pyridone unit.[41] The latter is known to form H-B dimers in nonpolar, nonprotic solvents and, though not strictly a template, serves to increase the encounter probability of two cinnamate chromophores. The directionality of H-B interactions readily allows nonsymmetrical systems to be tackled, as in, for example, the H-B-directed photocycloaddition of an alkene to cyanonaphthalene,[42] or in the intramolecular cycloaddition of an alkenyl-eneone.[43]

An example of cross-cycloaddition between thymine and coumarin relies on the use of Kemp's triacid as a U-shaped (covalent) template to juxtapose the reactive double bonds in proximity.[44] Irradiation of the resultant assembly thus favors the formation of a specific *syn* photoadduct (Fig. 2.11). Derivatives of Kemp's triacid have also been used by Bach and coworkers to control the diastereoselectivity of both inter- and intramolecular cycloadditions and of Paternò–Büchi cycloadditions (see the secton on "Templated Paternò–Büchi Reactions in Solution"). In the former case, the H-B interactions are used to place a sterically demanding residue so as to cover one of the faces of the reaction site (molecular shield principle). In this fashion, prochiral substrates may be coerced to undergo enantioselective photocylization. In the example shown in Figure 2.11, the intramolecular cycloaddition proceeds with modest enantionselectivity at room temperature (10–30% enantiomeric excess [ee]) but can be significantly increased by irradiating at low temperatures in the presence of an excess of template (>80% ee at –60°C).[45] This approach is also useful for intermolecular cycloadditons.[46] Both six- and five-membered lactams are potential substrates for undergoing templated intramolecular cyclization, but, in the case of simple enones, the short irradiation wavelengths required for excitation also results in decomposition of the template.[47]

A template possessing more than one binding site can, in principle, be used to bind two chromophores and control their ensuing photodimerization. Ideally, the binding sites should be strong and nonself-complementary to promote efficient formation of a photoreactive ternary complex. This can be accomplished through the combination of multiple H-B interactions, as is observed in natural nucleic acid derivatives. Bassani and coworkers used bar-

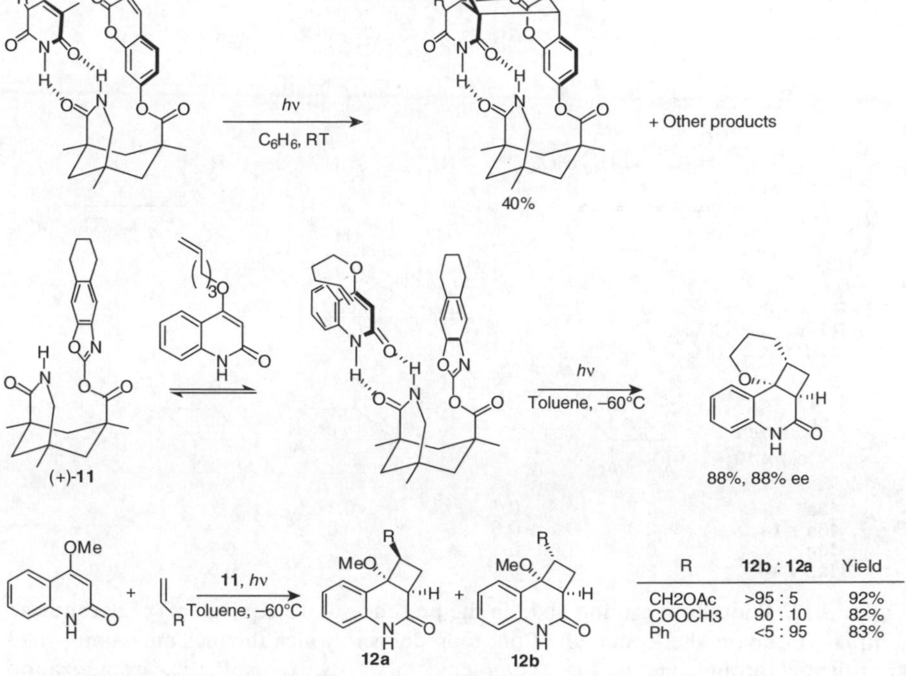

Figure 2.11. Examples of Kemp's triacid derivatives used to exert stereocontrol over intra- and intermolecular cycloaddition reactions.[44–46]

bituric acid derivatives as templates possessing two non-self-complementary binding sites (Fig. 2.12). The photodimerization of cinnamates,[48] stilbenes,[49] or styrenes[50] appended with diaminopyrimidine moieties that are complementary to barbituric acid is enhanced, and the authors proposed that the geometry of the photodimers obtained reflect the relative orientation of the exocyclic double bonds upon excitation. This is supported by the isolation of the elusive ε-truxillate photodimer upon dimerization of cinnamate **13b** in the presence of template **14**.[48] Interestingly, the photoproducts obtained retain the geometrical constraints imparted by the template and are therefore good receptors for barbiturates, as well as for analogous thymine and uracil derivatives.[49] More extended supramolecular templates possessing two rigid H-B sites linked through a xanthene scaffold were reported by Lehn and coworkers to assist the photodimerization of coumarin and thymine derivatives.[51,52]

Particularly noteworthy is the observation that the use of a supramolecular H-B template can induce the photodimerization of compounds not ordinarily known to undergo photoinduced dimerization under homogeneous conditions. This is partially the case for cinnamates and stilbenes, whose intermolecular photodimerization is inefficient in fluid solutions due to competing *E*,*Z* isomerization. Fullerene-C_{60} is an aromatic compound whose polymerization

13a R = Me, X = H
13b R = Me, X = CO_2Me
13c R = Pri, X = C_6H_5

$h\nu$, CH_2Cl_2, RT

Quantum yield x 10^3					
13a	0.7	0.1	<0.1	-	-
13a + 14	2.3	0.6	0.8	-	-
13b	0.6	0.1	-	0.7	0.1
13b + 14	2.4	0.6	-	0.5	1.6

Figure 2.12. Photodimerization of **13** in the presence of 0.5 equivalents of barbiturate template **14** favors the formation of photoproducts in which the melamine units of **13** are oriented through binding **14** via complementary H-B. Competing *E,Z* isomerization increases the number of structurally different photoadducts obtainable.[50]

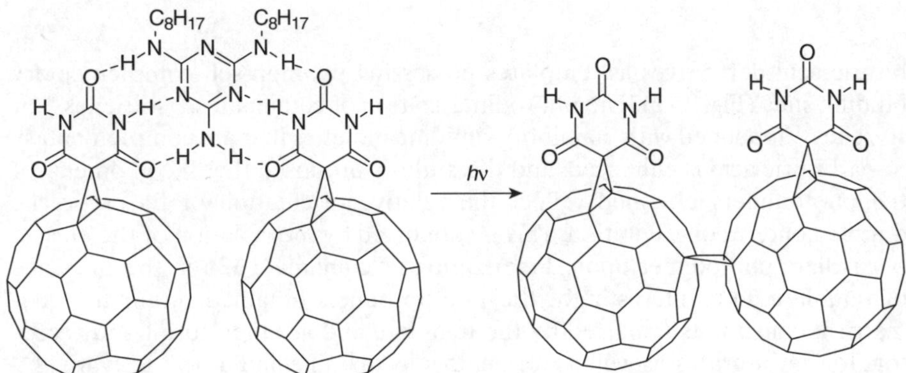

Figure 2.13. The use of a template can render otherwise improbable reactions possible, such as in the case of the intermolecular photodimerization of a C_{60} derivative in solution.[55]

has been observed to proceed efficiently in the solid and in solid clusters through sequential photoinduced [2 + 2] cycloadditions.[53,54] The use of a non-covalent H-B template can be used to assist the reaction in fluid media, where intermolecular photodimerization of C_{60} is not possible due to its short singlet lifetime and low solubility[55] (Fig. 2.13).

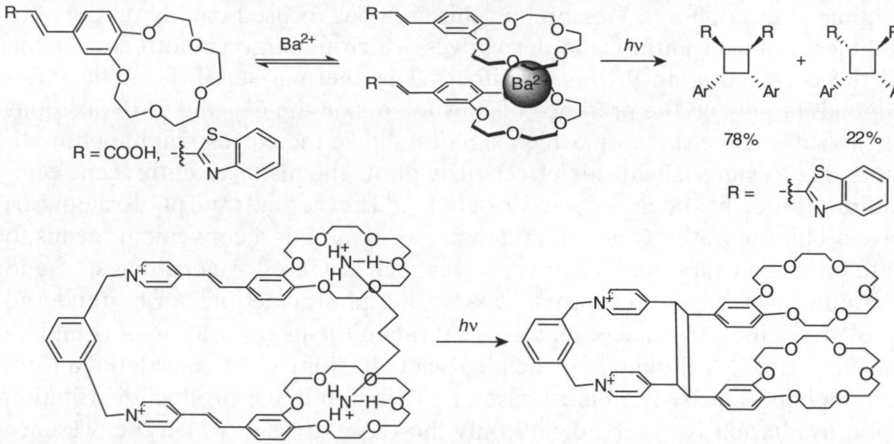

Figure 2.14. Inter- and intramolecular photocycloadditions assisted by crown ether complexation of metal ions or ammonium salts.[56,57]

Crown ethers are known to bind alkali metal ions via electrostatic interactions and ammonium ions via H-B. Supramolecular architectures based on these interactions can be designed to place two photoreactive chromophores in close proximity. Federova and coworkers[56] prepared 2-styrylbenzothiazole and cinnamic acid derivatives appended with 15-crown-5 moieties. The latter bind not only K^+ as 1:1 complexes but also Ba^{2+} in a 2:1 stoichiometry. In the presence of Ba^{2+}, two reactants are held in an HH geometry, as shown in Figure 2.14. Irradiation then leads to the formation of the corresponding HH photodimers. Interestingly, cross-photocycloaddition between the cinnamate and styryl derivatives could be realized owing to the intervention of an auxiliary H-B. Gromov et al.[57] showed that a diammonium salt can be used to template the population of a specific photoreactive conformer, which, upon irradiation, smoothly underwent intramolecular photodimerization of the two styrylpyridinium centers (Fig. 2.14). It is also possible to incorporate both H-B and metal ion interactions, as demonstrated by Bassani et al., since the two interactions are orthogonal and do not interfere with one another. The presence of both metal ion and H-B templates results in a synergistic effect, culminating in a 1000-fold enhancement in photodimerization quantum yield with respect to the untemplated reaction.[58]

TEMPLATED (4Π + 4Π) PHOTOCYCLIZATION REACTIONS IN SOLUTION

The topochemical aspects and other advantages of the [2 + 2] photocycloadditions described previously also apply to [4 + 4] photocycloaddition reactions. These include the formation of rigid photoproducts that retain the geometry of the starting reactants and that absorb at shorter wavelengths than the original chromophores. Research in this area has focused mainly on the photodimerization of anthracene derivatives, which undergo smooth conversion to the corresponding 9,10-photodimers. The thermal stability of the latter depends largely on the presence of substituents on the C-9 and C-10 positions of the anthracene skeleton, which can destabilize the corresponding photodimers due to stereoelectronic effects. The photochemistry of anthracene photodimerization has been reviewed in detail.[59,60] The thermal (and photochemical) reversibility of anthracene dimerization has made this a convenient means to gate binding in supramolecular receptors such as CDs,[61–63] calixarenes,[64,65] and Hamilton barbiturate receptors.[66,67] Also, the photoreaction can be rendered sensitive to the presence of metal ions through the incorporation of complexing side arms.[68] Although in principle such behavior can be considered a form to templated reactivity, in most cases, no variation in the product distribution upon irradiation is observed, and only those examples in which the presence of a guest can modify the course of the photoreaction will be covered in the following pages.

The ready availability of water-soluble anthracene derivatives, such as anthracene-1- or 2-sulfonic acid, and anthracene-2-carboxylic acid (**15**) greatly facilitates the investigation of the use of molecular encapsulation to template anthracene photodimerization, with the opportunity of inducing chiral selection in the photoproducts being a strong motivation. The acceleration of the photodimerization of **15** by γ-CD (Fig. 2.15) was reported by Tamaki et al. in 1984,[69,70] with application toward inducing regiocontrol in the dimerization process appearing soon thereafter.[71] Nakamura and Inoue[72] undertook detailed studies of the initial γ-CD/**15** system, and subsequently proposed a series of

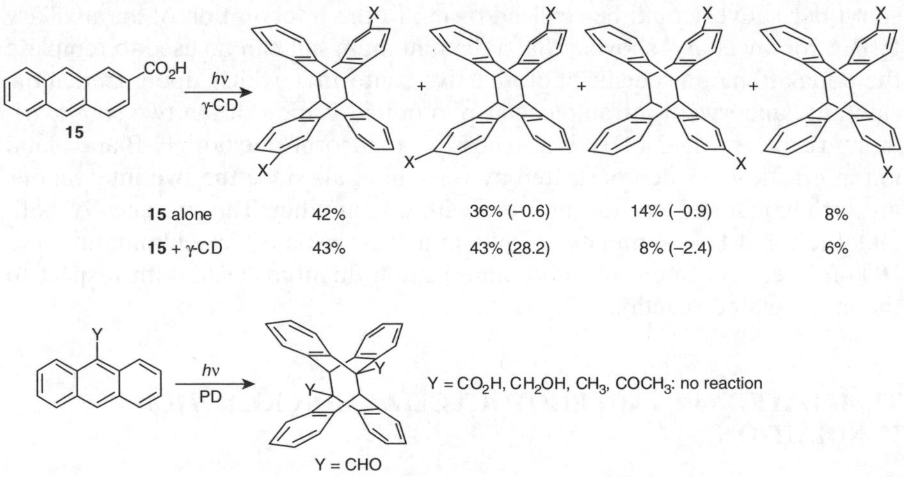

CO_2H — hν, γ-CD →

15

15 alone	42%	36% (–0.6)	14% (–0.9)	8%
15 + γ-CD	43%	43% (28.2)	8% (–2.4)	6%

hν, PD → Y = CHO

Y = CO_2H, CH_2OH, CH_3, $COCH_3$: no reaction

Figure 2.15. Photodimerization of anthracene derivatives encapsulated in CD and PD hosts. Product distribution and ee of chiral products (in parenthesis) given for the reaction of **15** in the presence of γ-CD (2:1 guest:host ratio).[40,72]

covalently modified CD hosts capable of steering the photodimerization of **15** toward the stereoselective formation of chiral adducts. Besides the modification of the glucosidic skeleton of the CD,[73,74] electrostatic interactions induced by pendant side chains[75–77] and capping the smaller opening of the CD[78] have proven successful. The palladium cage developed by Fujita et al.[33] (PD, Fig. 2.4) is of comparable size to γ-CD and can bind anthracene molecules in its cavity.[79] Karthikeyan and Ramamurthy examined the propensity of PD to template the photodimerization of 9-substituted anthracenes in aqueous media (Fig. 2.15).[40] Their results indicate that not all the anthracenes examined are encapsulated by PD, and that among those that are, photodimerizataion efficiency is limited. However, in view of the propensity of some 9-substituted anthracene HH photodimers to undergo rapid thermal retrodimerization, it is not impossible that, at least in some cases, photodimers may be formed transiently and go undetected.

Linking a chiral template to **15** via two-point H-B also proved a successful strategy to induce stereoselective photodimerization. Inoue and coworkers reported that the gastroprokinetic agent **16** (Fig. 2.16) forms 1:1 complexes with **15**, affording mixtures of photodimers that are enriched in the *anti* HT products, with ee's of 10–30% (up to 40% at –50°C).[80] Similarly, the chiral shield **11** developed by Bach blocks one of the two prochiral faces of a bound pyridone molecule to direct its ensuing [4 + 4] photocycloaddition to cyclopentadiene. The reaction is characterized by high conversion to a 2:3 mixture of the *endo* and *exo* products, both of which are obtained with high ee's (Fig. 2.16). Bovine and human serum albumin have also been used to induce

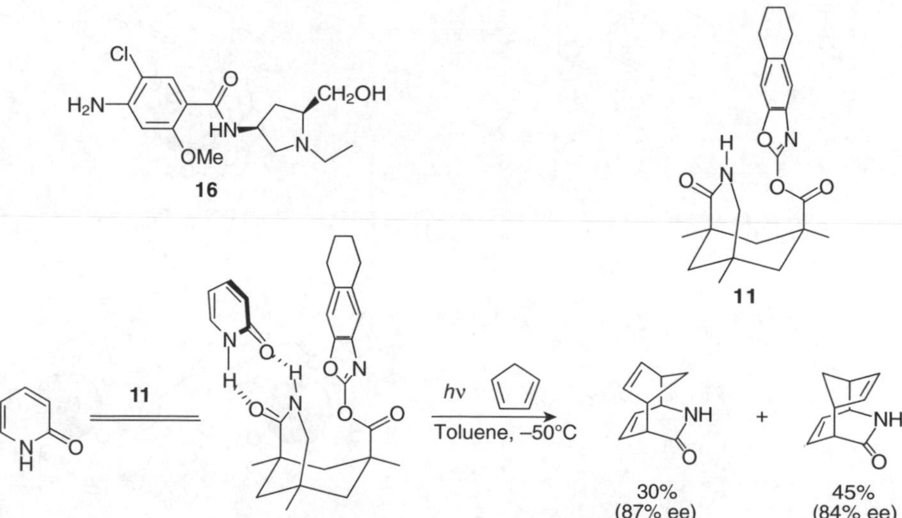

Figure 2.16. Chiral shields **11** and **16** bind anthracene-2-carboxylic acid and pyridone, respectively, furnishing stereoselectivity in the ensuing photocyclization reactions.[80,84]

chirogenesis in the photodimerization of **15**.[81–83] The protein environment presents hydrophobic pockets capable of hosting two or more chromophores in a chiral environment. While the exact environment offered by such hosts is difficult to apprehend, the use of biomolecules as templates for organic photoreaction is a promising venue.

Anthracenes substituted with crown ether complexing units are capable of complexing alkali metal ions. The resulting systems are interesting for sensing applications, and in some cases, the distribution of photoproducts is altered by the presence of complexed metal ions. Such systems thus represent examples of photoinduced [4 + 4] cycloadditions templated by metal ions. The effect of the metal ion template may at times be counterintuitive, as in the photoconversion of a crown ether to a cryptand through dimerization of pendant anthracenes, where the presence of bound metal ions induces a decrease in dimerization efficiency.[85] Instead, the intramolecular photodimerization of a macrocylic system incorporating two anthracence is sensitive to the presence of bound sodium ions, which direct the formation of the 9,10-photodimer (Fig. 2.17).[86]

Joining two anthracene moieties by a single polyethylene glycol chain represents a convenient means to photochromic compounds that are sensitive to metal ions. In the example by Nakamura and coworkers, the anthracenes are

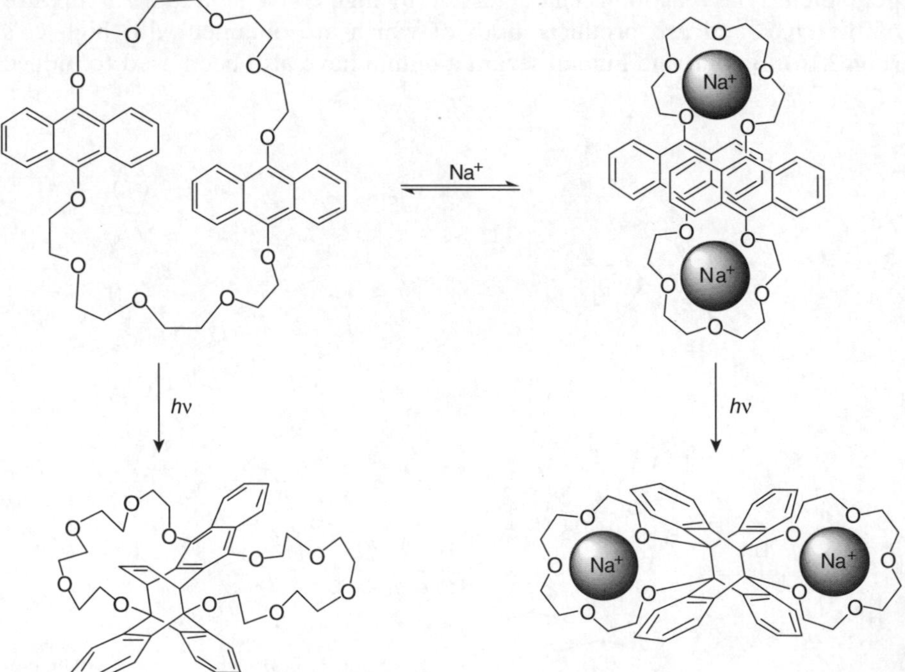

Figure 2.17. The presence of sodium ions templates the formation of the 9,10-photodimer. In its absence, a dissymmetrical photodimer is formed instead.[86]

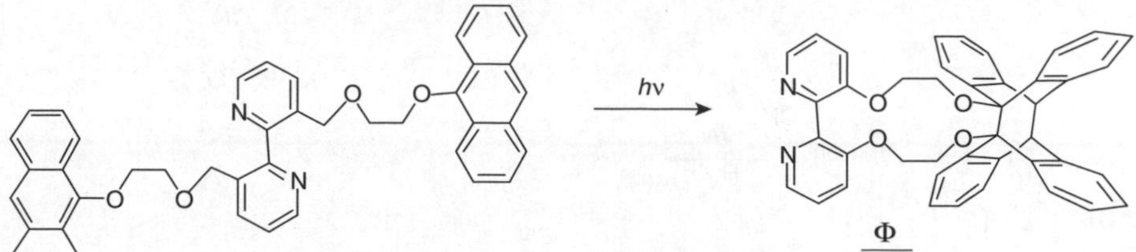

	Φ
17	0.19
17 + Na^+	0.29
17 + Hg^{2+}	0.26
17 + Hg^{2+} + Na^+	0.32

Figure 2.18. Photodimerization quantum yields for the reaction of **17** in the absence and presence of selected metal ions.[89]

joined through the C-1 or C-2 position, which leads to the formation of chiral photoproducts.[87,88] The C-1 linked system was shown to be sensitive to both monovalent and divalent akali metal ions, whereas the photoreactivity of the C-2 linked system proved to be relatively insensitive to metal ions. By combining binding sites adapted for different metal ions, Desvergne, Tucker, and coworkers constructed an assembly (compound **17**, Fig. 2.18) that is sensitive to the presence of two orthogonal stimuli. Transition metal ions preferentially bind the 2,2′-bipyridine site, whereas sodium ions are better bound by the podand moiety. Either ion will swing the anthracenes to a proximal position, accelerating the photodimerization reaction.[89]

Naphthalene and its derivatives are known to undergo [4 + 4] photodimerization to yield *trans* (major) and *cis* (minor) photoadducts, which, in a subsequent step, can give rise to cubane-like structures.[90,91] Little work has appeared on using supramolecular templates to control the mutual orientation of substituted naphthalenes, although micellar environments have been shown to exert influence over the final product distribution.[92] Wu et al. showed that encapsulation of two naphthalene moieties by CB[8] exerts stringent control of the inter- or intramolecular photodimerization of 2-naphthoates, strongly favoring formation of the *cis* photodimer.[93,94]

TEMPLATED PATERNÒ–BÜCHI REACTIONS IN SOLUTION

The [2 + 2] cycloaddition between an alkene and a ketone or aldehyde is known as the Paternò–Büchi reaction.[95,96] The oxetanes that are formed are versatile intermediates in organic synthesis, and this has generated considerable interest in understanding the regio- and stereoselectivity of the reaction.[97–102] The first example of using an exterior template to control oxetane formation was reported by Turro and coworkers.[103] They investigated the use of CDs to bind the carbonyl derivative, thus restricting its accessibility toward the alkene. Compared with the reaction in homogeneous solution in the

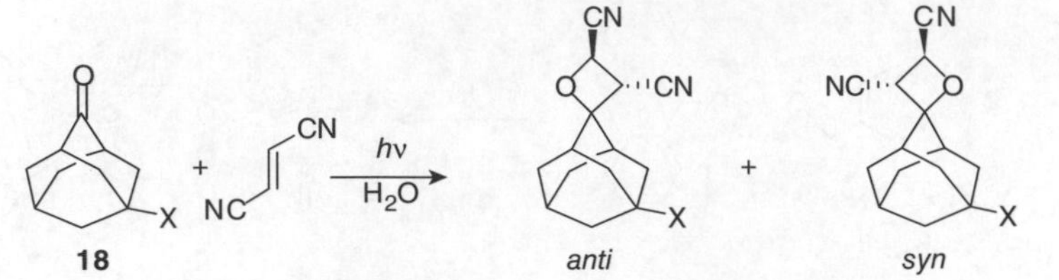

anti/syn	Br	OH	Ph	*t*-Bu
18	56/44	57/43	62/38	62/38
18 + β-CD (1:3)	20/80	38/62	23/77	14/86

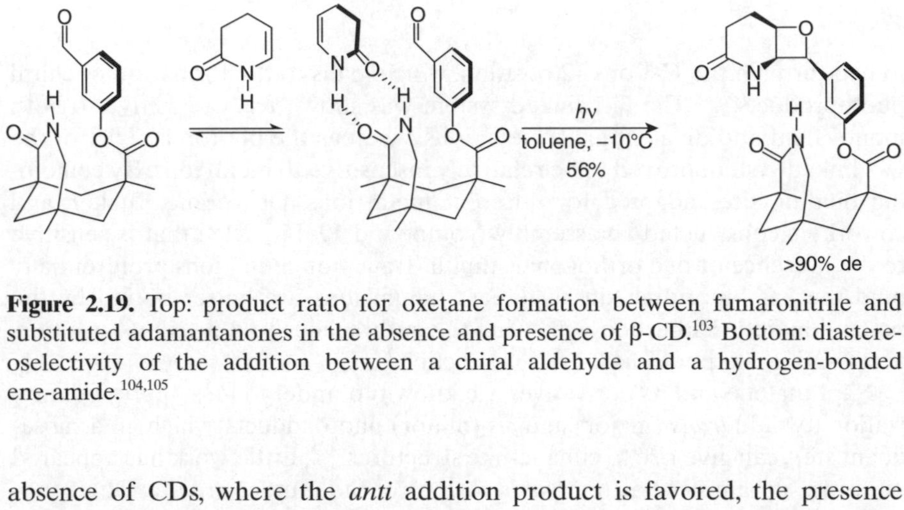

Figure 2.19. Top: product ratio for oxetane formation between fumaronitrile and substituted adamantanones in the absence and presence of β-CD.[103] Bottom: diastereoselectivity of the addition between a chiral aldehyde and a hydrogen-bonded ene-amide.[104,105]

absence of CDs, where the *anti* addition product is favored, the presence of β-CD caused a switch in stereoselectivity toward the *syn* addition product (Fig. 2.19). Smaller (α-CD) or larger (γ-CD) CDs had little or no effect. Bach et al. showed that H-B interactions between the alkene component and a chiral aldehyde could exert stereocontrol over the ensuing cycloaddition (Fig. 2.19).[104,105] Although not exactly a template, as is the case for the related molecular shield **15**, one may assimilate the Kemp's triacid derivative as a labile covalent template. This approach proved very successful, affording photoadducts in >90% diastereomeric excess (de) (>95% ee).

TEMPLATED PHOTOISOMERIZATION REACTIONS IN SOLUTION

Photoinduced *E*,*Z* isomerization of alkenes and related diazo derivatives has been extensively used as a probe for the local microenvironment. It is therefore natural to contemplate tailoring the surrounding environment of the chromophore so as to favor a specific isomer. The photoisomerization of cinnamates and stilbenes in CD hosts was examined by Syamala et al.[106] Wenz and coworkers[16] reported that encapsulation of water-soluble stilbenes in α- or β-CD, whose size is too small to allow encapsulation of two guests, suppresses $E \rightarrow Z$ isomerization along with the ensuing formation of phenanthrene, and a similar case was reported by Parthasarathy et al.[32] for excitation of 4,4′-dimethylstilbene in an OA host.

The isomerization of chiral or prochiral alkenes presents a particular interest due to its potential for inducing enantiodifferentiation and photochirogenesis.[107,108] CDs naturally posses a chiral environment and have been used as a starting point to develop hosts for the stereoselective isomerization of cyclooctene. Whereas the *Z*-cyclooctene is not chiral, the *E*-cyclooctene isomer can exist as one of two possible enantiomers. Despite their inherently chiral hydrophobic pocket, native CDs do not induce enantioselective photoisomerization of *Z*-cyclooctene.[109,110] However, Inoue et al. reported a variety of covalently modified CDs capable of inducing ee. The strategy relies on the covalent linkage of benzoyl sensitizers to the rim (O-6 position) of α-, β-, or γ-CDs.[111,112] Particularly promising results have been reported for β-CD, which presents the best cavity size match, possessing a 6-*O*-(3-methoxybenzoyl) substituent, which gives ee's approaching 50% (Fig. 2.20).[113] The results are dependent on the solvent, temperature, and pressure of the reaction conditions,[112,114] which indicates that the simple inclusion model is not sufficient to explain the chiral induction and suggest that entropic factors sometimes play a large role.[115] More recently, the same group has reported the isomerization of cyclooctadiene using a naphthalene-modified CD sensitizer.[116]

Cyclopropanes are another class of compounds whose isomerization can be brought about photochemically with induction of chirality, as demonstrated by early studies by Hammond and Cole on 1,2-diphenylcyclopropane using chiral sensitizers.[117] Koodanjeri and Ramamurthy later examined the photoisomerization of diphenylcyclopropane derivatives in the presence of β-CD.

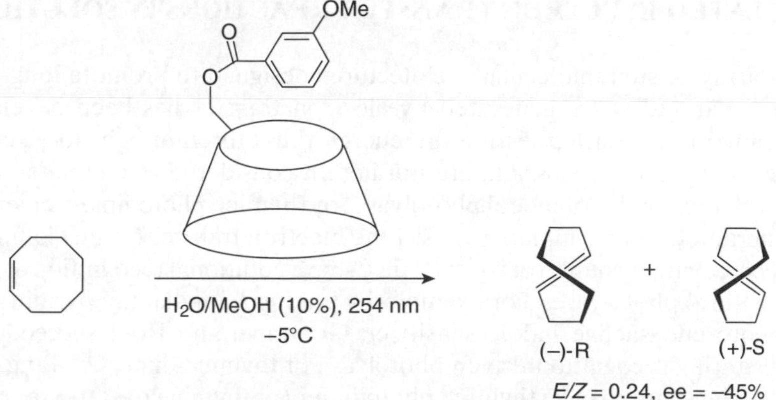

Figure 2.20. Enantiodifferentiation of *E*-cyclooctene by isomerization in a chiral host (β-CD)-sensitizer assembly.[113]

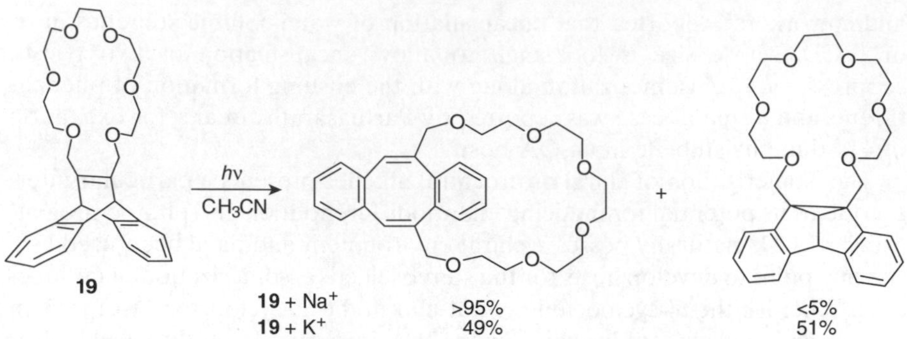

$h\nu$, CH_3CN

19

19 + Na^+	>95%	<5%
19 + K^+	49%	51%

Figure 2.21. Product distribution obtained upon irradiation of acetonitrile solutions of **19** in the presence of excess Na^+ or K^+ ions. Overall conversions are >95% and 79%, respectively.[120]

While the latter provides a chiral environment for the encapsulated cyclopropane, the ee's obtained are modest (13%).[118] More recently, the isomerization of chiral ene-carbamates included in γ-CD has been reported to proceed with moderate chiral induction in the solid, but almost no ee is observed upon irradiation of the host–guest complex in fluid solution.[119]

In contrast to template effects on *E*,*Z* isomerization of double bonds, template effects on valence or structural isomerizations are much more rare. An interesting example is the isomerization of a dibenzobarrelene derivative to either a cyclooctatetraene or a semibullvalene rearrangement photoproduct. Derivatives presenting a crown-ether-like cavity for binding metal ions (Fig. 2.21) exhibit product distributions that are dependent on the nature of the bound metal ion.[120]

TEMPLATED ELECTRON TRANSFER REACTIONS IN SOLUTION

A vast array of supramolecular architectures designed to promote long-lived charge separated states generated by electron transfer has been developed principally in view of harnessing the energy thus stored in light-to-electrical energy conversion devices, and are not herein considered as templated reactions. A few examples of model photolyase for thymine photodimer repair rely on a supramolecular template to assist the electron transfer event. Compared with their covalent counterparts,[121–123] they share additional recognition aspects of the natural photolyases. For example, by linking a 2,6-diaminopyridine unit to a photoredox-active indole sensitizer, Goodman and Rose succeeded in demonstrating recognition-driven photolyase of thymine dimers.[124] Tang et al. reported the synthesis of a thymine photodimer (or thymine oxetane, prepared from the photoreaction of thymine and benzopheneone) covalently linked to the amine of a modified β-CD (6-deoxy-6-amino-cyclodextrin).[125] The CD

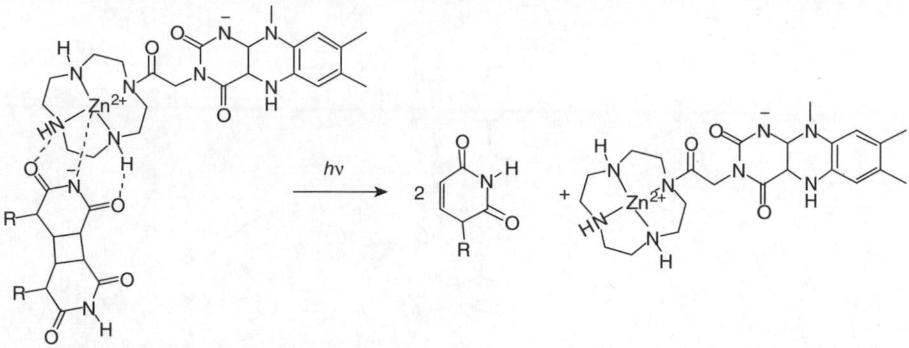

Figure 2.22. A zinc(II)cyclam-appended flavinium allies molecular recognition and photoinduced reductive cleavage of thymine dimers.[126]

moiety was designed to bind free sensitizer (dimethylaniline or indole) present in solution, thereby facilitating the electron transfer-mediated scission of the dimer or oxetane. Irradiation of the dimer–CD assembly in the presence of sensitizer afforded the cleaved photoproducts, and the results are in agreement with the intervention of a supramolecular sensitizer/reactant complex. König and coworkers developed an artificial photolyase model based on the use of a zinc cyclam moiety, known to bind thymine derivatives, to accelerate the reductive cleavage of thymine dimers (Fig. 2.22).[126] Despite their high association constant for thymine dimers, zinc cyclams present the drawback of binding thymine more strongly than the thymine dimer, which results in competitive inhibition by the product. A similar approach was used by Chibulka et al. to oxidize alcohols catalytically and is discussed in the section on "Templated Photooxydation Reactions in Solution."[5]

Bach and coworkers developed a chiral electron transfer sensitizer possessing an H-B binding site.[6] Modeled on their previous chiral shield (compound **11**), it possesses a benozophenone moiety, which, upon excitation, can undergo reductive electron transfer to oxidize a nearby substrate. The PET catalysis was shown to be particularly effective for substrates possessing H-B sites complementary to those of the sensitizer. By using a prochiral substrate (Fig. 2.23), it was demonstrated that the sensitizer could induce up to 70% ee (64% yield), while sustaining a catalytic cycle. Maximum turnover numbers >10 (at 5 mol% of sensitizer) could be achieved.

TEMPLATED REACTIONS OF PHOTOGENERATED RADICALS IN SOLUTION

Upon excitation, carbonyl compounds may undergo homolytic α-cleavage or γ-hydrogen abstraction (Norrish I and II reactions, respectively). The biradical intermediates formed then undergo secondary ground-state reactions to yield, in the case of Norrish I reactions, an alkene and an aldehyde (when the alkyl

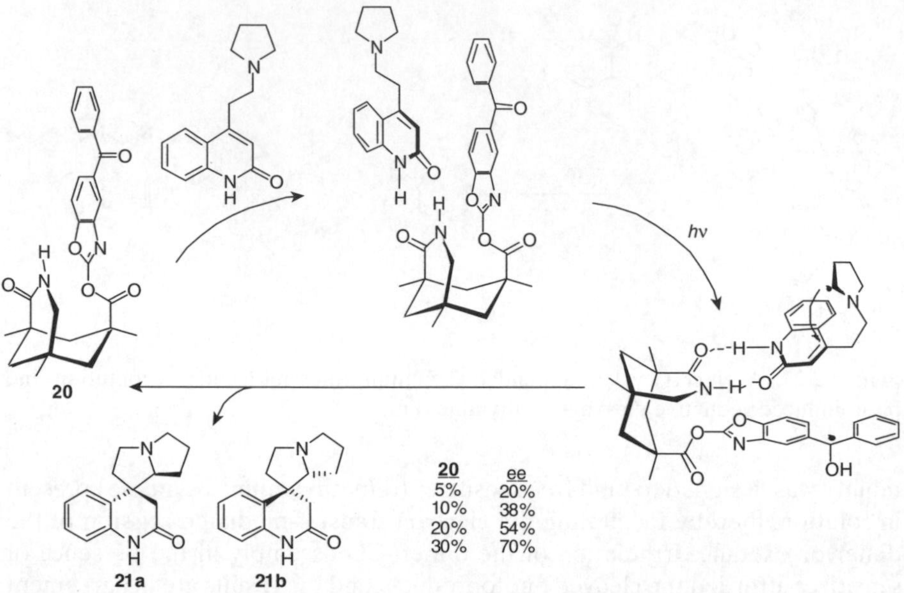

Figure 2.23. Example of a catalytic PET sensitizer capable of inducing good enantioselectivity (enantiomeric excess as a function of catalyst loading, toluene, –60°C). The catalyst doubles as a photosensitizer and as a molecular shield to direct the attack of the α-aminoalkyl radical formed by sequential electron and proton transfer to the benzophenone sensitizer.[6]

fragment has a β-proton), recombination products (with or without extrusion of carbon monoxide), or, when an α-proton is available on the carbonyl fragment, a ketene. Norrish II reactions generate a 1,4-biradical, which can cyclize to form an oxetane (Norrish–Yang cyclization), or undergo cleavage to give the corresponding alkene and an enol. Understandably, if the molecular structure of the chromophore determines the available reaction channels, the medium plays an important role in shaping the overall product distribution, particularly between in- and out-of-cage products.

In a seminal paper, Ramamurthy and coworkers examined the competition between Norrish I and II reactions in benzoin ethers included in β- and γ-CD hosts.[127] In solution, benzoin ethers preferentially undergo α-cleavage to give benzil and pinnacol ethers, whereas inclusion in β-CD markedly increases the formation of the oxetane product, resulting from Norrish II reaction, presumably by stabilizing the cisoid conformation in the ground state (Fig. 2.24). Temperature plays an important role in the formation and ensuing reactivity of the host–guest complex, and it is interesting to note that a switch in reactivity toward the benzil products is observed upon irradiation of the encapsulated benzoin ether in the solid. Nearly complete control over the competing I and II reactions of benzoin ethers was later obtained in solution by encapsula-

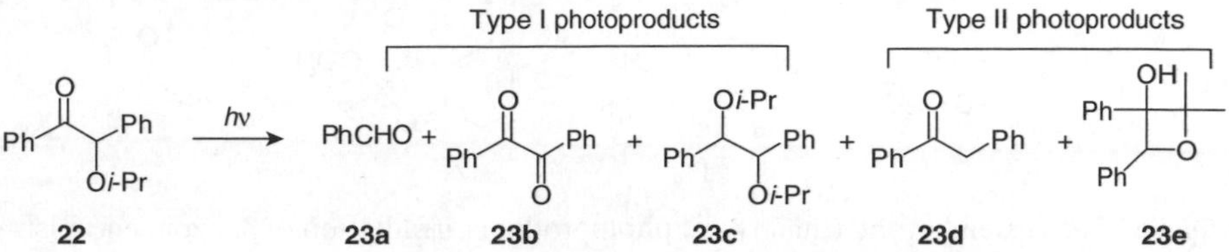

	23a	23b	23c	23d	23e
22 (MeOH)	27	11	62	–	–
22 + β-CD (H_2O, 5°C)	43	5	38	1	15
22 + β-CD (solid)	7	–	–	78	15

Figure 2.24. Partial control over type I versus type II photochemistry of benzoin alkyl ethers by CD hosts.[127]

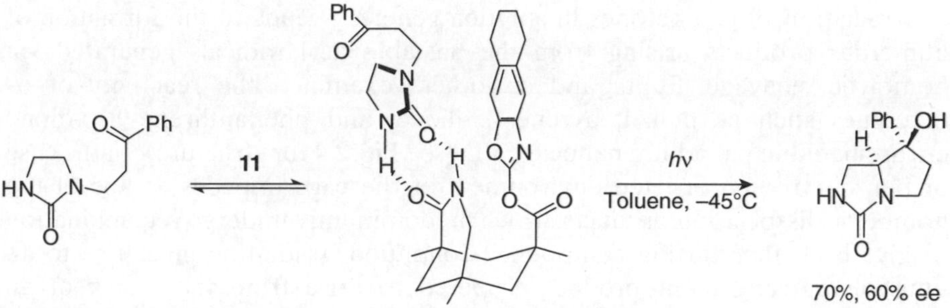

Figure 2.25. Example of an H-B template to direct intramolecular Norrish–Yang cyclization.[135]

tion in water-soluble calixarine hosts.[128] Nuclear magnetic resonance (NMR) studies suggest that the host forces the guest into a cisoid conformation that is particularly favorable for γ-hydrogen abstraction, thus templating the formation of deoxybenzoin photoproducts.[129]

OA, like CDs, can encapsulate aromatic ketones and induce directionality in the reaction trajectory. Gibb and coworkers examined the product distribution of various α-alkyl dibenzylketones included in OA hosts.[130–132] Efficient control of type I versus type II reaction is observed to augment with increasing alkyl chain length (C1 to C8), with longer chains favoring type II reactivity. In fluid solutions, exclusive type I reactivity is observed and, in the most favorable case (i.e., C8), the course of the reaction is switched from 85:15 to 10:90 (for type I to type II products) on going from hexane to OA capsules, respectively. A detailed mechanistic study of the partitioning of 1,4-biradicals due to conformational constraints (in diastereomers) was conducted by Moorthy et al. and provides a compelling background on which to rationalize template effects by molecular capsules on these reactions.[133] Promising stereoselectivity (ee's of 40–60%) was obtained in the Norrish–Yang cyclization of substituted imidazolininones using an H-B template (**11**) as a shield to direct the intramolecular cyclization of the intermediate 1,4-biradical (Fig. 2.25). As in applications

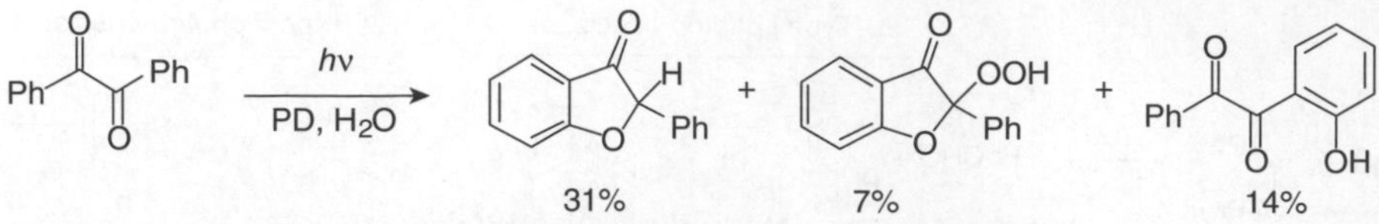

Ph–CO–CO–Ph $\xrightarrow[\text{PD, } H_2O]{h\nu}$ (benzofuranone, H, Ph) 31% + (benzofuranone, OOH, Ph) 7% + (Ph–CO–CO–C$_6$H$_4$OH) 14%

Figure 2.26. Instead of the usual type I photoproducts, irradiation of benzoin encaplsulated in a PD cage (see Fig. 2.4 for structure) gives cyclization of the intermediate oxygen-centered radical. Structural analyses indicates that two molecules of benzoin are encapsulated in one PD host.[136]

of noncovalent molecular shields, low temperatures and excess template were found to be necessary to enhance the observed stereoselectivity.[134,135]

Irradiation of α-diketones in solution generally leads to the formation of numerous products arising from the unstable acyl radicals generated via homolytic cleavage. Fujita and coworkers examined the reactivity of α-diketones such as benzil, pyrene-4,5-dione, and phenanthrene-9,10-dione encapsulated in palladium nanocage PD (see Fig. 2.4 for structure). In the case of benzoin, the constricted environment of the cage impedes escape of the homolytic dissociation products, which predominantly undergo recombination to give back the starting compound. Continued irradiation gives rise to an unusual rearrangement product proposed to arise from the attack of an oxygen-centered radical onto the adjacent benzoyl ring through an addition–elimination pathway (Fig. 2.26).[136] Not unexpectedly, irradiation of 1:2 inclusion complexes of an α-diketone and a compound with abstractable protons leads to coupling products.[137]

The photoreactivity of alkyl benzoate esters encapsulated in CD hosts was investigated by Annalakshmi and Pitchumani.[138] Benzyl alcohol is often the major product that is formed, though alkyl benzenes can become favored for longer alkyl chains, which presumably remain trapped in the CD. Similarly, 1-naphthyl benzoates undergo photo-Fries rearrangement inside a hydrophobic OA capsule to give product distributions that are somewhat simplified with respect to what is observed in fluid solution.[139]

TEMPLATED PHOTOOXIDATION REACTIONS IN SOLUTION

The confinement offered by the inclusion of reactants in a molecular capsule restricts their reactivity toward photoinduced oxidation. The reverse is also possible (i.e., confining a sensitizer within a molecular capsule), but this generally does not lead to a template effect though some degree of enantiomeric differentiation has been observed.[140] Addition of externally formed singlet oxygen to alkenes enclosed in an inert OA capsule has been investigated by Ramamurthy and coworkers. Briefly, singlet oxygen, generated through sensitization by dimetylbenzil or Rose Bengal, is used to oxidize cycloalkenes that

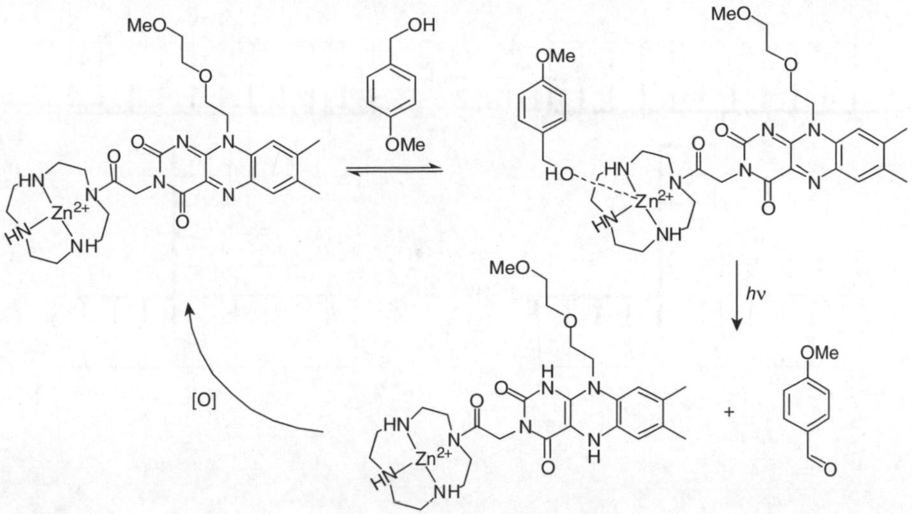

Figure 2.27. Catalytic cycle of flavin–salen complex for the oxidation of anisol to anisaldehyde.[5]

are contained in OA. By directing the approach of $^1O_2$ to the alkene, the capsule can provide control over the reaction. For this to work, the alkene guest must not be distributed anisotropically within the molecular container. In this example, an exocyclic methyl group serves the purpose of "molecular anchor" to restrict freedom of the guest. Although the outcome is variable, favorable cases can lead to a switch in product distribution in favor of a single oxidation product.[141] In at least one example, the molecular container was able to sensitize the oxidation of an alkane guest.[142] In this case the products obtained were those thermodynamically favored, making any directing ability of the template difficult to establish. It is in principle possible to conceive templates capable of both sensitizing and directing oxidation of bound substrates.

The model flavin-based photolyase developed by König is equally suitable to bind and catalyze the oxidation of benzylic alcohols into benzaldehydes (Fig. 2.27).[4,5] The complex binds the substrate through H-B interactions to the salen subunit. Following oxidation of the alcohol, the reduced flavin is regenerated by oxidation by atmospheric oxygen, thereby completing the catalytic cycle. The system is very efficient and boasts turnover numbers of 16–20.

PHOTOLIGATION OF OLIGONUCLEOTIDES IN SOLUTION

Insomuch as the joining of two oligonucleotide strands (chemical ligation) is accelerated by the presence of their complementary strand, the latter can be considered to act as a template for the ligation process. Indeed, this approach

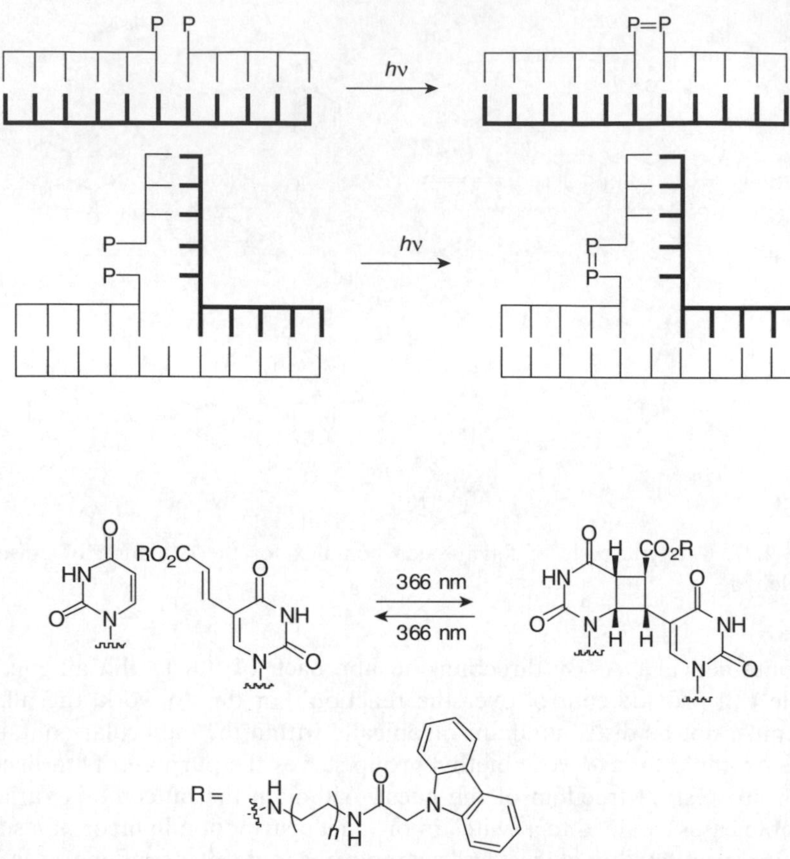

Figure 2.28. Use of photoligation to connect two strands in the presence of a complementary sequence (top), or to introduce a branching point (center). The action of the complementary strand (shown in bold) is that of a template. Bottom: photodimerization between uridine and a carboxyvinyluridine used to effect reversible ligation using monochromatic radiation.[153]

has been frequently used as an assay for the presence of specific oligonucleotide targets and forms the basis for many lab-on-chip devices. For some applications, photochemical ligation is an interesting alternative to chemical or enzymatic protocols. For example, it can be used to reversibly ligate three strands into a duplex, or to generate a branching point (Fig. 2.28).[143–146] Much work in the field is based on the [2 + 2] cycloaddition between uridine derivatives developed by Fujimoto and coworkers.[147–149] The method relies on a modified uridine in which the chromophore's absorption envelope is extended into the visible by conjugation and is applicable to circular DNA[150] and RNA.[151,152] A particularly interesting aspect is that the photoligation can be made reversible under the irradiation conditions by using a tethered carbazole PET sensitizer.[153] Thus, irradiation in the presence of template leads to ligation, whereas irradiation in the absence of the complementary sequence leads to cleavage of photoligated strands. Such behavior may be of interest to photochemically resolve combinatorial libraries and for screening oligonucleotide strands for matching sequences. Photochemical ligation can also be effected through anthracene [4 + 4] photodimerization, as shown by Ihara et al.[154] One advantage is the longer irradiation wavelength that does not damage DNA. The ligated oligonucleotides are expected to be sensitive to temperature in view of the anthracene dimer's propensity to undergo thermal cleavage.

CONCLUSION

The numerous examples of templated photochemical transformations described above offer compelling evidence that controlling photoinduced events in homogeneous solution is indeed possible. However, they also underpin the difficulties inherent to designing systems based on assumptions of the geometrical requirements of the reaction and the relative geometrical constraints offered by the template. On the other hand, it is quite clear that all photochemical transformations, even unimolecular ones, can be gated by an exterior template. The large body of work exerted on templating photoinduced [2 + 2] and [4 + 4] cycloadditions can serve as a basis to extrapolate toward less frequently explored photoreactions. The numerous parallels of template effects to biological processes will certainly provide fertile ground for innovation. Recent examples of enzyme-like behavior from artificial templates, the use of proteins as matrices, and photoligation of oligonucleotides attests to this.

REFERENCES

1. Schmidt, G.M.J. *Pure Appl. Chem.* **1971**, *27*, 647.
2. Damen, J.; Neckers, D.C. *J. Am. Chem. Soc.* **1980**, *102*, 3265.
3. Breslow, R.; Scholl, P.C. *J. Am. Chem. Soc.* **1971**, *93*, 2331.
4. Ritter, S.C.; König, B. *Chem. Commun.* **2006**, 4694.
5. Cibulka, R.; Vasold, R.; König, B. *Chem. Eur. J.* **2004**, *10*, 6223.
6. Bauer, A.; Westkamper, F.; Grimme, S.; Bach, T. *Nature* **2005**, *436*, 1139.
7. Svoboda, J.; König, B. *Chem. Rev.* **2006**, *106*, 5413.
8. Zitt, H.; Dix, I.; Hopf, H.; Jones, P.G. *Eur. J. Org. Chem.* **2002**, 2298.
9. Fleck, M.; Yang, C.; Wada, T.; Inoue, Y.; Bach, T. *Chem. Commun.* **2007**, 822.
10. Mecozzi, S.; Rebek, J. *Chem. Eur. J.* **1998**, *4*, 1016.
11. Hapiot, F.; Tilloy, S.; Monflier, E. *Chem. Rev.* **2006**, *106*, 767.
12. Szejtli, J. *Chem. Rev.* **1998**, *98*, 1743.
13. Takahashi, K. *Chem. Rev.* **1998**, *98*, 2013.
14. Tanaka, Y.; Sasaki, S.; Kobayashi, A. *J. Incl. Phenom.* **1984**, *2*, 851.
15. Moorthy, J.N.; Venkatesan, K.; Weiss, R.G. *J. Org. Chem.* **1992**, *57*, 3292.

16. Herrmann, W.; Wehrle, S.; Wenz, G. *Chem. Commun.* **1997**, 1709.
17. Syamala, M.S.; Ramamurthy, V. *J. Org. Chem.* **1986**, *51*, 3712.
18. Rao, K.S.S.P.; Hubig, S.M.; Moorthy, J.N.; Kochi, J.K. *J. Org. Chem.* **1999**, *64*, 8098.
19. Kaliappan, R.; Maddipatla, M.; Kaanumalle, L.S.; Ramamurthy, V. *Photochem. Photobiol. Sci.* **2007**, *6*, 737.
20. Banu, H.S.; Lalitha, A.; Pitchumani, K.; Srinivasan, C. *Chem. Commun.* **1999**, 607.
21. Pattabiraman, M.; Kaanumalle, L.S.; Natarajan, A.; Ramamurthy, V. *Langmuir* **2006**, *22*, 7605.
22. Pattabiraman, M.; Natarajan, A.; Kaanumalle, L.S.; Ramamurthy, V. *Org. Lett.* **2005**, *7*, 529.
23. Bassani, D.M. In: Horsepool, W.M.; Lenci, F., editors. *CRC Handbook of Photochemistry and Photobiology*, 2nd edition. Boca Raton, FL: CRC Press, **2003**.
24. Kim, J.; Jung, I.S.; Kim, S.Y.; Lee, E.; Kang, J.K.; Sakamoto, S.; Yamaguchi, K.; Kim, K. *J. Am. Chem. Soc.* **2000**, *122*, 540.
25. Jon, S.Y.; Ko, Y.H.; Park, S.H.; Kim, H.J.; Kim, K. *Chem. Commun.* **2001**, 1938.
26. Maddipatla, M.; Kaanumalle, L.S.; Natarajan, A.; Pattabiraman, M.; Ramamurthy, V. *Langmuir* **2007**, *23*, 7545.
27. Pattabiraman, M.; Natarajan, A.; Kaliappan, R.; Mague, J.T.; Ramamurthy, V. *Chem. Commun.* **2005**, 4542.
28. Ananchenko, G.S.; Udachin, K.A.; Ripmeester, J.A.; Perrier, T.; Coleman, A.W. *Chem. Eur. J.* **2006**, *12*, 2441.
29. Kaliappan, R.; Kaanumalle, L.S.; Natarajan, A.; Ramamurthy, V. *Photochem. Photobiol. Sci.* **2006**, *5*, 925.
30. Gibb, C.L.D.; Gibb, B.C. *J. Am. Chem. Soc.* **2004**, *126*, 11408.
31. Kaanumalle, L.S.; Ramamurthy, V. *Chem. Commun.* **2007**, 1062.
32. Parthasarathy, A.; Kaanumalle, L.S.; Ramamurthy, V. *Org. Lett.* **2007**, *9*, 5059.
33. Fujita, M.; Oguro, D.; Miyazawa, M.; Oka, H.; Yamaguchi, K.; Ogura, K. *Nature* **1995**, *378*, 469.
34. Yoshizawa, M.; Takeyama, Y.; Kusukawa, T.; Fujita, M. *Angew. Chem. Int. Ed. Engl.* **2002**, *41*, 1347.
35. Yoshizawa, M.; Takeyama, Y.; Okano, T.; Fujita, M. *J. Am. Chem. Soc.* **2003**, *125*, 3243.
36. Karthikeyan, S.; Ramamurthy, V. *J. Org. Chem.* **2007**, *72*, 452.
37. Karthikeyan, S.; Ramamurthy, V. *J. Org. Chem.* **2006**, *71*, 6409.
38. Nishioka, Y.; Yamaguchi, T.; Kawano, M.; Fujita, M. *J. Am. Chem. Soc.* **2008**, *130*, 8160.
39. Nishioka, Y.; Yamaguchi, T.; Yoshizawa, M.; Fujita, M. *J. Am. Chem. Soc.* **2007**, *129*, 7000.
40. Karthikeyan, S.; Ramamurthy, V. *Tetrahedron Lett.* **2005**, *46*, 4495.
41. Beak, P.; Zeigler, J.M. *J. Org. Chem.* **1981**, *46*, 619.
42. Yokoyama, A.; Mizuno, K. *Org. Lett.* **2000**, *2*, 3457.
43. Crimmins, M.T.; Choy, A.L. *J. Am. Chem. Soc.* **1997**, *119*, 10237.
44. Mori, K.; Murai, O.; Hashimoto, S.; Nakamura, Y. *Tetrahedron Lett.* **1996**, *37*, 8523.
45. Bach, T.; Bergmann, H.; Harms, K. *Angew. Chem. Int. Ed. Engl.* **2000**, *39*, 2302.

46. Bach, T.; Bergmann, H. *J. Am. Chem. Soc.* **2000**, *122*, 11525.
47. Albrecht, D.; Basler, B.; Bach, T. *J. Org. Chem.* **2008**, *73*, 2345.
48. Bassani, D.M.; Darcos, V.; Mahony, S.; Desvergne, J.-P. *J. Am. Chem. Soc.* **2000**, *122*, 8795.
49. Bassani, D.M.; Sallenave, X.; Darcos, V.; Desvergne, J.P. *Chem. Commun.* **2001**, 1446.
50. Darcos, V.; Griffith, K.; Sallenave, X.; Desvergne, J.-P.; Guyard-Duhayon, C.; Hasenknopf, B.; Bassani, D.M. *Photochem. Photobiol. Sci.* **2003**, *2*, 1152.
51. Skene, W.G.; Berl, V.; Risler, H.; Khoury, R.; Lehn, J.M. *Org. Biomol. Chem.* **2006**, *4*, 3652.
52. Skene, W.G.; Couzigne, E.; Lehn, J.-M. *Chem. Eur. J.* **2003**, *9*, 5560.
53. Sun, Y.-P.; Ma, B.; Bunker, C.E.; Liu, B. *J. Am. Chem. Soc.* **1995**, *117*, 12705.
54. Zhou, P.; Dong, Z.-H.; Rao, A.M.; Eklund, P.C. *Chem. Phys. Lett.* **1993**, *211*, 337.
55. McClenaghan, N.D.; Absalon, C.; Bassani, D.M. *J. Am. Chem. Soc.* **2003**, *125*, 13004.
56. Fedorova, O.; Fedorov, Y.V.; Gulakova, E.; Schepel, N.; Alfimov, M.; Goli, U.; Saltiel, J. *Photochem. Photobiol. Sci.* **2007**, *6*, 1097.
57. Vedernikov, A.I.; Lobova, N.A.; Ushakov, E.N.; Alfimov, M.V.; Gromov, S.P. *Mendeleev Commun.* **2005**, 173.
58. Pol, Y.V.; Suau, R.; Perez-Inestrosa, E.; Bassani, D.M. *Chem. Commun.* **2004**, 1270.
59. Bouas-Laurent, H.; Castellan, A.; Desvergne, J.P.; Lapouyade, R. *Chem. Soc. Rev.* **2001**, *30*, 248.
60. Bouas-Laurent, H.; Desvergne, J.-P.; Castellan, A.; Lapouyade, R. *Chem. Soc. Rev.* **2000**, *29*, 43.
61. Moriwaki, F.; Ueno, A.; Osa, T.; Hamada, F.; Murai, K. *Chem. Lett.* **1986**, 1865.
62. Ueno, A.; Moriwaki, F.; Azuma, A.; Osa, T. *J. Org. Chem.* **1989**, *54*, 295.
63. Ueno, A.; Moriwaki, F.; Osa, T.; Hamada, F.; Murai, K. *J. Am. Chem. Soc.* **1988**, *110*, 4323.
64. Schafer, C.; Eckel, R.; Ros, R.; Mattay, J.; Anselmetti, D. *J. Am. Chem. Soc.* **2007**, *129*, 1488.
65. Schafer, C.; Mattay, J. *Photochem. Photobiol. Sci.* **2004**, *3*, 331.
66. Molard, Y.; Bassani, D.M.; Desvergne, J.-P.; Horton, P.N.; Hursthouse, M.B.; Tucker, J.H.R. *Angew. Chem. Int. Ed. Engl.* **2005**, *44*, 1072.
67. Molard, Y.; Bassani, D.M.; Desvergne, J.P.; Moran, N.; Tucker, J.H.R. *J. Org. Chem.* **2006**, *71*, 8523.
68. Desvergne, J.-P.; Bouas-Laurent, H.; Perez-Inestrosa, E.; Marsau, P.; Cotrait, M. *Coord. Chem. Rev.* **1999**, *185–186*, 357.
69. Tamaki, T.; Kokubu, T. *J. Incl. Phenom.* **1984**, *2*, 815.
70. Tamaki, T. *Chem. Lett.* **1984**, 53.
71. Tamaki, T.; Kokubu, T.; Ichimura, K. *Tetrahedron* **1987**, *43*, 1485.
72. Nakamura, A.; Inoue, Y. *J. Am. Chem. Soc.* **2003**, *125*, 966.
73. Yan, C.; Nishijima, M.; Nakamura, A.; Mori, T.; Wada, T.; Inoue, Y. *Tetrahedron Lett.* **2007**, *48*, 4357.
74. Yang, C.; Nakamura, A.; Fukuhara, G.; Origane, Y.; Mori, T.; Wada, T.; Inoue, Y. *J. Org. Chem.* **2006**, *71*, 3126.

75. Ikeda, H.; Nihei, T.; Ueno, A. *J. Org. Chem.* **2005**, *70*, 1237.
76. Nakamura, A.; Inoue, Y. *J. Am. Chem. Soc.* **2005**, *127*, 5338.
77. Yang, C.; Fukuhara, G.; Nakamura, A.; Origane, Y.; Fujita, K.; Yuan, D.-Q.; Mori, T.; Wada, T.; Inoue, Y. *J. Photochem. Photobiol. A* **2005**, *173*, 375.
78. Yang, C.; Nakamura, A.; Wada, T.; Inoue, Y. *Org. Lett.* **2006**, *8*, 3005.
79. Yoshizawa, M.; Tamura, M.; Fujita, M. *Science* **2006**, *312*, 251.
80. Mizoguchi, J.-I.; Kawanami, Y.; Wada, T.; Kodama, K.; Anzai, K.; Yanagi, T.; Inoue, Y. *Org. Lett.* **2006**, *8*, 6051.
81. Nishijima, M.; Pace, T.C.S.; Nakamura, A.; Mori, T.; Wada, T.; Bohne, C.; Inoue, Y. *J. Org. Chem.* **2007**, *72*, 2707.
82. Nishijima, M.; Wada, T.; Mori, T.; Pace, T.C.S.; Bohne, C.; Inoue, Y. *J. Am. Chem. Soc.* **2007**, *129*, 3478.
83. Wada, T.; Nishijima, M.; Fujisawa, T.; Sugahara, N.; Mori, T.; Nakamura, A.; Inoue, Y. *J. Am. Chem. Soc.* **2003**, *125*, 7492.
84. Bach, T.; Bergmann, H.; Harms, K. *Org. Lett.* **2001**, *3*, 601.
85. McSkimming, G.; Tucker, J.H.R.; Bouas-Laurent, H.; Desvergne, J.-P.; Coles, S.J.; Hursthouse, M.B.; Light, M.E. *Chem. Eur. J.* **2002**, *8*, 3331.
86. Marquis, D.; Desvergne, J.-P.; Bouas-Laurent, H. *J. Org. Chem.* **1995**, *60*, 7984.
87. Hiraga, H.; Morozumi, T.; Nakamura, H. *Tetrahedron Lett.* **2002**, *43*, 9093.
88. Hiraga, H.; Morozumi, T.; Nakamura, H. *Eur. J. Org. Chem.* **2004**, 4680.
89. McSkimming, G.; Tucker, J.H.R.; Bouas-Laurent, H.; Desvergne, J.-P. *Angew. Chem. Int. Ed. Engl.* **2000**, *39*, 2167.
90. Bradshaw, J.S.; Hammond, G.S. *J. Am. Chem. Soc.* **1963**, *85*, 3953.
91. Chandross, E.A.; Dempster, C.J. *J. Am. Chem. Soc.* **1970**, *92*, 703.
92. Ramesh, V.; Ramamurthy, V. *J. Org. Chem.* **1984**, *49*, 536.
93. Lei, L.; Luo, L.; Wu, X.L.; Liao, G.H.; Wu, L.Z.; Tung, C.H. *Tetrahedron Lett.* **2008**, *49*, 1502.
94. Wu, X.-L.; Luo, L.; Lei, L.; Liao, G.-H.; Wu, L.-Z.; Tung, C.-H. *J. Org. Chem.* **2008**, *73*, 491.
95. Paternò, E.; Chieffi, G. *Gazz. Chim. Ital.* **1909**, *39*, 341.
96. Buchi, G.; Inman, C.G.; Lipinsky, E.S. *J. Am. Chem. Soc.* **1954**, *76*, 4327.
97. Hoffmann, N. *Chem. Rev.* **2008**, *108*, 1052.
98. Bach, T. *Synlett* **2000**, 1699.
99. Chung, W.S.; Turro, N.J.; Srivastava, S.; Li, H.; Le Noble, W.J. *J. Am. Chem. Soc.* **1988**, *110*, 7882.
100. Turro, N.J.; Chung, W.S.; Okamoto, M. *J. Photochem. Photobiol. A* **1988**, *45*, 17.
101. Buschmann, H.; Scharf, H.D.; Hoffmann, N.; Esser, P. *Angew. Chem. Int. Ed. Engl.* **1991**, *30*, 477.
102. Buschmann, H.; Scharf, H.D.; Hoffmann, N.; Plath, M.W.; Runsink, J. *J. Am. Chem. Soc.* **1989**, *111*, 5367.
103. Chung, W.S.; Turro, N.J.; Silver, J.; Le Noble, W.J. *J. Am. Chem. Soc.* **1990**, *112*, 1202.
104. Bach, T.; Bergmann, H.; Brummerhop, H.; Lewis, W.; Harms, K. *Chem. Eur. J.* **2001**, 7, 4512.

105. Bach, T.; Bergmann, H.; Harms, K. *J. Am. Chem. Soc.* **1999**, *121*, 10650.
106. Syamala, M.S.; Devanathan, S.; Ramamurthy, V. *J Photochem.* **1986**, *34*, 219.
107. Inoue, Y. *Chem. Rev.* **1992**, *92*, 741.
108. Inoue, Y. *Mol. Supramol. Photochem.* **2004**, *11*, 129.
109. Inoue, Y.; Kosaka, S.; Matsumoto, K.; Tsuneishi, H.; Hakushi, T.; Tai, A.; Nakagawa, K.; Tong, L.-H. *J. Photochem. Photobiol. A* **1993**, *71*, 61.
110. Tong, L.H.; Lu, R.H.; Inoue, Y. *Prog. Chem.* **2006**, *18*, 533.
111. Inoue, Y.; Dong, F.; Yamamoto, K.; Tong, L.-H.; Tsuneishi, H.; Hakushi, T.; Tai, A. *J. Am. Chem. Soc.* **1995**, *117*, 11033.
112. Inoue, Y.; Wada, T.; Sugahara, N.; Yamamoto, K.; Kimura, K.; Tong, L.-H.; Gao, X.-M.; Hou, Z.-J.; Liu, Y. *J. Org. Chem.* **2000**, *65*, 8041.
113. Lu, R.H.; Yang, C.; Cao, Y.J.; Wang, Z.Z.; Wada, T.; Jiao, W.; Mori, T.; Inoue, Y. *Chem. Commun.* **2008**, 374.
114. Inoue, Y.; Matsushima, E.; Wada, T. *J. Am. Chem. Soc.* **1998**, *120*, 10687.
115. Fukuhara, G.; Mori, T.; Wada, T.; Inoue, Y. *Chem. Commun.* **2005**, 4199.
116. Yang, C.; Mori, T.; Wada, T.; Inoue, Y. *New J. Chem.* **2007**, *31*, 697.
117. Hammond, G.S.; Cole, R.S. *J. Am. Chem. Soc.* **1965**, *87*, 3256.
118. Koodanjeri, S.; Ramamurthy, V. *Tetrahedron Lett.* **2002**, *43*, 9229.
119. Saito, H.; Sivaguru, J.; Jockusch, S.; Dyer, J.; Inoue, Y.; Adam, W.; Turro, N.J. *Chem. Commun.* **2007**, 819.
120. Ihmels, H.; Schneider, M.; Waidelich, M. *Org. Lett.* **2002**, *4*, 3247.
121. Butenandt, J.; Epple, R.; Wallenborn, E.-U.; Eker, A.P.M.; Gramlich, V.; Carell, T. *Chem. Eur. J.* **2000**, *6*, 62.
122. Carell, T.; Epple, R.; Gramlich, V. *Angew. Chem. Int. Ed. Engl.* **1996**, *35*, 620.
123. Epple, R.; Wallenborn, E.-U.; Carell, T. *J. Am. Chem. Soc.* **1997**, *119*, 7440.
124. Goodman, M.S.; Rose, S.D. *J. Org. Chem.* **1992**, *57*, 3268.
125. Tang, W.-J.; Song, Q.-H.; Wang, H.-B.; Yu, J.-Y.; Guo, Q.-X. *Org. Biomol. Chem.* **2006**, *4*, 2575.
126. Wiest, O.; Harrison, C.B.; Saettel, N.J.; Cibulka, R.; Sax, M.; König, B. *J. Org. Chem.* **2004**, *69*, 8183.
127. Reddy, G.D.; Usha, G.; Ramanathan, K.V.; Ramamurthy, V. *J. Org. Chem.* **1986**, *51*, 3085.
128. Kaliappan, R.; Kaanumalle, L.S.; Ramamurthy, V. *Chem. Commun.* **2005**, 4056.
129. Arumugam, S.; Kaanumalle, L.S.; Ramamurthy, V. *J. Photochem. Photobiol. A* **2007**, *185*, 364.
130. Gibb, C.L.D.; Sundaresan, A.K.; Ramamurthy, V.; Gibb, B.C. *J. Am. Chem. Soc.* **2008**, *130*, 4069.
131. Kaanumalle, L.S.; Gibb, C.L.D.; Gibb, B.C.; Ramamurthy, V. *J. Am. Chem. Soc.* **2004**, *126*, 14366.
132. Sundaresan, A.K.; Ramamurthy, V. *Org. Lett.* **2007**, *9*, 3575.
133. Moorthy, J.N.; Koner, A.L.; Samanta, S.; Singhal, N.; Nau, W.M.; Weiss, R.G. *Chem. Eur. J.* **2006**, *12*, 8744.
134. Bach, T.; Aechtner, T.; Neumuller, B. *Chem. Commun.* **2001**, 607.

135. Bach, T.; Aechtner, T.; Neumuller, B. *Chem. Eur. J.* **2002**, *8*, 2464.
136. Furusawa, T.; Kawano, M.; Fujita, M. *Angew. Chem. Int. Ed. Engl.* **2007**, *46*, 5717.
137. Yamaguchi, T.; Fujita, M. *Angew. Chem. Int. Ed. Engl.* **2008**, *47*, 2067.
138. Annalakshmi, S.; Pitchumani, K. *J. Photochem. Photobiol. A* **2006**, *184*, 34.
139. Kaanumalle, L.S.; Gibb, C.L.D.; Gibb, B.C.; Ramamurthy, V. *Org. Biomol. Chem.* **2007**, *5*, 236.
140. Weber, L.; Imiolczyk, I.; Haufe, G.; Rehored, D.; Hennig, H. *J. Chem. Soc. Chem. Commun.* **1992**, 301.
141. Natarajan, A.; Kaanumalle, L.S.; Jockusch, S.; Gibb, C.L.D.; Gibb, B.C.; Turro, N.J.; Ramamurthy, V. *J. Am. Chem. Soc.* **2007**, *129*, 4132.
142. Yoshizawa, M.; Miyagi, S.; Kawano, M.; Ishiguro, K.; Fujita, M. *J. Am. Chem. Soc.* **2004**, *126*, 9172.
143. Royer, G.P.; Cruickshank, K.A.; Morrison, L.E.; Amoco Corp., USA. Method and apparatus for template-directed photoligation and their use in assays. Patent 89-300258, **1989**, p. 29.
144. Ogasawara, S.; Fujimoto, K. *Chembiochem* **2005**, *6*, 1756.
145. Fujimoto, K.; Ogawa, N.; Hayashi, M.; Matsuda, S.; Saito, I. *Tetrahedron Lett.* **2000**, *41*, 9437.
146. Ogasawara, S.; Yoshimura, Y.; Hayashi, M.; Saito, I.; Fujimoto, K. *Bull. Chem. Soc. Jpn.* **2007**, *80*, 2124.
147. Fujimoto, K.; Yoshimura, Y.; Ikemoto, T.; Nakazawa, A.; Hayashi, M.; Saito, I. *Chem. Commun.* **2005**, 3177.
148. Ogino, M.; Yoshimura, Y.; Nakazawa, A.; Saito, I.; Fujimoto, K. *Org. Lett.* **2005**, *7*, 2853.
149. Yoshimura, Y.; Okamura, D.; Ogino, M.; Fujimoto, K. *Org. Lett.* **2006**, *8*, 5049.
150. Fujimoto, K.; Matsuda, S.; Yoshimura, Y.; Ami, T.; Saito, I. *Chem. Commun.* **2007**, 2968.
151. Yoshimura, Y.; Noguchi, Y.; Sato, H.; Fujimoto, K. *Chembiochem* **2006**, *7*, 598.
152. Yoshimura, Y.; Noguchi, Y.; Fujimoto, K. *Org. Biomol. Chem.* **2007**, *5*, 139.
153. Fujimoto, K.; Yoshino, H.; Ami, T.; Yoshimura, Y.; Saito, I. *Org. Lett.* **2008**, *10*, 397.
154. Ihara, T.; Fujii, T.; Mukae, M.; Kitamura, Y.; Jyo, A. *J. Am. Chem. Soc.* **2004**, *126*, 8880.

3

COMPLEXATION OF FLUORESCENT DYES BY MACROCYCLIC HOSTS

Roy N. Dsouza, Uwe Pischel, and Werner M. Nau

INTRODUCTION

The formation of host–guest complexes is a fundamental process in supramolecular chemistry. Varying spectroscopic techniques have been used to determine stability constants and binding stoichiometries, as well as complexation geometries. Among the various photophysical techniques used to determine such properties, fluorescence spectroscopy is exceptionally useful. First, many measurements can be performed by conventional steady-state fluorescence, where the change in fluorescence upon complexation provides a readily distinguishable feature to signal the formation of a supramolecular complex with a fluorescent dye. Second, the high sensitivity of fluorescence allows studies in more dilute solutions (μM to nM) than alternative techniques like nuclear magnetic resonance (NMR), UV-Vis spectrophotometry, and isothermal calorimetry (ITC).[1] This is particularly important when stability constants are high, and very dilute solutions are required to extract them accurately.

The structure of this chapter is such that we will provide a diagnostic compilation of fluorescence effects accompanying complexation with macrocyclic hosts, and the factors responsible for these photophysical effects will be summarized. We will subsequently focus on three different lines of applications: first, the use of dye–macrocycle complexes as model systems for fluorescence

Supramolecular Photochemistry: Controlling Photochemical Processes, First Edition. Edited by V. Ramamurthy and Yoshihisa Inoue.

resonance energy transfer (FRET) studies; second, the use of solvatochromic probes to sound out properties of the microenvironment inside the macrocyclic cavities (polarity and polarizability); and third, the application of fluorescent host–guest complexes for indicator displacement assays.

PHOTOPHYSICAL EFFECTS ON FLUORESCENT DYE ENCAPSULATION

The intermolecular complexes between macrocyclic host molecules and smaller guest molecules present prototypal supramolecular architectures. Owing to the fact that most macrocyclic hosts possess a concave inner cavity to include the guest, they are frequently considered as molecular containers with nanoscale dimensions.[2] The combination of a hydrophobic cavity with water solubility has further defined the enzyme-mimetic potential of water-soluble macrocyclic hosts.[3,4] Pertinent examples are cyclodextrins (CDs), calixarenes, and cucurbiturils, which are also available in different sizes to modulate their selectivity in guest binding (Fig. 3.1). CDs are naturally occurring macrocyclic compounds consisting of α-D-glucose subunits and were first encountered in the late 1800s.[5] They have been comprehensively studied[6] and have found numerous industrial applications.[5] *p*-Sulfonatocalix[*n*]arenes are water-soluble calixarene derivatives first synthesized by Shinkai and coworkers[7] and have consequently enabled the expansion of the aqueous chemistry of these aromatic macrocycles.[8] Cucurbit[*n*]urils are rigid glycoluril-based macrocycles that were first synthesized by Behrend et al. in 1905.[9] Their unique structure, however, was elucidated by Mock and coworkers in 1981.[10] Subsequent pioneering work by Mock,[11] Buschmann et al.,[12] Kim et al.,[13] and Day et al.[14] have significantly built upon the synthesis of their homologs and their applications in supramolecular chemistry.[15–18]

It is very interesting to investigate whether the inclusion of chromophoric guests into the various macrocycles, most importantly fluorescent dyes (Fig. 3.2), can affect their photophysical properties, and potentially lead to unprecedented effects and, ultimately, new applications.[2] In fact, numerous investigations have dealt with variations of fluorescent properties upon complexation.[19–22] We present a comparative compilation of selected host–guest systems where such effects have been documented (Table 3.1).

A wide spectrum of binding strengths (K_{ass}) coupled with varied fluorescence responses upon complexation provides a complete toolset of host–dye "reporter pairs" for indicator displacement applications (see section on Displacement sensor applications). Higher binding constants imply that these reporter pairs can be used in much smaller concentrations than those with moderate binding constants. Additionally, a large fluorescence response (I_{bound}/I_{free}) upon complexation, where a value greater than 1 indicates enhancement and a value less than 1 signals quenching, is highly desirable as this increases the sensitivity of the reporter pair.

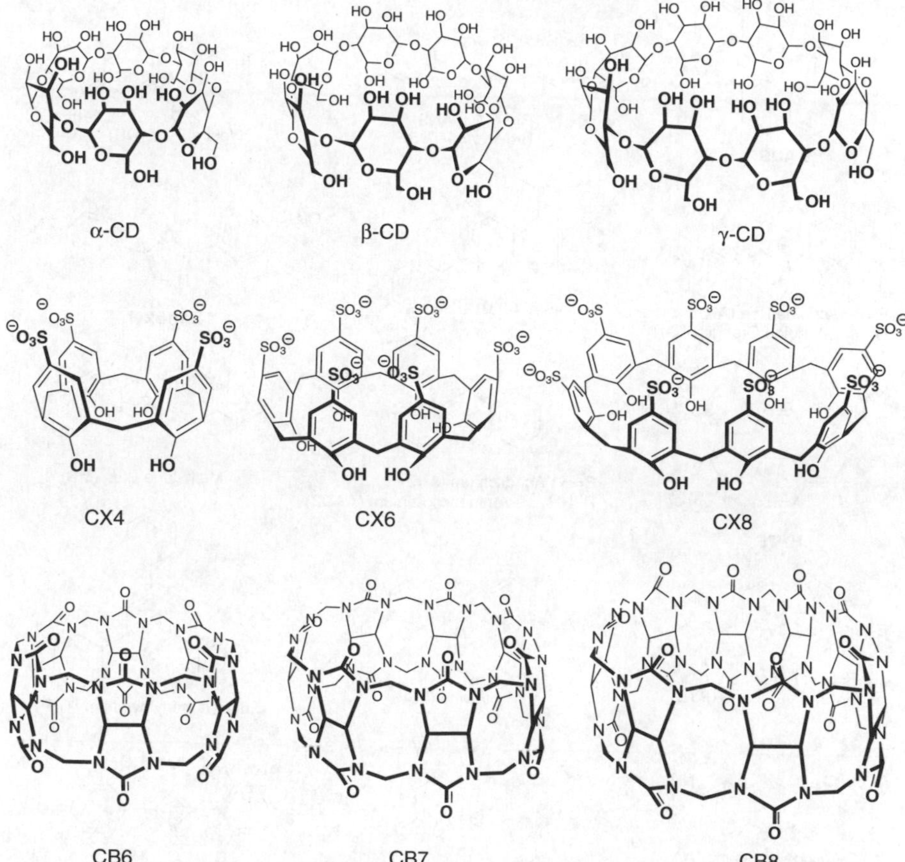

Figure 3.1. Water-soluble macrocyclic host molecules with varying size (from *top* to *bottom* and *left* to *right*): cyclodextrins (α, β, γ), *p*-sulfonatocalix[*n*]arenes (n = 4, 6, 8), and cucurbit[*n*]urils (n = 6, 7, 8).

Upon excitation to the first singlet excited state, several pathways of deactivation apply (Fig. 3.3), including radiative decay (k_r), fluorescence resonance energy transfer (k_{FRET}), photoproduct formation (k_P), internal conversion (k_{IC}), intersystem crossing (k_{ISC}), solvent-induced quenching, and quenching by additives or oxygen.[2] Several of these are being modified by macrocyclic encapsulation, and the observed fluorescence effects, either quenching or enhancement, can be traced back, in several cases, to predominant effects on individual rate constants. In most cases, the effects of inclusion complexation are satisfactorily understood in terms of effects on k_{IC}, related to either (1) the relocation of the fluorophore into the more hydrophobic environment of the host cavities or (2) the geometrical confinement of the chromophore within the host, which restricts rotational and vibrational freedom, thereby disfavoring nonradiative

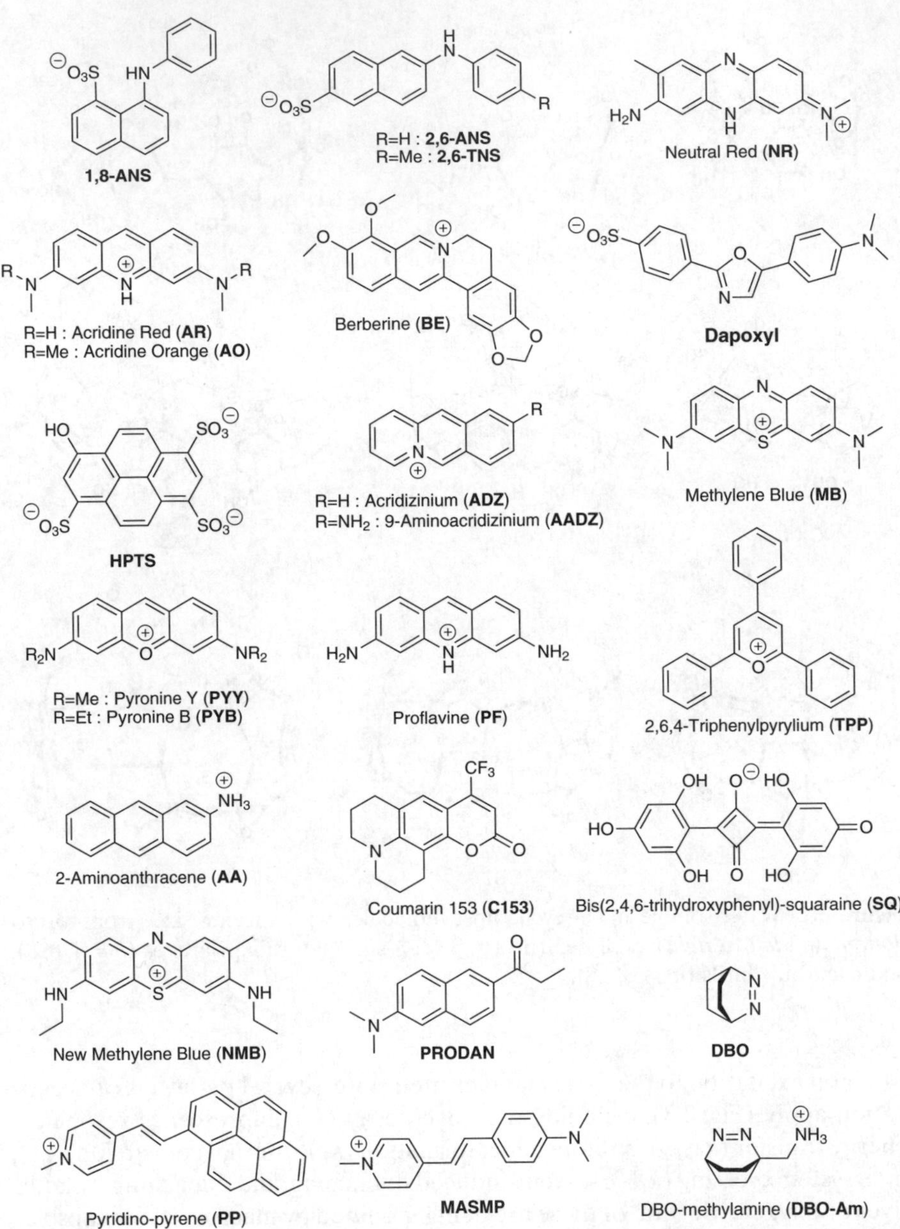

Figure 3.2. Examples of water-soluble fluorescent guest molecules that significantly alter their fluorescence response on host encapsulation. HPTS, 8-hydroxy-1,3,6-trisulfonic acid; MASMP, *trans*-4-[4-(dimethylamino)styryl]-1-methylpyridinium.

TABLE 3.1. A Comparative Study of the Photophysical Effects Exhibited by Different Dyes on Host Encapsulation in Aqueous Solution

Dye	Host	K_{ass} (M^{-1})	λ_{ex} (λ_{em})[a] (nm)	I_{bound}/I_{free}[b]	Reference
DBO	CB7	4×10^5	374 (427)	~0.73	23, 24
	β-CD	1100	358 (431)	<0.2	25
	CX4	1200	365 (420)	0.1	26
1,8-ANS	CB7	$K_1 = 19$ $K_2 = 160$	370 (469)	94	27
	β-CD	102	350 (510)	2.4	28
2,6-ANS	CB7	600	370 (452)	25	27
	β-CD	1350	325 (443)	32	21
2,6-TNS	β-CD	3700	350 (483)	16	28
NR	CB7	6.5×10^3	450 (600)	~8	29
	β-CD	$K_1 = 411$ $K_2 = 420$	444 (588)	~4.5	30
AO	CB7	2.0×10^5	465 (514)	3	31
	β-CD	1200	455 (548)	~0.75	31
AR	β-CD	2630	490 (553)	4	28
	CX6	9550	493 (560)	~0.33	32
BE	CB7	1.6×10^6	400 (508)	500	33
	CX4	2951	352 (546)	~40	34
	CX6	8192	352 (546)	~13	34
	CX8	$\sim 2.8 \times 10^5$	352 (546)	~40	34
Dapoxyl	CB7	2.0×10^4	336 (380)	200	35, 36
	β-CD	5488	353 (560)	7.8	37
HPTS	γ-CD	120	405 (435)	~6	38
ADZ	CB8	—	— (408)	0.05	39
AADZ	CB8	$4 \times 10^9\ M^{-2}$	— (508)	0.025	39
MB	CB7	1.3×10^7	665 (685)	~4.3	40
	CX4	1.41×10^5	640 (698)	~0.2	41
	CX6	6.61×10^5			
PYY	CB7	4.6×10^6	546 (565)	~1.8	40
	β-CD	380	515 (570)	0.5	42
PYB	CB7	—	556 (571)	~1.9	43
	β-CD	2000	515 (570)	0.53	42
PF	CB7	1.7×10^7	400 (500)	2.3	40
TPP	CB8	1.45×10^5	420 (590)	~5.5	44
AA	CB7	8×10^5	260 (412)	3.7	45
C153	β-CD	54	450 (555)	3.7	46
	γ-CD	—	405 (530)	2.1	47
NMB	β-CD	740	594 (647)	5.1	48
PRODAN	β-CD	862	360 (502)	~10	49
SQ	β-CD	$5.8 \times 10^5\ M^{-2}$	560 (608)	90	50
PP[c]	CX4	690	430 (580)	0.4	51
	CX6	5754		0.1	

(Continued)

TABLE 3.1. (*Continued*)

Dye	Host	K_{ass} (M^{-1})	λ_{ex} (λ_{em})[a] (nm)	I_{bound}/I_{free}[b]	Reference
MASMP	CX4	3.5×10^5	470 (604)	20	52
	CX6	1.0×10^5		30	
	CX8	$K_1 = 3 \times 10^6$		6	
		$K_2 = 8 \times 10^5$			
DBO-Am	CX4	2.3×10^4	365 (420)	~0.3	35, 53

[a]λ_{ex} represents the fluorescence excitation wavelength. λ_{em} represents the fluorescence maximum.
[b]I_{bound}/I_{free} represents the relative fluorescence of the host–dye complex (I_{bound}) compared with the free dye (I_{free}).
[c]Water/methanol mixture (1:1).
BE, berberine; ADZ, acridizinium; AADZ, 9-aminoacridizinium; PF, proflavine; TPP, 2,6,4-triphenylpyrylium; AA, 2-aminoanthracene; NMB, new methylene blue; SQ, bis(2,4,6-trihydroxyphenyl)-squaraine.

decay pathways.[21,27,43,54] Additionally, upon complexation, the fluorophore might also be protected from external (intermolecular) quenchers including the solvent and oxygen by the walls of the macrocycle (Fig. 3.3).[18,36]

For example, the polarity sensitive dyes **1,8-ANS** and **2,6-ANS** show a marked fluorescence enhancement on complexation with both CB7 as well as β-CD owing to a microenvironment of lower polarity inside these macrocyles.[21,27] Similarly, **Dapoxyl**, a known fluorescent intramolecular charge transfer (ICT) dye,[55,56] shows a pronounced increase in its charge transfer (CT) band on encapsulation.[36,37] It is interesting to note, however, that β-CD quenches the fluorescence of pyronine Y (**PYY**), pyronine B (**PYB**), and acridine orange (**AO**)—all similarly structured dye molecules.[31,57] Calixarenes generally quench the fluorescence of included guests (2,3-diazabicyclo[2.2.2]oct-2-ene (**DBO**), 1-methylamino-2,3-diazabicyclo[2.2.2]oct-2-ene (**DBOAm**), acridine red [**AR**], methylene blue [**MB**], and pyridine-pyrene [**PP**]) due to charge transfer-induced quenching by their electron-rich phenoxy rings.[26,32,41,51,53]

In the case of cucurbit[*n*]urils, although polarity and confinement effects contribute to fluorescence enhancement of included dyes, charge–dipole interactions play an unquestionably dominant role. In contrast to CDs, the carbonyl-lined portals of cucurbiturils (CBs) provide an ideal framework for host-assisted guest protonation[58]; that is, the dye becomes protonated when complexed. This accounts for the observation of a locally excited band in the case of **Dapoxyl**, which corresponds to the emission of the protonated form of the dye.[36] Furthermore, it also causes the relatively higher binding constant of the dye with CB7 as compared with β-CD. The same reasoning applies in the complexation of **AO** and neutral red (**NR**) with CB7 versus β-CD.

SUPRAMOLECULAR ENERGY TRANSFER (ET) IN HOST–GUEST COMPLEXES BETWEEN DYES AND CDS

Supramolecular host–guest complexes between dyes and macrocycles have also received special attention for the purpose of studying ET processes

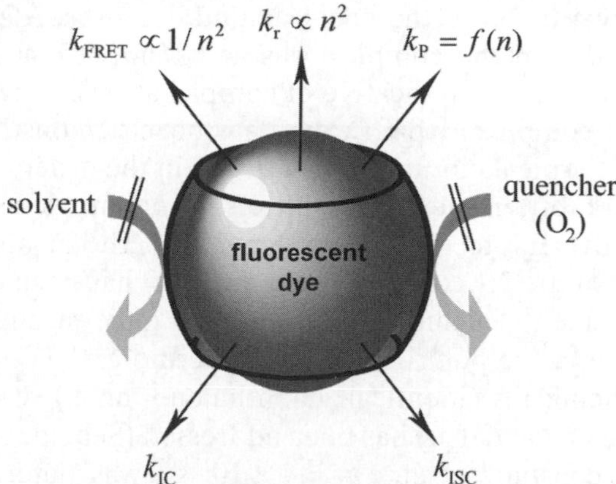

Figure 3.3. Deactivation of singlet excited states inside molecular containers illustrating the protection from quenchers and solvent-induced quenching as well as rate constants, which are being potentially modulated through the polarizability/refractive index.

between chromophores. In this regard, CDs have enjoyed preference as hosts, because of their convenient modification with chromophoric moieties via synthetic manipulation of free hydroxyl groups and the spectral neutrality due to absence of chromophores in the macrocycle itself. In this section, some selected recent examples for ET studies with dye–CD host–guest complexes will be reviewed, with special emphasis on their use as artificial supramolecular light-harvesting antennae, fluorescent chemosensors, and molecular switches. The focus will be on systems where all components are of organic nature. However, for the sake of completeness, it should also be mentioned that luminescent metal–organic complexes of Ln^{III} (Ln = Eu, Tb), Ru^{II}, or Os^{II} have been investigated as functional building blocks in CD-derived ET supramolecular devices.[59–64]

An early example for ET in CD-derived host–guest complexes involved a benzophenone-capped CD. Excitation of the aromatic ketone led to moderately efficient triplet–triplet ET (Φ_{ET} ca. 0.5–0.6) to encapsulated naphthalene derivatives, which was demonstrated by phosphorescence spectroscopy.[65] However, many dyes are involved in typical excited singlet state processes, which includes FRET. This notion is also based on often high extinction coefficients and elevated fluorescence quantum yields of many dyes, which are photophysical parameters favoring FRET.

Various modified β-CDs containing seven naphthalene chromophores (**CD1** and **CD2**) were described, which have been used to form supramolecular host–guest complexes with the rod-like shaped dyes oxazine 725 (**Ox 725**) and the merocyanine 4-dicyanomethylene-2-methyl-6-(*p*-diethanolaminostyryl)-4*H*-pyran (**DCM-OH**); see Scheme 3.1a, b.[66–69] For **CD1** and **DCM-OH**, a 1:1 complexation with a binding constant of $\log K_{11} = 5$ was determined.[68,69] Upon excitation in the naphthalene absorption band at 300 nm, an efficient ET (Φ_{ET} ca. 1) to the encapsulated dye proceeded. The emission maximum of the dye in the complex was at 604 nm (ca. 40 nm blue-shifted compared with the free dye, indicative of complexation). For **CD2**, the formation of 1:1 and 2:1 complexes (β-CD : dye) has been demonstrated for **Ox 725** and **DCM-OH**.[66,67] Typical binding constants are in the order of $\log K_{11} = 4$ and $\log K_{12} = 7$. Remarkably, in the 2:1 complexes, 14 naphthalene units are available for quantitative ET to the encapsulated dye. The arrangement of both **CD2** macrocycles in the 2:1 complex (Scheme 3.1c) has been corroborated by a series of experiments including variation of ionic strength, competitive binding, and fluorescence polarization.[67] Furthermore, the question of homo-ET (between chemically identical naphthalene units) versus hetero-ET (between naphthalene and dye) has been addressed (Scheme 3.1d).[66] For **CD2**, a homo-ET rate constant of $k_{\text{homoET}} = 4 \times 10^{11}\,\text{s}^{-1}$ was determined, which is significantly faster than hetero-ET to **Ox 725** (ca. 10 times) and **DCM-OH** (ca. three times). For both ET pathways, the main mechanism was assumed as FRET, albeit some minor contribution of contact ET was not excluded due to the spatial proximity between the chromophores in the assemblies. These investigations have shown that supramolecular host–guest complexes of modified CDs with covalently appended chromophores and encapsulated dyes may serve as efficient molecular functional units, operative through an absorption–ET–emission sequence. The molecular architecture (cyclic arrangement of the antenna chromophores), supramolecular organization, and high ET efficiency are reminiscent of natural light-harvesting systems in green plants. This renders these complexes as promising antenna mimics.

Photochromic compounds change absorption spectral properties depending on their switching state, for example, open versus closed form of spiropyran systems. This can be used to modulate the emission of signaling units in molecular dyads of photochrome–spacer–fluorophore type.[70,71] In many cases, the photochromic unit plays the role of the energy acceptor, which, depending on the spectral overlap of its absorption spectrum with the emission spectrum of the fluorophore, may engage in ET. Activation of this process leads to fluorescence quenching (ON–OFF switching).

Recently, this approach has been implemented in supramolecular assemblies containing CD as host for either the fluorophore[72] or a photochrome-modified anchor.[73] In Scheme 3.2, a spiropyran-modified β-CD is shown (**CD3**), which serves as host for rhodamine B (**RhB**).[72] The closed spiropyran form is not involved in ET, leaving the **RhB** emission at high intensity. Upon light-induced ($\lambda_{\text{exc}} = 365$ nm) conversion of the spiropyran to its open merocyanine form, ET from the encapsulated dye is activated. This led to efficient emission quenching (>85%) of **RhB**. The great value of supramolecular systems for this type of switches lies in their flexible architecture, which enables the use of the same photochrome-modified CD with a large variety of fluorophores, thereby leading to spectral fine-tuning of the output signal of the device. Another example used the inverse molecular situation: the fluorophore was bound covalently to β-CD and the photochrome supramolecularly assembled

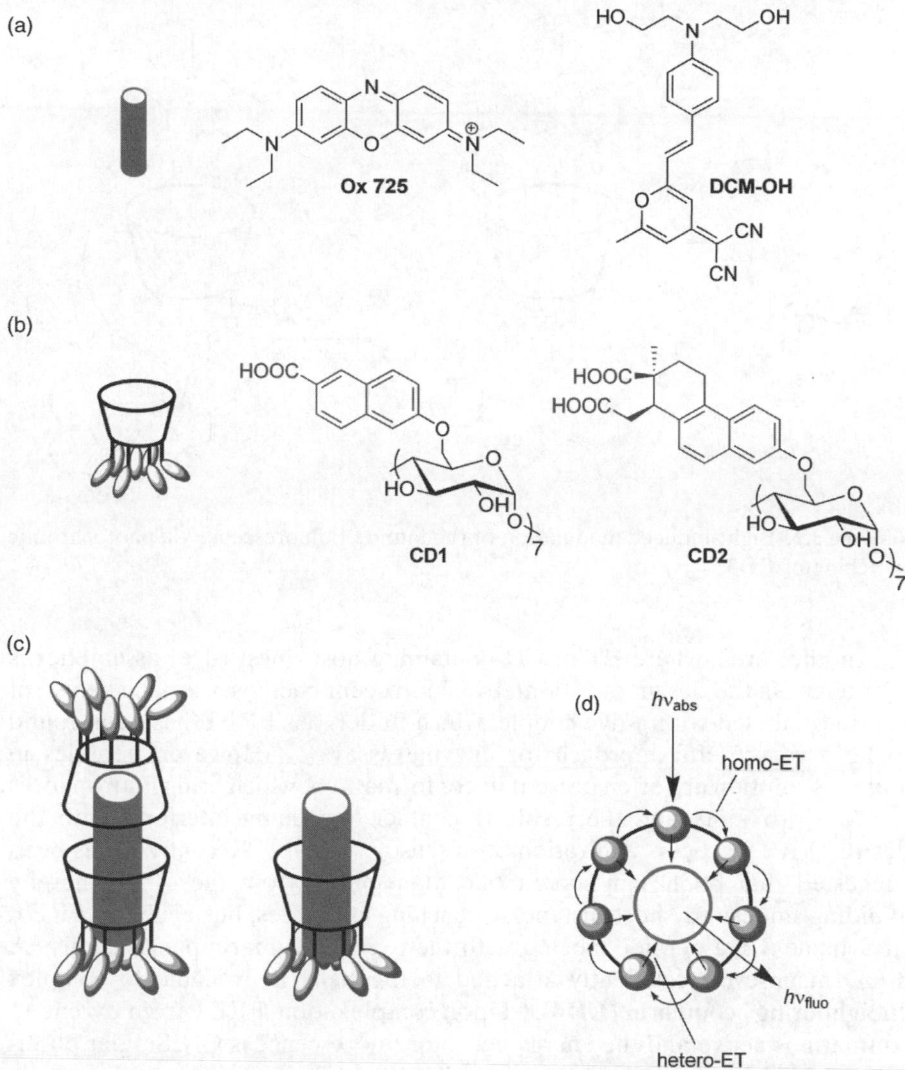

Scheme 3.1. (a) Structures of dyes, (b) structures of naphthalene-modified β-cyclodextrins, (c) representation of 2:1 (left) and 1:1 (right) β-CD : dye complexes, and (d) homo- and hetero-energy transfer processes in β-CD : dye complexes. Naphthalene unit peripheral spheres and dye as central circle. See ftp site for color: ftp://ftp.wiley.com/public/sci_tech_med/supramolecular_photochemistry.

via an adamantane anchor.[73] In the latter example, not only light was used as external stimulus of the switch, but also temperature, which affected the complexation of the anchor system with the CD. Thus, the distance between dye and photochrome was altered, which has impact on the efficiency of FRET.

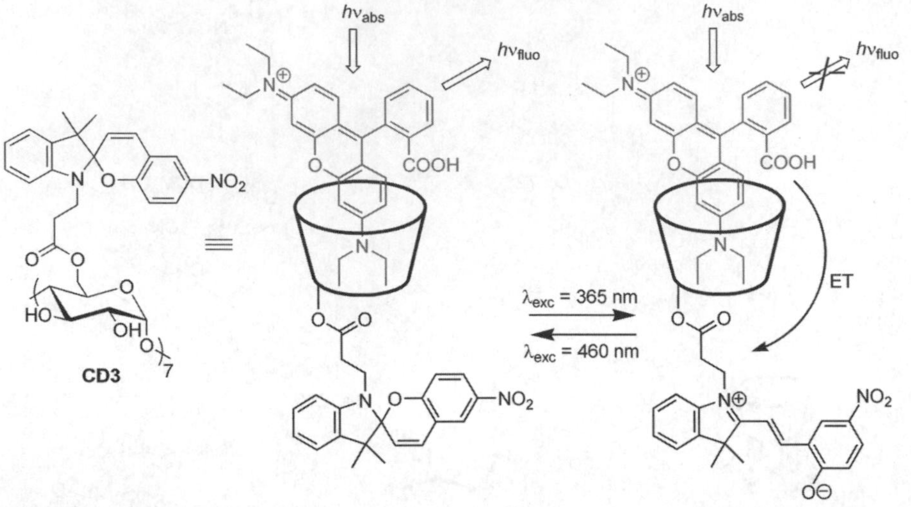

Scheme 3.2. Light-induced modulation of rhodamine B fluorescence via photochromic switching of **CD3**.

Another area, where ET in CD-containing host–guest (dye) assemblies is of interest, is the design of ratiometric fluorescent chemosensors.[74] The use of peptides labeled with a dye couple, which undergoes FRET, has been found to be a very useful approach for enzyme assays.[75–77] However, peptides in aqueous solution are often present in conformations, which bring fluorophores to close proximity and the resulting contact quenching interferes with the desired FRET process and ratiometric sensor response. Recently, it has been suggested that β-CD can screen one fluorophore from the other, thereby avoiding undesired short distance quenching processes, but enabling FRET. In Scheme 3.3, a peptide labeled with the pyrene/coumarin pair is shown. A β-CD macrocycle covalently attached to the helix is available to complex "neighboring" coumarin (**CD4**).[78] Upon complexation, FRET from pyrene to coumarin is active and the energy acceptor fluorescence is ON. Similar observations have been made for a coumarin–spacer–fluorescein dyad covalently linked to β-CD (**CD5**, see structure in Scheme 3.3).[79]

For **CD4**, it has been shown that the situation can be reverted by competitive binding of hyodeoxycholic acid by the macrocycle ($\log K_{11} = 5.7$), which leads to contact quenching in the dye pair and consequently the observation of fluorescence OFF switching. Thus, the system has been suggested as FRET peptide probe for cholesterol or other steroidal compounds.[78]

Finally, a remarkable system where all components, that is, energy donor dye, acceptor dye, and macrocycle, are organized noncovalently will be briefly introduced.[80] A [2]rotaxane made of γ-CD, a stilbene dye as thread, and terphenylene dicarboxylic acid stoppers, was shown to provide a supramolecular platform for enhanced binding of a second dye (a Cy3 dye derivative) in the

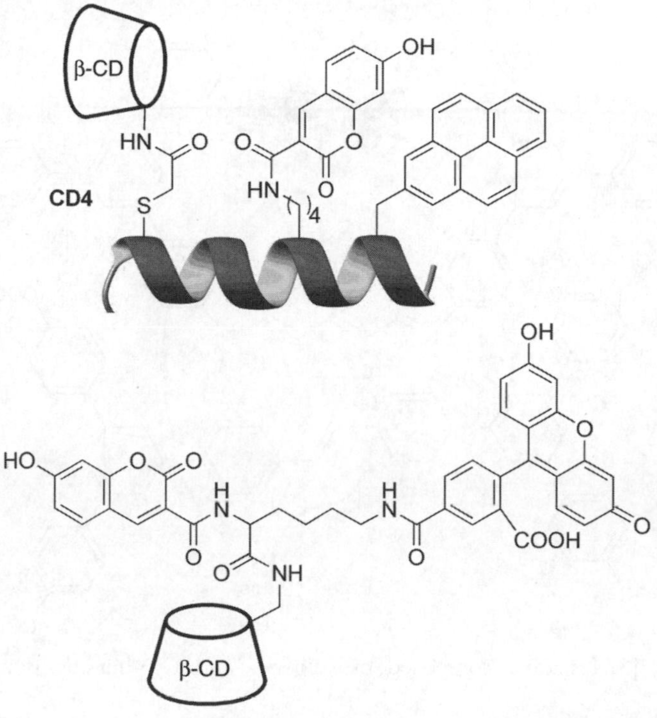

Scheme 3.3. Molecular structures of FRET-active cyclodextrins (CDs) **CD4** and **CD5**.

same macrocyclic cavity. It was reasoned that the stilbene π-system improves hydrophobic interactions with the Cy3 dye, which resulted in a three orders of magnitude higher binding constant than with γ-CD alone ($\log K_{11} = 5.0$ vs. 1.9, respectively). The thus enhanced preorganization favored the formation of a [3]rotaxane through locking of the Cy3 dye with the same bulky end groups as used for the stilbene dumbbell (Scheme 3.4). It was found that upon excitation of the stilbene chromophore at 350 nm, the Cy3 dye emits strongly ($\Phi_f = 0.56$) at a maximum of 603 nm. The quantitative quenching of stilbene fluorescence (430 nm) corroborates a very efficient ET between both encapsulated dyes ($\Phi_{ET} \sim 1$). The observed supramolecular binding features and efficient FRET process make this approach a potential candidate for sensing applications.

Interestingly, although this has not yet been experimentally tested, the rates of fluorescence resonance ET (k_{FRET}) should also show a systematic variation with the polarizability of the microenvironment offered by the macrocyclic interior (see Fig. 3.3). Indeed, the polarizability shows large variations for different macrocycles, as will be exposed in the following section.

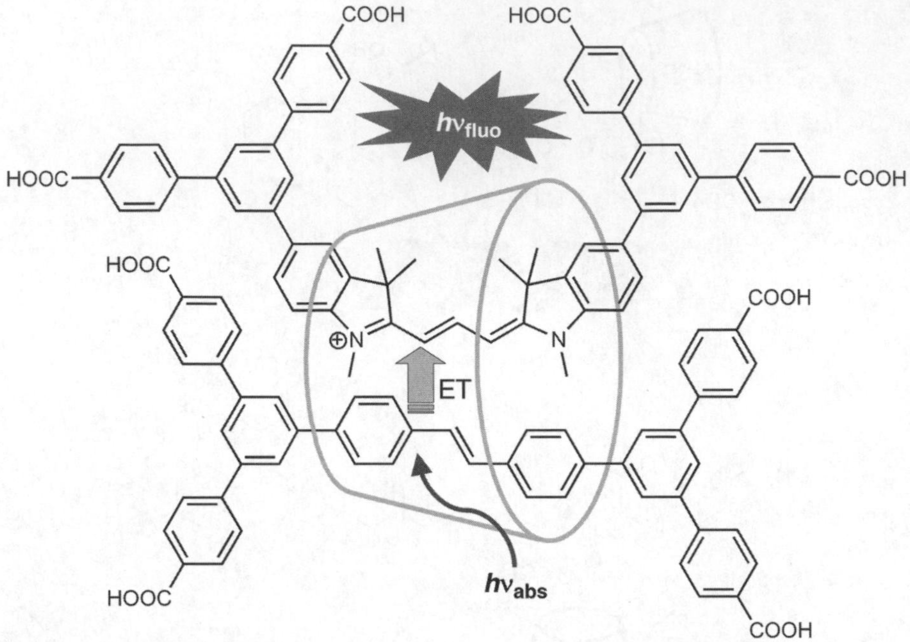

Scheme 3.4. [3]Rotaxane composed of stilbene- and Cy3-derived dumbbells and γ-CD.

POLARITY AND POLARIZABILITY OF MACROCYCLIC HOSTS

Cram postulated that the inner phase of supramolecules, for example, hemicarcerands, could behave as a new phase of matter.[81] Macrocyclic host molecules might potentially offer to included guests an extreme environment very different from common solvents, and thereby induce novel photophysical effects. In order to understand the properties of these microenvironments, several solvatochromic probes have been employed to investigate their polarity as well as polarizability.

Variations in the absorption and fluorescence spectra of solvatochromic probes can be principally employed to estimate the polarity of macrocyclic cavities, which cannot be assessed by direct spectroscopic methods. Wagner and coworkers developed a so-called polarity sensitivity factor (PSF) for a certain probe (e.g., cucurmin), in an attempt to standardize such measurements.[82] Table 3.2 presents a summary of the measured cavity polarity of different macrocycles compared with those of common organic solvents and solvent mixtures.

It can be easily seen that alcohol/water mixtures can successfully approximate the polarity of the interior cavity of the families of macrocycles discussed. However, the measured polarities are strongly dependent on the probe used as well as its orientation within the cavity. Presumably, the organic dyes are

TABLE 3.2. Polarities of Macrocyclic Cavities Compared with Common Solvents and Solvent Mixtures

Host	Probe	Equivalent Polarity	Reference
α-CD	**2,6-TNS**	52% EtOH	83
β-CD	**1,8-ANS**	66% MeOH	22
	1,8-ANS	55% EtOH	19
	2,6-TNS	59% EtOH	83
CX6[a]	**1,8-ANS**	80% EtOH	19
	2,6-TNS	60% EtOH	83
CX8[a]	Pyrene	1-Butanol	84
	2,6-TNS	85% EtOH	83
CB6	Cucurmin	87% EtOH	82
CB7	Rhodamine 6G	1-Octanol	54

[a]Modified macrocycles. See individual references.

not completely immersed in the cavities of any host, such that the chromophores remain at least partially exposed to the surrounding water. For example, studies have shown that α-, β-, and γ-CDs have quite similar polarities,[85–87] and observed differences are assumed to be attributed to differences in the degree of encapsulation of the probe into the CD cavity, or on the number of water molecules co-included with the guest.[22]

Aside from polarity effects experienced inside macrocycles, spectroscopic evidence has shown that the cavity of molecular containers can indeed impose extreme physical properties, namely an environment with extraordinary polarizability,[23] which can have a pronounced influence on photophysical properties, particularly on the fluorescence lifetimes.[24] To measure polarizabilities inside host molecules, new types of solvatochromic probes have been employed.

In particular, it has been observed that **DBO**, as well as other cyclic azoalkanes and some other n,π^*-excited chromophores like biacetyl, display environmentally dependent changes of the absorption intensity, that is, oscillator strength ($f = 4.32\times10^{-9}\int\varepsilon(\tilde{v})d\tilde{v}$).[23] The best correlation was observed in both cases (**DBO** and biacetyl) between the inverse oscillator strength and the (bulk) polarizability. Clear trends of the absorption and emission maxima with varying polarizability were also observed, but the changes of the oscillator strengths were most systematic in nature. Importantly, there was no correlation with the polarity or hydrogen-bond donor abilities of the solvent, such that polarizability effects could be reliably dissected from the more trivial, and less interesting, polarity effects.[88]

Consequently, when the oscillator strength of the near-UV absorption band of **DBO** was measured in different solvents as well as in the gas phase, a linear correlation between the inverse oscillator strength and the polarizability (P) of the solvent was found ($1/f = 3020–8320P$, $r = 0.979$, 11 data points),[23] which established **DBO** as a solvatochromic probe for the polarizability of the environment. Note that the bulk polarizability P is a simple function of the solvent refractive index $[P = (n^2 - 1)/(n^2 + 2)]$.

To extract the polarizabilities inside the various host molecules, **DBO** solutions with excess host were prepared, and the oscillator strengths of the resulting host–guest complexes were determined and corrected for the degree of complexation, where required. Subsequent correlation according to the established solvatochromic relationship provided the polarizabilities inside a variety of supramolecular container compounds (see Table 3.3). Interestingly, the polarizability inside CDs was found to be rather unspectacular, similar to that of water and alcohols.[24] The polarizability inside CX4, on the other hand, fell slightly below that of benzene, as would be expected from the partial exposure of the guest to the aqueous environment at the upper and lower opening of the calixarene.[26]

An exceptionally low polarizability was interpolated for cucurbit[7]uril, even lower than that in perfluorohexane and close to the gas phase. Furthermore, by employing biacetyl as a solvatochromic polarizability probe and using previously reported phosphorescence data for biacetyl,[89] an extremely high polarizability was extrapolated for hemicarcerands,[23] even higher than that in diiodomethane (Table 3.3). These results provided strong support for Cram's much debated "new phase of matter" hypothesis,[81] because they revealed refractive indices inside supramolecular architectures, which are unheard of in any solvent.

The observed trends, however, are not entirely unexpected, when the molecular structures are closely inspected. The inside of hemicarcerands is dominated by electron-rich π-systems pointing toward the interior, which are known to be highly polarizable; the guest is therefore exposed in all directions to a sea of π-electron density, even more pronounced than if it was immersed in benzene, where the tumbling of the solvent molecules reduces the effective polarizability exerted by the π-systems. In contrast, the inside of CB7 is the opposite of electron rich; all the surrounding bonds are polar, and there are no C–H bonds, π bonds, or lone pair orbitals pointing inward. Note, in particular, that the concave shape displaces the electron density of the ureido nitrogen lone pairs toward the outside of the cavity (see Fig. 3.1); the resonance interaction with the carbonyl groups further withdraws electron density toward the upper and lower portals, away from the inside of the cavity and the centrally positioned guest. The polarizabilities inside nanoscale containers cover therefore a very wide range and allow the exposure of guest molecules to environments with extremely low and high values.

The question regarding the relevance of a high or low environment polarizability on the photophysical properties of fluorescent dyes consequently arises. Several primary photophysical processes depend on the polarizability/refractive index of the environment (see Fig. 3.3).[90] The dependence of the decay rate constants on the refractive index can be analytically expressed for two pertinent rates: the FRET rate constant according to the Förster dipole–

TABLE 3.3. Refractive Index and Polarizability Inside Macrocyclic Host Molecules Determined by Using the DBO Chromophore as a Solvatochromic Probe, Relative to Those in Solvents and the Gas Phase[23]

Environment	Refractive Index (n)[a]	Polarizability (P)[b]
Gas phase	1.000	0.000
Cucurbit[7]uril	1.19	0.12
Perfluorohexane	1.252	0.159
β-CD[c]	1.33	0.20
H_2O	1.333	0.206
n-Hexane	1.375	0.229
CX4[d]	1.41	0.25
Benzene	1.501	0.295
Diiodomethane	1.742	0.404
Hemicarcerand[e]	1.86	0.45

[a]Refractive index, converted using the formula $P = (n^2 - 1)/(n^2 + 2)$.
[b]From the empirical relationship ($1/f = 3020–8320P$).
[c]See Reference 24.
[d]See Reference 26.
[e]Value determined using biacetyl as solvatochromic probe.[23]

dipole mechanism (k_{FRET}) and, more important since already experimentally confirmed, the radiative decay rate constant (k_r).

The Strickler–Berg equation predicts a dependence of the radiative decay rate on the square of the refractive index (Eq. 3.1)[91]:

$$k_r = 0.668\langle\tilde{v}\rangle_{av}^2 n^2 f. \tag{3.1}$$

The validity of this equation was previously tested for fluorophores with allowed electronic transitions; in these cases, the oscillator strength can be considered to remain constant in different environments.[92–96] A direct relationship between k_r and n^2 results in these limiting cases, which has been employed to estimate the polarizability in biological microenvironments like lipid membranes,[96] reverse micelles,[92–94] the interior of the green fluorescent protein,[95] and tryptophan residues in proteins.[97] The applicability of Equation 3.1 was also successfully tested for **DBO**, where the additional dependence of the oscillator strength on the polarizability needed to be explicitly taken into account.[24]

Assuming that other photophysical rate constants remain constant, one can project a concomitant change of the fluorescence lifetime τ_f and of the fluorescence quantum yield ϕ_f. As can be seen from Equations 3.2a, b, an increase in the fluorescence quantum yield at the expense of a reduced fluorescence lifetime is expected, when the radiative decay rate is increased and vice versa:

$$\text{(a) } \phi_f = \frac{k_r}{k_r + k_{nr}} \quad \text{and} \quad \text{(b) } \tau_f = \frac{1}{k_r + k_{nr}}. \tag{3.2}$$

The radiative decay rates of **DBO** in β-CD and CB7 provided an independent probe for the polarizabilities projected from the above solvatochromic method (effects on oscillator strength). Accordingly, if the polarizability inside CDs was similar to water and alcohols, the radiative decay rate should also be comparable, as was found to be the case.[24] If, however, the polarizability inside cucurbiturils was extremely low, in between perfluorohexane and the gas phase, the radiative decay rates should also fall within this uncommon region, therefore resulting in exceedingly slow radiative decay rates and, with other decay rates remaining similar, longer fluorescence lifetimes. Exactly this was observed: The fluorescence quantum yield in deaerated water is 0.26 and the fluorescence lifetime is 415 ns.[24] In the inclusion complex with CB7, a contrasting observation was made, because the fluorescence quantum yield decreased to 0.19, while the fluorescence lifetime increased to ca. 1 μs! This corresponds to a decrease in the radiative decay rate constant (simply calculated as ϕ_f/τ_f) by a factor of ca. 3! Noteworthy, this decrease in the radiative decay rate is only partially due to the decreased oscillator strength, but also due to a direct effect of the microenvironmental refractive index on the radiative decay rate, thereby demonstrating the influence of the critical n^2 term in Equation 3.1.

The fluorescence lifetime of DBO when included inside CB7 is the longest one known for an organic chromophore in solution.[18,98] Since then, radiative decay rates of fluorophores included in CB7 have been determined for numerous fluorescent dyes, including the commercially important rhodamine, pyronin, oxazine, coumarin, and cyanine dye types.[43,54] In all cases, increased fluorescence lifetimes were observed, often the longest ever observed for the respective dyes, and by calculating the radiative decay rates, several independent, but consistent, estimates for the refractive index inside this rigid molecular container could be obtained.

DISPLACEMENT SENSOR APPLICATIONS

The described photophysical effects and their use in determining the polarity or polarizability of macrocyclic cavities are complemented by promising applications of macrocyclic fluorescent dye complexes in the areas of analyte sensing and assaying. Such applications are based on the idea that a competitor (the analyte) is able to displace the dye from the host–guest complex,[26,35,51,99–104] thereby triggering a readily and accurately measurable fluorescence response, which reports on the absolute concentration of the analyte, or, in case of the recently developed tandem assays,[35,74,99] on its temporal evolution as a consequence of a catalytic (enzymatic) formation. Depending on whether the dye is more or less fluorescent in its uncomplexed form, one observes an increase or decrease in fluorescence upon analyte binding, which can be classified as either a "switch-on" or "switch-off" fluorescence response or sensor system (Scheme 3.5). Here, an interesting contrast between what is "desired" from a purely photophysical and application standpoint arises. For example, when

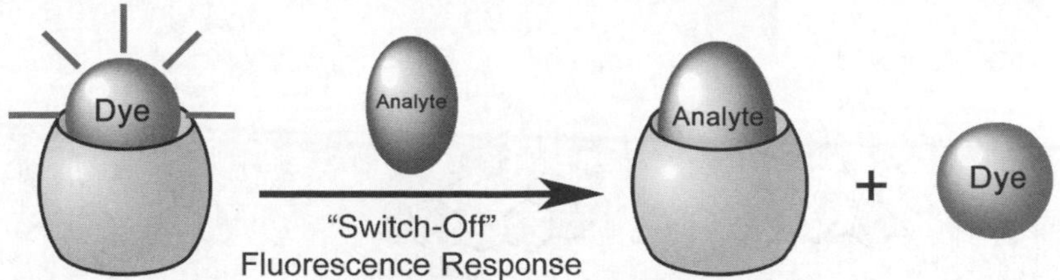

Scheme 3.5. A simple "switch-off" displacement assay.

searching for interesting fluorescent dyes, the quest has frequently been directed toward dyes, which show a fluorescence enhancement upon formation of the host–guest complex, corresponding to I_{bound}/I_{free} values ideally much larger than 1 (Table 3.1). One reason for this is that a reduction of fluorescence in the presence of an additive, as seen in many cases and particularly for calixarenes, corresponds to a common fluorescence quenching. Also, the PSF defined for solvatochromic fluorescent probes (see above)[82] is preferably based on a fluorescence enhancement upon binding. For displacement applications, however, the opposite is the case, because it is now advantageous to observe a switch-on fluorescence response upon addition of the analyte, which requires the fluorescent dye to be less fluorescent in its host–guest complex. As a consequence, quite a few displacement sensor systems have also been developed for calixarenes, which generally tend to quench fluorescence in their complexes.[26,35,53]

At the beginning of the development of a displacement application stands the identification of a suitable reporter pair composed of a fluorescent dye and a corresponding macrocycle. The fluorescent dye is selected such that it shows a large change in fluorescence, ideally a strong quenching, upon binding to the host. The macrocycle is chosen to display a high affinity to the dye, and in some cases, a highly selective binding and differentiation from structurally related analytes is considered essential. The binding constants of the fluorescent dye and the analyte, in its relevant concentration range, should also be comparable, such that the addition or formation of an analyte causes a sizable displacement and consequently fluorescence response. A particularly large amount of literature exists for calixarene-based assemblies tested for the sensing of neurotransmitters, particularly acetylcholine. The reporter pairs employed[26,51,52,100,105–107] are shown in Scheme 3.6, of which I–III do not operate in aqueous solution or neutral pH, but only in alcohol mixtures or in strongly alkaline solutions. The concept of all developed systems relies on a strong and competitive binding of the quarternary trimethylammonium group of acetylcholine to the calixarenes[26,51,52,105,106] (or resorcinarenes, in two studies),[100,107] which displaces the fluorescent dye from the complex and leads thereby to a (partial) recovery of the fluorescence characteristic for the uncomplexed dye. The spherical space-filling nature of the trimethylammonium group coupled

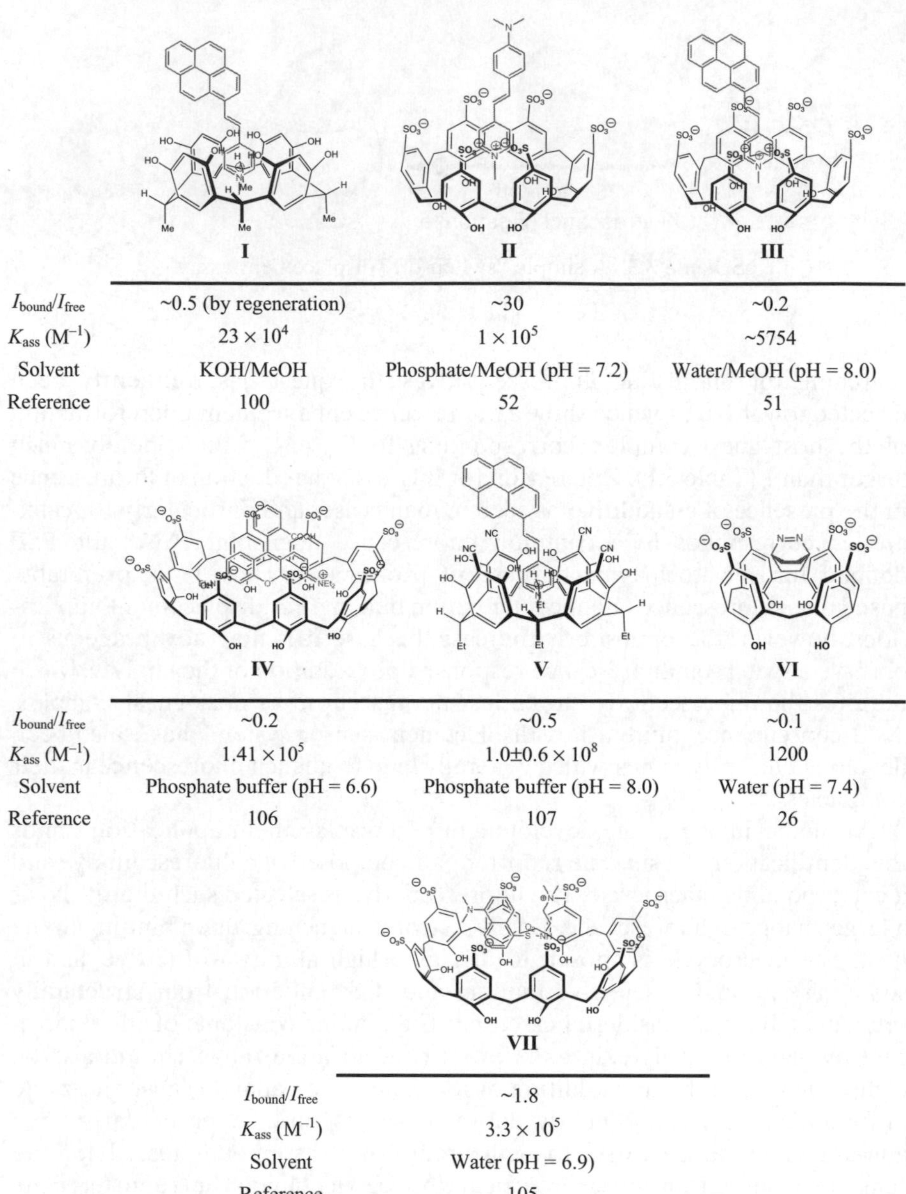

	I	II	III
I_{bound}/I_{free}	~0.5 (by regeneration)	~30	~0.2
K_{ass} (M^{-1})	23×10^4	1×10^5	~5754
Solvent	KOH/MeOH	Phosphate/MeOH (pH = 7.2)	Water/MeOH (pH = 8.0)
Reference	100	52	51

	IV	V	VI
I_{bound}/I_{free}	~0.2	~0.5	~0.1
K_{ass} (M^{-1})	1.41×10^5	$1.0 \pm 0.6 \times 10^8$	1200
Solvent	Phosphate buffer (pH = 6.6)	Phosphate buffer (pH = 8.0)	Water (pH = 7.4)
Reference	106	107	26

	VII
I_{bound}/I_{free}	~1.8
K_{ass} (M^{-1})	3.3×10^5
Solvent	Water (pH = 6.9)
Reference	105

Scheme 3.6. Reporter pairs used for acetylcholine sensing. See ftp site for color: ftp://ftp.wiley.com/public/sci_tech_med/supramolecular_photochemistry.

with its positive charge appears to provide an ideal match to ensure a high affinity with *p*-sulfonatocalixarenes, as well as a certain selectivity when compared with other neurotransmitters.

The displacement concept has also been used to sense other analytes. For example, the CX4•MB reporter pair (Table 3.1) has been successfully employed for sensing vitamin K_3,[41] while a CB8•2,7-dimethyldiazapyrenium reporter pair has been used to sound out catechol in solution.[108] It is also noteworthy to recognize reporter pairs that function not via displacement, but instead, by the formation of ternary complexes with the analyte. For example, the citrate anion was sensed by forming a ternary complex with a reporter pair consisting of an upper-rim functionalized calix[4]arene and a derivatized naphthalimide chromophore.[109]

One major disadvantage of using simple macrocycles for analyte sensing is their poor to mediocre selectivity. As a consequence, there are no real examples in which the examined systems, for example, for acetylcholine sensing, have been successfully applied in a direct biological context. Almost always will additional metabolites or even salts display some interaction with the macrocycle, which will contribute to the fluorescence variation and render the absolute concentration response of the sensor system unreliable. The moderate selectivity could, however, also present an advantage, because several structurally related analytes could be sensed with the same reporter pair. In addition, the undesirable effects due to the presence of other additives could be minimized, if only the target analyte changes with time (is formed or removed), such that the fluorescence change could be directly related to a concentration change. Exactly the latter application has recently been introduced as supramolecular tandem assays for monitoring enzymatic reactions.[35,99]

The detection of biomolecular analytes, and particularly of products of enzymatic reactions, is commonly achieved by using fluorescently or radioactively labeled analytes or substrates.[110,111] Alternatively, antibodies are broadly used, which allow a highly selective recognition.[112] Supramolecular tandem assays, which are based on the fluorescent reporter pairs discussed in this chapter,[35,99] allow a "label-free" detection of enzymatic reaction products, that is, devoid of covalent markers. The method allows the reliable and sensitive detection of various substrates for different enzymes.

The measurement is again based on the displacement principle and shown in Scheme 3.7. It requires the presence of a reporter pair consisting of a macrocycle and a fluorescent dye, for example, those compiled in Table 3.1. Although the macrocycles display a low selectivity, much lower than the alternatively used antibodies, it is relatively straightforward to identify several that show a differential binding between the substrate and product of an enzymatic reaction of interest. For example, such differentiation could be facilitated by varying size or increasing or decreasing charge upon enzymatic reaction, which is almost always fulfilled. In the examples investigated so far, the product was more tightly bound (strong competitor) than the substrate (weak competitor), although this is not compulsory for the enzyme assay to work. As can be seen from Scheme 3.7, the enzymatic reaction results then in the displacement of the dye and a switch-on or switch-off fluorescence response, which in turn depends on the photophysical response of the dye toward complexation. Thus, the enzymatic reaction can be conveniently followed, and the selective

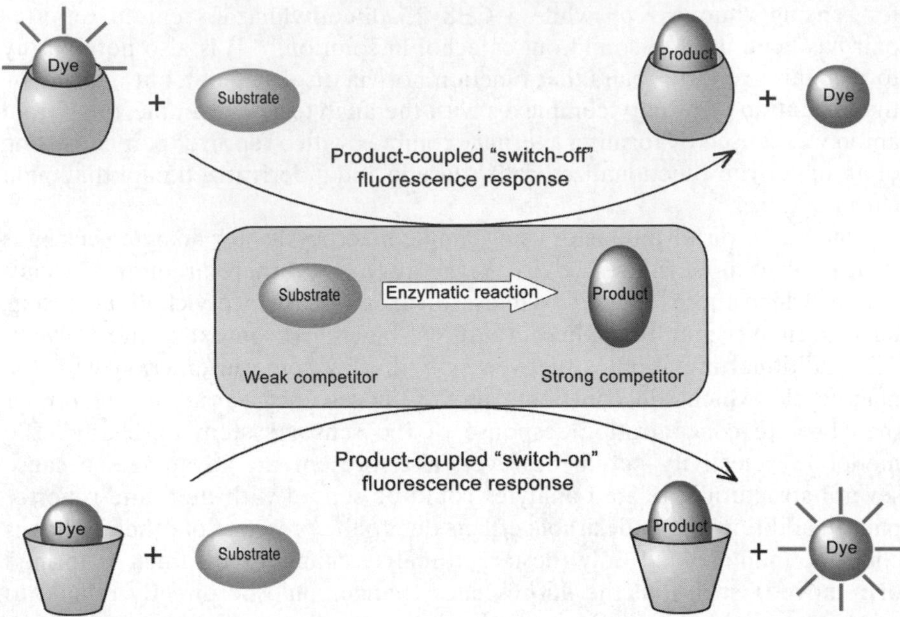

Scheme 3.7. Principle of product-coupled supramolecular tandem assays. See ftp site for color: ftp://ftp.wiley.com/public/sci_tech_med/supramolecular_photochemistry.

detection is ensured solely through the specificity of the enzyme, such that other components in the reaction mixture, even if they display some binding to the macrocycle, do not interfere.[35,99] Vice versa does the low selectivity of the macrocycle itself offer the possibility that entire substrate classes (e.g., amino acids), product classes (biogenic amines), and enzyme classes (e.g., decarboxylases) become accessible for assaying with a single reporter pair. This reduces time and cost dramatically, because the raising of several specific antibodies or the synthesis of covalently labeled antigens may easily consume several months to years, both in biochemical research laboratories as well as in the pharmaceutical industry.

As a real application, an example is shown in Figure 3.4—the decarboxylation of L-arginine by arginine decarboxylase. The cleavage of the negatively charged carboxylate group produces a biogenic amine (agmatine) as product, which exists as a doubly positively charged cation near neutral pH. Accordingly, macrocycles with cation receptor properties have been selected, namely cucurbit[7]uril (CB7) and *p*-sulfonatocalix[4]arene (CX4), which are both commercially available. CB7 and CX4 bind the amine product strongly ($1.1 \times 10^6\,M^{-1}$ and $7.1 \times 10^6\,M^{-1}$, respectively), but bind the amino acid substrate weakly ($310\,M^{-1}$ and $2.8 \times 10^3\,M^{-1}$, respectively).[35] As fluorescent dyes in the reporter pair, **Dapoxyl** and **DBO-Am** were selected to contrast the difference between a switch-off (for CB7•**Dapoxyl**) and switch-on (for CX4•**DBO-Am**)

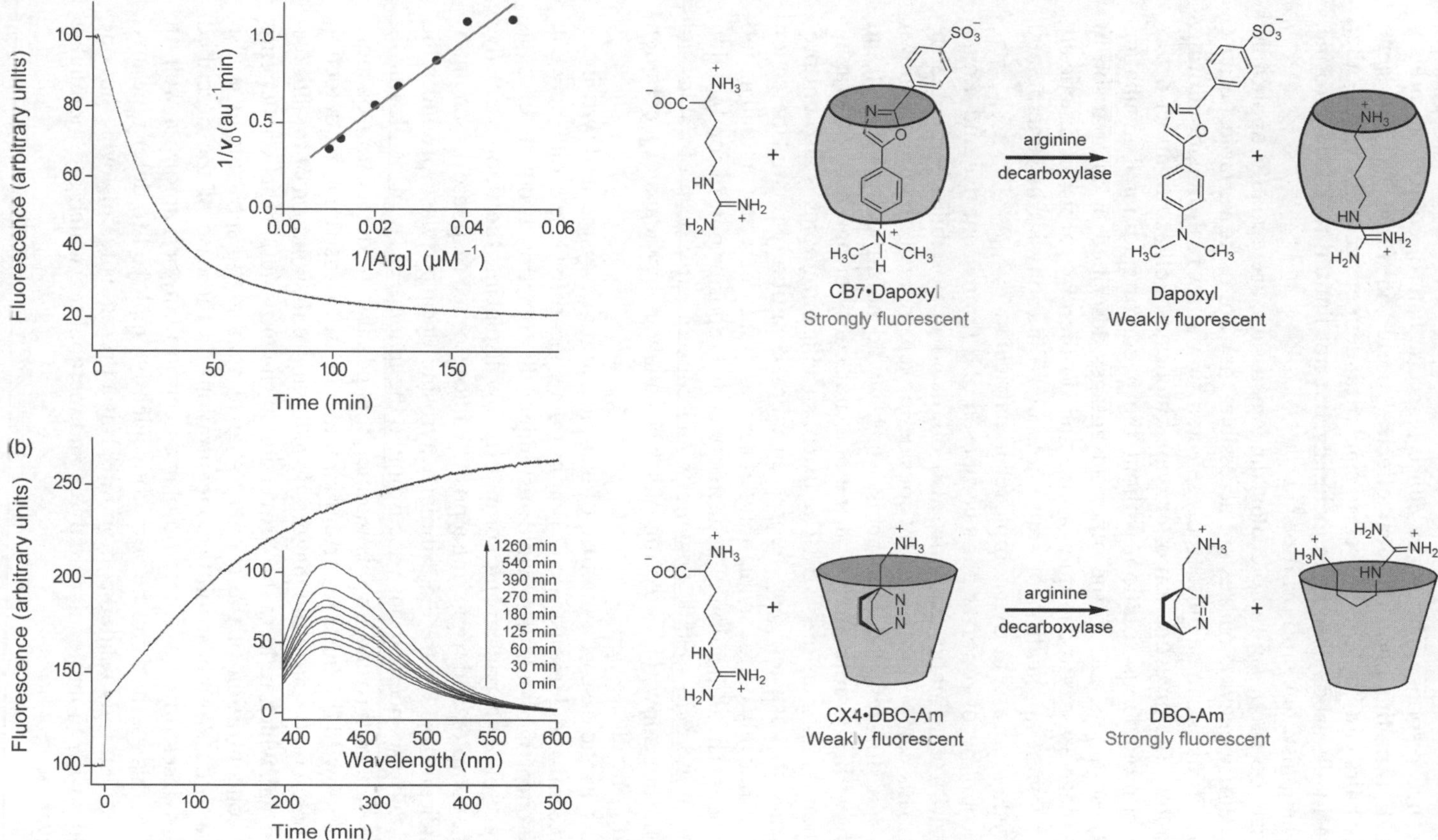

Figure 3.4. Continuous fluorescence enzyme assays for arginine decarboxylase in 10 mM NH_4OAc buffer at pH 6.0 by using (a) the CB7•Dapoxyl and (b) the CX4•DBO-Am reporter pair. The inset in (a) shows the Lineweaver–Burk plot for arginine decarboxylase. The inset in (b) depicts fluorescence spectra obtained at different reaction times in a noncontinuous aliquot assay. The principles are shown on the right.[35]

fluorescence response. The fluorescence response can, of course, be directly predicted from the photophysical effects reported in Table 3.1. The same reporter pairs can be flexibly used in similar tandem enzyme assays for other amino acid–decarboxylase pairs including, but not limited to, lysine, tyrosine, tryptophan, histidine, and ornitine.[35]

As can be seen, supramolecular tandem assays are broadly applicable, because entire enzyme classes are accessible; they are quite variable, because the reporter pairs can be quite freely selected; they allow real-time monitoring, which allows the investigation of enzyme kinetics as well as the direct detection in homogeneous solution, without the need for additional incubation steps. These assets add to the other advantages described in comparison to the alternative assay techniques above and do thereby furbish a practically highly relevant application of supramolecular photophysics in general, and macrocycle–fluorescent dye complexes in particular.

The coupling of reporter pairs composed of a fluorescent dye and a macrocycle with enzymatic reactions also offers novel possibilities for analyte sensing. This can be illustrated by considering a sample containing unknown amounts of different L-amino acids, which would be difficult to detect and quantify with a simple analytical method, not considering chromatographic separation techniques or NMR. The use of specific antibodies is also not preferred due to cost, and any fluorescence-based antibody-related detection would further rely on the availability of a series of fluorescently labeled amino acid analogs. The utilization of a macrocycle–dye reporter pair in combination with an amino acid-specific enzyme would consequently be preferable, and would further allow the detection of several analytes by means of a common readout principle.

In fact, in the case of a mixture of histidine, arginine, lysine, and tyrosine, it could recently be demonstrated[99] that histidine decarboxylase is only active in the presence of histidine to yield histamine; this reaction product serves as secondary analyte and strong competitor by displacing the fluorescent dye from the complex and triggering a positive fluorescence response, as schematically indicated in Figure 3.5. Similarly, tyrosine decarboxylase produces the desired fluorescence response according to the tandem assay principle only in the presence of tyrosine. In this manner, it is possible to detect four or more amino acids with the same reporter pair under identical instrumental detection conditions. And because the magnitude and temporal evolution of the fluorescence signal, which reflects the conversion in the enzymatic reaction, depends on the concentration of the primary analyte, that is, the substrate, it is also possible to determine absolute concentrations, in the case of the described decarboxylases down to a micromolar detection limit, or 1 nmol per well. In summary, this method can be very conveniently employed to determine mixtures of structurally related compounds. It is a prerequisite, however, that one is dealing with biomolecular analytes and metabolites, which are modifiable by enzymatic reactions.

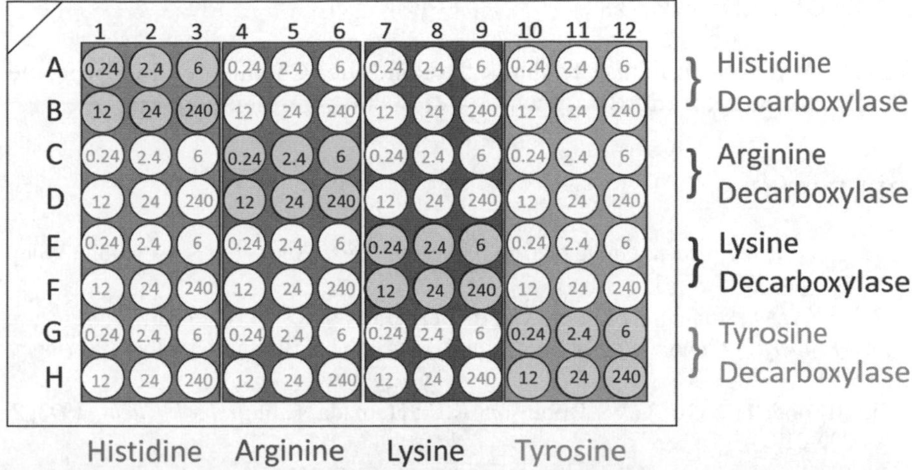

Figure 3.5. Microtiter plate setup employed for the multiparameter sensor array. The numbers within the wells refer to the amount of amino acid present, in nanomoles. Yellow wells mark expected positive responses. See ftp site for color: ftp://ftp.wiley.com/public/sci_tech_med/supramolecular_photochemistry.

CONCLUSIONS AND OUTLOOK

Since the original observation by Turro and coworkers on the phosphorescence enhancement of bromonaphthalene derivatives upon complexation by CDs,[113,114] photophysical effects upon dyes due to macrocyclic encapsulation have received considerable interest from the viewpoints of mechanistic understanding,[115] applications, and even chemical education.[116] Both a fluorescence switch-on as well as switch-off response have been documented for different combinations of macrocyclic hosts and fluorescent dye, and the two distinct responses are desirable from different viewpoints. A fluorescence enhancement accompanying complexation is desirable for a qualitative signaling that complexation has occurred while a fluorescence decrease concomitant with encapsulation is advantageous for monitoring the recovery of the fluorescence of the quenched complexed dye upon the addition of an analyte according to the indicator displacement principle. Particularly, with respect to sensor applications, there is an increasing demand for the development of refined reporter pairs in which (1) the binding constant between host and dye are matched with the range of binding constants for target analytes; (2) the fluorescence enhancement or quenching are sufficiently large to produce a significant fluorescence response in the relevant analyte concentration range; and (3) the excitation and emission wavelengths are matched to instrumental requirements, for example, laser wavelengths. The photophysical effects discussed in this chapter and the reporter pairs collected in Table 3.1 can provide design criteria for tailor-made reporter pairs.

ACKNOWLEDGMENTS

RND and WMN would like to thank the Fonds der Chemischen Industrie, the DFG (NA686/5-1), and BIG Bremen (VE0080B) for their support.

REFERENCES

1. Cragg, P. *A Practical Guide to Supramolecular Chemistry*. West Sussex: John Wiley & Sons, **2005**.
2. Nau, W.M.; Hennig, A.; Koner, A.L. In: Berberan-Santos, M.N., editor. *Fluorescence of Supermolecules, Polymers, and Nanosystems*, Vol. 4. Berlin, Heidelberg: Springer, **2008**, pp. 185–211.
3. Atwood, J.L.; Orr, G.W.; Robinson, K.D.; Hamada, F. *Supramol. Chem.* **1993**, *2*, 309–317.
4. Breslow, R.; Dong, S.D. *Chem. Rev.* **1998**, *98*, 1997–2011.
5. Szejtli, J. *Chem. Rev.* **1998**, *98*, 1743–1754.
6. D'Souza, V.T.; Lipkowitz, K.B. *Chem. Rev.* **1998**, *98*, 1741–1742.
7. Shinkai, S.; Araki, K.; Tsubaki, T.; Arimura, T.; Manabe, O. *J. Chem. Soc. Perkin Trans. 1* **1987**, 2297–2299.
8. Shinkai, S. In: Vicens, J.; Böhmer, V., editors. *Calixarenes: A Versatile Class of Macrocyclic Compounds*. Dordrecht, The Netherlands: Kluwer Academic Publishers, **1990**, pp. 173–198.
9. Behrend, R.; Meyer, E.; Rusche, F. *Justus Liebigs Ann. Chem.* **1905**, *339*, 1–37.
10. Freeman, W.A.; Mock, W.L.; Shih, N.Y. *J. Am. Chem. Soc.* **1981**, *103*, 7367–7368.
11. Mock, W.L. *Top. Curr. Chem.* **1995**, *175*, 1–24.
12. Hoffmann, R.; Knoche, W.; Fenn, C.; Buschmann, H.J. *J. Chem. Soc. Faraday Trans.* **1994**, *90*, 1507–1511.
13. Kim, J.; Jung, I.-S.; Kim, S.-Y.; Lee, E.; Kang, J.-K.; Sakamoto, S.; Yamaguchi, K.; Kim, K. *J. Am. Chem. Soc.* **2000**, *122*, 540–541.
14. Day, A.; Arnold, A.P.; Blanch, R.J.; Snushall, B. *J. Org. Chem.* **2001**, *66*, 8094–8100.
15. Kim, K. *Chem. Soc. Rev.* **2002**, *31*, 96–107.
16. Lagona, J.; Mukhopadhyay, P.; Chakrabarti, S.; Isaacs, L. *Angew. Chem. Int. Ed. Engl.* **2005**, *44*, 4844–4870.
17. Lee, J.W.; Samal, S.; Selvapalam, N.; Kim, H.-J.; Kim, K. *Acc. Chem. Res.* **2003**, *36*, 621–630.
18. Marquez, C.; Huang, F.; Nau, W.M. *IEEE Trans. Nanobioscience* **2004**, *3*, 39–45.
19. Arimura, T.; Nagasaki, T.; Shinkai, S.; Matsuda, T. *J. Org. Chem.* **1989**, *54*, 3766–3768.
20. Wagner, B.D. In: Nalwa, H.S., editor. *Handbook of Photochemistry and Photobiology*, Vol. 3. Stevenson Ranch, CA: American Scientific Publishers, **2003**, pp. 1–57.
21. Wagner, B.D.; Fitzpatrick, S.J. *J. Incl. Phenom. Macrocycl. Chem.* **2000**, *38*, 467–478.

22. Wagner, B.D.; MacDonald, P.J. *J. Photochem. Photobiol. A* **1998**, *114*, 151–157.
23. Marquez, C.; Nau, W.M. *Angew. Chem. Int. Ed.* **2001**, *40*, 4387–4390.
24. Mohanty, J.; Nau, W.M. *Photochem. Photobiol. Sci.* **2004**, *3*, 1026–1031.
25. Nau, W.M.; Zhang, X. *J. Am. Chem. Soc.* **1999**, *121*, 8022–8032.
26. Bakirci, H.; Nau, W.M. *Adv. Funct. Mater.* **2006**, *16*, 237–242.
27. Wagner, B.D.; Stojanovic, N.; Day, A.I.; Blanch, R.J. *J. Phys. Chem. B* **2003**, *107*, 10741–10746.
28. Liu, Y.; You, C.C. *J. Phys. Org. Chem.* **2001**, *14*, 11–16.
29. Mohanty, J.; Bhasikuttan, A.C.; Nau, W.M.; Pal, H. *J. Phys. Chem. B* **2006**, *110*, 5132–5138.
30. Singh, M.K.; Pal, H.; Koti, A.S.R.; Sapre, A.V. *J. Phys. Chem. A* **2004**, *108*, 1465–1474.
31. Shaikh, M.; Mohanty, J.; Singh, P.K.; Nau, W.M.; Pal, H. *Photochem. Photobiol. Sci.* **2008**, *7*, 408–414.
32. Liu, Y.; Han, B.-H.; Chen, Y.-T. *J. Org. Chem.* **2000**, *65*, 6227–6230.
33. Megyesi, M.; Biczok, L.; Jablonkai, I. *J. Phys. Chem. C* **2008**, *112*, 3410–3416.
34. Megyesi, M.; Biczok, L. *Chem. Phys. Lett.* **2006**, *424*, 71–76.
35. Hennig, A.; Bakirci, H.; Nau, W.M. *Nat. Methods* **2007**, *4*, 629–632.
36. Koner, A.L.; Nau, W.M. *Supramol. Chem.* **2007**, *19*, 55–66.
37. Diwu, Z.; Zhang, C.; Klaubert, D.H.; Haugland, R.P. *J. Photochem. Photobiol. A Chem.* **2000**, *131*, 95–100.
38. Mondal, S.K.; Sahu, K.; Sen, P.; Roy, D.; Ghosh, S.; Bhattacharyya, K. *Chem. Phys. Lett.* **2005**, *412*, 228–234.
39. Wang, R.; Yuan, L.; Ihmels, H.; Macartney, D.H. *Chem. Eur. J.* **2007**, *13*, 6468–6473.
40. Montes-Navajas, P.; Corma, A.; Garcia, H. *ChemPhysChem* **2008**, *9*, 713–720.
41. Lu, Q.; Gu, J.; Yu, H.; Liu, C.; Wang, L.; Zhou, Y. *Spectrochim. Acta A* **2007**, *68*, 15–20.
42. Reija, B.; Al-Soufi, W.; Novo, M.; Tato, J.V. *J. Phys. Chem. B* **2005**, *109*, 1364–1370.
43. Nau, W.M.; Mohanty, J. *Int. J. Photoenergy* **2005**, *7*, 133–141.
44. Montes-Navajas, P.; Teruel, L.; Corma, A.; Garcia, H. *Chem. Eur. J.* **2008**, *14*, 1762–1768.
45. Wang, R.; Yuan, L.; Macartney, D.H. *Chem. Commun.* **2005**, 5867–5869.
46. Scypinski, S.; Drake, J.M. *J. Phys. Chem.* **1985**, *89*, 2432–2435.
47. Roy, D.; Mondal, S.K.; Sahu, K.; Ghosh, S.; Sen, P.; Bhattacharyya, K. *J. Phys. Chem. A* **2005**, *109*, 7359–7364.
48. Lee, C.; Sung, Y.W.; Park, J.W. *J. Phys. Chem. B* **1999**, *103*, 893–898.
49. Banerjee, A.; Sengupta, B.; Chaudhuri, S.; Basu, K.; Sengupta, P.K. *J. Mol. Struct.* **2006**, *794*, 181–189.
50. Das, S.; Thomas, K.G.; George, M.V.; Kamat, P.V. *J. Chem. Soc. Faraday Trans.* **1992**, *88*, 3419–3422.

51. Koh, K.N.; Araki, K.; Ikeda, A.; Otsuka, H.; Shinkai, S. *J. Am. Chem. Soc.* **1996**, *118*, 755–758.
52. Korbakov, N.; Timmerman, P.; Lidich, N.; Urbach, B.; Sa'ar, A.; Yitzchaik, S. *Langmuir* **2008**, *24*, 2580–2587.
53. Bakirci, H.; Koner, A.L.; Dickman, M.H.; Kortz, U.; Nau, W.M. *Angew. Chem. Int. Ed.* **2006**, *45*, 7400–7404.
54. Mohanty, J.; Nau, W.M. *Angew. Chem. Int. Ed.* **2005**, *44*, 3750–3754.
55. Diwu, Z.; Lu, Y.; Zhang, C.; Klaubert, D.H.; Haugland, R.P. *Photochem. Photobiol.* **1997**, *66*, 424–431.
56. Zhu, Q.; Yoon, H.-S.; Parikh, P.B.; Chang, Y.-T.; Yao, S.Q. *Tetrahedron Lett.* **2002**, *43*, 5083–5086.
57. Al-Soufi, W.; Reija, B.; Novo, M.; Felekyan, S.; Kühnemuth, R.; Seidel, C.A.M. *J. Am. Chem. Soc.* **2005**, *127*, 8775–8784.
58. Bakirci, H.; Koner, A.L.; Schwarzlose, T.; Nau, W.M. *Chem. Eur. J.* **2006**, *12*, 4799–4807.
59. Faiz, J.A.; Williams, R.M.; Pereira Silva, M.J.J.; De Cola, L.; Pikramenou, Z. *J. Am. Chem. Soc.* **2006**, *128*, 4520–4521.
60. Haider, J.M.; Pikramenou, Z. *Chem. Soc. Rev.* **2005**, *34*, 120–132.
61. Haider, J.M.; Williams, R.M.; De Cola, L.; Pikramenou, Z. *Angew. Chem. Int. Ed.* **2003**, *42*, 1830–1833.
62. Mortellaro, M.A.; Nocera, D.G. *J. Am. Chem. Soc.* **1996**, *118*, 7414–7415.
63. Pikramenou, Z.; Nocera, D.G. *Inorg. Chem.* **1992**, *31*, 532–536.
64. Rudzinski, C.M.; Young, A.M.; Nocera, D.G. *J. Am. Chem. Soc.* **2002**, *124*, 1723–1727.
65. Tabushi, I.; Fujita, K.; Yuan, L.C. *Tetrahedron Lett.* **1977**, *18*, 2503–2506.
66. Berberan-Santos, M.N.; Choppinet, P.; Fedorov, A.; Jullien, L.; Valeur, B. *J. Am. Chem. Soc.* **2000**, *122*, 11876–11886.
67. Choppinet, P.; Jullien, L.; Valeur, B. *Chem. Eur. J.* **1999**, *5*, 3666–3678.
68. Jullien, L.; Canceill, J.; Valeur, B.; Bardez, E.; Lefèvre, J.-P.; Lehn, J.-M.; Marchi-Artzner, V.; Pansu, R. *J. Am. Chem. Soc.* **1996**, *118*, 5432–5442.
69. Jullien, L.; Canceill, J.; Valeur, B.; Bardez, E.; Lehn, J.-M. *Angew. Chem. Int. Ed. Engl.* **1994**, *33*, 2438–2439.
70. Gust, D.; Moore, T.A.; Moore, A.L. *Chem. Commun.* **2006**, 1169–1178.
71. Raymo, F.M.; Tomasulo, M. *J. Phys. Chem. A* **2005**, *109*, 7343–7352.
72. Wu, S.; Luo, Y.; Zeng, F.; Chen, J.; Chen, Y.; Tong, Z. *Angew. Chem. Int. Ed.* **2007**, *46*, 7015–7018.
73. Tan, W.; Zhang, D.; Wen, G.; Zhou, Y.; Zhu, D. *J. Photochem. Photobiol. A Chem.* **2008**, *200*, 83–89.
74. Praetorius, A.; Bailey, D.M.; Schwarzlose, T.; Nau, W.M. *Org. Lett.* **2008**, *10*, 4089–4092.
75. Sahoo, H.; Nau, W.M. *Chembiochem* **2007**, *8*, 567–573.
76. Sahoo, H.; Roccatano, D.; Hennig, A.; Nau, W.M. *J. Am. Chem. Soc.* **2007**, *129*, 9762–9772.

77. Sahoo, H.; Roccatano, D.; Zacharias, M.; Nau, W.M. *J. Am. Chem. Soc.* **2006**, *128*, 8118–8119.

78. Hossain, M.A.; Mihara, H.; Ueno, A. *J. Am. Chem. Soc.* **2003**, *125*, 11178–11179.

79. Takakusa, H.; Kikuchi, K.; Urano, Y.; Higuchi, T.; Nagano, T. *Anal. Chem.* **2001**, *73*, 939–942.

80. Klotz, E.J.F.; Claridge, T.D.W.; Anderson, H.L. *J. Am. Chem. Soc.* **2006**, *128*, 15374–15375.

81. Cram, D.J. *Nature* **1992**, *356*, 29–36.

82. Rankin, M.A.; Wagner, B.D. *Supramol. Chem.* **2004**, *16*, 513–519.

83. Shinkai, S.; Kawabata, H.; Matsuda, T.; Kawaguchi, H.; Manabe, O. *Bull. Chem. Soc. Jpn.* **1990**, *63*, 1272–1274.

84. Shi, Y.; Wang, D.; Zhang, Z. *J. Photochem. Photobiol. A* **1995**, *91*, 211–215.

85. Lichtenthaler, F.W.; Immel, S. *Liebigs Ann.* **1996**, 27–37.

86. Cox, G.S.; Hauptman, P.J.; Turro, N.J. *Photochem. Photobiol.* **1984**, *39*, 597–601.

87. Street, K.W.; Acree, W.E. *Appl. Spectrosc.* **1988**, *42*, 1315–1318.

88. Reichardt, C. *Solvents and Solvent Effects in Organic Chemistry*, 3rd edition. Weinheim: Wiley-VCH, **2003**.

89. Pina, F.; Parola, A.J.; Ferreira, E.; Maestri, M.; Armaroli, N.; Ballardini, R.; Balzani, V. *J. Phys. Chem.* **1995**, *99*, 12701–12703.

90. Lakowicz, J.R. *Principles of Fluorescence Spectroscopy*, 2nd edition. New York, London: Kluwer Academic/Plenum, **1999**.

91. Strickler, S.J.; Berg, R.A. *J. Chem. Phys.* **1962**, *37*, 814–822.

92. Lamouche, G.; Lavallard, P.; Gacoin, T. *J. Lumin.* **1998**, *76–77*, 662–665.

93. Lamouche, G.; Lavallard, P.; Gacoin, T. *Phys. Rev. A* **1999**, *59*, 4668–4674.

94. Lavallard, P.; Rosenbauer, M.; Gacoin, T. *Phys. Rev. A* **1996**, *54*, 5450–5453.

95. Suhling, K.; Siegel, J.; Phillips, D.; French, P.M.W.; Leveque-Fort, S.; Webb, S.E.D.; Davis, D.M. *Biophys. J.* **2002**, *83*, 3589–3595.

96. Toptygin, D.; Brand, L. *Biophys. Chem.* **1993**, *48*, 205–220.

97. Toptygin, D.; Savtchenko, R.S.; Meadow, N.D.; Roseman, S.; Brand, L. *J. Phys. Chem. B* **2002**, *106*, 3724–3734.

98. Nau, W.M.; Huang, F.; Wang, X.J.; Bakirci, H.; Gramlich, G.; Marquez, C. *Chimia* **2003**, *57*, 161–167.

99. Bailey, D.M.; Hennig, A.; Uzunova, V.D.; Nau, W.M. *Chem. Eur. J.* **2008**, *14*, 6069–6077.

100. Inouye, M.; Hashimoto, K.; Isagawa, K. *J. Am. Chem. Soc.* **1994**, *116*, 5517–5518.

101. Neelakandan, P.P.; Hariharan, M.; Ramaiah, D. *Org. Lett.* **2005**, *7*, 5765–5768.

102. Shi, Y.; Schneider, H.J. *J. Chem. Soc. Perkin Trans. 2* **1999**, 1797–1803.

103. Ueno, A.; Kuwabara, T.; Nakamura, A.; Toda, F. *Nature* **1992**, *356*, 136–137.

104. Wiskur, S.; Ait-Haddou, H.; Lavigne, J.; Anslyn, E.V. *Acc. Chem. Res.* **2001**, *34*, 963–972.

105. Jin, T. *J. Incl. Phenom. Macrocycl. Chem.* **2003**, *45*, 195–201.

106. Zhang, Y.J.; Cao, W.X.; Xu, J. *Chin. J. Chem.* **2002**, *20*, 322–326.

107. Tan, S.-D.; Chen, W.-H.; Satake, A.; Wang, B.; Xu, Z.-L.; Kobuke, Y. *Org. Biomol. Chem.* **2004**, *2*, 2719–2721.
108. Sindelar, V.; Cejas, M.A.; Raymo, F.M.; Chen, W.; Parker, S.E.; Kaifer, A.E. *Chem. Eur. J.* **2005**, *11*, 7054–7059.
109. Koner, A.L.; Schatz, J.; Nau, W.M.; Pischel, U. *J. Org. Chem.* **2007**, *72*, 3889–3895.
110. Gribbon, P.; Sewing, A. *Drug Discov. Today* **2003**, *8*, 1035–1043.
111. Matayoshi, E.D.; Wang, G.T.; Krafft, G.A.; Erickson, J. *Science* **1990**, *247*, 954–958.
112. Reymond, J.-L. *Enzyme Assays: High-Throughput Screening, Genetic Selection and Fingerprinting*. Weinheim: Wiley-VCH, **2006**.
113. Turro, N.J.; Bolt, J.D.; Kuroda, Y.; Tabushi, I. *Photochem. Photobiol.* **1982**, *35*, 69–72.
114. Turro, N.J.; Cox, G.S.; Li, X. *Photochem. Photobiol.* **1983**, *37*, 149–153.
115. Bortolus, P.; Monti, S. *Adv. Photochem.* **1996**, *21*, 1–133.
116. Wagner, B.D.; MacDonald, P.J.; Wagner, M. *J. Chem. Educ.* **2000**, *77*, 178–181.

4

SUPRAMOLECULAR PHOTOCHIROGENESIS

CHENG YANG AND YOSHIHISA INOUE

INTRODUCTION

Asymmetric synthesis is undoubtedly one of the most important research areas of modern organic chemistry. Studies on thermal asymmetric syntheses, both catalytic and enzymatic, have met great success in the last few decades.[1,2] In contrast, the photochemical counterpart, that is asymmetric synthesis via electronically excited states, has developed only recently.[3–7] Photochemical reaction is unique and attractive because molecules in the electronically excited state behave quite differently from those in the ground state to frequently give such products that are not readily obtainable in thermal reactions. On the other hand, there are some issues to be considered and overcome for attaining efficient control of a chiral photochemical reaction: First, an electronically excited molecule has only a very limited lifetime that is comparable to or even shorter than the diffusion timescale. It is thus a tough task to develop significant chiral interactions in such a short period of time. Also, a single excited-state interaction is not strong enough to efficiently manipulate the fate of the transient species involved in chirality transfer process. Furthermore, chiral photochemistry has a drawback inherent to excited-state reaction, that is, low activation energy, which makes the discrimination of a pair of diastereomeric intermediates or transition states challenging.

Supramolecular Photochemistry: *Controlling Photochemical Processes*, First Edition. Edited by V. Ramamurthy and Yoshihisa Inoue.

A variety of sources of chiral have hitherto been explored for the use in chiral photochemistry in solution. Circularly polarized light (CPL) is used as a physical source of chirality to selectively decompose or transform one of a pair of enantiomers in a racemic mixture to leave the antipodal one in excess. This strategy is called absolute asymmetric synthesis (AAS) because no chiral substance (reagent, catalyst, or substituent) is involved in the chirality differentiation process. However, the relative difference in extinction coefficient toward left- and right-handed CPL, on which the AAS relies, is small (typically ≪1%) for most chiral compounds, and hence an appreciable enantiomeric excess (ee) may be obtained only when most (>99%) of the starting material is consumed. This means that AAS is not suitable for practical use in asymmetric synthesis, but recent studies reveal a close connection between the CPL photochemistry and the extraterrestrial origin of the homochirality of natural amino acids on earth.[8–11]

Among the strategies using chemical sources of chirality, the chiral auxiliary method, in which an enantiopure substituent is temporarily introduced to a photosubstrate for intramolecular steric interactions, is known to be most effective in obtaining chiral photoproducts with high diastereoselectivities.[12] However, the use of an equimolar amount of chiral source and the tedious attachment/detachment of chiral auxiliary make this route less attractive. Noncovalently interacting chiral sources, such as optically active complexing agent and solvent that deliver chiral information through complexation or solvation, are less effective in general than a chiral auxiliary or require an excess amount of chiral source. Chiral photosensitization provides a unique approach to photochemical asymmetric synthesis that allows chirality amplification through the repeated use of a catalytic amount of chiral sensitizer. Thus, extensive efforts have been devoted to chiral photosensitization since the first study reported in 1965.[13] However, good enantioselectivities were achieved only recently by using chiral sensitizers with bulky chiral substituents under rather special conditions such as low temperature and/or high pressure,[14,15] probably due to the inherently weak sensitizer–substrate interactions in the excited state and also to the crucial roles played by the entropic factors.

More recently, a novel methodology has been invented to more efficiently control the stereochemical outcomes of chiral photoreaction by introducing a substrate into chiral supramolecular environment. This supramolecular photochirogenic method utilizes multiple noncovalent interactions operating in both ground and excited states, which allow the chiral source to intimately contact with the substrate in more strictly defined conformation for a longer period of time, eventually affording the photoproduct in better enantio- or diastereoselectivity. Indeed, recent studies have shown that supramolecular photochirogenesis is one of the most successful strategies hitherto examined for solution-phase chiral photochemistry. In this chapter, we will review and discuss the results reported in recent years in the field of supramolecular photochirogenesis.

SUPRAMOLECULAR PHOTOCHIROGENESIS WITH BIOMOLECULES

Biomolecules, such as protein, DNA, and antibody, are naturally occurring chiral macromolecules. Possessing high binding affinity and excellent molecular recognition ability for specific guest molecules, these biomolecules can be used as chiral supramolecular hosts for organic compounds that are practically insoluble in water but become soluble upon inclusion in the hydrophobic pocket(s) of a host. Hence, biomolecule-mediated photochirogeneses are usually carried out in aqueous solutions.

Indeed, biomolecules are more compatible with photochemical reactions, which can be done under mild conditions in aqueous solution at ambient temperature, rather than thermal reactions, which usually require organic media and thermal or catalytic activation and therefore potentially damage the shape and function of biomolecular hosts. The use of biomolecular hosts in chiral photochemistry needs some precautions: First, the UV-Vis spectrum of guest substrate should be carefully checked to avoid the absorption range of the biomolecular host to be used, unless the photosensitization by biomolecule is intended. Biomolecules absorb UV light in general. For example, the major absorption of proteins at longer wavelengths comes from tryptophan (λ_{max} 280 nm, ε 5600 $M^{-1}cm^{-1}$) and tyrosine (λ_{max} 274 nm, ε 1400 $M^{-1}cm^{-1}$) residues, while that of common nucleosides appears at ca. 260 nm. Although the aromatic residues of biomolecular host may function as built-in photosensitizer to activate the guest substrate included in its hydrophobic pocket, their photoreaction(s) and the possible structural changes derived therefrom should not be neglected. Second, protein hosts, such as serum albumin, often possess more than one binding site with a different shape, binding affinity, reactivity, and selectivity. This means that sophisticated photophysical and photochemical examinations under a variety of conditions are required to fully elucidate the binding behavior and the origin and mechanism of stereodifferentiation. Finally, as antipodal biomolecule is not available in general, only one of the enantiomer pair of chiral photoproduct is obtained by using the same biomolecular system.

Serum albumins, which are known to bind and transport specific endo- and exogenous (bio)organic molecules, are the most frequently used protein in chiral photochemistry. Zandomeneghi and coworkers reported the enantiomer-selective photodecomposition of racemic 1,1′-binaphthol (*rac*-**1**) (Scheme 4.1) mediated by bovine serum albumin (BSA) for the first time. (*S*)-(−)-Enantiomer of **1** is preferentially bound to BSA, developing a new absorption assignable to naphtholate anion at longer wavelengths, excitation of which leads to selective decomposition of (*S*)-(−)-**1** to enrich (*R*)-(+)-**1** in up to 99.5% ee after a 77% consumption of the starting material.[16,17]

Zandomeneghi and coworkers examined the enantioselective photodecomposition of *rac*-ketoprofen (**2**) (Scheme 4.1) in the presence of BSA to obtain 80% ee after 99.8% of the starting material was photodecomposed.[18] Miranda

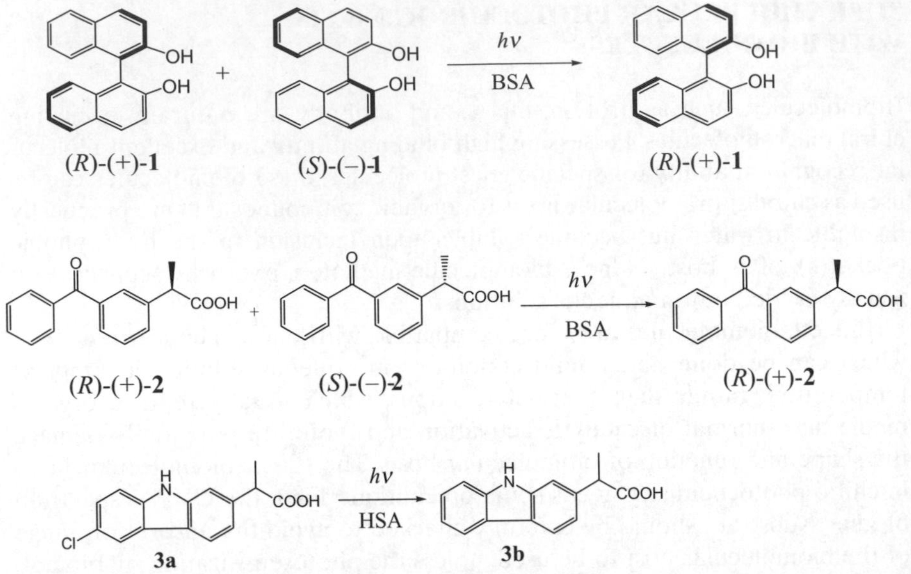

(*R*)-(+)-**1** + (*S*)-(−)-**1** —$h\nu$, BSA→ (*R*)-(+)-**1**

(*R*)-(+)-**2** + (*S*)-(−)-**2** —$h\nu$, BSA→ (*R*)-(+)-**2**

3a —$h\nu$, HSA→ **3b**

Scheme 4.1. Enantiodifferentiating photodecomposition and phototransformation with bovine (BSA) and human serum albumin (HSA).

and coworkers reported that the excited state of chiral ketoprofen **2** is quenched by thymidine and deoxyguanosine in an enantioselective manner, with accompanying Paternò–Büchi reaction and ketyl radical formation.[19] The binding site of human serum albumin (HSA) revealed only a subtle difference in affinity for the enantiomers of **2**, and the photolysis of *rac*-**2** mediated by HSA leads only to almost no enantioselectivity. Nevertheless, an addition of oleic acid to the system improved the ee to –4.8% after 46% photodecomposition of **2**.[18]

Carprofen **3a** (Scheme 4.1) is also included in the binding sites of HSA. The HSA-mediated photodechlorination of (*S*)-**3a** to give (*S*)-**3b** is faster than that of (*R*)-**3a** by a factor of 1.4 (equivalent to 58% ee).[20,21]

Since most of chiral substrates are bound to biomolecular hosts more or less enantioselectively, it is possible to conduct a photodecomposition in good enantioselectivity by appropriately choosing the irradiation wavelength, provided that the bound species absorbs light at appreciably different (preferably longer) wavelengths. In contrast, this advantage, intrinsic to chiral (racemic) substrates that originally possess stereogenic center(s), is not available for prochiral substrates, since the molecular chirality is created in a photoproduct right upon the photoreaction in the chiral environment of biomolecular host. This is more authentic photochirogenesis than the enantiodifferentiating photodecomposition mentioned above, but is much more difficult to attain with high enantioselectivity. Inoue and coworkers reported the enantiodifferentiating photoisomerization of (*Z*)-cyclooctene **4Z** to planar-chiral (*E*)-cyclooctene

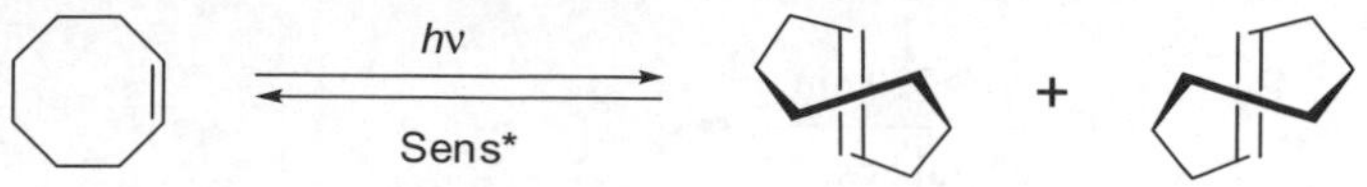

hν
Sens*

4Z (R)-(–)-4E (S)-(+)-4E

Scheme 4.2. Enantiodifferentiating geometrical photoisomerization of cyclooctene.

4E (Scheme 4.2) sensitized by nucleosides and nucleotides.[22] Pyrimidine nucleosides used as chiral sensitizers afford **4E** in up to 5.2% ee at the photostationary state, while the purine nucleosides are less effective in sensitizing the photoisomerization. Photoisomerization of **4Z** in the presence of calf thymus DNA (ctDNA) in aqueous solution gives **4E** in –15.2% ee with an *E*/*Z* ratio of 0.008. Raising temperature apparently reduces the enantioselectivity of the photoreaction, probably due to decomposition of the substrate from the hydrophobic minor groove of DNA. Photoisomerization of **4Z** in 50% aqueous methanol, where ctDNA keeps its B-form but the supramolecular interaction of **4Z** with ctDNA is significantly diminished, affords **4E** in only –0.9% ee, suggesting that the supramolecular interaction plays a crucial role in the formation of enantiomeric **4E**.

The [4 + 4] photocyclodimerization of anthracene derivatives is one of the oldest known photochemical reactions whose history can be traced back to the nineteenth century.[23] However, the chirogenic aspect of this well-documented photoreaction had not seriously been examined and discussed until the enantiodifferentiating photocyclodimerization of 2-anthracenecarboxylic acid (AC) was reported very recently by Inoue and coworkers.[24–36] It was found that BSA has four hydrophobic pockets that bind one, three, two, and three AC molecules, respectively, in the order of decreasing affinity.[25,37] Interestingly, AC bound to the first binding site exhibits sharp fluorescence similar to that in pentane and gives dual lifetimes of 4.8 and 2.1 ns, probably reflecting the positional or orientational difference in the first site. ACs bound to the second binding site of BSA have a lifetime of 13.2 ns, while those at third and fourth sites show broader fluorescence with an identical lifetime of 15.8 ns, which is indistinguishable from that of AC in bulk water. The fluorescence of free AC in water is significantly quenched by nitromethane, but those bound to the first and second sites are well protected from the attack of the quencher. AC molecules in the cavity of BSA are not located closely enough to show an exciton couplet in the circular dichroism spectra, and the quenching and photocyclodimerization of AC in the binding sites occur in a dynamic rather than static manner.

Photocyclodimerization of AC yields four stereoisomeric dimers, that is, *anti-* and *syn-head-to-tail* (HT) dimers **5a** and **5b** (Scheme 4.3) and *anti-* and *syn-head-to-head* (HH) dimers **5c** and **5d**, of which **5b** and **5c** are chiral. BSA-mediated photocyclodimerization of AC significantly improves the yield of HH dimers to afford **5c** in up to 40% yield. At an [AC]/[BSA] ratio of 3.6,

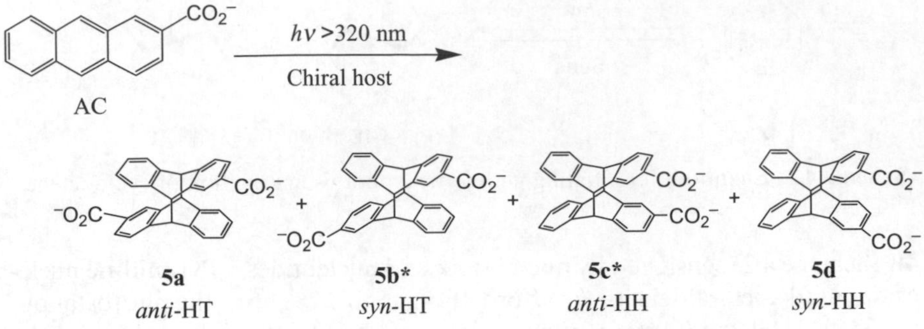

Scheme 4.3. Enantiodifferentiating photocyclodimerization of 2-anthracenecarboxylic acid (AC) with chiral host.

5b is given in 38% ee in the presence of 18 mM nitromethane added as a site-selective quencher, while **5c** is produced in 58% ee in the absence of nitromethane.[25,37]

HSA was also employed for mediating the photocyclodimerization of AC. It is known that HSA differs from BSA in only 26 amino acid residues out of ca. 600 residues, but both behave quite differently not only upon guest binding but also in the photophysical and photochemical behavior of bound guests.[38,39] Differing from BSA, HSA possesses five binding sites available for AC, which respectively accommodate one, one, three, five, and unspecified number (>10) of AC molecules. In sharp contrast to the BSA case, the HSA-mediated photocyclodimerization of AC affords the HT dimers as major products in much better ee's of 82% and 90% for **5b** and **5c**, respectively, at 5°C, for which site 3 is deduced to be responsible.[40]

SUPRAMOLECULAR PHOTOCHIROGENESIS WITH ZEOLITES

Zeolites are aluminosilicates composed of tetrahedral $[SiO_4]^{4-}$ and $[AlO_4]^{5-}$ units and are available as mesoporous crystalline solids with well-defined structures holding cavities called supercages.[41,42] The negative charge of $[AlO_4]^{5-}$ in zeolites is compensated by alkali or alkaline earth metal cations. Two kinds of zeolites, faujasites X and Y, which are characterized by their high cation exchange capacity, have been extensively investigated as host for mediating (chiral) photoreactions.[43,44] Faujasites X and Y hydrate can be represented by different formulae $M_{86}(AlO_2)_{86}(SiO_2)_{106}{\cdot}264H_2O$ and $M_{56}(AlO_2)_{56}(SiO_2)_{136}{\cdot}253H_2O$, respectively, but their framework topology is identical. Zeolites are transparent in the UV-Vis region since they are composed of Al-O-Si framework, like glass and quartz. The free volume in supercage is adjustable by the number and size of cations, and a variety of organic compounds can be immobilized on the inside walls as substrates and chiral inductors. These advantages make zeolite an attractive host for conducting chiral photoreactions. In this section, we will concentrate on the chiral aspect of photochemistry in zeolite supercage; for a full view, see a later chapter.

Diastereodifferentiating Photoreactions with Achiral Zeolites: The Confinement Effect

The chiral auxiliary method is one of the most successful strategies for obtaining good to excellent stereochemical outcomes in photochemical reactions. The covalent bonding of chiral substituent to photosubstrate ensures efficient intramolecular chirality transfer in general. The optical yield obtained critically depends on the chiral auxiliary introduced. Thus, the choice of "right" chiral auxiliary is an essential, but often laborious, task in this strategy. However, recent studies by Ramamurthy and coworkers showed that the inherently low or negligible diastereoselectivity obtained in solution can be significantly improved by confining the same substrate in the supercage of achiral zeolites.[45]

Photolysis of tropolone (*S*)-2-methylbutyl ether **6** (Scheme 4.4) in hexane affords a pair of diastereomers **7** as a 1:1 mixture, along with isomeric **8**. Strikingly, the same photoreaction in NaY zeolite at –20°C gave **7** in 68% diastereomeric excess (de).[46] The crucial role of confinement by supercage was confirmed by the fact that the photocyclization of **6** adsorbed on silica gel surface leads to **7** without any diastereodifferentiation.

2,2-Dimethyl-1-(2H)-naphthalenone-4-carboxylic acid esters **9b,c** (Scheme 4.5) gives oxa-di-π-methane rearrangement product **10b,c** in negligible de upon irradiation in hexane as well as on silica gel surface. In contrast, irradiation of the same compounds immobilized in supercage of NaY zeolite yields **10b** and **10c** in 57% and 47% de, respectively. Similarly, the diastereodifferentiating oxa-di-π-methane rearrangement of 2,4-cyclohexadienone-3-carboxylates **11b–e** gives **12b–e** in less than 5% de in trifluoroethanol solution, but the de's of **12b** and **12e** are significantly improved to 59% and 73%, respectively, upon irradiation within NaY zeolite.[47] Interestingly, the favored diastereomers of **12** can be switched by changing the counter cation (M) of MY

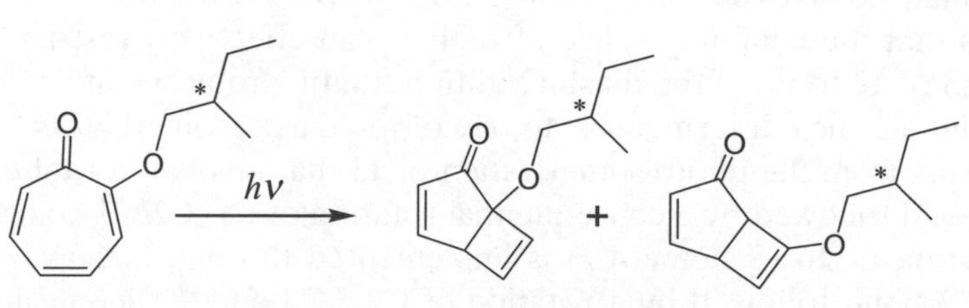

Scheme 4.4. Diastereodifferentiating photocyclization of enantiopure tropolone ether **6**.

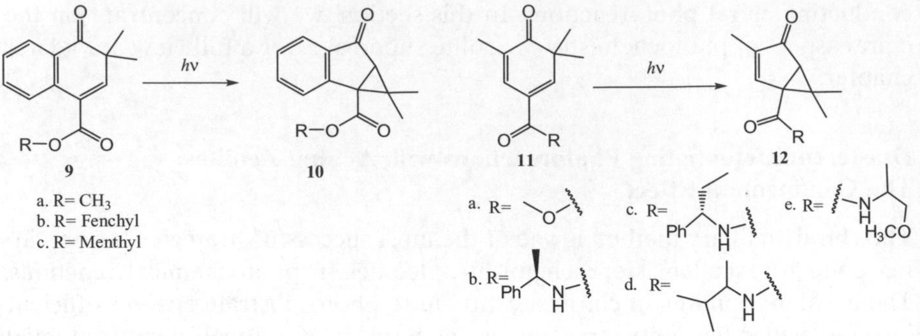

Scheme 4.5. Oxa-di-π-methane rearrangement of 2,2-dimethyl-1-(2H)-naphthalenone-3-carboxylates **9** and 2,4-cyclohexadienone-3-carboxylates **11**.

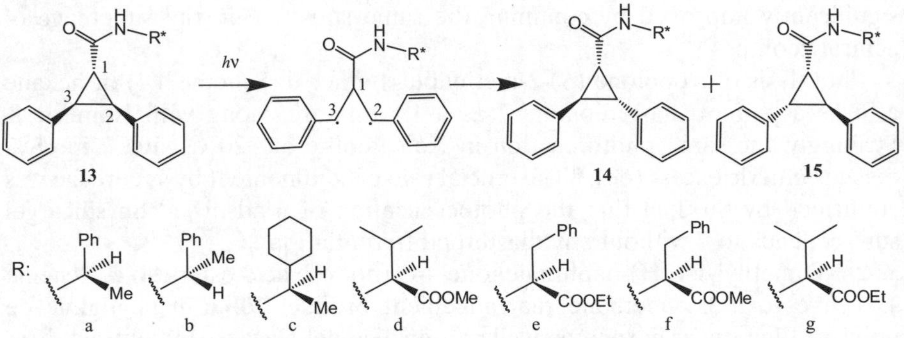

Scheme 4.6. Photoisomerization of 2,3-diphenylcyclopropane-1-carboxyamides **13**.

zeolite from Li and Na to larger K, Rb, and Cs, revealing that the cation in zeolites is able to critically influence the stereochemical outcomes.

Chiral amides of *cis*-2β,3β-diphenylcyclopropane-1α-carboxylic acid **13** (Scheme 4.6) photoisomerize to diastereomeric *trans* isomers **14** and **15** in only 0–5% de upon irradiation in homogeneous solution, which is however persistently enhanced up to 83% de by using LiY, NaY, and KY zeolites.[48] This result indicates that the confining space of zeolites can effectively restrict the conformation of **13** to drive the diastereodifferentiating rotation and recombination of the biradical intermediate to one of the diastereomeric *trans* isomers.

Differing from the photoisomerization of **13** that involves a rupture of the C2–C3 bond followed by a consequential rotation of C1–C2/C3 bond, photoisomerization of **16** (Scheme 4.7) is implemented through a cleavage of the C1–C2/C3 bond followed by a rotation of C2–C3 bond.[49] This mechanism is supported by the fact that the photolysis of *trans*-(2*S*,3*S*)-diphenyl-1-benzoylcyclopropane affords only *cis*-2,3-diphenyl-1-benzoylcyclopropane but no (2*R*,3*R*)-*trans*-diphenyl-1-benzoylcyclopropane.[50] Photolyses of chirally *meta*-

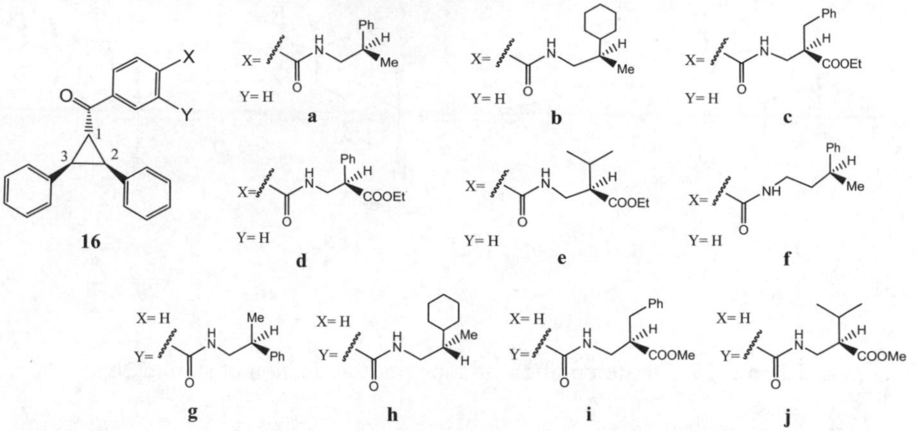

Scheme 4.7. 2,3-Diphenyl-1-aroylcyclopropanes **16** subjected to photoisomerization.

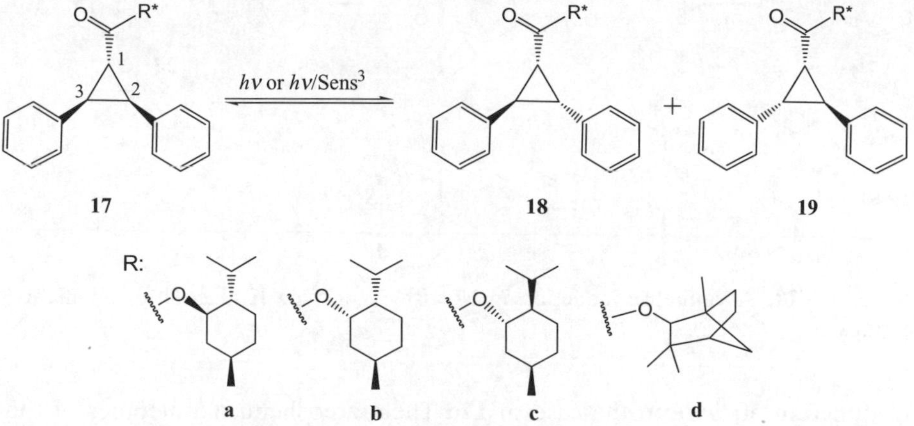

Scheme 4.8. Photoisomerization of 1,2-diphenylcyclopropane derivatives **17**.

substituted substrates **16g–j** in zeolite supercage consistently lead to higher de's of up to 71% (for **16g**) than *para*-substituted **16a–f**, regardless of the substituent employed.[44,48,51] Alternation of irradiation time or temperature does not affect the diastereoselectivity.

Photoisomerization of enantiopure menthyl, neomenthyl, isomenthyl, and fenchyl esters of 2β,3β-diphenylcyclopropane-1α-carboxylic acid **17** (Scheme 4.8) in the supercage of LiY, NaY, KY, RbY, and CsY zeolite has been examined by either direct irradiation or triplet sensitization with 4′-methoxyacetophenone.[52] Irradiation of **17a** or **17c** in homogeneous solution gives the first-eluted diastereomer of **18a/19a** or **18c/19c** pair in de <5%. In stark contrast, the photolysis of **17a** immobilized in zeolite supercage significantly improves the product's de to 55%, and interestingly, the antipodal product is

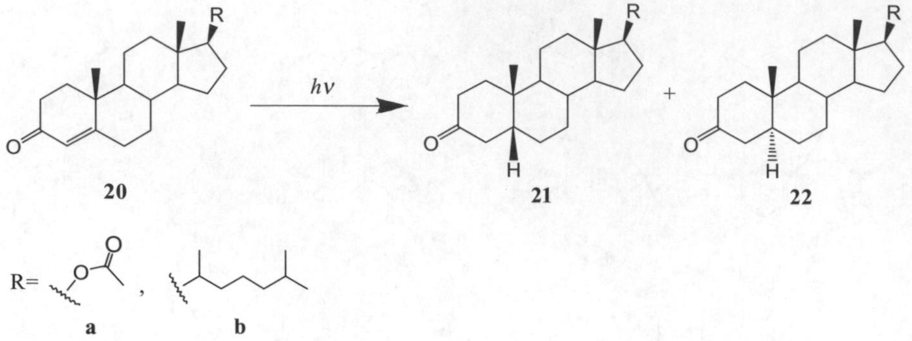

Scheme 4.9. Diastereodifferentiating photoreduction of steroid **20**.

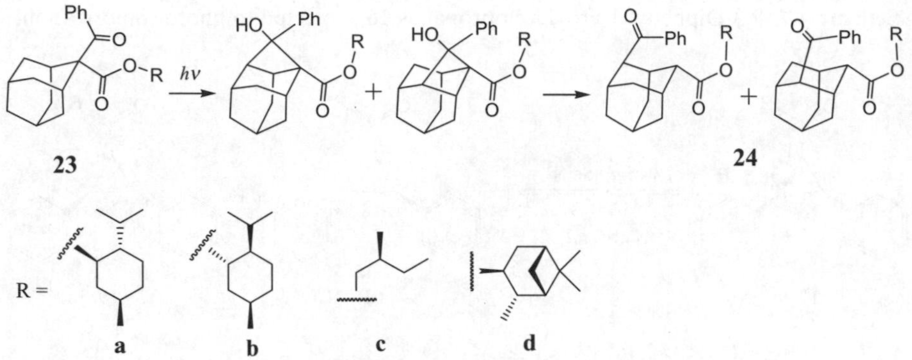

Scheme 4.10. Asymmetric induction in Norrish–Yang type II reaction of adamantyl ketoesters.

obtained in 50% de in the case of **17c**. The stereochemical outcomes of the sensitized photoisomerization considerably differ from those of the direct irradiation.[53]

Photoirradiation of steroids **20** (Scheme 4.9) affords reduction products **21** and **22**, respectively, through α- and β-face addition. Although irradiations of **20** in hexane, high-silica Y zeolite, or MCM-41 mesoporous silica give no reduction product, photolysis of **20** in 2-propanol yields the less-hindered α-face addition to give **22a** in 73% yield and **22b** in 65% yield. Unexpectedly, the photoreaction with NaY zeolite in hexane lead to a dramatic inversion of the diastereoselectivity to give **21a** and **21b** in 85% and 80% yield, respectively.[54]

As illustrated in Scheme 4.10, the diastereomeric cyclobutanols generated by the Norrish–Yang type II reaction of adamantly ketoesters **23** are readily converted to the corresponding δ-ketoesters **24** via retro-aldol reaction. The δ-ketoesters **24** are given in 4–22% de upon photolysis of **23** in isotropic solution of hexane and acetonitrile. The adamantane moiety was found to be too large in size to snugly fit into the supercage of zeolite, as only 60% of **23a** was adsorbed to zeolite after stirring a hexane solution containing **23a** in the presence of an excess amount of zeolite for 12h at 60°C. Photolysis of **23** loaded in LiY, NaY, and KY zeolite gives the photoproduct in 79% de.[47] The stereoselectivity of this zeolite-mediated photoreaction is guest-size dependent, and the lowest diastereoselectivity was observed for **24c** that possesses the most flexible and smallest chiral auxiliary.

Enantiodifferentiating Photoreactions with Chirally Modified Zeolites

Although some zeolites, such as zeolite β and titanosilicate ETS-10, are theoretically possible to exist in chiral form, so far no zeolite is separated in its enantiopure form.[42,55,56] Accordingly, photoreaction of a prochiral substrate mediated by native zeolite is anticipated to give a chiral product as a racemic mixture. For the use in enantiodifferentiating photochemical reactions, two approaches to chirally modified zeolites have been developed, that is, the adsorption of chiral organic molecules into zeolite supercages and the replacement of inorganic cations with chiral ammonium ions. These chiral modification strategies have been successfully applied to some thermal asymmetric reactions.[57–61]

Enantiodifferentiating photoreaction using chirally modified zeolites was first investigated by Ramamurthy and coworkers in the Norrish–Yang type II reaction of *cis*-4-*tert*-butylcyclohexyl ketones **25** (Scheme 4.11).[62] Chiral inductors, such as ephedrine, menthol, borneol and bornylamine, are embedded in zeolite with an average occupancy number of 0.3–2.6 per supercage. Among these chiral inductors, ephedrine was found to be most effective to give **26** in 25–30% ee. The same reaction performed in hexane in the presence of ephedrine gives **26** as a racemic mixture.

Photosensitized isomerization of *cis*-1,2-diphenylcyclopropane **27a** (Scheme 4.12), which is the very first example of photosensitized enantiodifferentiating reaction, has extensively been investigated since the report of Hammond and Cole in 1965.[13,63–65] However, the photoisomerization of **27a** sensitized by

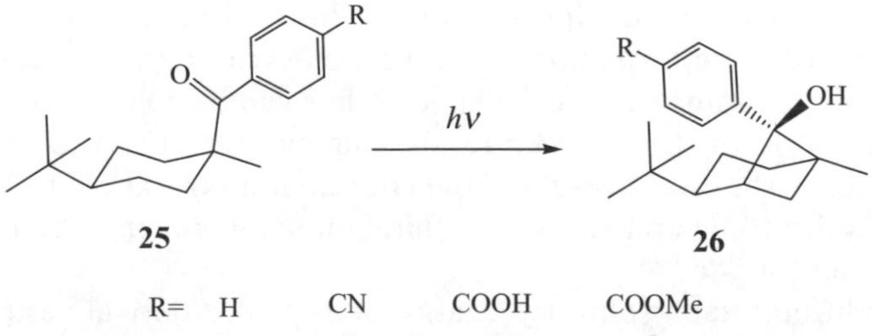

R
O
hν
R
OH
25
26

R=	H	CN	COOH	COOMe
	a	**b**	**c**	**d**

Scheme 4.11. Enantiodifferentiating Norrish–Yang type II reaction of *cis*-4-*tert*-butylcyclohexyl ketones.

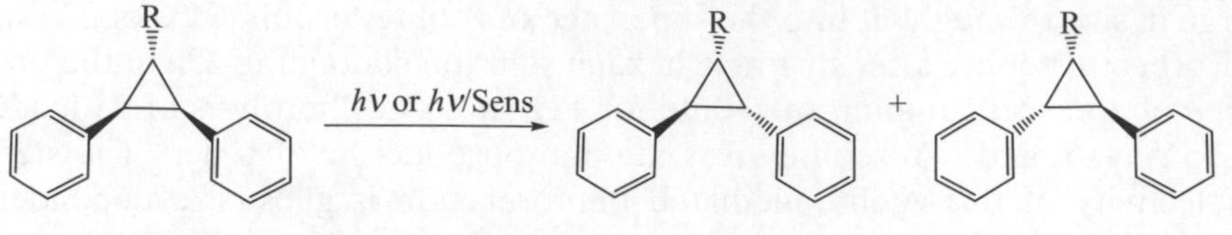

a: R = H **b**: R = COOH **c**: R = $CONHCH_2CH_2CH_2CH_3$ **d**: R = $CONHCH_2Ph$
d: R= $COOCH_3$ **e**: R= $COOCH_2CH_3$

Scheme 4.12. Enantiodifferentiating photoisomerization of *cis*-1,2-diphenylcyclopropane derivatives.

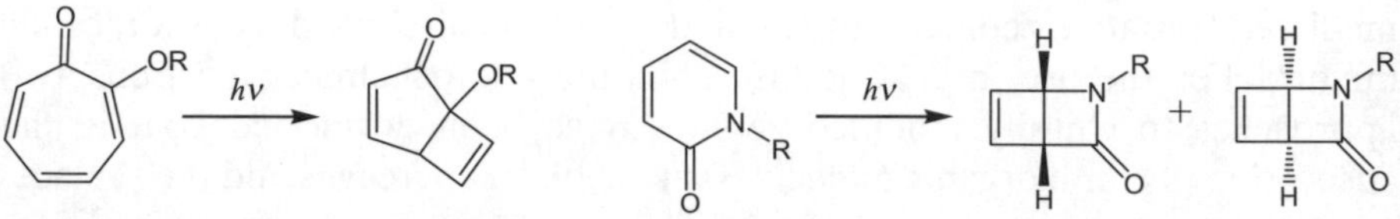

a: R=CH_3 **e**: R=$CH_2CH_2CH_3$ **a**: R= CH_3 **d**: R= CH_2Ph
b: R=CH_2CH_2Ph **f**: R= CH_2Ph **b**: R= CH_2CH_3 **e**: R= CH_2CH_2Ph
c: R=CH_2COOCH_3 **g**: R=$CH_2CH_2CH_2Ph$ **c**: R= $CH_2CH_2CH_3$ **f**: R= $CH_2CH_2CH_2Ph$
d: R=CH_2CH_3

Scheme 4.13. Enantiodifferentiating photocyclization of tropolone and pyridone derivatives.

conventional chiral sensitizer affords *trans*-1,2-diphenylcyclopropane **28** in only 10% ee at the best.[66] The quenching of the chiral sensitizer by the *trans* isomer rather than the decay from the intervening exciplex or radical ion pair was demonstrated to dominate the enantiodifferentiating process.[65] The enantio- and diastereodifferentiating photoisomerization of diphenylcyclopropane derivatives in chirally modified zeolites have been comprehensively studied in recent years.[48,51–53,67–75] No *cis*-to-*trans* isomerization of **27a** was observed in the supercage of MY zeolite either upon direct excitation or through triplet sensitization, for which the cation–π interaction that stabilizes the sandwich-type conformation of *cis*-diphenylcyclopropane was thought to be responsible. In contrast, its derivatives **27b–e** can be converted to the *trans* isomers smoothly in the *cis*/*trans* ratios ranging from 12:88 to 58:42 at the photostationary state, as a result of the coordination of cation with amide or ester moiety. By using ephedrine as a chiral inductor, a highest ee of 20% has been reported for **28c**.[53,67]

Enantiodifferentiating photocyclization of tropolone derivatives **29** (Scheme 4.13) to chiral bicycloheptadienones **30** has been considerably investigated in the solid state.[76–81] Ramamurthy and coworkers studied the photocyclization using zeolites chirally modified with norephedrine, ephedrine, or pseudo-

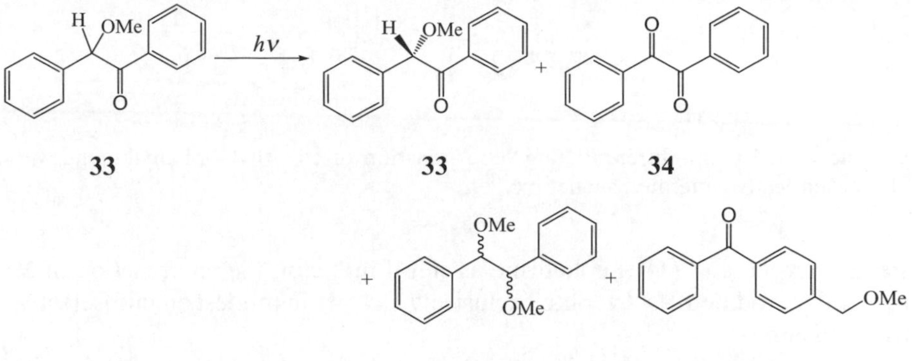

Scheme 4.14. Enantiodifferentiating radical recombination of benzoin.

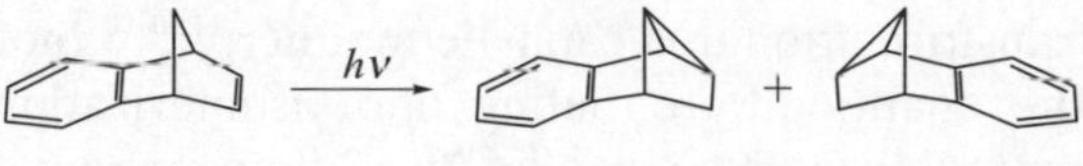

Scheme 4.15. Enantiodifferentiating photorearrangement of benzonorbornadiene.

ephedrine.[82,83] Chiral photoproduct **30c** was obtained in 69% ee by using ephedrine-modified zeolite.[82] These chiral inductors have also been applied to the enantiodifferentiating oxa-di-π-methane reaction of prochiral cyclohexadienone **11a** and naphthalenone **9a** (Scheme 4.5) to give the corresponding products **10a** and **12a** in 18% and 30% ee, respectively, in ephedrine-modified NaY zeolite.[44,75,84,85] Enantiodifferentiating photocyclization of pyridone derivative **31e** within norephedrine-modified KY zeolite gives cyclization product **32e** in 55% ee.

Turro and coworkers examined the effect of chirally modified zeolites on the photolysis of racemic benzoin methyl ether **33** (Scheme 4.14), in which the geminate radical pair formed upon the Norrish type I α-cleavage of **33** recombines to regenerate enantiomerically enriched photoproduct **33**. In solution, **34** and **35** are obtained as major products in 23% and 70% yield, respectively. However, **36** is obtained in 79% yield in zeolite at the expense of **33**–**35**. By using diethyl tartrate as chiral inductor, the photolysis in NaY zeolite gives **33** in 9.2% ee.[86]

Di-π-methane rearrangement of benzonorbornadiene **37** (Scheme 4.15) does not occur upon direct irradiation, but proceeds smoothly in the triplet manifold to give **38**.[87,88] Photolysis of **37** in TlY zeolite, where the intersystem crossing to reactive triplet state is accelerated due to the heavy atom effect of Tl, efficiently affords **38**.[89] Thus, the use of TlY zeolite modified with (+)-ephedrine hydrochloride leads to the formation of **38** in 14% ee.

The "ene" reaction of **39** with singlet oxygen (Scheme 4.16) in solution led to a 1:1 mixture of **40** and **41**. By using thionin as photosensitizer to generate

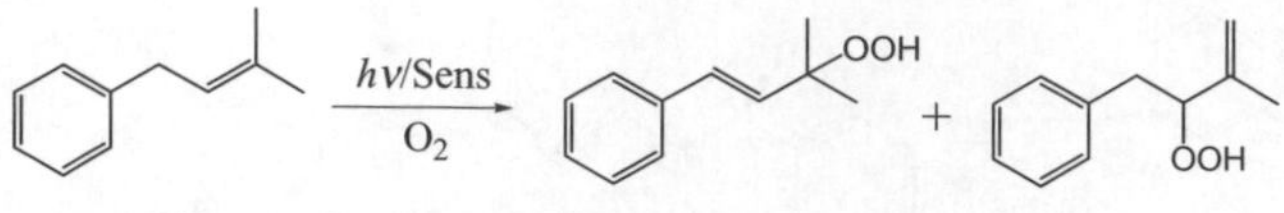

39 $\xrightarrow[O_2]{h\nu/\text{Sens}}$ **40** + **41**

Scheme 4.16. Enantiodifferentiating "ene" reaction of 1-methyl-4-phenylbutene with photochemically generated singlet oxygen.

singlet oxygen and (+)-ψ-ephedrine as chiral inductor, the ene reaction of **39** in chirally modified NaY zeolite exclusively gave **41** in modest enantioselectivity of 15% ee.[89]

In above-mentioned studies on supramolecular photochirogenesis with modified zeolite, each supercage is modified with at least one chiral inductor molecule for optimal chiral induction. As the enantiodifferentiating photosensitization has been demonstrated in solution to be a versatile strategy to effect photochemical transformation in a catalytic manner,[14,90–92] Inoue and coworkers attempted the enantiodifferentiating photoisomerization of **4Z** within chiral sensitizer-immobilized NaY zeolite.[93] In homogeneous solution, photosensitization with optically active 1-methylheptyl benzoate gives **4E** as a racemic mixture, whereas a modest ee of 4.5% was achieved in the photosensitization with 1-methylheptyl benzoate-immobilized NaY zeolite as a chiral sensitizing host.

Effect of Cation Species on the Stereoselectivity

In addition to the confinement by organic modifiers through van der Waals interactions, cation species in zeolites also plays a critical role in determining the stereoselectivity of photoreactions. Cations in X and Y zeolites are bound at three distinct sites in the zeolite pore system by electrostatic force, that is, the hexagonal prism faces between the sodalite nets, the open hexagonal faces, and the walls of the supercage. Cations can freely move in zeolite pores and exchange with different cationic species. The free volume accessible for adsorbing substrate relies on the number and nature of cations. It is thus deemed that cations have a key effect on adsorbing and positioning substrates in the supercage of zeolite.

The pivotal role of cations in the zeolite-mediated photochirogenic reactions has been implicated in various aspects. For example, in contrast to the greatly improved enantio- and diastereoselectivities found with a variety of native and modified zeolites, the photoreaction of substrates adsorbed on the surface of silica gel that contains no metal cations commonly affords negligible stereoselectivity. Addition of water, which causes hydration of cations and hence weakens the cation–substrate interaction, usually leads to a significant decrease of the stereoselectivity. Photoreactivity is critically governed by the cation–substrate interaction in zeolite. The *cis*-to-*trans* isomerization of 1,2-diphenylcyclopropane is completely inhibited due to the formation of stable sandwich-type complex between the phenyl groups and cations. However, the photoisomerization of its amide and ester derivatives, which form stronger carbonyl–cation complexes, takes place smoothly within the supercage of zeolite.[44,75,82]

The orientation of photosubstrate and chiral inductor in supercage varies greatly, depending on the nature and number of cation species. Changing cation species frequently leads to a dramatic change in stereochemical outcome even if the same chiral auxiliary is used.[44,57,82,84,94–98] For example, photolyses of **13a** and **13d** with LiY zeolite yield the corresponding *trans* isomer in 80% and 83% de, respectively, favoring the second-eluted isomer on gas chromatography (GC) analysis. However, the use of KY zeolite inverts the diastereoselectivity to prefer the first-eluted isomer in −14% and −80% de, respectively.

SUPRAMOLECULAR PHOTOCHIROGENESIS WITH CHIRAL TEMPLATES

Chiral templates that directly interact with target substrates through noncovalent interactions, such as hydrogen bonding, π–π, van der Waals, and electrostatic interactions, provide an effective and convenient tool for chiral photochemistry. Differing from other chiral supramolecular hosts, in which the inclusion in a three-dimensional cavity plays an important role, chiral templates are of lower molecular weights in general and transfer their chiral information to a bound substrate primarily through more clearly defined multiple noncovalent interactions. This strategy enables us to design the structure of host template to optimize the host–guest interactions.

Intramolecular Photoreaction with Chiral Templates

Hydrogen bond is one of the strongest and most commonly used noncovalent interactions for controlling supramolecular assembly and reaction in the ground and excited states.[99,100] Bach and coworkers reported the diastereodifferentiating Paternò–Büchi reaction of 3,4-dihydro-1H-pyridin-2-one in the presence of Kemp's triacid derivative **42a** (Scheme 4.17). The chiral template **42a** is able to choose one of the enantiofaces of 3,4-dihydro-1H-pyridin-2-one through the formation of dual hydrogen bonds to preferentially form one of the diastereomeric complexes, irradiation of which gives rise to oxetane **43a** in 90% de.[101,102] Although the template (**42a**) as well as the substrate (3,4-dihydro-1H-pyridin-2-one) may self-aggregate to yield hydrogen-bonded homodimers, the 1:1 complexation of **42a** with 3,4-dihydro-1H-pyridin-2-one is thermodynamically more favored ($K = 227\,M^{-1}$) over the self-dimerization of **42a** ($K = 24\,M^{-1}$) or 3,4-dihydro-1H-pyridin-2-one ($K = 85\,M^{-1}$).[101,103] In contrast to the high diastereoselectivity observed for **42a**, the *N*-methyl analog **42b** merely gives negligible de, lacking the binding ability toward 3,4-dihydro-1H-pyridin-2-one. In this particular case, **42** is not a simple template but also

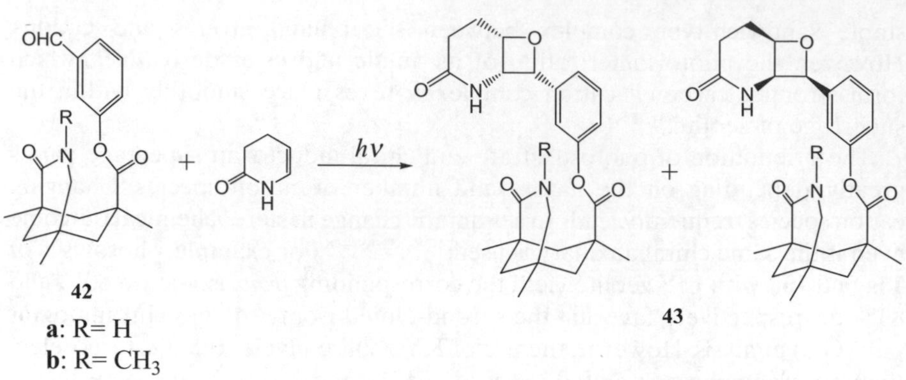

42

a: R= H
b: R= CH_3

43

Scheme 4.17. Enantioface-differentiating Paternò–Büchi reaction.

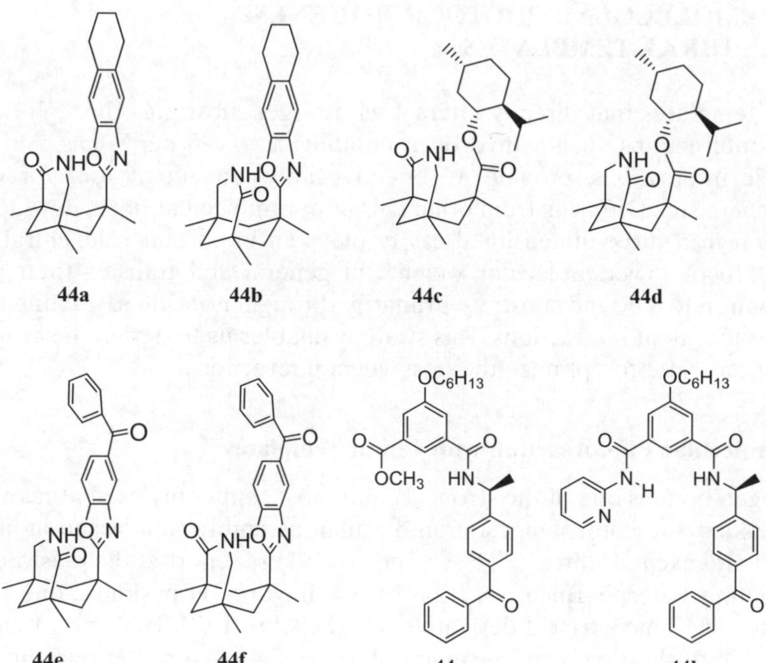

44a **44b** **44c** **44d**

44e **44f** **44g** **44h**

Scheme 4.18. Kemp's acid templates **44a–f** and isophthalamide template sensitizers **44g,h**.

a substrate, and hence the chiral moiety of **42** may be regarded as a functionalized chiral auxiliary that differentiates the diastereoface of the benzaldehyde moiety of enantiomeric **42** as well as the enantioface of prochiral 3,4-dihydro-1H-pyridin-2-one.

Bach and coworkers prepared a series of Kemp's triacid derivatives **44a–f** (Scheme 4.18) as "unreactive" or even "catalytic" chiral templates in order to achieve efficient chiral induction in various intra- and intermolecular photoreactions without consuming the templates. The enantiodifferentiating intramolecular photocycloaddition of prochiral 2-quinolone allyl ether **45** (Scheme 4.19) was carried out in the presence of chiral template **44a–c**. A great advantage of **44a,b** as hydrogen-bonding templates is that these chiral templates hardly form self-association dimer ($K = 0\,M^{-1}$) due to the steric hindrance of the bulky tetrahydronaphthalene wall. The amide moiety of these chiral templates forms dual hydrogen bonds with that of **45**. The bulky tetrahydronaphthalene wall shields one of the enantiofaces of substrate **45**, allowing the intramolecular attack of the allyl only from the open face. An excellent ee of 93% ee was achieved upon irradiation of **46** in the presence of **44b** at –60°C.[103,104] The same substrate complexed and triplet sensitized by chiral receptor **44h** at –70°C provides **46** in 22% ee, while template **44g**, lacking the

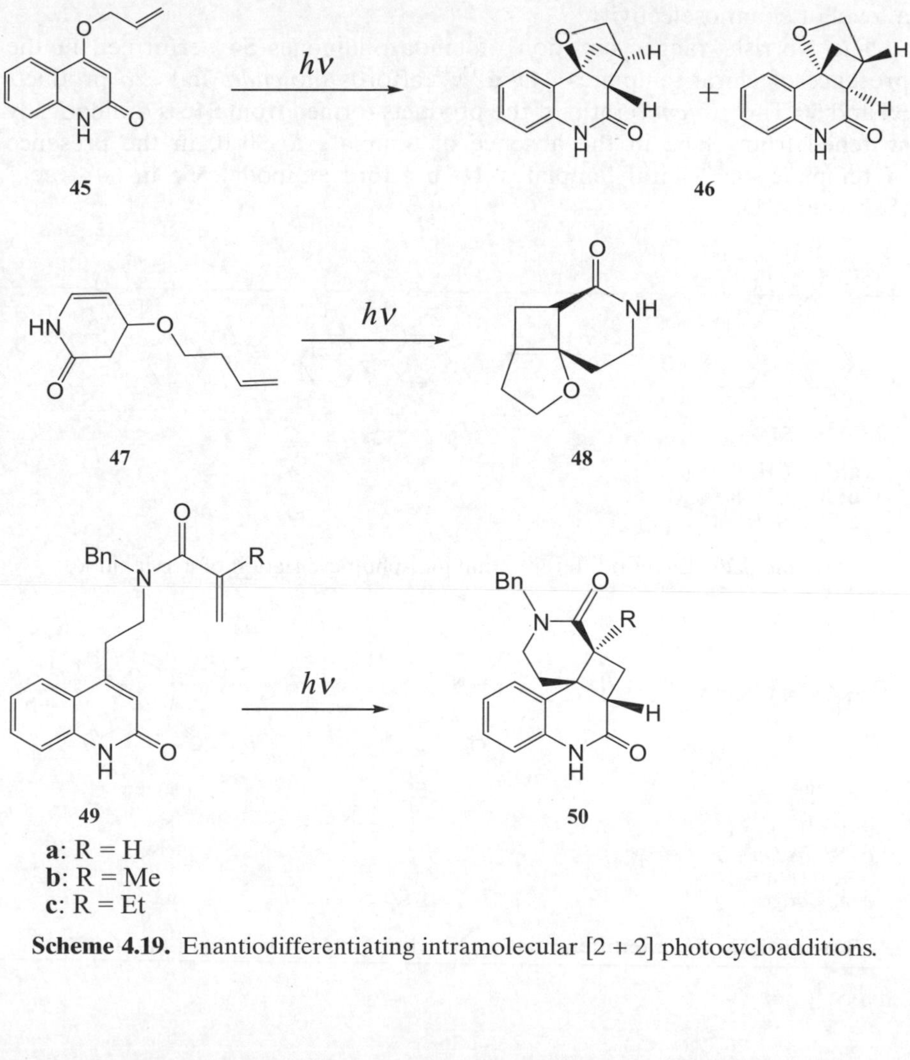

45 $\xrightarrow{h\nu}$ **46** (+ enantiomer)

47 $\xrightarrow{h\nu}$ **48**

49 $\xrightarrow{h\nu}$ **50**

a: R = H
b: R = Me
c: R = Et

Scheme 4.19. Enantiodifferentiating intramolecular [2 + 2] photocycloadditions.

hydrogen-bonding aminopyridine site, gives a racemic product.[105] Similar intramolecular [2 + 2] photocycloaddition reactions of **47** and **49** mediated by **44a** in toluene at –60°C give **48** in 59–75% ee and **50** in 74–92% ee.[106,107] The stereochemical outcomes in these enantiodifferentiating photocycloadditions can be improved by manipulating the reaction conditions, for example, by increasing the amount of template, lowering the temperature, and/or decreasing the polarity of solvent, all of which enhance the formation of template–substrate complex.

Enantiodifferentiating [6π]-photocyclization of acrylanilides **51a–c** (Scheme 4.20) in the presence of **44a** as a chiral template affords photoproducts **52** and **53** in ratios ranging from 48/52 to 27/73. Relatively high ee's of 57% and 39% are attained respectively for **52** and **53** through optimization of the template/guest ratio and the temperature.[55,108] Although similar reaction of furan-2-carboxyanilide derivatives carried out in cocrystals with a chiral inductor give much better ee's,[109,110] the same photoreaction in solution never reaches such a level of enantioselectivity.[111]

The Norrish–Yang cyclization of imidazolidinones **54** performed in the presence of chiral templates **44a–d**[112,113] affords the *endo* and *exo* products **55** and **56**. The *exo*/*endo* ratio of the products formed from **54c** is dramatically switched from 38/62 in the absence of template to 90/10 in the presence of template **44c**. Chiral templates **44a**,**b** afford antipodal **55c** in 60% ee[113] (Scheme 4.21).

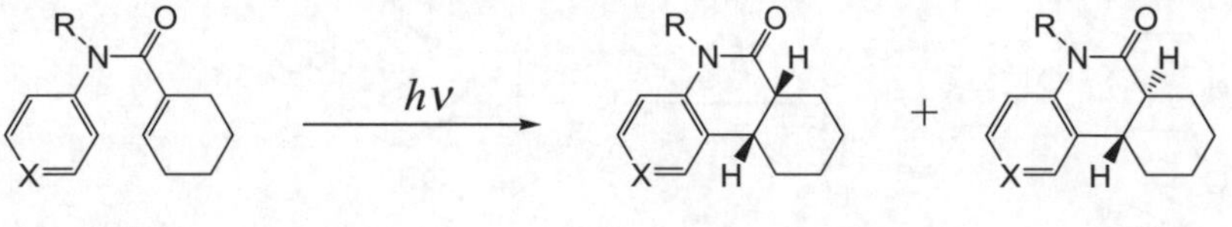

R–N(–C(=O)–cyclohexenyl)–pyridyl/phenyl (X) → $h\nu$ → **52** + **53**

51 **52** **53**

a: X = CH, R = H
b: X = CH, R = CH
c: X = N, R = H

Scheme 4.20. Enantiodifferentiating [6π]-photocyclization of acrylanilide.

54 → $h\nu$ → **55** + **56**

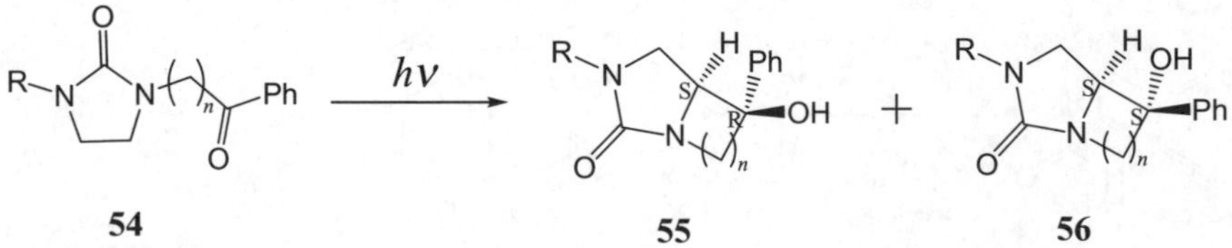

54 **55** **56**

a: R = H, n = 1
b: R = Ac, n = 1
c: R = H, n = 2
d: R = Ac, n = 2

Scheme 4.21. Enantiodifferentiating Norrish–Yang cyclization of ketones **54**.

Intermolecular Photoreactions Mediated by Chiral Templates

Precise control of intermolecular enantiodifferentiating photoreaction mediated by chiral template is more difficult to achieve in general, since one has to manipulate the termolecular interactions of the chiral host, guest substrate, and attacking reagent. A number of enantiodifferentiating intermolecular photoreactions have been examined with chiral templates **44**. Upon complexation with these chiral templates, one enantioface of the bound substrate is shielded by the template wall and free reagent can approach to the template–substrate complex from the unshielded face of substrate. Indeed, good to excellent ee's have been reported for the chiral intermolecular photoreactions mediated by **44**.

Photoirradiation of aromatic aldehyde **57** (Scheme 4.22) first affords (*Z*)- and (*E*)-dienols as reactive intermediates through the intramolecular γ-hydrogen abstraction. The resulting (*Z*)-dienol spontaneously returns to the starting material, while the (*E*)-isomer, possessing an inappropriate geometry for thc tautomcrism, survives long enough to undergo a Diels–Alder reaction with an alkene to afford two pairs of enantiomers **58** and **59**. This photoinduced cycloaddition reaction in the presence of **44a** favors the *exo* product **58** and furnished **58** in 92% ee and **59** in 94% ee.[114]

Intermolecular [4 + 4] photocycloaddition of pyridone **60** and cyclopentadiene in the absence of chiral template leads to a 2:3 mixture of cycloadducts **61** and **62** (Scheme 4.23). In the presence of chiral template **44a**, **61** and **62** are produced in remarkable ee's of 84–87%.[115]

Compound **63** (Scheme 4.24), which was originally developed as a gastroprokinetic agent, and its epimer **64** were incidentally found to strongly bind

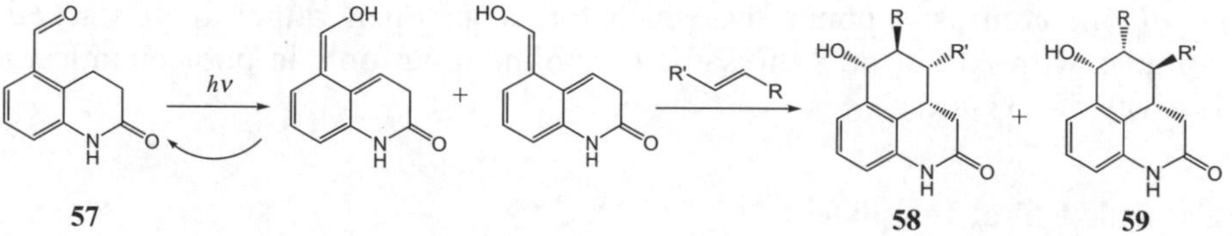

57 **58** **59**

a: R = CN, R' = H
b: R = COOMe, R' = H
c: R = R' = COOMe

Scheme 4.22. Enantiodifferentiating photoinduced [4 + 2] cycloaddition of **57** with dienophiles.

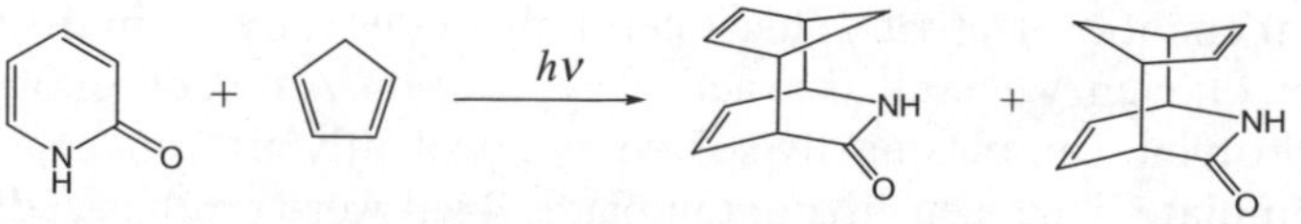

60 **61** **62**

Scheme 4.23. Enantiodifferentiating [4 + 4] photocycloaddition of pyridin-2(1H)-one with cyclopentadiene.

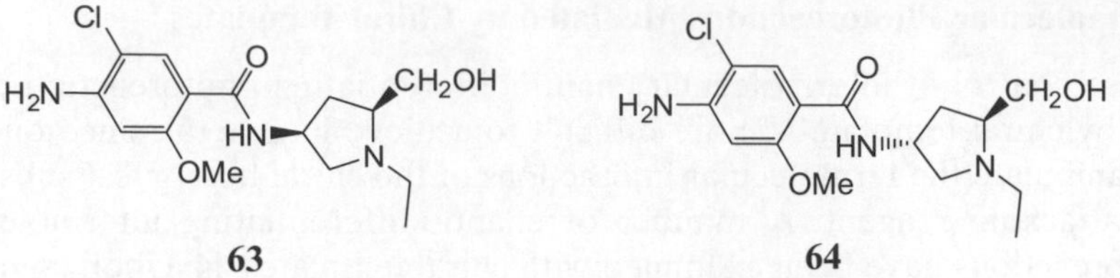

Cl H2N O HN OMe CH2OH N 63

Cl H2N O HN OMe CH2OH N 64

Scheme 4.24. Chiral template used for photocyclodimerization of 2-anthracenecarboxylic acid.

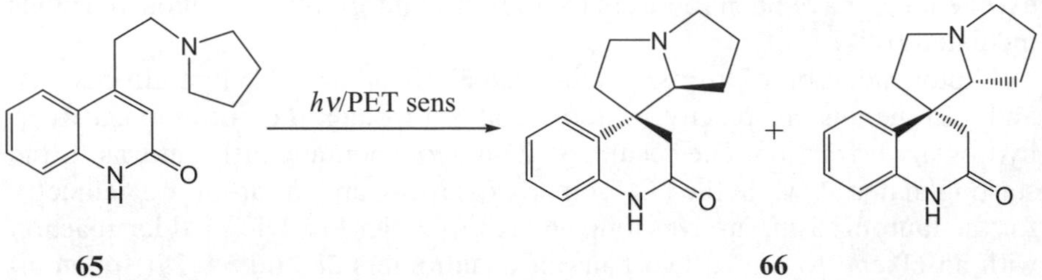

N N H O 65 *hν*/PET sens N N H O + N N H O 66

Scheme 4.25. Catalytic enantiodifferentiating photocycloaddition with PET sensitizing template (Fig. 4.1).

AC through a nine-membered dual hydrogen-bonding motif with the prolinol's amino and hydroxyl groups. Upon stacking complexation with AC, the 2-methoxybenzamide moiety of **63** shields one of the enantiofaces of AC, facilitating the attack from the opposite side of bound AC. Thus, the photocyclodimerization of AC (Scheme 4.3) in the presence of chiral template **63** affords HT dimer **5b** in –36% ee and HH dimer **5c** in –40% ee in CH_2Cl_2 at –50°C. In contrast, epimer **64**, which forms an open rather than stacked complex with AC, affords the same cyclodimers but only in poor enantioselectivity of <3% ee.

Catalytic Chiral Templates

The major drawback of using hydrogen-bonding templates is the formation of self-association dimers of both template and substrate. Photoirradiation of unbound substrate remaining in the bulk solution will give a racemic photoproduct. To suppress the contribution of free substrate to the product's ee, a stoichiometric or excess amount of chiral template is often required to achieve the optimal enantioselectivity. This is certainly undesirable from the viewpoint of chirogen efficiency as well as practical application. A recent report indicates that this problem can be smartly solved by covalently grafting a sensitizer to a chiral template. Thus, sensitizing templates **44e**,**f** were synthesized as photoinduced electron transfer (PET) sensitizers for chiral photocyclization of pyrrolidine-appended quinolone **65** (Scheme 4.25). Photoirradiation of **65** in the presence of 0.1 equivalent of **44e** affords cyclization product **66** in 52–64% yield with up to 70% ee.[116] As illustrated in Figure 4.1, substrate **65** is captured

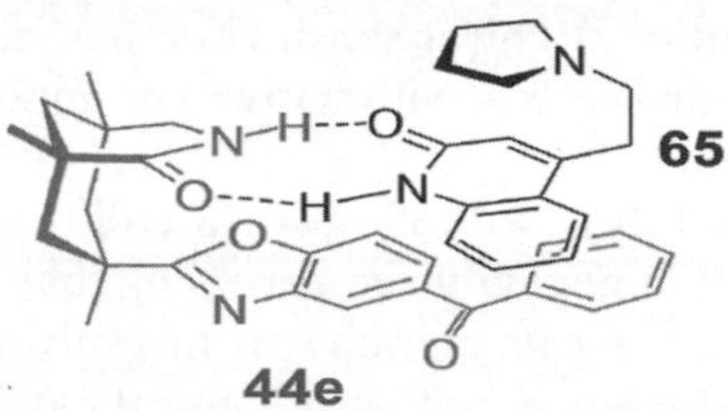

Figure 4.1. Mechanism for the catalytic chiral induction with sensitizer-bearing chiral template.

by chiral template **44e** through dual hydrogen bonds, with one of the enantiofaces of **65** being protected by the benzophenone moiety of **44e**. Although the photocyclization of **65** to chiral product **66** proceeds through rather complicated successive steps initiated by PET, it is to note that the intermediate pyrrolidyl radical attacks the *ipso*-carbon to form the spiro skeleton only from the open face of quinolone moiety (Fig. 4.1).[117]

Polar solvents are required in general to stabilize the radical ion pair formed in a PET process. However, solvation to the radical ion pair will cause a separation of the contact ion pair formed between substrate and chiral sensitizer, eventually leading to less efficient chirality transfer. Inoue and coworkers recently reported a potentially general methodology to circumvent this tradeoff by using a photosensitizer with protected saccharides in a nonpolar solvent. Under such circumstances, the PET process is accelerated by the local polarity around the sensitizer to produce a radical ion pair, which however cannot be separated by solvent molecules or diffuse to the bulk solution due to the low polarity of the nonpolar solvent.[37]

SUPRAMOLECULAR PHOTOCHIROGENESIS WITH CYCLODEXTRINS (CDS)

CDs are a series of cyclic oligosaccharides that possess truncated cone-shaped hydrophobic cavities. The commercially available α-, β-, and γ-CDs, which respectively consist of 6, 7, and 8 glucose units, can include a variety of organic compounds of different sizes in the central cavity in aqueous solution.[118] Hydrophobic interaction is the major driving force of the inclusion complexation of organic guests. CDs are water soluble, UV transparent, inherently chiral, feasible for chemical modification, and nontoxic in general. As a consequence of these advantages, CDs are so far the most extensively investigated chiral hosts for supramolecular photochirogenesis.[119,120]

CD-Mediated Chiral Intramolecular Photoreactions

The enantiodifferentiating photoisomerization of **4Z** to **4E** (Scheme 4.2), a benchmark chiral photoreaction frequently used for evaluating the performance of conventional chiral sensitizers, has also been employed in supramolecular photochirogenesis with native and modified CDs upon direct and sensitized excitation.

The vacuum UV photolysis at 185 nm of a solid-state complex of native β-CD with **4Z** affords **4E** in negligible ee, revealing that the native β-CD cavity is not efficient in transferring its chirality to the substrate.[65] This result is in line with the widely endorsed view that the chiral cavity of CD, in particular that of native CD, has only insignificant effect in chiral discrimination, presumably due to the round C_n symmetric structure with smooth inside walls as well as the nondirectional hydrophobic driving force.[121–123]

However, the originally low chirogenesis ability of native CD can be significantly improved through appropriate modifications. Inoue and coworkers reported the photoisomerization of **4Z** included and sensitized by a series of modified CDs **67** (Scheme 4.26).[124–131] The main idea is that the sensitizer moiety is self-included in its own cavity and unable to sensitize the unbound substrate in bulk solution, but an efficient energy/electron transfer occurs only upon inclusion of the substrate in the chiral cavity. Therefore, the sensitizing hosts **67** are expected to effectively transfer the excitation energy and chiral information from CD to bound **4Z**. Indeed, the ee of produced **4E** is significantly improved from nearly zero obtained with native β-CD[65] to 24% by using 6-*O*-(methyl phthaloyl)-β-CD **67b** as chiral sensitizer.[130] More recently, it was reported that the photoisomerization of **4Z** sensitized by 6-*O*-(*m*-methoxybenzoyl)-β-CD **67g** gives a much better ee of 46%, which is the highest ever reported for analogous CD-based sensitizers.[124,128] In contrast,

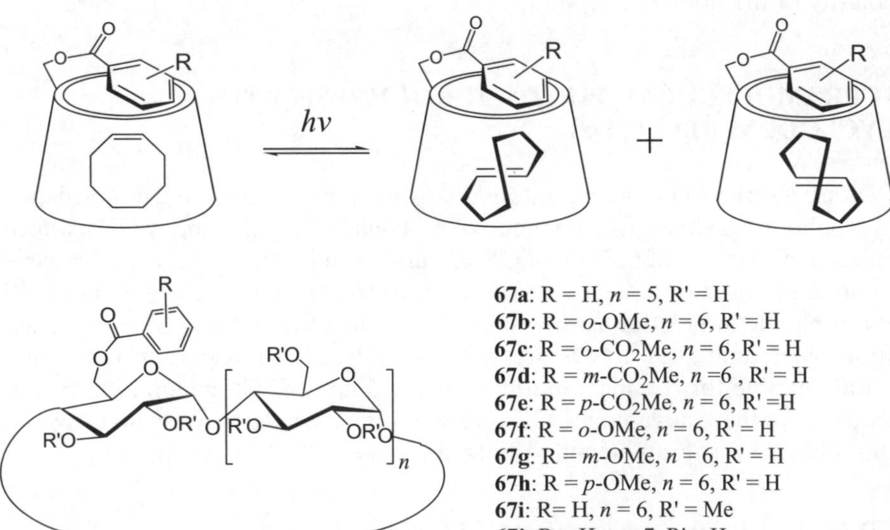

Scheme 4.26. Modified CDs as chiral sensitizing hosts **67**.

both *ortho* and *para* isomers **67f** and **67h** afford only poor enantioselectivity under the same reaction condition, revealing that a small difference in host structure can critically affect the optical yield and further that not only the choice of suitable host but also the fine-tuning of sensitizer structure are equally crucial for obtaining the best result in supramolecular photochirogenesis.[128]

In contrast to the significant effects of temperature and other entropy-related factors on the enantioselectivity obtained in conventional solution-phase chiral photosensitization, the entropy factor does not play a significant role in the supramolecular photoisomerization of **4Z** sensitized by 6-*O*-(methyl phthaloyl)-β-CD **67c**. Intriguingly, the ee obtained with *O*-permethylated 6-*O*-benzoyl-β-CD **67i**, in which the hydrogen-bonding network around the secondary rim of CD is broken, exhibits significant dependence on temperature. This contrasting behavior of ee obtained with native and permethylated CD is attributed to the difference in the rigidity of CD skeleton.[125,127] Hence, the dynamic ee control by entropy-related environmental factors is not feasible in most supramolecular photoreactions and thc original dcsign of chiral host becomes a crucial issue to obtain a significant ee.

The geometrical isomerization of (*Z*,*Z*)-1,3-cyclooctadiene **68ZZ**[132] (Scheme 4.27) to planar-chiral (*E*,*Z*)-isomer **68EZ** sensitized by naphthalene-appended CD **69**[133] affords modest ee of up to 4.6% ee. The enantioselectivities obtained with α- and β-CD-based hosts **69a,b** are only faintly affected by temperature, while the entropy factor plays a significant role when γ-CD-based **69c** is used, for which the more flexible framework of γ-CD is likely to be responsible.

The enantio- and diastereodifferentiating photoisomerization of 1,2-diphenylcyclopropane derivatives complexed with native β-CD have been investigated in both solution and solid state to give better ee's mostly in the solid state.[73] Photoirradiation of a solid-state mixture of β-CD complex of 1,2-diphenylcyclopropane **27a** (Scheme 4.12) and 4-methoxyacetophenone (added as a triplet sensitizer), which was prepared by grinding both in a mortar

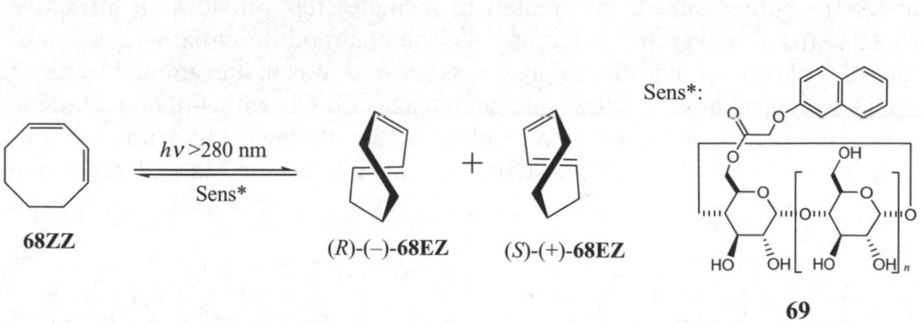

Scheme 4.27. Enantiodifferentiating photoisomerization of 1,3-cyclooctene included and sensitized by CD derivatives.

pestle, gives *trans* isomer **28a** in 13% ee. The enantiomer of **28a** that is preferentially included by β-CD was found to be enriched in the CD-mediated photoisomerization.[73] Direct irradiation of the solid-state complex of **13c** (Scheme 4.6) with native β-CD yields **14c** in 30% de.

Photolyses of *N*-methyl- and *N*-ethylpyridone **31a,b** (Scheme 4.13) in aqueous solution give **32a,b** in negligible ee even in the presence of an excess amount of β-CD. The solid-state complex can be prepared by mechanically mixing β-CD and **31a** or **31b**. Photoirradiation of the solid-state complex of **31b** with β-CD yields **32b** in 60% ee.[134] Interestingly, the enantioselectivity is sensitive to the water content of the solid-state complex. The product's ee dramatically decreases from 60% to 26% by just reducing the water content from 9% to 2%. This phenomenon is accounted for in terms of the hydrogen-bonding interactions of water molecules with the included substrate and/or host CD, since such a phenomenon is not caused by changing the content of nonhydrogen-bonding solvent like hexane.

Photoirradiation of tropolone ether **29** (Scheme 4.13) in the presence of CD in aqueous solution affords cyclization product **30** generally in low ee's (<5%). However, the photolysis of solid-state complex prepared from a mixture of **29** and CD in aqueous solution gives **30** in moderate ee's (up to 33%). The use of α-, β-, and γ-CD leads to the formation of **30a** in 28%, 5%, and 0% ee, respectively, indicating that the size matching between host and substrate is a most important factor for obtaining better enantiodifferentiation.[135]

Upon irradiation, *N,N*-dialkylpyruvamides **70** (Scheme 4.28) are converted to a mixture of the corresponding β-lactam and oxazolidin-4-one **71a–d** via the Norrish type II reaction.[136–138] The photoreaction performed in benzene gives **71c** as a major product, while **71a** is favored upon photolysis of β-CD-**70** complex in the solid state. The optical yield of **71d** obtained was determined to be 9% from the specific rotation.[136]

The intramolecular *meta* photocycloaddition of alkenoxybenzene, in which five stereogenic centers are created in a single step, provides an attractive route to fused polycyclic skeletons.[139,140] The enantiodifferentiating version of the *meta* photocycloaddition of alkenoxybenzene was first examined by using β-CD as a chiral host.[141] Solid-state complexes of β-CD with 4-phenoxybutene derivatives **72** (Scheme 4.29) were obtained by adding **72** to a hot aqueous solution of β-CD. The host:guest stoichiometry is 2:1 for **72a** and **72c–e** and

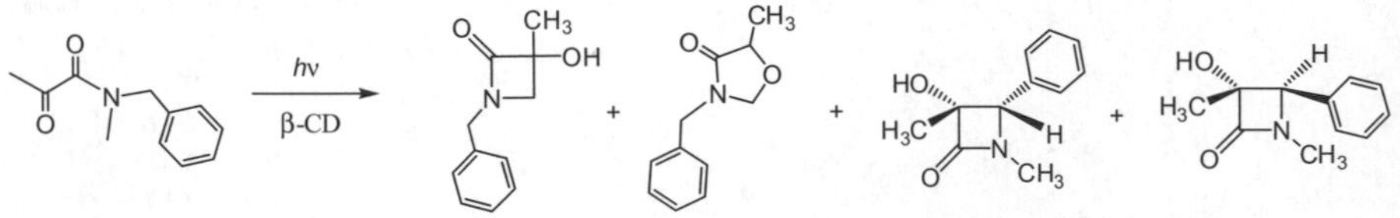

Scheme 4.28. Norrish type II reaction of *N,N*-dialkylpyruvamides.

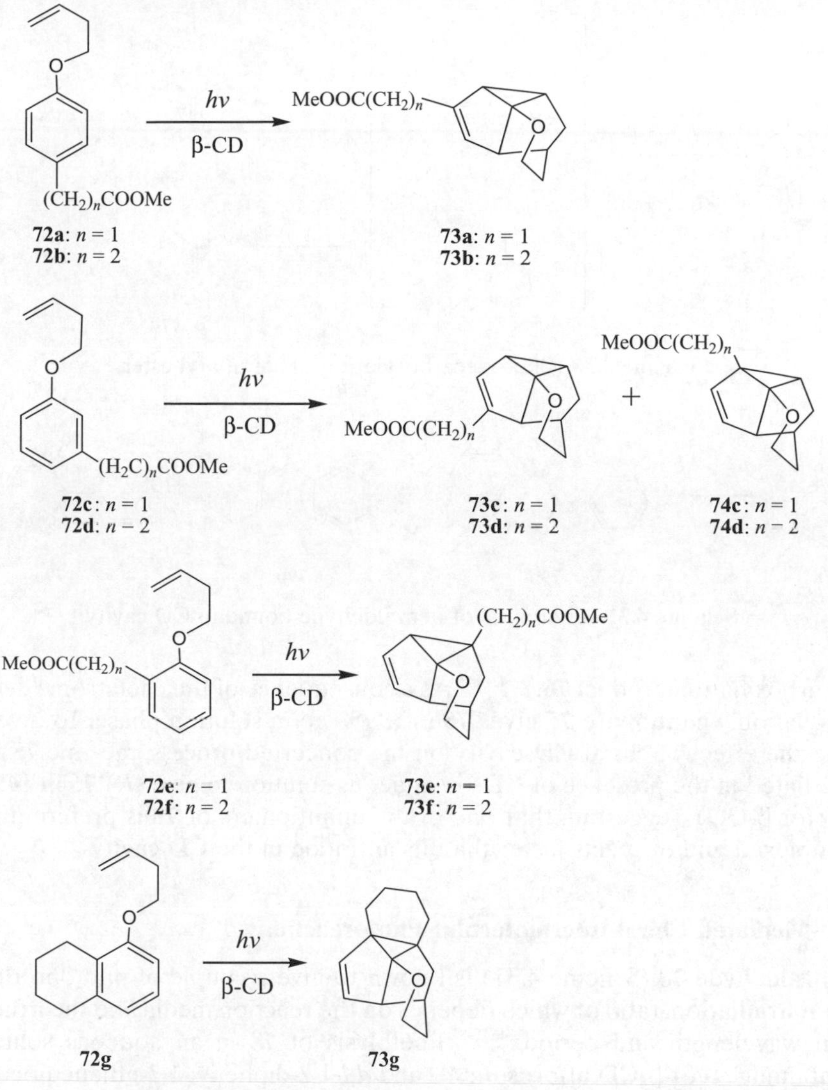

O

$(CH_2)_nCOOMe$

$h\nu$

β-CD

$MeOOC(CH_2)_n$

O

72a: $n = 1$
72b: $n = 2$

73a: $n = 1$
73b: $n = 2$

O

$(H_2C)_nCOOMe$

$h\nu$

β-CD

$MeOOC(CH_2)_n$

O

+

$MeOOC(CH_2)_n$

O

72c: $n = 1$
72d: $n = 2$

73c: $n = 1$
73d: $n = 2$

74c: $n = 1$
74d: $n = 2$

$MeOOC(CH_2)_n$

O

$h\nu$

β-CD

$(CH_2)_nCOOMe$

O

72e: $n = 1$
72f: $n = 2$

73e: $n = 1$
73f: $n = 2$

O

$h\nu$

β-CD

O

72g

73g

Scheme 4.29. Photocyclization of phenoxyalkenes **72**.

1:1 for the others. Upon irradiation, CD complexes of **72a**, **72b**, and **72e** give the corresponding cycloaddition products **73a,b,e** in only negligible ee, while the other complexes afford modest enantioselectivities of up to 17% ee (for **73c**). One of the enantiotopic faces of the benzene ring of **72** should be more hindered when included in the cavity of β-CD, and the tethered vinyl prefers to attack from the opposite side to yield enantiomeric **73** and **74**.

Chiral aryl ester **75** (Scheme 4.30) undergoes a concerted release of carbon dioxide via a spiro-lactonic transition state upon photoirradiation to afford

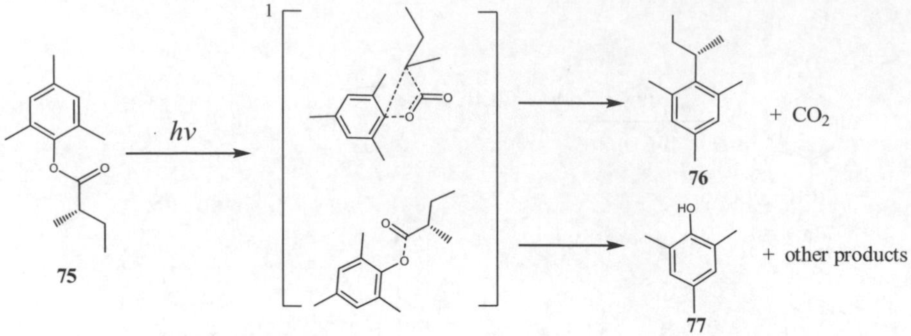

Scheme 4.30. Photodecarboxylation of chiral aryl ester.

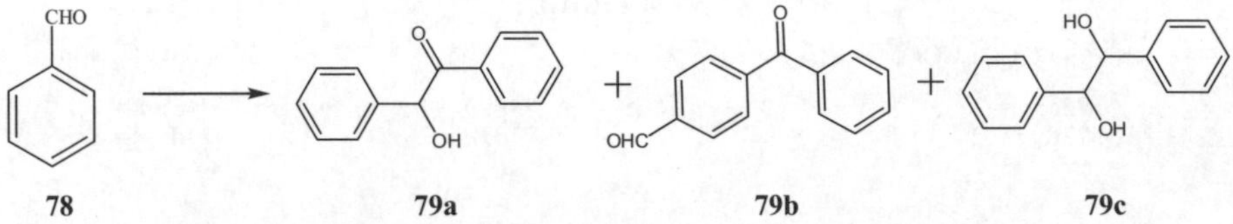

Scheme 4.31. Photolysis of benzaldehyde bound in CD cavity.

decarboxylation product **76**.[15,142–145] As a consequence of the chelatropic decarboxylation, enantiopure **75** gives **76** in >99% ee in solution phase. To investigate the effect of chiral CD cavity on the concerted process, racemic **75** was irradiated in the presence of CDs in aqueous solution to give (*R*)-**76** in 14.1% ee (for β-CD), suggesting that one of the enantiomers of **75** is preferentially complexed and/or reacts faster than its antipode in the CD cavity.

CD-Mediated Chiral Intermolecular Photoreactions

Benzaldehyde **78** (Scheme 4.31) is known to give a couple of photoproducts upon irradiation, ratio of which depends on the reaction media and the irradiation wavelength and period.[146–148] Photolysis of **78** in an aqueous solution containing α- or β-CD affords *meso*- and *d,l*-1,2-diphenyl-1,2-ethanediols **79c** in good 60–70% yield. Stirred with CD in aqueous solution, benzaldehyde **78** forms 1:1 and 1:2 solid-state complex with α- and γ-CD, respectively, while 2:2 complexation with β-CD is thought to occur in aqueous solution on the basis of the nuclear magnetic resonance (NMR) analysis of complexation behavior and the results of photoreaction. Irradiation of the solid-state complexes of **78** with β- and γ-CD gives **79a** and **79b** in different ratios of 70:30 (for β-CD) and 55:45 (for γ-CD). The enantioselectivity of obtained chiral product **79a** is 15% ee for β-CD, but almost racemic (<1% ee) for γ-CD, which is rationalized by the larger cavity of γ-CD that can only loosely bind the substrate.[149]

Photocyclodimerization of AC (Scheme 4.3) in the presence of native γ-CD was first investigated by Tamaki and coworkers in 1984.[150,151] It was found that the existence of γ-CD significantly improves the photocyclodimerization quantum yield from 0.05 to 0.4 with an accompanying increase of HT dimers at the expense of HH dimers. This photoreaction is inherently clean and the only by-product, 9,10-anthraquinone-2-carboxylic acid which arises from the incomplete deoxygenation of sample solution, is greatly suppressed in the presence of γ-CD. In contrast, β-CD forms a 2:2 complex with AC, and the photocyclodimerization of AC with β-CD affords exclusively achiral cyclodimer **5a**. Later in 2003, Nakamaura and Inoue reinvestigated this photoreaction in detail from the supramolecular photochirogenic point of view, since they found that enantiomers of **5b** and **5c** can be resolved by chiral high performance liquid chromatography (HPLC).[26] γ-CD forms stable 1:2 host–guest complex in a stepwise manner with association constants $K_1 = 161\,M^{-1}$ and $K_2 = 38{,}500\,M^{-1}$ in aqueous solution at 25°C. The much larger K_2 value is attributable thermodynamically to a greater enthalpy change ($-47.9\,kJ\,mol^{-1}$) for the inclusion of a second AC, most probably as a result of intimate contacts in the termolecular complex. Photocyclodimerization of AC with native γ-CD yields HT dimer **5b** in up to 41% ee and HH dimer **5c** in <5% ee.

Recently, a variety of peripheral and skeletal modifications have been applied to γ-CD for the purpose of improving the enantioselectivity of photocyclodimerization of AC. For example, secondary-rim-modified CDs **80a–d** (Scheme 4.32) give **5b** in varying enantioselectivity, which is comparable to or even smaller than the original ee obtained with native γ-CD. On the other hand, 3^A-amino-3^A-deoxy-*altro*-γ-CD **80e** greatly enhances the enantioselectivity, for which the distorted cavity and the electrostatic interaction between cationic ammonium and anionic carboxylate are jointly responsible.[30] The enantiodifferentiating photocyclodimerization of AC with native and modified γ-CD is sensitive to temperature, solvent, and hydrostatic pressure changes, and the use of **80e** leads to the best ee of 71% for **5b** in an aqueous solution at 210 MPa and −21°C.

γ-CD derivatives **80g**, **81**, and **82**, which possess two cationic groups on the primary rim, were designed and synthesized to improve the yield and enantioselectivity of HH dimers.[27,33] The attractive electrostatic interaction between the two ammonium cations on the primary rim and the carboxylate anion of included AC pair is expected to enhance the formation of HH-oriented precursor complexes. Indeed, **81e** and **82b** afforded chiral HH dimer **5c** in 15% and 35% ee, respectively, by optimizing the reaction temperature and solvent. The *anti*/*syn* ratio of HH dimers (**5c**/**5d**) is critically controlled by the distance between the two cationic sites on the rim of **82** to give more *anti*-HH dimer **5c** upon gradual elongation of the intercation distance from **82a** to **82d**.[129] Interestingly, photocyclodimerization of AC mediated by **80g** with one dicationic side arm affords **5c** in 41% ee, which is much higher than those obtained with its charge-equivalent analogs **82a–d** with two cationic substituents. This result indicates that the cations on a single flexible side arm are more effective in manipulating the enantiodifferentiation. Photolyses of solid-state complexes of AC with di-3,6-*anhydro*-γ-CDs **83a–d** with a remarkably distorted cavity

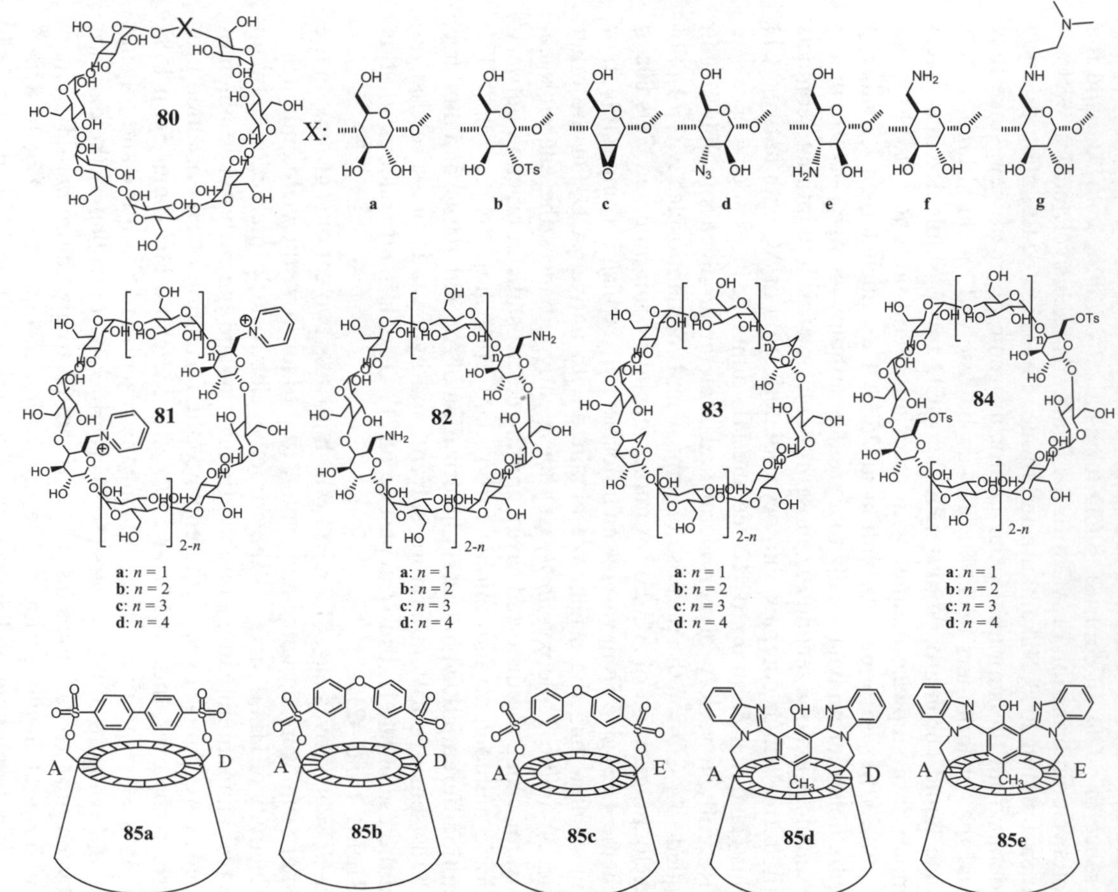

Scheme 4.32. Modified γ-CDs used for enantiodifferentiating [4 + 4] photocyclodimerization of 2-anthracenecarboxylic acid (AC).

give significantly improved combined yields (of up to 57%) and ee's (of up to 35% for **5c**) of HH dimers.[32]

A series of modified γ-CDs with flexible or rigid cap(s) have also been prepared for investigating the effect of aromatic cap on the photocyclodimerization of AC. Ditosylated γ-CD **84** gives HT dimer **5b** in yield and ee comparable to those observed with native γ-CD, indicating that aromatic substitution does not greatly alter the chiral environment of native γ-CD cavity at least for this photoreaction. The use of capped γ-CD **85b** and **85c** significantly diminish the enantioselectivity of **5b**, while the use of **85a** with a more rigid cap leads to the formation of antipodal **5b** in an improved ee of −56%. The opposite stereochemical consequences caused by the apparently trivial alteration in cap structure between **85a** and **85b** reveal the crucial role of cap rigidity in vitally controlling the supramolecular photochirogenic reactions in general.[30,128] γ-CDs **85d** and **85e**, possessing a bridging *p*-cresolbisbenzimidazole cap (Scheme 4.32), show obviously pH-dependent UV-Vis, circular dichroism, and fluorescence spectral changes. Furthermore, in the photocyclodimerization of AC mediated by **85d** and **85e**, a clear inversion of the enantioselectivity of HH dimer **5c** is induced by changing the solution pH, most probably due to the pH-responsive conformational change of the cap moiety.[128]

More recently, the supramolecular photocyclodimerizations of α-CD-appended AC **86** (Scheme 4.33) mediated by CD and cucurbituril hosts were examined to elucidate the effects of a bulky guest substituent (i.e., α-CD) located outside the binding site on the product ratio and ee. Photolysis of an aqueous solution containing native γ-CD and **86** leads to the formation of γ-CD-wheeled rotaxanes of almost exclusively HT dimers (HT:HH = 98:2), hydrolysis of which gives **5b** in 91% ee.[128] When achiral cucurbit[8]uril is used as a host, the product distribution is dramatically switched to the HH dimers (HT:HH = 1:99). These results demonstrate for the first time that a bulky group located outside the chiral host cavity can remotely but critically affect the complexation and the subsequent photoreaction occurring inside the cavity.

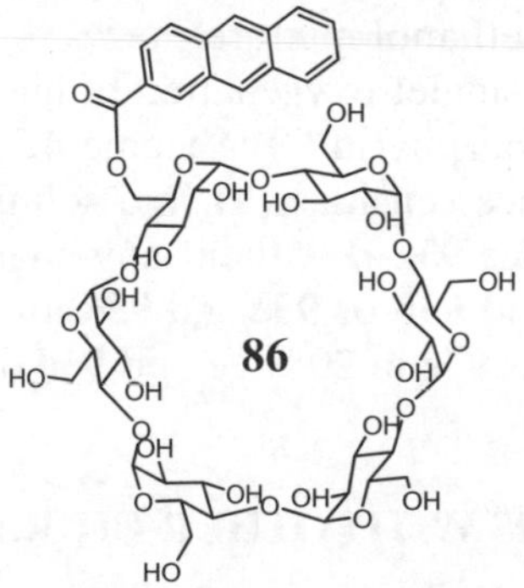

Scheme 4.33. α-CD-attached 2-anthracenecarboxylate.

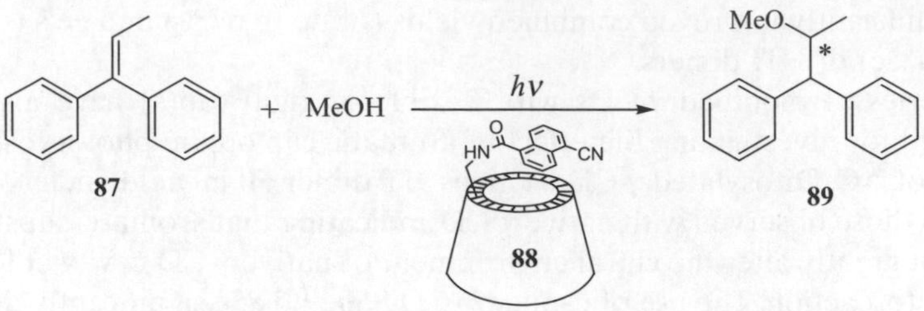

Scheme 4.34. Photosensitized anti-Markovnikov photoaddition of methanol to 1,1-diphenylpropene.

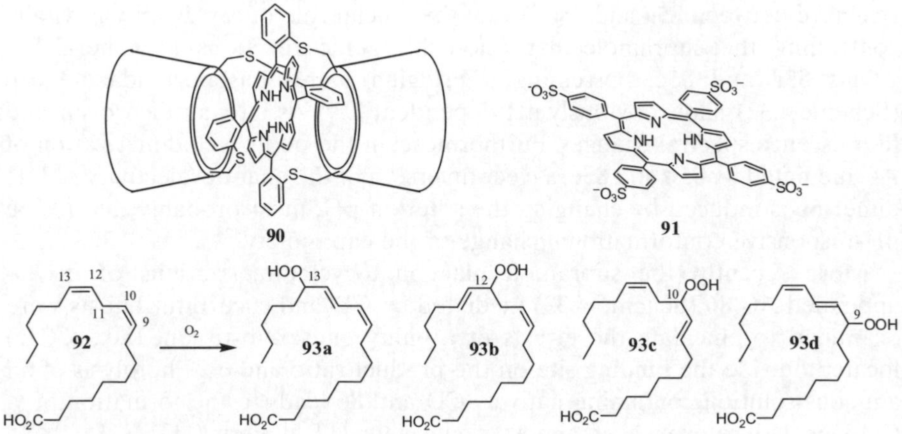

Scheme 4.35. Photosensitized singlet oxygenation of linoleic acid.

The anti-Markovnikov photoaddition of methanol to 1,1-diphenylpropene **87** (Scheme 4.34) can be mediated by 6-(5-cyanonaphthyl-1-carboamido)-6-deoxy-β-CD **88** as a PET sensitizing host.[125] Differing from the other supramolecular photochirogenic reactions with β-CD-based sensitizers, the enantioselectivity of photoproduct is highly sensitive to the temperature change, affording adduct **89** in –2.1% ee at 45°C but antipodal **89** in 5.8% ee at –40°C in a 1:1 water methanol mixture.

Enantiodifferentiating singlet oxygenation of linoleic acid **92** can be mediated by CD-sandwiched porphyrin **90** (Scheme 4.35).[152] Singlet oxygenation of **92** with achiral reference sensitizer **91** in a solution saturated with native β-CD gives hydroperoxides **93a–d** without any enantioselectivity or product selectivity between **93a** and **93b** or **93c** and **93d**. In contrast, photolysis of **92** in the presence of **90** gives **93c** in 20% ee and **93d** in 21% ee.

PHOTOCHIROGENESIS WITH OTHER CHIRAL HOSTS

Besides the supramolecular photochirogenesis systems described above, some new chiral hosts that possess attractive host properties have recently been

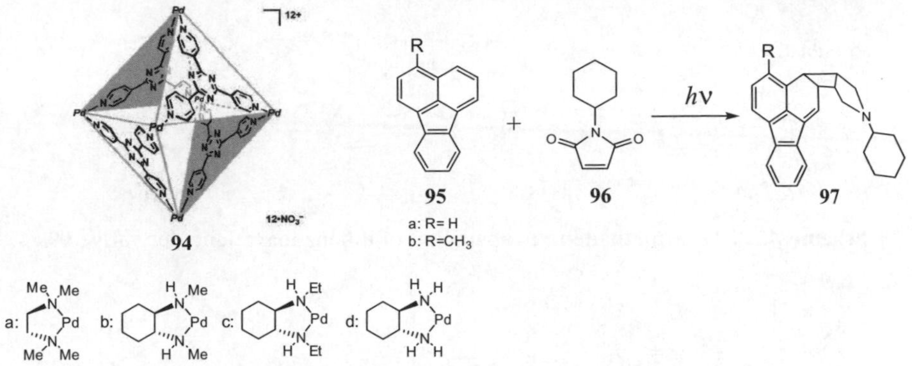

Scheme 4.36. Photocycloaddition with chiral self-assembled cage.

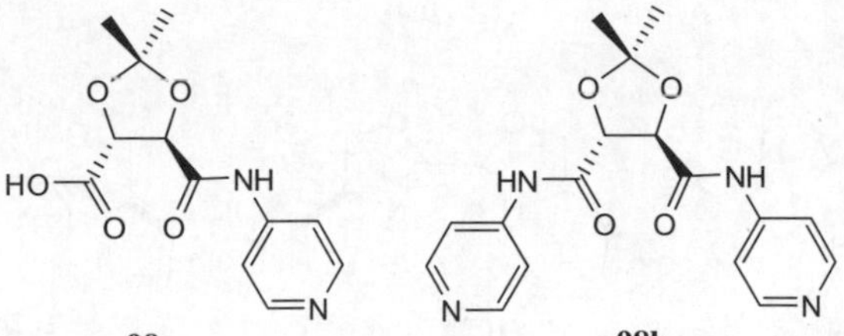

Scheme 4.37. Tartaric acid derivatives used for preparation of POST-1 and also for sensitized photoisomerization of (*Z*)-cyclooctene **1Z**.

reported. Fujita and coworkers reported an asymmetric photoaddition between **95** and **96** (Scheme 4.36) using self-assembled cages **94**. Like CDs, confined M_6L_4 cages accelerate the reaction rate and enhance the stereoselectivity of thermal and photochemical reactions.[153–156] Chiral ligands introduced to the periphery of cage **94** asymmetrically deform the cage frame, which enabled them to obtain **97b** in 50% ee upon photocycloaddition of **95–96** inside the cage of **94b**.[154] This result is encouraging, since even a chiral ligand coordinated to Pd from the outside of the cage can induce highly chiral environment to the originally achiral cavity.

A homochiral metal–organic porous material POST-1, which is composed of tartaric acid derivative **98a** (Scheme 4.37) coordinated to zinc, was used for mediating the enantiodifferentiating photoisomerization of **4Z**.[129,157] Supramolecular photosensitization of **4Z** with POST-1 gives **4E** in 5.4% ee, which is comparable to that observed in photoisomerization of **4Z** sensitized by **98b** (4.9% ee). The modest enantioselectivity may be attributed to the chiral channel that has a diameter (13 Å) much larger than the size of **4Z** and hence cannot efficiently confine the substrate **4Z**.

(1*S*,2*S*)-*trans*-1,2-Diaminocyclohexane can be immobilized in MCM-41 supercages to make a chiral mesoporous organosilica. The photoinduced di-π-methane rearrangement of 11-formyl-12-methyldibenzobarrelene **99**

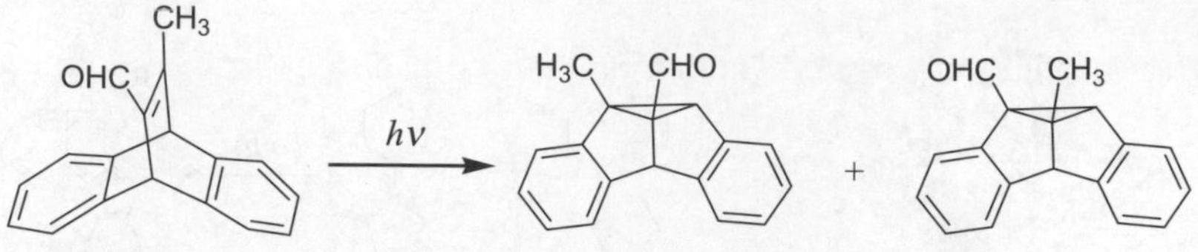

Scheme 4.38. Di-π-methane rearrangement of dibenzobarrelene derivative **99**.

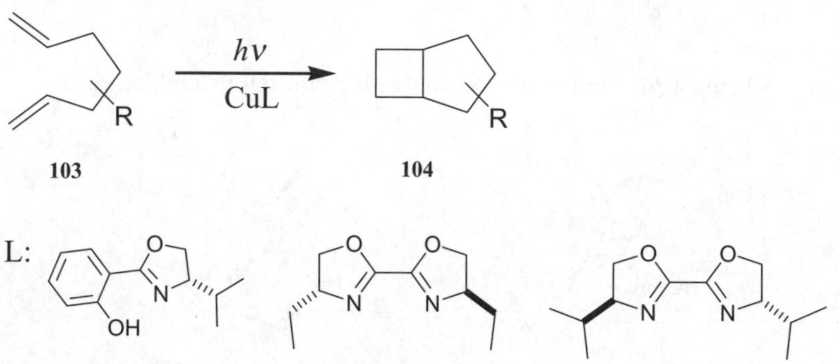

Scheme 4.39. Stereodifferentiating intramolecular copper(I)-catalyzed [2 + 2] photocycloaddition.

(Scheme 4.38) to **100** and **101** carried out in the chiral mesoporous organosilica gives **100** in 24% ee at a conversion of 11%.[158]

Langer and Mattay examined the intramolecular [2 + 2] photocycloaddition of 1,6-diene derivatives **103** catalyzed by copper(I) complexes with chiral ligands **102** (Scheme 4.39).[3] Despite that these chiral catalysts are successful in a variety of thermal asymmetric reactions, photocycloaddition of **103** with these chiral catalysts gives only low ee's of <5%. The coordination of a substrate to the chiral copper(I) catalyst leads to a decrease or even a complete suppression of the reactivity, and some chiral ligands are destroyed upon irradiation. The PET from the ligand rather than the olefin is likely to be responsible for the low reactivity and selectivity.

SUMMARY AND PERSPECTIVE

In the last decade, considerable progresses have been achieved in the field of supramolecular photochirogenesis. Many unique natures and new features that essentially differ from the conventional chiral photochemistry have been revealed for a variety of supramolecular photochirogenesis systems.

Complexation with a chiral supramolecular host enables efficient chirality transfer through noncovalent interactions in both ground and excited states to alter, and often enhance, the original reactivity and stereoselectivity and even open a new reaction channel. The confinement effect and the intimate supramolecular interactions in the cavity are the two most prominent factors that control the supramolecular photochirogenesis. Thus, the optical yield obtained in supramolecular photochirogenesis is a critical function of the host structure and modification and is controlled also by the external factors such as solvent, temperature, and pressure mostly through manipulation of the complexation in the ground state. Hence, the role of entropy-related factors is more or less reduced in the kinetic process of supramolecular photochirogenesis, unless a conformationally flexible host is used, and therefore the design of chiral host best fitted to the target photochirogenic reaction becomes highly crucial to achieve high stereoselectivity.

The chiral prearrangement of substrate(s) in the precursor complex with chiral host in the ground state is unique and essential to supramolecular photochirogenesis, and is expected to propagate to the excited state upon irradiation. This is particularly evident in the case of face-differentiating intermolecular photoreaction, because guest substrates accommodated in a binding site are not allowed to switch their position/orientation within the lifetime of excited substrate, which is much shorter than the time required for completing a decomplexation–recomplexation cycle. This means that the two inherent limitations or disadvantages of chiral photochemistry, that is, inherently weak interaction and short excited-state lifetime, are nicely overcome in supramolecular photochirogenesis by preorganizing the ground-state substrate(s) in an appropriate conformation/orientation (an entropic drive) and also by keeping the intimate chiral interactions of excited substrate(s) with supramolecular host for a longer period of time (an enthalpic drive).

In sensitized supramolecular photochirogenesis, the positional and conformational adjustment of substrate upon excited-state interaction with the sensitizer moiety should play a key role. The exciplex formed in a binding site is considered to be appreciably different in structure from the ground-state complex. In this context, theoretical and experimental studies on the supramolecular complex structure in both ground and excited states will afford important insights into the factors and mechanisms operating in supramolecular photosensitization and further improve the stereochemical outcomes.

While good to excellent diastereo- and enantioselectivities have been achieved in several supramolecular photochirogenesis systems, it is reasonable to expect that more significant stereochemical outcomes can be obtained through further explorations. The number of chiral hosts that have been examined for supramolecular photochirogenesis is still much smaller than that of chiral catalysts developed or enzymes examined for asymmetric syntheses. The optical yields hitherto reported are not necessarily satisfactory from the practical application point of view. On the other hand, the studies on chiral photochemistry have rather been limited to uni- or intramolecular photoreactions, and the intermolecular versions have not been extensively explored partly due to the difficulty in controlling the multimolecular interactions of chiral host with two or more prochiral guest substrates. In this context, chiral hosts with a larger binding site, which is suitable for inclusion of two or more substrates, may have to be used more frequently or newly developed in the next decade.

Pursuing photochirogenic reaction with a catalytic amount of supramolecular host is another important issue that deserves more attention of photochemists. So far, most of the supramolecular photochirogenesis systems require at least an equimolar, or even larger, amount of chiral host to obtain the optimal stereoselectivity, and catalytic photochirogenesis is achieved only for sensitized photoreaction. However, we have seen in several cases that even a nonsensitized photoreaction mediated by supramolecular host is remarkably accelerated, showing a distinctive feature of catalytic reaction. Thus, a catalytic photoreaction may also be implemented in nonsensitized supramolecular systems if the photoreaction occurring outside the binding site could be suppressed and the photoproduct could be removed timely from the system to avoid product inhibition.

Finally, we emphasize the green chemical aspect of supramolecular photochirogenesis. Most supramolecular hosts are inherently "organic" and environmentally benign, as they do not contain any hazardous heavy/transition or precious rare/noble metals, and are originally designed to work in aqueous solutions at ambient temperatures as "organic" catalysts, driven by clean light energy, and often biodegradable. These features may promote further development and practical application of the supramolecular photochirogenic strategies in the near future.

REFERENCES

1. Lough, W.J.; Wainer, I.W. *Chirality in Natural and Applied Science*. Boca Raton, FL: CRC Press, **2002**.
2. Ojima, I. *Catalytic Asymmetric Synthesis*. New York: Wiley-VCH, **2000**.
3. Langer, K.; Mattay, J. *J. Org. Chem.* **1995**, *60*, 7256.
4. Inoue, Y. *Chem. Rev.* **1992**, *92*, 741.
5. Rau, H. *Chem. Rev.* **1983**, *83*, 535.
6. Axel, G.; Griesbeck, U.J.M. *Angew. Chem. Int. Ed. Engl.* **2002**, *41*, 3147.
7. Mattay, J.; Dekker, M.; Kaanumalle, A.N.; Ramamurthy, V. *Synthetic Organic Photochemistry*. New York: Marcel Dekker, **2005**.
8. Hough, J.H.; Bailey, J.A.; Chrysostomou, A.; Gledhill, T.M.; Lucas, P.W.; Tamura, M.; Clark, S.; Yates, J.; Menard, F. *Adv. Space Res.* **2001**, *27*, 313.
9. Rikken, G.; Raupach, E. *Nature* **2000**, *405*, 932.
10. Nishino, H.; Kosaka, A.; Hembury, G.A.; Aoki, F.; Miyauchi, K.; Shitomi, H.; Onuki, H.; Inoue, Y. *J. Am. Chem. Soc.* **2002**, *124*, 11618.
11. Nishino, H.; Kosaka, A.; Hembury, G.A.; Shitomi, H.; Onuki, H.; Inoue, Y. *Org. Lett.* **2001**, *3*, 921.

12. Ruck-Braun, K.; Kunz, H. *Chiral Auxiliaries in Cycloadditions*. New York: Wiley-VCH, **1999**.
13. Hammond, G.S.; Cole, R.S. *J. Am. Chem. Soc.* **1965**, *87*, 3256.
14. Inoue, Y.; Matsushima, E.; Wada, T. *J. Am. Chem. Soc.* **1998**, *120*, 10687.
15. Mori, T.; Weiss, R.G.; Inoue, Y. *J. Am. Chem. Soc.* **2004**, *126*, 8961.
16. Levi-Minzi, N.; Zandomeneghi, M. *J. Am. Chem. Soc.* **1992**, *114*, 9300.
17. Ouchi, A.; Zandomeneghi, G.; Zandomeneghi, M. *Chirality* **2002**, *14*, 1.
18. Festa, C.; Levi-Minzi, N.; Zandomeneghi, M. *Gazz. Chim. Ital.* **1996**, *126*, 599.
19. Lhiaubet-Vallet, V.; Encinas, S.; Miranda, M.A. *J. Am. Chem. Soc.* **2005**, *127*, 12774.
20. Lhiaubet-Vallet, V.; Sarabia, Z.; Bosca, F.; Miranda, M.A. *J. Am. Chem. Soc.* **2004**, *126*, 9538.
21. Lhiaubet-Vallet, V.; Bosca, F.; Miranda, M.A. *J. Phys. Chem. B* **2007**, *111*, 423.
22. Wada, T.; Sugahara, N.; Kawano, M.; Inoue, Y. *Chem. Lett.* **2000**, 1174.
23. Fritzsche, I. *J. Prakt. Chem.* **1867**, *101*, 333.
24. Wada, T.; Nishijima, M.; Fujisawa, T.; Sugahara, N.; Mori, T.; Nakamura, A.; Inoue, Y. *J. Am. Chem. Soc.* **2003**, *125*, 7492.
25. Nishijima, M.; Pace, T.C.S.; Nakamura, A.; Mori, T.; Wada, T.; Bohne, C.; Inoue, Y. *J. Org. Chem.* **2007**, *72*, 2707.
26. Nakamura, A.; Inoue, Y. *J. Am. Chem. Soc.* **2003**, *125*, 966.
27. Nakamura, A.; Inoue, Y. *J. Am. Chem. Soc.* **2005**, *127*, 5338.
28. Mizoguchi, J.-I.; Kawanami, Y.; Wada, T.; Kodama, K.; Anzai, K.; Yanagi, T.; Inoue, Y. *Org. Lett.* **2006**, *8*, 6051.
29. Yang, C.; Fukuhara, G.; Nakamura, A.; Origane, Y.; Fujita, K.; Yuan, D.-Q.; Mori, T.; Wada, T.; Inoue, Y. *J. Photochem. Photobiol. A Chem.* **2005**, *173*, 375.
30. Yang, C.; Nakamura, A.; Fukuhara, G.; Origane, Y.; Mori, T.; Wada, T.; Inoue, Y. *J. Org. Chem.* **2006**, *71*, 3126.
31. Yang, C.; Nakamura, A.; Wada, T.; Inoue, Y. *Org. Lett.* **2006**, *8*, 3005.
32. Yang, C.; Nishijima, M.; Nakamura, A.; Mori, T.; Wada, T.; Inoue, Y. *Tetrahedron Lett.* **2007**, *48*, 4357.
33. Ikeda, H.; Nihei, T.; Ueno, A. *J. Org. Chem.* **2005**, *70*, 1237.
34. Yang, C.; Ke, C.; Fujita, K.; Yuan, D.-Q.; Mori, T.; Inoue, Y. *Aust. J. Chem.* **2008**, *61*, 1–4.
35. Yang, C.; Mori, T.; Inoue, Y. *J. Org. Chem.* **2008**, *73*, 5786.
36. Yang, C.; Mori, T.; Origane, Y.; Ko, Y.H.; Selvapalam, N.; Kim, K.; Inoue, Y. *J. Am. Chem. Soc.* **2008**, *130*, 8574.
37. Asaoka, S.; Wada, T.; Inoue, Y. *J. Am. Chem. Soc.* **2003**, *125*, 3008.
38. Kragh-Hansen, U. *Pharmacol. Rev.* **1981**, *33*, 17.
39. Peters, T. *All About Albumin: Biochemistry, Genetics, and Medical Applications*. San Diego, CA: Academic Press, **1996**.
40. Nishijima, M.; Wada, T.; Mori, T.; Pace, T.C.S.; Bohne, C.; Inoue, Y. *J. Am. Chem. Soc.* **2007**, *129*, 3478.
41. Breck, D.W. *Zeolite Molecular Sieves: Structure, Chemistry, and Use*. Malabar, FL: Krieger Pub Co, **1984**.
42. Davis, M.E.; Lobo, R.F. *Chem. Mater.* **1992**, *4*, 756.

43. Joy, A.; Ramamurthy, V. *Chem. Eur. J.* **2000**, *6*, 1287.
44. Kaanumalle, L.S.; Sivaguru, J.; Arunkumar, N.; Karthikeyan, S.; Ramamurthy, V. *Chem. Commun.* **2003**, 116.
45. Sivaguru, J.; Natarajan, A.; Kaanumalle, L.S.; Shailaja, J.; Uppili, S.; Joy, A.; Ramamurthy, V. *Acc. Chem. Res.* **2003**, *36*, 509.
46. Joy, A.; Uppili, S.; Netherton, M.R.; Scheffer, J.R.; Ramamurthy, V. *J. Am. Chem. Soc.* **2000**, *122*, 728.
47. Natarajan, A.; Joy, A.; Kaanumalle, L.S.; Scheffer, J.R.; Ramamurthy, V. *J. Org. Chem.* **2002**, *67*, 8339.
48. Chong, K.C.W.; Sivaguru, J.; Shichi, T.; Yoshimi, Y.; Ramamurthy, V.; Scheffer, J.R. *J. Am. Chem. Soc.* **2002**, *124*, 2858.
49. Zimmerman, H.E.; Flechtner, T.W. *J. Am. Chem. Soc.* **1970**, *92*, 6931.
50. Sivaguru, J.; Sunoj, R.B.; Wada, T.; Origane, Y.; Inoue, Y.; Ramamurthy, V. *J. Org. Chem.* **2004**, *69*, 5528.
51. Sivaguru, J.; Jockusch, S.; Turro, N.J.; Ramamurthy, V. *Photochem. Photobiol. Sci.* **2003**, *2*, 1101.
52. Sivaguru, J.; Sunoj, R.B.; Wada, T.; Origane, Y.; Inoue, Y.; Ramamurthy, V. *J. Org. Chem.* **2004**, *69*, 6533.
53. Kaanumalle, L.S.; Sivaguru, J.; Sunoj, R.B.; Lakshminarasimhan, P.H.; Chandrasekhar, J.; Ramamurthy, V. *J. Org. Chem.* **2002**, *67*, 8711.
54. Rao, V.J.; Uppili, S.R.; Corbin, D.R.; Schwarz, S.; Lustig, S.R.; Ramamurthy, V. *J. Am. Chem. Soc.* **1998**, *120*, 2480.
55. Auerbach, S.M.; Carrado, K.A.; Dutta, P.K. *Handbook of Zeolite Science and Technology*. Boca Raton, FL: CRC Press, **2003**.
56. van Bekkum, H. *Introduction to Zeolite Science and Practice*. Amsterdam: Elsevier Science, **2001**.
57. Auerbach, S.M.; Carrado, K.A.; Dutta, P.K., editors, *Handbook of Zeolite Science and Technology*. New York: Marcel Dekker, **2003**.
58. Carley, A.F.; Davies, P.R.; Hutchings, G.J.; Spencer, M.S., editors, *Surface Chemistry and Catalysis*. New York: Kluwer Academic/Plenum, **2002**.
59. Van de Velde, F.; Arends, I.W.C.E.; Sheldon, R.A. *Topics Catal.* **2000**, *13*, 259.
60. Shephard, D.S. *Stud. Surf. Sci. Cat.* **2000**, *129*, 789.
61. Hutchings, G.J. *Chem. Commun.* **1999**, 301.
62. Leibovitch, M.; Olovsson, G.; Sundarababu, G.; Ramamurthy, V.; Scheffer, J.R.; Trotter, J. *J. Am. Chem. Soc.* **1996**, *118*, 1219.
63. Murov, S.L.; Cole, R.S.; Hammond, G.S. *J. Am. Chem. Soc.* **1968**, *90*, 2957.
64. Ouannes, C.; Beugelmans, R.; Roussi, G. *J. Am. Chem. Soc.* **1973**, *95*, 8472.
65. Inoue, Y.; Kosaka, S.; Matsumoto, K.; Tsuneishi, H.; Hakushi, T.; Tai, A.; Nakagawa, K.; Tong, L. *J. Photochem. Photobiol. A Chem.* **1993**, *71*, 61.
66. Inoue, Y.; Yamasaki, N.; Shimoyama, H.; Tai, A. *J. Org. Chem.* **1993**, *58*, 1785.
67. Cheung, E.; Chong, K.C.W.; Jayaraman, S.; Ramamurthy, V.; Scheffer, J.R.; Trotter, J. *Org. Lett.* **2000**, *2*, 2801.
68. Lakshminarasimhan, P.; Sunoj, R.B.; Chandrasekhar, J.; Ramamurthy, V. *J. Am. Chem. Soc.* **2000**, *122*, 4815.

69. Sivaguru, J.; Scheffer, J.R.; Chandarasekhar, J.; Ramamurthy, V. *Chem. Commun.* **2002**, 830.
70. Koodanjeri, S.; Sivaguru, J.; Pradhan, A.; Ramamurthy, V. *Proc. Indian Natl. Sci. Acad. A* **2002**, *68*, 453.
71. Sivaguru, J.; Shichi, T.; Ramamurthy, V. *Org. Lett.* **2002**, *4*, 4221.
72. Sivaguru, J.; Wada, T.; Origane, Y.; Inoue, Y.; Ramamurthy, V. *Photochem. Photobiol. Sci.* **2005**, *4*, 119.
73. Koodanjeri, S.; Ramamurthy, V. *Tetrahedron Lett.* **2002**, *43*, 9229.
74. Poon, T.; Sivaguru, J.; Franz, R.; Jockusch, S.; Martinez, C.; Washington, I.; Adam, W.; Inoue, Y.; Turro, N.J. *J. Am. Chem. Soc.* **2004**, *126*, 10498.
75. Sivasubramanian, K.; Kaanumalle, L.S.; Uppili, S.; Ramamurthy, V. *Org. Biomol. Chem.* **2007**, *5*, 1569.
76. Kaftory, M.; Yagi, M.; Tanaka, K.; Toda, F. *J. Org. Chem.* **1988**, *53*, 4391.
77. Toda, F. *Mol. Crys. Liq. Crys.* **1988**, *161*, 355.
78. Toda, F.; Tanaka, K. *Chem. Commun.* **1986**, 1429.
79. Toda, F.; Tanaka, K.; Yagi, M. *Tetrahedron* **1987**, *43*, 1495.
80. Scheffer, J.R.; Wang, L. *J. Phys. Org. Chem.* **2000**, *13*, 531.
81. Ma, L.; Wu, L.-Z.; Zhang, L.-P.; Tung, C.-H. *Chin. J. Chem.* **2003**, *21*, 96.
82. Joy, A.; Kaanumalle, L.S.; Ramamurthy, V. *Org. Biomol. Chem.* **2005**, *3*, 3045.
83. Joy, A.; Scheffer, J.R.; Ramamurthy, V. *Org. Lett.* **2000**, *2*, 119.
84. Shailaja, J.; Lakshminarasimhan, P.H.; Pradhan, A.R.; Sunoj, R.B.; Jockusch, S.; Karthikeyan, S.; Uppili, S.; Chandrasekhar, J.; Turro, N.J.; Ramamurthy, V. *J. Phys. Chem. A* **2003**, *107*, 3187.
85. Uppili, S.; Ramamurthy, V. *Org. Lett.* **2002**, *4*, 87.
86. Kaprinidis, N.A.; Landis, M.S.; Turro, N.J. *Tetrahedron Lett.* **1997**, *38*, 2609.
87. Edman, J.R. *J. Am. Chem. Soc.* **1969**, *91*, 7103.
88. Hahn, R.C.; Johnson, R.P. *J. Am. Chem. Soc.* **1977**, *99*, 1508.
89. Joy, A.; Robbins, R.J.; Pitchumani, K.; Ramamurthy, V. *Tetrahedron Lett.* **1997**, *38*, 8825.
90. Inoue, Y.; Ikeda, H.; Kaneda, M.; Sumimura, T.; Everitt, S.R.L.; Wada, T. *J. Am. Chem. Soc.* **2000**, *122*, 406.
91. Nishiyama, Y.; Kaneda, M.; Saito, R.; Mori, T.; Wada, T.; Inoue, Y. *J. Am. Chem. Soc.* **2004**, *126*, 6568.
92. Nishiyama, Y.; Wada, T.; Asaoka, S.; Mori, T.; McCarty, T.A.; Kraut, N.D.; Bright, F.V.; Inoue, Y. *J. Am. Chem. Soc.* **2008**, *130*, 7526.
93. Wada, T.; Shikimi, M.; Inoue, Y.; Lem, G.; Turro, N.J. *Chem. Commun.* **2001**, 1864.
94. Sivaguru, J.; Saito, H.; Solomon, M.R.; Kaanumalle, L.S.; Poon, T.; Jockusch, S.; Adam, W.; Ramamurthy, V.; Inoue, Y.; Turro, N.J. *Photochem. Photobiol. Sci.* **2006**, *82*, 123.
95. Shailaja, J.; Kaanumalle, L.S.; Sivasubramanian, K.; Natarajan, A.; Ponchot, K.J.; Pradhan, A.; Ramamurthy, V. *Org. Biomol. Chem.* **2006**, *4*, 1561.
96. Natarajan, A.; Ramamurthy, V. *Org. Biomol. Chem.* **2006**, *4*, 4533.
97. Natarajan, A.; Ramamurthy, V.; Mague, J.T. *Mol. Crys. Liq. Crys.* **2006**, *456*, 71.

98. Warrier, M.; Kaanumalle, L.S.; Ramamurthy, V. *Can. J. Chem.* **2003**, *81*, 620.
99. Lehn, J.M. *Supramolecular Chemistry: Concepts and Perspectives*. Weinheim: VCH, **1995**.
100. Schneider, H.J.; Duerr, H., editors. *Frontiers in Supramolecular Organic Chemistry and Photochemistry*. New York: VCH, **1991**.
101. Bach, T.; Bergmann, H.; Harms, K. *J. Am. Chem. Soc.* **1999**, *121*, 10650.
102. Bach, T.; Bergmann, H.; Brummerhop, H.; Lewis, W.; Harms, K. *Chem. Eur. J.* **2001**, *7*, 4512.
103. Bach, T.; Bergmann, H.; Grosch, B.; Harms, K. *J. Am. Chem. Soc.* **2002**, *124*, 7982.
104. Bach, T.; Bergmann, H.; Harms, K. *Angew. Chem. Int. Ed. Engl.* **2000**, *39*, 2302.
105. Cauble, D.F.; Lynch, V.; Krische, M.J. *J. Org. Chem.* **2003**, *68*, 15.
106. Albrecht, D.; Basler, B.; Bach, T. *J. Org. Chem.* **2008**, *73*, 2345.
107. Selig, P.; Bach, T. *J. Org. Chem.* **2006**, *71*, 5662.
108. Bach, T.; Grosch, B.; Strassner, T.; Herdtweck, E. *J. Org. Chem.* **2003**, *68*, 1107.
109. Toda, F.; Miyamoto, H.; Kanemoto, K.; Tanaka, K.; Takahashi, Y.; Takenaka, Y. *J. Org. Chem.* **1999**, *64*, 2096.
110. Tanaka, K.; Kakinoki, O.; Toda, F. *Chem. Commun.* **1992**, 1053.
111. Naito, T.; Tada, Y.; Ninomiya, I. *Heterocycles* **1984**, *22*, 237.
112. Bach, T.; Aechtner, T.; Neumuller, B. *Chem. Commun.* **2001**, 607.
113. Bach, T.; Aechtner, T.; Neumuller, B. *Chem. Eur. J.* **2002**, *8*, 2464.
114. Grosch, B.; Orlebar, C.N.; Herdtweck, E.; Kaneda, M.; Wada, T.; Inoue, Y.; Bach, T. *Chem. Eur. J.* **2004**, *10*, 2179.
115. Bach, T.; Bergmann, H.; Harms, K. *Org. Lett.* **2001**, *3*, 601.
116. Bauer, A.; Westkaemper, F.; Grimme, S.; Bach, T. *Nature* **2005**, *436*, 1139.
117. Inoue, Y. *Nature* **2005**, *436*, 1099.
118. Rekharsky, M.V.; Inoue, Y. *Chem. Rev.* **1998**, *98*, 1875.
119. Breslow, R.; Dong, S.D. *Chem. Rev.* **1998**, *98*, 1997.
120. Takahashi, K. *Chem. Rev.* **1998**, *98*, 2013.
121. Kitae, T.; Nakayama, T.; Kano, K. *J. Chem. Soc. Perkin 2* **1998**, 207.
122. Rekharsky, M.V.; Inoue, Y. *J. Am. Chem. Soc.* **2002**, *124*, 813.
123. Rekharsky, M. *J. Am. Chem. Soc.* **2001**, *123*, 5360.
124. Inoue, Y.; Wada, T.; Sugahara, N.; Yamamoto, K.; Kimura, K.; Tong, L.-H.; Gao, X.-M.; Hou, Z.-J.; Liu, Y. *J. Org. Chem.* **2000**, *65*, 8041.
125. Fukuhara, G.; Mori, T.; Wada, T.; Inoue, Y. *J. Org. Chem.* **2006**, *71*, 8233.
126. Lu, R.; Yang, C.; Cao, Y.; Wang, Z.; Wada, T.; Jiao, W.; Mori, T.; Inoue, Y. *Chem. Commun.* **2008**, 374.
127. Fukuhara, G.; Mori, T.; Wada, T.; Inoue, Y. *Chem. Commun.* **2005**, 4199.
128. Lu, R.; Yang, C.; Cao, Y.; Wang, Z.; Wada, T.; Jiao, W.; Mori, T.; Inoue, Y. *J. Org. Chem.* **2008**, *73*, 7695.
129. Gao, Y.; Wada, T.; Yang, K.; Kim, K.; Inoue, Y. *Chirality* **2005**, *17*, S19–S23.
130. Inoue, Y.; Dong, S.F.; Yamamoto, K.; Tong, L.-H.; Tsuneishi, H.; Hakushi, T.; Tai, A. *J. Am. Chem. Soc.* **1995**, *113*, 2793.
131. Gao, Y.; Inoue, M.; Wada, T.; Inoue, Y. *J. Incl. Phenom. Macro.* **2004**, *50*, 111.

132. Inoue, Y.; Tsuneishi, H.; Hakushi, T.; Tai, A. *J. Am. Chem. Soc.* **1997**, *119*, 472.
133. Yang, C.; Mori, T.; Wada, T.; Inoue, Y. *New J. Chem.* **2007**, *31*, 697–702.
134. Shailaja, J.; Karthikeyan, S.; Ramamurthy, V. *Tetrahedron Lett.* **2002**, *43*, 9335.
135. Koodanjeri, S.; Joy, A.; Ramamurthy, V. *Tetrahedron* **2000**, *56*, 7003.
136. Aoyama, H.; Miyazaki, K.; Sakamoto, M.; Omote, Y. *Tetrahedron* **1987**, *43*, 1513.
137. Aoyama, H.; Hasegawa, T.; Omote, Y. *J. Am. Chem. Soc.* **1979**, *101*, 5343.
138. Aoyama, H.; Hasegawa, T.; Watabe, M.; Shiraishi, H.; Omote, Y. *J. Org. Chem.* **1978**, *43*, 419.
139. Wender, P.A.; Howbert, J.J. *J. Am. Chem. Soc.* **1981**, *103*, 688.
140. Vizvardi, K.; Toppet, S.; Hoornaert, G.J.; De Keukeleire, D.; Bak, P.; Van der Eycken, E. *J. Photochem. Photobiol. A Chem.* **2000**, *133*, 135.
141. Vizvardi, K.; Desmet, K.; Luyten, I.; Sandra, P.; Hoornaert, G.; Van der Eycken, E. *Org. Lett.* **2001**, *3*, 1173.
142. Mori, T.; Wada, T.; Inoue, Y. *Org. Lett.* **2000**, *2*, 3401.
143. Mori, T.; Inoue, Y.; Weiss, R.G. *Org. Lett.* **2003**, *5*, 4661.
144. Mori, T.; Saito, H.; Inoue, Y. *Chem. Commun.* **2003**, 2302.
145. Mori, T.; Takamoto, M.; Wada, T.; Inoue, Y. *Photochem. Photobiol. Sci.* **2003**, *2*, 1187–1199.
146. Closs, G.L.; Paulson, D.R. *J. Am. Chem. Soc.* **1970**, *92*, 7229.
147. Cocivera, M.; Trozzolo, A.M. *J. Am. Chem. Soc.* **1970**, *92*, 1772.
148. Berger, M.; Goldblatt, I.L.; Steel, C. *J. Am. Chem. Soc.* **1973**, *95*, 1717.
149. Rao, V.P.; Turro, N.J. *Tetrahedron Lett.* **1989**, *30*, 4641.
150. Tamaki, T.; Kokubu, T. *J. Incl. Phenom. Macro.* **1984**, *2*, 815.
151. Tamaki, T.; Kokubu, T.; Ichimura, K. *Tetrahedron Lett.* **1987**, *43*, 1485.
152. Kuroda, Y.; Sera, T.; Ogoshi, H. *J. Am. Chem. Soc.* **1991**, *113*, 2793.
153. Furusawa, T.; Kawano, M.; Fujita, M. *Angew. Chem. Int. Ed. Engl.* **2007**, *46*, 5717–5719.
154. Nishioka, Y.; Yamaguchi, T.; Kawano, M.; Fujita, M. *J. Am. Chem. Soc.* **2008**, *130*, 8160.
155. Yoshizawa, M.; Takeyama, Y.; Okano, T.; Fujita, M. *J. Am. Chem. Soc.* **2003**, *125*, 3243.
156. Yoshizawa, M.; Tamura, M.; Fujita, M. *Science* **2006**, *312*, 1472.
157. Seo, J.S.; Whang, D.; Lee, H.; Jun, S.I.; Oh, J.; Jeon, Y.J.; Kim, K. *Nature* **2000**, *404*, 982.
158. Benitez, M.; Bringmann, G.; Dreyer, M.; Garcia, H.; Ihmels, H.; Waidelich, M.; Wissel, K. *J. Org. Chem.* **2005**, *70*, 2315.

5

REAL-TIME CRYSTALLOGRAPHY OF PHOTOINDUCED PROCESSES IN SUPRAMOLECULAR FRAMEWORK SOLIDS

PHILIP COPPENS AND SHAO-LIANG ZHENG

INTRODUCTION

The dramatic increase in spectroscopic and crystallographic instrumentation during the last decade has opened up new vistas in the study of photochemical processes by diffraction methods that are only now beginning to be explored. The potential scope of the field is very large. By exploiting the cavities and channels in supramolecular framework solids to perform *in situ* chemical reactions triggered by light or thermal pulses, it is possible for the first time to observe the geometry of short-lived species using diffraction methods on fully periodic systems. The current challenge is to extend such methods to the examination of the progress of chemical reactions, in other words to make molecular movies that can revolutionize our insight in chemical processes and provide a direct test of theoretical methods, which are now increasingly capable of calculating transition states.

METHODS

Penetration of Light in Crystals: Molecular Dilution, One- and Two-Photon Processes, and Selection of Wavelength

Maximizing Penetration of a Light Beam in the Sample. A crucial component of the design of a photocrystallographic experiment is the evaluation of the

---

Supramolecular Photochemistry: Controlling Photochemical Processes, First Edition. Edited by V. Ramamurthy and Yoshihisa Inoue.

TABLE 5.1. Molecular Concentration in Some Photochemically Active Crystals

Compound	V/Z (Å^3)	Concentration (mol L^{-1})	Reference
α-Cinnamic acid	191	8.69	3
9-Methyl anthracene	260	6.40	4
2-Benzyl-5-benzylidene cyclopentanone	353	4.71	5
Sodium nitroprusside	283	5.87	6
CECR-HTA-2MeOH-1.5H_2O	1046	1.59	7

HTA, tiglic acid; CECR, C-ethylcalix[4]resorcinarene.

absorption of the beam in the sample, which can be surprisingly large even for the typically 100- to 200-μm crystals now typically used for crystallographic data collection at laboratory sources. The classic Beer–Lambert law is

$$I = I_0 e^{-\varepsilon c d}, \tag{5.1}$$

in which ε is the extinction coefficient in square meter per mole, c is the molar concentration in mole per cubic decimeter, and d is the path length in millimeter (SI units*). This means that for a 100-μm crystal with $c = 5$ and $\varepsilon = 10$, a modest value, the beam at the back of the crystal is attenuated by a factor $1/e^5$; in other words, the back of the crystal will not be illuminated. The situation can be much worse when the product is highly colored as is often the case (see e.g., References 1 and 2).

The supramolecular solid state offers some relief from this obstacle, as the active species can be significantly diluted by its incorporation into a guest framework. The effect is well demonstrated in Tables 5.1 and 5.2. In Table 5.1, the molecular concentrations in four neat crystals are compared with that of the small molecule tiglic acid (HTA) embedded in a C-ethylcalix[4]-resorcinarene (CECR) framework.† Table 5.2 shows the concentration of a larger molecule, the photosensitizer dye Cu bis(2,9-dimethyl-1,10-phenanthroline), in three of its salt and in four anionic frameworks. Given that each decrease of c by 1 mol L^{-1} improves I/I_0 by a factor $e^{-\varepsilon d}$, the improvements in using supramolecular solids are dramatic even for the ~0.1-mm crystals typically used in current diffraction experiments. But it is necessary to avoid frameworks of large conjugated systems, absorbing in the region needed for

*It should be noted that in the non-SI units of liter per mole per centimeter for ε and centimeter for d used in the older literature, ε is a factor of 10 larger.

†Abbreviations used in this chapter: CHTA, 1,3,5-cyclohexane tricarboxylic acid; THPE, tris(4-hydroxyphenyl)ethane; CMCR, C-methylcalix[4]resorcinarene; CECR, C-ethylcalix[4]-resorcinarene; HECR, hexaethyl calix[6]resorcinarenes; HTA, tiglic acid; TA, tiglic acid anion; HClA, 3-chloroacrylic acid; bpe, bis-pyridylethylene.

TABLE 5.2. Concentration of Salts of Cu Bis(2,9-Dimethyl-1,10-Phenanthroline)

Counterion	V/Z (Å^3)	Concentration (mol L^{-1})	Reference
NO_3	638	2.60	
PF_6	654	2.54	
CHTA	893	1.86	8
THPE	978	1.70	8
2 THPE	1376	1.21	8
CMCR	1406	1.18	8
CECR	1479	1.12	8

CHTA, 1,3,5-cyclohexane tricarboxylic acid; THPE, tris(4-hydroxyphenyl)ethane; CMCR, C-methylcalix[4]resorcinarene; CECR, C-ethylcalix[4]-resorcinarene.

photoactivation. The framework components included in Tables 5.1 and 5.2 fulfill this condition.

Selection of Wavelength. In early photocrystallographic experiments of Schmidt et al.,[9–11] single-crystal-to-single-crystal reactions were rarely encountered; typically, crystal lattices would break down after a few percent conversion. But in 1993, Enkelmann and Wegner showed convincingly that irradiation of a multicrystalline sample of randomly oriented crystals with a wavelength in the tail of the pertinent absorption band could preserve the crystal lattice to a very high degree of conversion.[12] This observation was challenged[13,14] but confirmed in subsequent detailed experiments.[3] An additional difficulty is encountered when the product of the photo process absorbs the exciting wavelength and thus prevents penetration of the beam in the crystal even when tail irradiation is used. Such a case was demonstrated by Harada et al.[2] in a study of the photochromic ring-closure reaction of fulgicides, during which the crystal color changes from yellow to dark red. But reaction throughout the crystal could be accomplished by irradiation with 742-nm light from a 5-ns pulse width 10-kHz Nd-YAG laser source, whereas 365- or 405-nm photon only converted a thin layer close to the incident surface. The 742-nm wavelength is well outside the absorption band of the starting material, and its activity can only be explained by two-photon absorption. This explanation has been confirmed in studies on α-cinnamic acid with 532-nm light from a 2- to 7-ns pulse width Nd-YAG laser,[15] which show a quadratic dependence on light intensity (Fig. 5.1).

With the increasing availability of high-powered tunable lasers, the two-photon technique may become the technique of choice in solid-state single-crystal photochemistry.

Tuning of the Absorption Spectrum by Chemical Substitution. It is often possible to tune the absorption spectrum to a desired wavelength by chemical substitution. Compilations can be found in the literature. For example, in

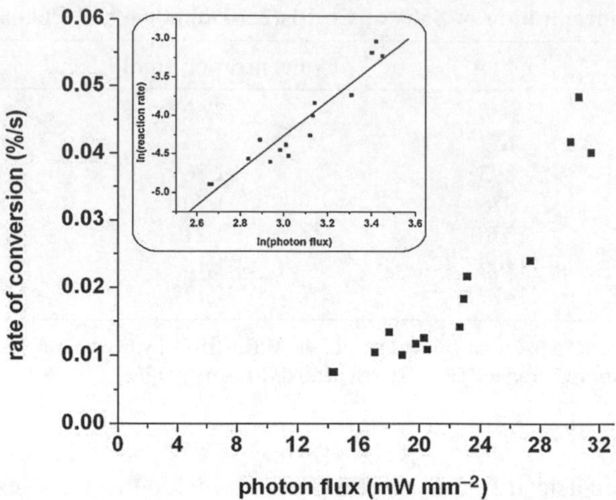

Figure 5.1. Rate of photo conversion of α-*trans*-cinnamic acid to α-truxillic acid as a function of 532-nm photon flux incident on (010). Inset: ln–ln plot; slope of the fitted line is 2.24 ± 0.18. From Reference 15.

Woodward's tabulation for conjugated carbonyl compounds, red shifts on the substitution of Cl, Br, and NR_2 are listed as 12–15, 30, and 95 nm, respectively.[16] The effect can be pronounced in coordination complexes when an innocent ligand, not participating in the photo reaction, is replaced by a conjugated ligand. *E*/*Z* photoisomerization of tiglic acid anion (TA) in the $[Zn(TA)_2(H_2O)_2]$ complex embedded in a CECR matrix (further discussed below) cannot be induced with 458-nm light from an Ar^+ laser. But on replacement of the two water molecules in the complex by the bidentate 2,2′-bipyridyl ligand, 325-nm radiation from a 48-mW He/Cd laser no longer leads to reaction, even after exposure for 24 h at 90 K. However, the reaction can be readily induced by exposure with 458-nm light from a 450-mW Ar^+ laser. After 4 h of irradiation (at 90 K), ~18% conversion to the *Z* isomer was achieved.[17] Theoretical calculations suggest that the initial photoinduced transition corresponds to electron transfer from a nonbonding orbital on one of the oxygens on TA to the bipyridyl ligand, which acts as an internal photosensitizer.

Measurement of the Extinction Coefficient ε. If thin crystals are available, the value of the extinction coefficient and its dependence on wavelength can be measured directly as shown for the $(n\text{-}Bu_4N)_4$ salt of the much investigated dinuclear Pt anion $Pt_2(P_2O_5H_2)^{4-}$, which has been measured at 15 K (Fig. 5.2).[18]

But this is often not the case, so that approximate methods may have to be resorted to. An estimate of the averaged isotropic extinction coefficient can be obtained by using solution-determined ε values with the concentration of the absorber encountered in the crystals. Such results are only approximate

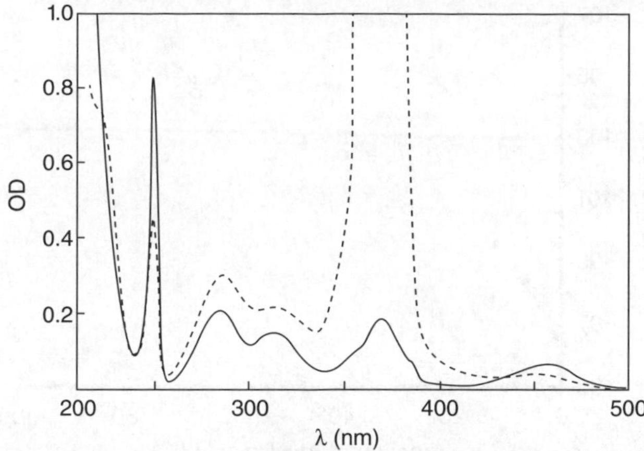

Figure 5.2. Polarized absorption spectra at 15 K of a thin single crystal of $(n\text{-}Bu_4N)_4$ $Pt_2(P_2O_5H_2)$; two different polarization directions are shown. From Reference 18.

as they neglect any changes in the UV-Vis absorption spectrum on crystallization.

Photoinduced Unit Cell Changes in Neat Crystals and in Supramolecular Solids

In neat crystals, the changes in cell dimensions are often indicative of the progress of a reaction. For several [2 + 2] solid-state photodimerization reactions surveyed by Turowska-Tyrk, unit cell contractions of 1–4% are reported.[19] Sometimes, as in the [4 + 4] dimerization of 9-methyl anthracene, an initial increase is followed by a decrease when the conversion exceeds 15%.[4] The unit cell change is often highly anisotropic, reflecting details of the reaction at the molecular level. In the photoinduced linkage isomerization of sodium nitroprusside, $Na_2[Fe(CN)_5NO]$,[6] two different photoproducts are formed depending on the conditions. They are an isonitrosyl isomer (MS_1) and an isomer with a sideways η^2-bound nitrosyl group (MS_2). The isomerization is accompanied by clearly identifiable changes in the cell dimensions of the orthorhombic crystals. The Fe-N-O vector in the virgin crystal is most closely aligned with the *a*-axis, which changes most on exposure. It lengthens by ~0.015 Å on the formation of ~37% MS_1 with 488-nm light, but it shortens by ~0.035 Å on subsequent exposure to 1064-nm light, which converts MS_1 into the side-bound, more compact, NO complex MS_2.

Even though the volume of the monoclinic unit cell of α-cinnamic acid changes little on [2 + 2] photodimerization of α-cinnamic acid to α-truxillic acid,[3] a pronounced change in the β-angle occurs. It changes from 96° in the neat crystal to 106° in the fully converted crystal.[3,15] The change is related to the offset of the double bonds before reaction. It causes the dimerization to

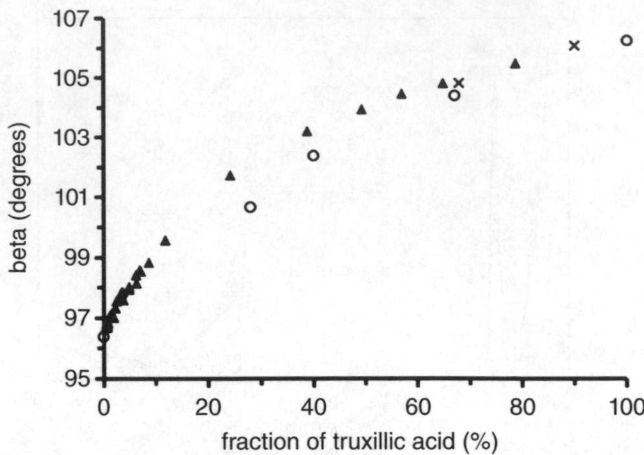

Figure 5.3. Plot of the fraction of α-truxillic acid versus β. Open circles: Reference 12; triangles: Reference 15; crosses: Reference 3. From Reference 15.

require a molecular sliding motion to juxtapose the molecules in a geometry suitable for the formation of the four-membered cyclobutane ring. The increase in the monoclinic angle correlates well with the conversion percentage determined by diffraction and can be used to monitor the progress of the reaction (Fig. 5.3). Similar effects can occur in multicomponent cocrystals.[20] Do such unit cell distortions occur in supramolecular framework solids?

Not unexpectedly, if the framework is reasonably rigid and its cavities are large enough to accommodate changes in the molecular guest, cell dimension changes on reaction or on excitation are much smaller in supramolecular framework solids. Only very minor (<1%) unit cell changes were observed at 40–60% conversion in the $Z \rightarrow E$ and $E \rightarrow Z$ isomerizations of chloroacrylic acid in a CECR matrix.[21] In the $E \rightarrow Z$ isomerization of HTA embedded in a CECR framework, the volume of the unit cell increases by ~1% on 30% conversion because of small increases in the *b*- and *c*-axes lengths. The small changes can be accounted for by an increase in volume of the reaction cavity as the isomerization proceeds and the reacting molecule shifts within the cavity.[7] In the time-resolved experiment on the excimer formation of xanthone embedded in hexaethylcalix[6]resorcinarene (HECR), further described below, a conversion percentage as high as 13% was reached, but unit cell changes were not significant and masked by the effect of a small temperature increase on repeated exposure to the laser pulses. A similar absence of unit cell change was observed on the contraction of the $[Cu(NH_3)_2]_2$ dimer in the strongly hydrogen-bonded framework of tris(4-hydroxyphenyl)ethane (THPE). Thus, the current evidence indicates that unit cell changes are much smaller when the reactive molecules are embedded in supramolecular solids, a result that can be attributed to the cohesiveness of many of the frameworks.

Multicomponent Analysis: Analyzing the Structures of the Photoproducts in a Two- or More Component Crystal

In time-resolved studies of molecular excited states, conversion percentages are always fractional. A fully converted "excited-state" crystal would unlikely be stable even for short periods. Its formation would also require an excessive amount of photons in a very short period, a flood that the crystal would be unlikely to survive long enough for measurement with currently available sources. On the other hand, chemical reactions in solids can and often do go to completion, depending on the matrix, the molecular changes, and the absence of a reverse reaction. Nevertheless, frequently crystals with incomplete conversion are studied. What is the best approach in these cases?

Although the partially converted crystal is strictly speaking a disordered one, there is the important difference as much is known about the initial component. This information can be exploited in both Fourier and least-squares methods. A *photodifference map*, defined as

$$\begin{aligned}\Delta\rho(\mathbf{r}) &= \rho_{\text{exposed}}(r) - \rho_{\text{initial}}(\mathbf{r})\\ &= \frac{1}{V}\sum[F_{\text{exposed}}(\mathbf{H})\exp i\phi_{\text{exposed}} - F_{\text{initial}}(\mathbf{H})\exp i\phi_{\text{initial}}]\exp{-2\pi\mathbf{H}\cdot\mathbf{r}},\end{aligned} \tag{5.2}$$

in which $\rho(\mathbf{r})$ represents the electron density at the point $\mathbf{r}$; and F, the amplitude of a reflection and Φ its phase, and $\mathbf{H}$, the corresponding reciprocal space vector, will show the product atoms as positive peaks and the reactant atoms as electron-deficient troughs. Shifts of the atoms of the reactant molecule will also be evident, as shown in the photodifference map in Figure 5.4.

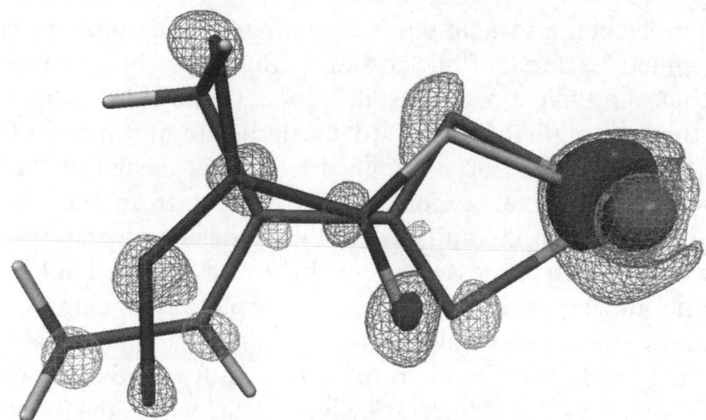

Figure 5.4. Photodifference map of a tiglic acid molecule in CECR·$[Zn(TA)_2(H_2O)_2]$ ·$4H_2O$ (2-h exposure). The displacement of the Zn atom and its changing coordination are clearly visible as are positive peaks for the product atoms, and negative peaks at the positions of the reactant atoms. Blue: 2.0; light blue: 1.0; orange: –1.0; red: –2.0 eÅ^{-3}. From Reference 22. See ftp site for color: ftp://ftp.wiley.com/public/sci_tech_med/supramolecular_photochemistry.

If the unit cell changes, the new cell can be used to calculate the photodifference map, as the parameters to be extracted are to be used in that cell. The subsequent least-squares refinement can start with the positional parameters thus obtained, but if unit cell changes are significant, two cells may have to be used as allowed in the program LASER05 used in our laboratory. Initially, all information on the reactant molecules should be kept fixed, apart from the molecular population that is set to 1-P, where P is a variable representing the product population. Restraints may have to be introduced to ensure that distances and thermal parameters in the photoinduced product are similar to those in the initial state. This is especially important when conversion percentages are low. In the final stages of the refinement, certain variables for the initial-state molecule may be refined, especially when conversion percentages are considerable. They include a rigid-body motion of the molecule and, for flexible molecules, parameters describing intramolecular deformations. What should not be done is ignore the multicomponent nature of the crystals in a refinement on the *average* atomic positions. This will invariably lead to an underestimate of the bond length and angle changes in the product and an artificial increase in the thermal parameters. Some further detail can be found in the literature.[7]

EXAMPLES OF SUPRAMOLECULAR PHOTOCRYSTALLOGRAPHY STUDIES

The Study of Excited States in Framework Supramolecular Crystals

To use supramolecular crystal in photocrystallography, a number of conditions must be attended to. First of all, crystal quality must be adequate and not inferior to that of a typical neat crystal; second, framework components must not absorb the exciting light; and third, excited-state lifetime-limiting energy or electron transfer should be minimized by proper design of the solid. The last condition is of special importance as evident from the luminescence quenching of benzophenone and benzil. In a series of supramolecular solids with frameworks of calixarenes connected by the unsaturated linker molecules 2,2′ bipyridine and bipyridyl ethylene,[23–25] complete quenching of the luminescence occurs. This is remarkable as benzophenone and benzil are prototype phosphorescing molecules with room-temperature emission lifetimes in neat crystals of 55 and 145 μs, respectively. But, when the nonconjugated 1,4-bis(imidazol-1yl-methyl)benzene molecule is used to link the C-methylcalix[4]resorcinarenes (CMCRs), a 77-K lifetime of 570 ns at 77 K is observed.[26] This is still considerably shorter than the neat crystal lifetime but illustrates the effect of avoiding conjugated linkers with their low unoccupied energy levels. The quenching can be explained by intermolecular energy transfer, which crucially depends on the relative spacing of the energy levels of the guest and framework molecules, as confirmed by theoretical analysis of the excited-state energy levels.[27]

Xanthone Embedded in a Calixarene Matrix.

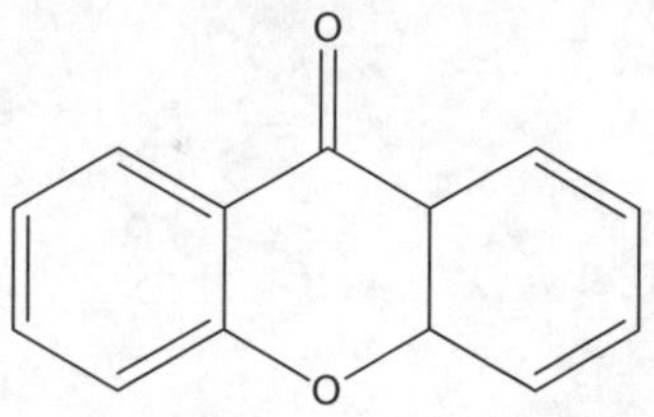

The phosphorescence of the xanthone molecule in rigid glasses has been extensively studied. Long phosphorescence lifetimes of 115 ms at 4.2 K and 2.5 ms at 77 K in hexane have been reported. The phosphorescence has been assigned as emissions from the closely spaced $T_1(\pi\pi^*)$ and $T_2(n\pi^*)$ states at 4.2 and 77 K, respectively,[28] and as due to an equilibrium between the two states.[29]

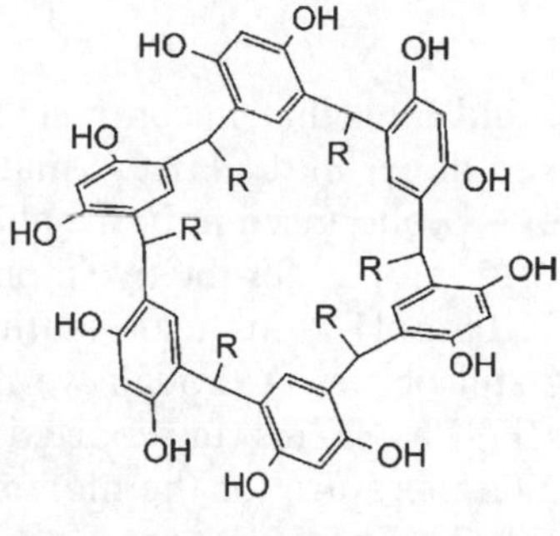

The first report on the phosphorescence of xanthone in supramolecular matrices appeared in 1998.[30] The room-temperature lifetime was found to be $5.2 \pm 0.5\,\mu s$ in a *p*-ethoxy *t*-butylcalix[6]arene framework, but shorter in a *p*-hydroxy *t*-butylcalix[6]arene matrix. Although no structure was reported, spectral evidence indicated a C=O/calixarene interaction in both matrices. Our subsequent measurement on crystalline samples gave emission lifetimes of 0.89 ms for neat xanthone and $0.22\,\mu s$ for xanthone in CECR, the latter less than even the room-temperature values of Barra and Agha, indicating a strong matrix dependence of the emission. Crystallographic analysis of CECR/xanthone shows a monomeric structure with composition CECR·xanthone·MeOH (Fig. 5.5). But a completely different structure is obtained when the calix ring is widened. When incorporated in a matrix of the larger HECR host, a solid of composition HECR·2xanthone·6MeOH is formed and the lifetime increases

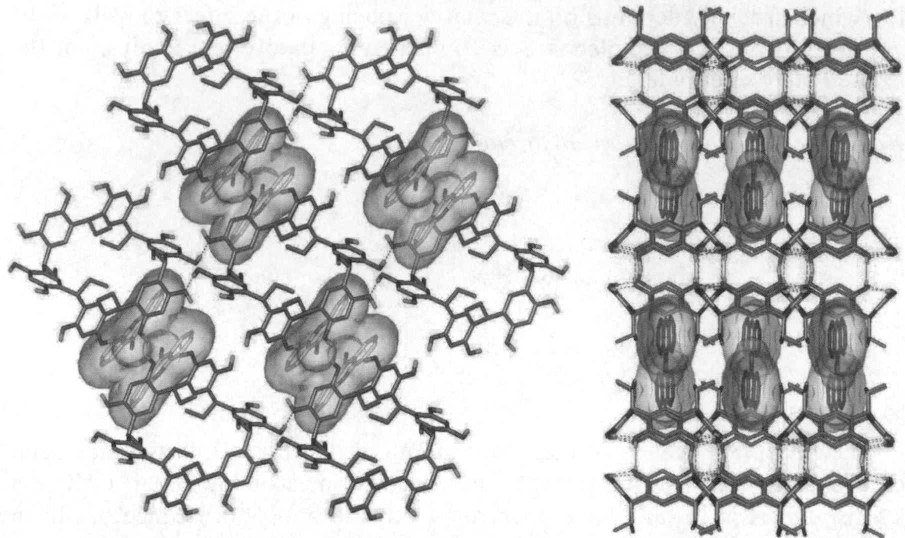

Figure 5.5. Three-dimensional view of HECR·2xanthone·6MeOH (left), containing dimeric xanthone, and of CECR·xanthone·MeOH (right) with monomeric xanthone molecules. Methanol molecules are omitted for clarity. From Reference 31.

to 5.56 μs.[31] X-ray analysis indicates that, in contrast to xanthone in CECR, the molecules form a π-stacked dimer in the HECR matrix (Fig. 5.5).

The molecular dilution is pronounced in both solids, the xanthone concentrations being 1.642 and 1.752 mol L^{-1} for the two monomer and dimer, respectively, compared with 7.106 mol L^{-1} in neat xanthone crystals. But some luminescence quenching still occurs. Although we measured the lifetime of neat xanthone crystals at 17 K as 887 ms (unreported results), the corresponding numbers at the same temperature for the monomer and dimer supramolecular crystals are only 0.22 and 5.56 ms, respectively, indicating significant luminescence quenching in the supramolecular solid. The luminescence maximum of xanthone in the CECR matrix is found at ca. 420 nm, while the emission maximum of the xanthone dimer is observed at 460 nm, in agreement with the red shifts commonly found in the luminescence spectra of the dimers in solution[32] and supported by our calculations of the energy level spacings.

The stroboscopic time-resolved experiments on HECR·2xanthone·6MeOH were performed at the 15-ID beamline at the Advanced Photon Source at a 10-kHz laser/X-ray pulse repeat frequency. The results indicate a contraction of the interplanar spacing by 0.25(3) Å, from 3.39 to 3.14 Å, as expected for excimer formation. Parallel theoretical calculations on the isolated dimer show a large dependence of the interplanar distance in the ground state on the basis set and the theoretical method, which included MP2 to DFT Hamiltonians and B3LYP, BP86, and SVWN functionals. The calculations of the excited state with

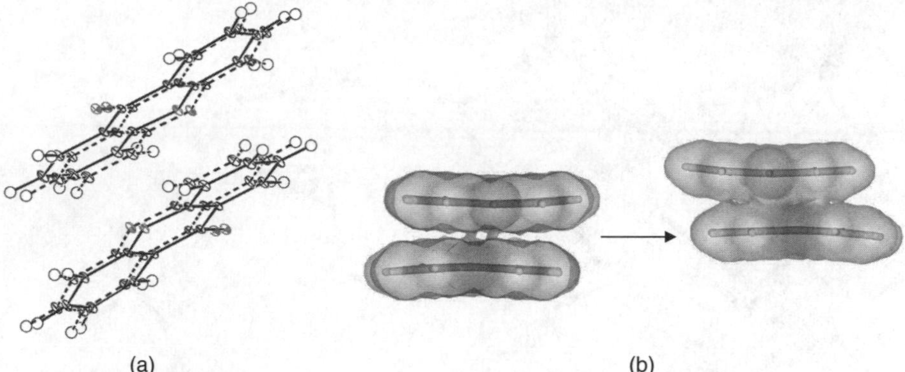

Figure 5.6. Change in the xanthone dimer on excimer formation. (a) Full lines: ground state; broken lines: excited state. (b) Sideways view. From Reference 33. See ftp site for color: ftp://ftp.wiley.com/public/sci_tech_med/supramolecular_photochemistry.

the SVWN functional and two different basis sets, selected because they gave reasonable agreement with the ground-state geometry, show a contraction of the isolated dimer by ~0.07 Å, less than what is observed experimentally in the crystal (Novozhilova and Coppens, unpublished). The contraction is accompanied by a lateral relative shift of the molecular planes of 0.24 Å (Fig. 5.6).

The experimental excited-state populations are found to be 8% and 13% in two successive experiments, the second with increased power in the laser beam. This is larger than the 3–6% excited-state populations achieved in our experiments with neat crystals, a result attributed to the improved photon/photoactive molecule ratio in the supramolecular crystals. Though the contraction and the magnitude of the lateral shift agree well in two separate experiments, the direction of the lateral shift is not exactly the same, indicating a possible dependence on crystal orientation effects, which must be explored in further work.

A $Cu(NH_3)_2$ Dimer Stabilized in an Anionic Framework and Its Stabilization on Excitation. A second example is the dimeric species $[Cu(NH_3)_2]_2^{2+}$. As shown by Carvajal et al., the isolated species is not stable, as the Coulombic repulsion between the positive copper atoms leads to an absence of a minimum as a function of the Cu–Cu distance in the molecular potential. That the ionic dimer nevertheless occurs in the solid state is a result of anion–cation interactions that stabilize the dimer. The effect has been described as *counterion-mediated bonding*.[34] The $[Cu(NH_3)_2]_2^{2+}$ bication is stabilized in a hydrogen-bonded host framework composed of photoinert THPE anions, to give a solid with the composition $[Cu(NH_3)_2]_2[THPE]_2{\cdot}3.25H_2O$ (Fig. 5.7).[35,36] Calculations on the isolated $[Cu(NH_3)_2]_2^{2+}$ dimer at the X-ray geometry indicate it to be less stable by 250–300 kJ mol^{-1} relative to the isolated monomers.

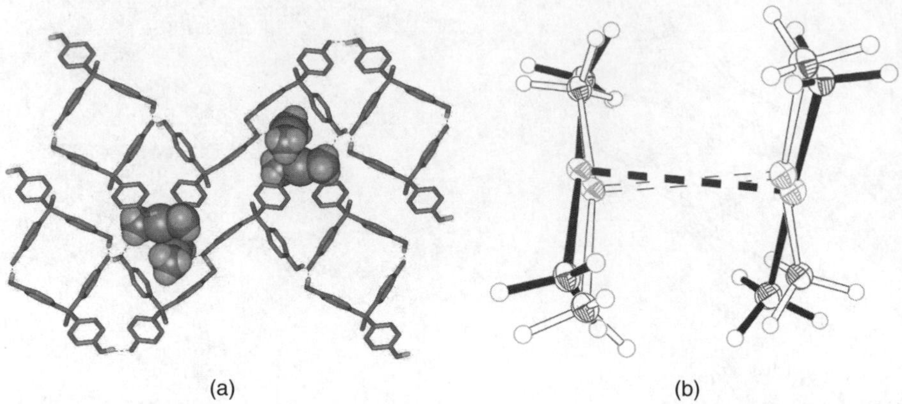

Figure 5.7. (a) Diagram of the molecular packing in $[Cu(NH_3)_2]_2[THPE]_2$ $3.25H_2O$. Copper atoms: purple; nitrogen: blue; hydrogen: green. The water molecules are located in the square channels. (b) The relation between the ground- and excited-state molecules in the crystal; full lines: ground state; open lines: excited state. From Reference 37. See ftp site for color: ftp://ftp.wiley.com/public/sci_tech_med/supramolecular_photochemistry.

Emission measurements show an intense photoluminescence at 17 K with an emission maximum at 495 nm, and a lifetime of ~4.2 μs, typical for a triplet excited state. Time-resolved diffraction data were collected on a sample embedded in a cold helium stream with a 0.49384-Å pulsed X-ray beam and 532-nm laser pulses preceding each excitation.[38] The Cu–Cu distance was found to contract by 0.30(1) Å from 3.0248(5) to 2.72(1) Å, a contraction accompanied by an 11.8° rotation of the of the Cu–Cu vector (Fig. 5.7). This is in agreement with other excited-state studies of binuclear complexes, which have shown that promotion from an electron in a full shell or subshell to a higher lying orbital with bonding character can lead to a pronounced shortening of the metal–metal distance.[39–41]

Theoretical optimization of the excited triplet state using the B88P86 functionel with a TZ2P basis set for Cu and a TZP basis set for nitrogen and hydrogen gives a reduced Cu(I)···Cu(I) distance of 2.61 Å, somewhat shorter than the experimental excited-state distance. This is perhaps not surprising. A similar discrepancy was found for the binuclear Rh complex $[Rh_2(1,8\text{-diisocyano-}p\text{-menthane})_4]^{2+}$ in its PF_6^- salt, in which case the metal–metal potential curve was found to be exceedingly shallow.[39] The hydrogen-bonded framework in the crystal appears unaffected, even though a conversion percentage of 9.5% was achieved in the experiment.

The theoretical calculations confirm that the HOMO to LUMO excitation corresponds to a transfer from an antibonding to a bonding orbital (Fig. 5.8). Calculations on the Ag and Au analogs show that even larger contractions on excitation, predicted as 0.43 and 0.79 Å, respectively, are to be expected for

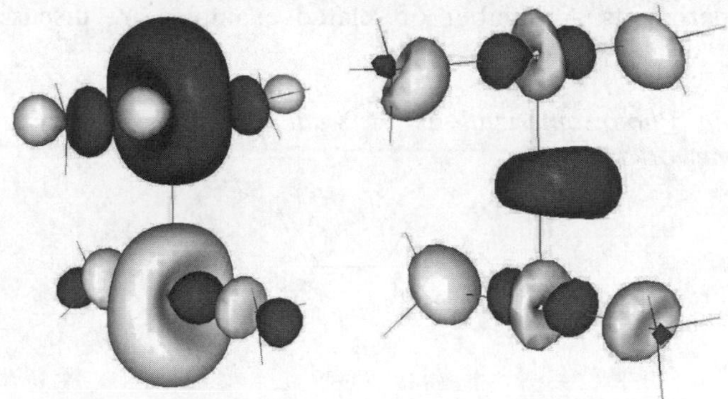

Figure 5.8. The HOMO and LUMO orbitals of the $[Cu(NH_3)_2]_2^{2+}$ dimer. Surfaces at 0.05 au (HOMO) and 0.07 au (LUMO).

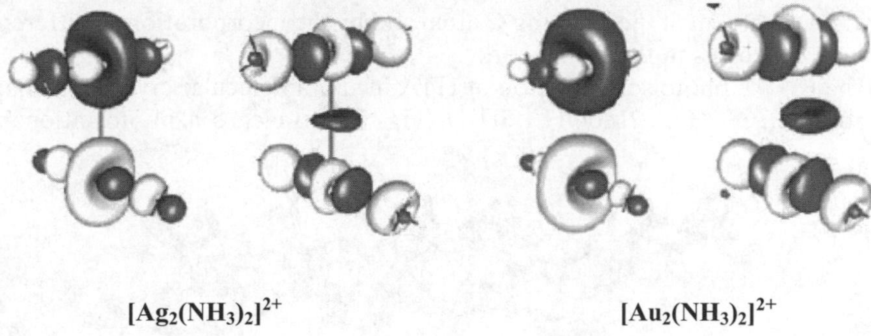

Figure 5.9. HOMO (left) and LUMO (right) for the $[Ag_2(NH_3)_2]^{2+}$ and $[Au_2(NH_3)_2]^{2+}$ cations. Surfaces at ±0.05 au.

these species (Novozhilova, Zheng, and Coppens, unpublished results). The HOMO and LUMO orbitals are shown in Fig. 5.9. However, no experimental data on these complexes are available as yet.

The Study of Photochemical Reactions in Supramolecular Solids

Supramolecular framework solids are the medium par excellence for the study of chemical processes. Their cavities can act as molecular flasks, as demonstrated in a series of experiments by Kawano and Fujita in which reactants embedded in the framework reacted with components of a solution diffusing into the solid without disturbing the crystal lattices.[42] In photochemical reactions, the trigger can be extremely sharp if pulsed lasers are used, opening the way for the study of chemical reactions on very short timescales. Such studies have been attempted in neat crystals[43] but have not yet produced definitive results. We have studied several solids by steady-state techniques as a prelude to such experiments. A number of related examples are discussed in the following.

Double-Bond Photoisomerizations of Small Olefinic Carboxylic Acids in CECR Frameworks.

Tiglic acid

For the same reason that C-substituted resorcinarene-based frameworks are suitable for time-resolved excited-state studies, they are favorable for the study of photochemical reactions. The absorption of the molecule is minor at wavelengths beyond 320 nm, and a great deal of variety is achieved by changing the substituent at the bridging C atom and by the incorporation of different solvent molecules in the framework.

The $E \rightarrow Z$ photoisomerization of HTA in supramolecular crystals of composition CECR·HTA·2MeOH·1.5H_2O (Fig. 5.10) reaches a photostationary

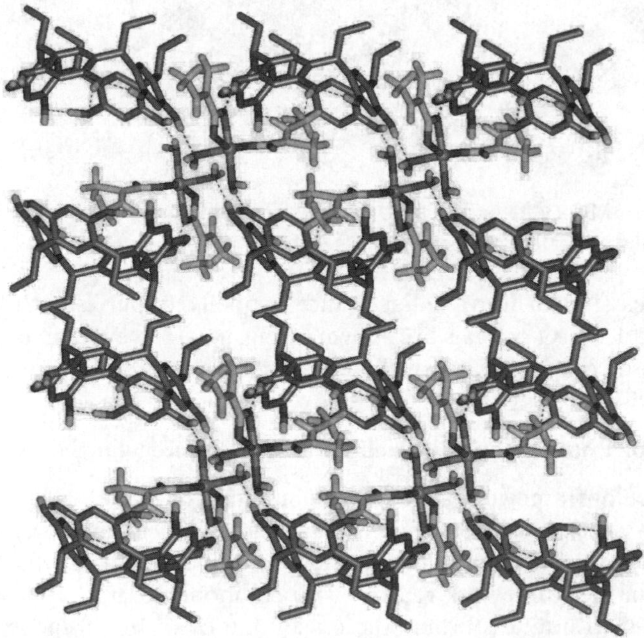

Figure 5.10. Supramolecular architecture of CECR·$[Zn(TA)_2(H_2O)_2]$·4H_2O. Zn: purple circles; red: oxygen atoms; green: tiglic acid carbon atoms; gray: CECR carbon atoms. From Reference 22. See ftp site for color: ftp://ftp.wiley.com/public/sci_tech_med/supramolecular_photochemistry.

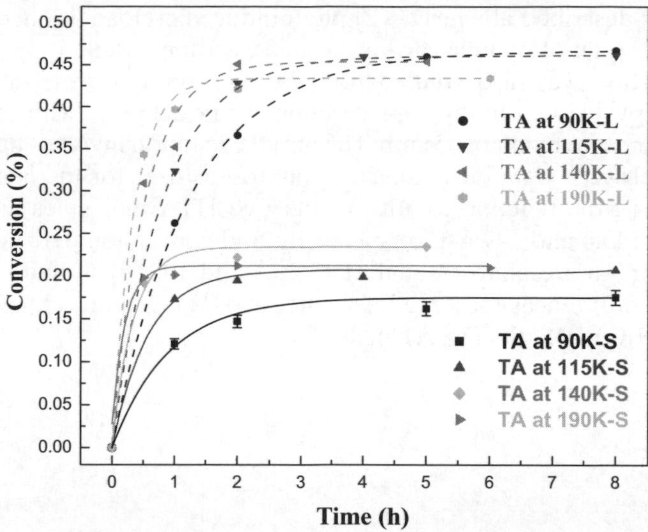

Figure 5.11. Percentage tiglic acid isomerization in CECR·$[Zn(TA)_2(H_2O)_2 \cdot 4H_2O$ as a function of exposure at different temperatures in the small cavity (S) and the large cavity (L). From Reference 22. See ftp site for color: ftp://ftp.wiley.com/public/sci_tech_med/supramolecular_photochemistry.

state of ca. 30% of the *Z* isomers, without perturbing the crystal lattice, by 6-h illumination with 325-nm light from a 45-mW He/Cd laser. In the photostationary state, the forward and the backward reactions are in equilibrium.[7] Analysis of the progress of the reaction as a function time indicates first-order kinetics.

HTA isomerization in the solid CECR·$[Zn(TA)_2(H_2O)_2]\cdot 4H_2O$ sheds light on the effect of the crystal environment on the photochemical reaction, as the two independent HTA molecules are embedded in cavities of different size with volumes of 92.6 and 153.8 Å^3, respectively, and isomerize at different rates (Fig. 5.11). The coordination of the Zn atoms changes from fivefold to fourfold in the reaction, as evident from the photodifference map (Fig. 5.4). Study of the temperature dependence of the reaction indicates small activation energies of 2.3(3) and 1.8(1) kJ mol^{-1} for the forward reaction, and 2.1(2) and 1.9(1) kJ mol^{-1} for the backward $Z \rightarrow E$ isomerization in the small and large cavities, respectively. Theoretical calculations on isolated ethylene give no evidence of such an activation barrier,[44] so the relatively small barriers may be a result of restrictions in the confined crystal space. On the other hand, activation energies of ~35 kJ mol^{-1} were measured by Lewis et al. for the $E \rightarrow Z$ isomerization of phenylpropene in hexane solution.[45] Such a barrier would not allow the reaction to proceed perceptibly at low temperatures at which the diffraction experiments have been conducted, suggesting a different mechanism in the two cases. It may be noted that in both HTA-containing frameworks described above, [2 + 2] photoinduced cycloaddition, common in many neat crystal of olefinic carboxylic acids, is suppressed.

The reaction was also studied for $[Zn(TA)_2(bpy)]$ in the solid CECR-$[Zn(TA)_2(bpy)]\cdot H_2O$.[17] In this case, reaction can be achieved with illumination at the longer wavelength of 458 nm. The initial excitation involves the bipyridyl ligand, which acts as an intramolecular photosensitizer for the isomerization process. Here, the reaction cavities of the two HTA molecules have similar volumes of 143.4 and 149.4 Å^3, respectively, and conversion percentages after exposure for 4 h are almost equal at 17.2(6) and 19.1(6)%, quite unlike the very large differences observed for the two HTA molecules in CECR-$[Zn(TA)_2(H_2O)_2]$-$4H_2O$ (Fig. 5.11).

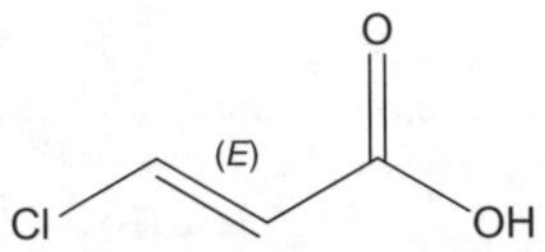

E-chloroacrylic acid

A fourth study in this series concerns the related molecule 3-chloroacrylic acid (HClA) embedded in CECR·HClA·2MeOH·1.5H_2O.[21] In this case, photostationary states could be approached from the two extremes, as both the *E* and the *Z* CECR inclusion solids could be synthesized. But the equilibria are different in the two stationary states. From the *Z* inclusion complex, a 1:1 *E*/*Z* stationary state was reached, whereas in the photostationary state reached by exposure of the *E*-containing framework crystals, the *E* structure is favored by a 3:2 ratio (Fig. 5.12) even though the two crystal structures are similar, but obviously not identical.

Can We Make a Movie of a Chemical Reaction?

The ability to monitor chemical reactions of molecules confined within frameworks raises the question if chemical reactions can be monitored on picosecond or even femtosecond timescales using pump-probe time-resolved diffraction methods. The time resolution of such experiments is limited by the longest of the width of the laser-pump or photon-pump pulse. As fs-width Ti sapphire and other very fast lasers are now available, the width of the X-ray (or electron pulse) becomes the limiting factor. Typical synchrotron sources produce pulses of 70 ps length. Time-slicing techniques have reduced this limit to fs length but at the cost of a very large decrease in intensity. Very bright X-ray free-electron lasers (XFELs) now becoming available produce very short fs pulses but require new experimental methods, which must be developed.

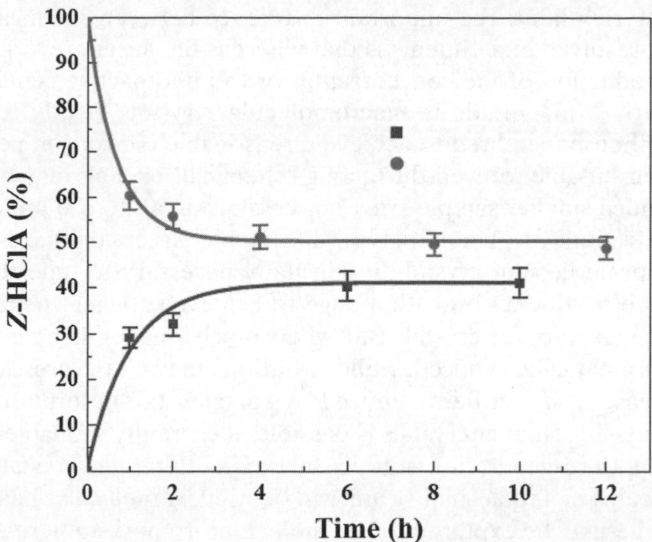

Figure 5.12. % *Z*-chloroacrylic acid starting from the *Z*-containing framework (red circles) from the *E*-guest-containing CECR framework (blue squares) as a function of exposure time. From Reference 21. See ftp site for color: ftp://ftp.wiley.com/public/sci_tech_med/supramolecular_photochemistry.

For a single-pulse experiment to be successful, $n_{incident}$, the number of photon incident on the sample during the pulse must be of the same order of magnitude as $N_{molecules}$, the number of photoactive molecules in the sample. More precisely, for a single-pulse experiment, f, the fraction of molecules in which the reaction is initiated, will be

$$f = n_{incident}(1 - 10^{-\varepsilon cd})\Phi / N_{molecules}, \tag{5.3}$$

in which ε is the extinction coefficient, c is the concentration of the photoactive molecules in the crystal, d is the path length of the beam in the crystal, and Φ is the quantum efficiency of the pertinent excitation process.* The challenge thus is to have a sufficient number of photon incident on the crystal to maximize f, but not such a large number as to lead to rapid deterioration of the crystal. Typical concentrations in supramolecular solids are 1–2 mol L^{-1} (Table 5.1) compared with 0.063 mol L^{-1} in photoactive yellow protein,[46] and 0.030 mol L^{-1} in myoglobin/CO,[47] on which time-resolved studies of the chemical reactions have been done at the lower spatial resolution imposed by the

*This equation ignores the reflection of the beam at the crystal interfaces, which will be sample dependent.

protein scattering limit. The important difference between chemical and biological time-resolved experiments is that whereas the latter are at lower resolution, the reduction of the concentration of the photoactive centers by one to two orders of magnitude in macromolecular crystals greatly reduces the number of photons required to achieve a reasonable conversion percentage.

A solution suitable for well-diffracting "chemical" crystals may be found in the use of much smaller sample sizes. For crystals of 50- to 100-μm dimension with concentrations of ~1 mol L^{-1}, typical for some larger complexes and well-designed supramolecular crystals, it is usually necessary to tune the laser to the tail of the pertinent absorption edge to achieve sufficient penetration of the pump beam into the crystal. But when much smaller samples are used, $N_{\text{molecules}}$ is dramatically reduced, while variations in $\varepsilon c d$ in expression (3) due to the decrease in d can be minimized by relaxing tail absorption and thus increasing ε. Using smaller crystals is possible at currently available undulator beamlines at third-generation source. In the XFELs, nanocrystals, or even single molecules in a molecular beam, can be used. A molecular beam diffraction technique may be explored for unimolecular isomerization reactions and ring closures. Although the nanocrystals are not expected to survive a single shot from the XFEL, so that angular scans on a single sample are no longer possible, techniques are now being developed to reconstruct the three-dimensional scattering patterns of identical multiple objects recorded in random orientations.[48,49]

With a typical solid-state concentration of 1 mol L^{-1}, a nanocrystal with a volume of 1000 nm^3 contains 6.02×10^8 molecules. This greatly relaxes the laser power requirements, even if the beam is not fully focused. With such small path lengths, much higher values of ε can be tolerated and are in fact required for absorption of a large fraction of the incident photons. With the nanocrystal of our example, an ε of 300 m^{-2} mol^{-1} would lead to 50% absorption, an ideal value giving decent absorption and reasonable homogeneous illumination throughout the crystallite.

We conclude that, although many hurdles are to be overcome, the prospects for further progress toward the ultimate goal of visualizing chemical reactions appear bright at this time.

REFERENCES

1. Rack, J.J.; Winkler, J.R.; Gray, H.B. *J. Am. Chem. Soc.* **2001**, *123*, 2432.
2. Harada, J.; Nakajima, R.; Ogawa, K. *J. Am. Chem. Soc.* **2008**, *130*, 7085.
3. Abdelmoty, I.; Buchholz, V.; Di, L.; Guo, C.; Kowitz, K.; Enkelmann, V.; Wegner, G.; Foxman, B.M. *Cryst. Growth Des.* **2005**, *5*, 2210.
4. Turowska-Tyrk, I.; Trzop, E. *Acta Cryst.* **2003**, *B59*, 779.
5. Turowska-Tyrk, I. *Chem. Phys. Lett.* **2002**, *361*, 115.
6. Carducci, M.; Pressprich, M.R.; Coppens, P. *J. Am. Chem. Soc.* **1997**, *119*, 2669.

7. Zheng, S.-L.; Messerschmidt, M.; Coppens, P. *Acta Cryst.* **2007**, *B63*, 644.
8. Zheng, S.-L.; Gembicky, M.; Messerschmidt, M.; Dominiak, P.M.; Coppens, P. *Inorg. Chem.* **2006**, *45*, 9281.
9. Schmidt, G.M.J. *J. Chem. Soc.* **1964**.
10. Cohen, M.D.; Schmidt, G.M.J.; Sonntag, F.I. *J. Chem. Soc.* **1964**, 2000.
11. Cohen, M.D.; Schmidt, G.M.J. *J. Chem. Soc.* **1964**, 1996.
12. Enkelmann, V.; Wegner, G. *J. Am. Chem. Soc.* **1993**, *115*, 10390.
13. Kaupp, G.; Haak, M. *Angew. Chem. Int. Ed. Engl.* **1996**, *35*, 2774.
14. Kaupp, G. *CrystEngComm* **2003**, *5(23)*, 117.
15. Benedict, J.B.; Coppens, P. *J. Phys. Chem. A* **2009**, *113*, 3116.
16. Hirayama, K. *Handbook of Ultraviolet and Visible Absorption Spectra of Organic Compounds*. Berlin: Springer-Verlag, **1967**.
17. Zheng, S.-L.; Gembicky, M.; Messerschmidt, M.; Coppens, P. *J. Chin. Chem. Soc.* **2009**, *56*, 16.
18. Stiegman, A.E.; Rice, S.F.; Gray, H.B.; Miskowski, V.M. *Inorg. Chem.* **1987**, *26*, 1112.
19. Turowska-Tyrk, I. *J. Phys. Org. Chem.* **2004**, *17*, 837.
20. Zheng, S.-L.; Pham, O.; Velde, C.M.L.; Gembicky, M.; Coppens, P. *Chem. Comm.* **2008**, 2538.
21. Zheng, S.-L.; Messerschmidt, M.; Coppens, P. *Chem. Comm.* **2007**, 2735.
22. Zheng, S.-L.; Velde, C.M.L.; Messerschmidt, M.; Volkov, A.; Gembicky, M.; Coppens, P. *Chem. Eur. J.* **2008**, *14*, 706.
23. Ma, B.; Coppens, P. *Cryst. Growth Des.* **2004**, *4*, 1377.
24. Ma, B.-Q.; Coppens, P. *J. Org. Chem.* **2003**, *68*, 9467.
25. Ma, B.-Q.; Zhang, Y.; Coppens, P. *Cryst. Growth Des.* **2001**, *1*, 271.
26. Ma, B.-Q.; Ferreira, L.F.V.; Coppens, P. *Org. Lett.* **2004**, *6*, 1087.
27. Zheng, S.-L.; Coppens, P. *Cryst. Growth Des.* **2005**, *5*, 2050.
28. Connors, R.E.; Christian, W.R. *J. Phys. Chem.* **1982**, *86*, 1524.
29. Vala, M.; Hurst, J.; Trabjerg, I. *Mol. Phys.* **1981**, *43*, 1219.
30. Barra, M.; Agha, K.A. *Supramol. Chem.* **1998**, *10*, 91.
31. Zheng, S.-L.; Coppens, P. *Chem. Eur. J.* **2005**, *11*, 3583.
32. Brouwer, F. In: Waluk, J., editor. *Conformational Analysis of Molecules in Excited States*. Weinheim: Wiley-VCH, **2000**, p. 177.
33. Coppens, P.; Zheng, S.-L.; Gembicky, M. *Zeitschr. Krist.* **2008**, *223*, 265.
34. Carvajal, M.A.; Alvarez, S.; Novoa, J.J. *Chem. Eur. J.* **2004**, *10*, 2117.
35. Zheng, S.-L.; Messerschmidt, M.; Coppens, P. *Angew. Chem. Int. Ed. Engl.* **2005**, *44*, 4614.
36. Stone, A.J. *The Theory of Intermolecular Forces*. Oxford: Clarendon, **1996**.
37. Coppens, P.; Zheng, S.-L.; Gembicky, M.; Messerschmidt, M.; Dominiak, P.M. *CrystEngComm.* **2006**, *8*, 735.
38. Coppens, P.; Row, T.N.G.; Leung, P.; Becker, P.J.; Yang, Y.W.; Stevens, E.D. *Acta Cryst.* **1979**, *A35*, 63.
39. Coppens, P.; Gerlits, O.; Vorontsov, I.I.; Kovalevsky, A.Y.; Chen, Y.-S.; Graber, T.; Novozhilova, I.V. *Chem. Comm.* **2004**, 2144.

40. Coppens, P.; Vorontsov, I.I.; Graber, T.; Gembicky, M.; Kovalevsky, A.Y. *Acta Cryst.* **2005**, *A61*, 162.
41. Vorontsov, I.I.; Kovalevsky, A.Y.; Chen, Y.-S.; Graber, T.; Gembicky, M.; Novozhilova, I.V.; Omary, M.A.; Coppens, P. *Phys. Rev. Lett.* **2005**, *94*, 193003/1.
42. Kawamichi, T.; Kodama, T.; Kawano, M.; Fujita, M. *Angew. Chem. Int. Ed. Engl.* **2008**, *47*, 1.
43. Davaasambuu, J.; Busse, G.; Techert, S. *J. Am. Chem. Soc.* **2006**, *110*, 3261.
44. Martinez, T.J. *Acc. Chem. Res.* **2006**, *39*, 119.
45. Lewis, F.D.; Bassani, D.M. *J. Am. Chem. Soc.* **1993**, *115*, 7523.
46. Ihee, H.; Rajagopal, S.; Srajer, V.; Pahl, R.; Anderson, S.; Schmidt, M.; Schotte, F.; Anfinrud, P.A.; Wulff, M.; Moffat, K. *Proc. Natl. Acad. Sci. U S A* **2005**, *102*, 7145.
47. Bourgeois, D.; Vallone, B.; Arcovito, A.; Sciara, G.; Schotte, F.; Anfinrud, P.A.; Brunori, M. *Proc. Natl. Acad. Sci. U S A* **2006**, *103*, 4924.
48. Shneerson, V.L.; Ourmazd, A.; Saldin, D.K. *Acta Cryst.* **2008**, *A64*, 303.
49. Fung, R.; Shneerson, V.; Saldin, D.K.; Ourmazd, A. *Nat. Phys.* **2009**, *5*, 64.

6

BIMOLECULAR PHOTOREACTIONS IN THE CRYSTALLINE STATE

ARUNKUMAR NATARAJAN AND BALAKRISHNA R. BHOGALA

INTRODUCTION

Bimolecular reactions in the crystalline state were discovered by Libermann during the [2 + 2] photodimerization of olefins to yield cyclobutanes as early as 1889.[1] Later, Kohlshutter proposed that the nature and properties of the products in the crystalline state reactions are governed by the fact that they take place within or on the surface of the solid.[2] In the early 1960s, Schmidt and his coworkers conducted the crystallographic investigations of cinnamic acid and a large number of its derivatives in order to understand the structure–reactivity correlation of bimolecular reactions. From these pioneering studies emerged the "topochemical postulates" that were used as ground rules in order to understand the outcome of a photochemical reaction.[3–5] In this chapter, we provide extensive coverage of bimolecular photodimerization leading to complex photoproducts in crystals that are not easily obtained via ground-state reactions.[6–21] Our emphasis in this chapter is to explore the growth in the field of solid-state bimolecular reactions from basic principles such as Schmidt's topochemical rules, the role of surrounding molecules in the crystalline state (reaction cavity concept), and the modern-day strategy of prealigning the reacting partners via templation (cocrystallization, ionic bonding, metal ion coordination, supramolecular nanocavities, etc.). The efficient synthesis of complex molecular targets such as cyclobutanes is one of the

Supramolecular Photochemistry: *Controlling Photochemical Processes*, First Edition. Edited by V. Ramamurthy and Yoshihisa Inoue.

central goals of synthetic chemistry. The solid state is an attractive medium for such a goal, since it offers a highly organized environment for stereoselective reactions. However, the solid state has remained largely unexploited as a medium for synthesizing molecular targets. We hope that the compilation of examples will encourage readers to investigate some of these systems in depth to better understand the mechanism of solid-state dimerization.

PHOTODIMERIZATION OF CINNAMIC ACIDS: EMERGENCE OF TOPOCHEMICAL POSTULATES

The reactions of cinnamic acids are examples of [2 + 2] photodimerization that have been investigated extensively. Some of these acids, on photolysis of the crystal, react to give dimeric products (Scheme 6.1), while in solution, *trans–cis* isomerization occurs but there is no dimerization. The acids are observed to crystallize in three polymorphic forms, namely, α, β, and γ, and show photochemical behavior, which is determined by this structure type. In all three modifications, cinnamic acid molecules pack in one-dimensional stacks, adjacent stacks being paired by hydrogen bonding across centers of symmetry. Within the stacks, the molecules lie parallel, with the normal distance between molecular planes being of the order of ~3.5 Å. The three structural types differ in the angle that the stack axis makes normal to the molecular planes. This is equivalent to a difference in the distance between equivalent points on the molecules, which is the crystallographic repeat distance, "d." In the β-type structure, the molecules are separated by a short repeat distance of 3.8–4.2 Å; thus, neighboring molecules up the stack are translationally equivalent and show considerable face-to-face overlap. The β-type packing arrangement in the case of a substituted cinnamic acid is shown in Figure 6.1. All cinnamic acids, which crystallize in this structure, react photochemically to give products

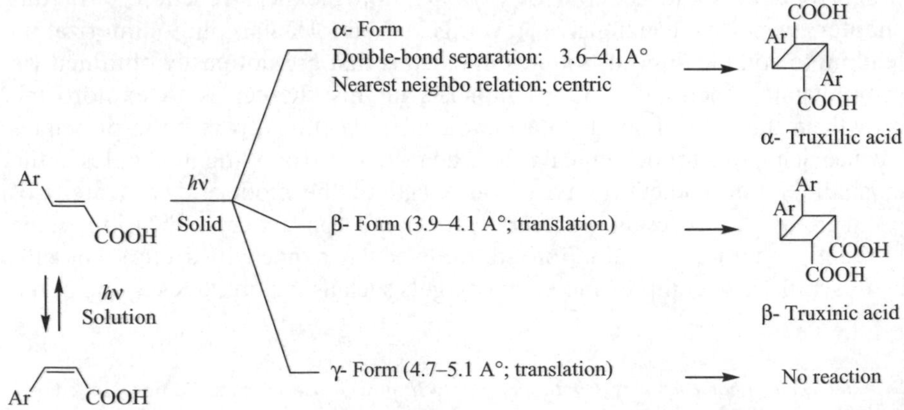

Scheme 6.1

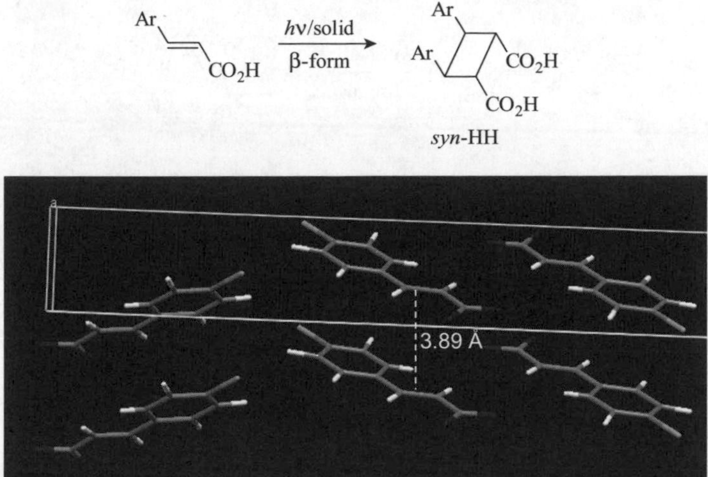

Figure 6.1. Packing arrangement of *p*-chloro cinnamic acid (β-packing).

of the same stereochemistry (mirror symmetric dimers). In the γ-type structure, adjacent molecules are offset so that the reactive double bonds do not overlap, and furthermore, the distance between them is large (4.8–5.2 Å). Crystals of this type are photostable. In the α-type, the double bond of a molecule in one stack overlaps with that of a centrosymmetrically related molecule in an adjacent stack. The distance between the equivalent double bonds is greater than 5.5 Å, but that between the overlapping double bonds is ~4.2 Å. This type of crystal upon irradiation produces centrosymmetric dimers. The α-type packing arrangement in the case of a substituted cinnamic acid is shown in Figure 6.2.

In Table 6.1, a list of most cinnamic acids whose behavior has been investigated in the solid state until 2008 is provided.[4,5,17,22–28] The important results obtained by analyzing the solid-state behavior of these are the following:

1. The product formed is governed by the environment rather than by the intrinsic reactivity of the reactive bonds in the crystalline state.
2. The proximity and degree of parallelism of the reacting centers are crucial for the dimerization.
3. There is a one-to-one relationship between the configuration and symmetry of the product with the symmetry between the reactants in the crystal.

Although there had been sporadic reports relating to solid-state photodimerization earlier, the systematic and thorough studies by Schmidt and coworkers on cinnamic acids laid the foundation for the growth of this field (Scheme 6.1).

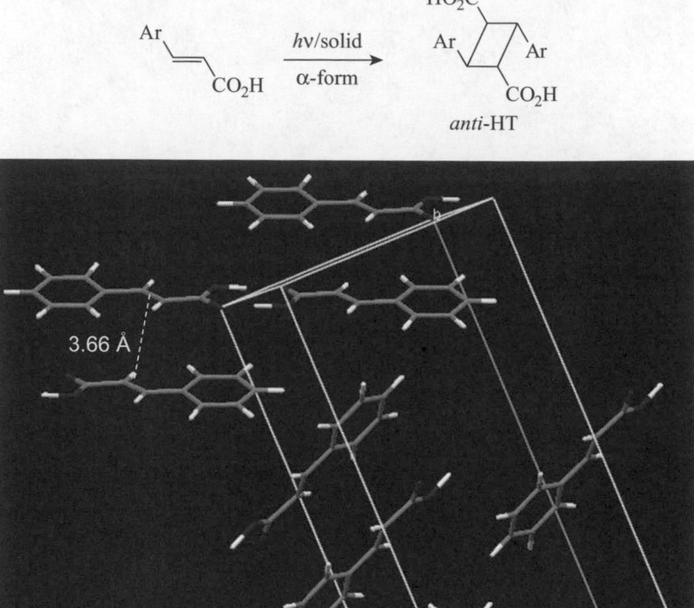

Figure 6.2. Packing arrangement of cinnamic acid (α-packing).

Schmidt has drawn attention to the fact that not only must the double bonds of the reacting monomers of cinnamic acid be within ~4.2 Å, but they must also be aligned parallel for cycloaddition to occur. A reaction that behaves in this way is said to be "topochemically controlled." Schmidt has drawn the geometrical criteria for dimerization only with the view of inferring how precisely the π-electron system of the reacting double bonds must be aligned in the crystal lattice for reaction to occur. However, investigation of several other systems as discussed below suggests that these concepts should be considered as guidelines rather than strict rules.

PHOTODIMERIZATION OF COUMARINS, STYRYLCOUMARINS, BENZYLIDENE CYCLOPENTANONES, AND BENZYLIDENE CYCLOPENTENONES

Following the pioneering studies of cinnamic acids by Schmidt, systematic investigations on the photodimerization of coumarins, styrylcoumarins, benzylidene cyclopentanones, and benzylidene cyclopentenones (Schemes 6.2–6.4) have been carried out during the last two decades.[29–57] The details of the investigation and the photochemical results are summarized in Tables 6.2–6.4. Most observations support the original topochemical postulate of Schmidt.

TABLE 6.1. Photodimerization of *trans*-Cinnamic Acids in the Crystalline State (See Scheme 6.1)

Cinnamic Acids (CAs)	Nature of Packing	% Yield	Nature of Dimer
CA[a]	α	74	*anti*-HT
	β	80	*syn*-HH
ortho-Hydroxy CA[a]	α	90	*anti*-HT
meta-Hydroxy CA[a]	α	76	*anti*-HT
para-Hydroxy CA[a]	α	78	*anti*-HT
ortho-Methoxy CA[a]	α	83	*anti*-HT
ortho-Ethoxy CA[a]	α	93	*anti*-HT
	β	90	*syn*-HH
ortho-Propoxy CA[a]	α	94	*anti*-HT
ortho-Isopropoxy CA[a]	α	97	*anti*-HT
ortho-Allylloxy CA[a]	α	93	*anti*-HT
ortho-Methyl CA[a]	α	—	None
para-Methyl CA[a]	α	95	*anti*-HT
ortho-Nitro CA[a]	β	27	*syn*-HH
meta-Nitro CA[a]	β	60	*syn*-HH
para-Nitro CA[a]	β	70	*syn*-HH
ortho-Chloro CA[a]	β	85	*syn*-HH
meta-Chloro CA[a]	β	70	*syn*-HH
para-Chloro CA[a]	β	71	*syn*-HH
ortho-Bromo CA[a]	β	82	*syn*-HH
meta-Bromo CA[a]	β	91	*syn*-HH
para-Bromo CA[a]	β	90	*syn*-HH
5-Bromo-2-hydroxy CA[a]	β	30	*syn*-HH
5-Chloro-2-methoxy CA[a]	β	85	*syn*-HH
5-Bromo-2-methoxy CA[a]	β	50	*syn*-HH
2,4-Dichloro CA[a]	β	78	*syn*-HH
2,6-Dichloro CA[a]	β	70	*syn*-HH
3,4-Dichloro CA[a]	β	60	*syn*-HH
3,4-Methylene dioxy CA[b]	β	74	*syn*-HH
3,4-Dimethoxy CA[b]	α	—	*anti*-HT
α-Acetylamino CA[c]	α	—	*anti*-HT
para-Formyl CA[d]	β	—	*syn*-HH
6-Chloro-3,4-methylenedioxy CA[e]	β	—	*syn*-HH
para-Cyano CA[e]	β	94	*syn*-HH
meta-Cyano CA[e]	β	80	*syn*-HH

[a]Cohen, M.D.; Schmidt,G.M.J. *J. Chem. Soc.* **1964**, 1996; Cohen, M.D.; Schmidt, G.M.J.; Sonntag, F.I. *J. Chem. Soc.* **1964**, 2000; Schmidt, G.M.J. *J. Chem. Soc.* **1964**, 2014.
[b]Desiraju, G.R.; Kamala, R.; Kumari, B.H.; Sarma, J.A.R.P. *J. Chem. Soc. Perkin 2* **1984**, 181.
[c]Iwamoto, T.; Kashio, S.; Haisa, M. *Acta Cryst.* **1989**, *C45*, 1753.
[d]Nakanishi, F.; Nakanishi, H.; Tsuchiya, M.; Hasegawa, M. *Bull. Chem. Soc. Jpn*. **1976**, *49*, 3096.
[e]Dhurjati, M.S.K.; Sarma, J.A.R.P.; Desiraju, G.R. *J. Chem. Soc. Chem. Commun*. **1991**, 1702.

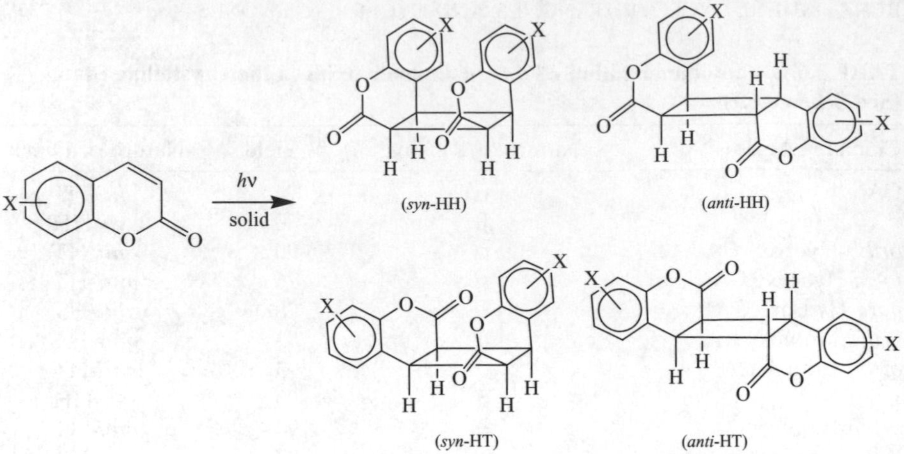

hν
solid
(syn-HH)
(anti-HH)
(syn-HT)
(anti-HT)

HH : head-to-head
HT : head-to-tail

Scheme 6.2

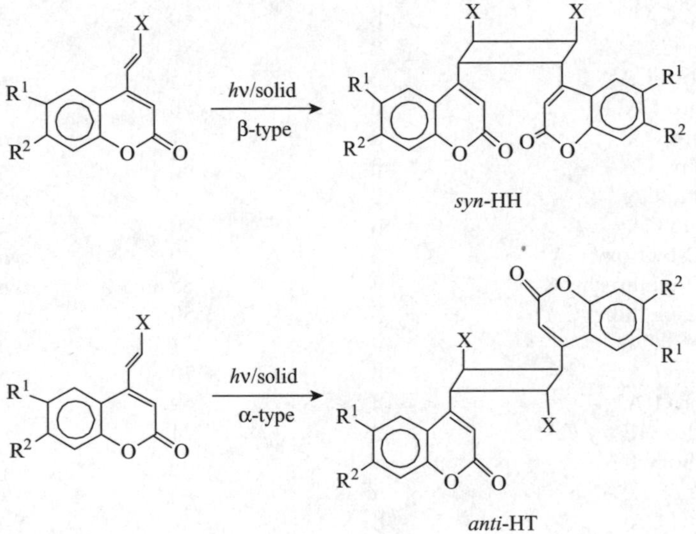

hν/solid
β-type
syn-HH

hν/solid
α-type
anti-HT

Scheme 6.3

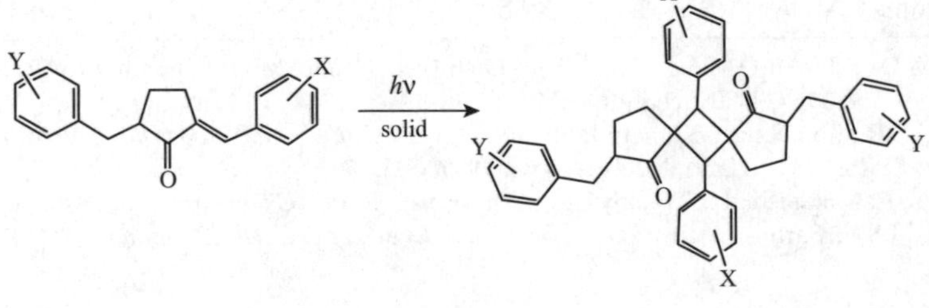

hν
solid

Scheme 6.4

TABLE 6.2. Photodimerization of Coumarins in the Crystalline State (Scheme 6.2)

Coumarins	Packing Behavior	% Yield	Nature of Dimer
Coumarin[b]	γ	Mixture	Three dimers
6-Chlorocoumarin[b]	β	100	*syn*-HH
7-Chlorocoumarin[b]	β	70	*syn*-HH
4-Methyl-6-chlorocoumarin[b]	β	50	*syn*-HH
4-Methyl-7-chlorocoumarin[b]	β	80	*syn*-HH
4-Chlorocoumarin[b]	–[a]	25	*anti*-HH
7-Methylcoumarin[b]	–[a]	65	*syn*-HH
6-Methoxycoumarin[b]	β	60	*syn*-HH
7-Methoxycoumarin[b]	β	90	*syn*-HT
8-Methoxycoumarin[b]	α	50	*anti*-HT
6-Acetoxycoumarin[b]	β	70	*syn*-HH
7-Acetoxycoumarin[b]	β	90	*syn*-HH
4-Methyl-7-acetoxycoumarin[b]	β	80	*syn*-HH
6-Bromocoumarin[c]	β	90	*syn*-HH
7-Bromocoumarin[c]	β	100	*syn*-HH
6-Fluorocoumarin[d]	β	100	*syn*-HH
7-Fluorocoumarin[d]	β	100	*syn*-HH
7-Fluoro-4-methylcoumarin[e]	β	25	*syn*-HH
6-Fluoro-4-methylcoumarin[e]	–[a]	30	*anti*-HT
6-Lodocoumarin[f]	β	40	*syn*-HH
Ethylcoumarin-3-carboxylate[g]	α	100	*anti*-HT

[a]Nontopochemical or defect initiated.
[b]Ramasubbu, N.; Gnanaguru, K.; Venkatesan, K.; Ramamurthy, V. *J. Org. Chem.* **1985**, *50*, 2337.
[c]Venugopalan, P.; Bharathi Rao, T.; Venkatesan, K. *J. Chem. Soc. Perkin 2* **1991**, 981.
[d]Amerendra Kumar, V.; Begum, N.S.; Venkatesan, K. *J. Chem. Soc. Perkin 2* **1993**, 463.
[e]Vishnumurthy, K.; Guru Row, T.N.; Venkatesan, K.; Ramamurthy, V.; Schanze, K., editors. *Molecular and Supramolecular Photochemistry*, Vol. 8. New York: Marcell Dekker, **2001**, p. 427.
[f]Desiraju, G.R. *Crystal Engineering: The Design of Organic Solids*. Amsterdam: Elsevier, **1989**.
[g]Ayala-Hurtado, S.; Flores-Larios, I.Y.; Padilla-Martinez, I.I.; Martinez-Martinez, F.J.; Garcia-Baez, E.V.; Cruz, A.; Hoepfl, H. *Supramol. Chem.* **2007**, 629.

α- and β-packing arrangements of coumarins and styrylcoumarins are shown in Figures 6.3 and 6.4. In majority of the examples, the structure of the dimer could be predicted based on the packing arrangement obtained through X-ray crystallographic investigations.

THE REACTION CAVITY CONCEPT: ANOMALY TO THE TOPOCHEMICAL POSTULATES

The topochemical postulate states that *reaction in the solid state is preferred and occurs with* a *minimum amount of atomic* or *molecular movement.* This implies that a certain amount of motion of various atoms in the crystal lattice is tolerable. Based on this, one could assume that for the formation of a

TABLE 6.3. Photodimerization of Styrylcoumarins in the Crystalline State (See Scheme 6.3)

Stryrylcoumarins	Nature of Packing	% Yield	Nature of Dimer
R^1=R^2=H, X=Ph[b]	α	46–48	*anti*-HT
R^1–F, R^2–H, X–Ph[c]	α	50	*anti*-HT
R^1=R^2=H, X=3-FC_6H_4[c]	α	78	*anti*-HT
R^1=R^2=H, X=3-FC_6H_4[c,f]	β	70	*syn*-HH
R^1–R^2–H, X–2-FC_6H_4[b]	β	80	*syn*-HH
R^1–R^2–H, X–4-FC_6H_4[b]	β	82	*syn*-HH
R^1=H, R^2=F, X=Ph[c]	γ	—	—
	β[d]	78–80	*syn*-HH
R^1=F, R^2=H, X=4-FC_6H_4[e]	β	80	*syn*-HH
R^1=F, R^2=H, X=2-FC_6H_4[e]	β	81	*syn*-HH
R^1–R^2–H, X–2,6-$F_2C_6H_3$[e]	γ	—	—
R^1=F, R^2=H, X=2,6-$F_2C_6H_3$[e]	β	85	*syn*-HH
R^1=H, R^2=F, X=2,6-$F_2C_6H_3$[e]	β	78	*syn*-HH
R^1–H, R^2–OH, X–Ph[a]	α	100	*anti*-HT
R^1=H, R^2=OMe, X=Ph[a]	α	—	—
R^1=H, R^2=Cl, X=Ph[a]	α	70–80	*anti*-HT
R^1=R^2=H, X=ClC_6H_4[a]	α	70–80	*anti*-HT
R^2=OH, R^1=H, X=ClC_6H_4[a]	α	70–80	*anti*-HT
R^1–R^2–H, X–$COCH_3$[a]	α	70–80	*anti*-HT
R^1=H, R^2=Cl, X=$COCH_3$[a]	α	70–80	*anti*-HT
R^1–R^2–H, X–$CSCH_3$[a]	α	70–80	*anti*-HT
R^1=H, R^2=H, X=ClC_6H_4[a]	α	70–80	*anti*-HT

[a]Moorthy, J.N.; Samant, S.D.; Venkatesan, K. *J. Chem. Soc. Perkin 2* **1994**, 1123.
[b]Vishnumurthy, K.; Row, T.N.G.; Venkatesan, K. *J. Chem. Soc. Perkin 2* **1996**, 1475.
[c]Vishnumurthy, K.; Row, T.N.G.; Venkatesan, K. *J. Chem. Soc. Perkin 2* **1997**, 615.
[d]Vishnumurthy, K.; Row, T.N.G.; Venkatesan, K. *Photochem. Photobiol. Sci.* **2002**, 799.
[e]Vishnumurthy, K.; Row, T.N.G.; Venkatesan, K. In: Ramamurthy, V.; Schanze, K., editors. *Molecular and Supramolecular Photochemistry*, Vol. 8. New York: Marcell Dekker, **2001**, p. 427.
[f]Coumarin was added as an additive during crystallization.

cyclobutane ring with C–C length of 1.56 Å, the double bonds can undergo a total displacement of about 2.64 Å toward each other from the original maximum distance of 4.2 Å. Even under ideal conditions, movement of double bonds toward each other is essential for dimerization to take place. The criterion of less than 4.2-Å separation implicitly assumes that such a motion would be accommodated by the molecules surrounding the reactant pair in the crystal. Thus, although the topochemical postulate focuses its attention essentially on the geometrical relationship of the reacting pairs, it seems to indirectly take into account the role of the surrounding molecules.

Once a compound has been crystallized, the template, either for good or otherwise, has been cast for the reaction. The topochemical postulate derives from this point. However, the postulate lacks precision in the following details:

TABLE 6.4. Photodimerization of Benzylidene Cyclopentanones in the Crystalline State (See Scheme 6.4)

Compound	Reactivity (Distance between Reactive Double Bonds)	Nature of Dimer
X=Y=H[a]	Yes (4.2 A°)	*anti*-HT
X=*p*-Br, Y=H[a]	Yes (3.8 A°)	*anti*-HT
X=H, Y=*p*-Cl[a,b]	Yes (4.0 A°)	*anti*-HT
X=*p*-Br, Y=*p*-Me[a,b]	Yes (3.92 A°)	*anti*-HT
X=H, Y=*p*-Br[a]	Yes	*anti*-HT
X=H, Y=*p*-Me[a,b]	Yes	*anti*-HT
X=*p*-Cl, Y=H[a,b]	No (5.03 A°)[c]	—
X=*p*-Br, Y=*p*-Cl[a,b]	No (4.7 A°)[c]	—
X=*m*-Br, Y=H[a,b]	No[c]	—
X=*o*-Br, Y=H[a,b]	No[c]	—
X=*o*-Cl, Y=H[a,b]	No[c]	—
X=*p*-Me, Y=H[a,b]	No[c]	—
X=*m*-Me, Y=H[a,b]	No[c]	—

[a]Theocharis, C.R. In: Patai, S.; Rappoport, Z. editors. *The Chemistry of Enones*. New York: John Wiley & Sons, **1989**, p. 1133; Jones, W.; Nakanishi, H.; Theocharis, C. R.; Thomas, J.M. *J. Chem. Soc. Chem. Commun.* **1980**, 610–611.

[b]Jones, W.; Ramdas, S.; Theocharis, C.R.; Thomas, J.M.; Thomas, N.W. *J. Phys. Chem.* **1981**, *85*, 2594–2597.

[c]Photostability arises from increased separation (>4.3 A°) of the reactive double bonds.

(1) Do the immediate neighbors of the reacting partners have any role to play? (2) Does the postulate consider the changes in the molecular geometry upon excitation? In order to take these into account at the phenomenological level, Cohen proposed the idea of the reaction cavity.[58,59] The cavity or cage is the space in the crystal occupied by the reacting partners. The reaction cavity by definition includes the space occupied by the reacting molecules and the void space surrounding them. The reaction cavity wall is made up of molecules adjacent to the reacting molecules. The atomic movements during a reaction would exert pressures on the cavity wall, which becomes distorted. However, the close packing works against large-scale changes in shape, so that only minimal change can occur (Fig. 6.5). This concept has been of help in qualitatively understanding the course of a variety of solid-state reactions.

The usefulness of the reaction cavity concept is readily apparent when applied to photostable crystals that would be expected to be otherwise on the basis of topochemical postulates. In compounds **6–12** listed in Scheme 6.5 and Table 6.5, the separation distances between reactive double bonds are less than 4.2 Å, yet they do not undergo dimerization upon photolysis.[23,49,50,55,60–63] The exceptional situations in all these cases can be understood qualitatively by invoking the "reaction cavity" concept.

Analysis of the systems listed in Table 6.5 and Scheme 6.5 highlights the role of the surrounding molecules in controlling the reactivity of olefins in

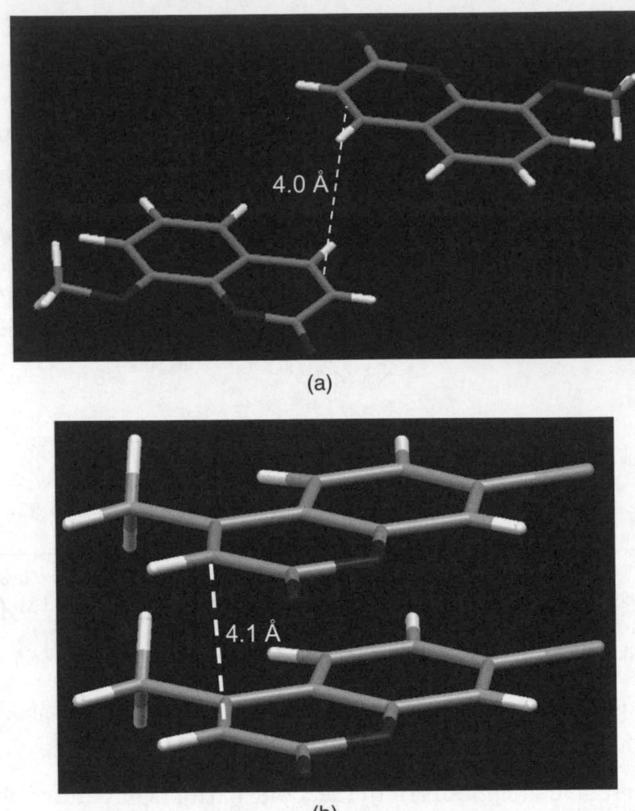

Figure 6.3. (a) 8-Methoxycoumarin (α-packing). (b) 4-Methyl-7-chlorocoumarin (β-packing).

crystals. One of the polymorphs[60,61] of distyryl pyrazine (**6**), where the potentially reactive double bonds are separated by 4.19 Å, is photostable.[60] The photostability of this compound has been ascribed to the layered structure that suppresses the molecular deformation necessary for the cycloaddition reaction. Another example where the molecular packing satisfies the topochemical criteria but yet is photostable is enone **7**.[62] The potentially reactive double bonds are parallel with a center-to-center distance of 3.79 Å. Nevertheless, **7** is photochemically inert when irradiated in the solid state. The attributed reason for the lack of solid-state reactivity of this enone is the steric compression experienced by the reacting molecules at the initial stages of photocycloaddition. In the crystal of 4-hydroxy-3-nitro methylcinnamate (**8**),[23] the neighboring molecules are related by a translation of 3.78 Å. But it has been observed that this compound is photostable in the solid state. In the crystal structure, the molecules are linked by hydrogen bonds to form a sheet-like structure close to the (102) plane. It is likely that the extensive intermo-

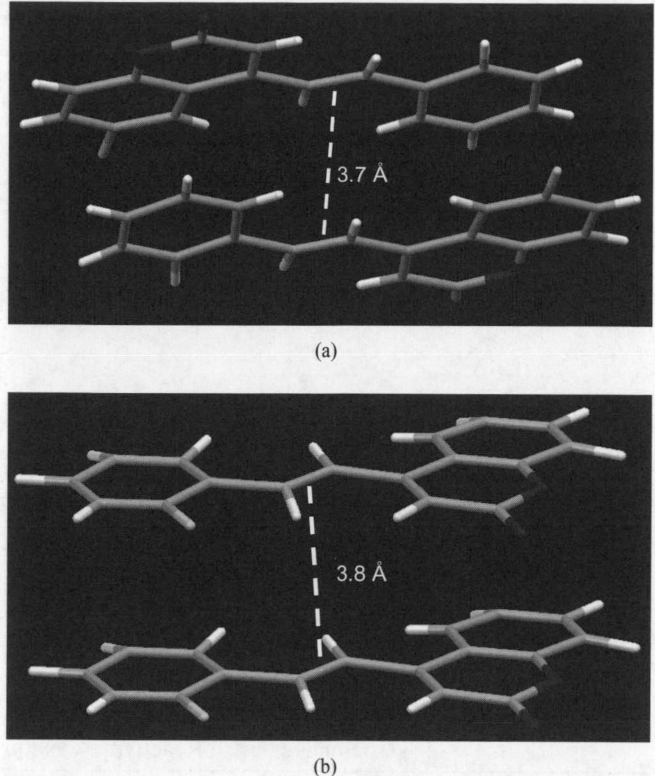

Figure 6.4. (a) Packing diagram of styryl-6-fluorocoumarin (α-packing). (b) Packing arrangement of 2-fluoro-styrylcoumarin (β-packing).

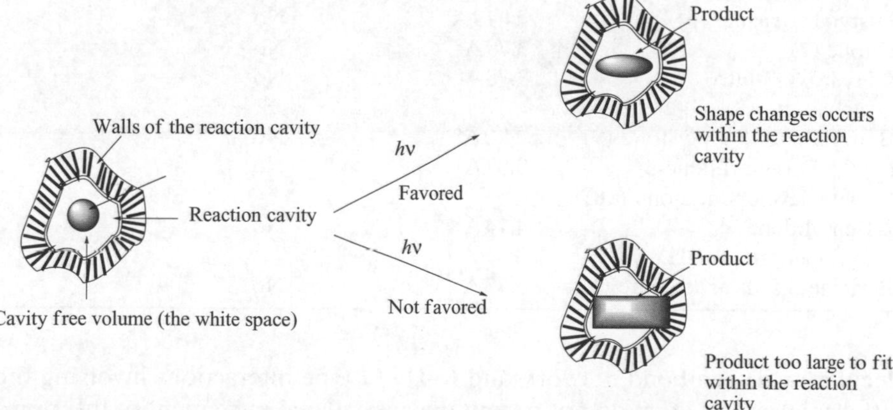

Figure 6.5. Concept of reaction cavity illustrated. Reaction cavity by definition includes the space occupied by the reactants and the empty space surrounding them.

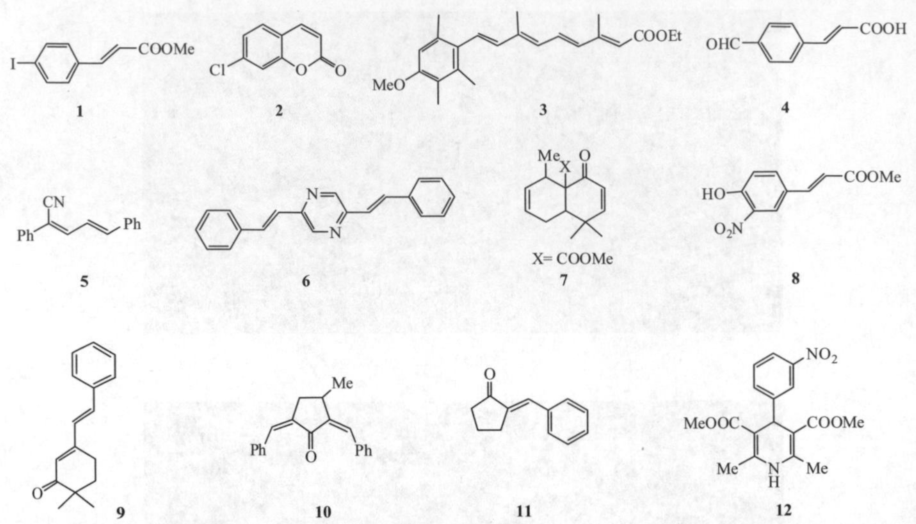

Scheme 6.5

TABLE 6.5. Examples of Exceptions to Original Topochemical Principles Regarding Distance (See Scheme 6.5)

Compound	Distance between Reactive Double Bonds	Reactivity	Nature of Dimer
Methyl-*p*-iodocinnamate (**1**)	β-Type, 4.3 A°	Yes	Mirror symmetric
7-Chlorocoumarin (**2**)	β-Type, 4.45 A°	Yes	*syn*-HH
Eteretinate (**3**)	4.4 A°	Yes	—
p-Formyl cinnamic acid (**4**)	β-Type, 4.825 A°	Yes	Mirror symmetric
(1*Z*,3*E*)-1-Cyano-1,4-diphenylbutadiene (**5**)	5.04 A°	Yes	—
Distyryl pyrazine (**6**)	<4.19 A°	No	—
Enone (**7**)	3.79 A°	No	—
4-Hydroxy-3-nitro cinnamate (**8**)	3.78 A°	No	—
Benzylidene-*dl*-piperitone (**9**)	<4.0 A°	No	—
(+)-2,5-Dibenzylidene-3-methylcyclopentanone (**10**)	3.87 A°	No	—
2-Benzylidene cyclopentanone (**11**)	4.14 A°	No	—
Pyridine-3,5-dicarboxylate (**12**)	3.73 A°	No	—

lecular hydrogen-bond network and C–H···O type interactions involving the ethylenic carbon atom do not permit the easy spatial movement of the atoms of the double bond in the lattice for the reaction to proceed. It has been reported that benzylidene-*dl*-pipertone (**9**)[55] is photostable in spite of the fact that there are two pairs of centrosymmetrically related double bonds that are parallel and at a distance of 3.92 and 3.98 Å, respectively. Crystalline (+)-2,5-dibenzylidene-3-methylcyclopentanone (**10**)[49] and 2-benzylidene cyclopentanone (**11**)[50] are photostable, while closely related molecules possessing similar packing arrangements undergo dimerization readily in the solid state. The distance between the centers of the olefinic bonds of the inversion-related pairs in the former and in the latter are 3.87 and 4.14 Å, respectively. The photostability is attributed to the reduced overlap between potentially reactive C=C bonds. Analyses of the above examples reveal that it is important to consider the arrangement of surrounding molecules with respect to the reacting pairs in addition to the relative orientation of the reacting molecules. As discussed below, the absence of photoreactivity in many of these cases can be understood by performing lattice energy calculations.

We discuss below the lattice energy calculations performed on one of these systems, 7-chlorocoumarin (**2**).[63] These calculations were performed using the computer program WMIN developed by Busing on a large number of photodimerizable olefins. It may be stressed that in these calculations, only the relative values within a series are meaningful in view of thc many approximations made. Although the calculations have been carried out using the ground-state geometry with the dispersion constants appropriate to the ground state, the results provide some insight. Irradiation of crystalline 7-chlorocoumarin yields a single dimer (*syn*-head-to-head [HH]).[32] The packing arrangement shown in Figure 6.6 reveals that there are two potentially reactive pairs of 7-chlorocoumarin molecules in a unit cell. One pair, being translationally related, has a center-to-center distance of 4.45 Å (favored to yield the *syn*-HH dimer). The other pair, being centrosymmetrically related, has a center-to-center distance of 4.12 Å (favored to yield the *anti*-head-to-tail [HT] dimer). Despite the favorable arrangement of the centrosymmetric pair, the dimer is obtained only from the translationally related pair. It has been calculated that

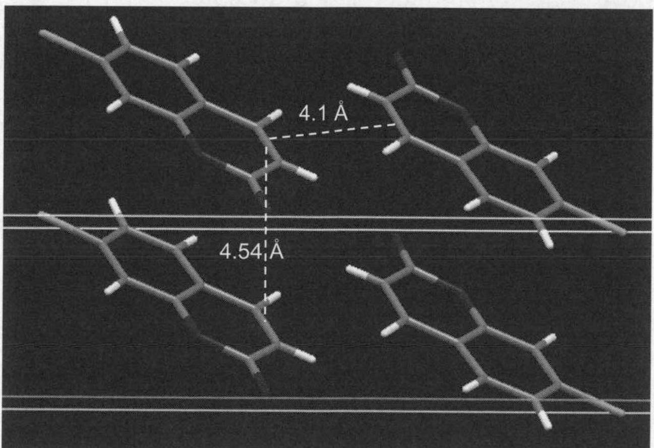

Figure 6.6. Packing diagram of 7-chlorocoumarin. Note that one pair has the olefins (α-packing) closely disposed than the other pair (β packing).

the rise in the lattice energy to achieve the ideal geometry for the translated pair (separated by 4.45 Å) is 177 kcal mol^{-1}, whereas for the centrosymmetric pair (separated by 4.12 Å), the energy increase is as large as 18,083 kcal mol^{-1}. This shows that the reaction pathway leading to the experimentally observed *syn*-HH dimer is energetically more favorable than the *anti*-HH isomer. In other words, the free volume around the translationally related pair is much larger than that near the centrosymmetrically related pair whose double bonds are initially closer. Lack of free volume in the most topochemically favored pair leads to no reaction, while presence of sufficient free volume allows dimerization of the less favored pair. The above example emphasizes the importance of void space around the reacting partners, the size of which may vary from system to system. Thus, the crystal reactivity requires the availability of free space around the reaction site.

FINE-TUNING OF TOPOCHEMICAL POSTULATES

According to the original topochemical postulates, the photodimerization in the solid state is likely to occur when separation between the reacting C=C π-bonds is less than 4.2 Å and the two C=C bonds are parallel to one another. These criteria were set based on extensive studies on cinnamic acids. These rules are still followed by a large number of molecules listed in Tables 6.2–6.4. With the exception of methyl-*p*-iodocinnamate (**1**), all the cinnamic acid derivatives that have adjacent double bonds separated by a distance of more than 4.2 Å in the crystalline phase are photostable. In the case of methyl-*p*-iodocinnamate (**1**), the molecules are arranged in a β-type packing with an inter-double-bond distance of 4.3 Å and yet react to yield the expected photodimer.[63] One should note that the upper limit of the critical distance for photodimerization in the solid state was set in the absence of experimental data in the range 4.2–4.7 Å, above which photodimerization does not occur.

Five examples in which photodimerization does occur even when the separation between reacting C=C bonds are more than 4.2 Å are presented in Table 6.5. Irradiation of crystalline 7-chlorocoumarin yielded a single dimer (*syn*-HH). The packing arrangement reveals that the two reactive 7-chlorocoumarin molecules are separated by 4.45 Å (Fig. 6.6). Since the only dimer obtained corresponds to *syn*-HH, it is clear that the reaction is between the pairs translated along the *a*-axis. It is noteworthy that the distance of 4.45 Å lies beyond the so far accepted limit of 3.5–4.2 Å for photodimerization in the solid state. Photodimerization of etretinate (**3**)[64] dimerizes in the solid state to yield two dimers. The center-to-center distance for the two sets of dimerizable bonds are 3.8 and 4.4 Å, the latter being outside the presently accepted limit. The most unusual case reported so far is *p*-formyl cinnamic acid (**4**).[24,25] This crystal, possessing a *b*-axis of 4.825 Å, dimerizes in the solid state to yield a mirror symmetric dimer. The above examples point out the need for a closer examination and modification of the distance criteria for photodimerization. These suggest that if the surrounding molecules can tolerate motions of the reacting pair, the reacting C=C bonds need not be within the initially stipulated distance of 4.2 Å. Apparently reacting pair can move much more than 2.64 Å than originally envisioned by Schmidt. Note that motion is along a plane perpendicular to the molecular plane.

A few cases (Scheme 6.6) have also been reported where exact parallelism between reactant double bonds has not been adhered to and yet photodimerization occurs.[63,65] Two most glaring examples are 7-methoxycoumarin (**16**)[32] and 2,5-dibenzylidene cyclopentanone (**17**).[49] In these two molecules, the reacting C=C pairs are crisscrossed (Figs. 6.7 and 6.8). In the crystals of 7-methoxycoumarin, the reactive double bonds are rotated by about 65° with

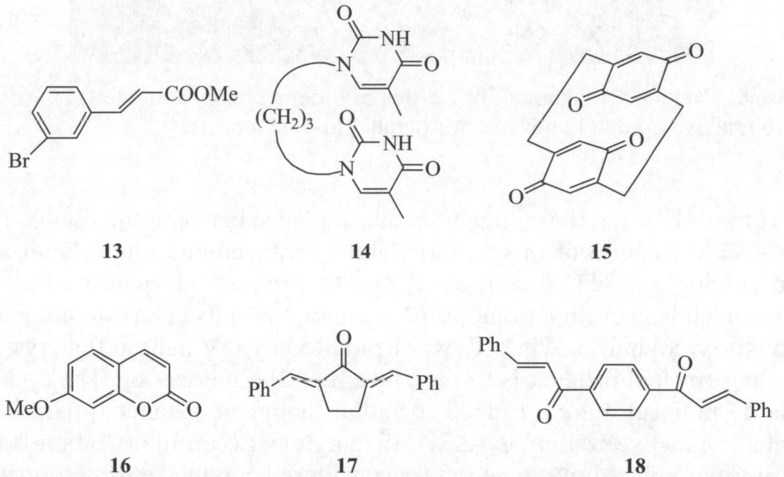

COOMe
Br
13

NH
O
N
(CH$_2$)$_3$
14

O
15

MeO
O
16

Ph
17

Ph
18

Scheme 6.6

3.6 Å

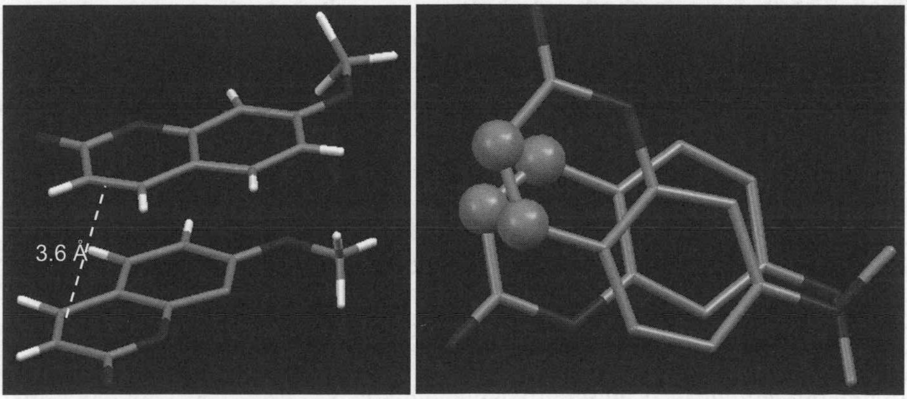

Figure 6.7. Packing arrangement of 7-methoxycoumarin. Note that the two reactive double bonds are not parallel to one another.

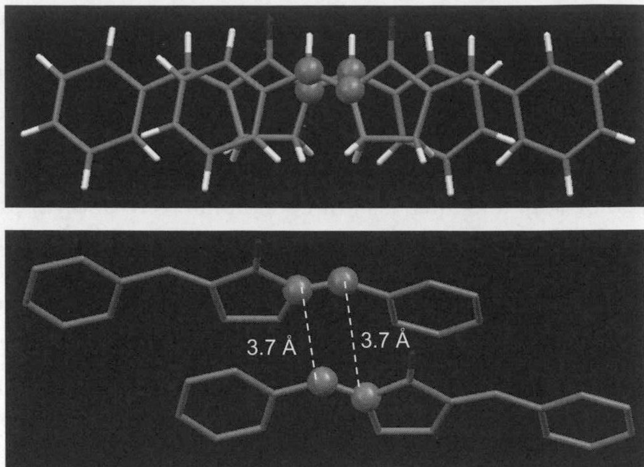

Figure 6.8. Packing arrangement of 2,5-dibenzylidene cyclopentanone (**11**). Note that the two reactive double bonds are not parallel to one another.

respect to each other, the center-to-center distance between the double bonds being 3.83 Å. In spite of this "unfavorable" arrangement, photodimerization occurs, giving *syn*-HT dimer as the only product in quantitative yield. 2,5-Dibenzylidene cyclopentanone **17** is analogous in its behavior and packing to 7-methoxycoumarin. When **17** was irradiated by UV light in the crystalline state, the principal product is formed by a [2 + 2] dimerization. The cyclopentanone **17** molecules are arranged such that the mean distance separating the potentially reactive centers is ~3.7 Å, the angle between the two bonds being 56°. Although this is not the geometry considered conducive for a topochemical reaction, dimerization does indeed take place in the solid state. It is remarkable that although the relevant olefinic π-orbitals are not overlapping in their ground-state geometry, both are photoreactive. These cases in which the nonparallel alignment of the π-orbitals does not inhibit photoreactivity indicate that there must be enough freedom for the reactive molecules to undergo the necessary movements to reorganize in their respective crystal lattices to allow dimerization to occur. In these two examples, the motion required to bring the two reactive C=C bonds one over the other is translation along the molecular plane (Table 6.6).

At this stage, one obvious question is if 7-methoxycoumarin and 2,5-dibenzylidene cyclopentanone that are not aligned properly react, why not *meta*-bromomethylcinnamate (**13**)[22] in which the two reactive C=C bonds make an angle of 28° (Fig. 6.9). Recognizition of the possible differences in the nature of reaction cavities in 7-methoxycoumarin, 2,5-dibenzylidene cyclopentanone and *meta*-bromomethylcinnamate leads to a better understanding. In *m*-bromomethylcinnamate similar to 7-methoxycoumarin, the double bonds are not ideally oriented for topochemical dimerization. Although the distance

TABLE 6.6. Examples of Exceptions to Original Topochemical Principles Regarding Parallelism of Double Bonds (See Scheme 6.6)

Compound	Rotational Angle of One Bond with Respect to the Other	Reactivity Dimerization
Methyl-*m*-bromocinnamate (**13**)	28°	No
1,1′-Trimethylene-bis-thymine (**14**)	6°	Yes
[2,2](2,5)-Benzoquinophane (**15**)	3°	Yes
7-Methoxycoumarin (**16**)	65°	Yes
2,5-Dibenzylidene cyclopentanone (**17**)	56°	Yes
1,4-Dicinnamoyl benzene (**18**)	6°	Yes

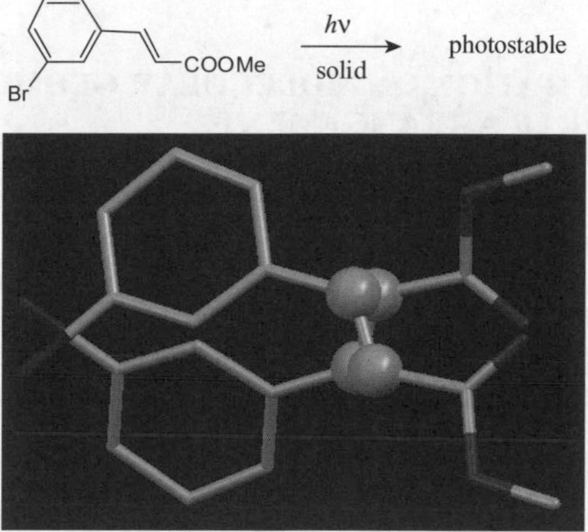

Figure 6.9. Packing arrangement of methyl-*m*-bromocinnamate (**13**). Note that the two reactive double bonds are crisscrossed.

between the centers of adjacent double bonds is 3.93 Å, the double bonds are not parallel. They make an angle of 28° when projected down the line joining the centers of the bonds. The energy increase needed to bring the two reactant molecules together to obtain the right isomer in 7-methoxycoumarin is about 200 kcal mol^{-1}, roughly the same order of magnitude as for many photoreactive crystals with favorably oriented pairs. On the other hand, in the case of *m*-bromomethylcinnamate, the energy increase to align the molecules parallel to each other in geometry suitable for dimerization is enormous (6726 kcal mol^{-1}). Such a large increase in the lattice energy probably does not favor reorientation of the molecule to result in photodimerization.

All these examples that appear anomalous in the light of topochemical postulates can be understood on a unified conceptual basis if one incorporates the reaction cavity concept of Cohen within the topochemical postulates due to Schmidt. Dimerization may be considered as taking place in a "nanocavity" within the bulk crystal, the latter being the host and the reactive pair the guest. The size and shape of the cavity and the interactions between the "guest reactants" and the host lattice will determine whether the nontopochemically arranged molecules would be permitted to undergo the motion necessary to reach a topochemical arrangement. Before assigning the reactivity of these anomalous pairs to bulk crystals, it is important to rule out defects being responsible for reactivity of these unusually placed pairs. In this chapter, we do not discuss defect-centered photodimerizations in the solid state.

It is clear that in addition to relative atomic positions, relative orientation of the reactive π-orbitals must be monitored to assess the feasibility of dimerization in the solid state. Less than ideal atomic and orbital orientations can still give rise to dimerization if the surrounding lattice can tolerate motions that would steer the molecules to proper mutual orientation.

IMPACT OF EXCITATION ON MOLECULAR GEOMETRY AND INTERMOLECULAR ARRANGEMENT

The static concept of preorganization does not correspond to reality in as much as it does not take into account the changes caused by molecular excitation. Excitation of molecules to higher electronic levels brings about changes, among other things, to the geometry and polarizability of molecules. For example, it is well known that formaldehyde undergoes pyramidalization upon excitation with a corresponding change in dipole moment. For olefins, the preferred minimum energy configuration in the excited state is the perpendicular (orthogonal π-orbitals) rather than planar form. It is also established that for some aromatics, dimeric complexes, that is, excimers, are stabilized with respect to monomers in the excited state. Such differences in geometry and polarizability between the ground and the reactive state (excited) are expected to have subtle consequences on the topochemical postulates based on ground-state properties. It is important to note that predictions concerning excited-state reactivity are made based on accurate ground-state geometries and packing arrangements obtained crystallographically. Accurate predictions are possible only if the difference in geometry between the ground and the reactive excited states is taken into account. In this context, it is of interest to note that work on obtaining X-ray crystal structures of molecules in excited states has already begun.

In the ground state, the crystal is expected to be homogeneous and the forces operating between molecules in the crystals are expected to be uniform. However, upon excitation, the crystal will contain two types of molecules, most in the ground state and a few in the excited state. The forces operating between an excited molecule and its neighbors differ from those operating between a ground-state molecule and its surroundings. The change in polarizability upon excitation increases the attractive part of the intermolecular force, while the repulsive part remains, initially, unchanged. *The localized excitation produces a particular type of local instability of the lattice configuration that may lead to large molecular displacements*. The displacements may favor the formation of excimers and photodimers in crystals.

Craig and coworkers[66–68] carried out an incisive theoretical investigation of this problem and have shown that a short-term lattice instability created upon excitation has the effect of driving one molecule close to a neighbor, thus promoting excimer or exciplex formation. The calculation for 9-cyanoanthracene showed that, for a short period after excitation, an excited molecule can be displaced away from its equilibrium crystal lattice position into an unsymmetrical local structure, with the excited molecule closer to one neighbor in the stack of molecules than to the other. In such a model, there is a transient preformation of an excimer not evident in the equilibrium local structure. The important message of the investigations by Craig and coworkers is that it is of the utmost importance to consider the dynamic properties of lattices (caused by photoexcitation) to understand the processes involved in photochemical reactions in crystals. This also implies that the dimerization may occur within a reaction cavity under conditions where the molcculcs are less than ideally oriented. The driving force to bring the pair into proper orientation will be provided by electronic excitation energy and the increased attractive interaction energy in the excited state.

Two recent examples provided in Figures 6.10 and 6.11 further highlight the role of excitation and flexibility of reactant molecules in the crystalline state. In these two examples, interaction between an amine (or an amide) and an acid is used to steer the olefinic chromophores within the reacting distance. In the example provided in Figure 6.10,[69,70] the hydrogen-bonded complex between *trans*-cinnamide and phthalic acid upon irradiation yields the β-truxinamide. However, the packing is not suitable for such a dimer formation.

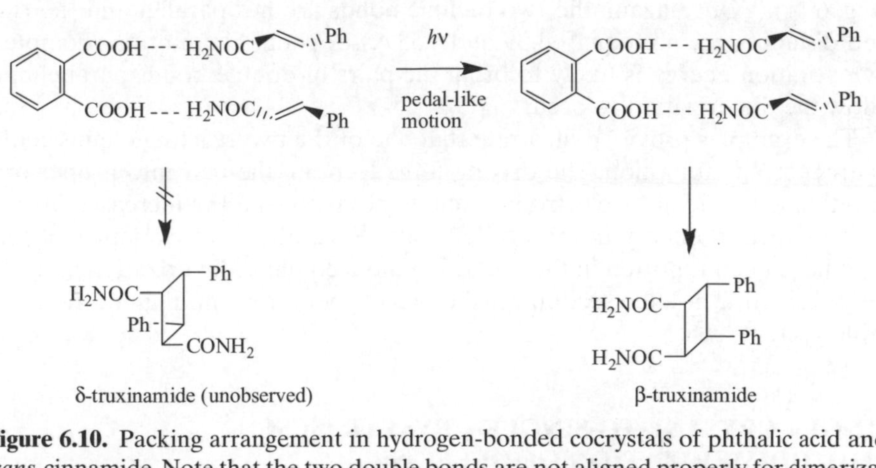

δ-truxinamide (unobserved)

β-truxinamide

Figure 6.10. Packing arrangement in hydrogen-bonded cocrystals of phthalic acid and *trans*-cinnamide. Note that the two double bonds are not aligned properly for dimerization. As shown in the scheme, pedal-like motion brings the two double bonds parallel and leads to β-dimer.

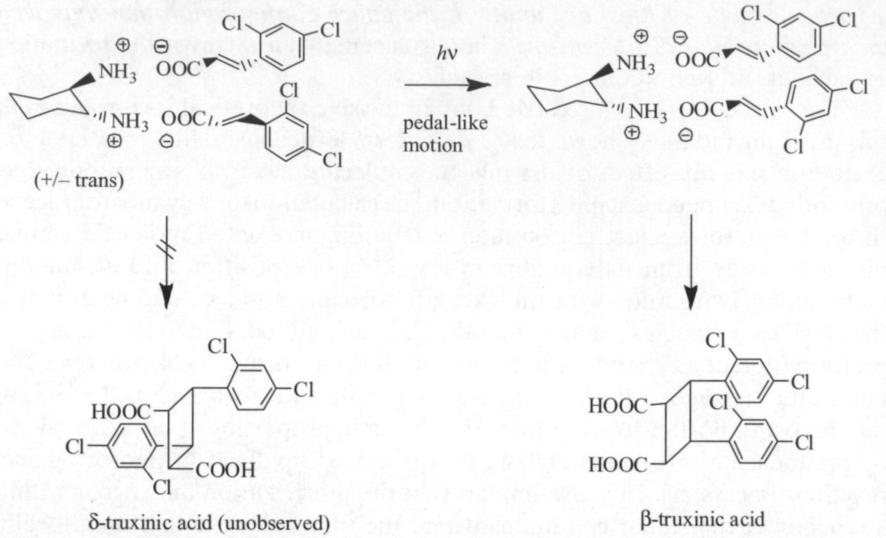

Figure 6.11. Packing arrangement in crystals of the double salt of 1*R*,2*R*-diaminocyclohcxanc and 2,4-dichlorocinnamic acid. Note that the two double bonds are not aligned properly for dimerization. As shown in the scheme, pedal-like motion brings the two double bonds parallel and leads to β-dimer.

As seen in Figure 6.10, the two olefinic bonds are not parallel and the two ends are separated by 3.8 and 4.8 Å. Based on the dimer (β-truxinamide) formed, it is speculated that one of the olefins perform a pedal-like motion prior to dimerization. A similar motion must also be involved during the irradiation of the salt between diaminocyclohexane and 2,4-dichlorocinnamic acid (Fig. 6.11).[71] Once again, the two olefinic bonds are not parallel and the two end distances are not equal (3.39 and 4.55 Å). A pedal-like motion prompted by excitation energy is likely to bring the pairs of double bonds parallel and allow the dimerization to occur.

The examples provided illustrate that one of the two reacting olefins could exert substantial motion. The driving force to bring the pair into proper orientation is provided by electronic excitation energy and the increased attractive interaction energy in the excited state. Recently, more examples of this type have been reported in the metal-ion-induced photodimerization and will be discussed in a later section on the use of metal ions in aligning reactant molecules.

SINGLE-CRYSTAL-TO-SINGLE-CRYSTAL (SCSC) PHOTODIMERIZATION OF OLEFINS

The overall phototransformation of olefins into cyclobutanes in the solid state can proceed by two pathways: single crystal to polycrystalline and SCSC. In the first case (single crystal to solid solution to polycrystalline), the product phase goes into solid solution in the lattice of the monomer, and then, as the dimer concentration rises, the solubility limit is exceeded and the new phase precipitates. Most of the dimerization examples presented in this chapter belong to this class. For example, X-ray powder diagrams in the case of coumarins show a gradual and complete loss of long-range order and an eventual appearance of an ordered product phase. There is no evidence yet as to whether the product phase separates out of the parent phase at specific or at random sites. Once the original monomer crystal breaks down due to contamination by the product dimer, the photodimerization may no longer be controlled by the initial packing. Since the solubility limit of the dimer in the monomer phase will vary with the reactant molecule, one might expect the maximum yield of topochemical dimer also to vary with the reactant molecule.

The second type of photoreactions namely SCSC transformation continues to be rare. There are at least three examples of photodimerization of olefins known to belong to this category (Figs. 6.12–6.14).[26,41,43–45,53,72] The chances of

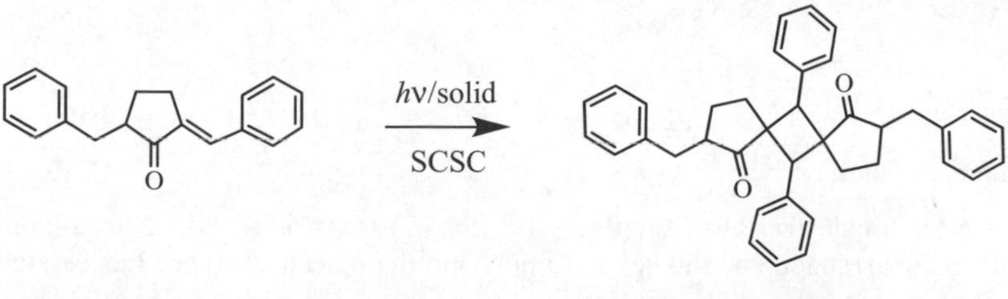

*h*ν/solid

SCSC

anti-HT dimer

4.15 Å

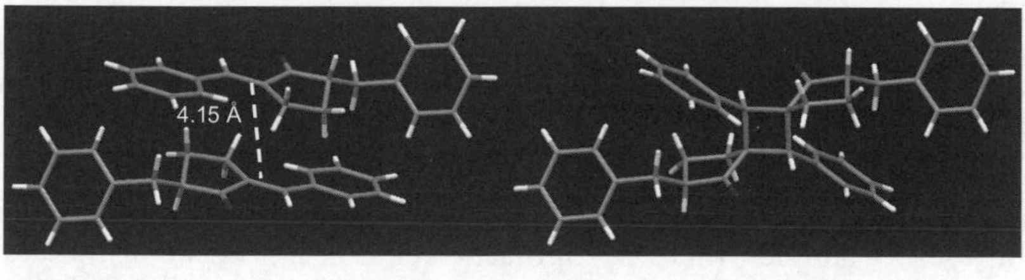

a/A°	31.10	31.32
b/A°	10.78	10.81
c/A°	8.69	8.63
Z	8	4
Space group	P_{bca}	P_{bca}
Unit cell volume A°3	2932	2922

Figure 6.12. Single-crystal-to-single-crystal (SCSC) transformation of 2-benzyl-5-benzylidene cyclopentanone. Packing arrangement and cell parameters in the reactant and product crystals are provided.

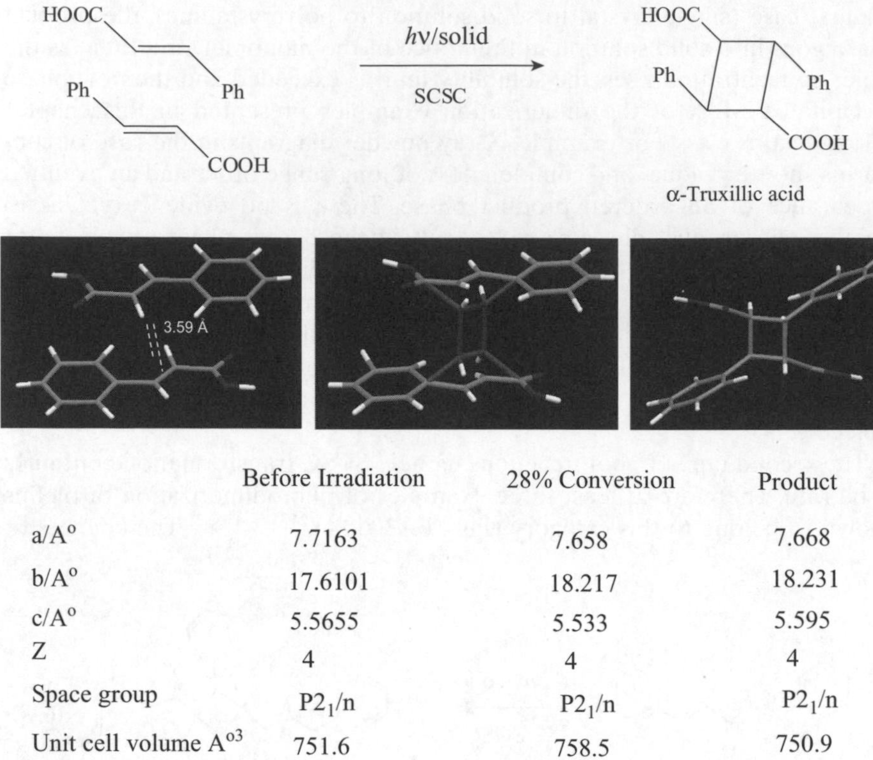

	Before Irradiation	28% Conversion	Product
a/A°	7.7163	7.658	7.668
b/A°	17.6101	18.217	18.231
c/A°	5.5655	5.533	5.595
Z	4	4	4
Space group	$P2_1/n$	$P2_1/n$	$P2_1/n$
Unit cell volume A^{o3}	751.6	758.5	750.9

Figure 6.13. Single-crystal-to-single-crystal (SCSC) transformation of *trans*-cinnamic acid. Packing arrangement and cell parameters in the reactant and product crystals are provided.

achieving SCSC transformation are much higher when the irradiation is conducted at the tail edge of the absorption of the olefin. This allows the molecules present at the surface as well as at the interior of the crystal to be uniformly excited. In these examples, the dimerization proceeds through a series of solid solutions of varying composition and is under topochemical control throughout. In these examples, there is topotactic relationship between these solid solution phases. Structure–photodimerization correlation studies on 2-benzyl-5-benzylidene cyclopentanone (Fig. 6.12)[41,44,45,53] reveal that it undergoes SCSC dimerization. Other examples in the one component crystals include the photodimerization of cinnamic acid (Fig. 6.13)[26] and styrylpyrilium salt (Fig. 6.14).[72] Crystal cell parameters of the reactant prior to and after irradiation are provided in the figures. There are only small changes in cell parameters even after total conversion of the reactant to the product suggesting that the phototransformation has occurred with very little changes in the atomic positions of the reactant molecule. We discuss one example below and the other two follow the same trend.

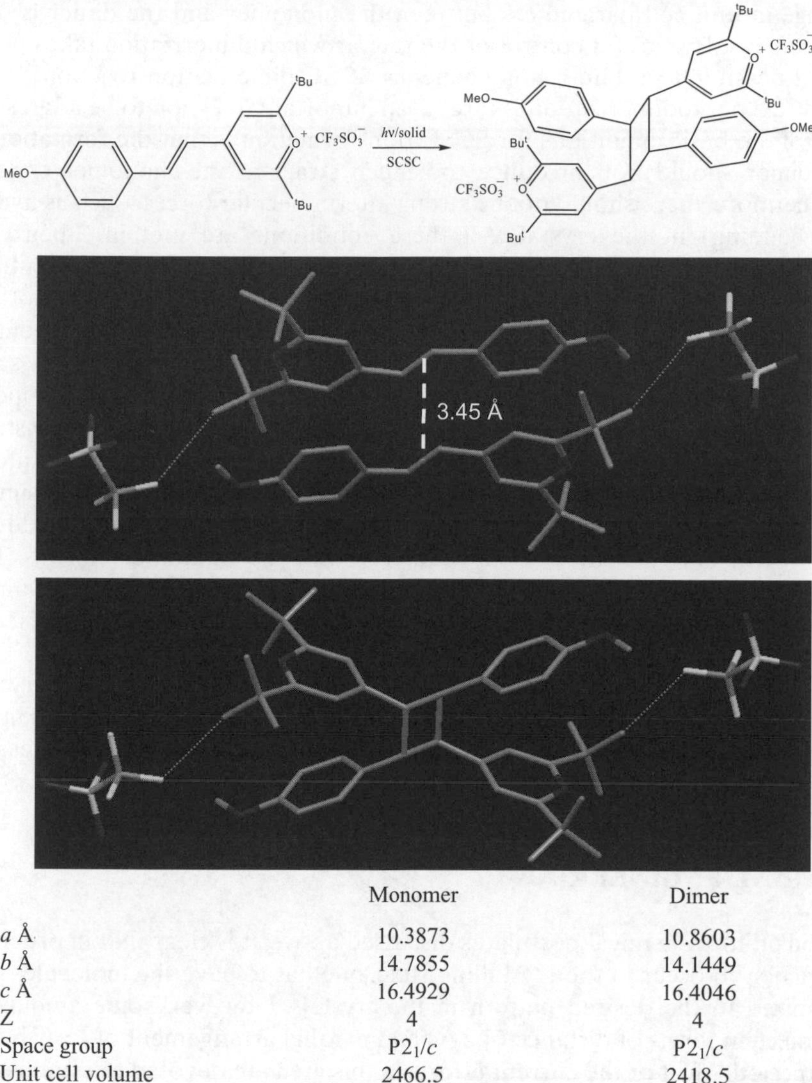

	Monomer	Dimer
a Å	10.3873	10.8603
b Å	14.7855	14.1449
c Å	16.4929	16.4046
Z	4	4
Space group	$P2_1/c$	$P2_1/c$
Unit cell volume	2466.5	2418.5

Figure 6.14. Single-crystal-to-single-crystal (SCSC) transformation of styrylpyrylium salt. Packing arrangements and cell parameters in the reactant and product crystals are provided.

In the crystals of 2-benzyl-5-benzylidene cyclopentanone, the neighboring molecules are related by a center of symmetry with the reactive double bonds separated by 4.1 Å. Photolysis of crystals of 2-benzyl-5-benzylidene cyclopentanone yields single crystals of its dimer (Fig. 6.12).[43,53] The fact that the product is crystalline indicates that there is a definite crystallographic relationship between the parent and the daughter phases. Indeed, the maximum change in unit cell parameters between the monomer and the dimer is only about 0.7%. By careful control of the rate at which dimerization takes place, it was possible to retain a homogeneous SCSC dimerization reaction. "Why the SCSC photodimerization is rare" is an important question to be addressed. One of the basic conditions for SCSC transformation is that the formation of the dimer should not introduce too much strain in the monomer crystals. Furthermore, there should not be strong intermolecular forces (such as hydrogen bonding) in the crystal. All these conditions are met in 2-benzyl-5-benzylidene cyclopentanone. In this case, the reactive double bond is essentially at the central part of the molecular framework. During the course of the dimerization, it is this part of the molecule that undergoes a large movement with the peripheral part of the molecule remaining essentially at the same position. In rigid molecular systems such as coumarins, one cannot hope to achieve this condition. Dimerization of the double bond in such rigid systems would result in large changes in the atomic positions of the peripheral atoms leading to disruption of the crystal. In the case of cinnamic acids and similar molecules, the presence of strong hydrogen bonding in the crystal would not allow sufficient relaxation of the dimer within the monomer crystals. This would result in the disruption of the crystal packing and formation of amorphous product. This is the case in 99+% of dimerization examples reported in this chapter.

There are other SCSC examples in the category of two-component crystals (cocrystals, inclusion complexes, and metal-ion coordination complexes) in the recent past and will be discussed in later sections on the use of templates and metal ions in aligning reactant molecules.

CRYSTAL ENGINEERING

Based on topochemical postulates discussed above, it is clear that in order for a reaction to occur in the crystalline state, one has to have the molecules preorganized in the desired pattern in the crystals. However, some amount of tolerance in terms of distance (4.2 Å) and parallel arrangement of C=C bonds is expected. Most of the current efforts in this area are devoted to establishing reliable strategies that would steer molecules so as to obtain an organic crystal structure of a predetermined form. Schmidt termed this operation "crystal engineering."[6] One of the major problems encountered here is lack of complete understanding of the intra- and intermolecular interactions leading to the observed crystal packing. If one had a complete understanding of the ways in which inter- and intramolecular interactions control packing of molecules in crystals, it would be feasible to design template groups, perhaps of temporary attachment, to the functional molecules to guide photochemically reactive groups into appropriate juxtaposition in crystals. In order to bring the reactive molecules into proper orientations, several distinct strategies have been employed: (1) intramolecular substitution, (2) templation with host structures,

(3) mixed-crystal formation, (4) generation of polymorphic forms, (5) steering crystallization through donor–acceptor and hydrogen-bonding strategies, and (6) structural isomorphism through groups of equivalent size. We briefly discuss the first two that have proven to be more general and reliable than the last three.

Substitution of Halogens

Schmidt and coworkers recognized quite early that monochloro substitution and especially dichloro substitution in aromatic molecules tend to steer molecules in crystal lattices with a short axis of ~4 , the so-called β-structure. A few examples are provided in Scheme 6.7.[6,30,40,73–78] It is remarkable to note in Scheme 6.7 that while the parent olefin in each case fails to photodimerize in the solid state, the dicholoro substitution leads to mirror symmetric dimers. The packing arrangement shown for 2,4-dichlorodiphenyl butadiene in Figure 6.15 reveals that the attractive interaction between the Cl—Cl steers the molecule in the correct orientation for dimcrization. Chloro substitution has bccn

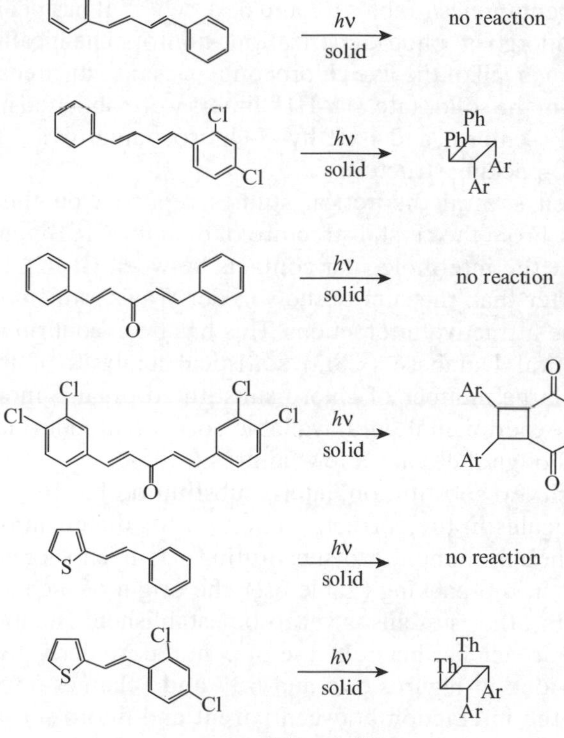

hν / solid → no reaction

Cl, Cl; hν / solid → Ph, Ph, Ar, Ar

O; hν / solid → no reaction

Cl, Cl, Cl, Cl, O; hν / solid → O, Ar, Ar, Ar, Ar, O

S; hν / solid → no reaction

S, Cl, Cl; hν / solid → Th, Th, Ar, Ar

Scheme 6.7

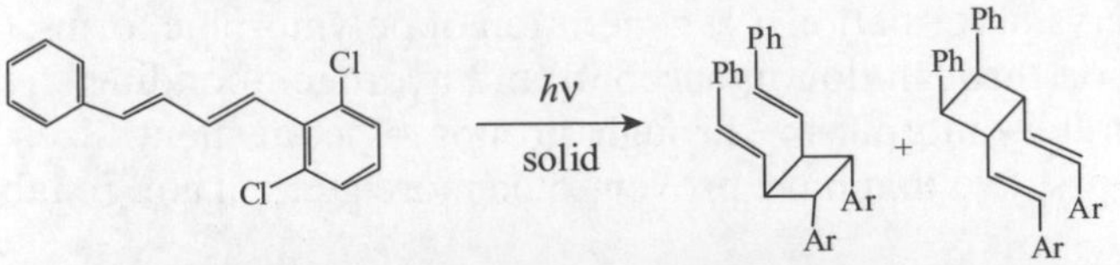

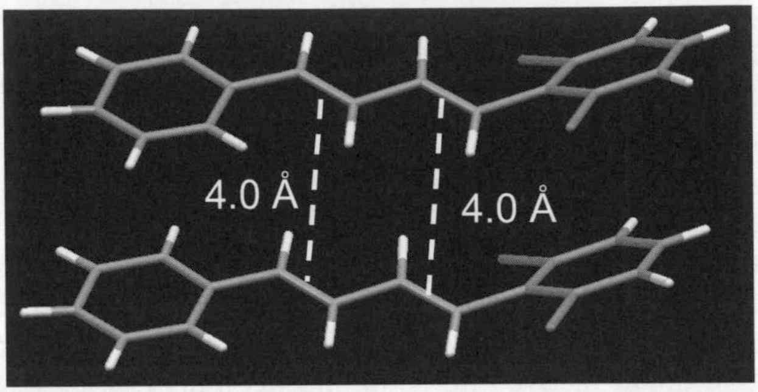

Figure 6.15. Packing arrangement in the crystals of (1*E*,3*E*)-1-phenyl-4-(2,6-dichlorophenyl)buta-1,3-diene. The diene substitution brings the two molecules within reactive distance and keeps them parallel.

successfully employed during dimerization of coumarins and 2-benzyl-5-benzylidene cyclopentanones (Tables 6.2 and 6.4).[42,45–48,52] It has been observed that coumarin undergoes photodimerization nontopochemically, yielding three dimers. However, all of the five chlorocoumarins investigated underwent clean dimerization in the solid state. *syn*-HH dimers were obtained in 6-chloro-, 7-chloro-, 4-methyl-6-chloro-, and 4-methyl-7-chlorocoumarins as a direct consequence of their β-packing structure.[32]

There have been several theoretical studies reported on the nature of Cl···Cl interactions. From the crystal structure of data for Cl^2, Br^2, and I^2, it has been observed that the intermolecular contacts between Cl···Cl, Br···Br, and I···I are much shorter than the sum of the van der Waals radii, indicating the presence of specific attractive interactions. This has been confirmed from the Cambridge Structural Database (CSD) statistical analyses of the packing arrangement of a large number of chloro-substituted organic molecules.[30] It appears from the experimental data available so far that chlorine is a good steering group, although there are a few failures.

In addition to chloro substitution, fluoro substitution has been effectively used to steer molecules in the correct orientation for dimerization (Scheme 6.8).[9,34,38,57,79–81] Although monofluoro substitution has been successfully used to steer coumarins into β-packing (Table 6.2), the origin of such an influence is unclear. Its use in other systems is yet to be established. The most reliable and predictable approach has been the use of pentafluoro derivatives. Several examples are provided in Figures 6.16 and 6.17 and Schemes 6.8 and 6.9.[79,81] In all these cases, the interaction between parent and fluoro substituted aryl groups drives the packing. The packing arrangements shown in Figures 6.16

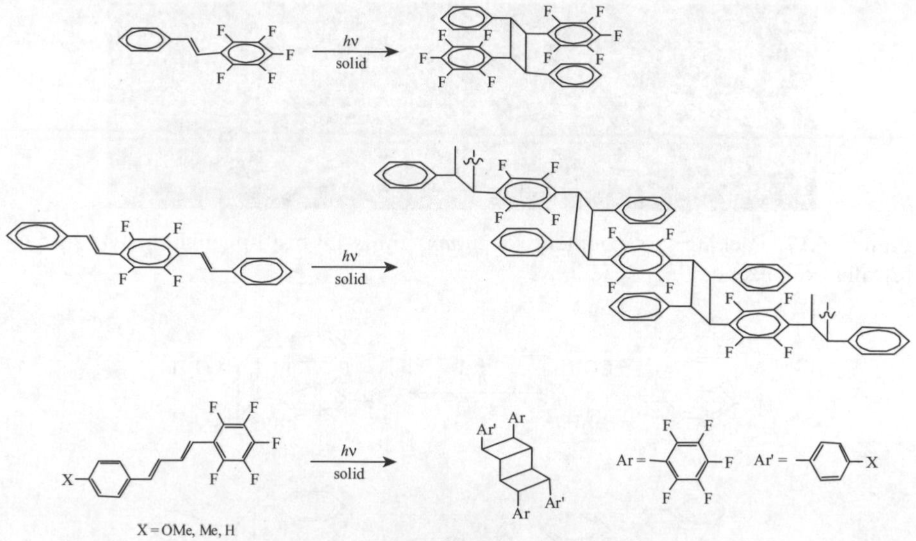

X = OMe, Me, H

Scheme 6.8

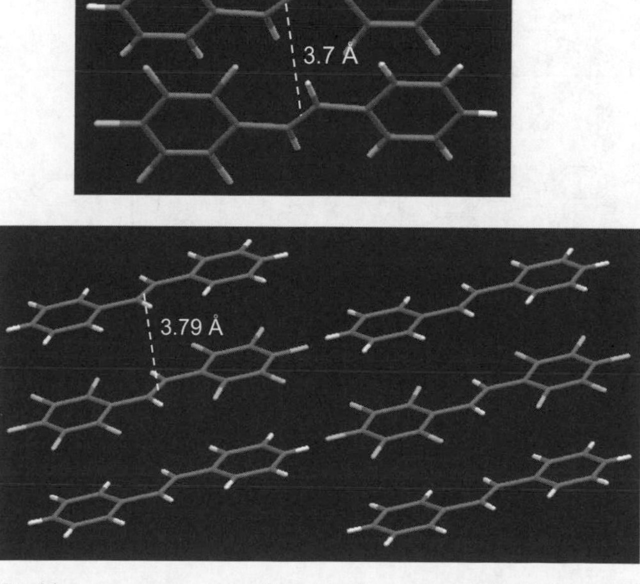

Figure 6.16. Packing arrangement in the crystals of *trans*-2,3,4,5,6-pentafluorostilbene (top) and mixed crystal of stilbene and decafluorostilbene (bottom).

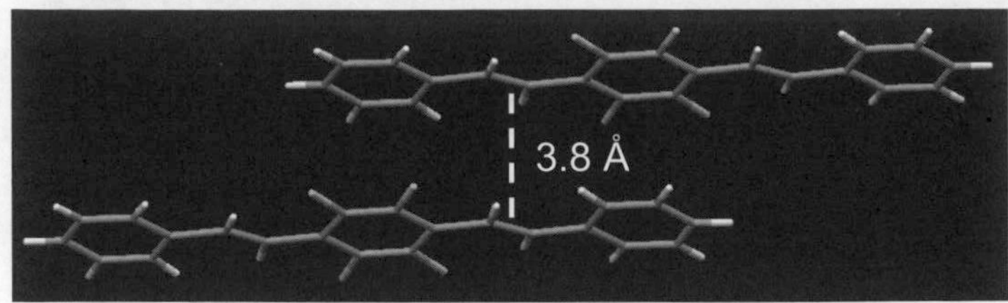

Figure 6.17. Packing arrangement of *trans*, *trans*-1,4-bis(2-phenylethenyl)-2,3,5,6-tetrafluorobenzene.

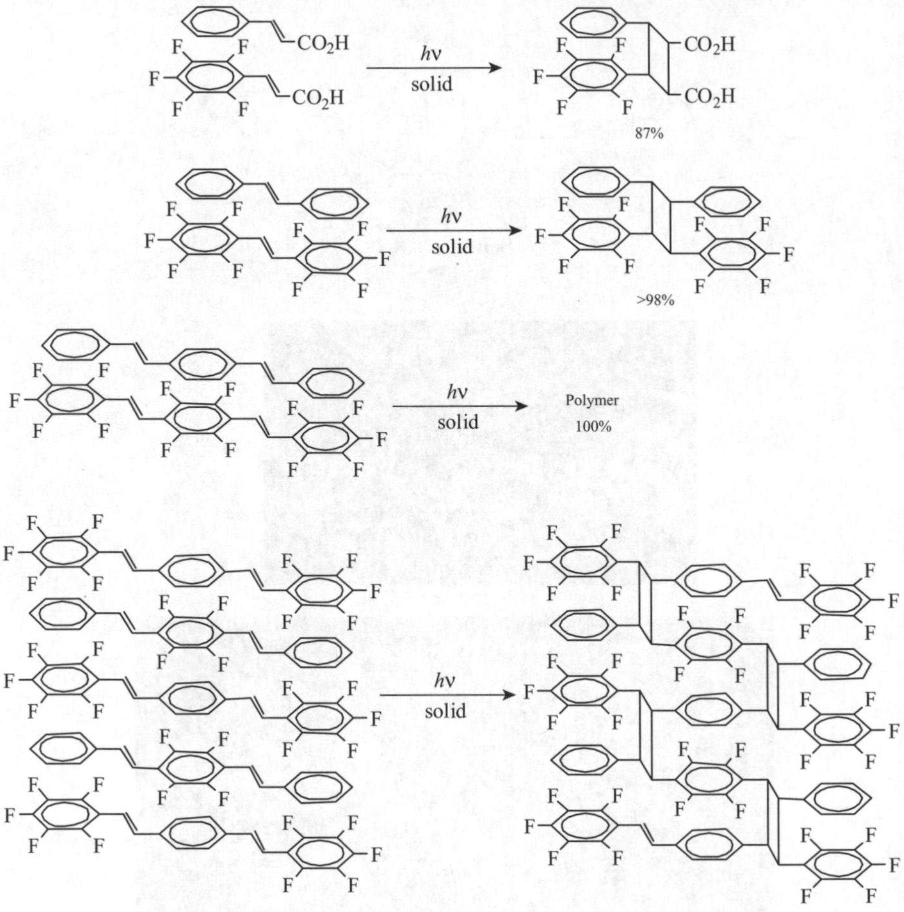

Scheme 6.9

and 6.17 bring out this feature. The nature of interaction may be of donor–acceptor or quadrupolar–quadrupolar type interaction between electron-rich phenyl and electron-deficient pentafluorophenyl groups. This strategy works well even during photodimerization between two different olefins (Scheme 6.9).

Use of Templates: Aligning Reactants through Hydrogen Bonding and Ionic Interactions

In this strategy, a template molecule is chosen such that the packing of the template/host molecules in the crystalline state will enable the potentially reactive guest molecules to pack in a manner that will facilitate photodimerization. In most examples thus far explored, hydrogen bonding between the template and the reactant molecules aligns the reacting pair.

One of the early examples involved the use of a diacetylene diol as a templating agent. Irradiation of powdered complexes of benzylidene acetophenone with the achiral diacetylene diol **19** gave a single photoproduct (>80% yield), which has been characterized as a *syn*-HT dimer (Scheme 6.10).[82] It is important to note that benzylidene acetophenone in the absence of template crystallizes in two polymorphic modifications, and the center-to-center distances between the double bonds are 5.2 and 4.8 Å in the two polymorphs. A remarkable effect of diacetylene diol **19** template is to bring the two reactive molecules closer. The molecules of the guest are packed in parallel pairs related by an inversion center. As a result, the planes of the double bonds are parallel and the center-to-center distance is 3.86 Å (Fig. 6.18). This arrangement enables the photodimerization to give the *syn*-HT dimer.[83–85] A number of coumarins have been successfully dimerized to *syn*-HH dimers with the help of a chiral diacetylene diol **20** template (Scheme 6.10).

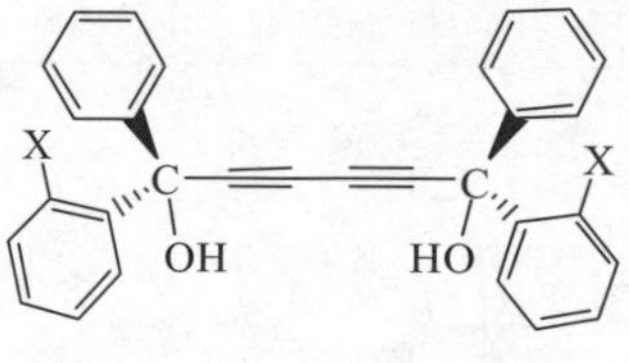

19 X = H
20 X = Cl

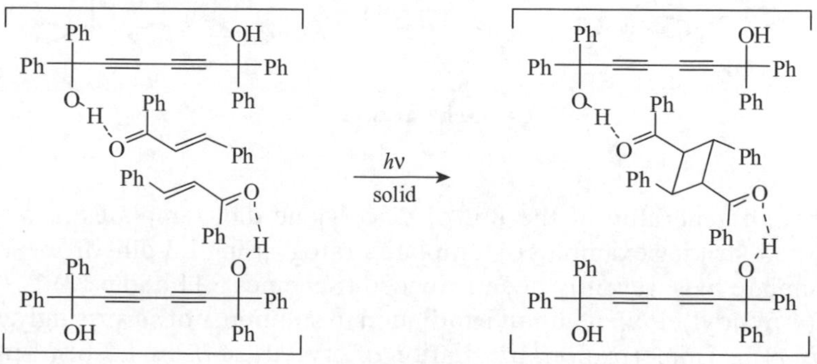

hν
solid

Scheme 6.10

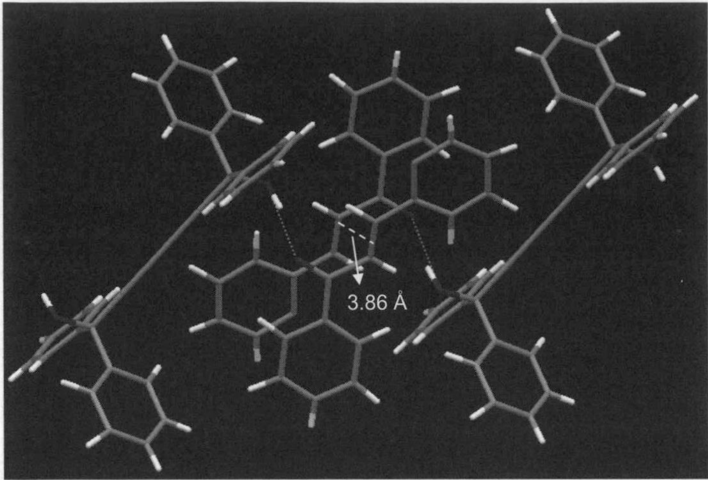

Figure 6.18. Packing arrangement of crystals of host–guest complex between 1,1,6,6-tetraphenyl-2,4-diyne-1,6-diol (host) and chalcone (guest).

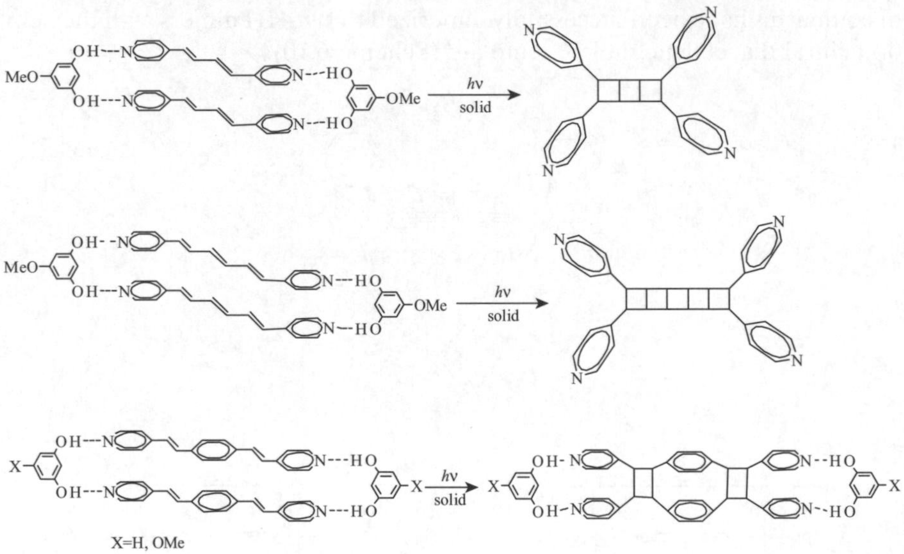

X=H, OMe

Scheme 6.11

Although generality of the use of diacetylene diol templates is yet to be established, striking examples of template strategy using 1,3-dihydroxybenzene as a template have recently been provided (Schemes 6.11 and 6.12).[86–91] *trans*-1,2-bis(4-pyridyl)ethylene upon irradiation in solution, not surprisingly, undergoes *cis–trans* isomerization. Irradiation of crystals of *trans*-1,2-bis(4-pyridyl)

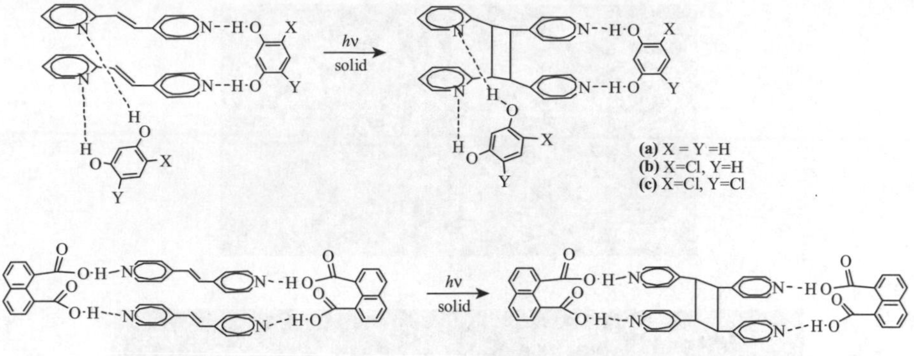

Scheme 6.12

ethylene does not give any products. Scrutiny of the crystal structure reveals that *trans*-1,2-bis(4-pyridyl)ethylene molecules crystallize in a layered structure in which olefins of neighboring molecules are separated by more than 6.52 Å (Fig. 6.19a).[92] Such a large distance does not allow dimerization. This molecule can be engineered to dimerize in the crystalline state if it is cocrystallized in the presence of 1,3-dihydroxybenzene. Irradiation of a mixture of 1,3-dihydroxybenzene and *trans*-1,2-bis(4-pyridyl)ethylene in solution resulted only in geometric isomerization. On the other hand, in the crystalline state, a single photodimer was obtained in quantitative yield. In this case, hydrogen bonding between 1,3-dihydroxybenzene and *trans*-1,2-bis(4-pyridyl)ethylene keeps the olefins parallel to each other and within 4.2 Å, thus facilitating the dimerization process (Fig. 6.19b).[86] As illustrated in Schemes 6.11, this approach has been extended to dienes and trienes.[91] The key realization in these studies is that 1,3-dihydroxybenzenes 1,8-disubstituted naphthalenes (Fig. 6.19c)[93–95] and Rebek's imide[96] (Scheme 6.15) can organize stacking of olefins at a distance of 4 Å. The main interaction used to organize is the hydrogen bonding.[97]

Double salt formation between diamines and acids has been used to align reactive C=C bonds within 4.2 Å. In these examples, if the amine is chosen properly, the two olefinic chromophores will stay within the reactive distance. The main force that holds the system together is the ionic interaction between the template and the olefin. Successful examples of this strategy are provided in Schemes 6.13 and 6.14.[71,98–101] One of the problems of this approach is that in a number of cases, the *cis* isomer (due to *trans–cis* isomerization) accompanies the dimer. This suggests that the packing must be loose enough to allow a large rotational motion of the reacting olefin.

While impressive examples to support the usefulness of the template strategy has been provided by Toda, Macgillivray, Thalladi, Christian Wolf, and others, there are likely to be equal number of examples that have not worked (Scheme 6.15). The key to success is the choice of the template as one template may not work well for all systems. For instance, if we were to investigate the choices of templates available for photodimerizations of 4,4; 3,3; 2,2; and other

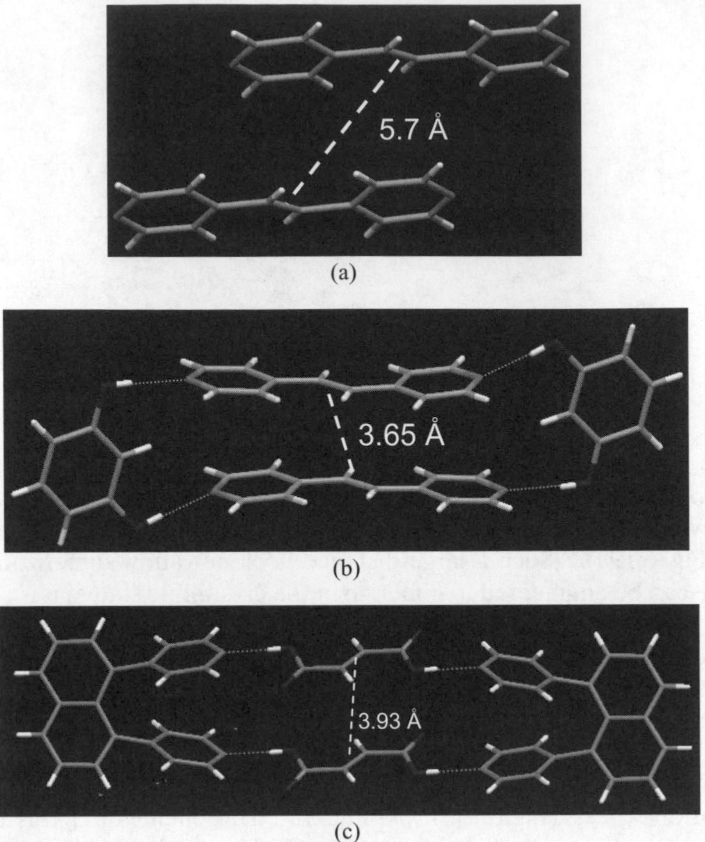

Figure 6.19. (a) Packing arrangement of *trans*-1,2-bis(4-pyridyl)ethylene, the two reactive bonds are far apart. (b) Packing arrangement in the cocrystal of *trans*-1,2-bis(4-pyridyl)ethylene and resorcinol; the two reactive bonds are brought closer. (c) Packing arrangement in the cocrystal of *trans*-1,2-bis(4-pyridyl)ethylene and 1,8-di(pyridine-4-yl)napthalene.

isomers of bispyridyl ethylenes, it is evident that more than one choice of template is needed for the successful alignment and outcome of the photodimerization products with the help of templates (Scheme 6.16). For instance, using resorcinol as template, Macgillivray showed that five of the seven isomers could be aligned in fashion favorable for dimerization but for [3,3] and [3,4] isomers. Similarly, the 1,8-napthalene was capable of aligning only three of the seven possible isomers.

In the venture of a versatile template that could align olefins to a greater degree of predictability, our group came across the template[102] "thiourea" after a detailed analysis of its cocrystals with pyridine bases available in CSD. Thiourea self-assembles into tapes via N–H···S hydrogen-bonded dimer

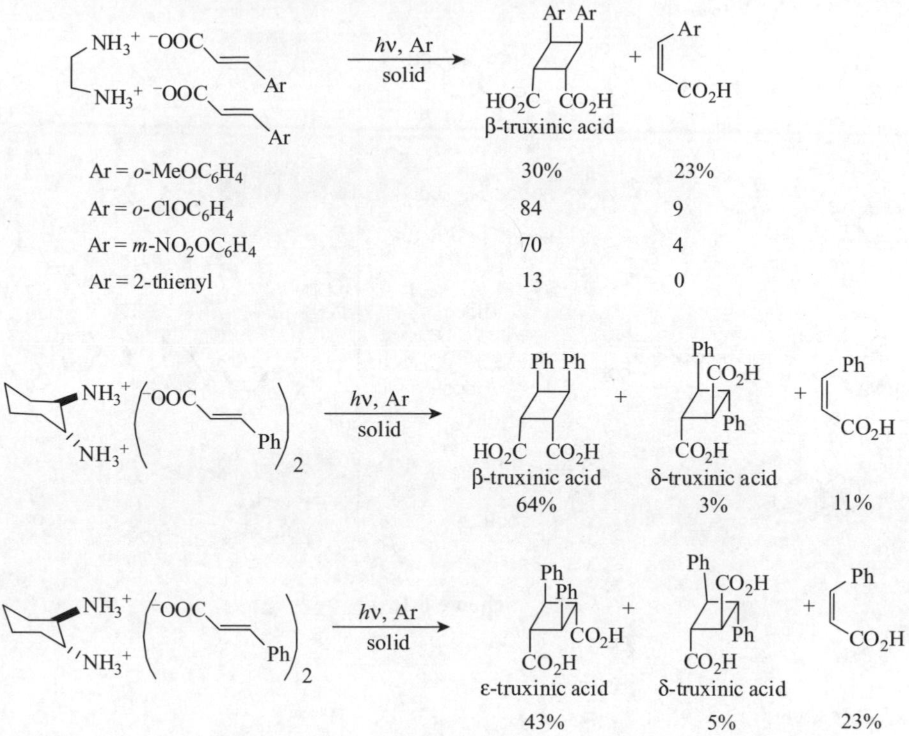

NH_3^+ ^-OOC—CH=CH—Ar / NH_3^+ ^-OOC—CH=CH—Ar —($h\nu$, Ar, solid)→ β-truxinic acid + (Z)-ArCH=CHCO$_2$H

	β-truxinic acid	cis acid
Ar = o-MeOC$_6$H$_4$	30%	23%
Ar = o-ClOC$_6$H$_4$	84	9
Ar = m-NO$_2$OC$_6$H$_4$	70	4
Ar = 2-thienyl	13	0

NH_3^+ / NH_3^+ (^-OOC—CH=CH—Ph)$_2$ —($h\nu$, Ar, solid)→ β-truxinic acid 64% + δ-truxinic acid 3% + (Z)-PhCH=CHCO$_2$H 11%

NH_3^+ / NH_3^+ (^-OOC—CH=CH—Ph)$_2$ —($h\nu$, Ar, solid)→ ε-truxinic acid 43% + δ-truxinic acid 5% + (Z)-PhCH=CHCO$_2$H 23%

Scheme 6.13

$\overset{+}{N}H_3$ / NH_3^+ $\bar{O}OC$—CH=CH—(2-ClC$_6$H$_4$) ×2 (+/– trans) —($h\nu$, solid)→ cyclobutane (2-ClC$_6$H$_4$)$_2$(COOH)$_2$ 7% + (Z)-2-ClC$_6$H$_4$CH=CHCOOH 26%

$\overset{+}{N}H_3$ / NH_3^+ $\bar{O}OC$—CH=CH—(2,4-Cl$_2$C$_6$H$_3$) ×2 (+/– trans) —($h\nu$, solid)→ cyclobutane (2,4-Cl$_2$C$_6$H$_3$)$_2$ (HOOC, COOH) 49% + (Z)-2,4-Cl$_2$C$_6$H$_3$CH=CHCOOH 6%

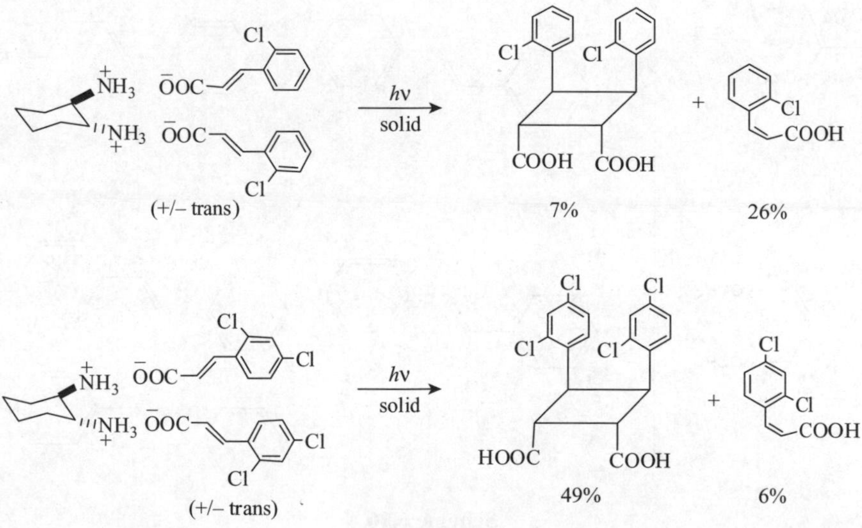

Scheme 6.14

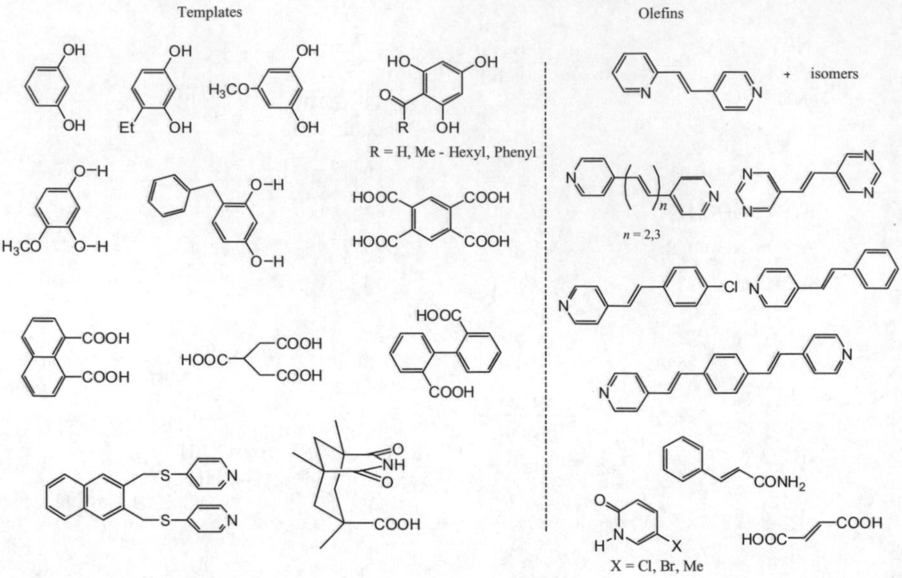

Templates

Olefins

+ isomers

R = H, Me - Hexyl, Phenyl

n = 2,3

X = Cl, Br, Me

Scheme 6.15

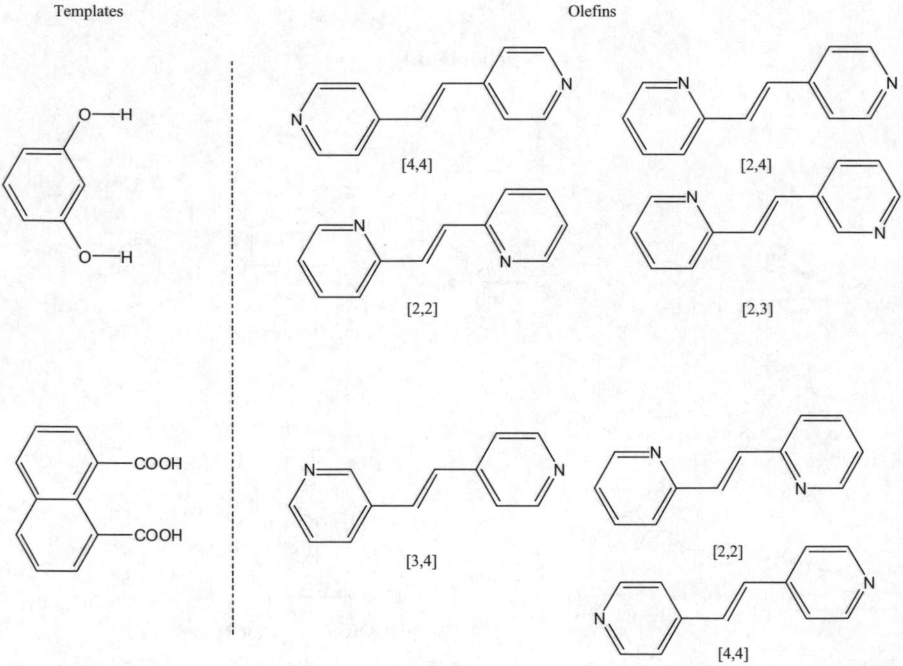

Templates

Olefins

[4,4]

[2,4]

[2,2]

[2,3]

[3,4]

[2,2]

[4,4]

Scheme 6.16

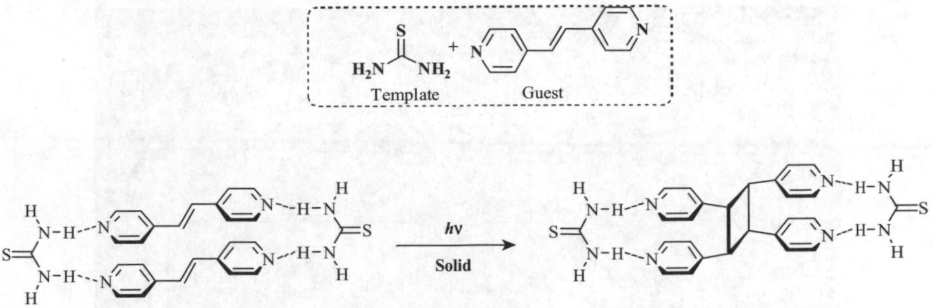

Scheme 6.17

synthon using *syn*-N–H atoms and the *anti*-N–H atoms are capable of holding pyridyl molecules in ~4 Å distance via $N\text{–}H_{(thiourea)}\cdots N_{(pyridyl)}$ hydrogen bond between the tapes. This crystal engineering strategy successfully yielded 1:1 cocrystals of thiourea with *trans*-1,2-bis(4-pyridyl)ethylene (4,4′-bpe). The pairs of olefins in this structure aligned alkene double bonds within 4.15 Å and leads to *syn* dimer in quantitative yield on photoexcitation (Scheme 6.17, Fig. 6.20a). Surprisingly thiourea forms 2:1 cocrystals instead of 1:1 with *trans*-1-(2-pyridyl)-2-(4-pyridyl)ethylene (2,4-bpe) where each thiourea molecule interacts with only one pyridyl nitrogen via an acceptor bifurcated $(N\text{–}H)_{2(thiourea)}\cdots N_{(pyridyl)}$ hydrogen bond (Fig. 6.20b). The reactant molecules here are preorganized in HT with the double bonds separated by 3.95 Å distance and gives *anti*-HT dimer in 100% yield upon photoexcitation. An inspection of this cocrystal structure gives a hint that the thiourea molecule at 2-pyridyl is just an auxiliary, and therefore, 4-stilbazole class of molecules would form a similar packing in 1:1 cocrystal with thiourea where the interdigitation of stilbazole molecules directs them to align in HT fashion. An experimental investigation of this hypothesis revealed the versatility of thiourea template (Scheme 6.18, Fig. 6.21). Six stilbazole derivatives were aligned in HT orientation in their 1:1 thiourea cocrystals and successfully yielded *anti*-HT dimer products in about 95% by photoirradiation. In case of thiourea and cyanostilbazole 1:1 cocrystal, the cyano group accepts hydrogen bond similar to pyridine nitrogen as in thiourea and 4-bpe 1:1 cocrystal and forms a similar packing. The cyanostilbazole molecules in this structure are preorganized in HH fashion, and the photoexcitation of this cocrystals leads to the formation of *syn*-HH dimer with 100% yield. Achieving similar packing patterns with different molecules is still a challenging task in solid state even in a family of compounds because very small changes in the geometrical and chemical properties of a functional group may lead to entirely different supramolecular arrangements. In spite of that, we were able to employ thiourea as a template for nine different pyridyl olefin compounds to yield single regio-isomer with >95% yield.

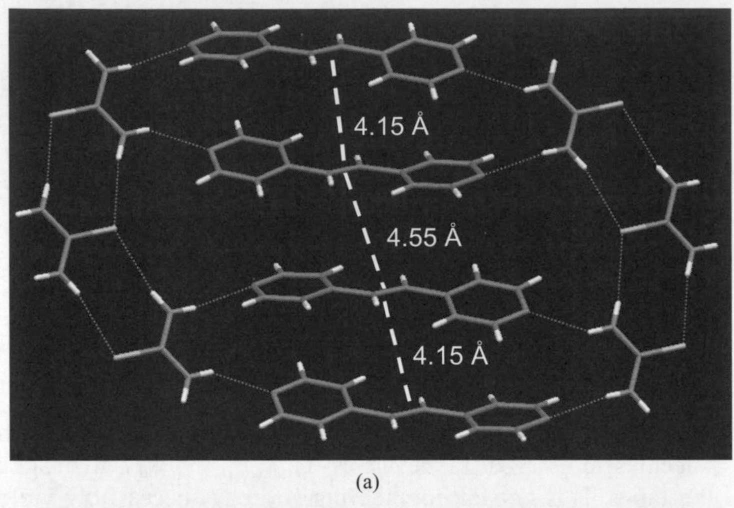

(a)

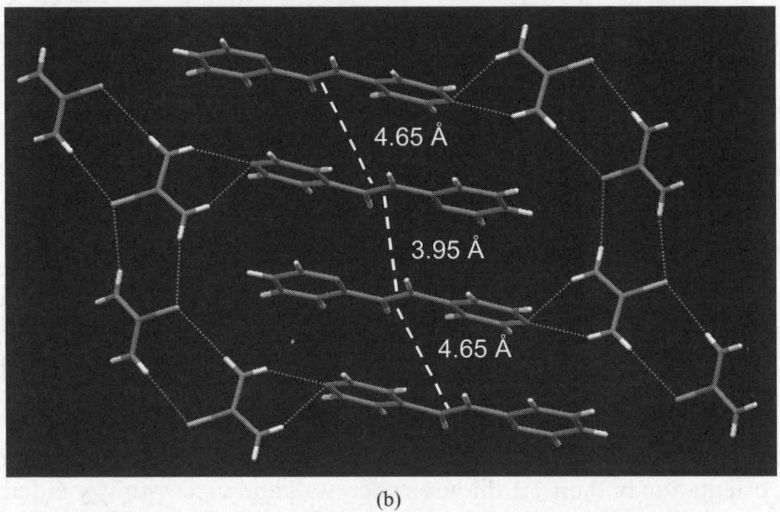

(b)

Figure 6.20. (a) Packing arrangement of thiourea and 4,4′-bpe 1:1 cocrystal. (b) Preorganization of 2,4-bpe molecules in head-to-tail fashion by interdigitation between thiourea tapes in 2:1 cocrystal of thiourea and 2,4-bpe. Thiourea molecules hydrogen bonded to 2-pyridyl nitrogen atoms are not shown for clarity.

Use of Metal Ions to Align the Reactant Molecules for Photodimerization

In the recent past, metal ion coordination has been cleverly exploited by chemists to align reacting partners to bring about a predictable photochemical outcome[21,103–109] In one example, a transition metal ion, in the form of AgI, is used to mediate a [2 + 2] photodimerization in the solid state.[104] Specifically, argentophilic forces, in the form of Ag···Ag interactions, are used to guide

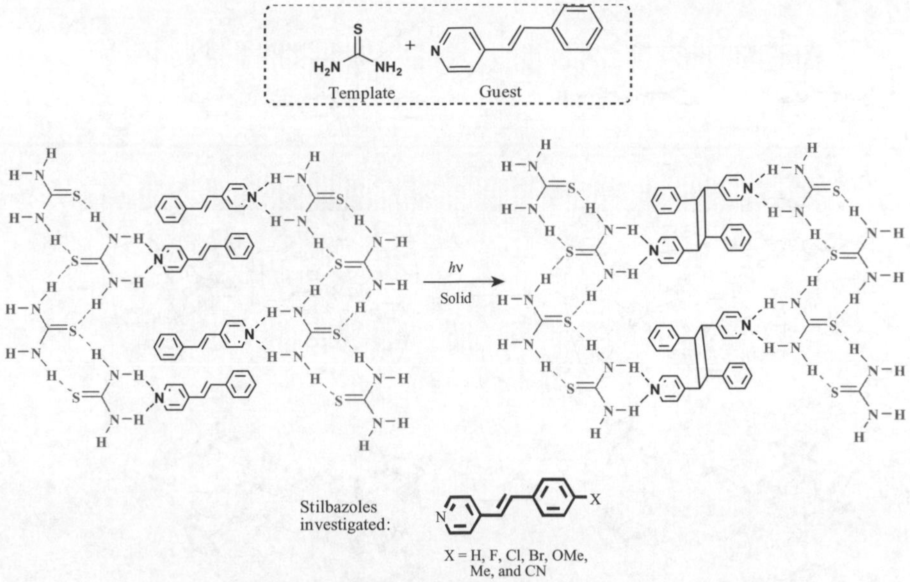

Template Guest

$h\nu$
Solid

Stilbazoles investigated:

X = H, F, Cl, Br, OMe, Me, and CN

Scheme 6.18

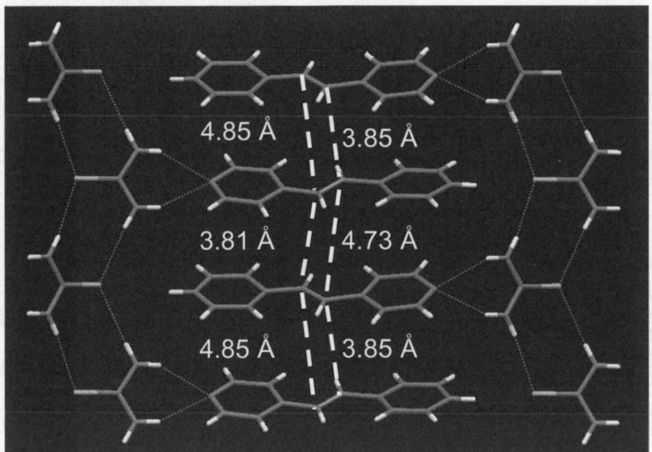

Figure 6.21. Packing arrangement of inclusion complex of stilbazole and thiourea.

stacking of olefins in the dinuclear complex $[Ag_2(4\text{-stilbz})_4][CO_2CF_3]_2$ (where 4-stilbz = *trans*-1-(4-pyridyl)-2-(phenyl)ethylene) **(21)** for a regiocontrolled HH [2 + 2] photodimerization (Scheme 6.19a). The metal and organic components have assembled to form a dinuclear complex sustained by an Ag···Ag force (Ag···Ag distance: 3.41 Å). The crystal structure of the complex (**21**) has two symmetry-independent 4-stilbz molecules preorganized in HH, and one

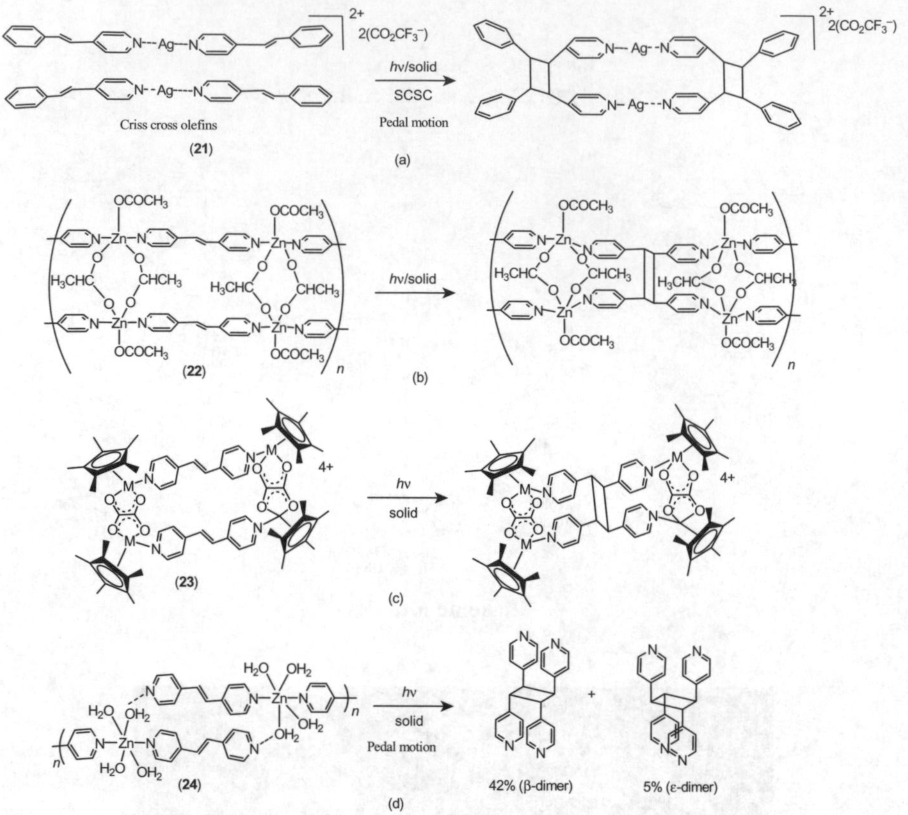

Scheme 6.19

of them is disordered where the olefins are positioned in two different ways in two distinct sites with occupancies (site A 0.89; site B 0.11) and adopt a crisscross arrangement within the molecule. The olefinic C=C of the ordered symmetry-independent molecule is aligned cross to the C=C at site A (89%) separated by 3.82 and 4.24 Å (C to C distances) and parallel to the C=C at site B (11%) separated by 3.82 and 3.86 Å (C to C distances), of disordered 4-stilbz molecule. Based on this structure, up to 11% of photoreaction is straightforward according to the Schmidt criteria. The remaining 89% of the dimerization could only proceed through pedal-like motion of one of the C=C prior to photoreaction yields *rctt*-1,2-bis(4-pyridyl)-3,4-bis(phenyl)cyclobutane product in 100%. The X-ray analysis of a photoreacted crystal shows that the reaction occurs in 100% yield by an SCSC transformation that, in addition to carbon–carbon single-bond making, involves breaking and formation of Ag···Ag and Ag···C interactions, respectively, accompanied by various changes in the crystal. The photoproduct adopts two orientations, with the cyclobutane ring exhibiting site occupancies different than that of the starting olefin's packing arrangement (occupancies: site A 0.60; site B 0.40). These occupancies are consistent with the C=C bonds of the olefins undergoing a pedal-like motion prior to the photoreaction. It is an unusual case because in general, the SCSC reactions occur in cases only when there is a very minimal amount of motion of atoms going from starting material to the photoproduct. The fact that this reaction occurs even when a larger motion of the reactants is necessary to align the reactants conducive for photodimerization shows that there is sufficient reaction cavity around the reaction site to accommodate the motion.

Vittal et al. published another example using a 1D metal coordination polymer, $[\{(H_3CCO_2)(\mu\text{-}O_2CCH_3)Zn\}_2(\mu\text{-bpe})_2]_n$ (**22**) (4,4′-bipyridyl ethylene [bpe]), which has a molecular ladder structure, underwent a [2 + 2] photochemical cycloaddition across the vertical strands into $[\{(H_3CCO_2)(\mu\text{-}O_2CCH_3)Zn\}_2\text{-}(\mu\text{-tpcb})_2]_n$ (tetrakis(4-pyridyl)-cyclobutane [tpcb]) in up to 100% yield.[105] The Zn(II) ions present in the two linear infinite polymers $[\{(O_2CCH_3)Zn\}\text{-}(\text{bpe})]_n$ in complex (**22**) are bridged by two acetate ligands to provide a molecular ladder polymeric structure, and hence, the ethylenic carbon atoms are separated by 3.66 (Scheme 6.19b). This complex did not undergo SCSC transformation as the crystal found to split into two on photoirradiation. However, an isostructural coordination polymer with different counterion $[\{(F_3CCO_2)(\mu\text{-}O_2CCH_3)Zn\}_2(\mu\text{-bpe})_2]_n$ underwent SCSC phototransformation (Fig. 6.22) 100% by UV irradiation in about 3 h. There is no change observed in single-crystal nature, morphology, transparency, or its diffracting power of X-rays.

Another interesting example in this section is the usage of μ-oxalato dinuclear iridium complexes as "organometallic clip" linear templates to direct [2 + 2] photodimerizations in the solid state (Scheme 6.19c).[108] The tetranuclear complexes bearing the 4,4′- bpe [4,4′-bpe = *trans*-1,2-bis(4-pyridyl)ethylene)] ligand, $[Cp^*_4M_4(\mu\text{-bpe})_2(\mu\text{-}\eta^2\text{-}\eta^2\text{-}C_2O_4)_2](OTf)_4$ (M = Ir or Rh) (**23**) were designed and synthesized with two equivalents of AgOTf and subsequent reaction with 4,4′-bpe in methanol at room temperature (ratio = 1:1). When single crystals or a powdered crystalline sample of iridium complex were subjected to UV irradiation, using an Hg lamp for a period of approximately 25 h resulted in dimerization of 4,4′-bpe to give *rctt*-tetrakis(4-pyridyl)cyclobutane in quantitative yield. As with the other examples, optical microscopy revealed that the transparency and shape of the crystals along with their single-crystal nature were retained during the photoreaction, which suggested that the reaction occurred via an SCSC transformation. Final example of metal-ion-induced photodimerization is the coordination complex $[Zn(bpe)_2(H_2O)_4](NO_3)_2{\bullet}8/3H_2O{\bullet}2/3bpe$ (**24**) formed by mixing, containing coordination complex cations $[Zn(bpe)_2{\bullet}(H_2O)_4]^{2+}$ reported by Peedikakkal and Vittal recently.[109] Interestingly, even this coordination complex forms parallel and crisscross double bonds, which undergoes photochemical [2 + 2] cycloaddition in the solid state and produces *rctt*-tetrakis(4-pyridyl)cyclobutane in up to 100% yield of photodimers with *rctt*-tpcb as major and *rtct*-tpcb as minor

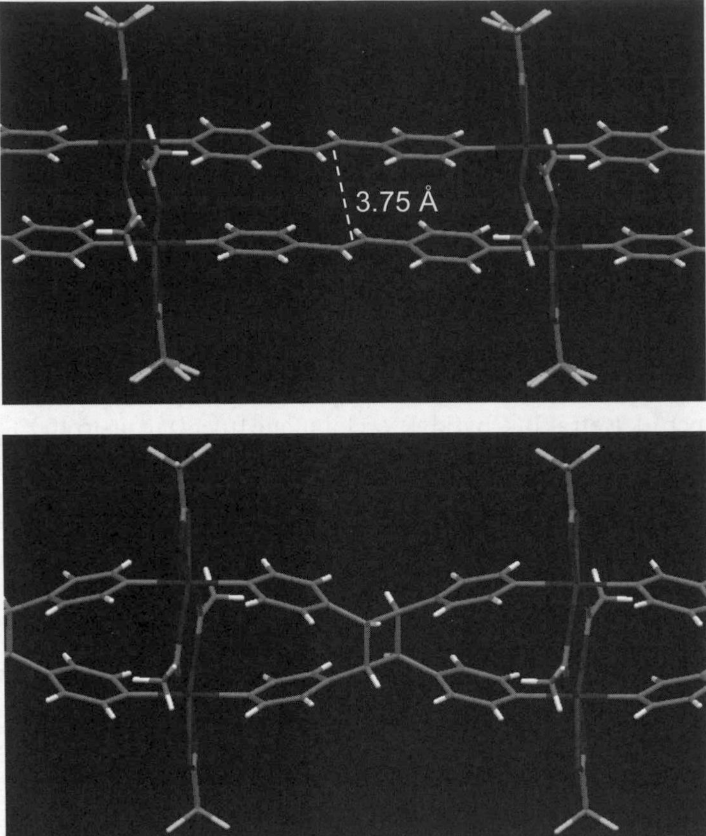

Figure 6.22. (Top) Packing arrangement of portion of the 1D molecular ladder polymeric structure of $[\{(F_3CCO2)(\mu\text{-}O_2CCH_3)Zn\}_2(\mu\text{-bpe})_2]_n$ and (bottom) its single-crystal-to-single-crystal-irradiated photoproduct $[\{(F_3CCO_2)(m\text{-}O_2CCH_3)Zn\}_2(m\text{-tpcb})_2]_n$.

products. Similar to the above examples, bpe ligands with crisscross conformation of C=C bonds appear to undergo pedal-like motion prior to photodimerization. Grinding single crystals to powder accelerates the pedal motion of crisscrossed olefins in the bpe ligands to parallel alignment and provides the *rctt*-cyclobutane stereoisomer quantitatively. When the single crystal of the complex was subjected to UV light for 25 h, 46% photochemical conversion was observed, of which 39% is due to *rctt*-tetrakis(4-pyridyl)-1,2,3,4-cyclobutane (*rctt*-tpcb, 2a) and 7% due to *rtct*-tpcb isomer (Scheme 6.19), and was monitored by ^{1}H nuclear magnetic resonance (NMR) spectroscopy. However, the percentage of minor isomer formed is much lower than the expected based on the orientation of C=C bonds and the C=C distances found. This was explained by assuming that C=C bonds of bpe ligands undergo pedal-like motion prior to photodimerization. Nevertheless, such motions are easier if the pair of molecules with C=C bonds are held on one side by weak interactions while the other side is left free. The sluggish reactivity of the complex was attributed to difficulty of pedal motion in the single crystals due to the presence of coordinative bonds on one side and hydrogen bonds on the other. Interestingly, when the single crystals were ground for 5 min to powder and subjected to UV irradiation for 4 h, 88% conversion of C=C bonds to cyclobutane rings was observed, and 100% conversion was achieved when UV irradiation was prolonged to 25 h, showing evidences that the grinding causes acceleration of the pedal motion.

Templation Using Hosts: Aligning the Olefins with Supramolecular (Organic and Inorganic) Nanocavities

Reactions within highly rigid medium such as crystalline state to a highly complex biological system (ex: protein) has inspired scientists over the years to conduct selective reactions using confined medium such as micelles, water-soluble polymers, dendrimers and synthetic hosts such as cyclodextrins (CDs), cucurbiturils, and octa acid. Bimolecular and unimolecular (not covered in this chapter) photoreactions done within some of these hosts have shown remarkable selectivities that are comparable to selectivity obtained in the crystalline state and are therefore worth covering under the templation approach. The nanocavities that are discussed in this section are tripalmitin (TP) monolayers, clay, faujasite Y zeolite, cucurbit-[*n*]-urils, and CDs. Unless and otherwise stated, the photoreactions in these cavities are conducted in water. The reaction vessels that we utilize to carry out chemistry are only slightly larger in size (nano) than those of the reactant molecules. One of the drawbacks of the crystalline state medium being a confined medium is the lack of flexibility of the reaction cavity, and we indeed discuss in the earlier sections with examples such as 4-hydroxy-3-nitromethylcinnamate whose neighboring molecules are aligned perfectly for photodimerization (alkene distance being 3.78 Å) but was photostable in the solid state. In the crystal structure, the molecules are linked by hydrogen bonds to form a sheet-like structure preventing easy spatial movement of the atoms of the double bond in the lattice for the reaction to proceed. These types of examples emphasize the need for void space around the reacting partners and for flexibility in intermolecular forces that hold the crystal. In crystals, the presence of strong intermolecular forces not only fails in giving directionality but also completely arrest the reaction in some cases. It is clear from the above examples that selectivity can be achieved in systems based on flexible rather than rigid reaction cavities and some of the supramolecular hosts have done a fantastic job on not only confining the reactant molecules but also producing excellent selectivity.

Heterogeneous Media: Monolayers, Clay, and Zeolite Materials. Over three decades ago, Quina and Whitten investigated photochemistry of surfactant

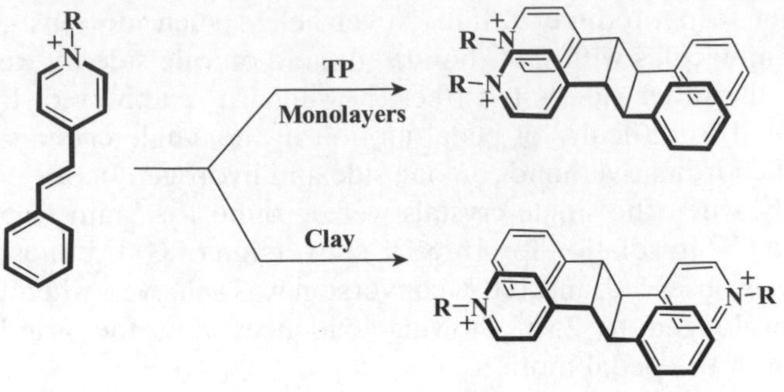

TP
Monolayers
Clay

Scheme 6.20

molecules in different media.[110] Since many of these compounds may be readily incorporated into monolayers, Whitten demonstrated that the oriented, semirigid environment of monolayer assemblies could be used to direct or modify their photochemistry. An example is the investigation on the photochemistry of the *p*-chlorobenzensulfonate and bromide salts of *N*-octadecyltrans-4-stilbazole. Irradiation of 4-stilbazole (1 g/100 mL) in 0.1 M HCl gave predominantly (ca. 80%) the *anti*-HT dimer (Scheme 6.20). However, the solid-state irradiation of *p*-chlorobenzene sulfonate leads to the formation of [2 + 2] dimer, and of the four possible dimers, only one photoproduct (*syn*-HH dimer) was obtained as the sole photoproduct. However, the bromide salt was rendered photostable and did not show any sign of photoproduct formation even at prolonged irradiations. Both of the above stilbazole salts form monolayers in mixtures with TP and are incorporated into monolayer assemblies. These incorporated assemblies have a long wavelength absorption maximum (345 nm) comparable to that in solution, but the band is quite broad on the long wavelength side, tailing well out past 430 nm. Analogous to the solid state, these assemblies exhibit a green fluorescence with a maximum of about 490 nm. In contrast to the solid state, the photoreaction of both the salts in TP assemblies resulted in a photodimer due to the reduced steric and like-charge repulsions. This result suggests that in the solid state, preferential association of hydrophobic groups may be important enough to offset unfavorable like-charge interactions in the hydrophilic zones. Similarly, Sawaki and coworkers[111,112] were interested in the photochemical behavior of intercalated substrates in clay layers. For example, smectite clays have anionic sites and sodium ions in the interlayer; the sodium ions are easily exchanged with organic cations. They observed that (*E*)-4-[2-(*p*-substituted pheny1)ethenylpyridinium ions intercalated in clay layers gave selectively *syn*-HT cyclodimers (Scheme 6.20).[111] These authors applied this technique to the mixture of stilbazolium salts to obtain crossbred cyclodimer products.[111,112] Addition of stilbazolium salts in amounts of over 20% equivalents on aqueous clay colloid resulted in the precipitation of a pale yellow gel. A suspended solution of a mixture of clay intercalated methyl- and cyano-stilbazolium salts, each 1 mmol dm^{-3} in

10% aqueous methanol, was irradiated with a UV lamp through a Pyrex filter (>300 nm). Homodimers and crossed cyclodimers were formed together with the corresponding (*Z*) isomers. However, the regioselectivity was still in favor of mainly *syn*-HT dimer.

Faujasite-type zeolite (MY), an inorganic nanoporous host, is a versatile host in controlling reactions of a large variety of organic guest molecules. The recognition of the nonpassive nature of the zeolite cavity has made it possible to control photoprocesses of included organic molecules. The faujasite Y zeolite is composed of alumino-silicate mineral with "open" structures called "supercage" that accommodates readily interchangeable alkali metal (Li-Cs) cations. Several chemists has made use of the confined cavity provided by the MY zeolite to study the motion and proximity of reactants and corresponding photoproducts with the supercage of the faujasite zeolite. Due to the presence of the three-dimensional network and cations the MY zeolites have been shown to confine the guest molecules within the supercage using weak interactions and thereby inducing selectivity on the photoproducts that are not easily accessible in the isotropic solvents. Ramamurthy et al. demonstrated that acenapthylenes could undergo bimolecular [2 + 2] photoprocesses in the supercage cavities of NaY.[113] They showed that a heavy atom perturber and an organic molecule could be "closeted" within a constrained environment, such as zeolites, and the interactions between them could be strengthened, leading to stronger heavy atom effects. Acenaphthylene upon excitation undergoes photodimerization to give both *cis* and *trans* dimers. Of the two, the *trans* dimer is derived only from the triplet state, while the *cis* dimer is from both the triplet and excited singlet states. Due to the light or heavy atom influencing the intersystem crossing rate, they found that only *cis* dimer was obtained in LiY, while in RbY, a mixture of *trans* and *cis* dimers was formed in nearly 1:1 ratio due to the heavy atom effect (Scheme 6.21). Consistent with

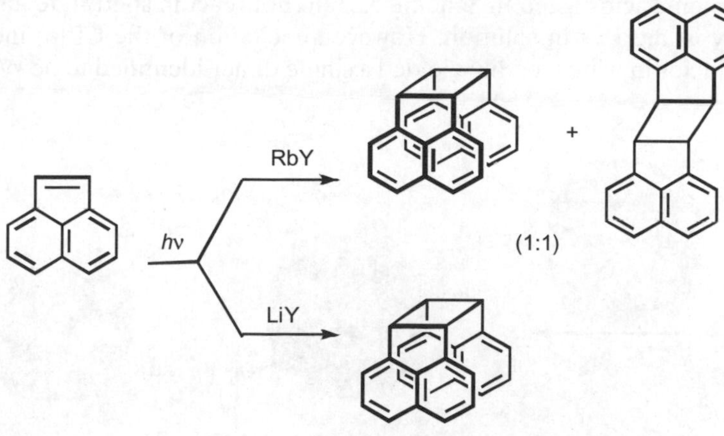

RbY

hν

(1:1)

LiY

Scheme 6.21

the conclusion that the *trans* dimer comes from the triplet state, irradiation of the RbY samples under oxygenated conditions gave only the *cis* dimer. The photoreactions were conducted in NaY as solid irradiations; reactant "loaded" zeolite samples were transferred to quartz/pyrex tubes and photolyzed directly as a powder or as slurry irradiations; and the loaded zeolite samples were transferred into fresh hexane and photolyzed as hexane slurry with continuous stirring. Similarly, Schuster and Turro demonstrated that a relatively high degree of regioselectivity in enone photodimerization can be achieved in zeolites, due to size constriction effects in the supercages of X- and Y-type faujasites as well as the complexing ability of the charge-compensating cation of the zeolite on included guests. In the case of cyclohexenone and cyclopentanone photodimerization in faujasites, the HH dimer was shown to be consistently preferred over the HT dimer immaterial of the different cation-exchanged MY zeolites. In all the examples discussed above in different reaction media, it is the flexibility of the media that provides enough motion to the reactant molecules so as to align in a favorable fashion for a bimolecular reaction to take place, but not as much as a solution media where the selectivity is compromised due to high degree of freedom available for the reactant molecules.

Cucurbiturils. Cucurbiturils that have reaction cavities similar to CDs form water-soluble host–guest complexes. Kim and our group have independently utilized the cavities of cucurbiturils to prealign aryl alkenes for photodimerization reactions. Kim and coworkers were the first to demonstrate that the cavity of cucurbituril [8] (CB[8]) can accommodate and orient the olefinic bonds of diamino stilbene hydrochloride to yield *syn* dimer in high yields (Scheme 6.22).[114] The selective formation of *syn* dimer within the host was proposed to be due to the arrangement of the olefins within the cavity that favors *syn* dimer formation. A few years ago, our group demonstrated the ability of cucurbiturils to dimerize alkenes that are completely photostable in the solid state.[115–118] The three cinnamic acids listed in Scheme 6.23 do not react in solid state and geometrically isomerizes in solution. However, excitation of the CB[8] included cinnamic acids in water readily yielded a single dimer identified to be *syn*-HH.

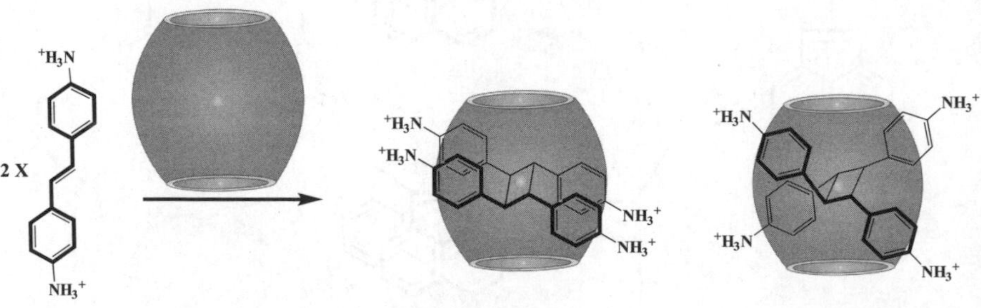

Scheme 6.22

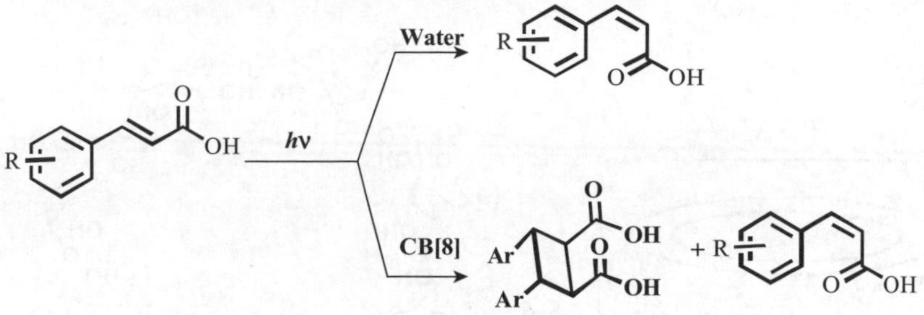

R	*syn*-HH Dimer	*cis*-Cinnamic Acid
3-MeO	72	28
3-Me	83	17
4-MeO	72	28

Scheme 6.23

While the confined cavity may help bring the two reactive olefin molecules together, it is the weak intermolecular forces that orient them in a proper geometry for dimerization. We believe that the inclusion of the two olefins in a HH fashion is a combination of π—π and hydrophobic interactions. In the HH arrangement, the two acid groups face water and the two phenyl groups would be forced to stay parallel favoring π—π interaction. Similarly, we demonstrated the ability of CB[8] to dimerize bispyridyl ethylenes with >85% yield of the photodimer product over the isomerization photoproduct. It was very well established by our group that CB[8] had ideal dimensions to align the olefins and promote photodimerizations very effectively, and even if the cavities with reduced dimensions such as CB[7] was used, only the corresponding isomerization photoproducts were obtained exclusively. The dimension of CB[8], which is comparable to γ-CD, is represented in Figure 6.23.

Just to illustrate the confining nature and flexibility provided by CB[8], the crystal structure of the inclusion complex of 4,4′-*trans*-bispryidyl ethylene and CB[8] (2:1 ratio) clearly confirmed the inclusion of the two guests within the cavity of CB[8]. And irradiation of these inclusion complex crystals for over 24 h resulted in no formation of photodimer. The crystals were completely photoinert. Analyzing closely, the crystal structure revealed that the olefinic bonds are placed at a distance of 4.8 Å and the olefins are aligned in "slipped

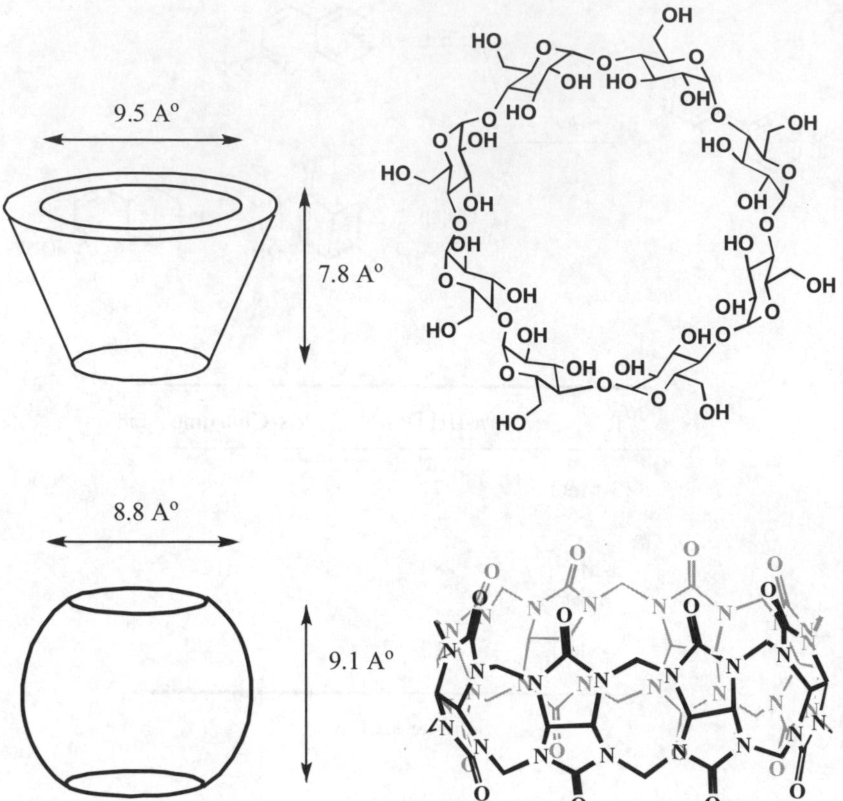

Figure 6.23. Pictorial and chemical representations of γ-cyclodextrin and cucurbit-[8]-uril.

fashion" (Fig. 6.24). This observation is in accordance with the "topochemical principle of solid-state reactivity" proposed by Schmidt et al., which states that solid-state photodimerizations can occur only if the olefinic bonds are close to each other by at least 4.2 Å. But when these crystals were dissolved in water and irradiated, this resulted in nearly 90% formation of *syn* dimer. Thus, the flexibility of the guest molecules in the water medium though confined within rigid cucurbituril walls underwent photodimerization effectively. We have also established that the CB[8] has the capability to align and dimerize, upon photoexcitation, the olefins such as stilbazoles, 1,4-bis(4-pyridyl)butadiene hydrochloride, and their corresponding diene analogs (Scheme 6.24).

CDs. CDs are doughnut-shaped cyclic oligosaccharides consisting of six, seven, or eight glucose units (α-, β-, and γ-CDs, respectively) with a hydrophobic inner cavity and a hydrophilic outside. These cavities have internal diameters varying from 4.7 to 8.3 Å, permitting them to form inclusion complexes

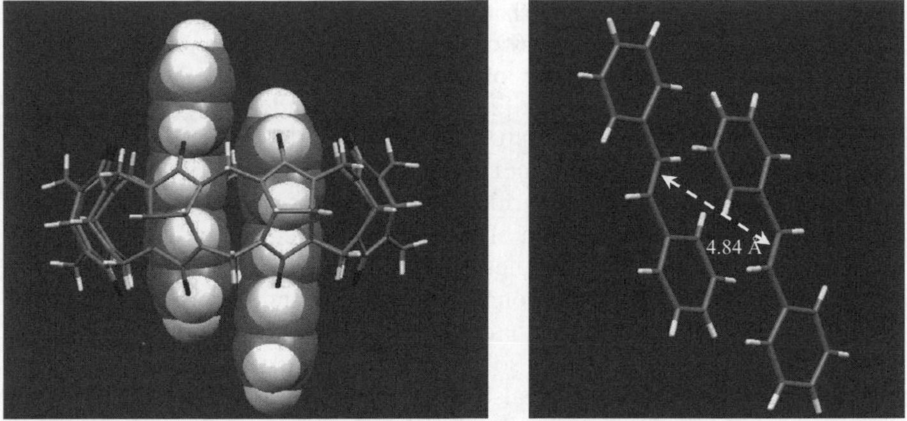

Figure 6.24. (Left) X-ray crystal structure of 2:1 inclusion complex of CB[8] and 4,4′-bpe and (right) the intermolecular alkene–alkene distance of the 4,4′-bpe within the inclusion complex.

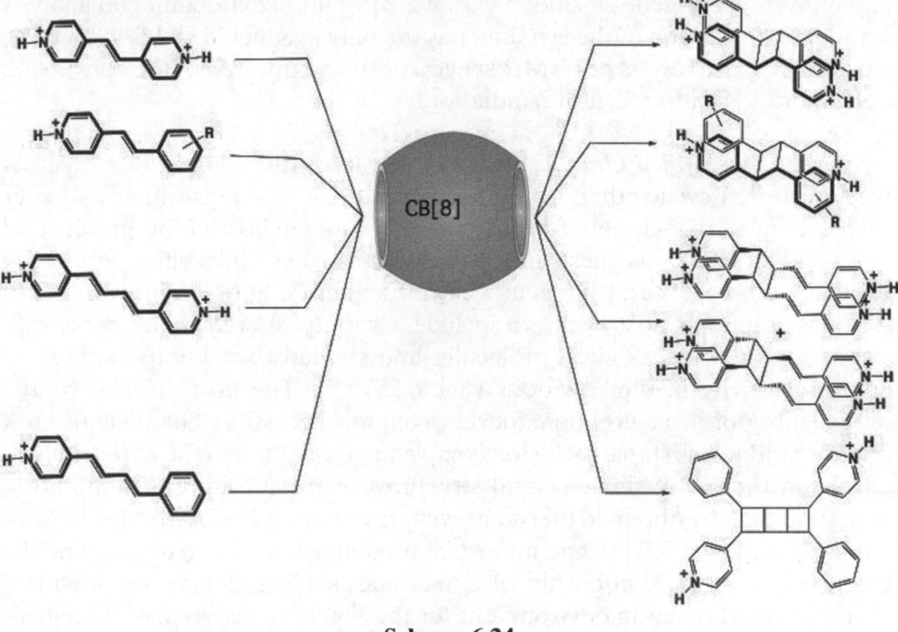

Scheme 6.24

with a variety of guest molecules. The depth of the binding cavity of all three CDs is almost the same (7.8 Å) (Fig. 6.23). Complexation between CDs and substrates in solution has been successfully exploited to induce selectivity in some unimolecular and bimolecular photochemical reactions. Tamaki and coworkers were the first to explore the utility of the CD cavity in effecting regio- and stereoselective dimerization of guest molecules like anthracene-2-sulfonate and anthracene-2-carboxylate.[119] Later, Moorthy et al.[120] demonstrated the photodimerization of acenaphthylene and coumarin upon complexation within γ-CD cavity. [2 + 2] Photocycloaddition of a symmetrical stilbene derivative to yield a mixture of dimers in the presence of γ-CD has also been reported by Herrmann et al.[121] Similarly, Pitchumani and coworkers showed that the *trans*-2-styrylpyridine, photolysis of a γ-CD complex in the solid state leads to the formation of the *syn*-HT dimer.[122] Our studies of [4,4], [3,3], and [2,2] bispyridyl ethylenes, included within CB[8] discussed in the earlier section, prompted us to conduct the same reaction within γ-CD as the cavity size of γ-CD is similar to that of CB[8].[115] Since the complex formation of guests with γ-CD is primarily solvatophobic interactions, unlike in case of CB[8] where dipolar interaction also plays a major role, we used the guests in their neutral forms. Stirring a solution of bispyridyl ethylenes dissolved in dichloromethane (5 mL of 20 mM solution) with an aqueous solution of γ-CD (5 mL of 10 mM) yielded the complex as a precipitate. The precipitate was dried, and the dry powder of inclusion complex was used for photochemical reactions. Irradiation of the powder of all the three complexes ([4,4], [3,3], and [2,2] bispyridyl ethylenes included within CB[8]) upon extraction and analysis showed the formation of the *syn* dimer as the only product in yields 98%, 95%, and 97%, respectively as per NMR spectra of the mixture. Also, the conversion was around 95% after 12 h of irradiation.

Porous Self-Assembled Organic Frameworks. Recently, Shimizu et al. synthesized a bis-urea cyclic ether, which forms columnar channels (inner diameter ~9–10 Å) via self-assembly of urea hydrogen bonding host in the presence of acetic acid molecules as guests in these columns in crystals.[123] This apohost can reversibly bind the guest molecules in self-assembly. Subsequently, they utilized this columnar host frame to include a variety of cyclic and acyclic α,β-unsaturated ketones as guest molecules and studied their photoreactions to obtain selectivity in dimerizations (Fig. 6.25).[124,125] The host, formed by the self-assembly of a bis-urea macrocycle, contains accessible channels of ~6 Å diameter and forms stable inclusion complexes. Even though the authors could not obtain the X-ray single-crystal structures of these bis-urea–enone host–guest systems, they obtained thermogravimetric analysis (TGA), NMR, powder X-ray diffraction (PXRD), and molecular modeling in order to understand the structure–reactivity relationship of enone guests. Host **25** crystals provide a robust confined reaction environment for the highly selective [2 + 2] cycloaddition of 3-methyl-2 cyclopentenone, 2-cyclohexenone, and 2-methyl-2-cyclopentenone, forming their respective *exo* HT dimers in high conversion (98%, 96%, and 80%, respectively). The products are readily extracted from the self-assembled host, and the crystalline host can be efficiently recovered and reused. Molecular modeling studies performed indicated that the origin of the observed selectivity is due to the excellent match between the size and shape of these guests to dimensions of the host channel and to the preorgani-

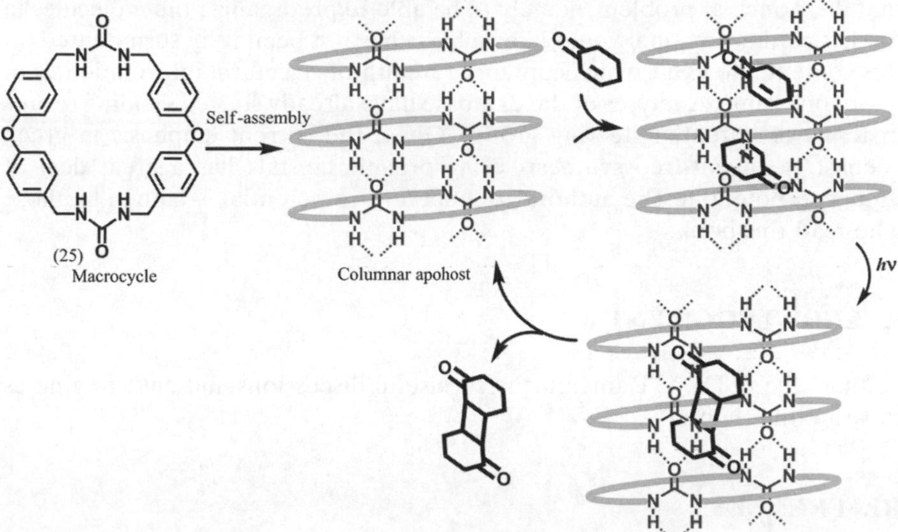

Figure 6.25. Bis-urea macrocycle **25** self-assembles into tubular structures that can bind suitable enone guest molecules. Irradiation of the enone in the confined environment selectively yields the head-to-tail photodimer.

zation of neighboring enones into favorable reaction geometries. Small substrates, such as acrylic acid and methylvinylketone, were bound by the host and were protected from photoreactions. Larger substrates, such as 4,4-dimethyl-2-cyclohexenone and mesityl oxide, do not undergo selective [2 + 2] cycloaddition reactions.

CONCLUDING REMARKS

Photodimerization of olefins in the solid state has been subjected to extensive and systematic study by several groups. Since the in-depth studies on cinnamic acids by Schmidt and coworkers, several groups have explored photodimerization of olefins in the solid state. These studies have repeatedly shown that the stereochemistry of dimer obtained in solid state is different from that in solution. These examples emphasize the importance of product selectivity in the solid state. Still, mechanistic studies are not that many, and this problem needs more attention. Recent progress in solving crystal structures of molecules present in the excited state should allow us gain a deeper understanding of the progress of photoreactions in the crystalline state. In spite of the active efforts put forth by several groups and some significant developments in solid-state photochemistry, it must be conceded that this area has not yet attracted the attention of mainstream organic chemists. This is mainly owing to the fact that the principal problem, namely to be able to preorganize the molecules in the lattice the way one would like to have, has not been fully surmounted. In this context, the CSD, which contains a wealth of structural information, is of enormous value. Analyses of the database have already started yielding results in terms of potential steering groups. Given the current emphasis in green chemistry, solvent-free synthesis using organic crystals has a great deal of untapped potential. The authors are hopeful this potential is tapped by those who read this book.

ACKNOWLEDGMENT

Authors thank Dr. V. Ramamurthy for useful discussions and encouraging us to write this chapter.

REFERENCES

1. Libermann, C. *Chem. Ber.* **1889**, *22*, 124, 782.
2. Kohlshutter, H.W.Z. *Anor. Allg. Chem.* **1918**, *105*, 121.
3. Cohen, M.D.; Schmidt, G.M.J. *J. Chem. Soc.* **1964**, 1996.
4. Cohen, M.D.; Schmidt, G.M.J.; Sonntag, F.I. *J. Chem. Soc.* **1964**, 2000.
5. Schmidt, G.M.J. *J. Chem. Soc.* **1964**, 2014.
6. Schmidt, G.M.J. *Pure Appl. Chem.* **1971**, *27*, 647.
7. Ramamurthy, V.; Venkatesan, K. *Chem. Rev.* **1987**, *87*, 433.
8. Natarajan, A.; Ramamurthy, V. In: Rappoport, Z.; Liebman, J.F., editors. *The Chemistry of Cyclobutanes*. Chichester: John Wiley & Sons, **2005**, p. 807.
9. Sonoda, Y. In: Horspool, W.; Lenci, F., editors. *CRC Handbook of Organic Photochemistry and Photobiology*, 2nd edition. Boca Raton, FL: CRC Press, **2003**, p. 73.
10. Venkatesan, K.; Ramamurthy, V. In: Ramamurthy, V., editor. *Photochemistry in Organized and Constrained Media*. New York: VCH, **1991**, p. 133.
11. Vishnumurthy, K.; Row, T.N.G.; Venkatesan, K. In: Ramamurthy, V.; Schanze, K., editors. *Molecular and Supramolecular Photochemistry*, Vol. 8. New York: Marcell Dekker, **2001**, p. 427.
12. MacGillivray, L.R. *CrystEngComm* **2002**, *4*, 37.
13. Toda, F. *CrytEngComm* **2002**, *4*, 215.
14. Keating, A.E.; Garcia-Garibay, M.A. In: Ramamurthy, V.; Schanze, K.S., editors. *Molecular and Supramolecular Photochemistry*. New York: Marcel Dekker, **1998**, p. 195.
15. Ito, Y. In: Ramamurthy, V.; Schanze, K.S., editors. *Molecular and Supramolecular Photochemistry*. New York: Marcel Dekker, **1999**, p. 1.
16. Ito, Y. *Synthesis* **1998**, 1.
17. Schmidt, G.M.J. *Reactivity of the Photoexcited Organic Molecule*. New York: Interscience, **1967**, p. 227.

18. Green, B.S.; Yellin, A.R.; Cohen, M.D. *Topics in Stereochemistry*, Vol. 16. Eliel, E.L.; Wilen, S.H.; Allinger, N.L., editors. New York: Wiley, **1986**, p. 131.
19. Theocharis, C.R. In: Patai, S.; Rappoport, Z., editors. *The Chemistry of Enones*. New York: John Wiley & Sons, **1989**, p. 1133.
20. MacGillivray, L.R. *Acc. Chem. Res.* **2008**, *41*, 280.
21. Nagarathinam, M.; Peedikakkal, A.M.P.; Vittal, J.J. *Chem. Commun.* **2008**, 5277.
22. Leiserowitz, L.; Schmidt, G.M.J. *Acta Cryst.* **1965**, *18*, 1058.
23. Hanson, A.W. *Acta Cryst.* **1975**, *B31*, 1963.
24. Nakanishi, F.; Nakanishi, H.; Tsuchiya, M.; Hasegawa, M. *Bull. Chem. Soc. Jpn.* **1976**, *49*, 3096.
25. Nakanishi, H.; Hasegawa, M.; Mori, T. *Acta Cryst.* **1985**, *C41*, 70.
26. Enkelmann, V.; Wegner, G.; Novak, K.; Wagener, K.B. *J. Am. Chem. Soc.* **1993**, *115*, 10390.
27. Fernandes, M.A.; Levendis, D.C.; Schoening, F.R.L. *Acta Crystallogr.* **2004**, *B60*, 300.
28. Fernandes, M.A.; Levendis, D.C. *Acta Crystallogr.* **2004**, *B60*, 315.
29. Ramasubbu, N.; Row, T.N.G.; Venkatesan, K.; Ramamurthy, V.; Rao, C.N.R. *J. Chem. Soc. Chem. Commun.* **1982**, 178.
30. Gnanaguru, K.; Murthy, G.S.; Venkatesan, K.; Ramamurthy, V. *Chem. Phys. Lett.* **1984**, *109*, 255.
31. Bhadbhade, M.M.; Murthy, G.S.; Venkatesan, K.; Ramamurthy, V. *Chem. Phys. Lett.* **1984**, *109*, 259.
32. Gnanaguru, K.; Ramasubbu, N.; Venkatesan, K.; Ramamurthy, V. *J. Org. Chem.* **1985**, *50*, 2337.
33. Arjunan, P.; Gnanaguru, K.; Ramamurthy, V.; Venkatesan, K. *Stud. Org. Chem.* **1985**, *20*, 347.
34. Kumar, V.A.; Begum, N.S.; Venkatesan, K. *J. Chem. Soc. Perkin 2* **1993**, 463.
35. Moorthy, J.N.; Samant, S.D.; Venkatesan, K. *J. Chem. Soc. Perkin 2* **1994**, 1123.
36. Vishnumurthy, K.; Row, T.N.G.; Venkatesan, K. *J. Chem. Soc. Perkin 2* **1996**, 1475.
37. Vishnumurthy, K.; Row, T.N.G.; Venkatesan, K. *J. Chem. Soc. Perkin 2* **1997**, 615.
38. Vishnumurthy, K.; Row, T.N.G.; Venkatesan, K. *Tetrahedron* **1999**, *55*, 4095.
39. Vishnumurthy, K.; Row, T.N.G.; Venkatesan, K. *Photochem. Photobiol. Sci.* **2002**, *1*, 799.
40. Green, B.S.; Schmidt, G.M.J. *Tetrahedron Lett.* **1970**, *11*, 4249.
41. Nakanishi, H.; Jones, W.; Thomas, J.M. *Chem. Phys. Lett.* **1980**, *71*, 44.
42. Jones, W.; Nakanishi, H.; Theocharis, C.R.; Thomas, J.M. *J. Chem. Soc. Chem. Commun.* **1980**, 610.
43. Nakanishi, H.; Jones, W.; Thomas, J.M.; Hursthouse, M.B.; Motevalli, M. *J. Chem. Soc. Chem. Commun.* **1980**, 611.
44. Thomas, J.M. *Nature* **1981**, *289*, 633.
45. Nakanishi, H.; Jones, W.; Thomas, J.M.; Hursthouse, M.B.; Motevalli, M. *J. Phys. Chem.* **1981**, *85*, 3636.
46. Jones, W.; Ramdas, S.; Theocharis, C.R.; Thomas, J.M.; Thomas, N.W. *J. Phys. Chem.* **1981**, *85*, 2594.

47. Jones, W.; Theocharis, C.R.; Thomas, J.M.; Desiraju, G.R. *J. Chem. Soc. Chem. Commun.* **1983**, 1443.
48. Theocharis, C.R.; Desiraju, G.R.; Jones, W. *J. Am. Chem. Soc.* **1984**, *106*, 3606.
49. Theocharis, C.R.; Jones, W.; Thomas, J.M.; Motevalli, M.; Hursthouse, M.B. *J. Chem. Soc. Perkin 2* **1984**, 71.
50. Kearsley, S.K.; Desiraju, G.R. *Proc. R. Soc. Lond. A* **1985**, *397*, 9.
51. Theocharis, C.R.; Clark, A.M.; Hopkin, S.E.; Jones, P.; Perryman, A.C.; Usanga, F. *Mol. Cryst. Liq. Cryst.* **1988**, *156*, 85.
52. Desiraju, G.R.; Bernstein, J.; Kishan, K.V.R.; Sarma, J.A.R.P. *Tetrahedron Lett.* **1989**, *30*, 3029.
53. Honda, K.; Nakanishi, F.; Feeder, N. *J. Am. Chem. Soc.* **1999**, *121*, 8246.
54. Honda, K. *Bull. Chem. Soc. Jpn.* **2002**, *75*, 2383.
55. Kanagapushpam, D.; Venkatesan, K.; Ramamurthy, V. *Acta Cryst.* **1987**, *C43*, 1128.
56. Venugopalan, P.; Venkatesan, K. *Acta Cryst.* **1990**, *B46*, 826.
57. Kumar, V.A.; Venkatesan, K. *J. Chem. Soc. Perkin 2* **1993**, 2429.
58. Cohen, M.D. *Angew. Chem. Int. Ed. Engl.* **1975**, *14*, 386.
59. Cohen, M.D. *Mol. Cryst. Liq. Cryst.* **1979**, *50*, 1.
60. Nakanishi, H.; Ueno, K. *Acta Cryst.* **1976**, *B32*, 3352.
61. Nakanishi, H.; Parkinson, G.M.; Jones, W.; Thomas, J.M.; Hasegawa, M. *Isr. J. Chem.* **1979**, *18*, 261.
62. Ariel, S.; Trotter, J. *Acta Cryst. C* **1984**, *C40*, 2084.
63. Murthy, G.S.; Arjunan, P.; Venkatesan, K.; Ramamurthy, V. *Tetrahedron* **1987**, *43*, 1225.
64. Pfoertner, K.H.; Englert, G.; Schoenholzer, P. *Tetrahedron* **1987**, *43*, 1321.
65. Kearsley, S.K. In: Desiraju, G.R., editor. *Organic Solid State Chemistry*. Amsterdam: Elsevier, **1987**, p. 69.
66. Craig, D.P.; Lindsay, R.N.; Mallet, C.P. *Chem. Phys.* **1984**, *89*, 187.
67. Craig, D.P.; Mallet, C.P. *Chem. Phys.* **1982**, *65*, 129.
68. Collins, M.A.; Craig, D.P. *Chem. Phys.* **1981**, *54*, 305.
69. Ito, Y.; Hosomi, H.; Ohba, S. *Tetrahedron* **2000**, *56*, 6833.
70. Ohba, S.; Hosomi, H.; Ito, Y. *J. Am. Chem. Soc.* **2001**, *123*, 6349.
71. Natarajan, A.; Mague, J.T.; Venkatesan, K.; Ramamurthy, V. *Org. Lett.* **2005**, *7*, 1895.
72. Novak, K.; Enkelmann, V.; Wegner, G.; Wagener, K.B. *Angew. Chem. Int. Ed. Engl.* **1993**, *32*, 1614.
73. Cohen, M.D.; Elgavi, A.; Green, B.S.; Ludmer, Z.; Schmidt, G.M.J. *J. Am. Chem. Soc.* **1972**, *94*, 6776.
74. Cohen, M.D.; Green, B.S.; Ludmer, Z.; Schmidt, G.M.J. *Chem. Phys. Lett.* **1970**, *7*, 486.
75. Desiraju, G.R.; Sarma, J.A.R.P. *Acc. Chem. Res.* **1986**, *19*, 222.
76. Ramasubbu, N.; Parthasarathy, R.; Rust, P.M. *J. Am. Chem. Soc.* **1986**, *108*, 4308.
77. Thomas, J.M.; Desiraju, G.R. *Chem. Phys. Lett.* **1984**, *110*, 99.
78. Sarma, J.A.R.P.; Desiraju, G.R. *Chem. Phys. Lett.* **1985**, *117*, 160.

79. Vishnumurthy, K.; Guru Row, T.N.; Venkatesan, K. *Photochem. Photobiol. Sci.* **2002**, *1*, 427.
80. Venkatesan, K. *Proc. Indian Natn. Sci. Acad. A Phys. Sci.* **1999**, *65*, 129.
81. Coates, G.W.; Dunn, A.R.; Henling, L.M.; Ziller, J.W.; Lobkovsky, E.B.; Grubbs, R.H. *J. Am. Chem. Soc.* **1998**, *120*, 3641.
82. Kaftory, M.; Tanaka, K.; Toda, F. *J. Org. Chem.* **1985**, *50*, 2154.
83. Tanaka, K.; Toda, F. *J. Chem. Soc. Perkin 1* **1992**, 943.
84. Tanaka, K.; Mochizuki, E.; Yasui, N.; Kai, Y.; Miyahara, I.; Hirotsu, K.; Toda, F. *Tetrahedron* **2000**, *56*, 6853.
85. Hatano, B.; Hirano, S.-Y.; Yanagihara, T.; Toyota, S.; Toda, F. *Synthesis* **2001**, 1181.
86. MacGillivray, L.R.; Reid, J.L.; Ripmeester, J.A. *J. Am. Chem. Soc.* **2000**, *122*, 7817.
87. Hamilton, T.D.; Papaefstathiou, G.S.; MacGillivray, L.R. *J. Am. Chem. Soc.* **2002**, *124*, 11606.
88. Papaefstathiou, G.S.; Friscic, T.; MacGillivray, L.R. *J. Supramol. Chem.* **2002**, *2*, 227.
89. MacGillivray, L.R.; Reid, J.L.; Ripmeester, J.A.; Papaefstathiou, G.S. *Ind. Eng. Chem. Res.* **2002**, *41*, 4494.
90. Friscic, T.; MacGillivray, L.R. *Chem. Commun.* **2003**, 1306.
91. Gao, X.; Friscic, T.; MacGillivray, L.R. *Angew.Chem. Int. Ed. Engl.* **2004**, *43*, 232.
92. Vansant, J.; Smets, G.; Declercq, J.P.; Germain, G.; Van Meerssche, M. *J. Org. Chem.* **1980**, *45*, 1557.
93. Papaefstathiou, G.S.; Kipp, A.J.; MacGillivray, L.R. *Chem. Commun.* **2001**, 2462.
94. Varshney, D.B.; Papaefstathiou, G.S.; MacGillivray, L.R. *Chem. Commun.* **2002**, 1964.
95. Mei, X.; Liu, S.; Wolf, C. *Org. Lett.* **2007**, *9*, 2729.
96. Varshney, D.B.; Gao, X.; Friscic, T.; MacGillivray, L.R. *Angew.Chem. Int. Ed. Engl.* **2006**, *45*, 646.
97. MacGillivray, L.R. *J. Org. Chem.* **2008**, *73*, 3311.
98. Ito, Y.; Kitada, T.; Horiguchi, M. *Tetrahedron* **2003**, *59*, 7323.
99. Ito, Y.; Borecka, B.; Trotter, M.; Scheffer, J.R. *Tetrahedron Lett.* **1995**, *36*, 6083.
100. Ito, Y.; Borecka, B.; Olovsson, G.; Trotter, J.; Scheffer, J.R. *Tetrahedron Lett.* **1995**, *36*, 6087.
101. Ito, Y.; Fujita, H. *Chem. Lett.* **2000**, 288.
102. Bhogala, B.R.; Captain, B.; Parthasarathy, A.; Ramamurthy, V. *J. Am. Chem. Soc.* **2010**, *132*, 13434.
103. Papaefstathiou, G.S.; Georgiev, I.G.; Friscic, T.; MacGillivray, L.R. *Chem. Commun.* **2005**, 3974.
104. Qianli, C.; Swenson, D.C.; MacGillivray, L.R. *Angew. Chem. Int. Ed. Engl.* **2005**, *44*, 3569.
105. Toh, N.L.; Nagarathinam, M.; Vittal, J.J. *Angew. Chem. Int. Ed. Engl.* **2005**, *44*, 2237.
106. Briceno, A.; Leal, D.; Atentio, R.; De Delgado, G.D. *Chem. Commun.* **2006**, 3534.
107. Hill, Y.; Briceno, A. *Chem. Commun.* **2007**, 3930.
108. Han, Y.-F.; Lin, Y.-J.; Jia, W.-G.; Wang, G.-L.; Jin, G.-X. *Chem. Commun.* **2008**, 1807.

109. Peedikakkal, A.M.P.; Vittal, J.J. *Chem. Eur. J.* **2008**, *14*, 5329.
110. Quina, F.H.; Whitten, D.G. *J. Am. Chem. Soc.* **1975**, *97*, 1602.
111. Usami, H.; Takagi, K.; Sawaki, A.Y. *J. Chem. Soc. Perkin 2* **1990**, 1723.
112. Usami, H.; Takagi, K.; Sawaki, A.Y. *J. Chem. Soc. Faraday Trans.* **1992**, *88*, 77.
113. Ramamurthy, V.; Corbin, D.R.; Kumar, C.V.; Turro, N.J. *Tetrahedron Lett.* **1990**, *31*, 47.
114. Jon, S.Y.; Ko, Y.H.; Park, S.H.; Kim, H.-J.; Kim, K. *Chem. Commun.* **2001**, 1938.
115. Pattabiraman, M.; Natarajan, A.; Kaanumalle, L.S.; Ramamurthy, V. *Org. Lett.* **2005**, *7*, 529.
116. Pattabiraman, M.; Natarajan, A.; Kaliappan, R.; Mague, J.T.; Ramamurthy, V. *Chem. Commun.* **2005**, 4542.
117. Kaanumalle, L.S.; Natarajan, A.; Sivasubramanian, K.; Kaliappan, R.; Pattabiraman, M.; Ramamurthy, V. *Spectrum* **2006**, *9*, 16.
118. Maddipatla, M.V.S.N.; Kaanumalle, L.S.; Natarajan, A.; Pattabiraman, M.; Ramamurthy, V. *Langmuir* **2007**, *23*, 7545.
119. Tamaki, T.; Kokubu, T.; Ichimura, K. *Tetrahedron* **1987**, *43*, 1485.
120. Moorthy, J.N.; Weiss, R.G.; Venkatesan, K. *J. Org. Chem.* **1992**, *57*, 3292.
121. Herrmann, W.; Wehrle, S.; Wenz, G. *Chem. Commun.* **1997**, 1709.
122. Banu, H.S.; Lalitha, A.; Pitchumani, K.; Srinivasan, C. *Chem. Commun.* **1999**, 607.
123. Shimizu, L.S.; Hughes, A.D.; Smith, M.D.; Davis, M.J.; Zhang, B.P.; Zur Loye, H.C.; Shimizu, K.D. *J. Am. Chem. Soc.* **2003**, *125*, 14972.
124. Yang, J.; Dewal, M.B.; Shimizu, L.S. *J. Am. Chem. Soc.* **2006**, *128*, 8122.
125. Yang, J.; Dewal, M.B.; Profeta, S. Jr.; Smith, M.D.; Li, Y.; Shimizu, L.S. *J. Am. Chem. Soc.* **2008**, *130*, 612.

7

STRUCTURAL ASPECTS AND TEMPLATION OF PHOTOCHEMISTRY IN SOLID-STATE SUPRAMOLECULAR SYSTEMS

MENAHEM KAFTORY

INTRODUCTION

Photochemical reactions in the solid state are highly dependent on the geometry of the reacting compounds. To understand the reaction mechanism, the course of the reaction, and the reaction control factors, one must be able to follow the structural changes during the reaction. Unfortunately, most solid-state reactions result in the destruction of the single-crystal integrity of the sample under study (heterogeneous reaction). Therefore, the mechanism and the resulting crystal structure can only be speculated about. In homogeneous solid-state reactions, on the other hand, in which the single-crystal integrity remains throughout the reaction, the crystal structure and its changes can be monitored; however, the chemical systems enabling such a monitoring are rare.[1–10] In some cases, the crystal structures of a solid solution containing both the reactant and the product were analyzed structurally.[4,5,10–13] In these rare cases, the structures of the substrate and the product can be monitored. In

Supramolecular Photochemistry: *Controlling Photochemical Processes*, First Edition. Edited by V. Ramamurthy and Yoshihisa Inoue.

other very exciting experiments, the structural changes in photochromic salicylidene aniline crystals using two-photon excitation were observed by X-ray diffraction of a single crystal,[14] and the structure of the light-induced radical pair from a hexaarylbiimidazolyl derivative together with its ground-state substrate was determined by X-ray diffraction of a single crystal.[15] Yet topotactic (single-crystal-to-single-crystal) reactions, where the transformation takes place with retention of the crystal lattice and enabling crystal structure determination, are rare. Most of those rare cases are limited to reactions with small atomic movements. The main reason for the limited number of examples undergoing topotactic photochemical reaction is that any substantial structural change that a reactive molecule undergoes in the crystal will affect the neighbor molecules undergoing the same structural variation and result in a local disorder. We may compare it to a crowd of people walking on a sidewalk in one direction, when suddenly a few pedestrians change direction, immediately causing a disorder in the crowd.

A general method by Enkelmann et al.[16] describes the induction of a homogeneous photochemical reaction by irradiating the crystal with wavelength corresponding to the chromophore's absorption tail. This method enables monitoring structural changes during photoreactions. It may also provide another means by which new polymorphs of the product compound can be obtained.[17] However, in most cases, if the photochemical reaction is executed on the neat light-sensitive compound, the reaction does not proceed to full conversion before the crystal disintegrates. Therefore, monitoring of the reaction process is very limited.

It was recently demonstrated[18–23] that in supramolecular systems composed of light-stable host molecules and light-sensitive guest molecules, the photochemical reaction may proceed to completion with the preservation of the single-crystal integrity. If the host molecules provide topochemical conditions required for mono- or bimolecular reactions and the guest molecules are photochemically active, regio- and stereoselective reactions are anticipated.[24–26] If the host molecule is chiral, it induces chirality into the space where the guest molecules react, and as a result enantioselective reactions take place.[27–30] Therefore, inducing photochemical reactions in supramolecular compounds proved to be a unique method for synthesizing a large variety of compounds.[31] The volume of space available for the guest molecules determines whether the reaction will be heterogeneous or homogeneous. In cases where the size of the reaction core is large enough to accommodate the substrate and the product (not simultaneously), the reaction is expected to be homogeneous. This volume is also called "reaction cavity," originally introduced and developed by Cohen to describe reactions in crystals.[24] This model was further developed by Weiss et al.[25] to accommodate supramolecular systems.

The advantages of performing photochemical reactions in solid supramolecular systems compared with carrying the photochemical reaction in the neat solid reactive compound are the following. (1) The dilution and the separation between potentially photochemical active molecules enable structural variation with no interference from molecules of the same kind, and this will lead to homogeneous reaction. (2) By changing the light-stable host molecules, it is possible to identify the system where the reaction will be topotactic; it also gives the opportunity to study the effect of different neighbor molecules on the reactivity of the light-sensitive guest molecule. (3) Replacing the host molecules may affect the conformation and the geometry of the reactive molecular target and may provide valuable data to the understanding of the reaction mechanism. (4) It reduces the number of photons required to achieve a certain amount of conversion.[23] (5) It provides the ability to engineer templates for carrying out special synthesis.

THE SCOPE AND LIMITATION OF THE REVIEW

The field of solid-state photochemistry was established in the 1960s by the group of Gerhards Schmidt at the Weizmann Institute of Science. The synthetic potential of carrying photochemical reactions in organized media was appreciated, and investigation in the field was revived in the 1980s. Attempts to control photochemical reactions have been made by the use of zeolite, micelles, vesicles, and other assemblies. This review will mainly discuss photochemical reactions in solid organic supramolecular systems from the structural point of view. It will provide information according to the type of the photochemical reactions with less information on photophysical events. Throughout the review, we will use the conventional notation of inclusion compounds or host–guest compounds. In all the examples that will be described, those compounds are composed of light-stable host molecule and light-sensitive guest molecule (unless stated otherwise).

There are many review articles that have been devoted to cover the progress in the field of solid-state photochemistry. A few of these articles are mentioned in this chapter.

REGIO- AND STEREOSELECTIVE [2 + 2] AND [4 + 4] PHOTOREACTIONS

At the beginning of the 1980s, Toda, his group, and his collaborators showed the advantages of carrying out [2 + 2] and [4 + 4] photodimerization in solid inclusion compounds. In each of the crystals, the host molecules were light stable and the guest molecules were light sensitive. The host molecules such as **1–13** are diols bearing hydroxyl groups that provide means to hold the guest molecules by hydrogen bonding. The host molecules are packed in such a way that there is space and isolation of one or two guest molecules, thus enabling the reaction.[26,32–35]

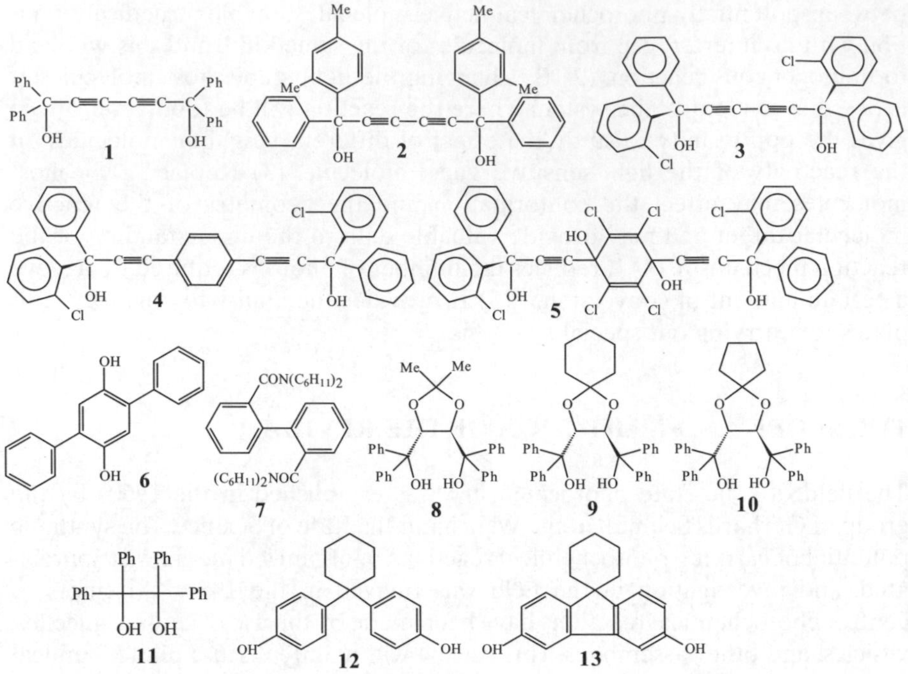

For example, theoretically [2 + 2] photocyclization of **14** may lead to one (or more) of the four different products, **15–18** (Scheme 7.1). **14a** photodimerized in EtOH solution with a very low yield (9%). No dimerization was observed when the neat solid **14a** was irradiated. However, irradiation of the solid host–guest compounds made of **14a** and **1** or **6** for 6 h provided **15** in 90% and 85% yield, respectively. The host molecules **1** as well as **6** have molecular centers of symmetry and reside on a crystallographic inversion center site, thus inducing the same symmetry between the guest molecules, as shown in Figure 7.1. The presence of the inversion center in the crystal between the substrate molecules (Fig. 7.1) ensures that the product will have an inversion center (**15**). The same symmetry is expected for the product of the [4 + 4] photodimerization of 9-anthraldehyde (Fig. 7.2). The yields of the reactions were over 70%.

PHOTOCHEMICAL [2 + 2] CYCLOADDITION

Photochemical [2 + 2] cycloaddition in solution can provide a mixture of products. Similar reactions in the solid state can yield a single product. Below are some unique examples of such a reaction in solid supramolecular systems.

A supramolecular complex with pseudorotaxene-like architecture, such as the crown ether bis-*p*-phenylene[34] crown-10 (**19**), was shown to host

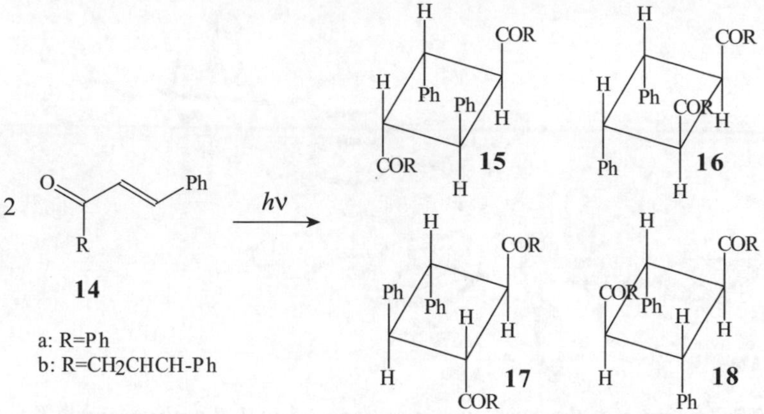

2 **14** $\xrightarrow{h\nu}$ **15** **16** **17** **18**

a: R=Ph
b: $R=CH_2CHCH\text{-}Ph$

Scheme 7.1.

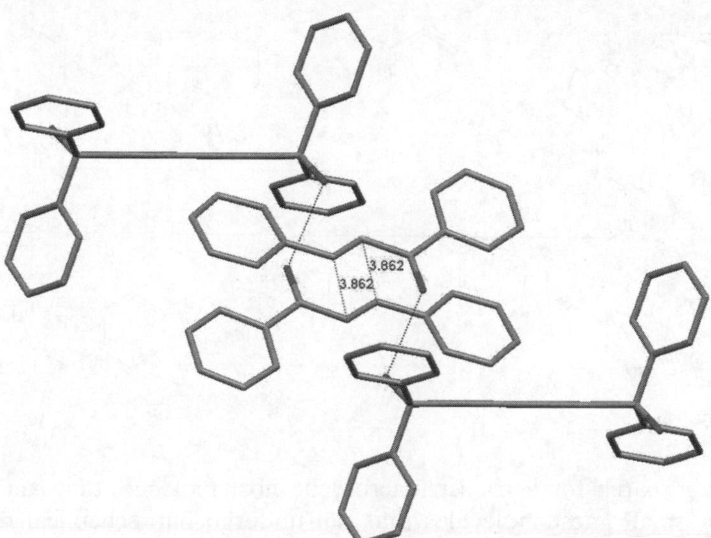

Figure 7.1. Example of the host–guest packing of **1** with benzylidene acetophenone (**14a**) (hydrogen atoms are omitted for clarity).

bis(dialkylammonium ion)-containing threadlike dication **20**·2(**PF**$_6$) in a crystalline host–guest compound. By replacing the *p*-phenylene in **20** by a *trans*-stilbenoid unit (**21**), a crystal was obtained with the two stilbenoid ions parallel to each other with a distance of 4.203 Å between potential reactive centers (Fig. 7.3). Irradiation for 30 h yielded the photodimer shown in Figure 7.4.[18]

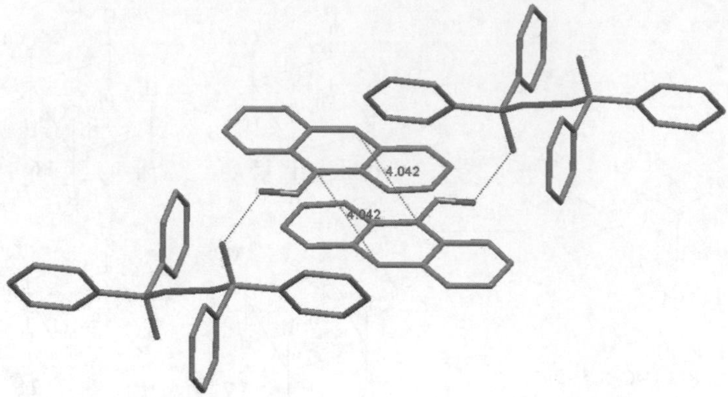

Figure 7.2. Example of the host–guest packing of **1** with 9-anthraldehyde (hydrogen atoms are omitted for clarity).

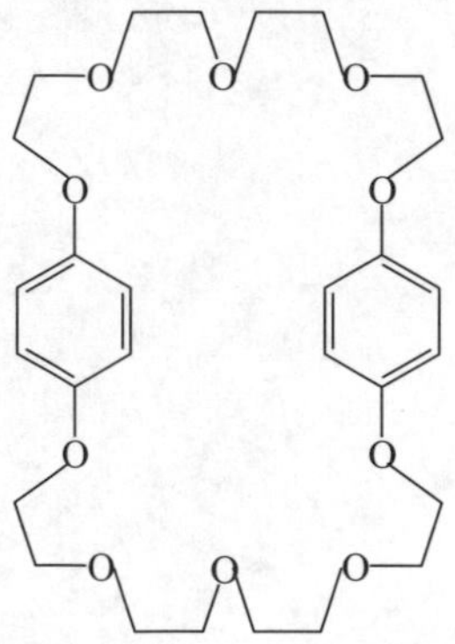

In their search for a robust macrocyclic host molecule that can provide pores for small guest molecules that can undergo photochemical reactions, Shimizu et al.[36,37] prepared the bis-urea macrocycle **22**. The macrocycle was crystallized with α,β-unsaturated ketone derivatives such as **23**[38,39] and irradiated to yield the head-to-tail [2 + 2] cycloaddition product, such as **27** (Scheme 7.2). Of the 12 host–guest compounds examined, four underwent photochemical cycloaddition (**23**–**26**), the first three in very good yields (80–100% conversion). Not only was the conversion higher than that observed in the [2 + 2] photocycloaddition in the neat substrate, but also the selectivity between the head-to-tail and head-to-head isomers (such as **27** and **28**) was much better. However, in the rest of the inclusion compounds they prepared, no reactions were observed while cycloaddition of the neat substrates took place, although to different levels of conversions and no specific selectivity.

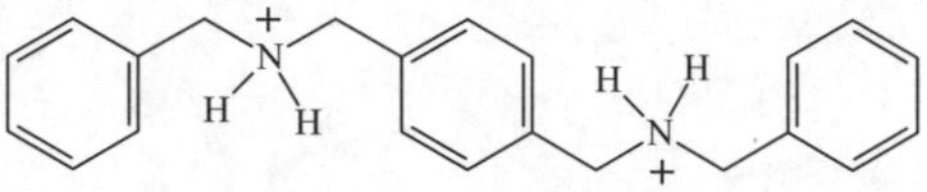

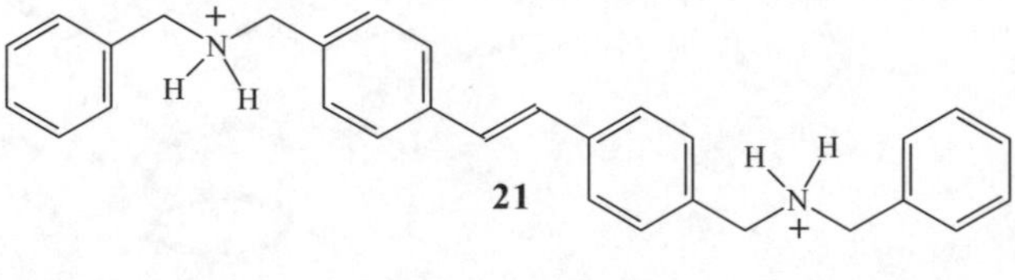

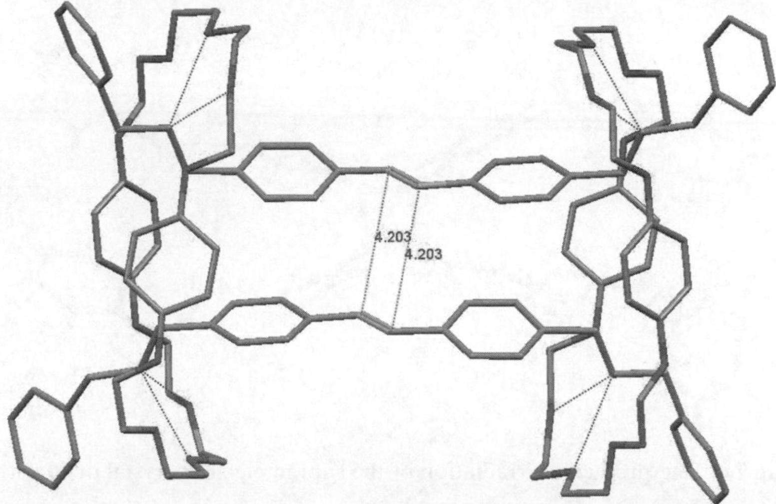

Figure 7.3. The packing of **20** and **21** in the supramolecular crystal (the reactive double bonds are in violet at the base of the dotted lines). See ftp site for color: ftp://ftp.wiley.com/public/sci_tech_med/supramolecular_photochemistry.

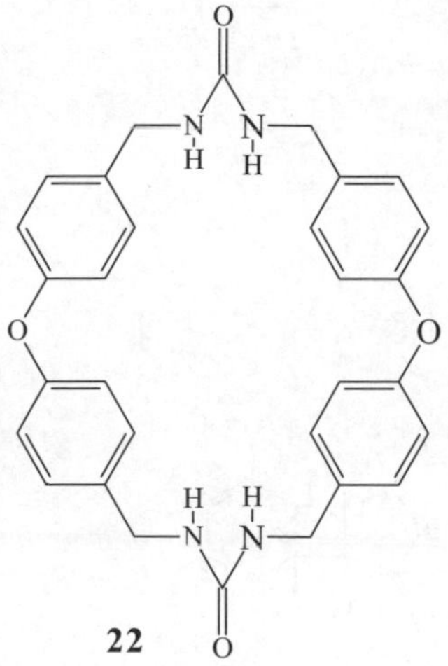

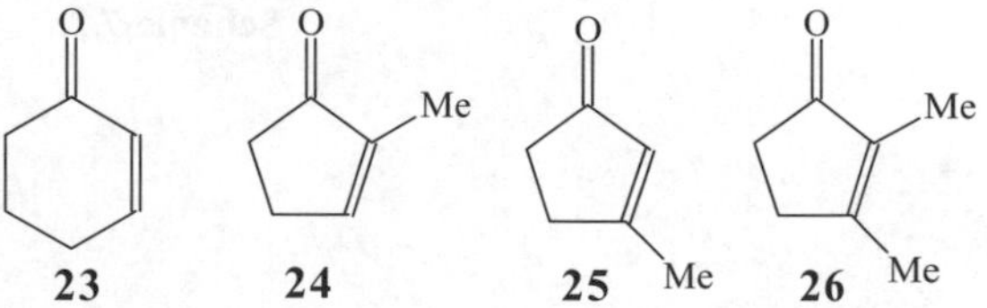

Sometimes the role of the host and guest are not determined, and in cases where both compounds are light sensitive, a photochemical reaction may take place between the two. A remarkable example is the photocyloaddition of 2-pyrones with maleimide[40] (Scheme 7.3). Photocycloaddition was not obtained from **29f·30** and **29g·30**. Because of the difficulties in crystallization of **29a**, **29b**, **29d**, **29f**, and **29g**, the two substrates (**29** and **30**) were dissolved, the solvent was evaporated, and the powder obtained was irradiated. The

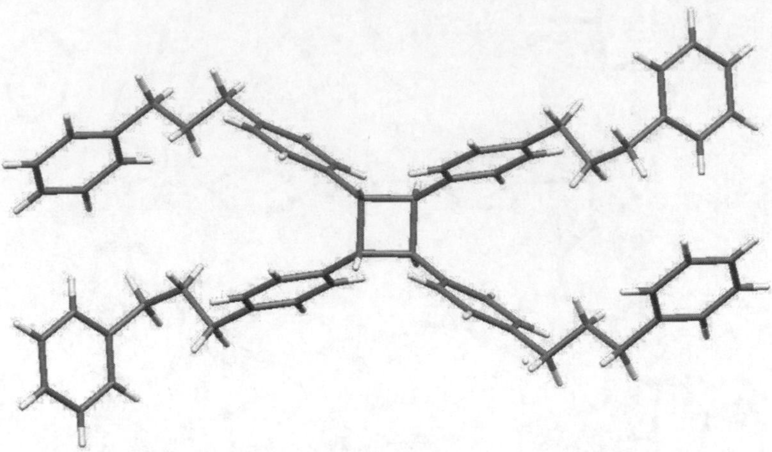

Figure 7.4. The product of irradiation of the supramolecular crystal of **20** with **21**.

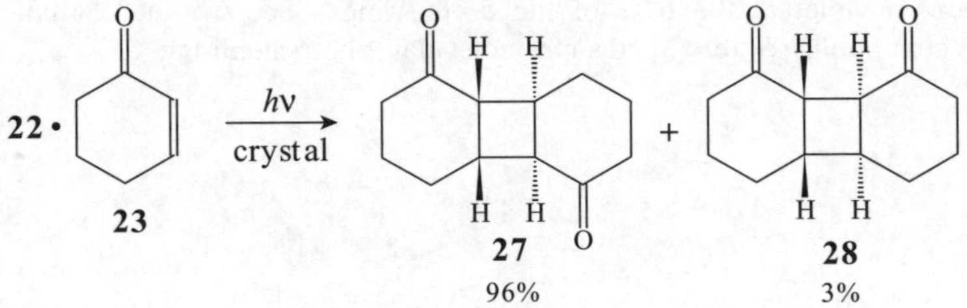

Scheme 7.2.

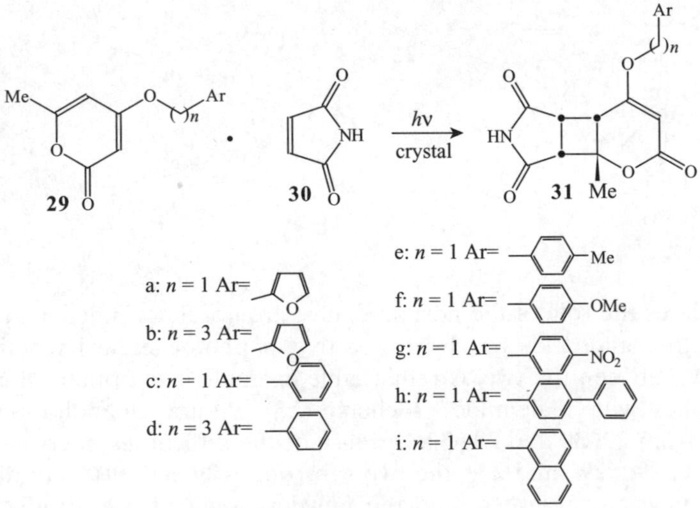

Scheme 7.3.

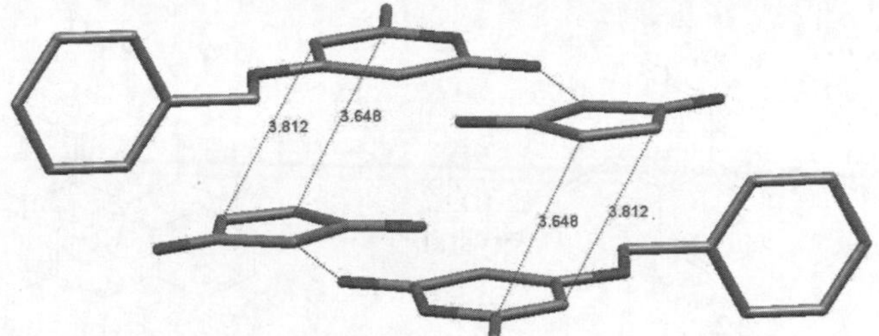

Figure 7.5. The relative geometry in **29c**.

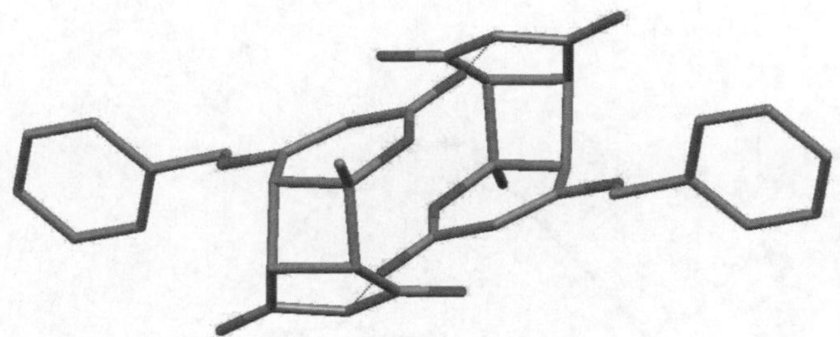

Figure 7.6. The structure of the product **31c**.

relative geometry of the host and guest molecules in the crystal of **29c** is shown in Figures 7.5 and 7.6 as well as the structure of the product **31c**. The distances between the reacting centers are 3.648 and 3.812 Å.

[4 + 4] PHOTODIMERIZATION

The examples of [4 + 4] photodimerization in host–guest solid compounds are mainly of anthracene derivatives. As the guest molecules, of the two possible photodimers that can be obtained from 9-anthraldehyde (**32a**, Scheme 7.4), 9-acetylanthracene (**32**b), or 9-(methoxycarbonyl)anthracene (**32c**), with 1,1,6,6-tetraphenyl-2,4-hexadiyne-1,6-diol as the host molecule (**1**),[22,26,32] only the corresponding *anti* dimer (**33**) is obtained. The arrangement in the first two is similar in the sense that the guest molecules are held by hydrogen bonds to the host molecules. The host molecules are packed around inversion centers in the first (**1**:2(**32a**)) and in the second (**1**:2(**32b**)), and as a result, the two guest molecules are also related by inversion centers, with distances of 4.042 and 3.805 Å in the first and the second compound, respectively, between the reactive centers, thus leading to the *anti* dimer (Fig. 7.7 as an example). The packing in the latter (**1**:2(**32c**)) is somewhat different (Fig. 7.8), and the distances between the reactive centers are 3.946 and 4.401 Å. Nevertheless, the guest molecules are related by pseudoinversion centers and form the dimer with the inversion center. The molecules of **32c** form infinite stacks (Fig. 7.9).

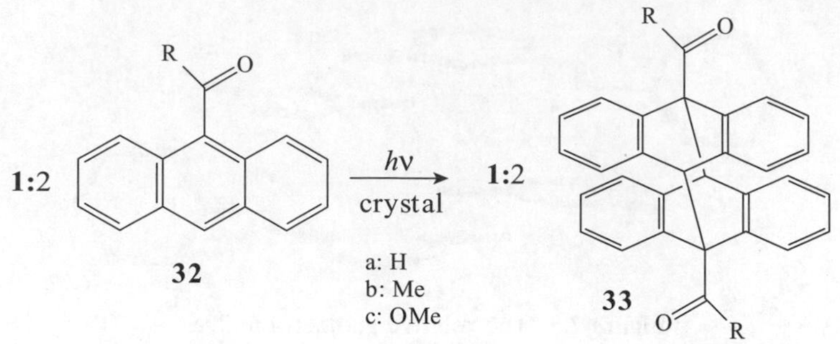

1:2 **32** $\xrightarrow[\text{crystal}]{h\nu}$ 1:2 **33**

a: H
b: Me
c: OMe

Scheme 7.4.

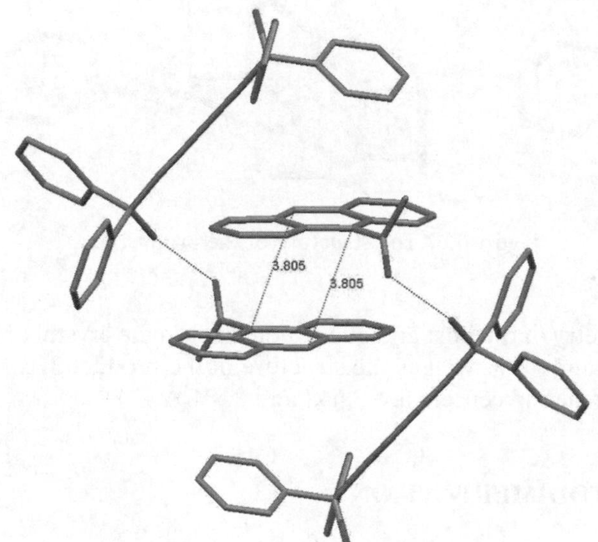

Figure 7.7. Host–guest relation in **1**:2(**32b**).

Pairs are not related by crystallographic inversion centers, and each molecule can photodimerize with a molecule either above or below in the same stack.

A unique example showing the great advantage of carrying photodimerization in solid supramolecular system is the asymmetrical [4 + 4] photodimerization of 9-aminomethylanthracene (**34**–**35**).[41] Because of the dynamic of molecules in solution, it would be improbable to even imagine such an outcome when the compound is irradiated in solution. Salts made up of 9-aminomethylanthracene (**34**) with various dicarboxylic acids were prepared. Irradiation of solid host–guest compounds obtained with muconic acid or acetylenedicarboxylic acid (with methanol included in the crystal) provided in

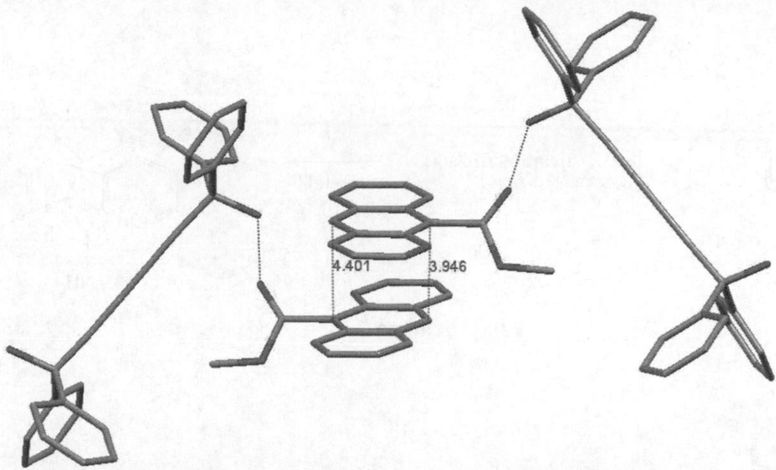

Figure 7.8. Host–guest relation in **1**:2(**32c**).

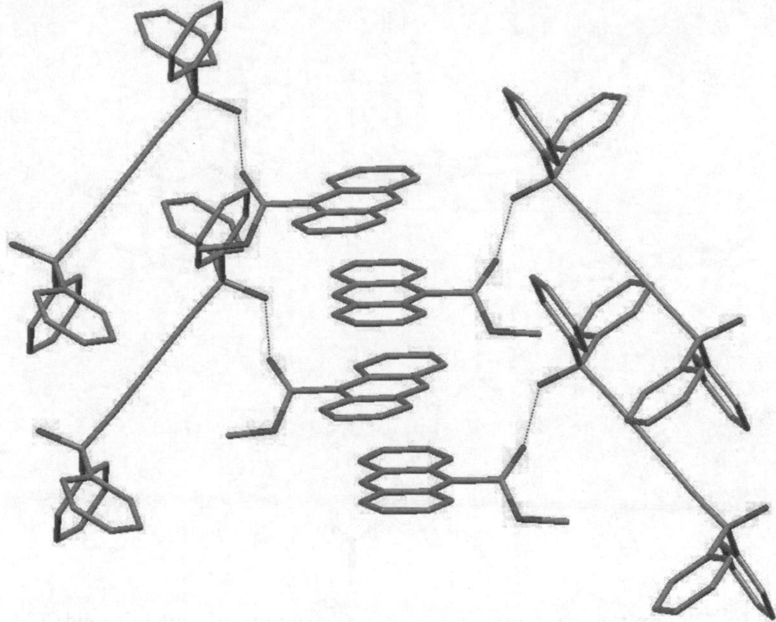

Figure 7.9. Infinite stacks in **1**:2(**32c**).

100% yield the surprisingly asymmetrical dimer **35** (see Scheme 7.5). The crystal structure of the salt composed of two molecules of 9-aminomethylanthracene, and muconic acid is shown in Figure 7.10. The distances between the reactive centers are 3.620 and 3.675 Å, short enough to facilitate the reaction.

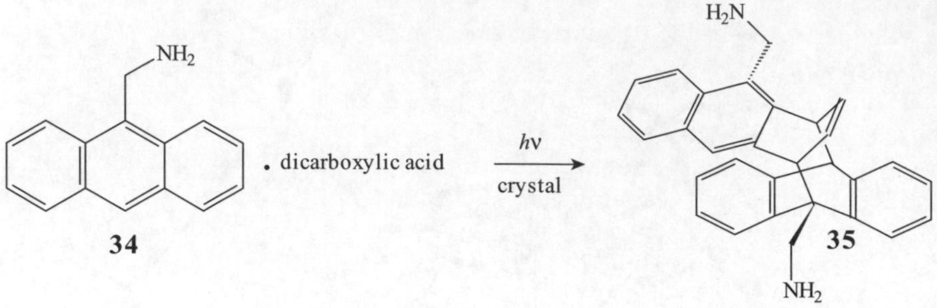

NH_2

34 . dicarboxylic acid $\xrightarrow[\text{crystal}]{h\nu}$ **35**

H_2N NH_2

Scheme 7.5.

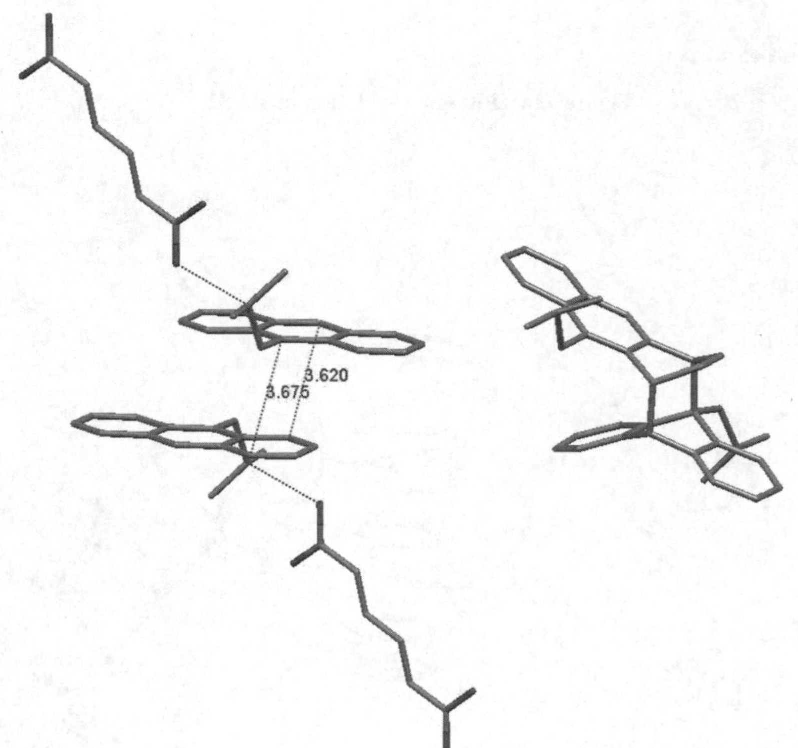

Figure 7.10. The mutual geometry in the crystal structure of **muconic acid:**2(**34**) (left) and the unsymmetric dimer (right).

Pyridone and its derivatives are another set of guest molecules that are found to undergo [4 + 4] photodimerization in solid host–guest compounds. Earlier work is presented in this section and later work is described in another section with regard to single-crystal-to-single-crystal transformation. In the group of Toda and collaborators,[35] different host–guest compounds of **1** as host and pyridone derivatives (**36**) as guest molecules have been prepared. Only two solid host–guest compounds underwent photodimerization: **36a** was photodimerized to yield **37a**, and **36b** was photodimerized in low yield to give **38b**. All other inclusion compounds were inert to irradiation. Lavy et al.[42–44] prepared three host–guest compounds—**12:**2(**36a**), **12:**2(**36c**), and **13:**2(**36e**)—and irradiated all of them. They were inert to irradiation. The crystal structure of the three compounds exemplified the weakness of the method. The strength of the method is that one can use different host molecules to prepare variable host–guest solid compounds that may undergo photodimerization; however, there is no control on the packing, and there is no guarantee that the structure will be such to enable photodimerization. In this particular case, irradiation of the neat compound in EtOH solution revealed photodimers (**37**) in better yields (except **36a**). When similar solid inclusion compounds with 1-methyl-2-pyridones (**36** with R^1=Me) were irradiated, photodimers of type **37** were obtained in good yield, except for **36d** (R^1=Me), **36f** (R^1=Me), and **36i** (R^1=Me). The reason is unknown because there is no crystal structure information; however, it should be a result of the crystal packing.

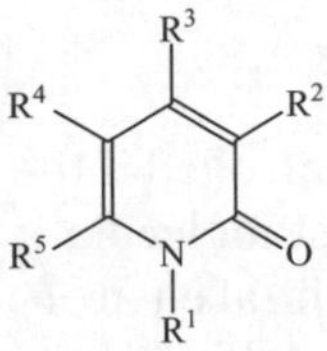

a: $R^1=R^2=R^3=R^4=R^5$=H
b: R^1=H; $R^3=R^4=R^5$=H; R^2=Me
c: R^1=H; $R^2=R^4=R^5$=H; R^3=Me
d: R^1=H; $R^2=R^3=R^5$=H; R^4=Me
e: R^1=H; $R^2=R^3=R^4$=H; R^5=Me
f: $R^1=R^3=R^5$=H; $R^2=R^4$=Me
g: $R^1=R^2=R^4$=H; $R^3=R^5$=Me
h: $R^1=R^2=R^4$=H; $R^3=R^5$=Me
i: $R^1=R^2$=H; $R^3=R^4=R^5$=Me
j: $R^1=R^2=R^3=R^4$=H; R^5=OMe

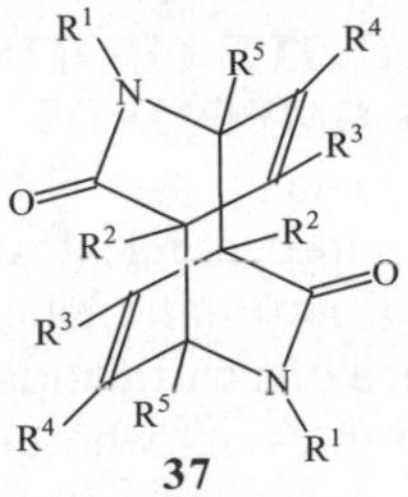

36 **37** 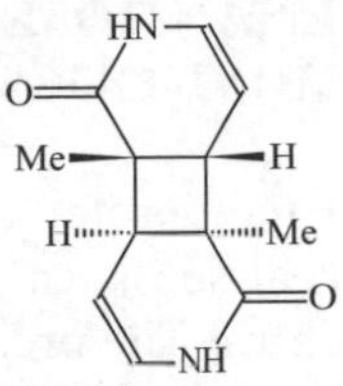 **38**

ASYMMETRIC PHOTOREACTIONS IN SOLID SUPRAMOLECULAR SYSTEMS

One of the major advantages of carrying out photochemical reactions in the solid state of a neat compound over photochemical reactions in solution is that in the solid, the potentially reactive molecules are fixed in a geometry that enables the reaction to adopt a single course of reaction. In a solution, on the other hand, the continuous changes of the molecular structure expressed by variation of the conformation may lead to different products. To impose enantioselective photochemical reaction of a solid compound of a prochiral neat compound, it should be crystallized in a specific way to enable such an enantioselective reaction. However, since the ability of crystal engineering is still very limited and there is no control on the packing of molecules in the solid state, it is not yet a useful method for carrying out an enantioselective photochemical reaction. The notion that enantioselective photochemical reactions will take place in a chiral environment attracted scientists to examine it by crystallizing the potential prochiral photoreactive species in a chiral environment. This approach was translated into a search for different chemical systems that can provide the chiral environment for the photoreactive species. The different approaches provided systems such as host–guest compounds composed of light-stable chiral host and light-sensitive prochiral guest molecules. An alternative approach was the use of ionic chiral auxiliaries and the formation of chiral solid salts. The use of zeolites with chiral auxiliaries and prochiral molecules has also been tested. The use of chiral molecular capsules including the prochiral reactive molecules is another approach. Yet another approach is to construct hybrid organic–inorganic materials; the cavities created in these frameworks may be chiral and may serve the purpose of enantioselective photochemical reactions. In this section, we describe mainly the use of two approaches: the use of host–guest systems with chiral host molecules, a method that was extensively studied and reviewed by Tanaka and Toda[31,45–47] and the ionic chiral auxiliaries method that was designed and thoroughly investigated by the group of Scheffer.[48,49] Many other examples are documented by Wada and Inoue[50] and by Koshima.[51]

ENANTIOSELECTIVE PHOTOCHEMICAL REACTIONS IN SOLID HOST–GUEST COMPOUNDS

Enantioselective photochemical reactions can be successfully controlled if the reaction is carried out in an inclusion compound where the host molecule is chiral. The presence of a chiral molecule in the crystal induces a chiral environment in the reaction cavity where the photochemical reaction is taking place. For example, disrotatory intramolecular photocyclization of tropolone alkyl ethers can be controlled by using optically active host molecules. Irradiation of a 1:1 inclusion compound of α-tropolone methyl ether **39a** and (*S*,*S*)-(−)-**3** in the solid state gave (1*S*,5*R*)-(−)-**40a** of 100% enantiomeric excess (ee) and (+)-**42a** of 91% ee in 11% and 26% yields, respectively[52,53] (Scheme 7.6). Similar irradiation of the inclusion compound of **39b** with (−)-**3** gave (1*S*,5*R*)-(−)-**41b** of 100% ee and (+)-**42b** of 72% ee in 12% and 14% yields, respectively. The interpretation is that the disrotatory photoreaction of **39** in the inclusion crystal with (−)-**3** occurs only in the A direction (see **39**) due to the steric hindrance of the host molecule. This interpretation is supported by the crystal structure of the inclusion compound.[27]

Another striking example is the stereo- and enantioselective photoreactions of the *N*,*N*-dialkylphenylglyoxylamides obtained as a result of irradiation solid inclusion compound of **43** in different optically active host compounds such as **3**, **5**, **9**, and **10**[27,53] (Scheme 7.7). When a crystal of **43a** and (−)-**3** was irradiated, optically pure β-lactam(−)-**44a** was obtained in 90% yield.

The crystal structure of the inclusion compound (Fig. 7.11) showed that the conformation of the photoactive molecule is fixed in the crystal by the two hydrogen bonds with the two hydroxyl groups of the same host molecule and

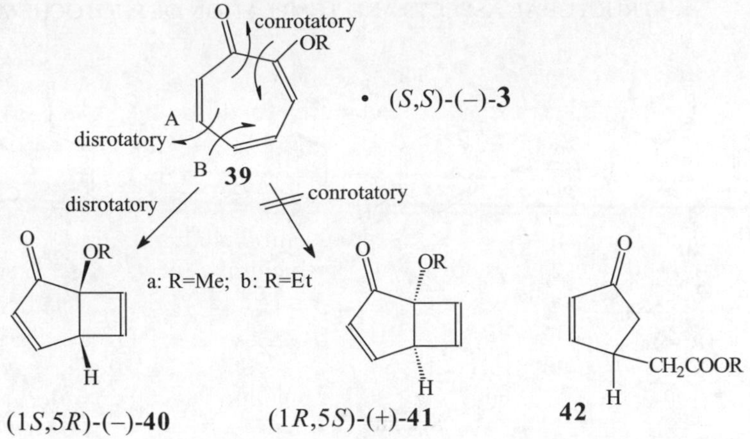

O conrotatory
OR

• (*S*,*S*)-(−)-**3**

A
disrotatory
B **39**
disrotatory
conrotatory

O OR
a: R=Me; b: R=Et
H
(1*S*,5*R*)-(−)-**40**

O OR
H
(1*R*,5*S*)-(+)-**41**

O
CH_2COOR
H
42

Scheme 7.6.

O CHR^1R^2
Ph C N
C CHR^3R^4
O
43

$h\nu$
solid

OH R^1
Ph R^2
O N
CHR^3R^4
44

+

R^1
O R^2
N
Ph
O
45

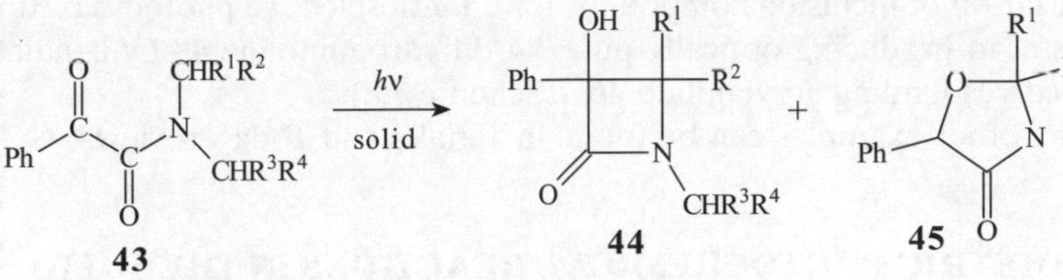

a: $R^1=R^2=R^3=R^4=H$
b: $R^1=R^3=H$; $R^2=R^4=Me$
c: $R^1=R^3=H$; $R^2=R^4=Et$
d: $R^1=R^2=H$; $R^3=R^4=Me$
e: $R^1=H$; $R^2=Me$; $R^3=R^4=Me$
f: $R^1=R^2=R^3=R^4=Me$
g: $R^1=R^3=H$; $R^2,R^4=-(CH_2)_3-$
h: $R^1=R^3=H$; $R^2,R^4=-(CH_2OCH_2)-$

Scheme 7.7.

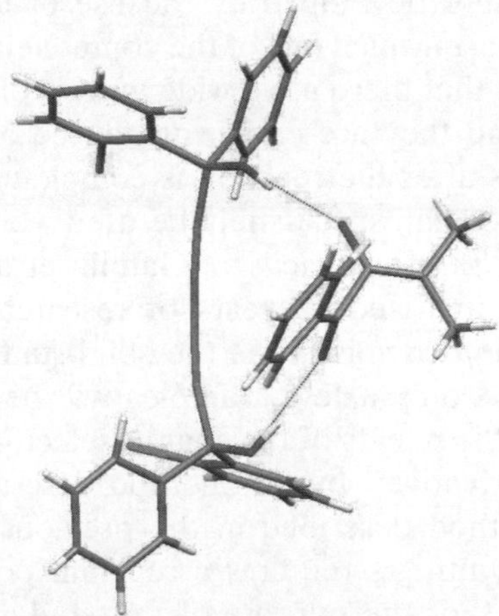

Figure 7.11. Host molecule **4** holding by two hydrogen bonds **43a**.

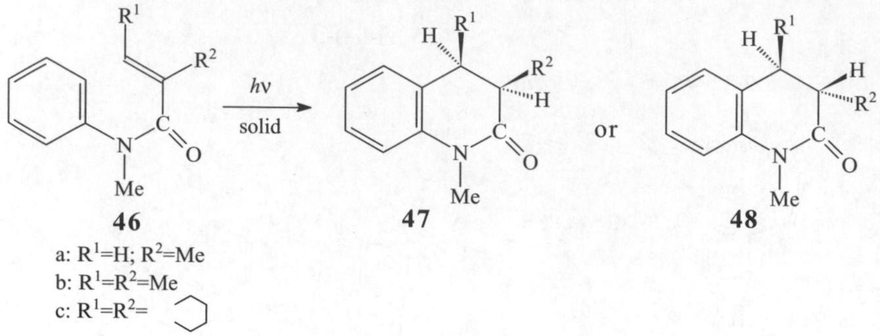

a: R^1=H; R^2=Me
b: R^1=R^2=Me
c: R^1=R^2=

Scheme 7.8.

in the absence of a mirror image molecule (the structure is in a chiral space group $P2_12_12_1$); only one enantiomer of the β-lactam could be produced.

Utilization of inclusion compounds for enantioselective photocyclization is successful in producing optically pure 3,4-dihydroquinoline **48** by irradiation of the corresponding acrylanilide **46**[30] (Scheme 7.8).

Many other examples can be found in Tanaka and Toda's review.[46]

ASYMMETRIC PHOTOCHEMICAL REACTIONS IN THE SOLID STATE BY THE ADDITION OF CHIRAL IONIC AUXILIARIES

As an alternative to the method of carrying out photochemical reactions on chiral host-prochiral guest organic systems, the group of Scheffer prepared organic salts made of carboxylic acids and organic base such as amines. The idea was that these compounds would have higher melting points and therefore might lead to higher degrees of conversions without crystal melting. This new approach was further developed by the use of ionic chiral auxiliaries, namely by the use of salts in which one of the components is chiral. The advantage of this method was that there are a wide variety of optically pure amines and carboxylic acids, and they are easily introduced and removed from the two-component systems after the reaction is complete. The first few demonstrations that this method can successfully be used were published in several publications[54–60] as well as in a review by Gamlin et al.[49b] In a later review, Scheffer and Xia[49a] summarized 15 years of research into the asymmetric induction in organic photochemistry via the solid-state ionic chiral auxiliary approach, and therefore, only a few examples will be mentioned here. It is interesting to compare the results of the enantioselective photoelectrocyclization of a tropolone derivative[61] by the method described in this paragraph compared with the method described in the previous one. The ionic chiral auxiliary approach to synthesis requires a reactant possessing an acidic (or basic) functional group. For the purpose of this study, the acid derivative of tropolone (**39c**) was prepared. The induction of the crystal chirality was via the use of commercially available, optically pure amines. The best results were obtained with 1-phenylmethylamine and 1-amino-2-indanol, which gave ee's in the 60–80% range. It is interesting to note that this same photoreaction was a subject for testing an alternative approach using chiral-modified zeolites for photochemical asymmetric synthesis. The approach of Joy and Ramamurthy[62] was to modify the zeolite so that it would be chiral. For this purpose, they used a "chiral inductor," which was a photochemically inert, optically pure compound such as ephedrine, which, when absorbed by the zeolite, renders it chiral. This methodology was applied also to tropolone derivatives. The best results (78% ee) were obtained with **39d**. Ramamurthy et al. pointed out[63] that the use of **3** gave very good results in terms of enantioselective photoreaction; however, the host molecule is very expensive. He suggested the use of other chiral organic host molecules, such as cyclodextrins, which are inexpensive. Although cyclodextrin has been one of the molecules of choice to separate chiral isomers, it was tested for its ability to impose enantioselective photoreaction in only a few examples, with very poor results. Koodanjeri et al.[63] used α-, β-, and γ-cyclodextrin to prepare host–guest compounds with a variety of tropolone derivatives. The photoreaction was carried out in solution and in the solid state. The good news is that no enantioselectivity was detected from the solution experiment, and enantioselectivity was detected from the solid-state experiment. The bad news is that the ee was quite low. The highest ee of 33% was found with **39d** as the guest molecule.

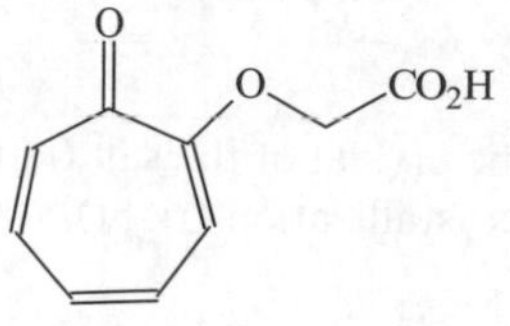

O O CO_2H O O Ph

39c 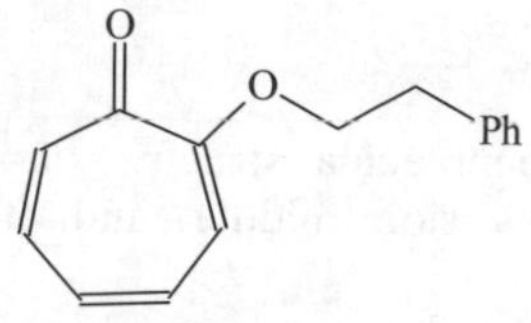 **39d**

The well-known Norrish–Yang type II reaction is the most thoroughly investigated reaction from the point of view of the ionic auxiliary approach. Perhaps the most exciting example shows single-crystal-to-single-crystal transformation. Reactions where the transformation is a single-crystal-to-single-crystal are very rewarding because of the ability to obtain X-ray crystal structural information at any stage of the reaction. 7-Methyl-7-benzoylnorbornane-*p*-carboxylic acid (**49a**)[64] was treated with a variety of optically pure amines to yield the corresponding 1:1 salts (**49b**) (Scheme 7.9). Irradiation of the solid salts, followed by workup by diazomethane, led to the corresponding cyclobutanols (**50c**). The results show that the product's ee ranged from 84% to 98% and the conversion from 88% to 100%. This is an enantioselective reaction with an exceptionally high yield. In many other cases, a slight decrease in ee with increasing conversion was observed. This is also a good example of the ability to receive different polymorphs by single-crystal-to-single-crystal transformation. **49** crystallizes with (*S*)-(−)

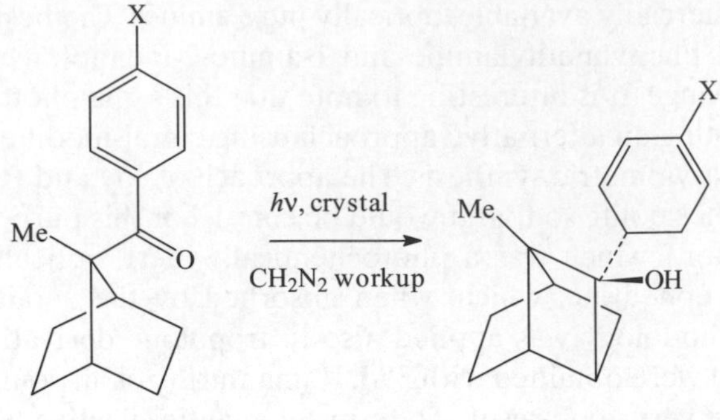

X Me O $h\nu$, crystal CH_2N_2 workup X Me OH

49 **50**

a: X=COOH; b: X=COO^- optically pure amine$^+$; c: X=COOMe

Scheme 7.9.

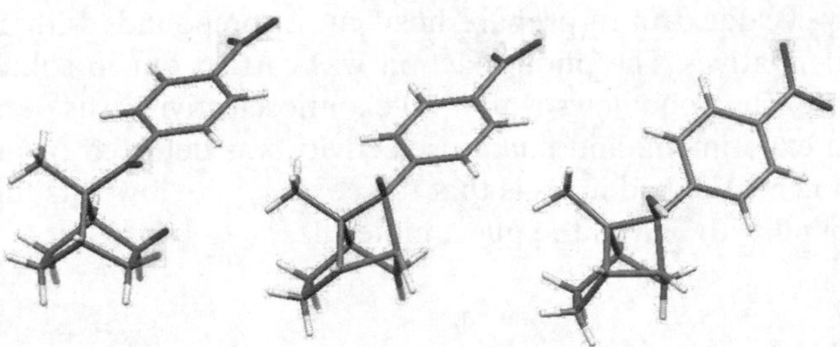

Figure 7.12. The molecular structure of **49** in the crystal of the salt (left), after irradiation to 93% conversion (middle), and after recrystallization (right).

-1-phenylethylamine in the $P2_12_12_1$ space group. After full conversion, it remains in the same space group; however, it recrystallizes from methanol in space group $P2_1$. The crystal structure (Fig. 7.12) shows very clearly that the carbonyl oxygen is much closer to the γ-hydrogen atom that undergoes abstraction, H_x (green, 2.66 Å) than to H_y (purple, 3.43 Å). In the absence of mirror symmetry in the crystal, the reaction should be enantioselective.

In a more recent publication, Scheffer and his collaborators[65] demonstrated that based on molecular mechanics calculations, and the knowledge gained from many examples on the geometric requirement for γ-hydrogen atom abstraction, the Yang photocyclization of α-1-norbornylacetophenone derivatives in the crystalline state can be engineered through methylation adjacent to the carbonyl group. As a consequence of the methylation, the conformation of the molecule in the crystal changes, leading to enhanced diastereo- and enantioselectivity. Molecular mechanics calculations of the minimum energy conformation of the α-1-norbornylacetophenone **51** (R=H) shows that the carbonyl oxygen atom is equidistant from the enantiotopic *pro-R* and *pro-S* *endo* hydrogens (Scheme 7.10). Such conformation will not lead to an enantioselective reaction. However, the minimum energy conformation of **52** (R=Me)

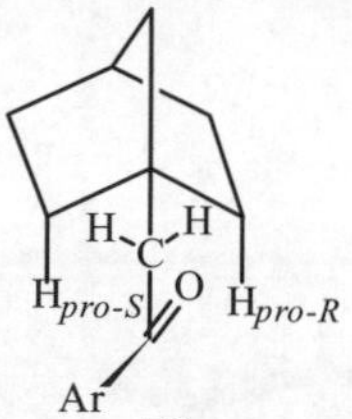

Scheme 7.10.

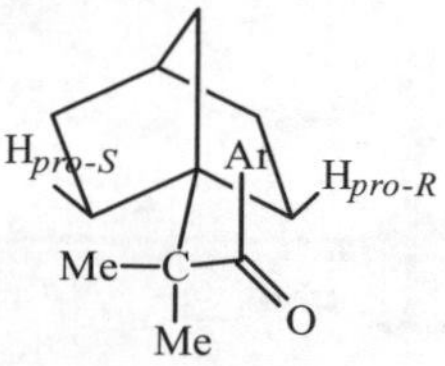

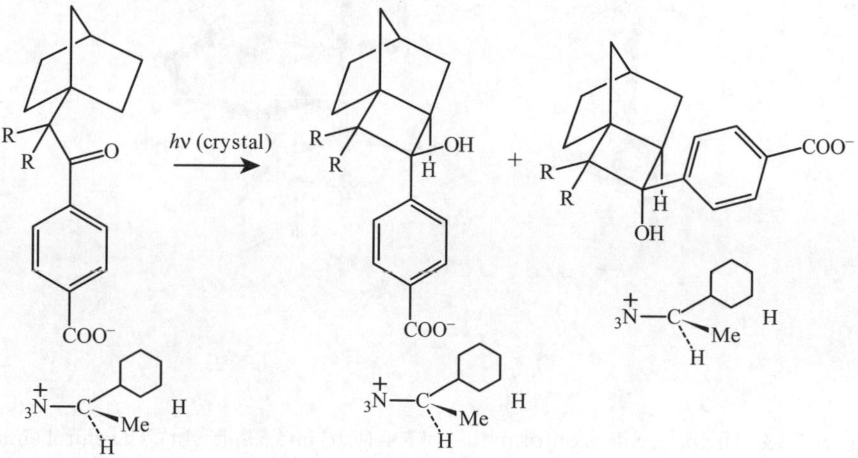

is different, and the ketone oxygen is much closer to the *pro-R exo* hydrogen than to the *pro-S exo* hydrogen. Therefore, an enantioselective reaction is anticipated. Indeed, irradiation of **53a** (Fig. 7.13, left) yielded 46% conversion with **54/55** ratio of 46/54 with ee of 42% (−) and 33% (+), respectively, while irradiation of **53b** (Fig. 7.13, right) yielded 100% conversion with **54/55** ratio of 4/96 and 99% (−) ee.

Koshima and his group[66a,b] used the cocrystal approach for asymmetric photocyclization. The chiral cocrystals are composed of two different achiral molecules. Koshima et al. found that although the cocrystal approach was efficient for the preparation of new chiral crystals, their asymmetric reactions failed. The failure was attributed either to unsuitable molecular arrangement enabling enantioselectivity or to the lack of photoreactivity.[66b] However, when chiral cocrystals of isopropylbenzophenone derivatives with achiral amines were irradiated, Norrish type II photocyclization took place. This group prepared cocrystals of **56** or **57** with many achiral amines as well as with substituted benzylamines; however, only three crystallized in chiral space groups (**56:58**, **57:59**, and **57:60**). For example, when cocrystal **56:58** (space group $P2_12_12_1$) was irradiated, three chiral products—cyclopentanol **61**, cyclopentanol **62**, and hydrol **63**—were obtained (Scheme 7.11) in 19%, 11%,

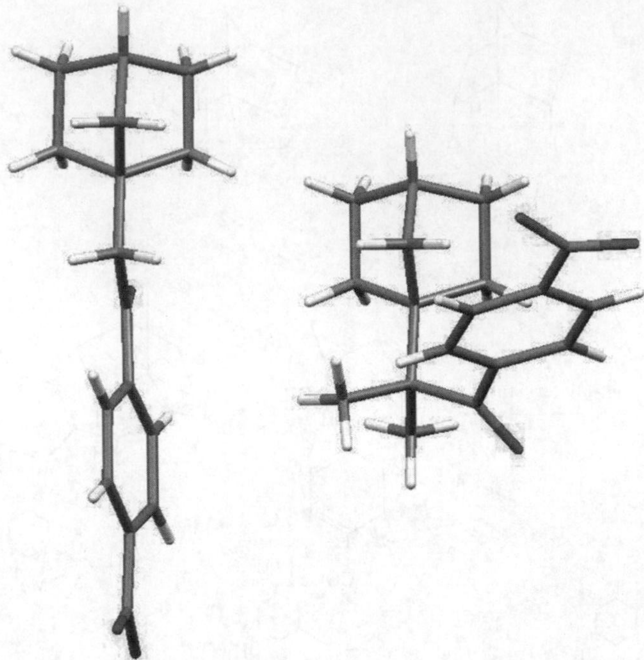

Figure 7.13. The molecular conformation of **53a** (left) and **53b** (right). The chiral amine is omitted.

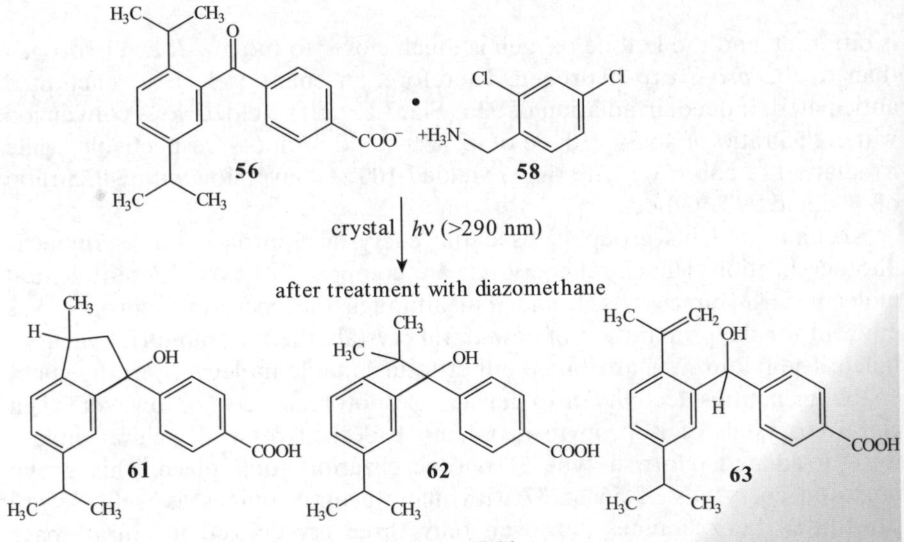

H_3C CH_3 O Cl Cl

COO^- ^+H_3N

56 • **58**

H_3C CH_3

crystal | $h\nu$ (>290 nm)

after treatment with diazomethane

CH_3 H OH | CH_3 H_3C OH | H_3C CH_2 OH H

COOH **61** | COOH **62** | COOH **63**

H_3C CH_3 | H_3C CH_3 | H_3C CH_3

Scheme 7.11.

and 3% yields, respectively, in 82%, 86%, and 76% ee. Irradiation of the other two cocrystals revealed the two enantiomers of a product similar to **62**.

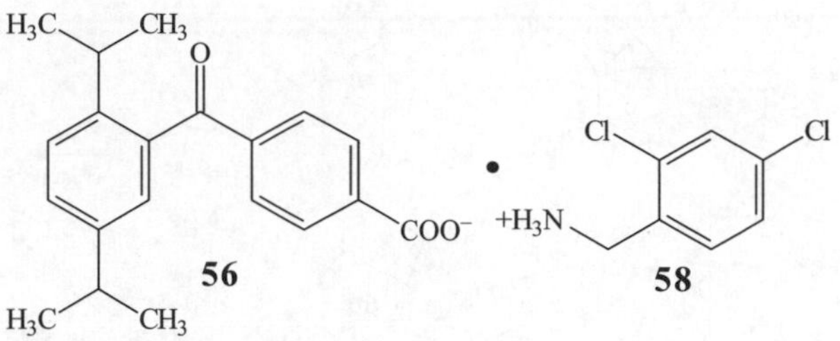

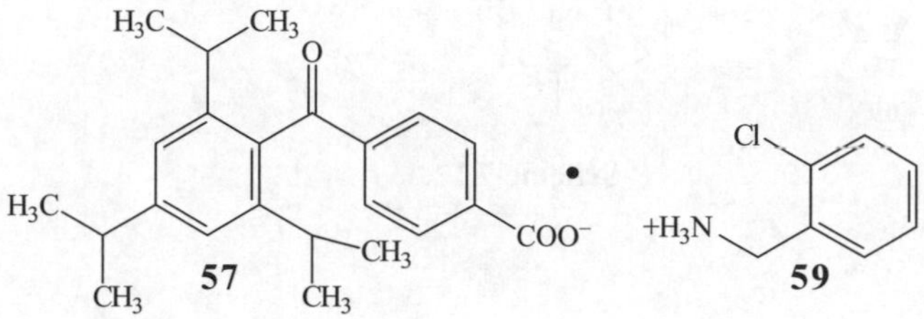

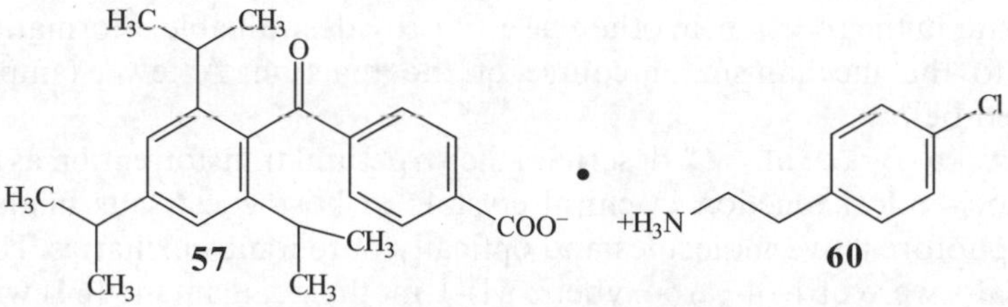

SINGLE-CRYSTAL-TO-SINGLE-CRYSTAL TRANSFORMATION

Topotactic reactions where the transformation from the reactant to the product is single-crystal-to-single-crystal offer an opportunity to follow structural changes in great detail. Such a structural insight during the reaction cannot be obtained when the reaction is carried out in any other medium. Single-crystal-to-single-crystal transformations are rarely found in the reactions of solid neat compounds. Many are described in review articles such as by Keating and Garcia-Garibay.[67] Some of them are involved with intramolecular cyclization or other reactions with minimal molecular migrations.[10,13,68–70] On the other hand, there are recently many publications describing examples of reactions in solid supramolecular systems that undergo single-crystal-to-single-crystal transformation, such as intramolecular photocyclizations[71–73] or photodimerization and polymerization reactions, including $[2 + 2]$[29,46,74,75a,b] and $[4 + 4]$

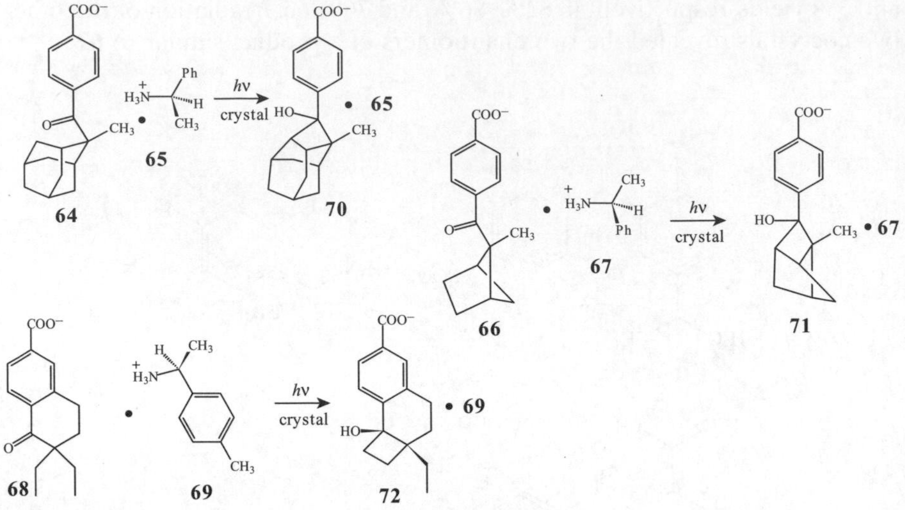

Scheme 7.12.

photodimerization.[44] In some cases, the study of these reactions provides information about the structural effect on the neighboring molecules that do not participate in the reaction; in other cases, it provides valuable information with regard to the mechanism or course of the reaction. A few examples are described below.

Turowska-Tyrk et al.[72,73,76] described the structural transformation as a result of photochemical reaction in chiral crystals of host–guest salts made up of achiral photoreactive molecules and optically pure ionic auxiliaries. The crystalline salts were of 1-(4-carboxybenzoyl)-1-methyladamantane (**64**) with (*R*)-(+)-1-phenylethylamine (**65**), 2-(4-carboxybenzoyl)-2-methyl-*endo*-bicyclo[2.1.1]hexyl (**66**) with (*S*)-(−)-1-phenylethylamine (**67**), and 6,6-diethyl-5-oxo-5,6,7,8-tetrahydronaphthalene-2-carboxylate (**68**) with (1*S*)-1-(4-methyl)phenylamine (**69**) during the Yang photocyclization, yielding **70**, **71**, and **72**, respectively (Scheme 7.12). The first step in the Yang photocyclization is γ-hydrogen abstraction, thus forming a biradical followed by ring closure. The crystal structures were determined after exposing the crystal to irradiation for different periods of time. It should be noted that since the timescale of X-ray diffraction data collection is much longer than the reaction time, the structure observed is an average structure of the reactant and the product at these intervals. However, the variation in the lattice parameters, as well as other geometric data, shows the effect of the photochemical reaction on the crystal structure.

The group of Kaftory[22,71,77,78] studied the possible homogeneity in the photochemical reactions of three different reactions: electrocyclic ring closure of α-oxoamides, [4 + 4] photodimerization of pyridone derivatives, and [4 + 4] photodimerization of anthracene derivatives. Only the first two are described in this section.

Solution photochemistry of α-oxoamides showed[79] that the primary step in the reaction is γ-hydrogen abstraction by the carbonyl oxygen to form a biradical, which can then proceed in one of three types of reactions: type II cyclization to yield β-lactam; 1,4-hydrogen shift followed by cyclization to yield oxazolidine-4-one; and type II elimination followed by addition of a nucleophile to give a mandelic acid derivative. In the solid of neat compounds, it was shown[80] that the studied compounds undergo selective type II reactions to yield β-lactams. The photochemistry of 1-(phenylglyoxyl) piperidine (**73a**) and of 4-oxo-(phenylacetyl)-morpholine (**73b**) embedded in various host molecules (**1, 3, 6, 9, 11**) was studied by Lavy et al.[78] It was found that the irradiation of the solid host–guest compounds leads selectively to the β-lactam or to the oxazolidine-4-one. The solid compounds were examined for possible single-crystal-to-single-crystal transformation, with the aim of understanding the reasons for adoption of different reaction courses. Four of the host–guest crystals underwent single-crystal-to-single-crystal transformation on irradiation, and the crystal structures could be monitored. Two of the compounds, **9:73a** and **9:73b**, adopted the course of reaction leading to α-oxoamide of type **74a** (tetrahydro-2-phenyl-oxazolo[2,3-c][1,4]oxazine-3(2H)-one), and the other two, **1:73a** and **11:73b**, led to the β-lactam of type **75a** (4-oxa-I-azabicyclo[4.2.0]octan-8-one,7-hydroxy-7-phenyl) (Scheme 7.13). The reason for the different behavior is attributed either to the different reaction cavity shape or to the different conformations adopted by the reactive molecules (Fig. 7.14).

The [4 + 4] solid-state photodimerization of pyridone and various derivatives of pyridone was a subject of many investigations. 2-Pyridone exists in solution as an equilibrium mixture with 2-hydroxypiridine, and therefore, its

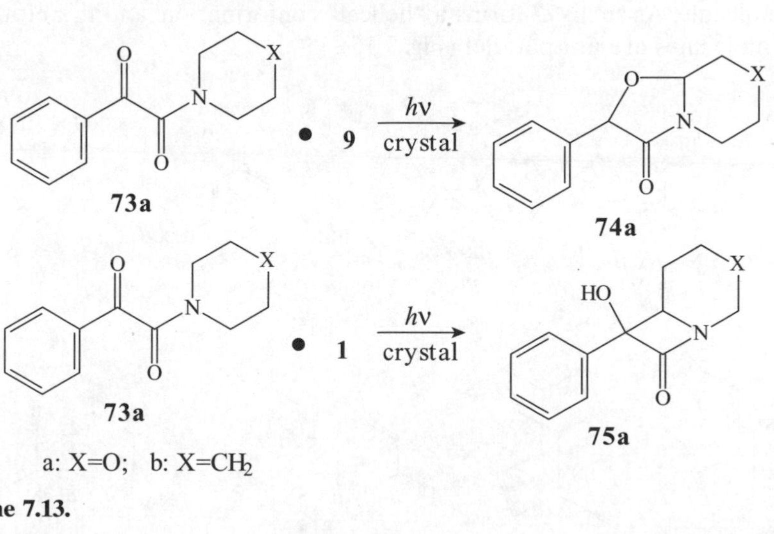

73a • 9 —$h\nu$, crystal→ 74a

73a • 1 —$h\nu$, crystal→ 75a

a: X=O; b: X=CH_2

Scheme 7.13.

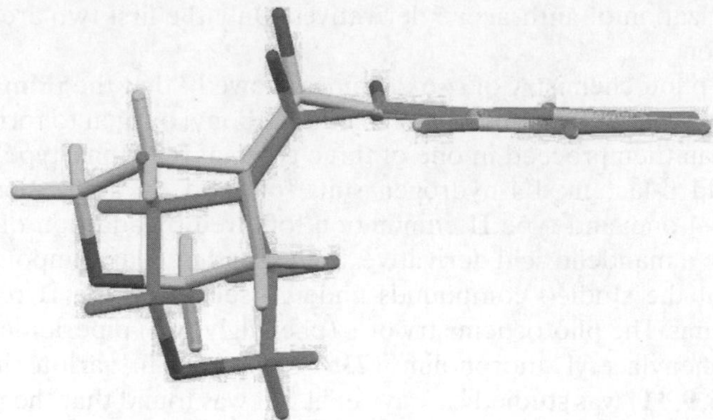

Figure 7.14. Overlay of **73a** as depicted from **9:73a** (yellow or light gray) and **1:73a** (green or darker gray). See ftp site for color: ftp://ftp.wiley.com/public/sci_tech_med/supramolecular_photochemistry.

solution photodimerization led to very poor yields. Irradiation of solid inclusion compounds of 2-pyridone with various hosts, on the other hand, gives the photodimer in very good yields. Recently, the group of Toda[81–84] demonstrated that irradiation of solid host–guest compounds composed of 2-pyridone (**76a**), 5-chloro-2-pyridone (**76b**), and 5-methyl-2-pyridone (**76c**) as the guest molecule and 1,1′-biphenyl-2,2′-dicarboxylic acid (**77**), 1,2,3-benzene tricarboxylic acid (**78**), or 1,2,4,5-benzene tetracarboxylic acid (**32**), may lead to either [4 + 4] photodimerization (**80a** or **80b**) or [2 + 2] photodimerization (**81a** or **81b**), depending on the packing of the guest molecules in the crystal. The host–guest compounds of **77**:2(**76b**) and **77**:2(**76c**) show similar structural relation between the pairs of guest molecules. Each host molecule holds, via hydrogen bonding, two guest molecules; the mutual orientation is determined by the host molecule. As such, **77** forms a "helical" conformation, and therefore, the guest molecules are not parallel (Fig. 7.15).

76 a: X=H b: X=Cl c: X=Me **77** **78** **79**

80a **80b** **81a** **81b**

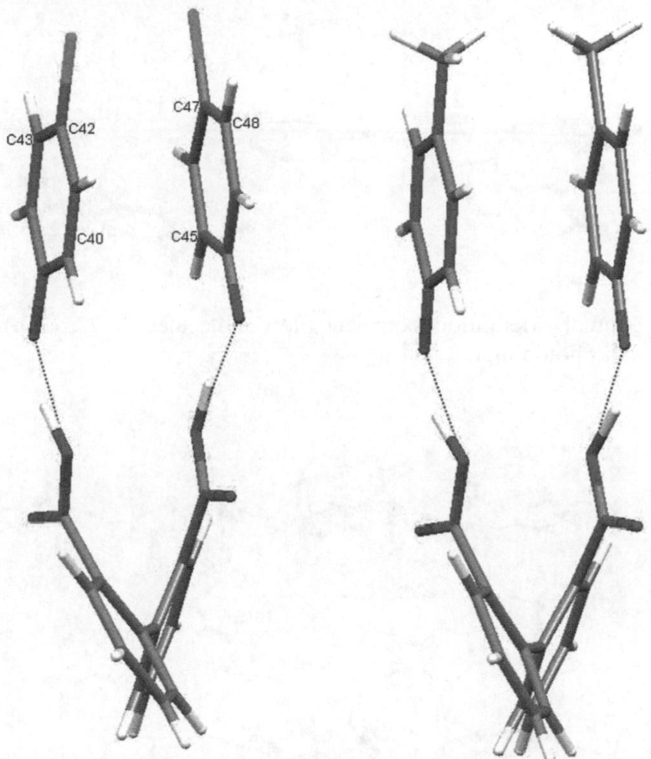

Figure 7.15. Mutual geometry between pairs of guest molecules in **77:**2(**76b**) (left) and **77:**2(**76c**) (right).

The distances between C43–C47 and C42–C48 (3.432 and 3.325 Å, respectively) to form the [2 + 2] photodimer are much shorter than C43–C48 and C40–C45 (3.372 and 4.511 Å, respectively) to form the [4 + 4] photodimer. The relative geometry between the guest molecules in **77:**2(**76c**) is very similar. It seems that the shorter distances leading to the [2 + 2] dimer dictate the photoreaction. Irradiation of the inclusion compound of **79:**2(**76b**) resulted in the *meso-cis-syn* [4 + 4] photodimer, of type **80b** (Fig. 7.16). In the case of **79:**2(**76a**), each guest molecule is held by another host molecule; the control imposed by the conformation of the host molecule is lost, and a [4 + 4] photodimer *trans–anti* of type **80a** is produced. It is interesting to note that in some cases where the crystals were inert to photoirradiation, heating or exposing the crystals to acetonitrile vapor initiated the photoreaction.[84]

Lavy and Kaftory[77] showed that 2-pyridone and 6-methyl-2-pyridone as inclusion compounds with **1**, undergo single-crystal-to-single-crystal transformation to yield [4 + 4] photodimer of type **80a**. This fact raises an interesting question regarding the space required for solid-state topotactic reactions. It is known that if the reaction cavity is too small, the reaction will be

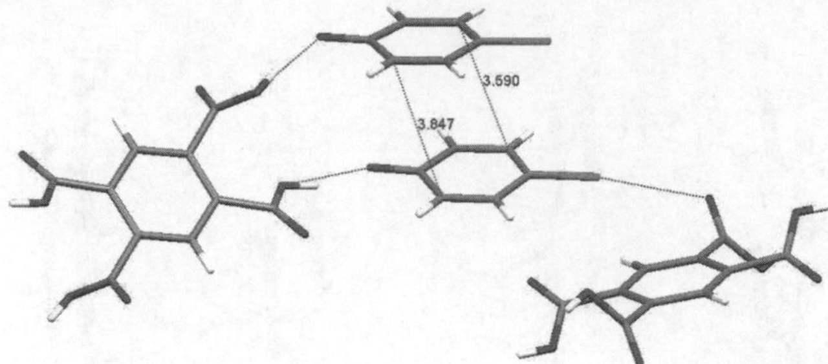

Figure 7.16. Mutual orientation between guest molecules in **77**:2(**76b**) leading to [4 + 4] *trans–anti* photodimer.

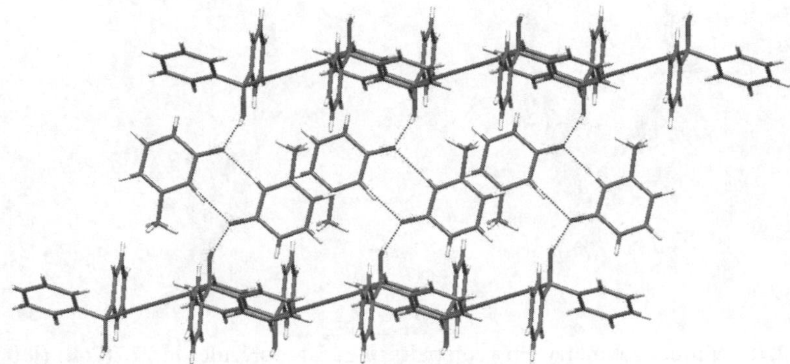

Figure 7.17. The structure of 6-methyl-2-pyridone embedded within the host molecule **1** before irradiation.

heterogeneous and the crystal will disintegrate. On the other hand, if the reaction cavity is sufficient to accommodate the reactant and the product (not at the same time), the reaction can be homogeneous and the integrity of the crystalline will be preserved. What happens if the volume of the product is smaller than that of the substrate and therefore smaller than the reaction cavity? In this particular case, water molecules penetrate into the crystal and reside in an ordered manner, forming hydrogen bonding with the guest or with the host molecules (Figs. 7.17 and 7.18).

Moreover, in these particular examples, it was found that molecules of the monomer (or dimer) flip upside down during the transformation. In the absence of moisture during irradiation, the crystal disintegrates.

Other examples of molecular or atomic reorientation or rotation in the crystal without affecting the crystal integrity were shown to occur in solid host–guest compounds. The most thoroughly investigated event is the *cis–trans* isomerization that has been studied by single-crystal-to-single-crystal transfor-

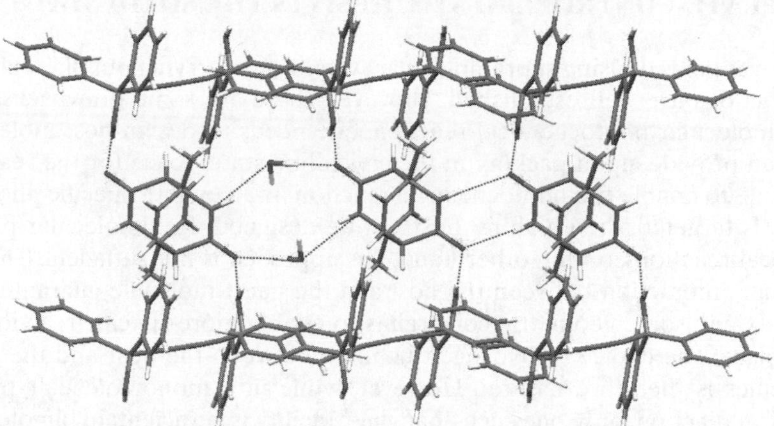

Figure 7.18. The structure of the host–guest crystal in Figure 7.17 after irradiation and water penetration.

mation. Ananchenko et al.[20] demonstrated the phototransformation of stilbene within *para*-hexa-nonylcalix[4]arene. The *cis*-stilbene stacks as a π–π dimer located at the center of the capsule of the calix[4]arene. The *trans*-stilbene is not packed in dimers. Each crystalline inclusion compound was irradiated and the other isomer was found. After prolonged irradiation, the monomers of stilbene photodimerized. The group of Coppens[85–87] showed the single-crystal-to-single-crystal *E* to *Z* and *Z* to *E* isomerization of 3-chloroacrylic acid (**82**) and *E* to *Z* isomerization of tiglic acid (**83**) in supramolecular framework. The host in both examples was *C*-ethylcalix[4]resorcinarene (**86**). Upon stepwise exposure of a single crystal to 325-nm light from He/Cd laser, the *Z* isomer of tiglic acid underwent isomerization to the *E* isomer. An equilibrium was reached at 30% of the *Z* isomer. The kinetic measurement of the isomerization indicates first-order kinetics. The group[87] also described a single-crystal-to-single-crystal isomerization of two Zn-coordinated tiglic acid molecules.

82(*E*) **83(*Z*)** **84(*E*)** **85(*Z*)** **86**

TEMPLATE-CONTROLLED SYNTHESIS IN THE SOLID STATE

The advantages of using supramolecular systems for carrying out photochemical reactions are well established. However, there are some drawbacks. In a monomolecular photochemical reaction, one needs to design host molecules that can provide, upon packing in the crystalline state, space for the reactant molecule to enable the photochemical reaction. If an enantiospecific photoreaction is targeted, then a chiral host can be designed. In bimolecular photochemical reactions, on the other hand, this approach is not sufficient. The use of strong interaction between the host and the guest molecule guarantees, to some extent, some geometric constraints to enable more specific reactions to take place. The choice of hydrogen bonding between the host and the guest molecules is therefore natural. However, while in a monomolecular photochemical reaction, only one such "hanging" facility is sufficient; in bimolecular reactions, the circumstances are different. To ensure that bimolecular reactions take place, one should control the packing so that the two reacting molecules will be in close proximity and with the right mutual alignment. To enhance the control on relative geometry and the proximity of two potentially reacting molecules, it is better to have them bound (by hydrogen bonds or any other strong intermolecular interactions) to a single molecule, as seen schematically in Scheme 7.14. A rational design of template is needed in order to execute the plan shown in Scheme 7.13.

The geometric requirements for a [2 + 2] photodimerization have been well established by the extensive work of the Schmidt group.[88] Therefore, it was only natural that this photoreaction was used for testing the template concept. The first attempt was made using the double salt formation by diamines (**87**) with *trans*-cinnamic acid (**88**) to yield truxilic acid (**89**), as shown in Scheme 7.15.[89,90] An example of the crystal structure of such a pair of reactive mole-

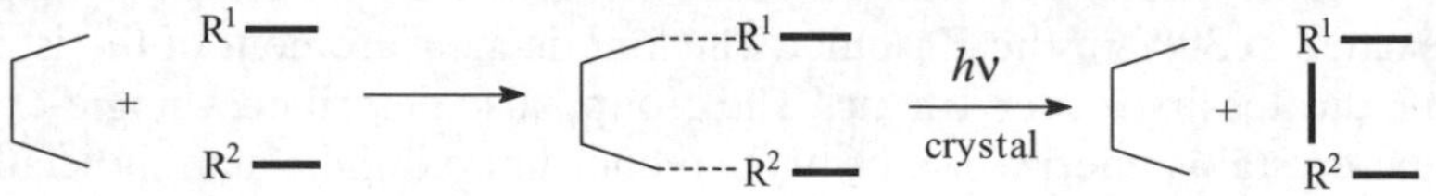

R^1 —— + R^2 —— → ----R^1 —— ----R^2 —— $\xrightarrow[\text{crystal}]{h\nu}$ + R^1 —— R^2 ——

Scheme 7.14.

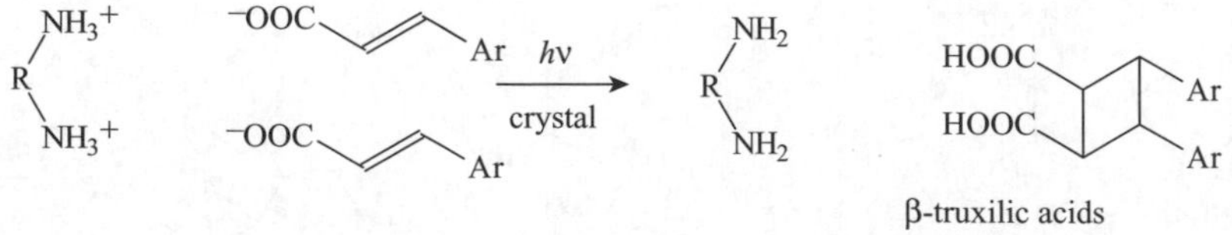

NH_3^+ R NH_3^+ (**87**) + ^-OOC–CH=CH–Ar / ^-OOC–CH=CH–Ar (**88**) $\xrightarrow[\text{crystal}]{h\nu}$ NH_2 R NH_2 + HOOC, HOOC, Ar, Ar — β-truxilic acids (**89**)

Scheme 7.15.

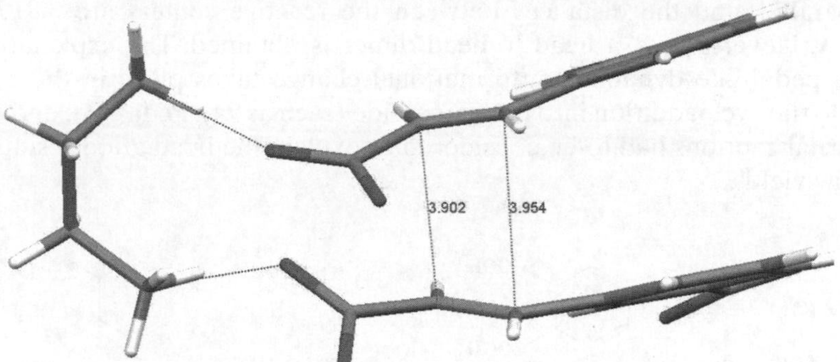

Figure 7.19. The template of diamine (**87**) with *trans*-cinnamic acid (**88**).

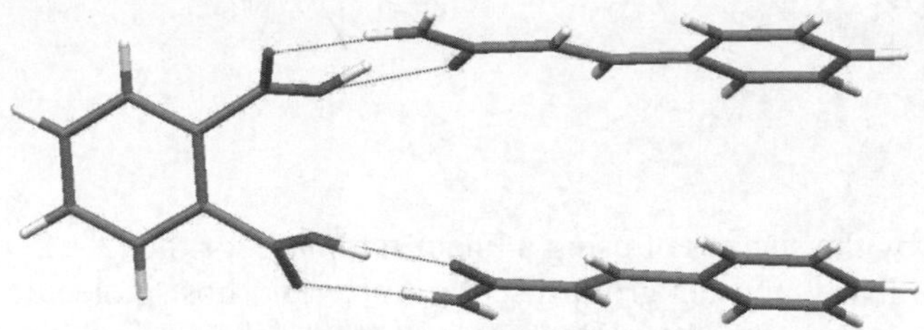

Figure 7.20. The structure of **91:93a** before irradiation (the photoactive double bond is marked in violet). See ftp site for color: ftp://ftp.wiley.com/public/sci_tech_med/supramolecular_photochemistry.

cules is shown in Figure 7.19. This demonstrates the concept of using a template for bimolecular reactions.

Ito et al.[91] continued their efforts to show the advantage of using the double salt to orient potential reactive molecules with double bonds to undergo [2 + 2] photodimerization. In that publication, the role of the cation–anion is interchanged so that the light-stable species is a dicarboxylic acid, such as oxalic acid, succinic acid, phthalic acid, or fumaric acid (**90–92**), and the light-sensitive molecules are positively charged amides, such as *trans*-cinnamamaides (**93**). It was shown that dimeric-like hydrogen bonds hold the molecules in a double layer arrangement that enables, in some of the supramolecular systems, [2 + 2] photodimerization to occur. Of the eight different cocrystals, four (**90a:93a**, **91:93a**, **90a:93b**, and **90a:93c**) gave significant yields (42%, 37%, 43%, and 86%, respectively) of the corresponding β-truxinamide (**94**). Irradiation of crystals of **90c:93a** led to the α-truxillamide (**95**). An example of the template arrangement of **91:93a** is shown in Figure 7.20. The phthalic acid holds two guest molecules in a head-to-head orientation. The double bonds that should undergo [2 + 2] photodimerization (marked in violet in Fig. 7.20) are not parallel, and the distances between the reactive centers are 3.819 and 4.849 Å. Nevertheless, a head-to-head dimer is obtained. The explanation is that a pedal-like dynamic conformational change takes place in the crystal prior to the cycloaddition into β-truxinamide (such as **94**). Pedal-like and other molecular motions had to be considered to explain the irradiation results and the low yields.

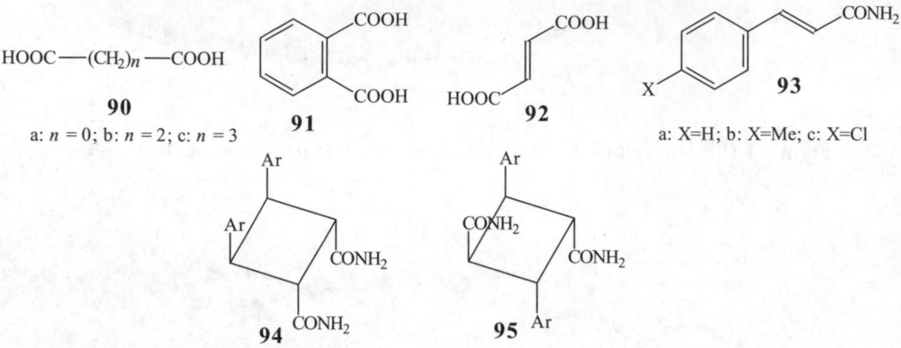

HOOC—$(CH_2)n$—COOH

90
a: $n = 0$; b: $n = 2$; c: $n = 3$

91 **92** **93**
a: X=H; b: X=Me; c: X=Cl

94 **95**

Inspired by the success of using a linear template for the [2 + 2] photoreaction, MacGillivray and his group used a more rigid host molecule (aromatic ring) with two hydrogen bond donor groups that will ensure the organization of the guest molecules independent of the crystal packing. The linear template was based on 1,3-dihydroxybenzene (resorcinol) (**96**), and the photoreactive guest molecule was *trans*-1,2-bis(4-pyridiyl)ethylene (**97**), as shown in Scheme 7.16 and Figures 7.21 and 7.22.

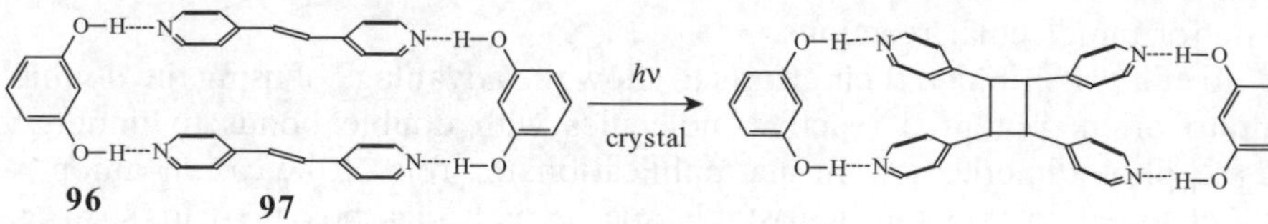

96 **97** $h\nu$ crystal

Scheme 7.16.

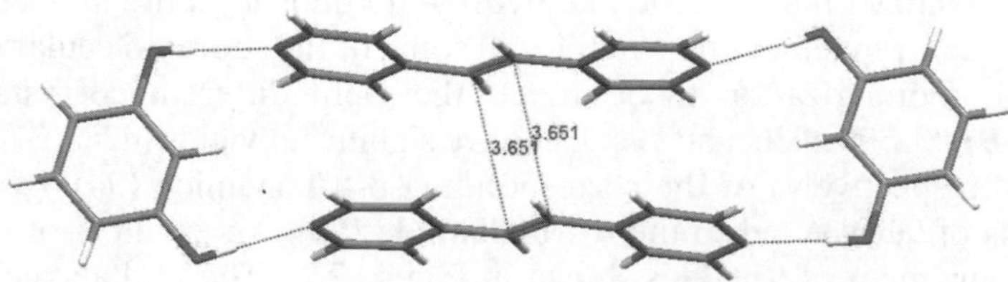

Figure 7.21. The template structure of resorcinol and *trans*-1,2-bis(4-pyridyl)ethylene before irradiation.

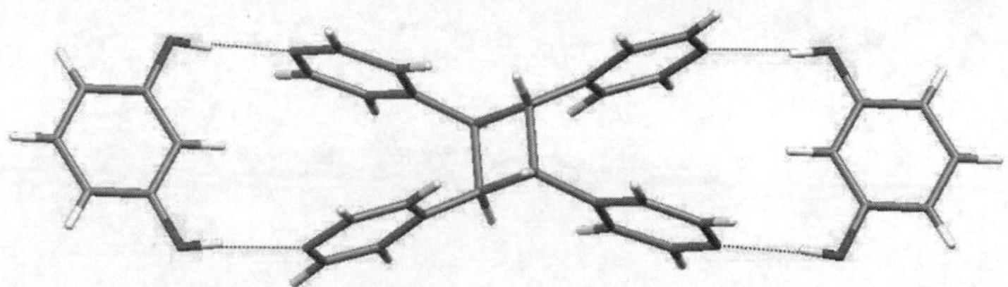

Figure 7.22. The template structure of resorcinol and the dimer of *trans*-1,2-bis(4-pyridyl)ethylene obtained after irradiation.

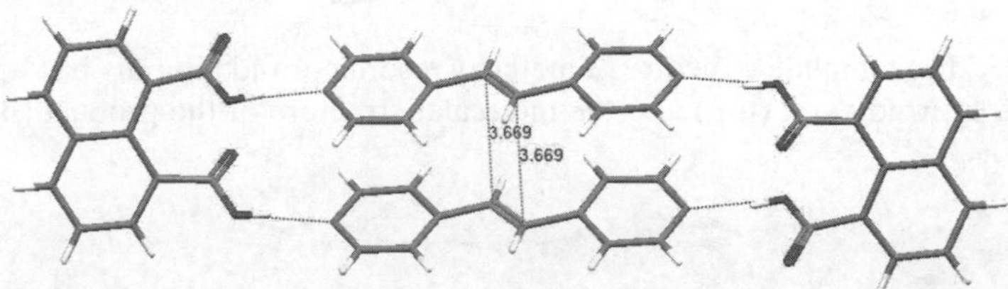

Figure 7.23. The template structure of 1,8-naphthalenedicarboxylic acid and *trans*-1,2-bis(4-pyridyl)ethylene before irradiation.

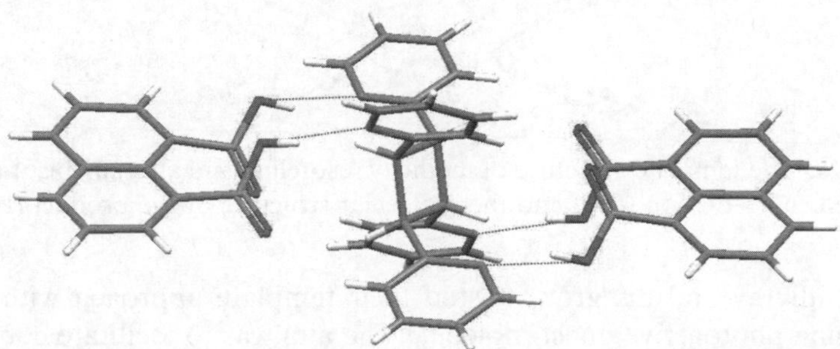

Figure 7.24. The template structure of 1,8-naphthalenedicarboxylic acid and *trans*-1,2-bis(4-pyridyl)ethylene after irradiation.

Following the report of MacGillivray et al.,[92] Garcia-Garibay et al.[93] showed that a flexible template based on a crown ether can organize compounds with C=C double bond to facilitate [2 + 2] photodimerization in the solid state with 80% yield. Further development of the linear template for employing [2 + 2] photodimerization was carried out by the group of MacGillivray by modifying the template by different substituents. Also, 1,8-naphthalenedicarboxylic acid was tested as a template[94] (Figs. 7.23 and 7.24). Similarly, Shan and Jones[95] used tricarboxylic acid and 1,2,4,5-benzene tetracarboxylic acid as templates for the alignment in the solid state of 1,2-bis(4-pyridyl)ethylene and demonstrated that irradiation yields the photodimer. Other examples of the use of similar linear templates can be found in the review article by MacGillivray et al.[96]

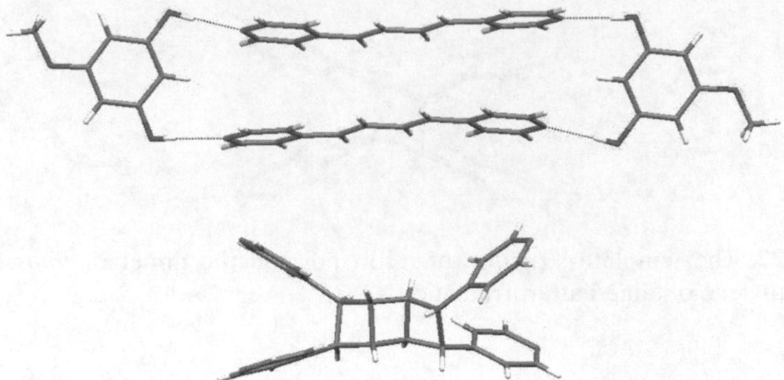

Figure 7.25. The template structure of methoxy resorcinol and all-*trans*-bis(4-pyr-poly-2-ene) before irradiation (top) and the molecular structure of the product (bottom).

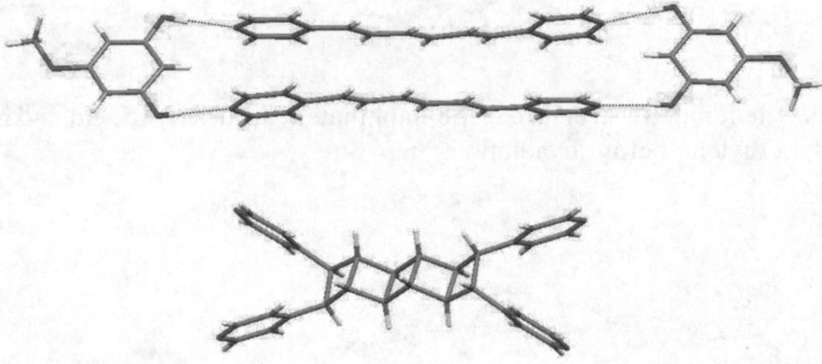

Figure 7.26. The template structure of methoxy resorcinol and all-*trans*-bis(4-pyr-poly-3-ene) before irradiation (top) and the molecular structure of the product (bottom).

MacGillivray and his group tested their template approach with a more challenging photoactive guest molecule. The aim was to facilitate consecutive [2 + 2] photodimerization to form ladderanes (Gao et al.[74]). They used the methoxy resorcinol, all-*trans*-bis(4-pyr-poly-*m*-ene) where $m = 2$ or 3 (**98** and **99**) as the guest molecules. Irradiation over a period of 120 h revealed the ladderanes, as confirmed by their X-ray crystal structures (Figs. 7.25 and 7.26). Template-controlled photodimerization by using metal ions and other inorganic frameworks is a promising new topic of research. This subject is beyond the scope of this review; however, the interested reader may find the review of Buĉar et al.[97] interesting.

98 **99**

Polymerization of diacetylenes in the solid state was extensively studied and the geometric requirement was specified (see e.g., Enkelmann[98]). Fowler et al.[99] described a strategy based on host molecules that serve as templates to orient the diacetylenes to enable thermal polymerization. Their approach is a wonderful example of the clever use of data from known crystal structures for crystal engineering. The first successful design was the host–guest compound of **100** and **101**.[100] Unfortunately, ultraviolet (UV) irradiation did not initiate polymerization. However, exposure of a single crystal to ^{60}Co γ-radiation resulted in the crystals becoming dark red. Raman spectroscopy has proven to be an excellent method for characterizing conjugated polymer, and it was shown that polymerization had indeed taken place. The crystal structure before irradiation is shown in Figure 7.27. In their effort to design a better host–guest system to enable polymerization of diacetylenes, they were able to prepare eight cocrystals with oxalamides **102a**, **102b**, and **102c** as the host molecule and the pyridyldiacetylenes **103** and **104**.[101] The crystals of **102a:104** (Fig. 7.28) were quite sensitive to UV irradiation. A color change was observed and the crystal was turned red. The yield of the conversion was estimated to be 61%. This cocrystal was also reactive when exposed to γ-ray irradiation.

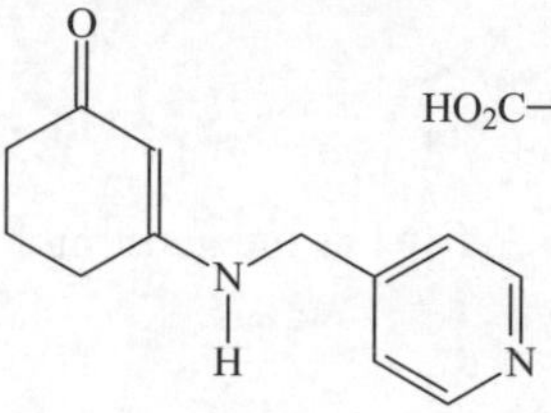

O N H N **100**

HO_2C CO_2H **101**

H O $(CH_2)n$ N H O C O H N O $(CH_2)n$ O H O

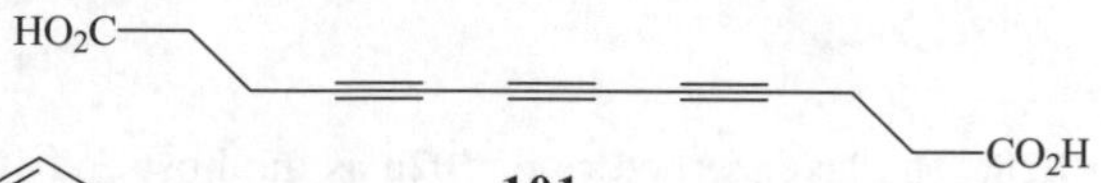

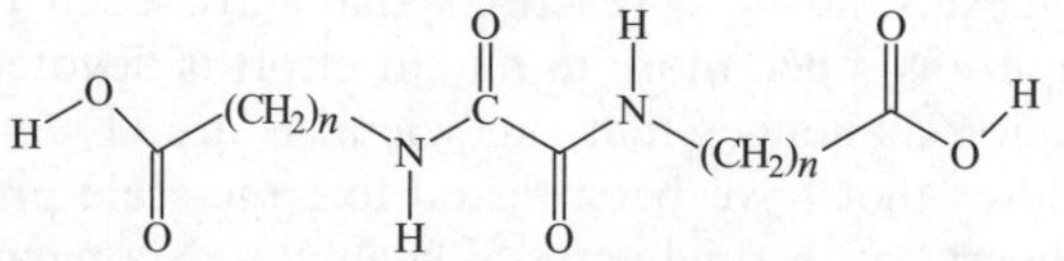

a: n = 1; b: n = 2; c: n = 4 **102**

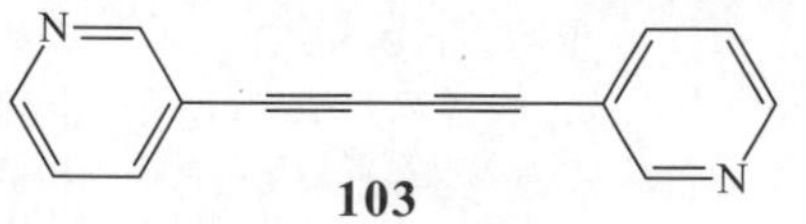

N N **103** N N **104**

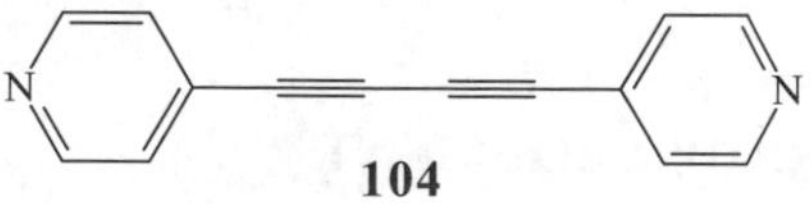

CONCLUDING REMARKS

This chapter did not cover all the published work on the structural aspects and templating in the photochemistry of solid host–guest compounds. However, it provides the flavor of the subject by presenting interesting examples showing how far the scientific community has come and the advantages of working in

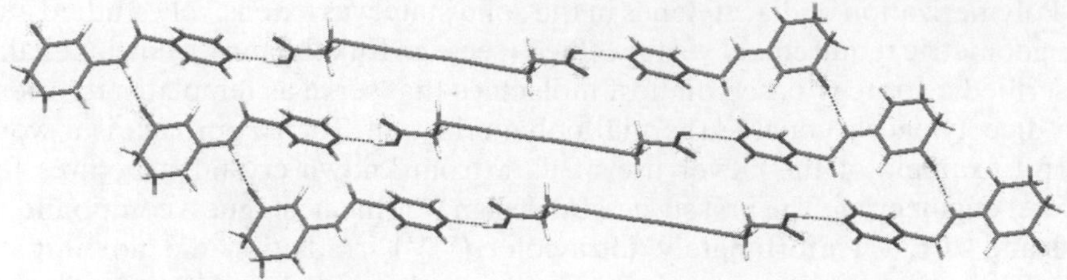

Figure 7.27. The effect of the template on the alignment of the triacetylenic moiety.

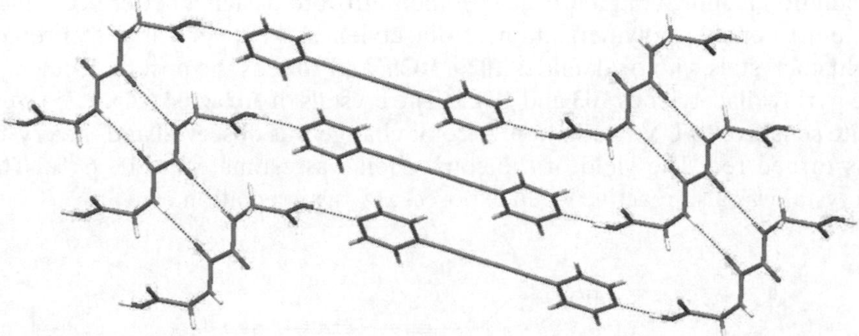

Figure 7.28. The layer structure of **102a** as the host and **104** as the guest molecules.

this challenging field of chemistry. There are many examples of photochemical reactions in host–guest solid-state reactions that have been discovered in recent years. And there is much more to find, if effort is devoted to research in the field. Two major challenges that call scientists are (1) to translate the in-laboratory syntheses that have been tested to large-scale production and (2) to use the single-crystal-to-single-crystal examples as a means to explore reaction mechanisms. Hopefully, the scientific community will do its best to reply to this call.

ACKNOWLEDGMENT

The work was supported by the Israel Science Foundation, #499/08.

REFERENCES

1. Wegner, G.Z. *Naturforsch* **1969**, *24b*, 824.
2. Osaki, K.; Schmidt, G.M.J. *Isr. J. Chem.* **1972**, *10*, 189.

3. Cheng, K.; Foxman, B. *J. Am. Chem. Soc.* **1977**, *99*, 8102.
4. Nakanishi, H.; Jones, W.; Thomas, J.M.; Hursthouse, M.B.; Motevalli, M. *J. Phys. Chem.* **1981**, *85*, 3636.
5. Chang, H.C.; Popovitz-Biro, R.; Lahav, M.; Leiserowitz, L. *J. Am. Chem. Soc.* **1982**, *104*, 614.
6. Ohashi, Y.; Yanagi, K.; Kurihara, T.; Sasada, Y.; Ohgo, Y. *J. Am. Chem. Soc.* **1982**, *104*, 6353.
7. Braun, H.-G.; Wegner, G. *Makromol. Chem.* **1983**, *184*, 1103.
8. Tieke, B.; Chapuis, G. *J. Polym. Sci. Polym. Chem. Ed.* **1984**, *22*, 2895.
9. Wang, W.-N.; Jones, W. *Tetrahedron* **1987**, *43*, 1273.
10. Leibovitch, M.; Olovsson, G.; Scheffer, J.R.; Trotter, J. *J. Am. Chem. Soc.* **1998**, *120*, 12755.
11. Theocharis, C.R.; Desiraju, G.R. *J. Am. Chem. Soc.* **1984**, *106*, 3606.
12. Turowska-Tyrk, I. *Acta Crystallogr.* **2003**, *B59*, 670.
13. Turowska-Tyrk, I.; Trzop, E. *Acta Crystallogr.* **2003**, *B59*, 779.
14. Harada, J.; Uekusa, H.; Ohashi, Y. *J. Am. Chem. Soc.* **1999**, *121*, 5809.
15. Kawano, M.; Sano, T.; Abe, J.; Ohashi, Y. *J. Am. Chem. Soc.* **1999**, *121*, 8106.
16. Enkelmann, V.; Wegner, G.; Novak, K.; Vagener, K.B. *J. Am. Chem. Soc.* **1993**, *115*, 10390.
17. Novak, K.; Enkelmann, V.; Wegner, G.; Wagener, K.B. *Angew. Chem. Int. Ed. Engl.* **1993**, *32*, 1614.
18. Amirsakis, D.G.; Elizarov, A.M.; Garcia-Garibay, M.A.; Glink, P.T.; Stoddart, J.F.; White, A.J.; William, D.J. *Angew. Chem. Int. Ed. Engl.* **2003**, *42*, 1126.
19. Toda, F.; Bishop, B. *Separation and Reactions in Organic Supramolecular Chemistry: Perspectives in Supramolecular Chemistry*, Vol. 8. New York: John Wiley & Sons, **2004**.
20. Ananchenko, G.S.; Udachin, K.A.; Ripmeester, J.A.; Perrier, T.; Coleman, A.W. *Chem. Eur. J.* **2006**, *12*, 2441.
21. Halder, G.; Kepert, C. *Aust. J. Chem.* **2006**, *59*, 597.
22. Zouev, I.; Lavy, T.; Kaftory, M. *Eur. J. Org. Chem.* **2006**, 4164.
23. Coppens, P.; Zheng, S.-L.; Gembicky, M.; Messerschmidt, M.; Dominiak, P.M. *CrystEngComm* **2006**, *8*, 735.
24. Cohen, M.D. *Angew. Chem. Int. Ed. Engl.* **1975**, *14*, 386.
25. Weiss, R.G.; Ramamurthy, V.; Hammond, G. *Acc. Chem. Res.* **1993**, *26*, 530.
26. Kaftory, M.; Tanaka, K.; Toda, F.J. *Org. Chem.* **1985**, *50*, 2154.
27. Kaftory, M.; Yagi, M.; Tanaka, K.; Toda, F.J. *Org. Chem.* **1988**, *53*, 4391.
28. Tanaka, K.; Mizutani, H.; Miyahara, I.; Hirotsu, K.; Toda, F. *CrystEngComm* **1999**, *3*, 8.
29. Tanaka, K.; Toda, F.; Mochizuki, E.; Yasui, N.; Kai, Y.; Miyahara, I.; Hirotsu, K. *Angew. Chem. Int. Ed. Engl.* **1999**, *38*, 3523.
30. Ohba, S.; Hosomi, H.; Tanaka, K.; Miyamoto, H.; Toda, F. *Bull. Chem. Soc. Jpn.* **2000**, *73*, 2075.
31. Tanaka, K.; Toda, F. In: Toda, F., editor. *Organic Solid-State Reactions*. Dordrecht, the Netherlands: Kluwer Academic Publishers, **2002**, pp. 109–158.

32. Tanaka, K.; Toda, F. *J. Chem. Soc. Chem. Commun.* **1983**, 593.
33. Tanaka, K.; Toda, F. *Nippon Kagakukaishi* **1984**, 141.
34. Kaftory, M. *Tetrahedron* **1987**, *43*, 1503.
35. Fujiwara, T.; Tanaka, K.; Toda, F. *J. Chem. Soc. Perkin 1* **1989**, 663.
36. Shimizu, L.S.; Hughes, A.D.; Smith, M.D.; Davis, M.J.; Zhang, B.P.; Zur Loye, H.C.; Shimizu, K.D. *J. Am. Chem. Soc.* **2003**, *125*, 14972.
37. Shimizu, L.S.; Hughes, A.D.; Smith, M.D.; Samuel, S.A.; Ciurtin-Smith, D. *Supramol. Chem.* **2005**, *17*, 27.
38. Yang, J.; Dewal, M.B.; Shimizu, L.S. *J. Am. Chem. Soc.* **2006**, *128*, 8122.
39. Yang, J.; Dewal, M.B.; Profeta, S.; Smith, M.D.; Li, Y.; Shimizu, L.S. *J. Am. Chem. Soc.* **2008**, *130*, 612.
40. Obata, T.; Shimo, T.; Yasutake, M.; Shinmyozu, T.; Kawaminami, M.; Yoshida, R.; Somekawa, K. *Tetrahedron* **2001**, *57*, 1531 and references therein.
41. Horiguchi, M.; Ito, Y.J. *Org. Chem.* **2006**, *71*, 3608.
42. Lavy, T.; Kaftory, M. *Acta Cryst.* **2006**, *E62*, 3977.
43. Lavy, T.; Kaganovich, M.; Kaftory, M. *Acta Cryst.* **2006**, *E62*, 3979.
44. Lavy, T.; Meirovich, N.; Sparkes, H.A.; Howard, J.A.K.; Kaftory, M. *Acta Cryst.* **2007**, *C63*, 89.
45. (*a*) Toda, F. *Top. Curr. Chem.* **1988**, *149*, 211; (*b*) Toda, F. *Acc. Chem. Res.* **1995**, *28*, 480.
46. Tanaka, K.; Toda, F. *Chem. Rev.* **2000**, *100*, 1025.
47. Toda, F. *Top. Curr. Chem.* **2005**, *254*, 1.
48. Gamlin, J.N.; Jones, R.; Leibovitch, M.; Patrick, B.; Scheffer, J.R.; Trotter, J. *Acc. Chem. Res.* **1996**, *29*, 203.
49. (*a*) Scheffer, J.R.; Xia, W. *Top. Curr. Chem.* **2005**, *254*, 233; (*b*) Gamlin, J.N.; Jones, R.; Leibovitch, M.; Patrick, B.; Scheffer, J.R.; Trotter, J. *Acc. Chem. Res.* **1996**, *29*, 203.
50. Wada, T.; Inoue, Y. In: Inoue, Y.; Ramamurthy, V., editors. *Chiral Photochemistry*. New York: Marcel Dekker, **2004**, pp. 341–384.
51. Koshima, H. In: Inoue, Y.; Ramamurthy, V., editors. *Chiral Photochemistry*. New York: Marcel Dekker, **2004**, pp. 485–532.
52. Toda, F.; Tanaka, K. *J. Chem. Soc. Chem. Commun.* **1986**, 1429.
53. Toda, F.; Tanaka, K.; Yagi, M. *Tetrahedron* **1987**, *43*, 1493.
54. Gudmundsdottir, A.D.; Scheffer, J.R. *Tetrahedron Lett.* **1990**, *31*, 6807.
55. Gudmundsdottir, A.D.; Scheffer, J.R. *Photochem. Photobiol.* **1991**, *54*, 535.
56. Jones, R.; Scheffer, J.R.; Trotter, J.; Yang, J. *Tetrahedron Lett.* **1992**, *33*, 5481.
57. Gudmundsdottir, A.D.; Scheffer, J.R.; Trotter, J. *Tetrahedron Lett.* **1994**, *35*, 1397.
58. Gudmundsdottir, A.D.; Li, W.; Scheffer, J.R.; Rettig, S.; Trotter, J. *Mol. Cryst. Liq. Cryst.* **1994**, *240*, 81.
59. Jones, R.; Scheffer, J.R.; Trotter, J.; Yang, J. *Acta Crystallogr.* **1994**, *B50*, 601.
60. Koshima, H.; Maeda, A.; Masuda, N.; Matsuura, T.; Hirotsu, K.; Okada, K.; Mitzutani, H.; Ito, Y.; Fu, T.Y.; Scheffer, J.R.; Trotter, J. *Tetrahedron Asymmetry* **1994**, *5*, 1415.
61. Scheffer, J.R.; Wang, L. *J. Phys. Org. Chem.* **2000**, *13*, 531.

62. Joy, A.; Ramamurthy, V. *Chem. Eur. J.* **2000**, *6*, 1287.
63. Koodanjeri, S.; Joy, A.; Ramamurthy, V. *Tetrahedron* **2000**, *56*, 7003.
64. Patrick, B.O.; Scheffer, J.R.; Scott, C. *Angew. Chem. Int. Ed. Engl.* **2003**, *42*, 3775.
65. Botoshansky, M.; Braga, D.; Kaftory, M.; Maini, L.; Patrick, B.O.; Scheffer, J.R.; Wang, K. *Tetrahedron Lett.* **2005**, *46*, 1141.
66. (*a*) Koshima, H.; Miyamoto, H.M.; Yagi, I.; Uosaki, K. *Cryst. Growth Des.* **2004**, *4*, 807 and references therein; (*b*) Koshima, H.; Kawanishi, H.; Nagano, M.; Yu, H.; Shiro, M.; Hosoya, T.; Uekusa, H.; Ohashi, Y. *J. Org. Chem.* **2005**, *70*, 4490.
67. Keating, A.E.; Garcia-Garibay, M.A. In: Ramamurthy, V.; Schanze, K.S., editors. *Photochemical Solid-to-Solid Reactions*. New York: Marcel Dekker, **1998**, 195–248.
68. Leibovitch, M.; Olovsson, G.; Scheffer, J.R.; Trotter, J. *J. Am. Chem. Soc.* **1997**, *119*, 1462.
69. Hosomi, H.; Ohba, S.; Tanaka, K.; Toda, F. *J. Am. Chem. Soc.* **2000**, *122*, 1818.
70. Turowska-Tyrk, I.; Grzesniak, K.; Trzop, E.; Zych, T. *J. Solid State Chem.* **2003**, *174*, 459.
71. Lavy, T.; Sheynin, Y.; Kaftory, M. *Eur. J. Org. Chem.* **2004**, 4802.
72. Turowska-Tyrk, I.; Trzop, E.; Scheffer, J.R. *Acta Cryst.* **2006**, *B62*, 128.
73. Turowska-Tyrk, I.; Bakowicz, J.; Scheffer, J.R. *CrystEngComm* **2006**, *8*, 616.
74. Gao, X.; Friščič, T.; MacGillivray, L.R. *Angew. Chem. Int. Ed. Engl.* **2004**, *43*, 232.
75. (*a*) Friščič, T.; MacGillivray, L.R. *Chem. Commun.* **2003**, 1306–1307; (*b*) Friščič, T.; MacGillivray, L.R. *Z. Kristallogr.* **2005**, *220*, 351.
76. Turowska-Tyrk, I.; Bakowicz, J.; Scheffer, J.R. *Acta Cryst.* **2007**, *B63*, 933.
77. Lavy, T.; Kaftory, M. *CrystEngComm* **2007**, *9*, 123.
78. Lavy, T.; Sheynin, Y.; Sparkes, H.A.; Howard, J.A.K.; Kaftory, M. *CrystEngComm* **2008**, *10*, 734.
79. Aoyama, H.; Hasegawa, T.; Omote, Y. *J. Am. Chem. Soc.* **1979**, *101*, 5343.
80. Aoyama, H.; Sakamoto, M.; Kuwabara, K.; Yoshida, K.; Omote, Y. *J. Am. Chem. Soc.* **1983**, *105*, 1958.
81. Hirano, S.; Toyota, S.; Toda, F. *Mendeleev Commun.* **2004**, *14*, 247.
82. Hirano, S.; Toyota, S.; Toda, F. *Heterocycles* **2004**, *64*, 383.
83. Hirano, S.; Toyota, S.; Toda, F. *Chem. Commun.* **2005**, 643.
84. Hirano, S.; Toyota, S.; Toda, F.; Fujii, K.; Uekuasa, H. *Angew. Chem. Int. Ed. Engl.* **2006**, *45*, 6013.
85. Zheng, S.-L.; Messerschmidt, M.; Coppens, P. *Acta Cryst.* **2007**, *B63*, 644.
86. Zheng, S.-L.; Messerschmidt, M.; Coppens, P. *Chem. Commun.* **2007**, 2735.
87. Zheng, S.-L.; Vande Velde, C.M.L.; Messerschmidt, M.; Volkov, A.; Gembicky, M.; Coppens, P. *Chem. Eur. J.* **2008**, *14*, 706 and references therein.
88. Schmidt, G.M.J. *Pure Appl. Chem.* **1971**, *27*, 647.
89. Ito, Y.; Borecka, B.; Trotter, J.; Scheffer, J.R. *Tetrahedron Lett.* **1995**, *36*, 6083.
90. Ito, Y.; Borecka, B.; Trotter, J.; Scheffer, J.R. *Tetrahedron Lett.* **1995**, *36*, 6087.
91. Ito, Y.; Hosomi, H.; Ohba, S. *Tetrahedron* **2000**, *56*, 6833.
92. MacGillivray, L.R.; Reid, J.L.; Ripmeester, J.A. *J. Am. Chem. Soc.* **2000**, *122*, 7817.

93. Amirsakis, D.G.; Garcia-Garibay, M.A.; Rowan, S.J.; Stoddart, J.F.; White, A.J.P.; Williams, D.J. *Angew. Chem. Int. Ed. Engl.* **2001**, *40*, 4256 and references therein.
94. Papaefstathiou, G.S.; Kipp, A.J.; MacGillivray, L.R. *Chem. Commun.* **2001**, 2462.
95. Shan, N.; Jones, W. *Tetrahedron Lett.* **2003**, *44*, 3687.
96. MacGillivray, L.R.; Papaefstathiou, G.S.; Friščič, T.; Varshney, D.B.; Hmilton, T.D. *Top. Curr. Chem.* **2004**, *248*, 201.
97. Bucar, D.-J.; Papaefstathiou, G.S.; Hamilton, T.D.; Chu, Q.L.; Georgiev, I.G.; MacGillivray, L.R. *Eur. J. Inorg. Chem.* **2007**, 4559.
98. Enkelmann, V. *Adv. Polym. Sci.* **1984**, *63*, 91.
99. Kane, J.J.; Liao, R.F.; Lauher, J.W.; Fowler, F.W. *J. Am. Chem. Soc.* **1995**, *117*, 12003.
100. Xiao, J.; Yang, M.; Lauher, J.M.; Fowler, F.W. *Angew. Chem. Int. Ed. Engl.* **2000**, *39*, 2132.
101. Curtis, S.M.; Le, N.; Fowler, F.W.; Lauher, W. *Cryst. Growth Des.* **2005**, *5*, 2313.

8

PHOTOCHROMISM OF MULTICOMPONENT DIARYLETHENE CRYSTALS

MASAKAZU MORIMOTO AND MASAHIRO IRIE

INTRODUCTION

Photochromism is defined as a photoinduced reversible transformation of a chemical species between two isomers having different absorption spectra.[1] Photochromic molecules change not only their colors but also other chemical and physical properties, such as oxidation/reduction potentials, dielectric constants, and refractive indices. These changes are ascribed to the difference in electronic structures of the two isomers. So far, various types of photochromic molecules have been reported. Some of them are shown in Scheme 8.1. Azobenzene and spirobenzopyran undergo thermally reversible photochromic reactions. The photogenerated colored isomers are thermally unstable and return to the colorless ones in the dark. On the other hand, photochromic reactions of furylfulgide and diarylethene are thermally irreversible. The photogenerated colored isomers are thermally stable in the dark and hardly return to the colorless ones at room temperature. The decoloration process is induced by visible irradiation.

Photochromic molecules, which undergo photochromic reactions in the crystalline state, are rare, because in crystals, large geometrical structure change is prohibited. For example, azobenzene and most of spirobenzopyran derivatives cannot undergo photochromic reactions in the single crystals. The reason is that they require large geometrical structure changes during the

Supramolecular Photochemistry: Controlling Photochemical Processes, First Edition. Edited by V. Ramamurthy and Yoshihisa Inoue.

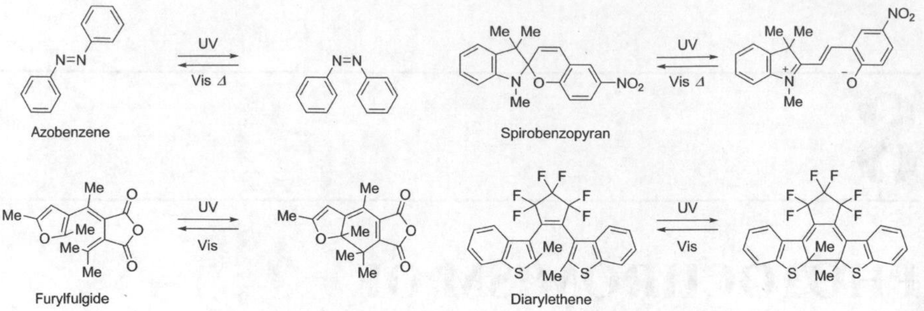

Azobenzene

Spirobenzopyran

Furylfulgide

Diarylethene

Scheme 8.1. Typical photochromic molecules.

photochromic reactions. Photochromic molecules that can undergo photochromism in the crystalline phase are paracyclophanes,[2] *N*-salicylidene aniline,[3] aziridine,[4] triarylimidazole dimer,[5] diphenylmaleronitrile,[6] 2-(2,4-dinitrobenzyl) pyridine,[7] and triazene.[8] However, most of the photogenerated colored states are thermally unstable at room temperature. In contrast to these molecules, diarylethene derivatives having heterocyclic aryl groups undergo thermally irreversible and fatigue-resistant photochromic reactions in the single-crystalline phase.[9] This chapter focuses on the photochromism of diarylethene multicomponent single crystals. Preparation of mixed crystals or cocrystals composed of different kinds of diarylethene derivatives and various molecules creates special photofunctions, which cannot be realized by usual single-component crystals. First, general photochromic performance of single-component diarylethene single crystals will be introduced. Subsequently, characteristic photochromic performances of multicomponent crystals, such as multicolor photochromism, supramolecular structures, asymmetric photoreactions, and nanostructure fabrication, will be described.

SINGLE-COMPONENT SINGLE CRYSTALS

Photochromic reactions of diarylethene derivatives are based on photoreversible 6π electrocyclic reactions of the central hexatriene moieties. The molecules reversibly convert between open- and closed-ring isomers by cyclization and cycloreversion reactions upon irradiation with ultraviolet (UV) and visible light. In most cases, the open-ring isomers are colorless, while the closed-ring isomers are colored. Diarylethene derivatives undergo thermally irreversible and fatigue-resistant photochromic reactions. Both isomers are thermally stable at room temperature,[10] and the coloration/decoloration cycles can be repeated more than 10,000 times.[11] The cyclization and cycloreversion reactions complete in less than 10 ps after photoexcitation.[12] The cyclization quantum yield in solution is moderate ($\Phi = 0.1 \sim 0.5$).[13] The cycloreversion quantum yield is strongly dependent on the substituents. When methoxy groups are introduced at the central reactive carbons, the cycloreversion quantum yield is strongly suppressed to less than 10^{-4}, while it increases to 0.41 when the substituents are cyano groups.[14]

Diarylethene derivatives undergo thermally stable and fatigue-resistant photochromic reactions not only in solution but also in the single-crystalline phase.[9] Figure 8.1 shows photographs of diarylethene single crystals. Upon irradiation with UV light, the colorless crystals exhibit various colors depending on the chemical structures of the molecules. The anisotropy of polarized absorption spectra of the colors indicates that the photoreactions proceed in the crystal lattice.[15] The colored states remain stable in the dark at room temperature. Upon irradiation with visible light, the photogenerated closed-ring isomers in the crystals undergo cycloreversion reactions and the colored crystals return to the initial colorless ones.

Photochromic reactivity of diarylethene single crystals is controlled by the molecular conformation in the crystals. The open-ring isomer of diarylethene adopts two types of conformations with the aryl rings in mirror symmetry (parallel) and C_2 symmetry (antiparallel) as shown in Scheme 8.2.[16] According to the Woodward–Hoffmann rule, photochemical electrocyclic reactions of 6π systems proceed in a conrotatory mode.[17] The conrotatory photocyclization of diarylethene can proceed only from the antiparallel conformation. The requirement for diarylethene molecules to undergo photocyclization reactions in crystals is that the molecules should be fixed in the antiparallel conformation and that the distance between the reactive carbons is less than 4 Å.[18] If these requirements are fulfilled, the cyclization quantum yields become close to unity (100%).

The photocyclization process was directly followed by X-ray crystallographic analysis.[19] The molecular structures of diarylethene **1** before and after the photocyclization reaction are shown in Figure 8.2. The open-ring isomer undergoes a cyclization reaction in a conrotatory mode to form the closed-ring isomer. Almost all atoms remain the same positions except the sulfur and the reacting carbon atoms. The small structural change allows the molecule to undergo the photochromic reaction in the crystal.

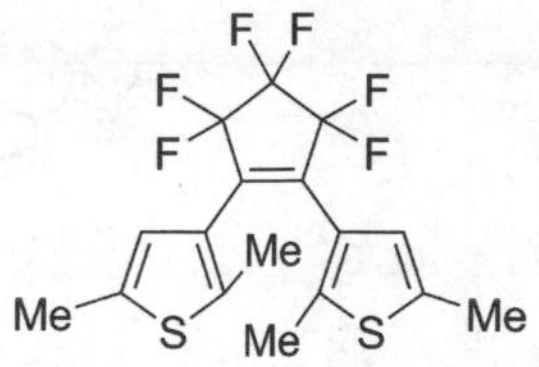

F F
F F
F F
Me
Me S Me S Me

1

One of the characteristic properties of the crystalline-state reactions is a topochemical reaction.[20,21] A crystal lattice can provide a space for reactions with stereo-, regio-, diastereo-, and enantiospecificity. Photocyclization of diarylethene derivatives produces two enantiomers of the closed-ring isomers

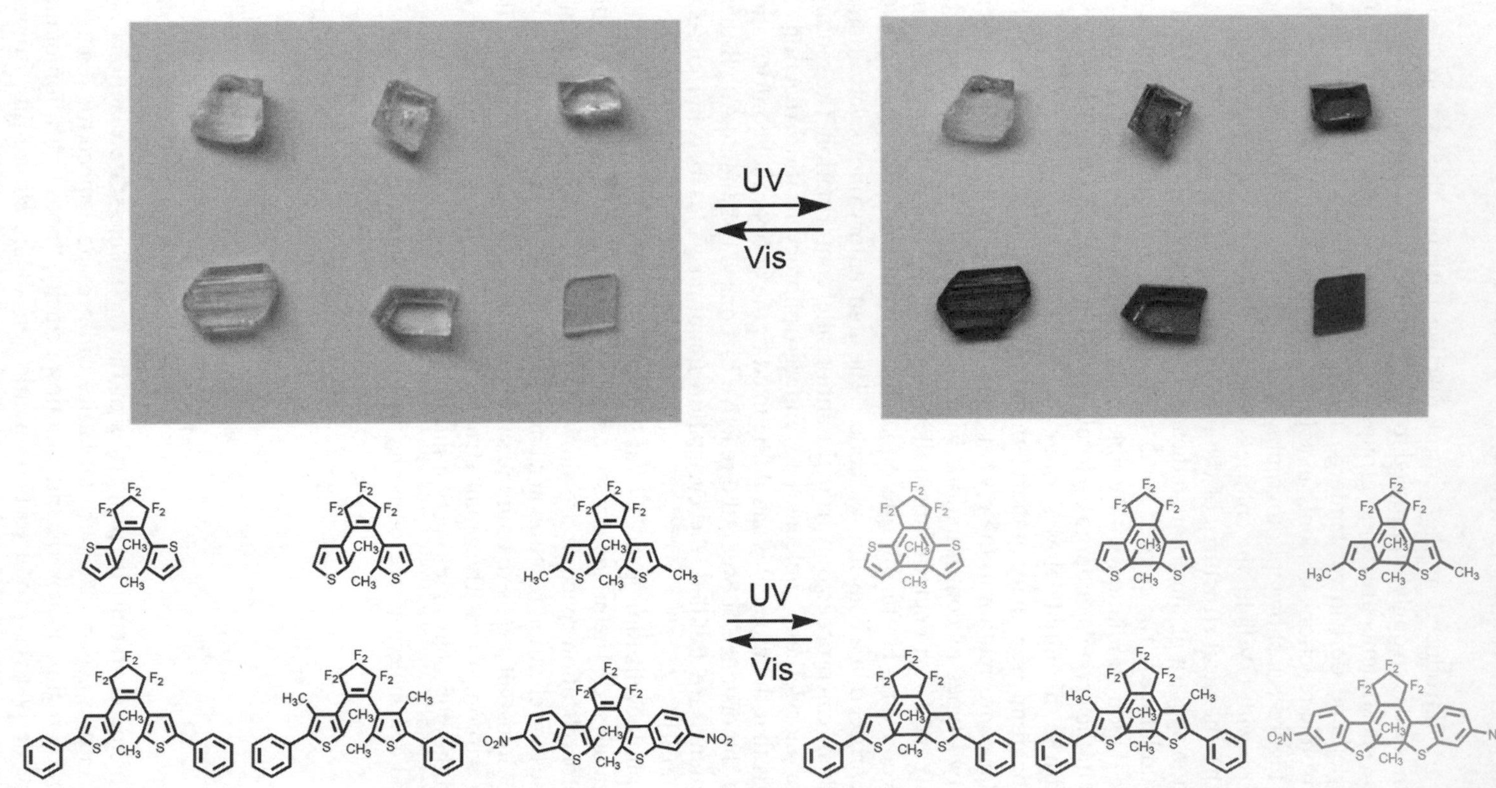

Figure 8.1. Photographs of photochromic diarylethene single crystals. See ftp site for color: ftp://ftp.wiley.com/public/sci_tech_med/supramolecular_photochemistry.

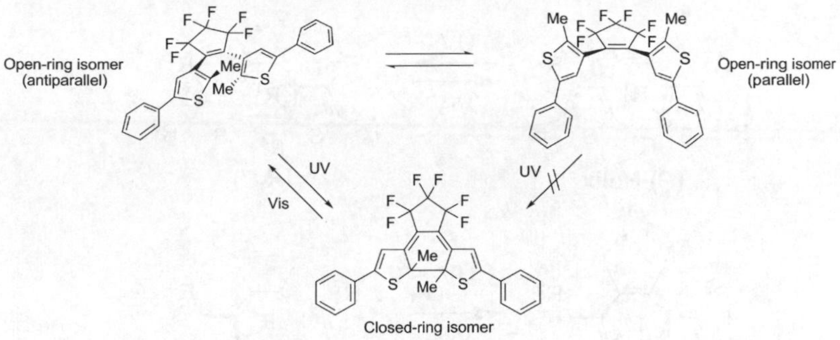

Open-ring isomer (antiparallel)

Open-ring isomer (parallel)

UV

Vis

UV

Closed-ring isomer

Scheme 8.2. Photochromic reaction of diarylethene.

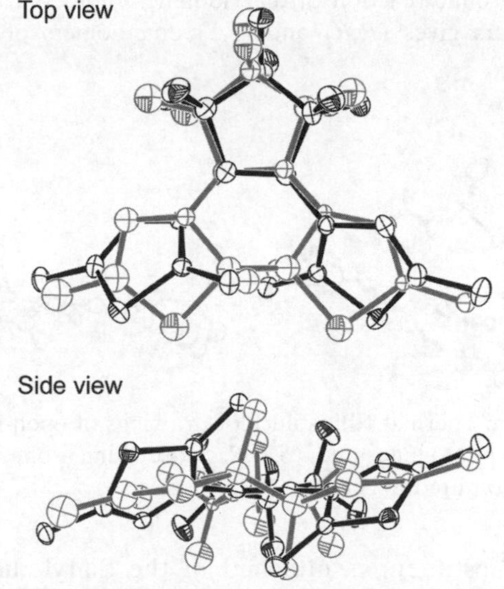

Figure 8.2. Oak Ridge Thermal Ellipsoid Plot drawing of UV-irradiated crystal of **1**. Black and red molecules are open-ring and photogenerated closed-ring isomers, respectively. Hydrogen atoms are omitted for clarity. See ftp site for color: ftp://ftp.wiley.com/public/sci_tech_med/supramolecular_photochemistry.

with (*R*,*R*) and (*S*,*S*) absolute configurations, which originate from two asymmetric carbon atoms at the reacting points, as shown in Scheme 8.3. The absolute configuration is derived from the conformation of the central hexatriene moiety in the open-ring isomer. Conrotatory cyclization reactions from (*P*)-helical (right-handed) and (*M*)-helical (left-handed) conformers of the open-ring isomer yield the (*R*,*R*) and (*S*,*S*) enantiomers of the closed-ring isomer, respectively. In general, the photocyclization in solution results in the formation of the two enantiomers in equal amounts, that is a racemic mixture.

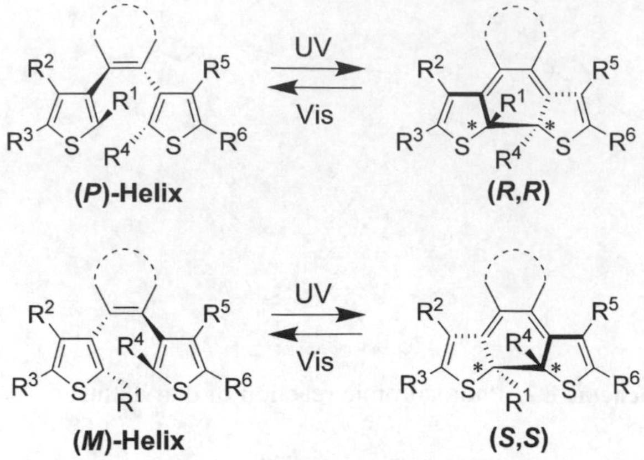

Scheme 8.3. Photochromic reaction of diarylethene. Photocyclization from (*P*)- and (*M*)-helical conformers gives (*R*,*R*) and (*S*,*S*) enantiomers of closed-ring isomer, respectively.

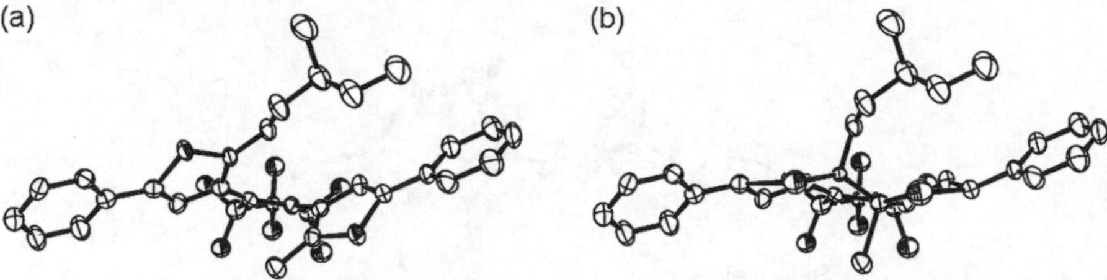

Figure 8.3. Oak Ridge Thermal Ellipsoid Plot drawings of open-ring isomer in chiral crystal of **2** (a) and photogenerated (*S*,*R*,*R*) closed-ring isomer in the crystal (b). Hydrogen atoms are omitted for clarity.

Even if a chiral substituent is attached in the diarylethene molecule, the enrichment of one of diastereomers is hardly observed.[22]

Asymmetric photoreactions can take place in diarylethene single crystals. A diastereoselective photoreaction was observed in a single crystal of diarylethene **2** having a chiral (*S*)-3-methyl-1-penten-1-yl substituent.[23] The diarylethene crystallized into a chiral crystal with an orthorhombic space group $P2_12_12_1$. All diarylethene molecules in the chiral crystal adopt a (*P*)-helical conformation, as shown in Figure 8.3a. The chiral (*S*)-3-methyl-1-penten-1-yl substituent provides a chiral environment in the crystal. Photocyclization reaction of this molecule in solution gave a mixture of equal amounts of the two diastereomers. On the other hand, the reaction in the chiral crystal was highly diastereoselective, and one of the two diastereomers was predominantly produced with the diastereomeric excess value of >95%. *In situ* X-ray crystallographic analysis of the photocyclization process revealed that the photoproduct was the diastereomer (*S*,*R*,*R*), as shown in Figure 8.3b.

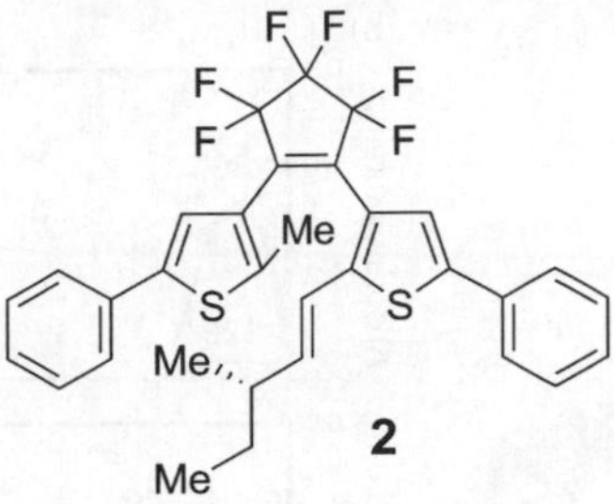

If the open-ring isomers of diarylethenes having no chiral substituents would crystallize to a chiral packing structure, absolute asymmetric photoreactions, in which achiral molecules produce chiral products in the absence of any external chiral agents, could be observed. It was found that achiral diarylethene **3** formed chiral single crystals with a monoclinic chiral space group $P2_1$.[24] Two kinds of crystals, **(*P*)-3** and **(*M*)-3**, which are mirror images of each other, were obtained from the same crystallization batch. All molecules in **(*P*)-3** are fixed in a (*P*)-helical conformation, while all molecules in **(*M*)-3** are fixed in an (*M*)-helical conformation. (*P*)- and (*M*)-helical conformers are spontaneously separated during the crystallization process. Owing to topochemical reactions in the chiral crystals, **(*P*)-3** and **(*M*)-3** gave (*R*,*R*) and (*S*,*S*) enantiomers, respectively, with the enantiomeric excess (ee) value of >94%.

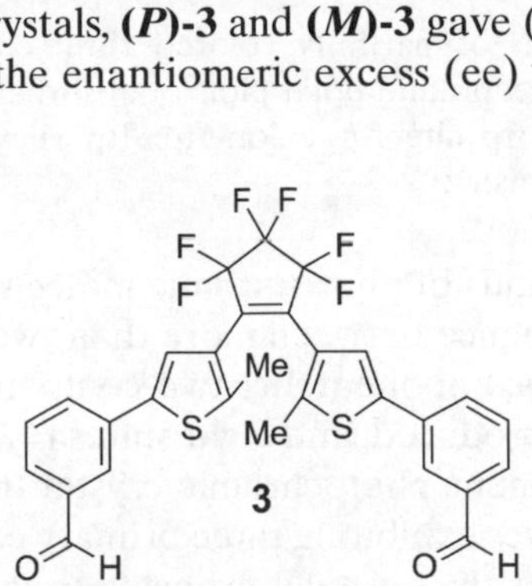

MULTICOMPONENT SINGLE CRYSTALS

Characteristic photochromic performances, such as multicolor photochromism, supramolecular structures, asymmetric photoreactions, and nanostructure fabrication, can be given by the preparation of multicomponent crystals, including mixed crystals of different diarylethene molecules and stoichiometric cocrystals formed by intermolecular aryl–perfluoroaryl interactions.

Multicolor Photochromism of Mixed Crystals

Usual single-component photochromic systems interconvert between only two states: "colorless" and "colored." In contrast, multicomponent systems

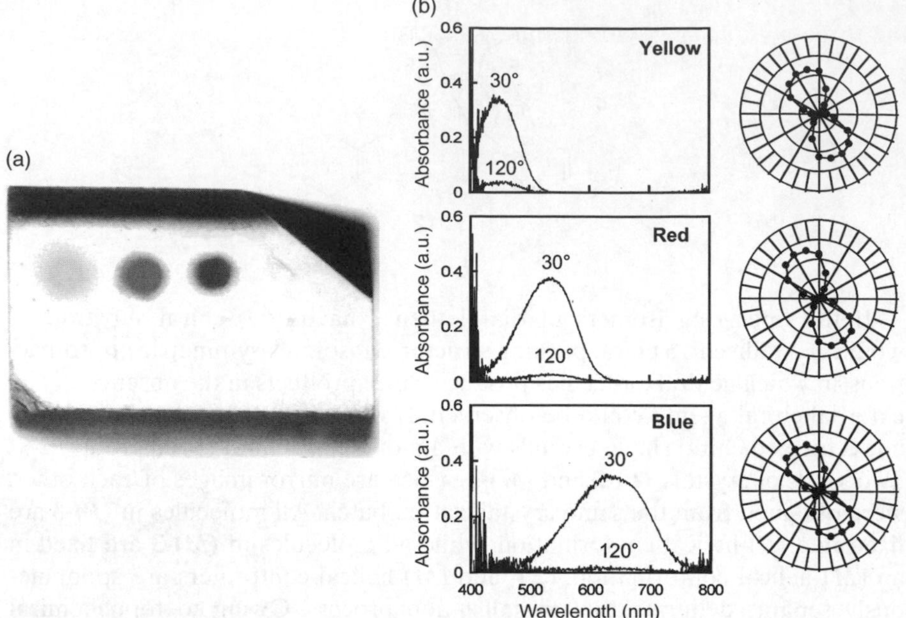

Figure 8.4. (a) Photograph of partially colored three-component crystal of **1·4·5**. (b) Polarized absorption spectra and polar plots of absorbance of yellow, red, and blue colors in the crystal. See ftp site for color: ftp://ftp.wiley.com/public/sci_tech_med/supramolecular_photochemistry.

composed of different kinds of photochromic molecules show reversible multistate switching performance between more than two states by the combination of two states of each component. In a two-component system, for example, four states (=2^2) can be produced, and eight states (=2^3) in a three-component system. If a multicomponent photochromic crystal that contains three kinds of photochromic molecules exhibiting three primary colors, such as yellow, red, and blue, can be prepared, the crystal is expected to show a full range of colors upon photoirradiation.

A full-color photochromic crystal was prepared by introducing three kinds of diarylethene molecules, **1**, **4**, and **5**, exhibiting red, yellow, and blue, in a single crystal.[25] Recrystallization of a mixture of **1**, **4**, and **5** (molar ratio **1**:**4**:**5** = 1:0.4:0.5) from acetonitrile gave a single crystal composed of **1**, **4**, and **5** in the molar ratio of 97.4:2.4:0.2. The space group and the cell parameters of the three-component crystal was the same as those of a single-component crystal of **1**. The crystal lattice of **1** provides the sites where **4** and **5** can occupy. Figure 8.4a shows a photograph of the three-component crystal. Upon irradiation with appropriate wavelength of light, the colorless crystal turned to yellow, red, and blue. These colors are due to the generation of closed-ring isomers of **1** (red), **4** (yellow), and **5** (blue) in the crystal. Figure 8.4b shows polarized absorption spectra of the three colors. Although the molecular structures of **4** and **5** were not discerned by X-ray crystallographic analysis because of their small contents in the crystal, the polarized absorption anisotropy of the three colors suggests that the three kinds of diarylethene molecules are packed in a similar manner and undergo photochromic reactions in the single-crystalline phase.[15] All the colors were thermally stable and did not fade in the dark at room temperature. Upon irradiation with visible light ($\lambda > 450\,\text{nm}$), the colors were completely bleached.

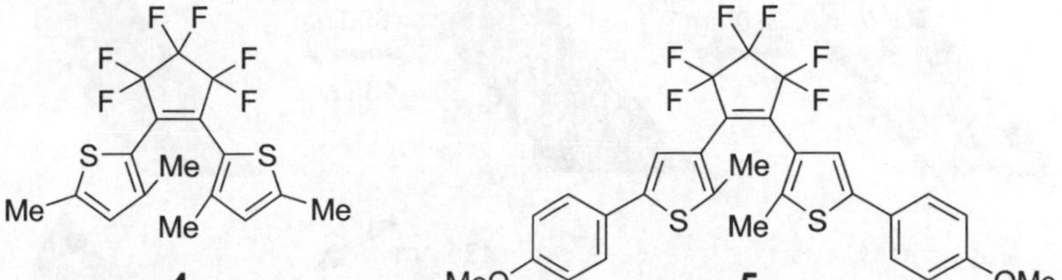

F F F F F F S Me S Me Me Me 4

F F F F F F Me MeO S Me S OMe 5

Another full-color photochromic crystal was prepared from diarylethenes **6**, **7**, and **8**, which have oxazole, thiazole, and thiophene rings and turn yellow, red, and blue, respectively, upon UV irradiation.[26] The geometrical structures of these three molecules are quite similar to each other. Therefore, the molecules can form a mixed crystal composed of almost equal amounts of the three components. Recrystallization of a mixture of **6**, **7**, and **8** (molar ratio **6**:**7**:**8** = 22:32:46) from ethanol gave a three-component mixed crystal with the composition of **6**:**7**:**8** = 32:53:15. This crystal exhibited yellow, red, and blue upon irradiation with appropriate wavelength of light, as shown in Figure 8.5.

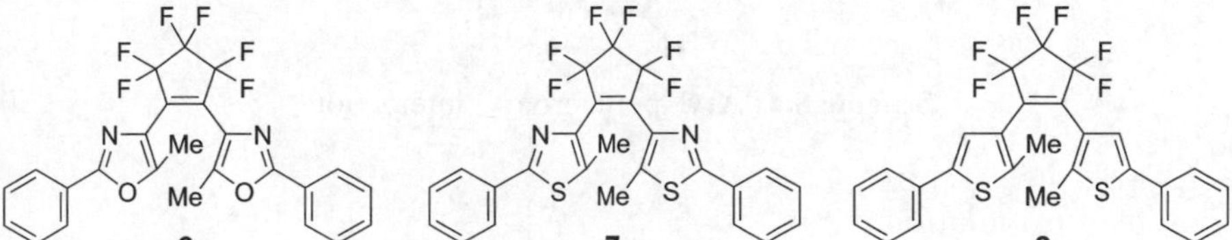

F F F F F F N Me N O Me O 6

F F F F F F N Me N S Me S 7

F F F F F F Me S Me S 8

Crystal Engineerng of Cocrystals by Aryl–Perfluoroaryl Interaction

In the field of crystal engineering, intermolecular noncovalent interactions, such as hydrogen-bonding, metal-coordinating, and π–π interactions, have been widely used for the control of structures and functions of molecular crystals.[27] A stacking interaction between aromatic hydrocarbons and aromatic perfluorocarbons, namely aryl–perfluoroaryl interaction (Scheme 8.4), is a special type of aromatic–aromatic interaction[28] and has been used for the preparation of organic two-component cocrystals and the favorable prearrangement of reactive molecules for crystalline-state reactions.[29] Here, several

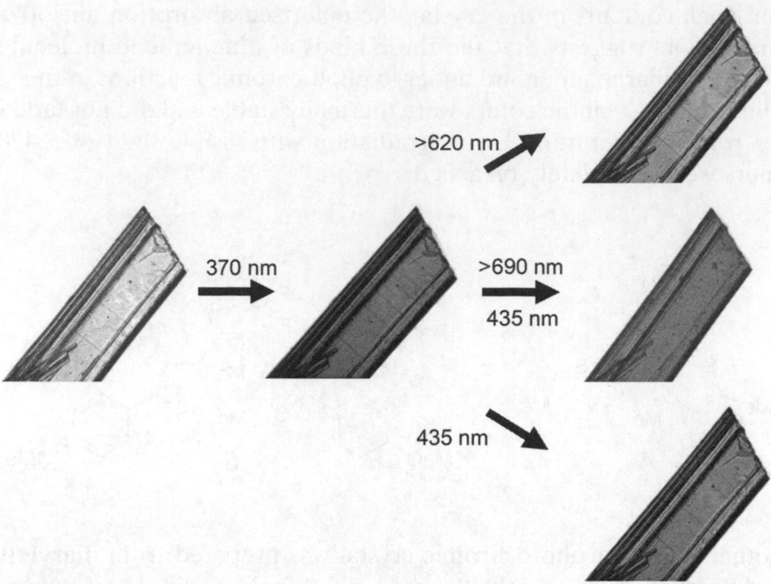

Figure 8.5. Color changes of three-component crystal of **6·7·8**. See ftp site for color: ftp://ftp.wiley.com/public/sci_tech_med/supramolecular_photochemistry.

Scheme 8.4. Aryl–perfluoroaryl interaction.

examples of crystal engineering of photochromic diarylethenes by utilizing aryl–perfluoroaryl interactions will be described.

Supramolecular Structures in Cocrystals of Diarylethene and Aromatic Molecules. Diarylethene **9** having two pentafluorophenyl groups was synthesized, and cocrystals of **9** with aromatic molecules, such as benzene (**Bz**) and naphthalene (**Np**), were prepared.[30] **9** cocrystallized with **Bz** and **Np** in the stoichiometric ratio of 2:1 and 1:1, respectively, upon recrystallization of the mixtures from hexane. Figure 8.6 shows supramolecular structures formed by aryl–perfluoroaryl interactions in the cocrystals of **9·Bz** and **9·Np**. In **9·Bz**, a one-dimensional chain structure, which is composed of **9** and **Bz** in the ratio of 2:1, is formed (Fig. 8.6a), while in **9·Np**, a discrete sandwiched structure

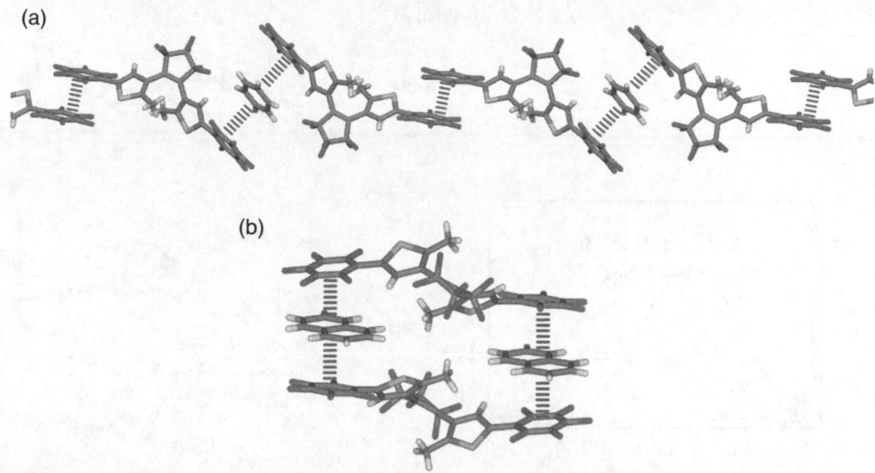

Figure 8.6. X-ray structures of a one-dimensional chain in cocrystal of **9·Bz** (a) and a discrete sandwiched structure in cocrystal of **9·Np** (b).

composed of **9** and **Np** molecules is formed (Fig. 8.6b). In both crystals, aryl–perfluoroaryl stacking structures between the pentafluorophenyl groups of **9** and **Bz** or **Np** molecules are clearly observed.

9

The cocrystals and a homocrystal of **9** underwent photochromic reactions. Figure 8.7a shows absorption spectra of the colored crystals. The absorption maxima were located at 555, 620, and 630 nm for **9·Bz**, **9·Np**, and **9**, respectively. The absorption spectra of the colored crystals showed a significant spectral shift (by as much as 75 nm), despite the colors are due to the same closed-ring isomer of **9**. The shift is ascribed to the difference in the molecular conformation, especially the torsion angles between the thienyl and pentafluorophenyl rings, induced by the intermolecular aryl–perfluoroaryl interactions. Figure 8.7b shows the structures of the diarylethene molecules in the crystals. The absorption spectra reflect the π-conjugation length of the photogenerated closed-ring isomers in the crystals. As can be seen from Figures

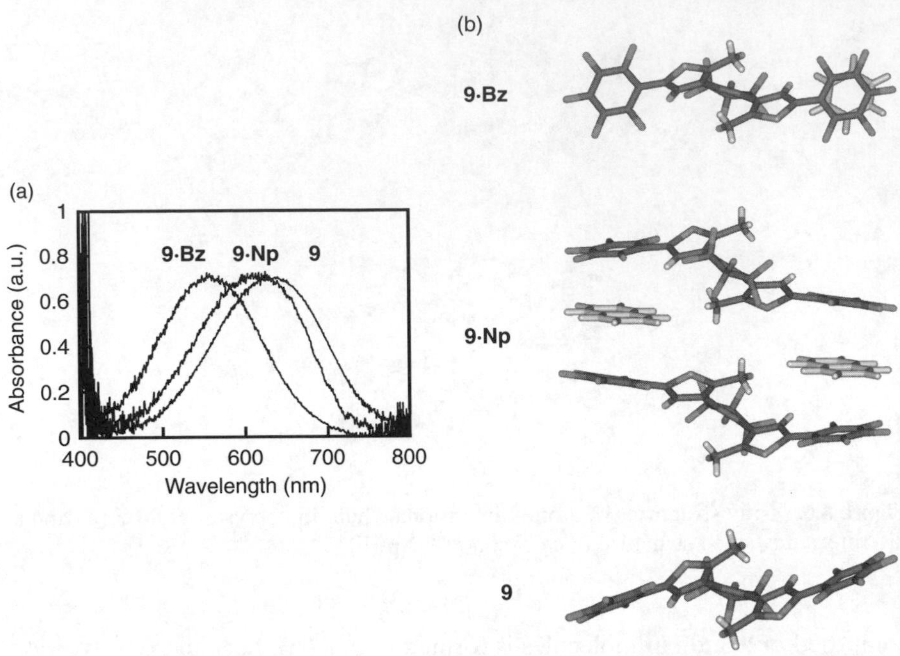

Figure 8.7. Absorption spectra of colored crystals of **9·Bz**, **9·Np**, and **9** (a) and structures of the diarylethene molecules in the crystals (b).

8.7b, pentafluorophenyl rings in the crystals of **9·Np** and **9** are almost parallel to the molecular plane. This means that the photogenerated closed-ring isomers have coplanar structures and the π-conjugation extends over the entire molecule. On the other hand, in the crystal of **9·Bz**, the planes of the phenyl rings are almost perpendicular to the molecular plane. The twisted phenyl rings do not enter the π-conjugation of the central part of the closed-ring isomer. The photogenerated closed-ring isomers in **9·Bz**, in which the π-conjugation is limited, show absorption at a shorter wavelength than those in **9·Np** and **9**, in which π-electrons are delocalized over the entire molecule. Cocrystallization with aromatic molecules changed the molecular conformation in the crystal and affected the absorption properties of the photogenerated closed-ring isomer.

Asymmetric Photoreaction in Cocrystal of Diarylethene and Aromatic Molecules. Another absolute photochromic reaction was found in a cocrystal of diarylethene **10** having naphtyl groups and octafluoronaphthalene (**Np^F**) formed by aryl–perfluoroaryl interaction.[31] Examples of an absolute asymmetric reaction in a chiral two-component cocrystal composed of achiral molecules are very rare.[32] **10** and **Np^F** cocrystallized to form a 1:2 cocrystal with a monoclinic chiral space group *C*2, despite both of the molecules are achiral

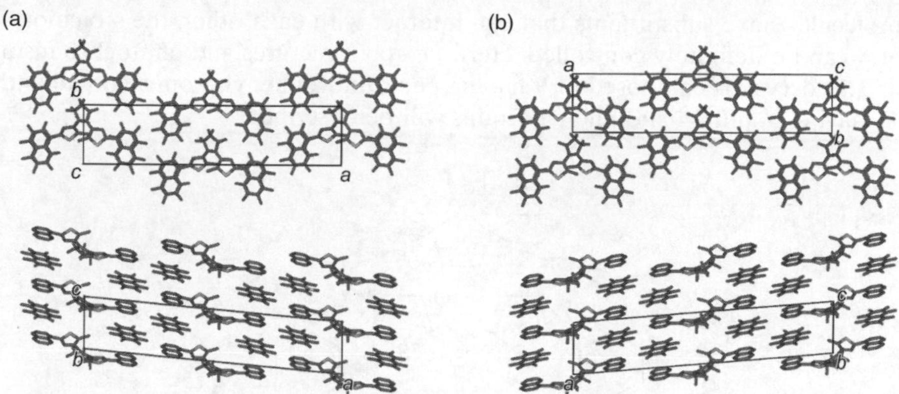

Figure 8.8. Molecular packing diagrams of cocrystals of **(*P*)-10·Np^F** (a) and **(*M*)-10·Np^F** (b).

and both of the homocrystals are also achiral. As in the case of diarylethene **3**, two kinds of crystals, **(*P*)-10·Np^F** and **(*M*)-10·Np^F**, were obtained, in which the diarylethene molecules adopt (*P*)-helical and (*M*)-helical conformations, respectively. Packing diagrams in Figure 8.8 shows the existence of aryl–perfluoroaryl stacking interactions between the naphtyl groups of **10** and **Np^F** molecules. There is no enantioselectivity in photocyclization in a hexane solution and the achiral homocrystal of **10**. In the chiral cocrystals, on the other hand, highly enantioselective reactions took place with the ee values of >99%. The photocyclization with such a high enantioselectivity in the cocrystals can be attributed to the rigid stacking of the naphthalene-octafluoronaphthalene piles, which should not waver largely during the photochemical transformation.

10

Nanostructure Fabrication in Cocrystals of Two Kinds of Diarylethene. In the full-color photochromic crystals described above, the dopant molecules were substitutionally incorporated into the host crystal lattice. Although the crystals showed full-color photochromism, the composition ratio in the crystals is hard to control because of the absence of strong intermolecular interactions. When molecules have substituents that can interact with each other, the stoichiometry can be definitely controlled. Here, crystal structures and photochromism of 1:1 cocrystals composed of **9** having pentafluorophenyl groups and **10** and **11** having naphthyl and phenyl groups will be described.[33]

11

Recrystallization of a 1:1 (molar ratio) mixture of **9** and **10** from hexane gave colorless single crystals. Figure 8.9a shows molecular packing diagrams in a cocrystal of **9·10**. The crystal contains **9** and **10** in the molar ratio of 1:1. The pentafluorophenyl groups of **9** and the naphthyl groups of **10** are stacked by intermolecular aryl–perfluoroaryl interactions. The diagrams viewed from the *a*-, *b*-, and *c*-axes indicate that **9** and **10** molecules are packed in a three-dimensional alternating arrangement to form a mosaic-like structure. A cocrystal of **9** and **11** was also prepared by recrystallization of a 1:1 mixture of the compounds. Molecular packing diagrams in **9·11** are shown in Figure 8.9b. The composition ratio in the crystal was 1:1. The crystal has a layered structure in which unimolecular layers of **9** and **11** are alternately stacked. The thickness of each layer is about 0.65 nm. The pentafluorophenyl rings of **9** and the phenyl rings of **11** and are not stacked with each other. This indicates that aryl–perfluoroaryl interaction does not exist between **9** and **11**.

The cocrystals showed photochromism. Upon irradiation with UV light, the colorless crystals of **9·10** and **9·11** turned green and blue, respectively. The colors

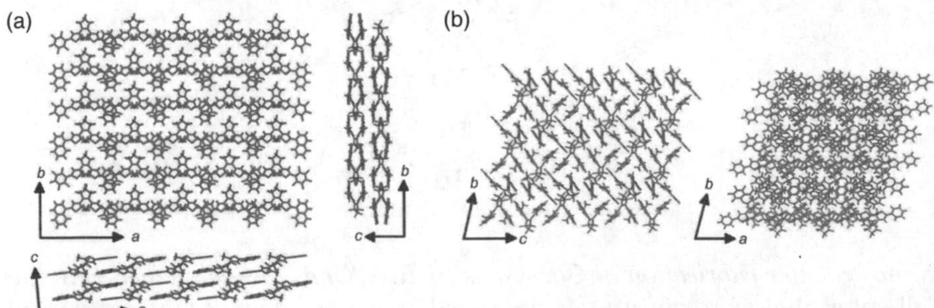

Figure 8.9. Molecular packing diagrams of cocrystals of **9·10** (a) and **9·11** (b). Red, green, and blue molecules indicate **9**, **10**, and **11**, respectively. See ftp site for color: ftp://ftp.wiley.com/public/sci_tech_med/supramolecular_photochemistry.

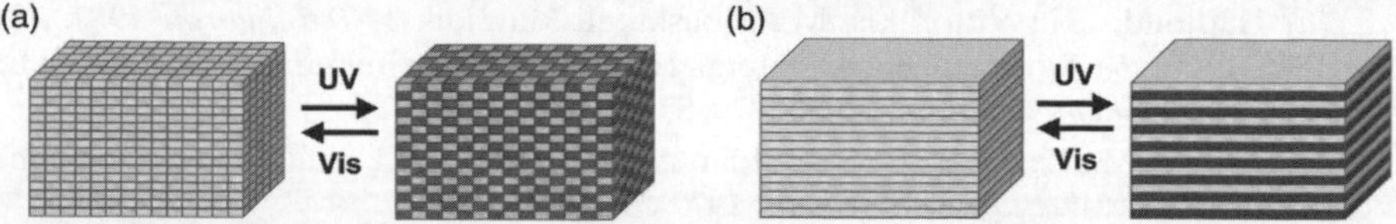

Figure 8.10. Schematic illustrations of photochromic reactions in cocrystals **9·10** (a) and **9·11** (b). Red, green, and blue areas indicate **9**, **10**, and **11**, respectively. See ftp site for color: ftp://ftp.wiley.com/public/sci_tech_med/supramolecular_photochemistry.

were thermally stable in the dark at room temperature, and bleached upon irradiation with visible light ($\lambda > 450$ nm). The ratio of the reacted two components was monitored by high performance liquid chromatography. In the cocrystals of **9·10** and **9·11**, **10** and **11** selectively underwent photocyclization to yield the closed-ring isomers, while photocyclization of **9** was significantly suppressed. In homocrystals of the three compounds, efficient photocyclization reactions with quantum yields close to 1 took place. The highly selective photoreaction of **10** and **11** was also confirmed by *in situ* X-ray analysis.

Such a dramatic difference in the photoreactivity between the homocrystals and the cocrystals can be explained by intermolecular excited-energy transfer in the cocrystals. Absorption edges of **10** and **11** in hexane are located at longer wavelengths than that of **9**. This means that the excited S_1 energy levels of **10** and **11** are lower than that of **9**. In stoichiometric cocrystals, the two different diarylethene molecules are closely packed, and the intermolecular separation between the molecules is less than 1 nm. Although photocyclization reactions of diarylethene derivatives in the single-crystalline phase usually proceed very fast (in less than 10 ps),[34] the close intermolecular contact in the cocrystals allows the excited energy to transfer from **9** to **10** or **11**. The excited energy absorbed by **9** is efficiently transferred to **10** or **11**, and consequently the highly selective photocyclization reactions of **10** or **11** took place in the cocrystals.

Figure 8.10 shows schematic illustrations of photochromic reactions in the cocrystals of **9·10** and **9·11** with well-controlled nanostructures. As a result of the selective photocyclizations, the colored and colorless molecules, which have different refractive indices, are arranged periodically at the molecular level in the UV-irradiated crystals. Such photoreversible periodic changes in refractive indices in the crystalline nanostructures have potential application as a new type of photonic device.[35]

REFERENCES

1. (*a*) Brown, G.H. *Photochromism*. New York: Wiley-Interscience, **1971**; (*b*) Dürr, H.; Bouas-Laurent, H. *Photochromism: Molecules and Systems*. Amsterdam: Elsevier, **2003**; (*c*) Bouas-Laurent, H.; Dürr, H. *Pure Appl. Chem.* **2001**, *73*, 639–665.
2. Golden, J.H. *J. Chem. Soc.* **1961**, 3741–3748.

3. (*a*) Hadjoudis, E.; Vittorakis, M.; Moustakali-Mavridis, I. *Tetrahedron* **1987**, *43*, 1345–1360; (*b*) Amimoto, K.; Kanatomi, H.; Nagakari, A.; Fukuda, H.; Koyama, H.; Kawato, T. *Chem. Commun.* **2003**, 870–871.
4. Trozzolo, A.M.; Leslie, T.M.; Sarpotdar, A.S.; Small, R.D.; Ferraudi, G.J.; DoMinh, T.; Hartless, R.L. *Pure Appl. Chem.* **1979**, *51*, 261–270.
5. (*a*) Maeda, K.; Hayashi, T. *Bull. Chem. Soc. Jpn.* **1970**, *43*, 429–438; (*b*) Kawano, M.; Sano, T.; Abe, J.; Ohashi, Y. *J. Am. Chem. Soc.* **1999**, *121*, 8106–8107.
6. Ichimura, K.; Watanabe, S. *Bull. Chem. Soc. Jpn.* **1976**, *49*, 2220–2223.
7. (*a*) Sixl, H.; Warta, R. *Chem. Phys.* **1985**, *94*, 147–155; (*b*) Schmidt, A.; Kababya, S.; Appel, M.; Khatib, S.; Botoshansky, M.; Eichen, Y. *J. Am. Chem. Soc.* **1999**, *121*, 11291–11299.
8. Mori, Y.; Ohashi, Y.; Maeda, K. *Bull. Chem. Soc. Jpn.* **1989**, *62*, 3171–3176.
9. (*a*) Irie, M. *Chem. Rev.* **2000**, *100*, 1685–1716; (*b*) Kobatake, S.; Irie, M. *Bull. Chem. Soc. Jpn.* **2004**, *77*, 195–210; (*c*) Morimoto, M.; Irie, M. *Chem. Commun.* **2005**, 3895–3905; (*d*) Irie, M. *Bull. Chem. Soc. Jpn.* **2008**, *81*, 917–926.
10. (*a*) Irie, M.; Lifka, T.; Kobatake, S.; Kato, N. *J. Am. Chem. Soc.* **2000**, *122*, 4871–4876; (*b*) Takami, S.; Kobatake, S.; Kawai, T.; Irie, M. *Chem. Lett.* **2003**, 892–893.
11. (*a*) Hanazawa, M.; Sumiya, R.; Horikawa, Y.; Irie, M. *J. Chem. Soc. Chem. Commun.* **1990**, 206–207; (*b*) Uchida, K.; Nakayama, Y.; Irie, M. *Bull. Chem. Soc. Jpn.* **1990**, *63*, 1311–1315.
12. (*a*) Miyasaka, H.; Araki, S.; Tabata, A.; Nobuto, T.; Mataga, N.; Irie, M. *Chem. Phys. Lett.* **1994**, *230*, 249–254; (*b*) Tamai, N.; Saika, T.; Shimidzu, T.; Irie, M. *J. Phys. Chem.* **1996**, *100*, 4689–4692; (*c*) Ern, J.; Bens, A.T.; Martin, H.-D.; Mukamel, S.; Schmid, D.; Tretiak, S.; Tsiper, E; Kryschi, C. *Chem. Phys.* **1999**, *246*, 115–125; (*d*) Okabe, C.; Nakabayashi, T.; Nishi, N.; Fukaminato, T.; Kawai, T.; Irie, M.; Sekiya, H. *J. Phys. Chem. A* **2003**, *107*, 5384–5390.
13. Irie, M.; Sakemura, K.; Okinaka, M.; Uchida, K. *J. Org. Chem.* **1995**, *60*, 8305–8309.
14. (*a*) Shibata, K.; Kobatake, S.; Irie, M. *Chem. Lett.* **2002**, 618–619; (*b*) Morimitsu, K.; Shibata, K.; Kobatake, S.; Irie, M. *J. Org. Chem.* **2002**, *67*, 4574–4578; (*c*) Morimitsu, K.; Kobatake, S.; Nakamura, S.; Irie, M. *Chem. Lett.* **2002**, 858–859.
15. (*a*) Kobatake, S.; Yamada, T.; Uchida, K.; Kato, N.; Irie, M. *J. Am. Chem. Soc.* **1999**, *121*, 2380–2386; (*b*) Kobatake, S.; Yamada, M.; Yamada, T.; Irie, M. *J. Am. Chem. Soc.* **1999**, *121*, 8450–8456; (*c*) Kobatake, S.; Shibata, K.; Uchida, K.; Irie, M. *J. Am. Chem. Soc.* **2000**, *122*, 12135–12141.
16. (*a*) Uchida, K.; Nakayama, Y.; Irie, M. *Bull. Chem. Soc. Jpn.* **1990**, *63*, 1311–1315; (*b*) Irie, M.; Miyatake, O.; Uchida, K. *J. Am. Chem. Soc.* **1992**, *114*, 8715–8716.
17. (*a*) Woodward, R.B.; Hoffmann, R. *The Conservation of Orbital Symmetry*. Weinheim: Verlag Chemie, **1970**; (*b*) Nakamura, S.; Irie, M. *J. Org. Chem.* **1988**, *53*, 6136–6138.
18. (*a*) Shibata, K.; Muto, K.; Kobatake, S.; Irie, M. *J. Phys. Chem. A* **2002**, *106*, 209–214; (*b*) Yamada, T.; Muto, K.; Kobatake, S.; Irie, M. *J. Org. Chem.* **2001**, *66*, 6164–6168;; (*c*) Kobatake, S.; Uchida, K.; Tsuchida, E.; Irie, M. *Chem. Commun.* **2002**, 2804–2805.
19. (*a*) Yamada, T.; Kobatake, S.; Muto, K.; Irie, M. *J. Am. Chem. Soc.* **2000**, *122*, 1589–1592; (*b*) Yamada, T.; Kobatake, S.; Irie, M. *Bull. Chem. Soc. Jpn.* **2000**, *73*, 2179–2184.

20. (*a*) Venkatesan, K.; Ramamurthy, V. In: Ramamurthy, V., editor. *Photochemistry in Organized and Constrained Media*. Weinheim: VCH, **1991**, pp. 133–184; (*b*) Ramamurthy, V.; Venkatesan, K. *Chem. Rev.* **1987**, *87*, 433–481; (*c*) Gamlin, J.N.; Jones, R.; Leibovitch, M.; Patrick, B.; Scheffer, J.R.; Trotter, J. *Acc. Chem. Res.* **1996**, *29*, 203–209; (*d*) Tanaka, K.; Toda, F. *Chem. Rev.* **2000**, *100*, 1025–1074.

21. (*a*) Hosomi, H.; Ohba, S.; Tanaka, K.; Toda, F. *J. Am. Chem. Soc.* **2000**, *122*, 1818–1819; (*b*) Ito, Y.; Kano, G.; Nakamura, N. *J. Org. Chem.* **1998**, *63*, 5643–5647.

22. Yamaguchi, T.; Uchida, K.; Irie, M. *J. Am. Chem. Soc.* **1997**, *119*, 6066–6071.

23. Kodani, T.; Matsuda, K.; Yamada, T.; Kobatake, S.; Irie, M. *J. Am. Chem. Soc.* **2000**, *122*, 9631–9637.

24. Yamamoto, S.; Matsuda, K.; Irie, M. *Angew. Chem. Int. Ed. Engl.* **2003**, *42*, 1636–1639.

25. (*a*) Morimoto, M.; Kobatake, S.; Irie, M. *Adv. Mater.* **2002**, *14*, 1027–1029; (*b*) Morimoto, M.; Kobatake, S.; Irie, M. *J. Am. Chem. Soc.* **2003**, *125*, 11080–11087.

26. (*a*) Kuroki, L.; Takami, S.; Shibata, K.; Irie, M. *Chem. Commun.* **2005**, 6005–6007; (*b*) Takami, S.; Kuroki, L.; Irie, M. *J. Am. Chem. Soc.* **2007**, *129*, 7319–7326.

27. Desiraju, G.R. *Angew. Chem. Int. Ed. Engl.* **1995**, *34*, 2311–2327.

28. Patrick, C.R.; Prosser, G.S. *Nature* **2001**, *187*, 1021.

29. (*a*) Coates, G.W.; Dunn, A.R.; Henling, L.M.; Dougherty, D.A.; Grubbs, R.H. *Angew. Chem. Int. Ed. Engl.* **1997**, *36*, 248–251; (*b*) Dai, C.; Nguyen, P.; Marder, T.B.; Scott, A.J.; Clegg, W.; Viney, C. *Chem. Commun.* **1999**, 2493–2494.

30. Morimoto, M.; Kobatake, S.; Irie, M. *Cryst. Growth Des.* **2003**, *3*, 847–854.

31. Morimoto, M.; Kobatake, S.; Irie, M. *Chem. Commun.* **2008**, 335–337.

32. (*a*) Suzuki, T.; Fukushima, T.; Yamashita, Y.; Miyashi, T. *J. Am. Chem. Soc.* **1994**, *116*, 2793; (*b*) Koshima, H.; Ding, K.; Chisaka, Y.; Matsuura, T. *J. Am. Chem. Soc.* **1996**, *118*, 12059; (*c*) Koshima, H.; Kawanishi, H.; Nagano, M.; Yu, H.; Shiro, M.; Hosoya, T.; Uekusa, H.; Ohashi, Y. *J. Org. Chem.* **2005**, *70*, 4490.

33. (*a*) Morimoto, M.; Kobatake, S.; Irie, M. *Photochem. Photobiol. Sci.* **2003**, *2*, 1088–1094; (*b*) Morimoto, M.; Kobatake, S.; Irie, M. *Chem. Rec.* **2004**, *4*, 23–38.

34. Miyasaka, H.; Nobuto, T.; Itaya, A.; Tamai, N.; Irie, M. *Chem. Phys. Lett.* **1997**, *269*, 281–285.

35. (*a*) Joannopoulos, J.D.; Meade, R.D.; Winn, J.N. *Photonic Crystals*. Princeton, NJ: Princeton University Press, **1995**; (*b*) Joannopoulos, J.D.; Villeneuve, P.R.; Fan, S. *Nature* **1997**, *386*, 143–149.

9

MANIPULATION OF ENERGY TRANSFER PROCESSES WITHIN THE CHANNELS OF L-ZEOLITE

GION CALZAFERRI AND ANDRÉ DEVAUX

INTRODUCTION

Green plants have developed very sophisticated tools for trapping and transporting electronic excitation energy in their antenna system. The detailed structure of the antenna system of purple bacteria has been resolved. It consists of regular arrangements of chlorophyll molecules held at fixed positions by means of proteins.[1] Light absorbed by any of these chlorophyll molecules is transported to the reaction center, providing the energy necessary for chemical processes to be initiated. A green leaf consists of millions of such well-organized antenna devices. Recreating this system in the laboratory would be a hopeless task—at least regarding the current possibilities of chemical synthesis.

In order to realize systems with similar light harvesting properties, we should first understand the basic principles that govern the transport of electronic excitation energy. Fortunately, this understanding is very advanced and goes back to the pioneering work of Theodor Förster.[2] A chlorophyll molecule consists essentially of a positively charged backbone with some delocalized electrons. The energy of an absorbed photon is transformed into kinetic energy of one of these delocalized electrons. This fast-moving electron causes an oscillating electromagnetic field that can interact with a neighboring acceptor

Supramolecular Photochemistry: *Controlling Photochemical Processes*, First Edition. Edited by V. Ramamurthy and Yoshihisa Inoue.

molecule A, if it bears states that are in resonance with the excited state of the donor D^*. Thus, the excitation energy can be transferred from one molecule to the other. The radiationless electronic excitation energy transfer (EnT) is caused by the very weak near-field interaction between excited configurations of the initial state $(D^* \ldots A_i)$ and of the final state $(D \ldots A_i^*)$. The Förster mechanism involves no orbital overlap between the donor and acceptor molecules, and thus no electron transfer occurs:

$$(D^* \ldots A_i) \xrightarrow{k_{EnT(i)}} \left(D \ldots A_i^*\right). \tag{9.1}$$

In such a system, the "optical electrons" associated with individual component molecules (or chromophoric units) preserve essentially their individual characteristic. The donor D and the acceptor A_i can be the same kind of molecules or they can be different. Förster observed that the rate constant k_{EnT} for the transfer from one electronic configuration to the other can be expressed as a product of three terms:

$$k_{EnT} \propto G \cdot DA \cdot S. \tag{9.2}$$

The geometrical expression G describes the dependence of the rate constant on the distance and angle between the electronic transition dipole moments (ETDMs) of the donor and the acceptor. The term DA specifies the chromophores involved, by taking into account the resonance condition as well as the photophysical properties of the donor; while the factor S takes into account the environment of the pair.

Our design of a model mimicking the key functionality of the green plants antenna system was inspired by the experience we had with different zeolite materials.[3,4] The properties of molecules, complexes, and clusters inside cavities and channels—outside the field of catalysis—have been investigated by several authors.[5–7] A one-dimensional channel system has the advantage of being the simplest possible choice, as is illustrated in Figure 9.1. The donor molecules are represented in light gray and the acceptors in dark gray. A donor, being excited by photon absorption, transfers its electronic excitation to an unexcited neighbor. After series of such steps, the electronic excitation reaches a luminescence trap (acceptor molecule) and is then released, for example, as fluorescence. The acceptors are thought to mimic the "entrance to the reaction center" of the natural antenna. The dimensions given in Figure 9.1 correspond to the pore opening and the center-to-center distance between two neighboring channels in zeolite L (ZL). According to the Förster theory, the largest EnT rate constant is observed if the ETDMs are oriented parallel to the channel axis. Electronic excitation energy transport can be extremely fast in such systems because of its low dimensionality.

Different materials bearing one-dimensional channels can be envisaged for realizing the situation sketched in Figure 9.1. We found that ZL is an excellent host for the supramolecular organization of organic dyes. The synthesis of ZL

Figure 9.1. Scheme of an artificial photonic antenna. The chromophores are embedded in the channels of the host. The dyes shown in light gray act as donor molecules that absorb the incoming light and transport the excitation via Förster resonance energy transfer (FRET) to the acceptors shown at the channel ends on the right. The process can be analyzed by measuring the emission of the acceptors and comparing it with that of the donors. The double arrows indicate the orientation of the ETDM.

crystals of different morphologies in the size range of 30 nm up to about 10,000 nm is well established. It is possible to cover about seven orders of magnitude in terms of volume.[8–10] We focus on systems based on ZL as a host but the theoretical reasoning is, however, also valid for other host materials with similar properties. The structure and morphology of ZL is depicted in Figure 9.2. The primary building unit of the framework consists of TO_4 tetrahedrons where T represents either Al or Si. The secondary building unit, the cancrinite cage, is made up of 18 corner-sharing tetrahedrons. These cages are stacked into columns. Connecting the latter by means of oxygen bridges in the *a*,*b* plane, one obtains a one-dimensional channel system running parallel to the crystals *c*-axis. The channel system exhibits hexagonal symmetry. The molar composition of ZL is $(M^+)_9([AlO_2]_9[SiO_2]_{27}) \times nH_2O$, where M^+ are monovalent cations, compensating the negative charge resulting from the aluminum atoms, and the number of water molecules n is 21 in fully hydrated materials, and 16 for crystals equilibrated at about 22% relative humidity.[11]

It is useful to imagine ZL as consisting of a bunch of strictly parallel channels as shown in Figure 9.2d.[12,13] The channels have a smallest free diameter of about 0.71 nm, the largest diameter inside is 1.26 nm. The distance between the centers of two neighboring channels is 1.84 nm. Each ZL crystal consists of a large number of channels (n_{ch}), which can be estimated as follows:

$$n_{ch} = 0.267(d_Z)^2, \tag{9.3}$$

where d_Z is the diameter of the crystal in nanometer. For example, a crystal with a diameter of 600 nm features nearly 100,000 strictly parallel channels. The ratio of void space available in the channels with respect to the total volume of a crystal is about 26%. An important consequence is that ZL allows,

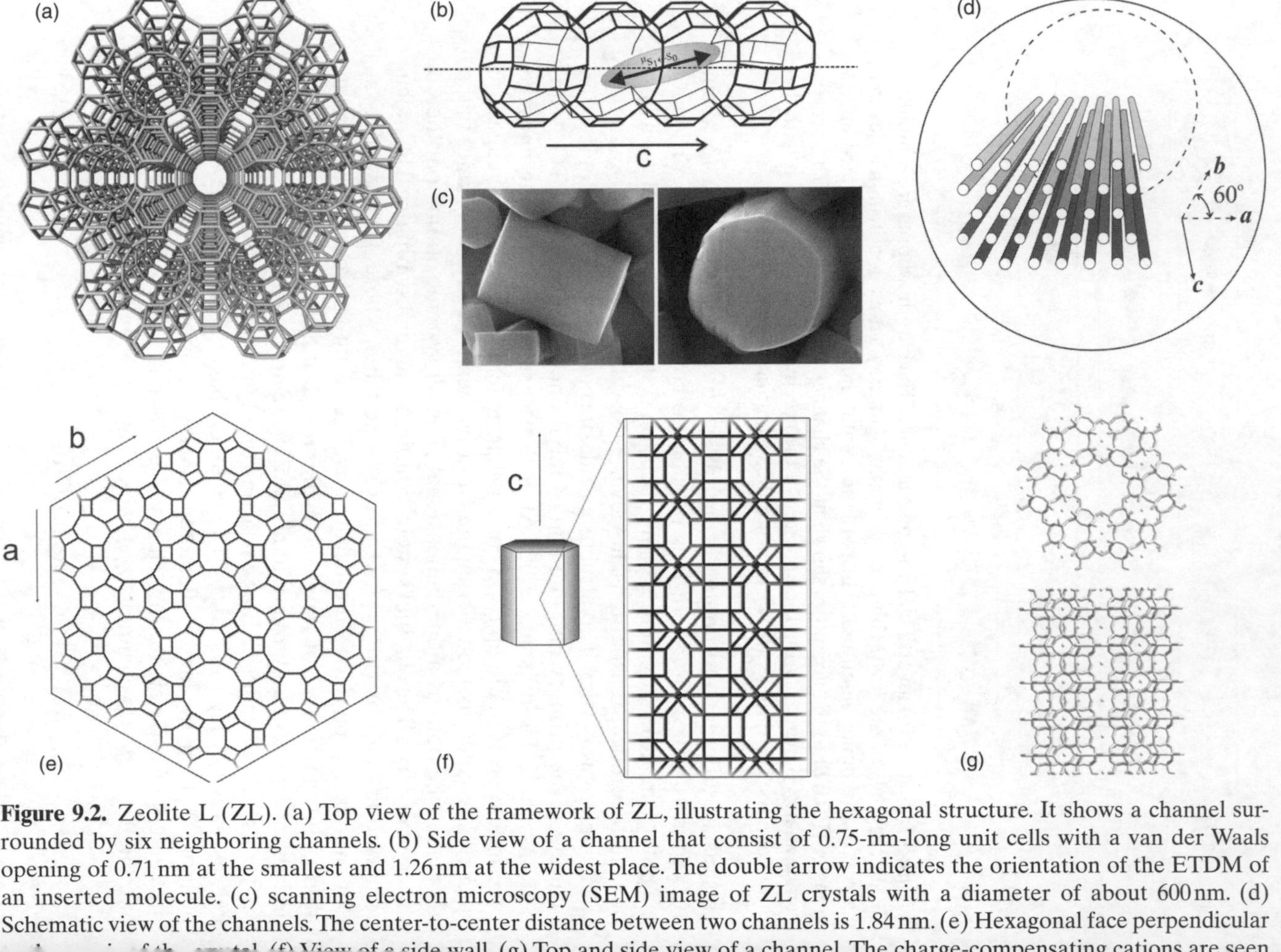

Figure 9.2. Zeolite L (ZL). (a) Top view of the framework of ZL, illustrating the hexagonal structure. It shows a channel surrounded by six neighboring channels. (b) Side view of a channel that consist of 0.75-nm-long unit cells with a van der Waals opening of 0.71 nm at the smallest and 1.26 nm at the widest place. The double arrow indicates the orientation of the ETDM of an inserted molecule. (c) scanning electron microscopy (SEM) image of ZL crystals with a diameter of about 600 nm. (d) Schematic view of the channels. The center-to-center distance between two channels is 1.84 nm. (e) Hexagonal face perpendicular to the c axis of the crystal. (f) View of a side wall. (g) Top and side view of a channel. The charge-compensating cations are seen

through geometrical constraints, the realization of extremely high concentrations of well-oriented molecules that behave essentially as monomers. A 30×30 nm crystal can take up to nearly 5000 dye molecules that occupy two unit cells, while a 60×60 nm crystal can host nearly 40,000. Many different molecules and clusters have been inserted into the channels of ZL.[13–15] The different surfaces of a ZL crystal, illustrated in Figure 9.2e–g, do not have the same chemical reactivity. This property can be used for obtaining objects with selectively functionalized surfaces. The base surfaces have significantly different reactivity compared with the coat, mainly due to the fact that the channel entrances are exclusively located at the base surface. This difference forms the basis for the stopcock principle.[16]

Here we focus on situations where molecules behave in the states of interest as individuals. This means that the "optical electrons" associated with the chromophoric units preserve essentially their individual characteristics. The chromophores can, however, communicate with each other. We are interested in molecules that are so large that they can neither pass each other inside of the channels nor sit on top of each other, as illustrated in Figure 9.3. The geometrical constraints imposed by the host determine the orientation of the dye molecules inside the channels. Exciton splitting becomes important at sufficiently short ETDM distances of neighboring chromophores. Small molecules such as naphthalene and anthracene can stack as shown on the right side of Figure 9.3a. Larger ones can stack in order to form J-aggregates (Figure 9.3c, e). In both cases, considerable concentration quenching of the fluorescence can result. The latter is completely avoided if the molecules are sufficiently bulky so that they have no other option to arrange (Figure 9.3b, d).

From the whole ZL volume, only a part, namely the large channels, is available for the guest species. Therefore, it is convenient to introduce a parameter bearing the information on dye concentration but based on purely geometrical (space-filling) properties of ZL as a host, that is, showing to what extent the ZL channels are filled with dye molecules. The *loading*, or occupation probability p, of a dye–ZL material is defined as follows:

$$p = \frac{\text{Number of occupied sites}}{\text{Total amount of sites}}, \tag{9.4}$$

where the *site* n_s represents the number of unit cells occupied by a dye molecule. It can, for example, be equal to 1, 2, or 3. n_s must not necessarily be an integer number. The loading ranges from 0 for an empty ZL to 1 for a fully loaded one. The dye concentration of a dye–ZL material $c(p)$ can be expressed as a function of the occupation probability as follows:

$$c(p) = 0.752 \frac{p}{n_s} \left(\frac{\text{mol}}{\text{L}} \right). \tag{9.5}$$

Förster's theory leads to a simple expression for the EnT rate constant that contains only experimentally accessible parameters. This is of great value. The

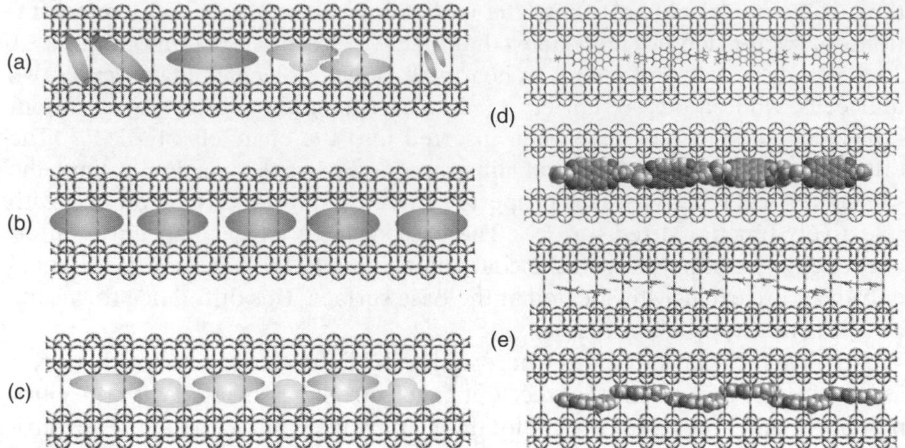

Figure 9.3. Packing of dyes in the channel; simplified view of different orientations and arrangements of molecules. (a) The molecule on the left is small enough to fit into one unit cell and its shape is such that it is oriented nearly perpendicular to the channel axis. Then, we see a molecule that occupies two unit cells and is oriented at an angle of about 45°. The next molecule is so large that it can only align parallel to the channel. Next, we illustrate a situation where two molecules come so close that their orientation and their optical properties are influenced by the packing. The molecules shown on the right are so small that they can form dimers and eximers. (b) Large molecules that align parallel to the channel axis because of their size and their shape. (c) Stacking of molecules of appropriate size and shape, leading, for example, to excitonic states. (d) Dense packing of the perylene dye PR149. The shape of the molecule is such that dense packing corresponds to scheme (b). (e) The shape of the cyanine dye $PC21^{+}$ allows a packing that corresponds to scheme (c).

theory has been adapted and generalized in order to meet more complex situations.[17,18] Here we explain the rate equation for Förster resonance energy transfer (FRET) by following the arguments given in the original work of Förster, and we describe exciton coupling based on Davydov's theory.[2,19,20] The results have been found to be very useful for designing and understanding organized systems based on nanochannel materials.[4,13,21–25] The discussion is restricted to one-photon processes under ambient temperature conditions. This means that processes in which two or more electronically excited molecules interact with each other—leading not only to phenomena such as annihilation processes, super radiance, lasing, and others, but also to very low temperature situations where shallow traps may be important—are not discussed. Energy transfer is abbreviated as EnT in order to distinguish from electron transfer, for which ET has often been used.

The FRET process can be extended from the inside of a crystal to its environment (or vice versa) by means of the so-called stopcock molecules.[16,21,24] Such stopcock molecules consist of a head group that is too large to pass the pore opening and a tail that fits into the channel. These chromophores are

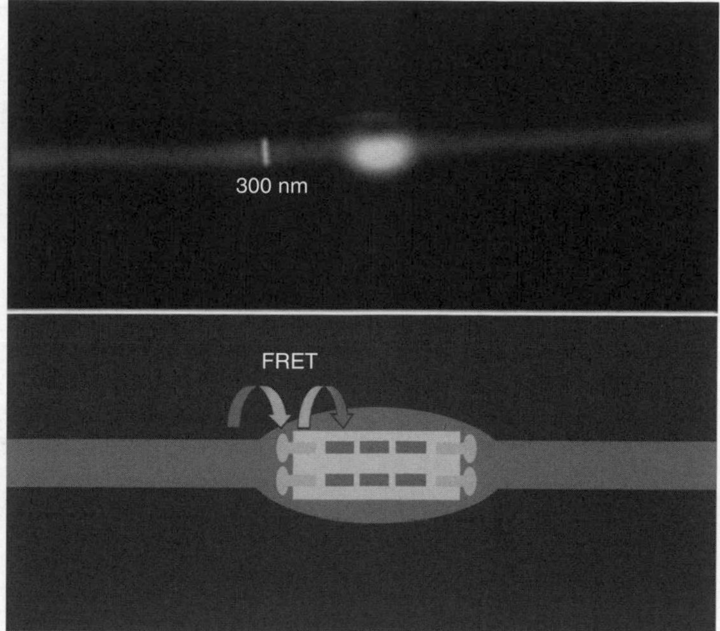

Figure 9.4. Example of FRET communication from the environment (here an electroluminescent polymer fiber) to the chromophores located inside of the ZL crystals. Top: fluorescence microscopy image of a dye-loaded, stopcock-plugged ZL crystal embedded in a poly (phenylene vinylene) (PPV) fiber. Bottom: Schematic representation of the above situation.

therefore located at the ZL basal surfaces and can promote the communication between dyes inside the crystal and the outside world. An example for such a system is given in Figure 9.4, where a dye-loaded, stopcock-modified ZL crystal has been embedded into electroluminescent polymer fiber.[25] Electronic excitation energy from the polymer is first transferred to the stopcock molecules, which in turn pass it to the chromophores located deeper in the channels.

A wide variety of molecules has been successfully incorporated into ZL or used as stopcocks. Tables 9.1–9.3 give examples of cationic or neutral dyes as well as stopcock molecules. The preparation and properties of supramolecularly organized dye–ZL composites are discussed in the section on "Supramolecularly Organized Dye-ZL Composites." Electronic excitation EnT in these materials is reported in the section on "FRET in Supramolecularly Organized Dye-ZL Composites." The theoretical basis for understanding these properties are explained in detail in the Appendix.

SUPRAMOLECULARLY ORGANIZED DYE–ZL COMPOSITES

ZL is an excellent host for the supramolecular organization of guest species, be they organic molecules, complexes, or clusters.[12–16,21–25] Such guests can be

TABLE 9.1. Examples of Cationic and Zwitterionic Dyes That Have Been Inserted into the Channels of ZL

Dye	Structure	Dye	Structure	Dye	Structure
$ProFl^+$		Th^+		MC^+	
Py^+		$MeAcr^+$		$PC20^+$	
$PyGY^+$		AY		$PC21^+$	
PyB^+		AQ		$BTMPI^+$	
Ox^+		MV^{2+}		DSM^+ (n = 1)	
$DEOx^+$		BDP^+		SQ III $R = CH_3$ SQ(JG9) $R = (CH_2)_3CH_3$	
$Ox1^+$				$Ln^{3+}L_n$	$Ln^{3+} = Eu^{3+}, Tb^{3+}, Er^{3+}$ L = bpy, benzophenone

inserted either by ion exchange or by adsorption from the gas phase or a solution, depending on whether they are charged or neutral species, respectively. Smaller complexes can be brought into the channel system as they are, while larger ones have to be prepared by means of a "ship-in-a-bottle" synthesis. This kind of synthesis was pioneered by Lunsford et al. in the early 1980s[26] and has since been used successfully for preparing many different materials.[27–30] The name is derived from artistic bottles containing a ship that is larger than the bottle neck. The principle is illustrated in Figure 9.5: the first step involves the insertion of the metal centers by ion exchange. The next step is to let potential ligands diffuse into the channels where they will form the desired metal complex. The reaction is easier and faster close to the channel entrances, since spatial restrictions make it more difficult for the ligand to reach the ions lying deeper inside. Several luminescent rare earth complexes have been prepared in ZL channels via this method.[14b, 28–30]

Our discussion here will be focused on organic dyes as guests. The geometrical constraint imposed by the host material leads to a highly ordered and well-defined arrangement of the guests inside the channels, as is illustrated in Figure 9.3. Organic dye molecules that can pass the 0.71-nm pore openings of ZL are usually too large to overtake or stack with molecules already present in the channels. Thus, one can create materials with two or more defined domains, each containing only one type of guests. This geometrical confine-

TABLE 9.2. Examples of Neutral Dyes That Have Been Inserted into the Channels of ZL

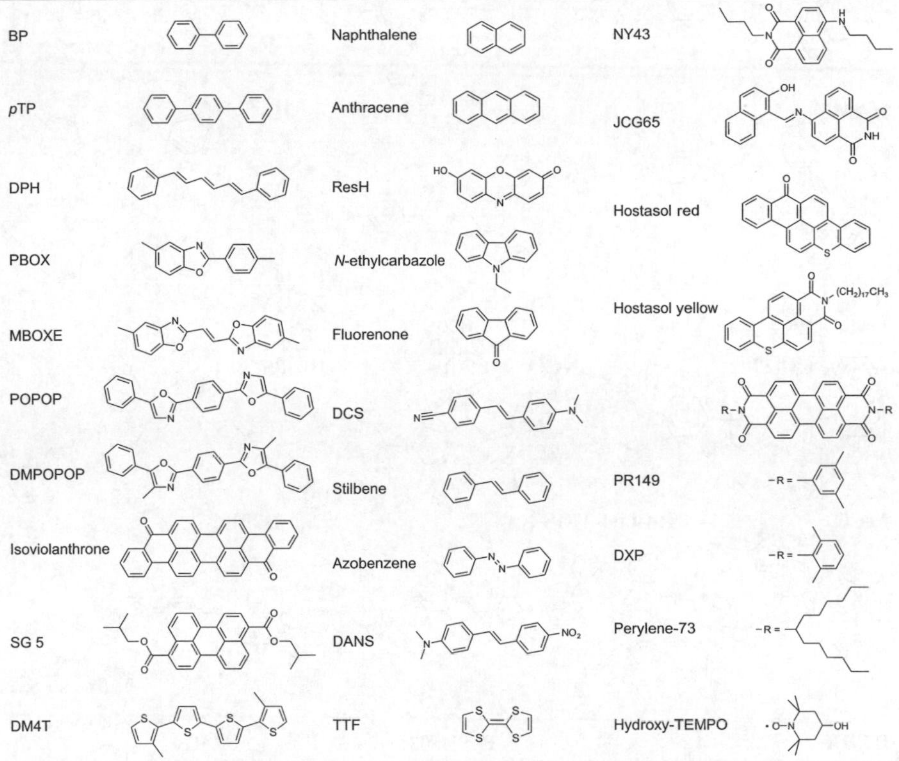

BP	Naphthalene	NY43
*p*TP	Anthracene	JCG65
DPH	ResH	Hostasol red
PBOX	*N*-ethylcarbazole	Hostasol yellow
MBOXE	Fluorenone	
POPOP	DCS	
DMPOPOP	Stilbene	PR149
Isoviolanthrone	Azobenzene	DXP
SG 5	DANS	Perylene-73
DM4T	TTF	Hydroxy-TEMPO

ment also makes it possible to prepare systems with very high dye concentrations where the molecules are still acting as monomers. However, small dyes, such as naphthalene, can form dimers inside of the channel system.[31] We focus on molecules that are too large for forming aggregates of this type. Another point is that the molecules' size and shape can allow them to come close enough so that coupling of the ETDM becomes significant.[21–23] $PC21^+$-ZL materials are examples where such a situation can be observed. Since the molar extinction coefficient of $PC21^+$ is very high (165,900 $M^{-1}\,cm^{-1}$) and its oscillator strength is 1.4, the J-coupling can be considerable. At the top left of Figure 9.6, the absorption spectrum of a 10^{-6} M $PC21^+$ solution in 1-butanol is compared with the excitation spectra of ZL samples with different dye loadings ($p = 0.068$ and $p = 0.18$, respectively). With increasing loading, the excitation band splits into two components, one at longer (492 nm) and one at shorter wavelength (452 nm). Also, the emission spectra shown at the top right of Figure 9.6 feature two additional bands at 540 and 572 nm. The long wavelength component is more prominent for the material with higher dye

TABLE 9.3. Examples of Stopcock Molecules That Have Been Attached to the Channel Entrances of ZL

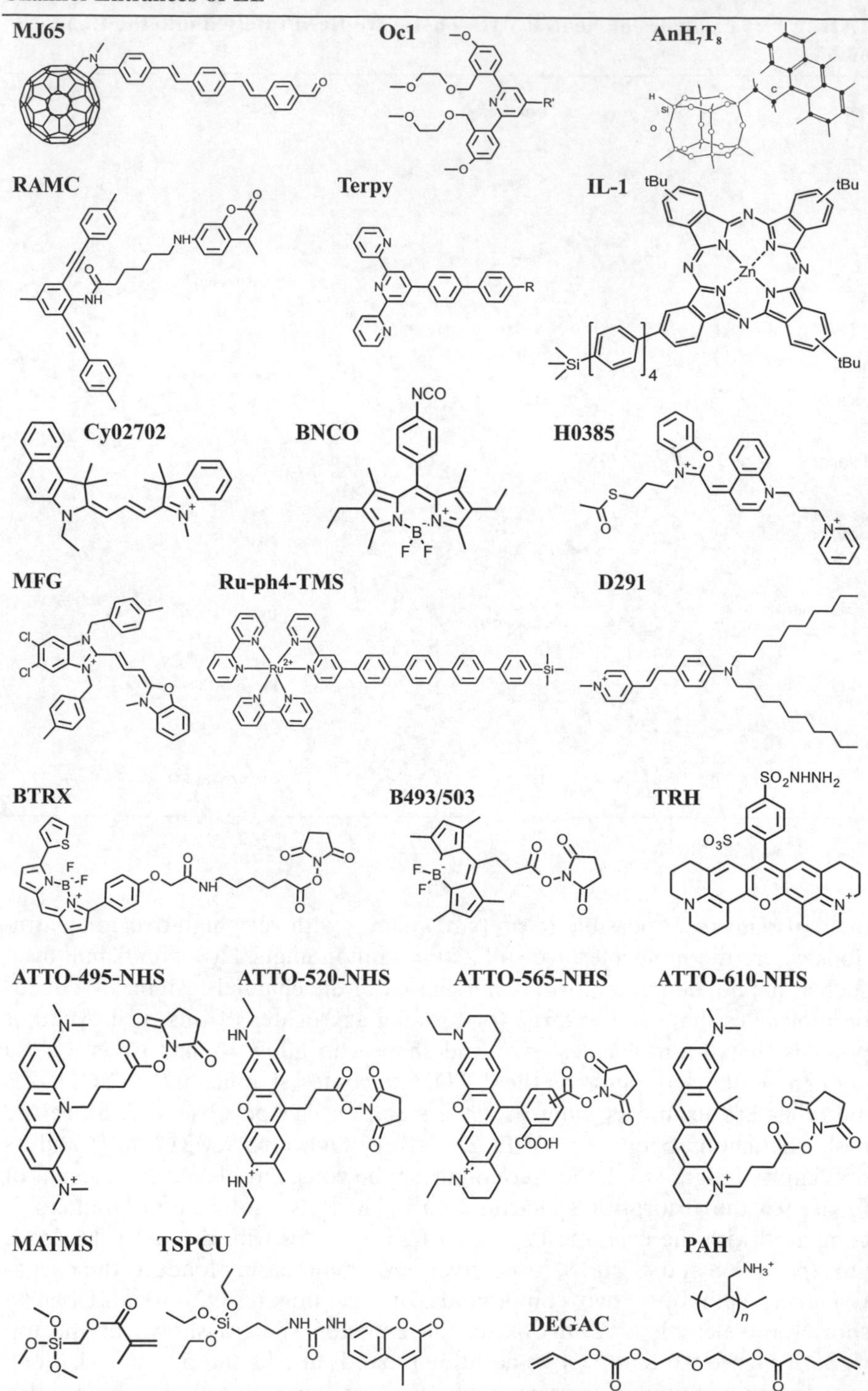

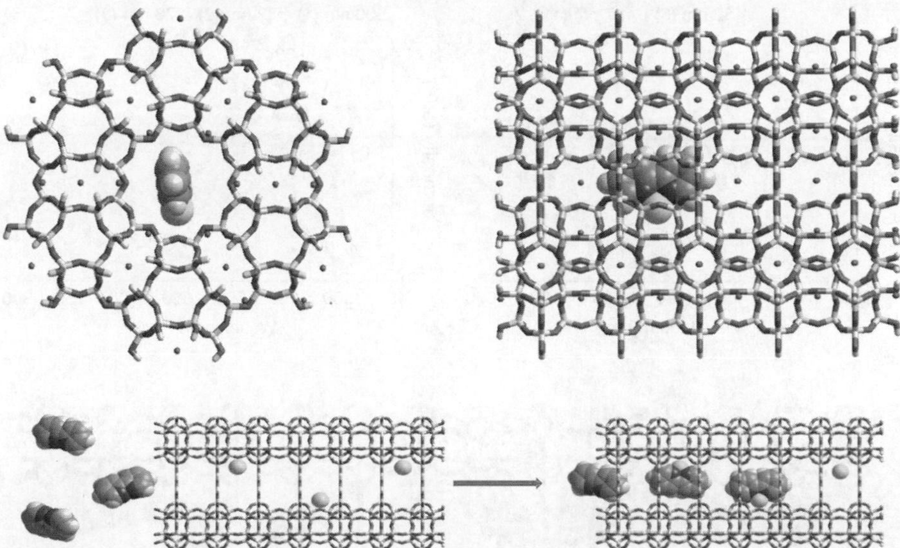

Figure 9.5. "Ship-in-a-bottle" synthesis in the channels of ZL. The upper part of this scheme shows a front and side view of the structure of ZL containing Ln^{3+}(bpy); the Ln^{3+} ions are shown as light gray balls. The small spheres represent potassium cations belonging to ZL. Monovalent cations (3.6) per unit cell can be exchanged. Water molecules are not shown. The lower part illustrates the synthesis procedure. Bpy enters the channels, shown in a simplified way, to form the complex. The scheme also illustrates that the reaction is easier and faster close to the channel entrance. Space restriction makes it more difficult for the ligand to reach the Ln^{3+} ions lying deeper inside.

concentration and is slightly red shifted, from 572 to 578 nm. Based on these spectroscopic observations, polarized fluorescence microscopy images and the fact that $PC21^+$ enters the channels only reluctantly, we can deduce the situation depicted at the lower part of the figure. The noninteracting arrangement shown on the left side of the channel correspond to samples with low dye concentrations, while the stacking indicated on the right side is expected for higher loadings.[21] A nice demonstration of exciton coupling in a dye–ZL composite has recently been reported for Py^+ in ZL in sophisticated time, space, and spectrally resolved confocal microscopy experiments;[23a–c] in experiments with a furo-furanone;[23d] and with perylene bisimide dyes.[23e]

Another factor to keep in mind is that the spectral behavior and stability of molecules embedded in the channels of ZL can be much influenced by the charge compensating ions.[7,32] Out of the 9 monovalent cation positions in each unit cell, 3.6 are located along the wall of the main channel and can therefore interact with the guests. These positions are usually occupied by monoatomic cations such as protons, alkali, or earth alkali metals. An alternative would be to replace these exchangeable cations with small organic cations

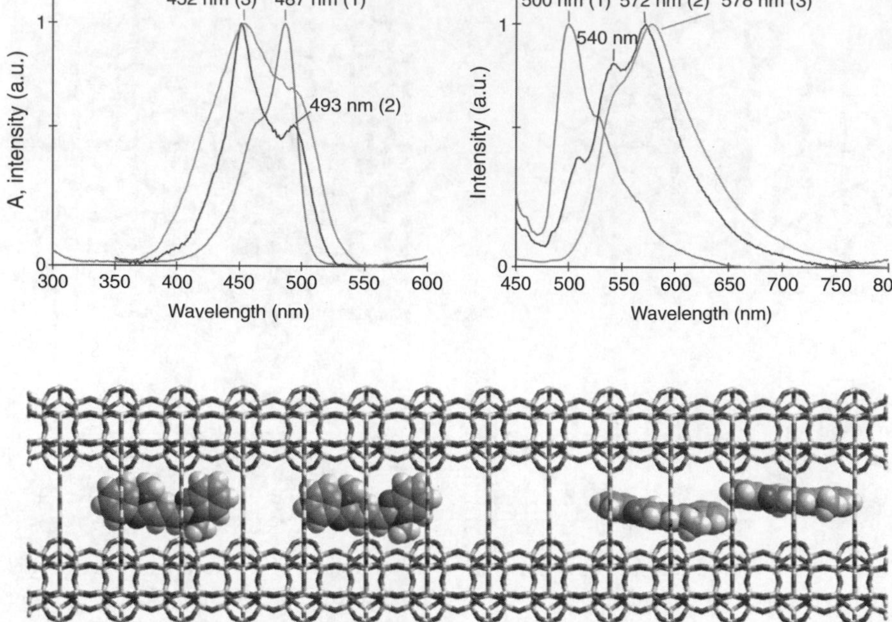

Figure 9.6. Example of J-coupling in the channels of ZL. Absorption, excitation, and emission spectra of $PC21^+$ and scheme of molecular arrangement in a channel of ZL. Upper left: absorption spectrum of $PC21^+$ in 10^{-6} M 1-butanol solution (1); excitation spectra of $PC21^+$-ZL at $p = 0.068$ (2), emission observed at 540 nm; and at $p = 0.18$ (3), emission observed at 640 nm. Upper right: emission spectra of $PC21^+$-ZL at $p = 0.045$ (1), at $p = 0.068$ (2), and at $p = 0.18$ (3). Lower: $PC21^+$ molecules in a channel of ZL; left: noninteracting molecules; right: stacking occurring at higher loading.

such as imidazolium derivatives. To our knowledge, this variant has not yet been explored.

ZL as a host material allows for the design and preparation of a large variety of highly organized host–guest systems. Some relevant types are illustrated in Figure 9.7. Important steps in the development of these composite materials were the sequential filling with different dyes that lead to sandwich structures, the invention of the stopcock principle, the discovery of quasi-1D EnT, the preparation of unidirectional EnT material, and finding ways to create hybrid materials fully transparent in the visible range.[13,16,33–41] The latter is important for spectroscopic investigations, as well as for optical and electro-optical applications, since the small ZL crystals exhibit considerable light scattering due to their size and refractive index between 1.4 and 1.5. ZL is the only currently available microporous material allowing the realization of the full range of organizational patterns presented here.

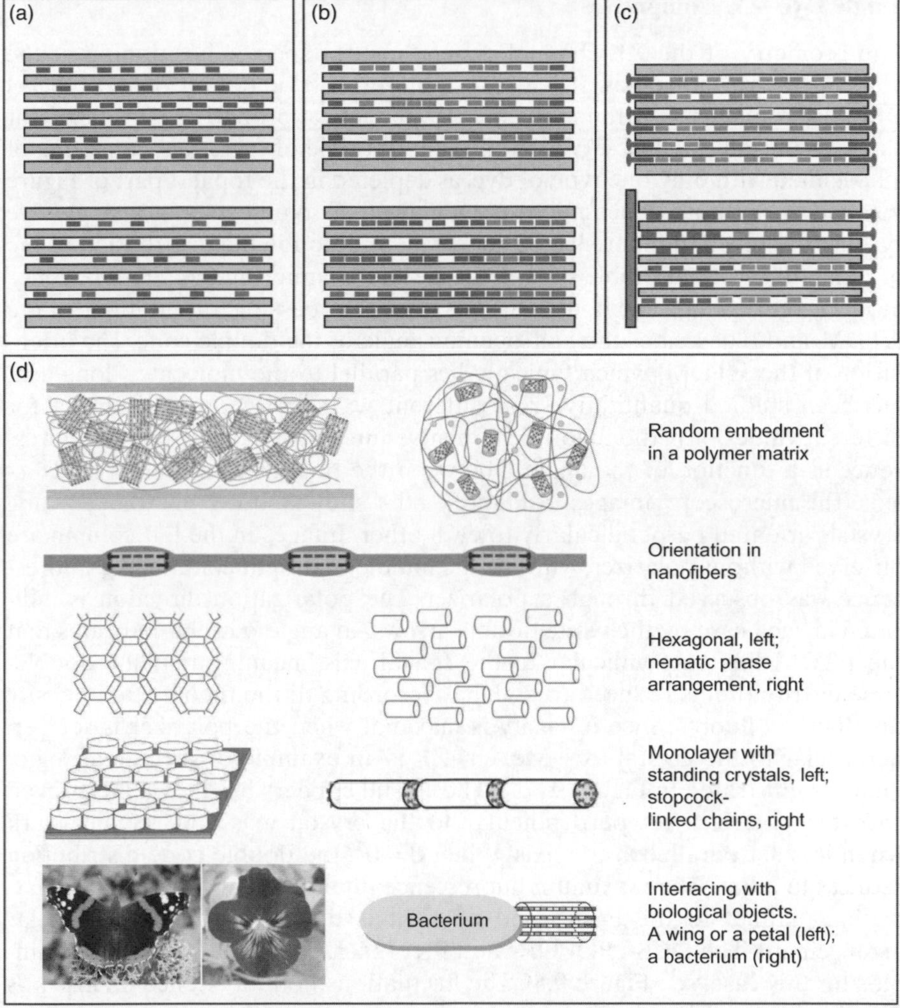

Figure 9.7. Schematic representations of supramolecularly organized functional dye–ZL composites. (a) *Single- and mixed-dye materials* are obtained by either loading ZL crystals with one kind of dye (top) or by simultaneous insertion of different dye molecules (bottom). (b) *Antenna materials* can be prepared by the sequential insertion of different dyes. (c) *Stopcock-plugged antenna materials* are obtained by modifying either bidirectional (top) or monodirectional (bottom) antenna materials with specific closure molecules, called stopcocks. (d) *Organization of ZL crystals.* Arrangement of functional dye–ZL crystals into different patterns and interfacing with biological objects. See ftp site for color: ftp://ftp.wiley.com/public/sci_tech_med/supramolecular_photochemistry.

Single Dye–ZL Composites

The geometry of the ZL channel system ensures specific positioning of the embedded dye molecules. It leads to a preferred orientation of the ETDM and thus to materials with highly anisotropic optical properties. This behavior can be especially well observed in the simplest type of materials, consisting of a host filled with only one type of dye, as depicted in the top left part of Figure 9.7. Polarized fluorescence microscopy images of some exemplary single-dye systems are given in Figure 9.8. A good way to describe the distribution of the orientation of an ensemble of dyes inside the channels is the "double cone" model (Figure 9.8a). The double arrows represent possible orientations of the ETDM, and β describes the half-opening angle of the double cone. The orientation of the ETDM, which typically lies parallel to the molecules' long axis, has been studied quantitatively in different ways.[21,37] Figure 9.8c shows, for different values of β, the theoretical relative intensity of the observed fluorescence as a function of the angle ε between the polarizer and the crystals' *c*-axis. The microscopy images in Figure 9.8d–f show about 2-micrometer-long crystals oriented perpendicularly to each other. Images in the left column are observed without polarizer, while in the middle and right column, the fluorescence was observed through a polarizer. The polarization direction is indicated in each case by the white double arrow. An angle α of 90° indicates that the ETDM lies perpendicular to the crystal axis, meaning that the double-cone distribution is reduced to a plane. According to the theoretical intensity distribution, fluorescence intensity is maximal when the polarizer is set perpendicular to the crystal axis. $MeAcr^+$-ZL is an example of such an arrangement, as illustrated in Figure 9.8d.[38] The crystal appears bright when observed with the polarizer set perpendicular to the crystal axis and remains dark when it is set parallel to the axis. When $\beta = 0°$, the double cone distribution reduces to a line. The maximum fluorescence intensity is observed with polarization parallel to the *c*-axis, and no light is detected in a perpendicular arrangement. The DCS-, POPOP-, and PR149-ZL composites are good examples for this case; see Figure 9.8f. The fluctuations in fluorescence intensity as function of the polarizer angle are less extreme for the three examples shown in Figure 9.8e. The orientation of the Py^+ ETDM in ZL is 72°.[37] In the case of a longer molecule, $PyGY^+$, we hardly observe any angle dependence. This means that the orientation of the ETDM is close to the magic angle (52.1°). The fluorescence intensity of PyB^+-ZL is highest when the polarizer is oriented parallel to the *c*-axis, and the difference between the two extremes is quite large.

Composites Consisting of Two or More Different Dyes

A more complex material can be prepared by loading two dyes simultaneously into ZL. If the dyes have similar insertion kinetics, as is the case for Py^+ and Ox^+, one obtains a host filled with a random mixture of both guests, as sketched

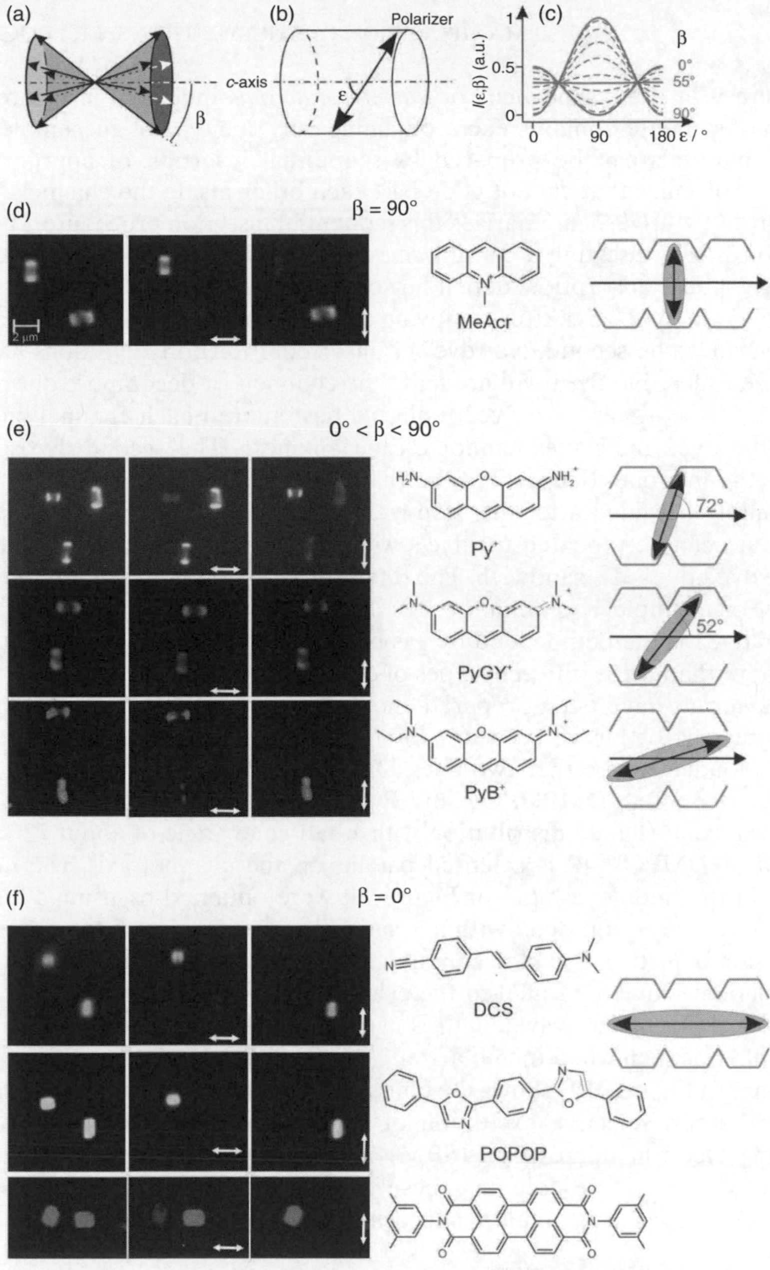

Figure 9.8. Orientation of ETDM of dye molecules inside the channel. (a) Distribution of the ETDM on a double cone with a half-opening angle β. (b) Polarization of emission observed when a single crystal is examined by means of a polarizer. (c) Relative intensity of the observed fluorescence as a function of the observation angle ε with respect to the c-axis, for different half cone angles β. (d–f) Different orientations of the ETDM of dye molecules. (d) Perpendicular to the channel axis, $\beta = 90°$. The double cone distribution of the dye molecules reduces to a plane. (e): $0° < \beta < 90°$. (f) The orientation is parallel to the channel axis, $\beta = 0°$. The double cone distribution reduces to a line. See ftp site for color: ftp://ftp.wiley.com/public/sci_tech_med/supramolecular_photochemistry.

in Figure 9.7a. The "sandwich" or *antenna materials* indicated in Figure 9.7b consist of separate domains, each containing only one type of guest molecules. These materials can be prepared by sequential insertion of appropriately chosen molecules that cannot glide past each other inside the channels.[39] The left part of Figure 9.9 summarizes this sequential insertion procedure. The first step consists of inserting a certain amount of a dye1 either by ion exchange or by gas-phase adsorption, depending on its properties. This leads to a material we name dye1-ZL. After removing dyes, which might be adsorbed at the outer surface, the second dye (dye2) is inserted. Insertion conditions have to be chosen such that dye1 will not leave the channels or decompose during the process. Once at least one dye2 molecule has entered each channel on both sides, the dye1 molecules cannot escape anymore. The second dye species pushes the first ones deeper into the channels, as they cannot pass each other. The sample obtained after this step is described as dye2,dye1-ZL. The same procedure can be repeated for dye3, with the resulting material being called a dye3,dye2,dye1-ZL sandwich. The different insertion methods can be combined: For example, a cationic dye1 is inserted by means of ion exchange. A neutral dye2 is then embedded by gas-phase adsorption and so on. Materials with more than three different types of dyes can be prepared if desired. Many such materials have been prepared and characterized.[13,21,22,34,39–41] The middle and right part of Figure 9.9 show fluorescence microscopy images of single crystals loaded with either two dyes, Ox^+ and DMPOPOP (middle column), or three dyes, Ox^+, DMPOPOP, and Py^+ (right column). The ETDM of Ox^+ exhibits a cone-shaped distribution with a half cone angle of about 72°, while ETDM of DMPOPOP is oriented parallel to the channel axis. The images shown in the middle column of Figure 9.9 were obtained by using a fluorescence microscope equipped with a polarizer, an appropriate filter set, and an immersion objective: The blue emission seen in Figure 9.9A was recorded with the polarizer aligned parallel to the cylinder axis and with specific excitation of DMPOPOP in the wavelength range of 330–385 nm; the red emission in Figure 9.9C is seen when the polarizer is oriented perpendicular to the crystal; the image in Figure 9.9B shows the emission from the system without polarization and upon selective excitation of Ox^+ in the range of 545–580 nm. A scanning electron microscopy (SEM) image of the crystal is given at the bottom of the middle column. These data show that the dyes are indeed organized inside the channels as suggested by the scheme for the second insertion step.

Stopcock-Plugged Composites

Stopcock-plugged antenna materials (Figure 9.7c) are obtained by modifying the channel ends of dye-loaded ZL by specific closure molecules, which can only partially enter the channels. Such stopcock molecules consist typically of a head and a tail, as indicated in Figure 9.10A. Due to size restrictions, only

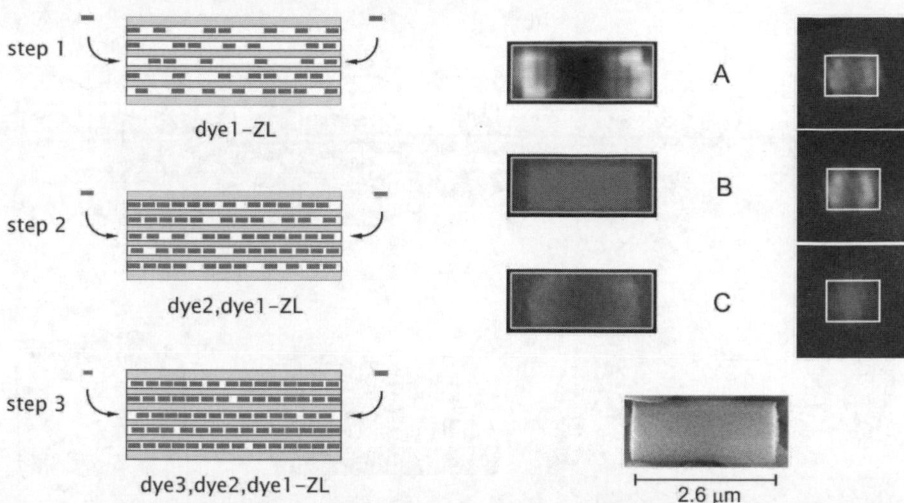

Figure 9.9. Preparation and fluorescence microscopy images of sandwich materials. Left column: scheme for sequential insertion (steps 1, 2, and 3) of three different dyes into the channels of a ZL crystal to form a sandwich material.[39] Middle column: DMPOPOP,Ox^+-ZL crystals. The blue emission (A) is observed through a polarizer parallel to the cylinder axis when specifically exciting DMPOPOP at $\lambda = 330$–385 nm. The red emission (C) is seen when the polarizer is turned into vertical position. Image (B) is observed without polarizer upon excitation of Ox^+ at $\lambda = 545$–580 nm. At the bottom, we show the SEM image of a crystal used in these experiments.[40] Right column: Py^+,DMPOPOP,Ox^+-ZL crystal. The outline of the about 1.5-micrometer-long ZL crystal we have used is indicated in white. The green Py^+ emission (A) was visualized by exciting the crystal at 470–490 nm. The emission was detected through a broadband interference filter at 550 nm. The blue DMPOPOP emission (B) was recorded by exciting the DMPOPOP at 330–385 nm and observing the emission polarized in the direction of the crystal axis. The red Ox^+ emission (C) was detected after excitation at 545–580 nm.[13] See ftp site for color: ftp://ftp.wiley.com/public/sci_tech_med/supramolecular_photochemistry.

the tail can enter the channel.[12,13,16,21] Depending on their nature, stopcocks can be bound either by physisorption, by electrostatic interaction, or by covalent bonding. Examples of stopcocks are given in Table 9.3. Since these molecules are located at the interface between the interior of a ZL crystal and the surrounding, they can be considered as mediators for communication between molecules inside the nanochannels and objects outside of the crystals. Stopcock molecules can also be used to prevent penetration of small molecules such as oxygen and water or to hinder encapsulated dye molecules from leaving the channels.[32] A nice example for the sealing properties of stopcock molecules is presented in Figure 9.10b. The Ru-ph4-TMS complex has a tail fitting

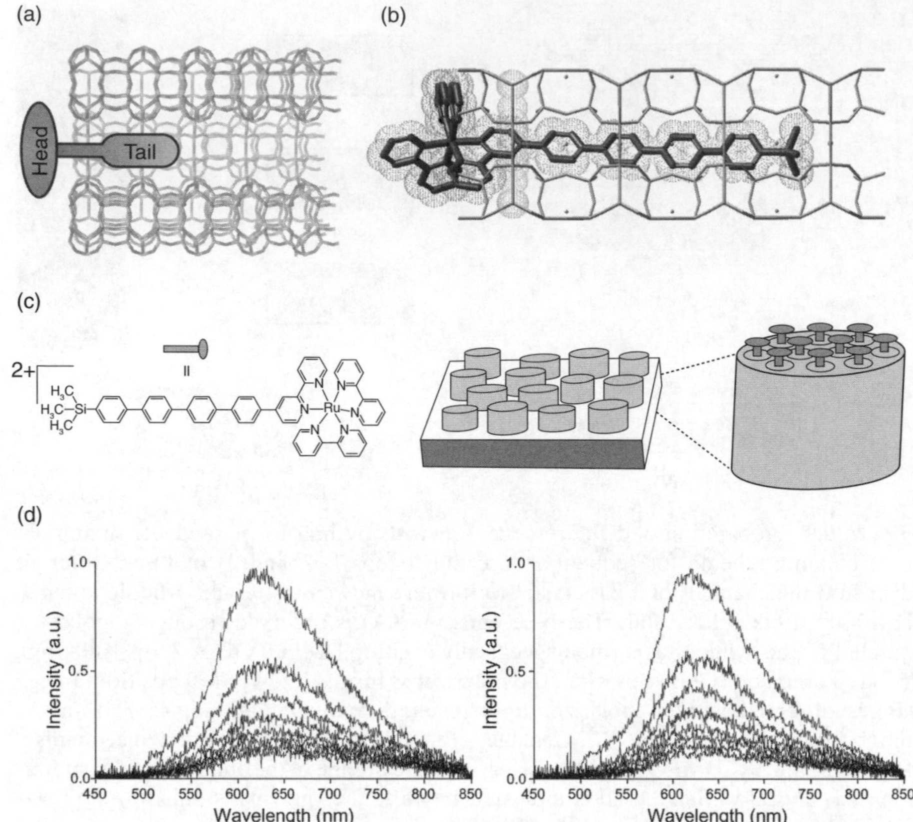

Figure 9.10. Stopcock principle and (Ru-ph4-TMS)-ZL assembly. (a) Schematic representation of one channel entrance with a stopcock molecule. (b) Side view of the (Ru-ph4-TMS)-ZL assembly. The stopcock molecule is electrostatically bound to the channel entrance. The dotted clouds outline the van der Waals surface of the molecule, showing the plugging capability of the stopcock. (c, left) Structure of the Ru-ph4-TMS stopcock. (c, right) A pictorial view of a ZL monolayer on a glass substrate with an enlarged view of the (Ru-ph4-TMS)-ZL assembly. The ZL channels are oriented perpendicular to the substrate plane. (d) Time-resolved emission spectra of a (Ru-ph4-TMS)-ZL monolayer in toluene under O_2 (left) and under N_2 (right) atmosphere. The spectra were excited at 460 nm and recorded at room temperature. The delay increment between consecutive spectra is 200 ns.[47]

nicely into the channel. Due to the positively charged head, the stopcock is electrostatically bound to the channel entrance.[42] The van der Waals model indicates that the head group is bulky enough to fully block the pore opening. Ru-ph4-TMS is a triplet emitter and, as many other long lived species, is quenched by oxygen via a diffusion-controlled collision.[43–45] We have observed that the luminescence quenching of the Ru-ph4-TMS stopcock by oxygen nearly disappears when the former is bound to ZL. This is due to the delocalization of the emitting ^{3}MLCT state over the bpy-ph4 ligand. When bound to ZL, this part of the molecule is shielded from O_2 collisions by the framework.[46] Time-resolved luminescence spectra of Ru-ph4-TMS-plugged ZL monolayers under O_2 or N_2 atmosphere are shown in Figure 9.10c, with a time delay of 200 ns between each curve. Both sets of spectra are remarkably similar, and the lifetimes lie in the same range as for the free complex in solution under N_2 atmosphere, at about 1200 ns.[47] This supports the idea of excellent shielding from O_2 collisions offered by the ZL channel to the bound Ru-ph4-TMS tail.

We wish the communication between exterior and interior to be efficient and spatially well controlled. Therefore, the method for sequential modification of the channel entrances[21,48] is a very effective and flexible way to realize such an interfacing. As an example, all *N*-hydroxysuccinimidyl ester (NHS) derivatives shown in Table 9.3 were covalently bound to ZL by first selectively modifying the channel entrances with alkoxysilanes featuring a protected amino group. After the removal of the protecting group, the amino groups at the ZL surface can bind the NHS derivatives. Stopcock molecules bearing a positive charge either on head, tail, or even both groups are also of great interest. The electrostatic binding of such molecules to the negatively charged channels of ZL is quite strong but still reversible under the right conditions, thus allowing flexibility in the plugging of ZL channel entrances.

The location of the stopcock can be verified experimentally by means of fluorescence microscopy.[16,21,25,48] An example is given in Figure 9.11a, where ATTO-680 has been covalently attached to both ends of a ZL crystal.[49] The red luminescence on the crystal bases stems from the attached ATTO-680. The shape of the ZL crystal is outlined by a white rectangle. Another example for a molecular stopcock is octasilasesquioxane ($H_7Si_8O_{12}$) substituted with one anthracene group, indicated as AnH_7T_8 in Table 9.3.[21,50] The van der Waals model given in Figure 9.11b illustrates that the $H_7Si_8O_{12}$ cage fits tightly into the channel, completely sealing it, while the anthracene part remains at the outside. This interpretation is supported by the confocal fluorescence microscopy images in Figure 9.11c of AnH_7T_8-plugged ZL crystals. They exhibit blue emission only from the basal surfaces. The white rectangles show the outline of the crystal. Superior stability of the (stopcock)-ZL assembly is obtained after covalent attachment to the channel entrances. It is also possible to work with a polymer bearing specific side groups, which can plug the channel entrances through self-organization. Combination of selective channel entrance functionalization by the stopcock principle and of the selective coat surface modification leads to challenging new materials. The different options for attaching stopcock molecules to the channel entrances of ZL make this principle very attractive. The van der Waals attachment is reversible, depending on a variety of parameters such as the surrounding solvent, pH, or ionic strength. The same is partly true for the electrostatic binding that exploits the negative charge of the ZL framework: A positive charge of the tail or head

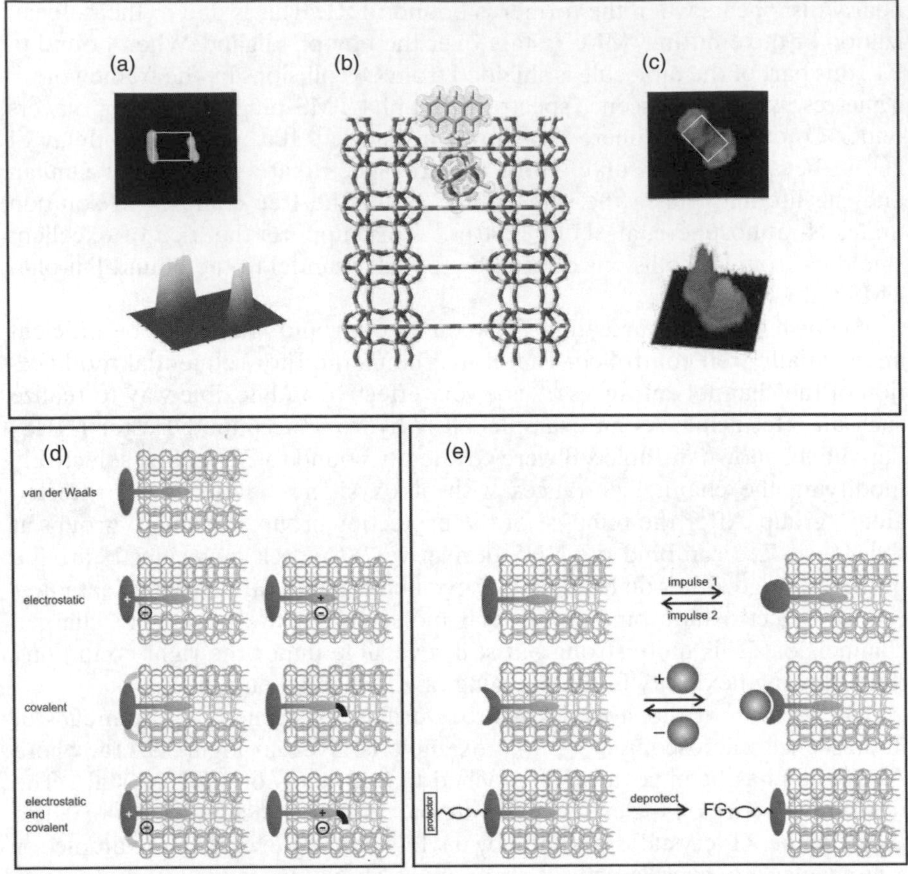

Figure 9.11. Stopcocks. (a) Confocal fluorescence microscopy image and intensity map of an (ATTO680)-ZL crystal. The image indicates that only the base of the crystal is modified with ATTO680. (b) van der Waals model of the stopcock AnH_7T_8 adsorbed to the entrance of a ZL channel. The molecule fully blocks the channel entrance. (c) Confocal fluorescence microscopy image and intensity map of a (AnH_7T_8)-ZL crystal, which show that the stopcocks are only located at the basal surface. (d) Different ways of attaching stopcock molecules. (e) Functionalities of stopcock heads. See ftp site for color: ftp://ftp.wiley.com/public/sci_tech_med/supramolecular_photochemistry.

can be used to substitute a charge-compensating cation. Both van der Waals and electrostatic attachments can be used to build a precursor for covalent binding. Stopcocks can be chosen such that they have the desired reactivity as schematized in Figure 9.11e. If a photochromic moiety such as an azobenzene or a spiropyran is used, the stopcocks can have switching properties, which can be triggered by illumination. In such a case, the channel would be partially open in one situation and completely blocked in another. Similar switching behavior can also be realized by using heads reacting on stimuli such as pH changes or redox processes.

Highly Organized Systems

A higher degree of supramolecular organization can be reached by arranging the dye–ZL composites discussed so far into larger structures. Such hierarchically organized structures, presenting successive ordering from the molecular up to macroscopic scale, are of great interest for nanotechnology, due to the relationship between molecular arrangements and macroscopic properties.[51–53] The organization of quantum-sized particles, nanotubes, and microporous materials has been studied on different surfaces. It is used in science, technology, diagnostics, and medicine.[6,54–62] Size, shape, and surface composition of the objects and also the properties of the surface on which they should be organized play a decisive role and, in some cases, determine not only the quality of the self-assembly but also its macroscopic properties. Self-assembly strategies for the organization of matter make hierarchical ordering attractive by avoiding expensive techniques such as photolithography. An overview of hierarchically ordered materials that were created by self-assembly from dye–ZL composites is given in Figure 9.7d and in Figure 9.12.

Hybrid (polymer),dye-ZL materials with randomly oriented crystals, as shown in Figure 9.12a, are obtained by dispersing surface-modified ZL in the precursor monomer, followed by a specific polymerization procedure.[36] Surface modification with alkoxysilane derivates improves the miscibility of the crystals with the organic monomers, preventing microphase separation and even enabling copolymerization when an appropriate organic substituent is used. In this type of material, the usually strong light scattering from ZL can be suppressed by refractive index matching and avoidance of microphase separation. This property is of great value, among others, for spectroscopic investigations, as will be discussed later on.

Dye–ZL composites usually feature high optical anisotropy. Generating a similar anisotropy on the macroscopic scale is useful for many applications and also for spectroscopic analysis. Materials containing crystals oriented as shown in the leftmost sketch of Figure 9.12b can be obtained by using elastic polymers as substrate. After appropriate treatment of the polymer substrate, such as stretching it with or without gentle heating, the crystals form the desired nematic phase. The mechanical force of the stretching influences the position of the ZL crystals and most will align in the pulling direction. The SEM and fluorescence microscopy images of Py^+-ZL embedded into a poly(vinyl chloride) (PVC) polymer film, shown in the middle of Figure 9.12b, illustrate this well. The polarized fluorescence microscopy images shown present another example of this nematic arrangement. The sample was prepared by drop coating ZL loaded with either Ox^+ or MC^+ onto a poly(isobutylene) (PIB) plastic film.[36] The film was then stretched to its sixfold size upon gentle heating. The ETDM of the red fluorescent dye Ox^+, when inserted into ZL, is

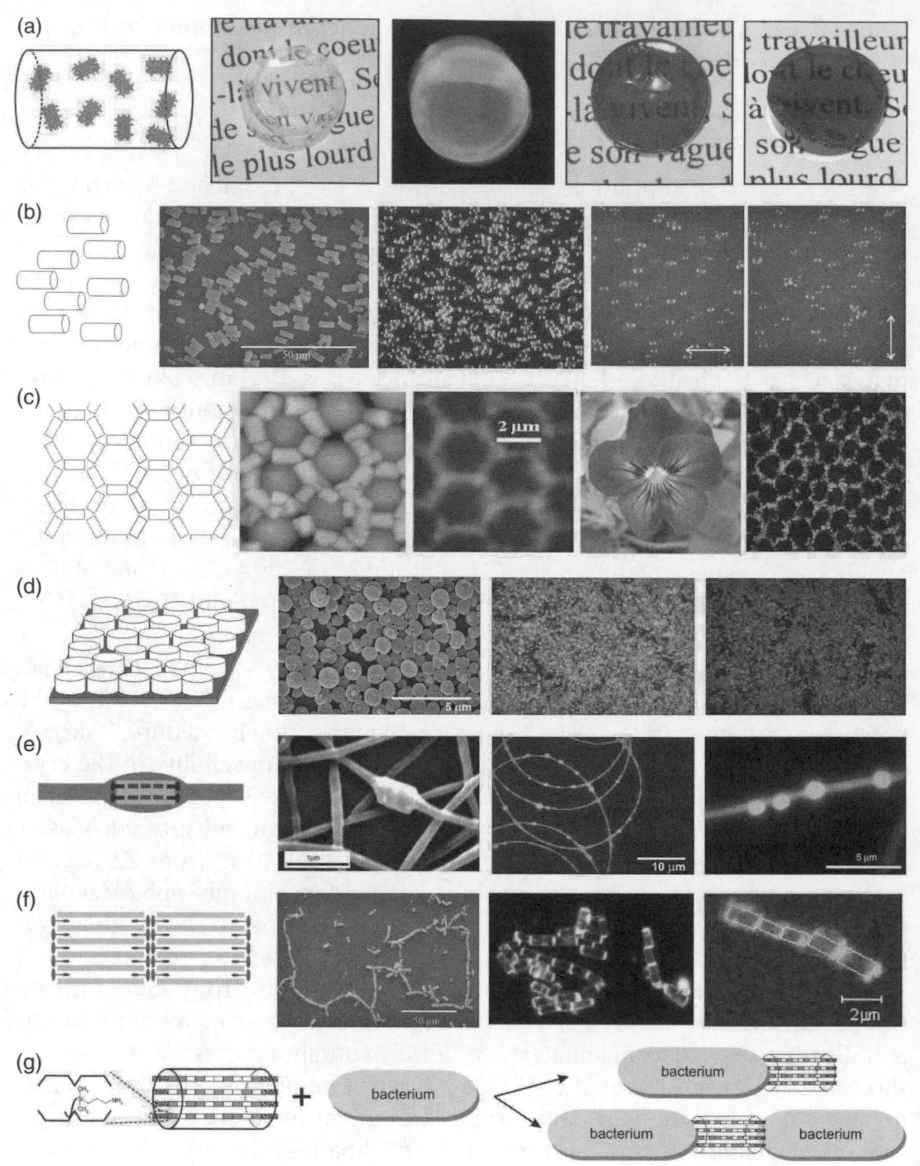
(a)
dont le coeu
-là vivent. Se
le plus lourd
travailleur
(b)
50 μm
(c)
2 μm
(d)
5 μm
(e)
1μm
10 μm
5 μm
(f)
50 μm
2μm
(g)
+
bacterium
bacterium
bacterium
bacterium

Figure 9.12. Overview of supramolecular organizations of ZL crystals. (a) Randomly oriented ZL crystals in a polymer. Photographic images of ZL–polymer hybrid materials prepared from DMPOPOP-ZL (observed under normal and UV illumination), from hostasol-red-GG-loaded crystals, or from Ox^+-ZL (from left to right). (b) Nematic phase arrangement. Left: SEM and fluorescence microscopy image of Py^+-ZL crystals aligned in a stretched PVC plastic film. The Py^+ molecules are located at the channel ends due to the short insertion time.[83] (c) Hexagonal arrangement. Left: dye–ZL crystals organized by a surface-tension-driven self-assembly process on a PDMS/PS film (SEM and fluorescence microscopy image).[64] Right: Fluorescence microscope image of Ox^+-ZL aligned on the petal of a horned pansy (*Viola cornuta*).[65] (d) Monolayers of standing ZL crystals. Left: SEM image of a monolayer. Right: fluorescence microscopy images of monolayers loaded with Py^+ and Ox^+, respectively.[34] (e) Nanofiber. Left: Atomic force microscopy (AFM) image of nanofiber with embedded dye–ZL crystals. The crystals are oriented along the fiber axis. Right: fluorescence image PEO/DTT (dithieno[3, 2-b:2′3′]-thiophene-S,S dioxide) nanofiber (blue) with embedded Ox^+-ZL crystals (red sparks).[68] (f) Chains. Right: SEM and fluorescence microscope images of ZL crystals linked together by modifying the channel entrances with the cationic dye D291.[83] Right: fluorescence microscopy image of a chain formed by (AnH_7T_8)-ZL crystals; see Figure 9.11b. (g) Assembly of ZL-bacterium in phosphate buffered saline (PBS) solution and self-assembly of two bacteria with functionalized 1-μm ZL as the junction.[73a] See ftp site for color: ftp://ftp.wiley.com/public/sci_tech_med/supramolecular_photochemistry.

aligned with an angle of about 72° with respect to the channel axis, while the blue emitting MC^+ is longer and thus oriented parallel to the channels. The polarization analysis now shows only crystals with blue fluorescent MC^+ when light parallel to the channels is detected and only red fluorescent Ox^+ when the analyzer is turned by 90° (left and right images, respectively). It should be noted, that the same result can be obtained by loading both dyes into the same ZL and preparing the stretched PIB film with only this type of crystals. Strategies based on the minimization of interfacial free energy for self-assembly have been applied to position small objects at the micrometric scale[63] and novel approaches toward micropatterning employing surface tension engineering have been realized.[64] It was shown that dye–ZL crystals can be organized into hexagonal motifs on a patterned polydimethylsiloxane (PDMS) surface. We show in Figure 9.12c an SEM and fluorescence microscopy image of such a material. The assembly process here is driven by surface tension interactions. This principle is a powerful tool for manipulations at the micrometer scale, allowing us to hierarchically organize molecular dyes on a macroscopic level. A similar type of arrangement can be observed by drop coating dye–ZL crystals on the petals of a horned pansy (*Viola cornuta*), as shown in the photograph and microscopy image in right part of Figure 9.12c.[65] This alignment is not determined by topological forces.

Compact-oriented monolayers on glass substrates of either standing or lying ZL crystals, as sketched in Figure 9.7d, can be prepared in a variety of ways.[34,62,66] Methods for the preparation of large-scale 2D ZL crystal arrays have been recently reported.[30,66,67] The monolayer quality depends strongly on the type and surface roughness of the substrate. The determining factor here is the availability of accessible surface hydroxyl groups. The best results can be obtained by using glass plates with a low melting point, quartz or a material with similar surface properties. In the first step, the OH groups on the substrate surface are functionalized with linker molecules, such as 3-chloropropyltrimethoxysilane. ZL crystals are then coupled to the linkers through their surface hydroxyl groups by sonication at room temperature. The difference in silanol group density, and thus in chemical reactivity, between the ZL base and coat surfaces leads to the preferred "standing crystal" orientation, as can be seen from the SEM image on the left of Figure 9.12d. Using disk-shaped ZL crystals improves the quality of this arrangement. Such two-dimensional arrays can be realized with either empty or dye-filled ZL. The latter is illustrated by the rightmost microscopy images, showing the luminescence from a monolayer filled with either green or red emitting dyes.

The successful assembly of zeolite crystals largely depends on the availability of crystals with narrow particle size distribution and well-defined morphology. Subsequent insertion of guests into the channels and addition of stopcocks is only possible if the free channel openings are not blocked or damaged during the preparation of the monolayer. If successful, the procedure leads to materials with exciting properties such as monodirectional transfer of electronic excitation energy. First materials bearing such unidirectional electronic excitation energy transport properties have been realized,[34] and improvement of the methods for preparing corresponding ZL monolayers on a substrate, which we call c-oriented open-channel monolayers (c-ocMLs), has been demonstrated.[30,66] A c-ocML prepared from empty ZL crystals and filled in a second step with dyes, thus leading to asymmetric or monodirectional loading patterns, is schematically depicted in Figure 9.13a. Fluorescence microscopy images of such a material are presented in Figure 9.13b. It was prepared by sequential insertion of $MeAcr^+$ followed by the red emitting DTCI into the channels of ZL crystals arranged as a monolayer; DTCI has the same structure as $PC20^+$, but with N-Et instead of N-Me. Some crystals were removed from the monolayer by scratching before investigation by means of optical microscopy. A representative amount of the crystals exhibited fluorescence only from one end, indicating that, in a monolayer, dye molecules can enter the channels from only one side.

ZL crystals, with diameter around 700 nm and length in the range of 1000 nm, have been successfully embedded into electrospun polymer fibers.[68] Electrospinning is a very efficient process for the preparation of fibers with diameters ranging from the nano- to the micrometer scale.[69,70] Polymer wires as thin as 150 nm are still able to enclose ZL. The crystals are aligned parallel to the fiber axis, as can be seen from the SEM image in Figure 9.12e. Nanowires prepared by incorporating dye–ZL act as very bright, polarized light sources (Figure 9.12e, left part) and are currently investigated for advanced nanopho-

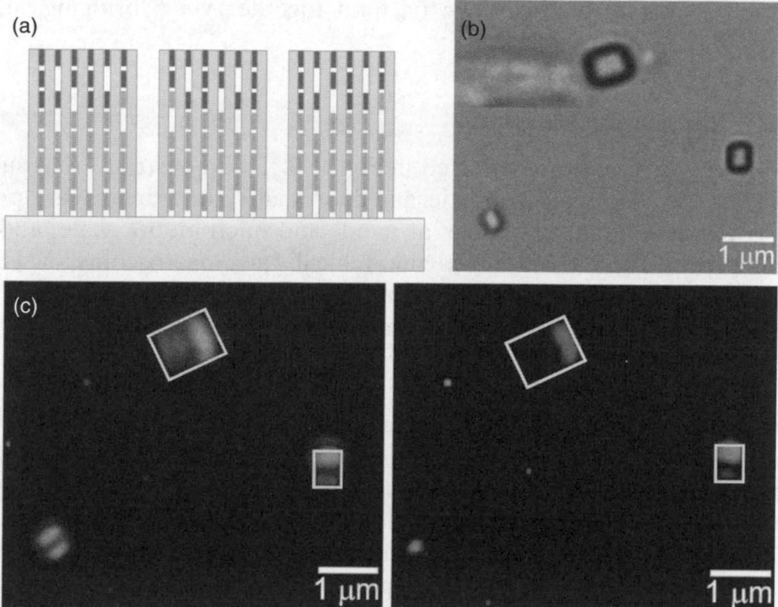

Figure 9.13. Monodirectional antenna on glass substrate. (a) Schematic representation of a c-ocML with sequential dye loading. (b) White light microscopy image of DTCI,$MeAcr^+$-ZL crystals removed from a monolayer. (c) Fluorescence microscopy image with specific excitation of $MeAcr^+$ at 470–490 nm (left) or of DTCI at 545–580 nm (right).[38] See ftp site for color: ftp://ftp.wiley.com/public/sci_tech_med/supramolecular_photochemistry.

tonic applications. Stopcock-plugged ZL crystals can be arranged into chain-like one-dimensional assemblies by exploiting coordinative interactions between the head groups of the stopper molecules, as illustrated in the scheme of Figure 9.12f. The reversible nature of such coordinative bonds makes it possible to maximize the base-to-base interaction between crystals.[71] The chain-like arrangement of ZL crystals seen in the SEM image of Figure 9.12f is due to the van der Waals interactions between the long alkyl tails of the used stopcock D291. The fluorescence microscopy image in the middle part of Figure 9.12f shows similar assemblies of D291-plugged ZL. The chain seen on the right is composed of crystals to which AnH_7T_8 has been attached; see also Figure 9.11b.[21] The enhanced luminescence at the ZL basal surfaces shows the presence of the stopcock molecules. Such micro-barcode structures may be of interest for tagging or bio-imaging purposes.[72] Attaching dye–ZL composites by means of amino-functionalized surfaces to living entities, such as nonpathogenic *Escherichia coli* (*E. coli*) bacteria, is very appealing. The situation schematically depicted in Figure 9.12g was investigated by means of optical and scanning electron microscopy. An interesting observation was that the amino-functionalized crystals seem to always bind to one of the poles of *E. coli*. It was also possible to connect two bacteria together via a bridging, amino-functionalized dye–ZL crystal.[73]

Optically Transparent Materials

The organizational pattern reported in Figure 9.12 demonstrates that surface tension, van der Waals, electrostatic, and coordinative interactions are powerful tools for manipulating objects at nano- and micrometric scale, and that molecular species can be organized hierarchically at a macroscopic level based on self-assembling processes. As a summary of this section, we can say that dye–ZL composites offer very high flexibility in the realization of supramolecularly organized materials.

The properties of dye–ZL composites outlined above promote them for the development of optically useful materials, such as lenses, infrared plastic light-emitting diodes for use in telecommunication, nanostructured materials for optical data storage, or for improvement of a polymer's chemical–physical properties.[74] For many optical applications, the host–guest systems have to be inserted into a polymer matrix while maintaining transparency in the visible range.[75] Based on experiments on dispersion and refractive index matching with dye–ZL crystals of different size, a synthesis procedure for transparent polymer–ZL materials has been developed.[36,76] Photographic images of a series of hybrid materials, prepared from dye–ZL crystals and the commercial polymer CR39, are given in Figure 9.14. All materials are transparent and exhibit the characteristic coloring of the inserted dyes. ZL monolayers strongly scatter visible light, as can be seen from the sample in the upper part of Figure 9.14a. This phenomenon disappears completely when the dense monolayer is covered with a polymer film, giving rise to a perfectly transparent material. The hybrid polymer shown in Figure 9.14c contains an EnT host–guest material based on the dyes Ox^+, hostasol yellow 3G (HY3G), and DMPOPOP. The UV-Vis absorption and luminescence spectra of this material are given in Figure 9.14d, e. The material exhibits three prominent absorption bands: one at 600 nm, corresponding to Ox^+; a broad band centered on 482 nm caused HY3G; and a structured band system at 395 nm originating from DMPOPOP. The light scattering is weak and the material fully transparent in the visible range. The functionality of the EnT system can be demonstrated by selectively exciting HY3G at 420 nm. The emission bands of both donor at 529 nm and the acceptor at 605 nm can be seen in this case. The excitation spectrum detected at 620 nm, where only Ox^+ emits, exhibits bands for all three dyes.

Light scattering arising from uncoated monolayers complicates spectroscopic investigations, especially for the measurement of absorption spectra. On the other hand, the homogeneous and highly ordered arrangement of ZL crystals found in these monolayers is ideal for determining the ETDM orientation of dyes embedded in the channels. The suppression of light scattering through polymer coating allows measuring angle-dependent absorption spectra of a ZL monolayer. A Py^+,Ox^+-ZL sample will be used as a demonstra-

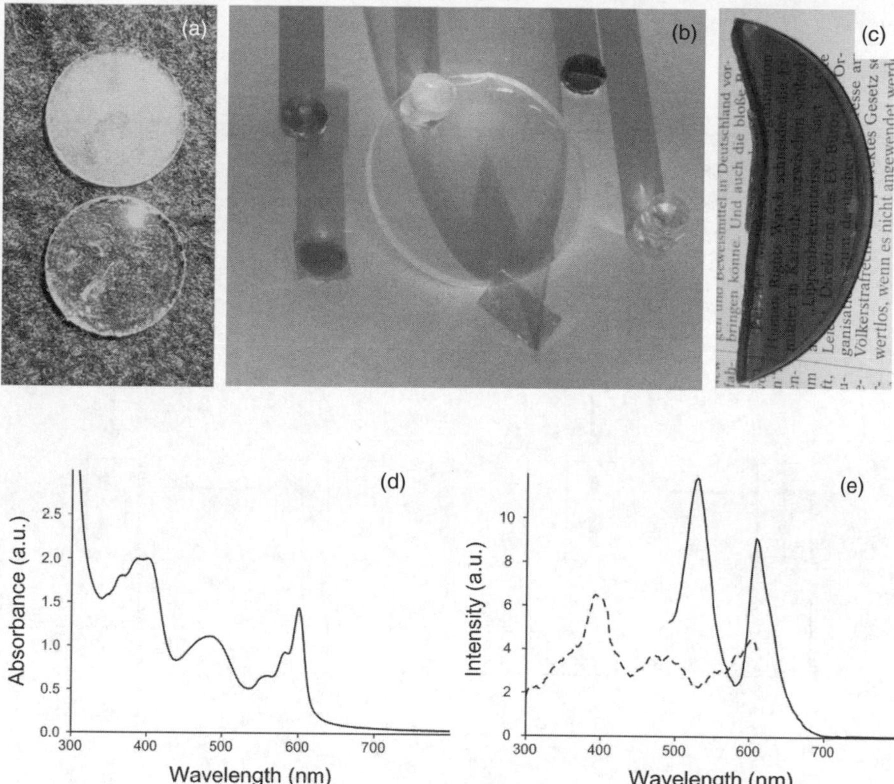

Figure 9.14. Photographic images and spectra of transparent polymer–ZL hybrid materials. (a) Unloaded ZL monolayer (top) and similar monolayer covered with a polymer film (bottom). (b) Selection of transparent dye-loaded and unloaded ZL–polymer hybrid materials. (c) Hybrid material (1% w/w) prepared from Ox^+,HY3G,DMPOPOP-ZL and CR39.[100] (d) Absorption spectrum of the hybrid material (CR39)–Ox^+,HY3G, DMPOPOP-ZL (1% w/w). (e) Excitation (dashed, λ_{det} = 620 nm) and emission (solid, λ_{ex} = 420 nm) spectra of the same material. The Ox^+ emission at 605 nm is mainly due to energy transfer from HY3G to Ox^+, while the excitation spectrum detected at 620 nm, where only Ox^+ emits, exhibits bands for all three dyes. Seeftp site for color: ftp://ftp.wiley.com/public/sci_tech_med/supramolecular_photochemistry.

tion. The scheme in Figure 9.15a explains the measurement method, where the dye-loaded monolayer is represented by two single crystals standing on a substrate. The absorptivity depends on the angle α defined as the angle between the incident light beam and the monolayer support, and on the orientation angle β of the inserted dyes ETDM with respect to the crystal axis. The cases for three different β values are illustrated in Figure 9.15b–d. When the molecules are oriented perpendicular to the channel axis (i.e., $\beta = 90°$), no absorption will occur if the incident light beam is parallel to the monolayer support

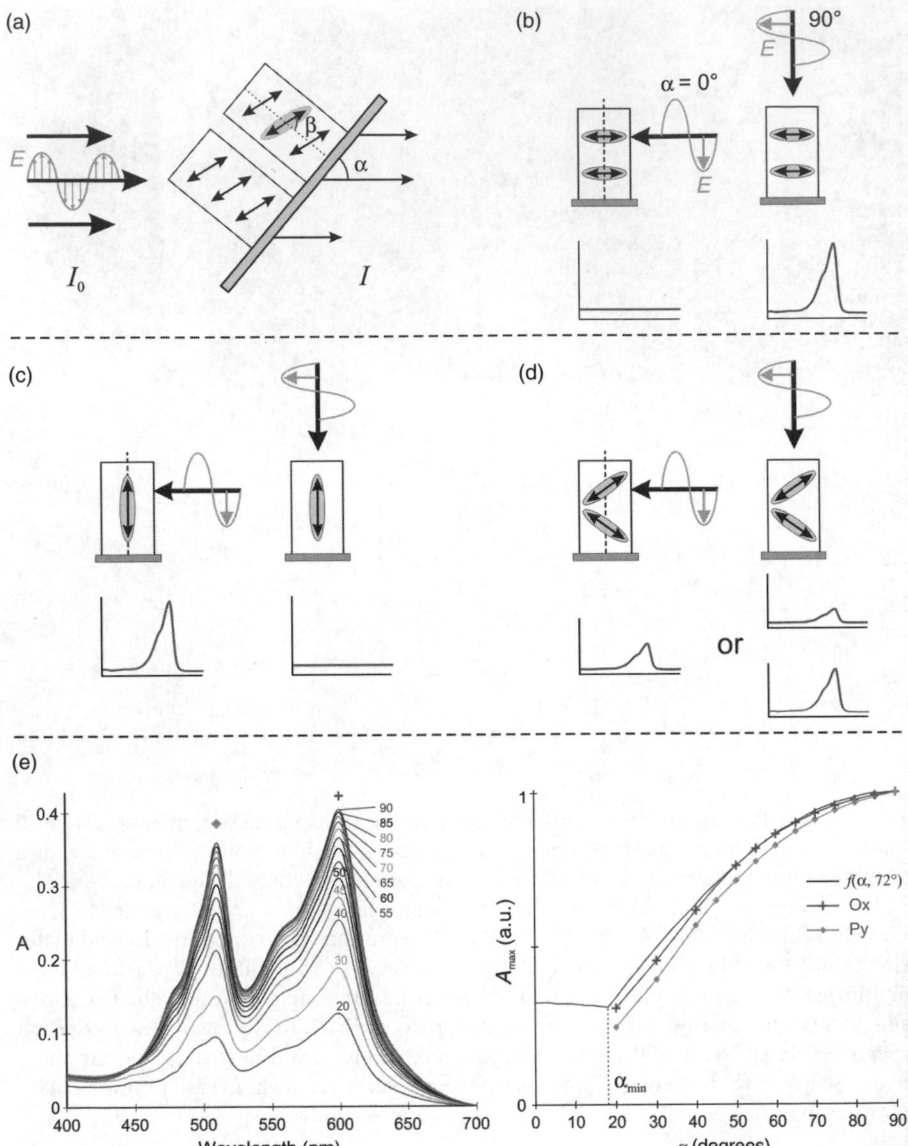

Figure 9.15. Angle-dependent absorption spectra of a Py^+,Ox^+-ZL monolayer. (a) Schematic representation of the measurement situation. (b–d) Dependence of the monolayer absorbance on the orientation of molecules inside the channels (β) and on the angle of the incident light (α). (e) Absorption spectra of a Py^+,Ox^+-ZL monolayer measured at angle α from 20° to 90°. The numbers correspond to values of angle α (left). The chart on the right show a comparison of the theoretical function $f(\alpha,72°)$ with the scaled experimental data. Points have been connected by lines for better visualization.[21]

($\alpha = 0°$). The molecules cannot be excited, since in this case, their ETDM lie perpendicular to the lights' electric vector. The observed absorbance will be maximal when the light falls perpendicular to the substrate (i.e., $\alpha = 90°$), as the molecules ETDM and the lights' electric vector will be coincident. The situation is exactly reversed for molecules oriented parallel to the channel axis (Figure 9.15c; $\beta = 0°$): Light absorption is maximal when the incident beam is parallel to the substrate and minimal when it falls perpendicularly on it. For molecules oriented at an angle $0° < \beta < 90°$, the absorbance will depend on the value of β. Expressing the angle dependence is easy for the two-dimensional case and goes as follows[21]:

$$f(\alpha,\beta) = |\cos(\alpha+\beta)| + |\cos(\alpha-\beta)|. \tag{9.6}$$

The function $f(\alpha,\beta)$ can be correlated to the maximum of the absorption spectrum A_{max} at the corresponding angles. One also has to consider that the light path through the monolayer is lengthened for incidence angles different from 90°. This lengthening leads to an increase in absorbance that has to be corrected. Correction to the same path length is obtained by multiplying the absorption spectrum at a given angle a by the sine of this angle[38]:

$$A_{corr}(\lambda) = A_{\alpha}(\lambda)\sin(\alpha). \tag{9.7}$$

The measured angle-dependent absorption spectra for a Py^+,Ox^+-ZL monolayer are shown in Figure 9.15e. Absorptivity is highest for α close to 90°, since the ETDM of both dyes are organized in a cone-shaped distribution with a half-opening angle of about 72°.[37] A comparison of these measured data points with the expected values for $f(\alpha,72°)$ is given at the right part of Figure 9.15e. From this plot, we can conclude that the optical anisotropy of the Py^+,Ox^+-ZL can be described by Equation 9.6.

FRET IN SUPRAMOLECULARLY ORGANIZED DYE–ZL COMPOSITES

The supramolecular organization of dyes inside the channels of ZL, as presented in the previous section, offers the opportunity to create materials in which electronic excitation energy can be transported by means of resonance EnT. The well-defined geometric ordering of the included guest species and the high optical anisotropy, as well as the great preparative flexibility, make dye–ZL composites ideal systems for the study of FRET processes. These processes are not limited to energy migration inside of a single crystal. Using stopcock molecules to interface the ZL to its environment makes it also possible to out-couple the electronic excitation energy.[21] The same principle can be used to transfer energy from an external component, like a luminescent or electroluminescent polymer in which the crystals are embedded, to the dyes located in the channels.[25] In this section, we will discuss different types of EnT materials and review examples thereof. In-depth theoretical background on FRET processes is discussed in the section on "Theoretical Background."

FRET between Chromophores inside of Dye–ZL Antenna Materials

The first two types of EnT systems that will be discussed are randomly mixed-dye systems and antenna materials. In both cases, the excitation energy transport process is limited to the interior of a single crystal.

Mixed-Dye Systems. ZL crystals filled with a random mixture of two or more dyes as shown at the left part of Figure 9.16 are called *mixed-dye systems*. Such materials can be prepared by simultaneous insertion of different dye types, as long as they all enter the channel at a similar rate. If the absorption spectrum of one dye (acceptor) overlaps well with the emission spectrum of a donor, and if their ETDMs are not oriented perpendicular to each other, FRET can occur between them. The processes that can take place are schematically depicted in Figure 9.16. Upon selective excitation of the donor (light gray rectangle), the electronic excitation energy can be either transferred to a neighboring acceptor (dark gray rectangle) or reemitted by the donor. The EnT is also in competition with radiationless decay, quenching, and photochemically induced degradation. FRET efficiency is strongly distance dependent. The mean distance between donors and acceptors can be controlled by varying the overall dye concentration. Mixed-dye systems are therefore well suited for studying this dependence.

A nice case study is the mixed Py^+,Ox^+-ZL composite. The insertion kinetics of the strongly fluorescent dyes Py^+ and Ox^+ into ZL are very similar, providing

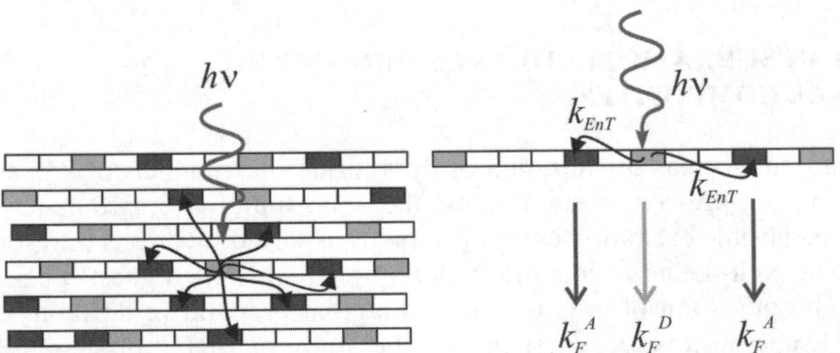

Figure 9.16. Scheme of a mixed dye–ZL composite. Left: schematic representation of a few ZL channels. Each rectangle marks a site. White boxes stand for empty sites, while donors *D* and acceptors *A* are shown in light gray and dark gray, respectively. Right: main processes taking place after selective excitation of a donor. The rate constant for energy transfer is k_{EnT}, while the rate constants for fluorescence of donors and acceptors are labeled k_F^D and k_F^A, respectively.

a statistically homogeneous distribution of both dyes over the whole crystal. The two dyes form an excellent donor–acceptor pair, due to the large overlap between the Py^+ fluorescence and Ox^+ absorption spectra; see figure 1 in Reference 77. The EnT process can be well observed in this system. Figure 9.17a shows a photographic image of seven fluorescent samples. ZL crystals with an average length of 300 nm were used for the preparation of all samples. The two references Py^+ and Ox^+ were correspondingly loaded with $p = 0.007$ of either Py^+ or Ox^+. Samples A–E were filled with a 1:1 mixture of both dyes. The loading of each component is denoted as $p_{1/2}$, indicating that the total dye loading is $2{\cdot}p_{1/2}$. The concentrations for samples shown in Figure 9.17a are as follows: A, $2 \times (5 \times 10^{-4}$ M or $p_{1/2} = 0.0014)$; B, $2 \times (1.25 \times 10^{-3}$ M or $p_{1/2} = 0.0035)$; C, $2 \times (2.5 \times 10^{-3}$ M or $p_{1/2} = 0.007)$; D, $2 \times (5.0 \times 10^{-3}$ M or $p_{1/2} = 0.014)$; and E, $2 \times (1.0 \times 10^{-2}$ M or $p_{1/2} = 0.028)$. The dye concentration increases from A to E, which in turn leads to a decrease in the mean donor–acceptor distance. All samples were excited at 485 nm, corresponding to strong absorption for Py^+ and low Ox^+ absorbance. Sample A exhibits mainly the green fluorescence of Py^+, indicating that EnT is insignificant. This is schematically shown by the donor–acceptor configuration on the left of the photograph. The yellow color of sample B is due to a mixture of green and red fluorescence, meaning that EnT becomes significant in this case. The process becomes more and more important with increasing concentration so that from sample C to E, the red fluorescence stemming from Ox^+ is dominant. The donor–acceptor scheme on the right of the photograph illustrates this for case E.[77]

The same effect can be observed on the single crystal scale, by following the diffusion of the random dye mixture as a function of time.[78] The situation sketched at the top of Figure 9.17b represents the (idealized) initial state where the dye mixture is still tightly clustered at the channel entrances. The dark and light gray boxes stand for Ox^+ and Py^+, respectively. Letting the dyes move deeper into the channels by thermal diffusion leads to an increase in the statistical average distance between neighboring molecules. This is shown at the lower part of Figure 9.17b. Thus, the EnT efficiency from donors to acceptors is expected to be lower for crystals having been equilibrated for a longer time. The fluorescence microscopy images of a mixed Py^+,Ox^+-ZL sample given in Figure 9.17c illustrate this very nicely. The images were taken after loading the crystals for 20 min (1), 60 min (2), 470 min (3a, 3b), and 162 h (4), respectively. For all images, Py^+ was selectively excited, with the exception of (3b), where Ox^+ was selectively excited. The samples with short equilibrium times exhibit an orange to yellow luminescence, due to the overlap of green Py^+ and red Ox^+ emissions, which is an indication for efficient FRET. After a diffusion time of 162 h, the dyes are so far apart from each other that, upon selective excitation of Py^+, only its green emission can be observed.

Antenna. Antenna systems are supramolecular arrangements in which electronic excitation of molecules occurs in a given volume and the electronic excitation energy is then transported by FRET to a well-defined location.

(a)

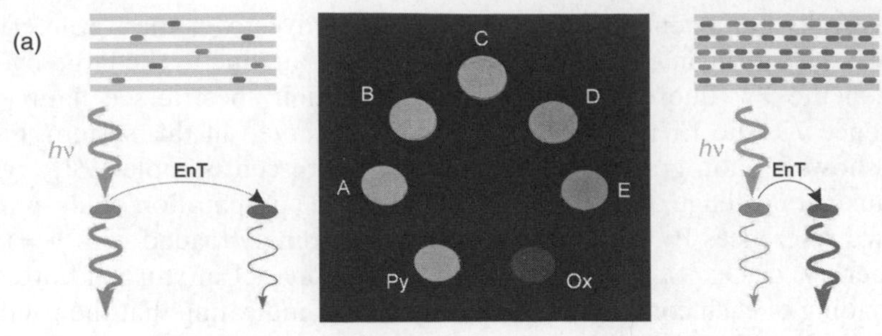
hν
EnT
C
B
D
A
E
Py
Ox
hν
EnT

(b)

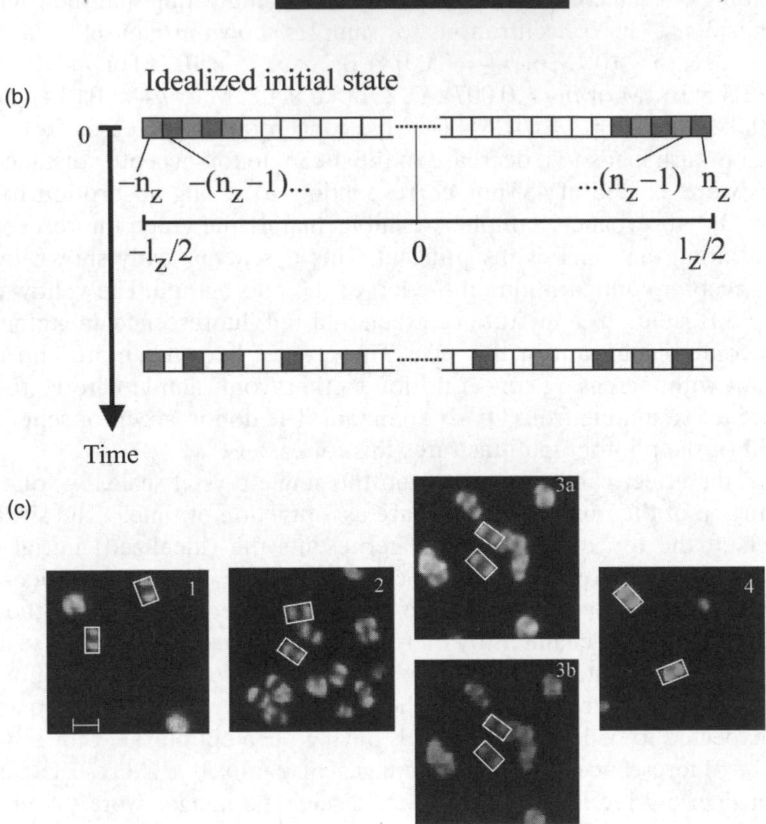
Idealized initial state
0
$-n_z$
$-(n_z-1)$...
...(n_z-1)
n_z
$-l_z/2$
0
$l_z/2$
Time
(c)
1
2
3a
3b
4

Figure 9.17. Visual demonstration of FRET. (a, middle) Photographic image of the fluorescence of dye–ZL layers, upon monochromatic irradiation at 485 nm, through an interference filter and observation through a 500-nm cutoff filter. The two samples indicated as -Py and Ox are references loaded with pure Py^+ and Ox^+, respectively. Samples A–E contain a 1:1 mixture of Py^+ and Ox^+ with the following loading for each dye: A, $p_{1/2} = 0.0014$; B, $p_{1/2} = 0.0035$; C, $p_{1/2} = 0.007$; D, $p_{1/2} = 0.014$; and E, $p_{1/2} = 0.028$. (a, left) Scheme of a crystal with a donor–acceptor pair describing the situation in sample A. The dye loading is low; hence, the donor–acceptor distance is large. When the donor is selectively excited, its fluorescence intensity is dominant (green light), while the energy transfer process to an acceptor is insignificant, and the acceptor flo-rescence is very weak. (a, right) Scheme describing the situation in sample E. High dye loading results in a short donor–acceptor distance. The energy transfer process is very efficient. Donor fluorescence is weak, while fluorescence of the acceptor is strong. (b) Diffusion of dye molecules in the channels of ZL. Idealized initial state of a channel and state after diffusion has occurred for some time. (c) Fluorescence microscopy images, visualizing the diffusion of Ox^+ and Py^+ in ZL, taken after 20 min (1), 60 min (2), 470 min (3a, 3b), and 162 h (4), respectively. Py^+ was selectively excited, with the exception of 3b where Ox^+ was selectively excited. Two crystals in each image are framed. The scale given in 1 corresponds to a length of 1500 nm.[77,78] See ftp site for color: ftp://ftp.wiley.com/public/sci_tech_med/supramolecular_photochemistry.

Sequential insertion of dyes into the channels of ZL has been shown to be an excellent tool for preparing such systems.[39] Since the geometrical constrain imposed by the ZL host prevents the molecules from gliding past each other, the crystal will be divided into compartments where the density of one dye type is dominant. This section focuses on the discussion of EnT in antenna materials and on the dependence of its efficiency on several parameters.

A schematic view of the photonic antenna invented by us is illustrated in Figure 9.18.[4,13,21] The monomeric dye molecules are represented by rectangles. The dye molecule that has been excited by absorbing a photon transfers its electronic excitation energy to another molecule. After a series of such steps, the electronic excitation reaches luminescent traps, which we have pictured as dark rectangles. The energy migration is in competition with spontaneous emission, radiationless decay, quenching, and photochemically induced degradation. Fast energy migration is therefore crucial if a trap should be reached before other processes can take place. Due to the symmetry of the systems shown in Figure 9.18, the excitation energy can be transported in two directions: either to or from each channel entrance (left and right scheme, respectively). These materials are thus called *bidirectional antenna*.

As an example for such antenna materials, we show at the bottom part of Figure 9.19 confocal fluorescence intensity maps of a ZL crystal sequentially loaded first with Py^+ and then with Ox^+.[79] The upper part of Figure 9.19 sketches the experimental situation: The green emitting Py^+ domain, located in the middle of the crystal, is selectively excited at 485 nm. The excitation energy can then travel randomly, as indicated by the black arrows, until it

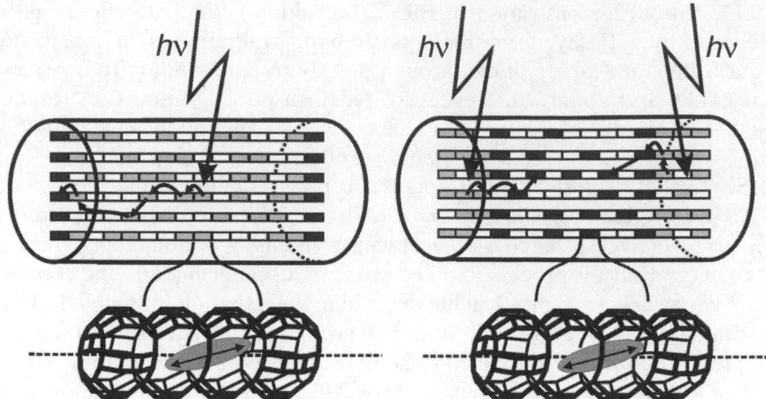

Figure 9.18. Schematic representation of a bidirectional antenna, consisting of organized dye molecules acting as donors (light gray rectangles) and acceptors (dark gray rectangles). Left: Donors are in the middle part of the crystal, and the acceptors are at the front as well as the back of each channel. Right: Donors are located at each channel entrance, and the acceptors are in the middle part. The enlargements show details of a channel containing a dye and the orientation of its ETDM (double arrow), which lies parallel with respect to the channel axis for long molecules and is tilted for shorter ones.

reaches an Ox^+ molecule. The Ox^+, with its lower lying electronic transition, acts as a trap. Thus, the energy cannot be transferred back to a Py^+ molecule. Confocal microscopy images of suitably large crystals show that the dyes are indeed arranged as proposed. The green donor fluorescence is observed in the middle part of the crystal upon selective excitation and use of an appropriate high-pass filter (Figure 9.19, lower left). The acceptors' red fluorescence can be observed by either selectively exciting them, as shown at the lower right part of Figure 9.19, or by exciting the donor and applying a suitable optical filter.

In these antenna materials, the FRET efficiency, that is, the efficiency of electronic excitation delivery to the acceptors, depends mainly on two parameters: the donor loading and the crystal length. The efficiency will be lowest for a low donor loading in long crystals. The top part of Figure 9.20 summarizes an experiment where 600-nm-long ZL crystals were loaded with varying donor amounts, while the acceptor concentration was kept constant (i.e., two Ox^+ molecules on average per channel entrance). The donor loading p_{Py} increases from samples 1 to 6 as follows: 1, 0.007; 2, 0.012; 3, 0.031; 4, 0.057; 5, 0.106; and 6, 0.182. The fluorescence spectra, shown in Figure 9.20a, have been recorded after selective excitation of Py^+ at 470 nm. All spectra are scaled to the same height at the emission maximum of Py^+ at 520 nm. The Ox^+ emission intensity, and hence the FRET efficiency, increases with increasing donor loading. The reason for this is that the mean distance between molecules decreases, causing an increase in probability for electronic excitation to reach a neighboring

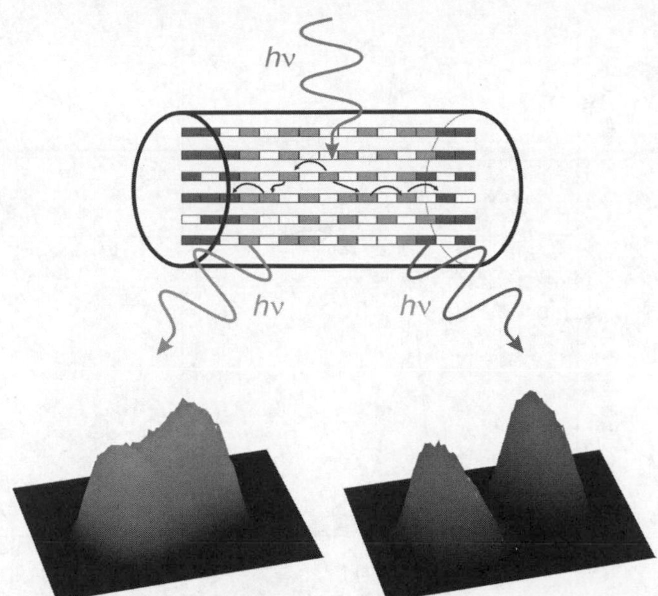

Figure 9.19. Fluorescence behavior of an Ox^+,Py^+-ZL antenna. Top: Scheme of an antenna crystal and processes taking place upon excitation of a donor molecule: absorption of incident light, energy migration from an excited donor molecule to a neighboring unexcited one and trapping by an acceptor (black arrows), fluorescence of both donor and acceptor. Lower: Confocal microscopy intensity map of an antenna crystal upon selective excitation of the donor and observation trough an appropriate filter (left) and similar intensity map obtained upon selective excitation of the acceptor (right).

molecule before decaying. The acceptor-to-donor intensity ratios versus donor loading p_{Py} are plotted in Figure 9.20b. We observe that the EnT efficiency in antenna materials increases faster and nonlinearly with donor loading than in mixed-dye materials where this dependence is linear. The reason for this different behavior is assumed to be due to the influence of staggered donor–acceptor interface profiles, which will be discussed later in more detail. There are always some favorable molecular arrangements at such an interface, where donor and acceptor are close enough for efficient FRET. These favorable cases considerably increase the mean FRET rate constant at low donor loadings.

Another way to manipulate the mean donor–acceptor distance is illustrated at the lower part of Figure 9.20. In this experiment, the donor and acceptor loadings are kept constant ($p_{Py} = 0.11$ and two Ox^+ molecules on average per channel entrance), while the average crystal length is varied.[8] The results obtained for crystals with lengths of 300 nm (1), 500 nm (2), 850 nm (3), 1400 nm (4), and 2400 nm (5) are summarized in Figure 9.20c, d. The fluorescence spectra, observed after selective excitation of Py^+ at 460 nm, are scaled to the

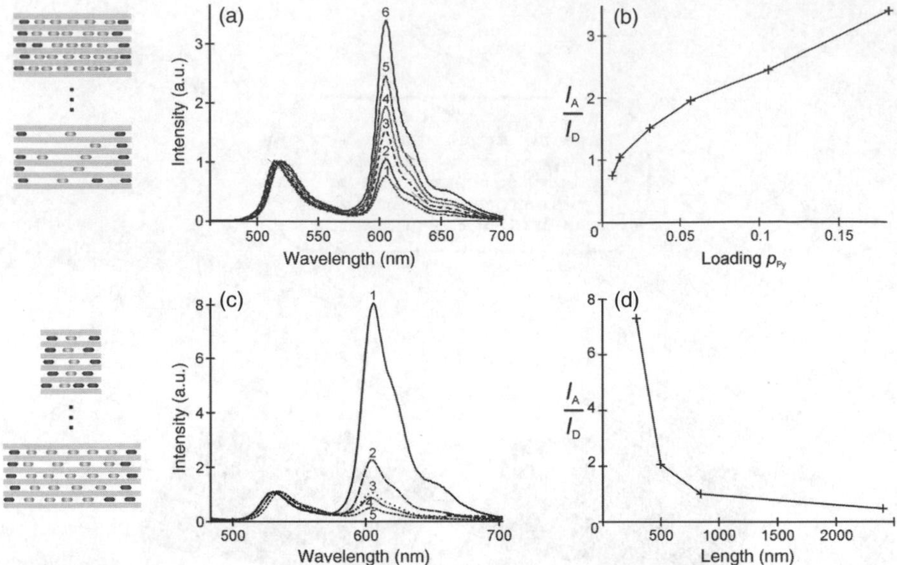

Figure 9.20. FRET efficiency dependence on donor loading and crystal length. (a, b) Experiments with different donor loading. Crystal length and acceptor concentration (dark rectangles) were kept constant, while donor (gray rectangles) loading p_{Py} varied: 0.007 (1); 0.012 (2); 0.031 (3); 0.057 (4); 0.106 (5); 0.182 (6). (a) Fluorescence spectra upon excitation of Py^+ at 470 nm. The FRET efficiency increases with increasing donor loading. (b) Acceptor-to-donor emission maxima ratio plotted versus p_{Py}. (c, d) Experiments with different crystal lengths. The loading of donors was kept constant, $p_{Py} = 0.11$, and all samples were modified with two acceptor molecules at each channel end on average, while the crystal length varied: 300 nm (1), 500 nm (2), 850 nm (3), 1400 nm (4), and 2400 nm (5). (c) Fluorescence spectra upon excitation of Py^+ at 460 nm. The FRET efficiency decreases with increasing crystal length. (d) Acceptor-to-donor emission maxima ratio plotted versus crystal length.

same height at the Py^+ emission maximum. A strong increase in FRET efficiency is observed with decreasing crystal length. Due to the constant donor concentration, the efficiency of each individual energy migration step is the same in short and long crystals. However, fewer steps are needed to reach an acceptor in the case of a short crystal. The probability for an excitation to reach an acceptor is therefore higher, as there is less time to find a nonluminescent trap, to decay radiationless, or to emit spontaneously. About 90% of the light emitted from the 300-nm crystals is due to energy migration with subsequent transfer to the luminescent Ox^+ traps.[8] Such a result is expected for FRET as the main process. The small red shift in Py^+ emission, observed for high loadings and also in longer crystals, is due to self-absorption. This phenomenon becomes more prominent as the absorptivity of individual crystals increases.[21]

Time-resolved confocal investigations are helpful in understanding and characterizing FRET processes. An Ox^+,DMPOPOP-ZL antenna material is

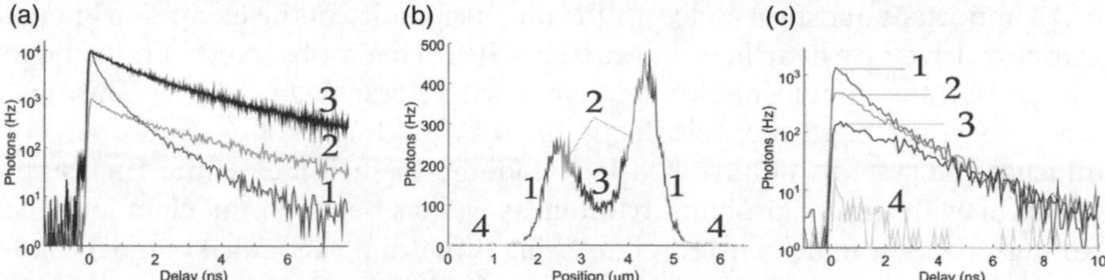

Figure 9.21. Luminescence decay measured on single dye–ZL crystals in a confocal microscope. (a) Decay observed as an average over the whole crystal. (3) Decay of DMPOPOP when a DMPOPOP-ZL crystal is excited at 320 nm and the luminescence is observed at 470 nm. (2) Decay of Ox^+ when a DMPOPOP,Ox^+-ZL crystal is excited at 495 nm and the luminescence is observed at 600 nm. (1) Decay of DMPOPOP when a DMPOPOP,Ox^+-ZL crystal is excited at 320 nm and the luminescence observed at 470 nm. (b, c) Time- and space-resolved confocal microscopy results. (b) Confocal intensity map of such a crystal with differently numbered regions: (1)–(4). (c) Decay curves of Ox^+ obtained from a DMPOPOP,Ox^+-ZL crystal excited at 320 nm and the luminescence observed above 660 nm at the different positions indicated in (b).[40]

used as an example.[40] Curve (3) in Figure 9.21a shows the luminescence decay for the donor (DMPOPOP) upon selective excitation at 320 nm and in the absence of any acceptors, while curve (1) is observed in the presence of Ox^+ (acceptor). A shortening of the donor lifetime in the presence of acceptor molecules is expected to occur due to FRET processes. The EnT gives an additional pathway for the depopulation of the donor excited states. Curve (2) indicates the Ox^+ fluorescence decay in the Ox^+,DMPOPOP-ZL system when the acceptor is directly excited at 495 nm. Figure 9.21b contains a luminescence intensity cross section over a single crystal obtained by means of confocal microscopy. The different regions are indicated by numbers: (4) represents luminescence detected outside of the crystal, (1) is used for the channel entrances, while (2) and (3) stand for the donor–acceptor interface and the pure donor domain, respectively. Lifetime measurements were made in each of these sections and the resulting decay curves are given in Figure 9.21c, with corresponding numbers. In this measurement, donors were selectively excited at 320 nm and detection took place above 660 nm, so that only luminescence from the acceptor would be observed. As expected, no significant luminescence intensity can be found outside of the crystal. The acceptor fluorescence is strongest at the channel entrances and gets weaker the closer one gets to the central donor region. The intensity dynamics of the acceptors, as plotted in Figure 9.21c, exhibit a fast rise before the decay starts. This is due to the fact that the acceptors are not directly excited but are "pumped" by the donors via FRET. As the rise time gets shorter, the more efficient the EnT is. This means that it is often very difficult to see this rise time for molecules with luminescence lifetimes in the range of a few nanoseconds. Nevertheless, nice observations of this phenomenon have been reported for both antenna materials[79] and mixed-dye systems.[77]

An important question concerns the dimensionality of the excitation energy transport driven by near-field interactions. It has been observed in mixed-dye systems that the dimensionality decreases with increasing loading.[77] This phenomenon arises from the anisotropy imposed by the ZL host and does not influence the systems behavior at low loadings. In such a case, the EnT steps will occur with similar probability along as well as between the channels. The transfer process can be considered as being three-dimensional. At higher loadings, however, the geometrical constrains of the host play a more prominent role. EnT steps along the channel will be more and more preferred, while transfers between channels will become less likely. In this case, the transfer process can be described as one-dimensional.

To further investigate this, ZL crystals were first loaded with red acceptor molecules (Ox^+), then spacers (DMPOPOP), and finally green-emitting donors (Py^+). The material is schematically presented in Figure 9.22a. The idea behind

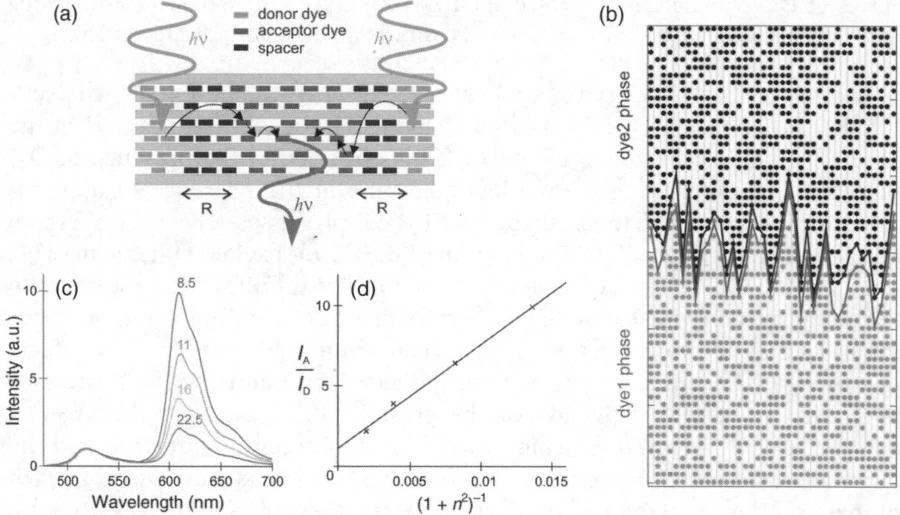

Figure 9.22. Dimensionality of the FRET process. (a) Scheme of an individual crystal of the material used for the experiment. Channels contain red-emitting dyes in the middle (dark gray), followed by spacer layers of different average thickness (black), and finally green-emitting molecules at the ends (light gray). (b) Phase boundary of dyes illustrated by an image of a ZL crystal consisting of 40 channels, each containing 32 sites, as obtained by a Monte Carlo simulation. The sites were first randomly filled with one kind of dye (black dots) and then, after virtually sealing the upper channel entrances, with a second kind of dye molecules (light gray) from the still open channels' entrances located on the opposite side. The total dye loading is 0.6. The calculation indicates that the phase boundary extends over about 10 sites. (c) Fluorescence spectra scaled to the same peak height of Py^+ emission upon selective excitation at 460 nm. The numbers correspond to the average number of spacer molecules per layer. (d) Intensity I_{rel} of the acceptor versus $1/(1 + n^2)$, where n is the average number of spacer molecules confirming the value of $\alpha = 2$.[33]

this experiment was to investigate the distance dependence of the EnT probability from donors to acceptors for varying thickness R of the spacer layer. The spacer DMPOPOP absorbs light at higher energy than Py^+, and thus cannot participate in EnT processes initiated by specific excitation of the donor. The spacer serves only to spatially separate Py^+ and Ox^+. The fluorescence spectra shown in Figure 9.22c were recorded after excitation at 460 nm. The numbers 8.5, 11, 16, and 22.5 indicate the average number of DMPOPOP molecules in the spacer layer. As one can see, the EnT efficiency decreases with increasing thickness of the spacer layer. This clearly indicates that the process is controlled by near-field interaction. We do not have direct access to the mean value of the donor–acceptor distance. It is, however, proportional to the average number n of spacer molecules. The mean distance cannot be obtained by simply multiplying the number of spacer molecules by their length, because the phase boundaries of the domains are staggered. This is illustrated in Figure 9.22b, where the phase boundary between dye1 and dye2 was obtained by means of a Monte Carlo simulation. In neighboring channels, the domains of dye1 and dye2 can overlap over several sites. In this case, the distance between dyes situated in neighboring channels may be shorter than between adjacent molecules in the same channel. Therefore, in Figure 9.22d, the acceptor fluorescence intensity was plotted versus $1/(1 + n^{\alpha})$. Fitting the data gives a value of 2 for α, which is characteristic for one-dimensional transport. In the case of the investigated dye–ZL materials, however, it is more correct to use the term "quasi-1D EnT": The statistical mean of the EnT process is directed along the c-axis of the crystal, but some individual EnT steps to neighboring channels can still occur. This result seems to be the first experimental proof for quasi-1D electronic excitation energy transport given in the literature.[33]

FRET to or from a Stopcock

The efficiency of a FRET process depends on several parameters, among them the spectral overlap integral between the donor emission and acceptor absorption band, as well as the distance and the relative orientation of the ETDM of the pair. We consider FRET between a dye molecule inside and a stopcock located at the entrance of a ZL channel. As long as other parameters remain constant (i.e., type of dyes and spatial separation), the EnT rate will be highest when both ETDM are collinear to each other and minimal in the perpendicular case. This is described by the orientation factor $\kappa^2_{D^*A}$ (see also the section on "Theoretical Background" and the Appendix), which takes values from 0 for perpendicular, to 1 for parallel, and up to 4 for collinear ETDM arrangements, as depicted in Figure 9.23a. Five possible configurations for stopcock ETDM orientations are sketched in Figure 9.23, along with a typical cone-shaped ETDM distribution of guest molecules. The stopcocks ETDM can be arranged perpendicular to the channel axis in one (b) or two (c) dimensions, parallel to the channel axis in one (d) or two dimensions (e), and in a sphere-

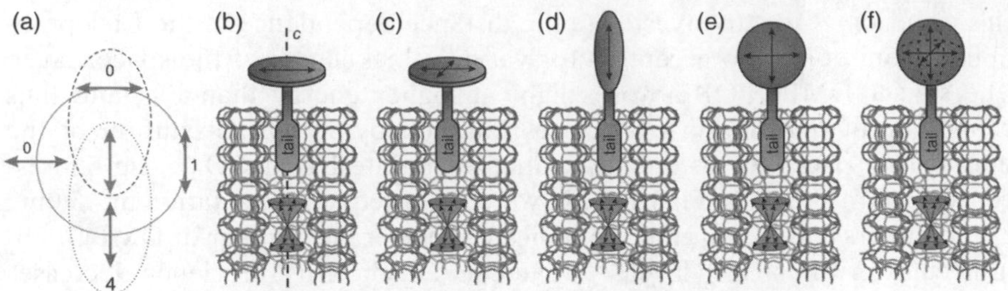

Figure 9.23. Orientation of ETDMs. (a) Selected values for the orientation factor κ^2 of the donor–acceptor pairs for $\varphi = 0°$. (b–f) Examples of orientation of the transition dipole moment of stopcock molecule: (b) perpendicular to the crystals' *c*-axis, (c) in plane, (d) parallel to the channel axis, (e) parallel and perpendicular orientations, and (f) spherical orientations. The typical cone-shaped distribution of a guest molecule's ETDM is indicated as dashed double arrows.

like fashion (f). In the first two cases, maximum coupling will be obtained when the dyes ETDM is lying perpendicularly to the channel axis, while a parallel orientation will be best for the third situation. The last two cases, (e) and (f), are somewhat special in so far as coupling between dye and stopcock will be possible no matter how the dyes' ETDM are placed. Systems corresponding to all five arrangements have been prepared. There is plenty of room, however, for synthetic chemists to explore the many other possibilities.

We now present some materials corresponding to the guest–stopcock ETDM arrangements sketched above. Two examples for stopcock molecules with ETDMs oriented perpendicular to the *c*-axis (situation (b) in Fig. 9.23) are shown in Figure 9.24b, e. The first system was realized by loading ZL with red-luminescent Ox^+ and subsequently modifying the channel entrances with the green-emitting stopcock B493/503.[16] The excitation and emission spectra of the individual components are summarized in Figure 9.24a. Exciting the B493/503 stopcock at 460 nm leads to a considerable Ox^+ fluorescence, as shown by the spectrum in Figure 9.24c. This indicates that EnT from the stopcock to the acceptor inside the channels takes place. The stopcock-plugged antenna material given in Figure 9.24e works in the reverse way, transferring excitation energy from the inside to the outside of the crystal. This was achieved by embedding the blue-emitting MC^+ as donor and adding the green-emitting stopcock MFG as acceptor to the channel entrances.[80] The excitation and emission spectra of each single component for this system are given in Figure 9.24d. The EnT from the molecules inside to the stopcock molecules can be observed in the fluorescence spectrum of the (MFG),MC^+-ZL composite shown in Figure 9.24f. For this measurement, the donor MC^+ was selectively excited at 350 nm. The two examples demonstrate that FRET can be realized in both directions: from the outside of the crystal to the inside, and reverse, from the inside of the crystal to the outside.

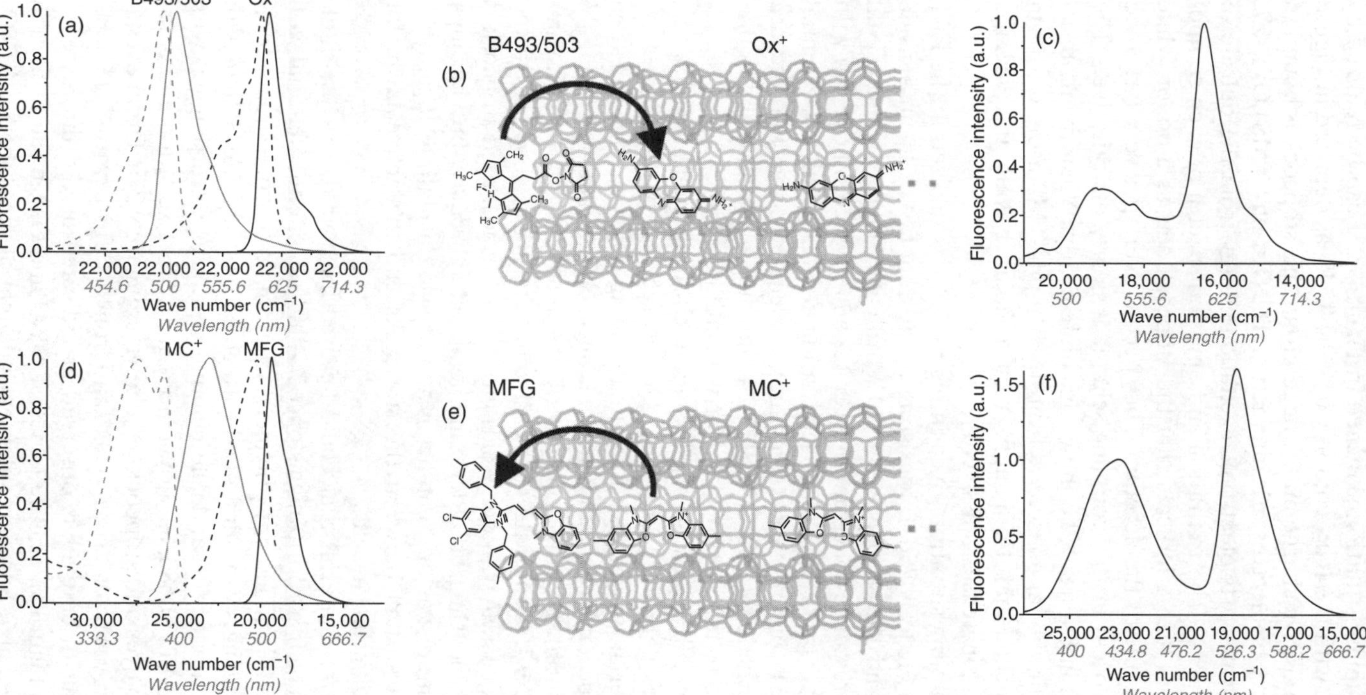

Figure 9.24. FRET from the outside to the inside of a crystal (top) or in the reverse direction (bottom). (a) Excitation (dashed) and emission (solid) spectra of (B493/503)-ZL and Ox^+-ZL. (b) Scheme of a (B493/503),Ox^+-ZL channel, where electronic excitation energy is injected from the outside to the inside of the crystal. (c) Fluorescence spectra of the (B493/503),Ox^+-ZL stopcock-plugged antenna material upon selective excitation of B493/503 at 460 nm.[16] (d) Excitation (dashed) and emission (solid) spectra of (MFG)-ZL and MC^+-ZL. (e) Scheme of an (MFG),MC^+-ZL channel, where electronic excitation energy is transported from the inside to the outside of the crystal. (f) Fluorescence spectra of the (MFG),MC^+-ZL stopcock-plugged antenna material upon selective excitation of MC^+ at 350 nm.[16,80]

The ETDM coupling situation sketched in Figure 9.23f can be realized by plugging $Ox1^+$-ZL with the Ru^{2+} complex Ru-ph4-TMS, as depicted in Figure 9.25b. Due to the positive charge on the head group, this stopcock is bound to the channel entrance by electrostatic interaction. The (Ru-ph4-TMS),$Ox1^+$-ZL antenna material transports electronic excitation energy from the outside to the inside of the crystal. The excitation and emission spectra of each component given in Figure 9.25A indicate that the stopcock acts as a donor, while the $Ox1^+$ takes the role of the acceptor. The spectral overlap is very large and thus the FRET process in this coupling scheme is nearly quantitative.[42] The emission spectrum in Figure 9.25c shows this quite neatly: When the material is excited at 460 nm, the luminescence from the Ru-ph4-TMS around 610 nm is so strongly quenched that it only appears as a shoulder, while the emission intensity from $Ox1^+$ increases massively.

The ETDM coupling situation given in Figure 9.23c is realized in the EnT material shown in Figure 9.25d–f, consisting of the donor SG5, located inside of the channels, and the stopcock BNCO, covalently bound to the crystal entrance. The fluorescence spectrum of SG5 has a large spectral overlap with the excitation spectrum of BNCO, as seen in Figure 9.25d, which is essential for efficient FRET. Figure 9.25f summarizes excitation and fluorescence spectra of the bidirectional (BNCO),SG5-ZL antenna material. It shows that the EnT from SG5 to the stopcock is nearly quantitative.[22]

A stopcock molecule based on phtalocyanine, allowing the realization of a coupling scheme as depicted in Figure 9.23e, has recently been reported.[24,81] The antenna material was prepared by first loading ZL with Ox^+ and then adsorbing the stopcock IL-1 to the channel entrances, as sketched in Figure 9.26b. The absorption and emission spectra of the single components, summarized in Figure 9.26a, indicate that EnT will take place from Ox^+ to the phtalocyanine stopcock. A van der Waals model for the adsorbed stopcock is shown in Figure 9.26c. The phthalocyanine head bears three *tert*-butyl groups, which is of great interest as they can act as an intrinsic "insulating" layer (preventing electron transfer) between the stopcocks and the active layer, for example, in an organic solar cell.[82]

We present now a material that could be used for obtaining white-light-emitting materials by incorporating a stopcock-plugged antenna material with multiple dyes into a luminescent polymer. An example has been realized as follows: The ZL channels were first filled with green-emitting NY43, then with red-emitting Ox^+, and at the end, the channel entrances were plugged with deep-red-emitting ATTO-680, as shown in Figure 9.27b. Excitation and emission spectra of the dyes and stopcock are shown in Figure 9.27a. Upon excitation of NY43, energy is transferred to Ox^+, and from Ox^+ further to ATTO-680. The emission of all three dyes is observed as indicated by the fluorescence spectrum given in Figure 9.27c. The bands of NY43 and Ox^+ in the excitation spectrum, measured at 725 nm, are intense, while that of the stopcock ATTO-680 is weak, due to its low amount. The presented example is not an exception; similar systems can be prepared in many different ways.[21,24] Recently,

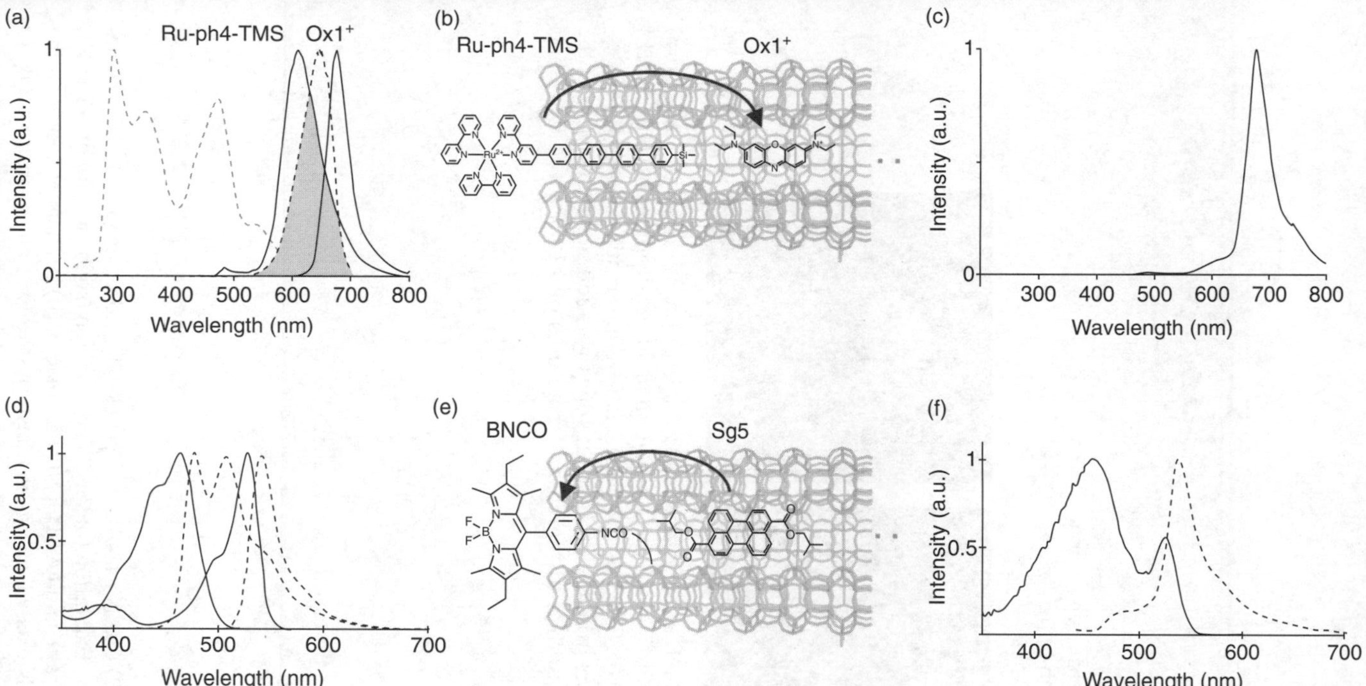

Figure 9.25. FRET from and to stopcocks. (a–c) (Ru-ph4-TMS),$Ox1^+$-ZL stopcock-plugged antenna material. (a) Excitation (dashed) and emission (solid) spectra of (Ru-ph4-TMS)-ZL and $Ox1^+$-ZL. The spectral overlap of the Ru^{2+} emission and $Ox1^+$ absorption is shaded. (b) Schematic drawing of a (Ru-ph4-TMS),$Ox1^+$-ZL channel. (c) Fluorescence spectrum of the stopcock-plugged antenna material upon excitation at 460 nm. The shoulder at 610 nm is the remaining weak emission of the Ru^{2+} complex. (d–f) (BNCO),SG5-ZL stopcock-plugged antenna material. (d) Excitation (solid) and emission (dashed) spectra of SG5 in ethanol and of BNCO in CH_2Cl_2. (e) Scheme of the (BNCO),SG5-ZL antenna material. The BNCO stopcock is covalently bound to the ZL framework. (f) Excitation (solid) and emission (dashed) spectra of (BNCO),SG5-ZL. The emission spectrum was recorded after selective excitation of SG5 at 430 nm. The excitation spectrum was detected at 640 nm.

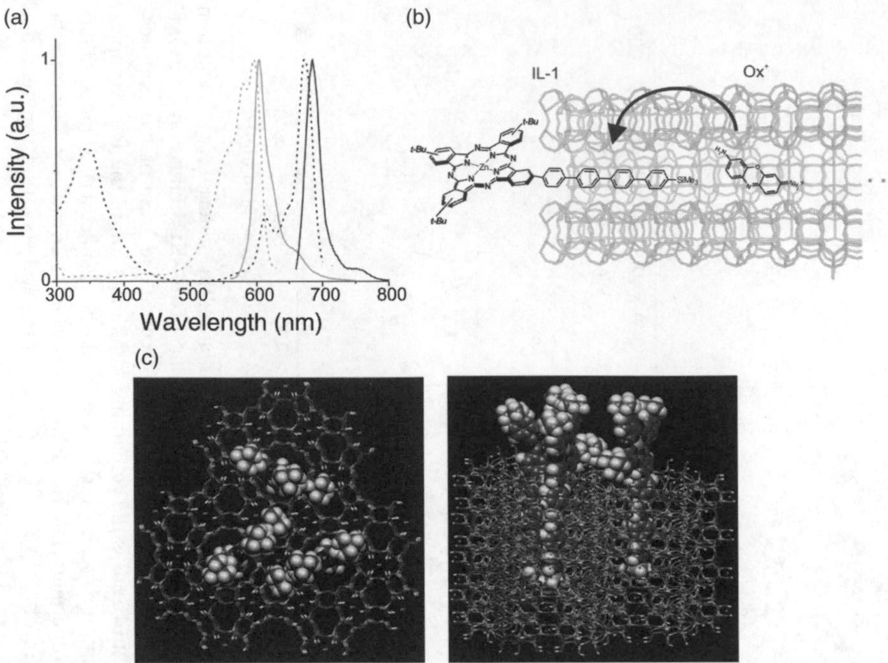

Figure 9.26. (IL-1),Ox$^+$-ZL energy transfer material. (a) Emission (solid) and absorption (dashed) spectra of a 7×10^{-7} M IL-1 solution in tetrahydrofurane (THF) (black) and of Ox$^+$ in ZL ($p = 0.05$, light gray). (b) Schematic representation of the (IL-1),Ox$^+$-ZL system. FRET occurs from Ox$^+$ to IL-1. (c) Top (right) and side (left) views of the phthalocyanine IL-1 adsorbed to the channel entrances of ZL as a van der Waals model.[81]

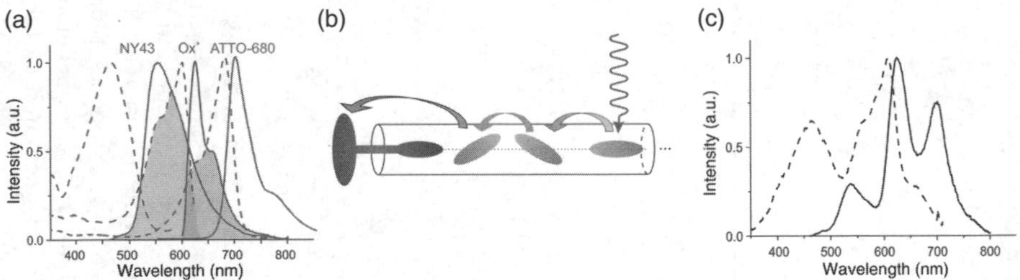

Figure 9.27. (ATTO680),Ox$^+$,NY43-ZL, a stopcock-plugged antenna material containing two kinds of dyes. (a) Excitation (dashed) and emission (solid) spectra of NY43-ZL (light gray), Ox$^+$-ZL (gray), and (ATTO680)-ZL (black). (b) A channel ending of the material. When NY43 is excited, energy is transferred to Ox$^+$, and then from Ox$^+$ to ATTO680. (c) Excitation spectra observed at 725 nm (dashed) and emission spectra upon excitation at 450 nm (solid) of the material. In the emission spectrum, bands of all three dyes are observed.[83]

considerable progress on white light emission from dye-loaded ZL materials has been reported.[25]

It is essential to control the interaction of an individual crystal with its environment. Therefore, appropriate functionalization of the external crystal surface is important. This is possible because base and coat of the crystals differ in their chemical reactivity. Full coverage of the external crystal surface can, for example, be achieved by reacting the crystals with triethoxysilane derivatives. We show in Figure 9.28a a fluorescence microscopy image of an

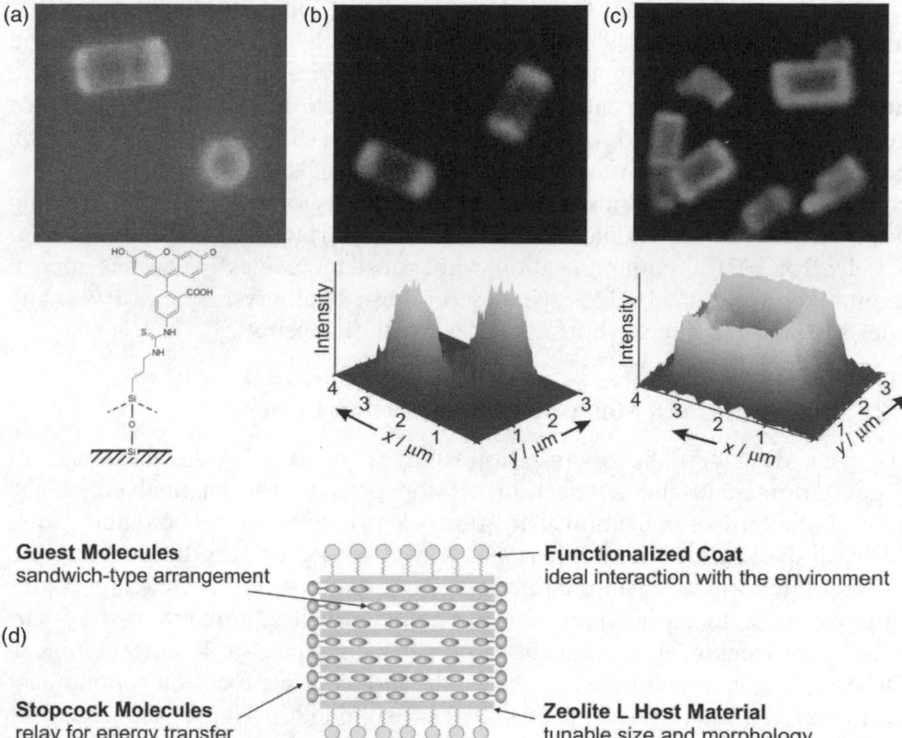

Figure 9.28. Full and selective functionalization of the external crystal surface. (a–c) Fluorescence microscopy images of functionalized ZL crystals. (a) Fluorescence microscopy image of ZL with fluorescein covalently bound to the external surface. The length of the cylindrically shaped crystals is approximately 5000 nm. The crystal on the right is standing on its base. Scheme of the covalent binding of the fluorescein to the ZL surface via an aminopropyl methoxy silane (APMS) linker (bottom). (b) Selective functionalization of the base. Two APMS-functionalized ZL crystals, marked with ATTO-610-NHS, (top) and the corresponding relative intensity distribution of one of these crystals (bottom). (c) Several ZL crystals with base and coat aminopropyl triethoxy silane (APTES) functionalization and marking ATTO-610-NHS (top), and the corresponding relative intensity distribution of one of these crystals (bottom). (d) Scheme for complex functionalization of ZL crystals. See ftp site for color: ftp://ftp.wiley.com/public/sci_tech_med/supramolecular_photochemistry.

aminopropyl triethoxy silane (APTS)-modified crystal marked with fluorescein. It shows that the whole crystal surface is covered with the dye. We have discussed that it is often desirable to have a selective functionalization of the channel entrances. An elegant and general method for selective covalent attachment of stopcocks is to first attach a molecule containing a protected functional group (NH_2, SH, or COOH) to the channel entrance. The protective group is then cleaved off, leaving behind the desired functional group that can react with an appropriate reagent. This has been reported for the first time in Reference 48 and later been used for different purposes,[21,24,73a] such as to realize complex functionalities on dye–ZL crystals. Figure 9.28d illustrates that the desired structure inside of the crystals is built up first, followed by the modification of the channel entrances in the desired way and then by the functionalization of the coat.[15h] The individual functionalization steps must not necessarily be carried out with fluorescent dyes. Some steps may also be directed toward appropriate sealing of the channels, influencing the solvent compatibility of the crystals, changing the reactivity, or modifications pursuing other targets. Full functionalization of the outer surface with a donor or acceptor dye for FRET communication with guest molecules inside has special importance for nanosized ZL, because in these small crystals, access to many guests is possible due to short distances in all dimensions.[25b, 73b, 83]

FRET between Guest Molecules and a Reaction Center

We have discussed the organization of dyes inside ZL—the first stage of organization—and the connection of stopcocks to the channel ends—the second stage of organization. The stopcock principle allows communication between dyes inside the channels and a reaction center outside of the crystal. This requires what we call the third stage of organization. The reaction center can be a molecule, a polymer matrix, a semiconductor, a quantum-sized particle, a molecular or nanomagnet, or a biochemical or biological object. Different kinds of communications can be induced. We focus on communication mediated by a FRET process. This is explained in Figure 9.29 where we distinguish between stopcocks that do not take an active part in the FRET process and may act as a protecting chemical linker, as closure, or as insulator molecules, (a) and (b), and those that are actively involved in FRET, (c) and (d). The distance between the donor and the acceptor should in both cases be shorter or at least not longer than the Förster radius. Chemistry provides an enormous variety of options for realizing such highly organized structures. We discuss only a few of them.

Randomly Oriented Dye–ZL Composites. We choose, as a first example, the use of stopcock-plugged dye–ZL materials as luminescent labels for analytical purposes. The working principle of a functional EnT label is sketched in Figure 9.30a, which is related to schemes (a) and (b) in Figure 9.29. The channel entrances of dye–ZL crystals are terminated with stopcocks featuring an

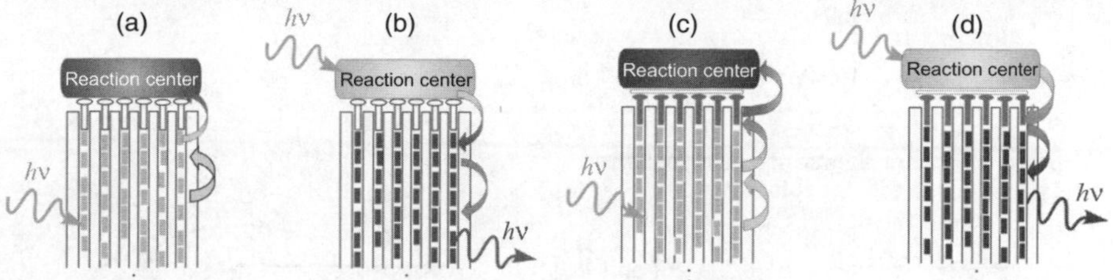

Figure 9.29. Communication of guests with an external reaction center via FRET. (a, b) The stopcock molecules must not be fluorescent if they act as insulating or protecting chemical linkers between dyes inside the channels and chromophores located in the reaction center. We illustrate in (c) and (d) that a thin insulating part between the luminescent stopcock and the reaction center is usually desirable if the stopcocks are luminescent molecules in order to avoid direct contact, which might lead to electron transfer or other undesirable reactions. The insulating part can, for example, be covalently bound to either the stopcock or the reaction center. The distance between donors and acceptors should not exceed the Förster radius if efficient FRET is desired.[21]

appropriate receptor head that can bind to selected molecules or ions. Upon binding, the distance between dyes inside of the ZL (dye1) and the bound analyte becomes short enough for FRET. After selective excitation of dye1, the luminescence from the analyte is observed. An example is given in Figure 9.30b, c. Biotin has been attached to dye1-ZL crystals by sequential functionalization of the channel entrances. DMPOPOP has been used as dye1 and ZL nanodisks as host. The composites excitation and fluorescence spectra are shown in Figure 9.30c. A weak-binding antigen modified with ATTO-488 (dye2) (anti-biotin-IgG-ATTO488) was added to the biotin-plugged dye1-ZL. The resulting interaction was studied by measuring the fluorescence spectra in a buffer solution of pH 7. The emission spectrum upon excitation at 340 nm and the excitation spectrum, recorded at 600 nm, of pure biotin-plugged dye1-ZL (gray) features only one band, that of dye1. Reaction with antigen-dye2 (black) leads to an additional band, that of dye2, in the excitation spectrum. The emission spectrum, also recorded upon excitation at 340 nm (specific excitation of dye1), shows that the dye1 emission is considerably quenched, while the intensity of dye2 emission increases significantly—due to FRET, as the distance between the guest (dye1) and the reaction center (dye2) molecule is sufficiently short. This example shows that functional EnT biolabels based on dye–ZL systems are feasible.[21] In other experiments, it was shown that dye–ZL materials may have interesting properties as phototherapeutic agents.[73]

A simple way to transfer electronic excitation energy according to the scheme in Figure 9.29d, consist in coupling stopcock-plugged antenna crystals with a luminescent polymer, so that energy injection takes place from the polymer via the stopcock to the guest molecules. The polymer can be excited by radiative or electrical excitation. This is illustrated in Figure 9.31a for a random arrangement of dye–ZL objects. In the example presented, not only

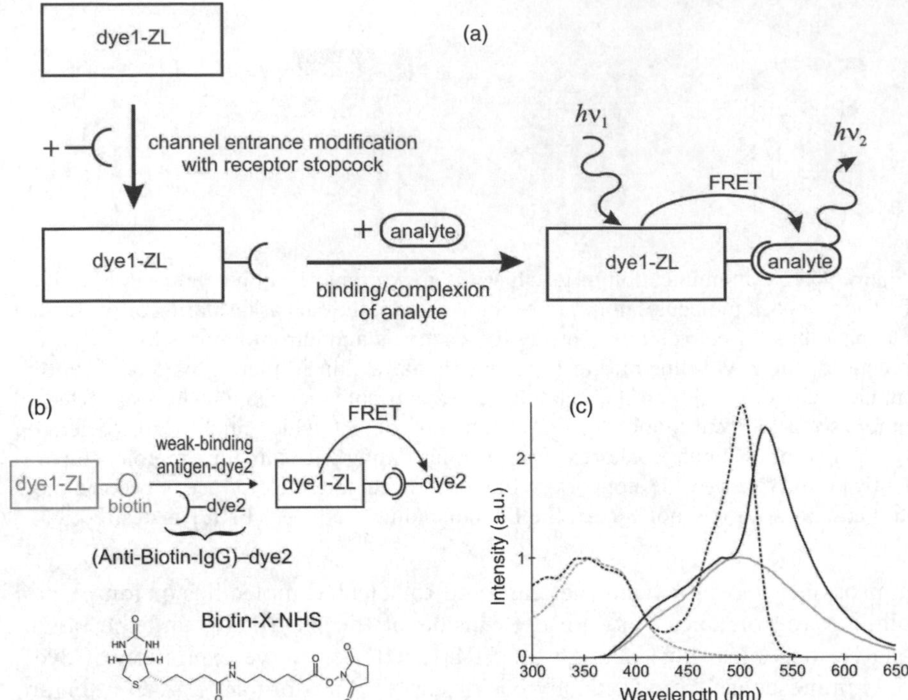

Figure 9.30. Functional FRET biolabel. (a) Working principle of a *functional FRET label*. In the first step, the channel entrances of ZL crystals, loaded with dye1, are modified with a receptor stopcock. The resulting material can now bind to a compatible analyte, so that the distance between dye1 and the analyte is short enough for FRET to occur. (b) Demonstration experiment carried out with Biotin-X-NHS as functional receptor. (c) Excitation (emission at 600 nm, dashed line) and emission (excitation at 340 nm, solid line) spectra before reaction (gray lines) and after reaction (black lines) of biotin-plugged dye1-ZL with antigen-dye2, measured in a pH 7 buffer solution. The spectra are scaled to same peak height at the dye1 absorption/emission maxima.[21,100]

the channel entrances but the whole external surface of the crystals was covered with stopcock molecules. The luminescent polymer in which the dye–ZL crystals were embedded was poly(*N*-vinyl carbazole) (PVK). PVK is a photoconductive polymer emitting in blue.[84] The absorption maximum of ATTO-495 in ethanol lies at 493 nm and the emission maximum at 525 nm. The spectral overlap regions between PVK and ATTO-495 and between ATTO-495 and Ox^+ are large. ATTO-495 is thus expected to harvest electronic excitation energy from PVK and pass it on to Ox^+.[80]

The (ATTO-495),Ox^+-ZL crystals were embedded in PVK by preparing a polymer/dye–ZL suspension in dichloromethane. Layers were prepared from this suspension by solvent evaporation. Figure 9.31b shows the emission spectra of ATTO495,Ox^+-ZL embedded in PVK (solid), and of pure (ATTO-

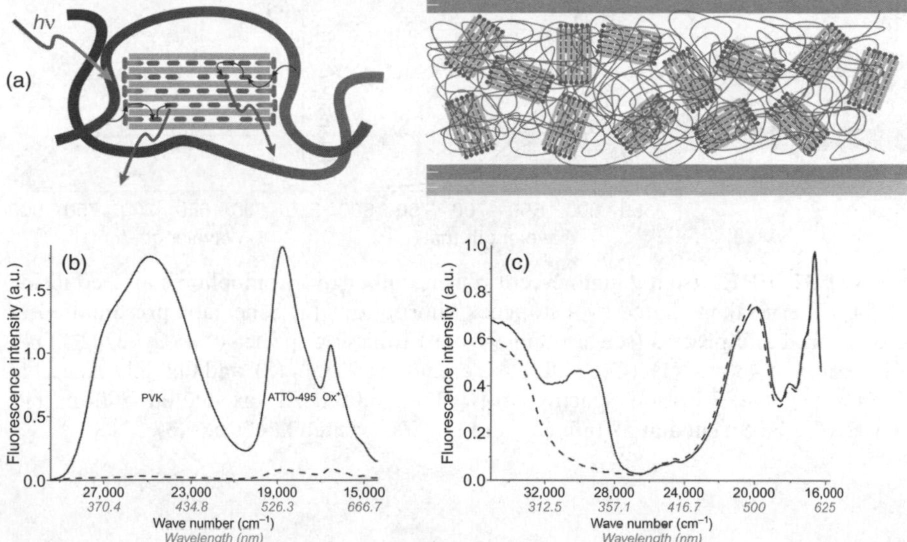

Figure 9.31. Coupling of stopcock-plugged dye–ZL antenna with a luminescent polymer. (a) Scheme of the system. Upon excitation of the polymer, energy is transferred to a stopcock and from there further to guest molecules. The polymer is drawn as a ribbon. (b) Fluorescence spectra of (ATTO-495),Ox^+-ZL embedded in PVK (solid) and of an (ATTO-495),Ox^+-ZL sample (dashed). Both samples contained an equal amount of (ATTO-495),Ox^+-ZL and were excited at 330 nm. The emission bands of ATTO-495 and Ox^+ are very weak in the sample without PVK, while they are intense in the polymer embedded sample. (c) Excitation spectra of both samples detected at 615 nm.[80]

495),Ox^+-ZL (dashed) when the layers are excited at 330 nm, where the absorption of the polymer is strong. The emission of ATTO-495 and of Ox^+ is very weak in the sample without polymer; since both dyes have very low absorptivity at 330 nm, the direct excitation is, as expected, very weak. Their emission is intense, however, for the polymer embedded sample. This indicates that electronic excitation energy is transferred from the photoconductive polymer PVK to the ATTO-495 stopcocks and from there to the Ox^+ molecules inside the ZL channels. The excitation spectra shown in Figure 9.31c were recorded at 615 nm, where Ox^+ emission is detected, and do support this conclusion. The spectrum of the PVK-containing sample shows a typical band belonging to the polymer around 333 nm. This proves that the polymer is able to generate Ox^+ emission by EnT via the intermediate ATTO-495 stopcocks.

A more advanced application of this has been realized by embedding dye–ZL into an electroluminescent poly(ethylene oxide) (PEO) fiber doped with dithieno[3, 2-b:2′3′]-thiophene-*S*,*S* dioxide by electrospinning.[68] PEO is widely used for the fabrication of light-emitting electrochemical cells, organic batteries, and supercapacitors.[85] Electrospinning is a well-known technique

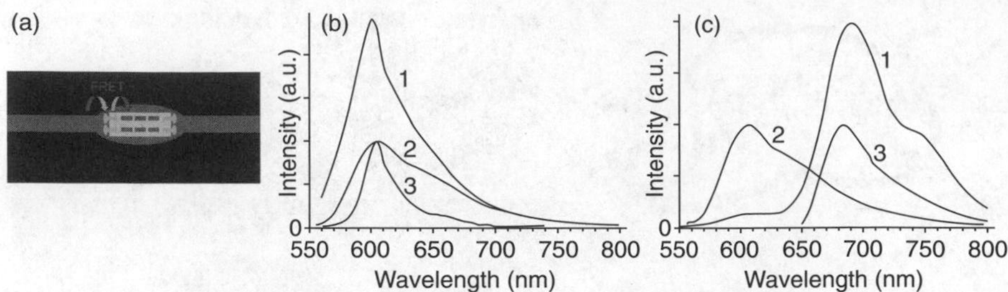

Figure 9.32. FRET from a luminescent polymer fiber to chromophores located inside of the ZL crystal, mediated by a stopcock fluorophore. (a) Schematic presentation of the material and process (see also Fig. 9.4). (b) Emission spectra of a ($Cy02702^+$),Ox^+-ZL excited at 480 nm (1), ($Cy02702^+$)-ZL excited at 520 nm (2), and Ox^+-ZL excited at 540 nm (3). (c) Emission spectra of ($Cy02702^+$),$Ox1^+$-ZL excited at 500 nm (1), ($Cy02702^+$)-ZL excited at 520 nm (2), and Ox^+-ZL excited at 620 nm (3).[25]

that allows obtaining polymeric fibers both on micro- and nanoscale by using a high-voltage electric field to spin a polymer melt or a polymer solution into oriented fibers, which can be collected on a metallic screen.[69] Recently, it has been shown that FRET from the polymer to dyes inside of ZL can be promoted via a stopcock molecule, provided that the spectral overlap between partners is sufficiently large. The conjugated polymer used in the fiber was poly[(9,9-dioctylfluorenyl-2,7-diyl)-alt-co-(1,4-benzo-{2,1′,3}-thiadiazole)] blended with polystyrene (PS). The results, summarized in Figure 9.32, show that efficient FRET occurs for ($Cy02702^+$),$Ox1^+$-ZL, while little transfer is observed if $Ox1^+$ is replaced by Ox^+, for which the spectral overlap with the $Cy02702^+$ stopcock is poor.[25]

The basic working principle of organic solar cells is the dissociation of photogenerated excitons at the interface between electron donor and acceptor phases by a photoinduced charge transfer process. Subsequently, the charge carriers have to be transported through the respective phase to the electrodes. Critical parameters for photocurrent generation are therefore the active layer absorptivity, the efficiency of the charge transfer, and the charge carrier transport properties of the materials involved.[86] A major problem of organic solar cells is the low overall power conversion efficiency compared with established inorganic solar cell techniques. This is due in a large part to the fact that the good organic semiconducting materials available have an absorption edge at comparatively high energy, usually above 2 eV.[87] Therefore, a large part of the solar spectrum is not used by current organic solar cells. An exception is zinc-phthalocyanine (ZnPc), which is used as a low band gap material. The absorption reaches to over 800 nm, leading to a much higher potential for photon harvesting. ZnPc shows almost no absorption in the region between the Q- and the Soret-band around 500 nm. This has severe consequences on the spectral

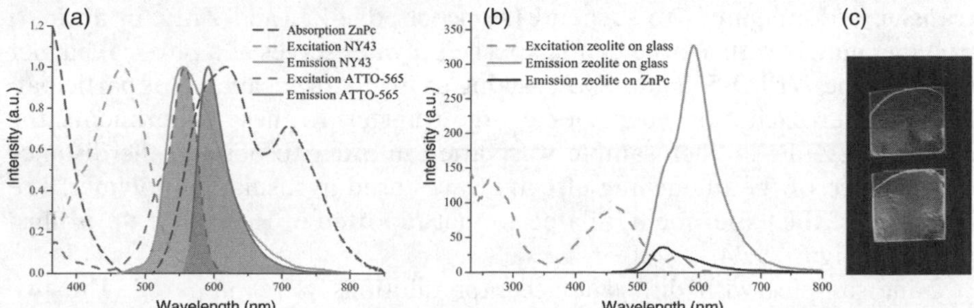

Figure 9.33. Sensitization of ZnPc by (ATTO-565),NY43-ZL composites. (a) The absorption spectrum of ZnPc (blue) exhibits a gap between 400 and 550 nm. The excitation (dashed) and emission (solid) spectra of NY43 (light gray) and ATTO-565 (dark gray) show that light from 400 to 550 nm can be absorbed and transferred onto ZnPc. The shaded areas correspond to the overlap integrals between the donor and acceptor pairs. (b) Excitation (dashed, λ_{det} 645 nm) and emission spectra (solid, λ_{ex} 450 nm) of (ATTO-565),NY43-ZL on glass (gray) and on ZnPc (black). A strong quenching of the emission is observed when the ZL is coated on a ZnPc substrate. (c) Photographic image of (ATTO-565),NY43-ZL on a glass substrate (lower) and on a ZnPc-coated substrate (upper) under UV irradiation. A color change from orange to yellow-green can be observed.[82,83] See ftp site for color: ftp://ftp.wiley.com/public/sci_tech_med/supramolecular_photochemistry.

efficiency. A way to solve this problem is to use FRET from a suitable chromophore to the low-band-gap solar cell material. Results from simple experiments prove that such an approach works. Dye-loaded and stopcock-modified disk-shaped ZL composites were drop coated onto ZnPc layers. For these experiments, NY43 was inserted into disk-shaped ZL, of 50- to 80-nm length and 300- to 500-nm diameter, via gas-phase adsorption. A loading of $p_{NY43} = 0.1$ was used. The channel entrances were then covalently modified with ATTO-565. Approximately 10 mg (ATTO-565),NY43-ZL were suspended in 1.0 mL *n*-butanol, and 100 mL of this suspension was drop coated onto a ZnPc-coated glass substrate. As a reference, the same amount was drop coated onto a plain glass substrate. An additional excitation spectrum of the reference sample on glass was recorded to confirm that the ATTO-565 emission at 590 nm is due to FRET from the donor NY43. We observe that the emission of the composite on a ZnPc-coated glass substrate is strongly quenched with respect to the reference, as reported in Figure 9.33. The ATTO-565 emission at 590 nm is even more strongly quenched than that of NY43 appearing at 540 nm. This change in the peak ratio leads to a color change that can also be observed by eye when the samples are irradiated with UV light. A photographic image of the two samples under UV illumination is shown in Figure 9.33c.

This strong quenching of the dye–ZL emission when in contact with ZnPc evidently indicates an interaction between the ATTO-565 and the ZnPc. A way to find out whether the quenching due to excitation energy or to electron transfer is dominant is to suspend (stopcock),dye-ZL and ZnPc in an inert polymer matrix with a large band gap. The polymer serves as a physical barrier between the ATTO-565 and ZnPc, acting as insulator incapable of participating in either excitation energy or charge transfer. In these experiments, the amount of ZnPc in each sample was varied in order to observe the distance dependence of the quenching effect. PS was used as insulating polymer. The principle of the experiment, related to that reported in Reference 88, is illustrated in Figure 9.34.

Nine samples with different acceptor dilutions were prepared. The two reference samples 0 and 8 contained either no PS or no ZnPc. The samples were drop coated onto glass plates and dried at room temperature. The luminescence spectra of these samples recorded upon excitation at 450 nm are shown in Figure 9.35a.

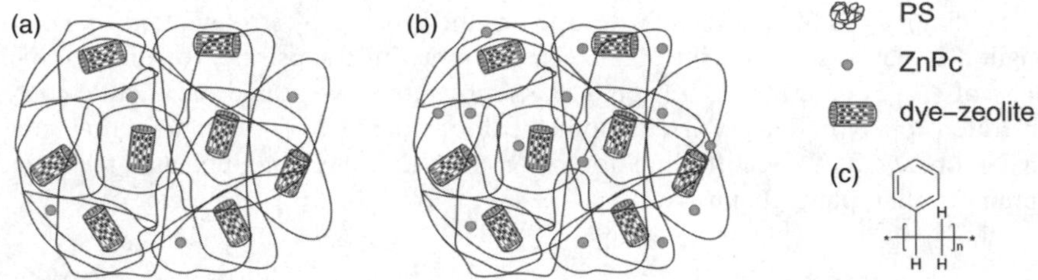

Figure 9.34. Principle of the quenching experiment, where the (stopcock),dye-ZL composite is suspended in PS. A varying amount of ZnPc is added to each sample, as illustrated in (a) and (b), to examine the distance dependence of the quenching. (c) Structure of the insulating PS.

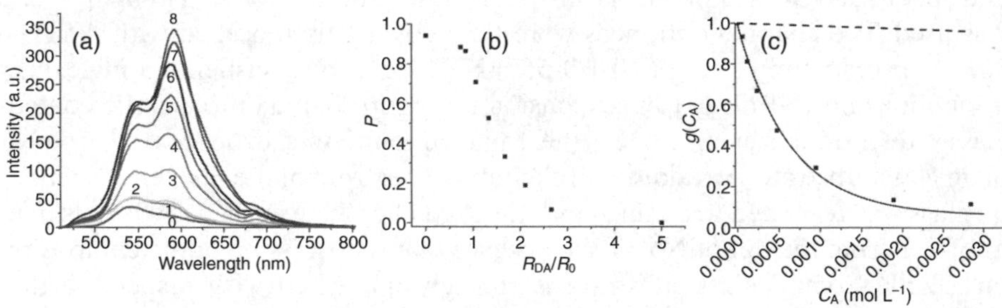

Figure 9.35. Quenching of (ATTO-565),NY43-ZL by ZnPc. (a) Luminescence spectra recorded upon excitation at 450 nm. The mean distance between the (ATTO-565),NY43-ZL composite and the acceptor ZnPc increases from samples 0 to 8. (b) Energy transfer probability P as a function of the relative distance R_{DA}/R_0. (c) Emission intensity of samples 1–8 plotted against the ZnPc concentration C_A (■). The dashed line is the calculated quenching assuming electron transfer and setting R_{eff} to 1.5 nm. The solid curve represents the fitted function of $g(C_A)$, Equation 9.11 with $\alpha = 447$, assuming FRET.[83]

The total sample volume was in each case 2.75×10^{-7} L. From this the molar ZnPc concentrations C_A in each sample can be calculated. The mean donor–acceptor distance R_{DA} can be estimated according to Equation 9.8 assuming isotropic conditions, where N_A is Avogadro's number[13]:

$$R_{DA} = \left[\frac{3}{4\pi}\frac{1}{C_A N_A}\right]^{\frac{1}{3}}. \tag{9.8}$$

To evaluate the distance dependence of the quenching effect, the probability P for FRET was calculated according to Equation 9.9, where I_D and I_{DA} represent the donor emission in the absence and presence of acceptors, respectively:

$$P = 1 - \frac{I_{DA}}{I_D}. \tag{9.9}$$

The probability P was plotted against the relative distance R_{DA}/R_0, where R_0 is the Förster radius, as defined in Equations (9.71)–(9.75). The R_{DA}/R_0 value for sample 9 was arbitrarily set to 5. The Förster radius R_0 for the ATTO-565 to ZnPc pair is approximately 7.1 nm. We observe that the probability for EnT shows the distance dependence expected for FRET. It approaches unity at small distances and decreases rapidly after a certain donor–acceptor distance has been reached; see also Figure 9.48. The donor–acceptor distance at which the EnT probability is 50% is not exactly at $R_{DA}/R_0 = 1$. This could well be a consequence of inhomogeneities in the PS layer. The graph, however, clearly indicates that the observed quenching effect on ZnPc is mainly due to FRET. Electron transfer shows a different distance dependence. It occurs with a probability of 1 until the critical radius R_{eff} is reached and then drops off rapidly. The quenching ratio I_{DA}/I_D for electron transfer is given in Equation 9.10[88]:

$$\frac{I_{DA}}{I_D} = e^{-\frac{C_A}{c_0}}, \quad \text{where} \quad c_0 = \frac{3}{4\pi}\frac{1}{R_{eff}^3}. \tag{9.10}$$

The quenching ratio I_{DA}/I_D, which we express as $g(\gamma)$ for excitation EnT, however, is calculated using Equation 9.11, which is derived in the next section, Equation 9.123:

$$g(\gamma) = 1 - \sqrt{\pi}\,\gamma e^{\gamma^2}[1 - erf(\gamma)], \quad \text{where} \quad \gamma = \frac{C_A}{c_0} \text{ and } c_0 = \frac{\alpha}{R_0^3}. \tag{9.11}$$

The term c_0 is called critical concentration and represents an acceptor concentration that results in 76% EnT. Considering these equations, the data of the previous quenching experiment can be interpreted in more detail. The relative emission intensities were plotted against the corresponding acceptor concentrations and the data fitted as a function of R_0 using Equation 9.11. The resulting curve with $\alpha = 447$ is shown in Figure 9.35c. The function corresponding to electron transfer was added to the plot assuming a critical radius R_{eff} of 1.5 nm, which is a reasonable boundary for electron transfer to occur. From this, it is obvious that FRET is the dominant quenching mechanism.

Dye–ZL composites can be considered as a new hybrid material that has a great chance for solving an old problem of luminescent solar concentrators (LSCs). An LSC is a transparent plate containing luminescent chromophores.[89] Light entering the face of the plate is absorbed and reemitted by these centers. A fraction of the emitted light is trapped by total internal reflection and guided to the edges of the plate where it can, for example, be converted to electricity by a photovoltaic device. As the edge area of the plate is much smaller than the face area, the LSC operates as a light concentrator. A major drawback in conventional LSC is the overlap between absorption and emission spectra, which is, as a basic principle, considerable in strongly luminescent molecules. Self-absorption therefore becomes an important loss mechanism. As a solution to this problem, antenna materials can be used.[22] In such host–guest materials, the absorption and emission processes can be separated by employing an absorbing dye present in large excess and a monolayer of an emitting dye. A further advantage of using these materials is that the photostability of many dyes can be considerably improved by including them into the channels of ZL. The photophysical and photochemical behavior of included molecules depend on the loading, co-cations, and co-adsorbed solvent present in the channels, like water (as is the case for all materials presented so far). We show in Figure 9.36 two simple experiments in which the active layer, while being only a few micrometers thick, is embedded between two glass plates. Light scattering has been suppressed by a procedure reported in Reference 36. Sample (a) was prepared by inserting DXP as donor into the channels of ZL, which was then modified with ATTO 565, acting as an emitter. In sample (b), $PC21^+$ was used as absorbing dye and Ox^+ as emitter with a ratio of about 15:2 between donor and emitter. Emission and excitation spectra for all components as well as for the device are summarized in Figure 9.36c, d. A donor–acceptor ratio of more than 50 has been successfully used in experiments with other dye combinations.[101]

Higher Organization Trough Anisotropic Arrangement. Many interesting materials can be made based on randomly arranged dye–ZL composites, some of which are useful in practical applications. However, more sophisticated devices can be envisaged by using options to arrange the composite crystals in specific patterns as explained in Figures 9.7 and 9.12. Here we focus on functional ZL crystals that are vertically oriented on a surface. We distinguish between systems with centrosymmetric and non-centrosymmetric organization of dyes because they lead to different macroscopic properties. This is sketched in Figure 9.37 where we show in the upper part an SEM image of a ZL monolayer on a glass substrate, as well as two schematic representations

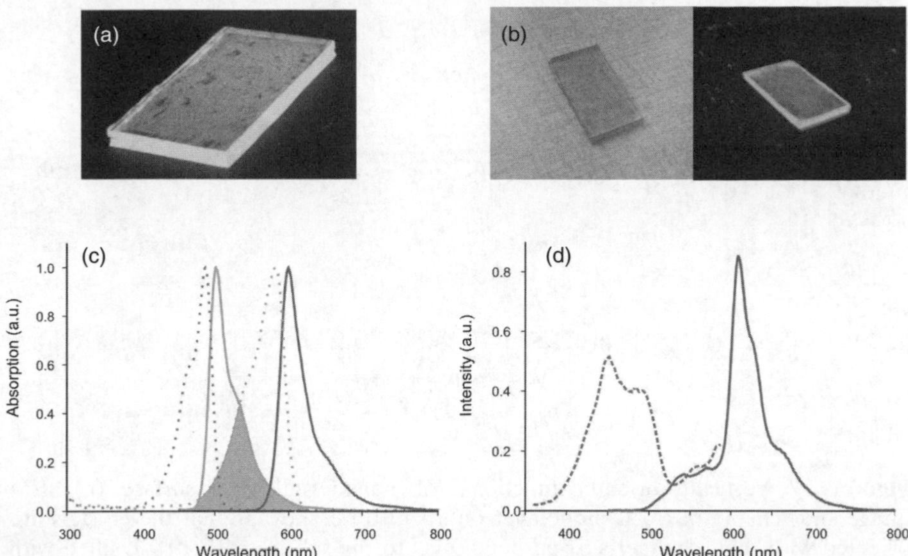

Figure 9.36. Luminescent solar concentrators (LSCs). (a) Fluorescence of a small concentrator (2×1 cm) under near-UV illumination with DXP as donor and ATTO 565 as acceptor. (b) Small fluorescent concentrator with $PC21^+$ as donor and Ox^+ as acceptor. Left: appearance of the device. Right: fluorescence under near-UV illumination. (c) Absorption (dotted) and fluorescence spectra (solid) of $PC21^+$ and Ox^+ in ZL. The spectral overlap between the fluorescence spectrum of the donor and the acceptor is shaded. (d) Excitation (dotted) and fluorescence spectrum of the Ox^+,$PC21^+$-ZL material. The excitation spectrum was measured at 640 nm. The fluorescence spectrum was measured by excitation at 425 nm where Ox^+ does not absorb. FRET from $PC21^+$ to Ox^+ is considerable. See ftp site for color: ftp://ftp.wiley.com/public/sci_tech_med/supramolecular_photochemistry.

of it. The scheme on the right shows a side view of ZL. The gray bars indicate the channel walls and the empty part the channels that can be filled with dye molecules. This is a very simple cartoon, similar to that used in Figure 9.7. One has to keep in mind that a crystal of 600-nm diameter consists of roughly 100,000 of such channels. In the middle part, we show different arrangements with centrosymmetrically loaded crystals. Figure 9.37b represents crystals filled with one dye only, while (c) and (d) show centrosymmetric simple and stopcock type antenna, respectively. The lower end of the figure shows non-centrosymmetrically filled crystals (e–g). The synthesis of layers consisting of centrosymmetrically modified crystals is easier as they allow more options for crystal surface modification.[61,67] For obtaining arrays consisting of non-centrosymmetric crystals, c-ocMLs are necessary. This means to use a monolayer preparation technique that allows insertion of dyes and/or the addition of stopcock molecules after monolayer preparation, as has been reported in References 34 and 66.

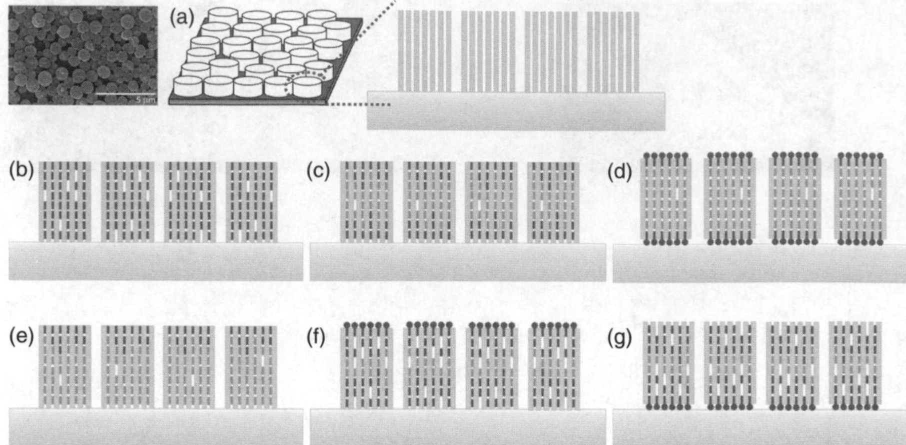

Figure 9.37. Vertically oriented functional ZL nanocrystals on a surface. (a) SEM image and scheme of a ZL monolayer on a substrate, showing that the crystals are oriented with their channel axis perpendicular to the substrate. (b–d) ZL filled with dye molecules in a centrosymmetric fashion arranged on a substrate. (e–g) Non-centrosymmetric dye–ZL crystals arranged as a c-oriented open-channel monolayer (c-ocML) on a substrate.

An antenna material, which absorbs all light in a specific wavelength range and transfers the electronic excitation energy by means of FRET to well-organized acceptors, offers unique possibilities for developing dye-sensitized solar cells, LSCs, a new generation of organic light-emitting diodes (OLEDs), energy downconverters, color changing media, and as arrays for analytical purposes. In some of these applications, both non- and centrosymmetrically modified crystals can be used. The non-centrosymmetric arrangement allows, however, for the realization of the most advanced devices. We report in Figure 9.38 EnT experiments in oriented (stopcock),dye2-ZL monolayers, where dye1 is a molecule located on one side of the crystals only. It can be either inside of the channels or attached as a stopcock molecule. On the top left of the figure, we show an SEM image of a layer and next to it a cartoon of the arrangement for a situation with three dyes (stopcock),dye2,dye3-ZL, where dye1 is a stopcock molecule. Below we show luminescence spectra of different systems. They prove that any combination of unidirectional EnT materials can be made, namely between dyes organized inside the channels (b), between a donor stopcock to acceptors inside (c) and (e), and from donor molecules inside to an acceptor stopcock (d). These systems can be considered as being the first unidirectional antennae realized on a macroscopic scale.[21,30,34,38,66] Many fascinating experiments are waiting to be performed with this kind of materials.

We have shown that supramolecular organization of dyes inside ZL allows light harvesting within the volume of a dye–ZL crystal and radiationless

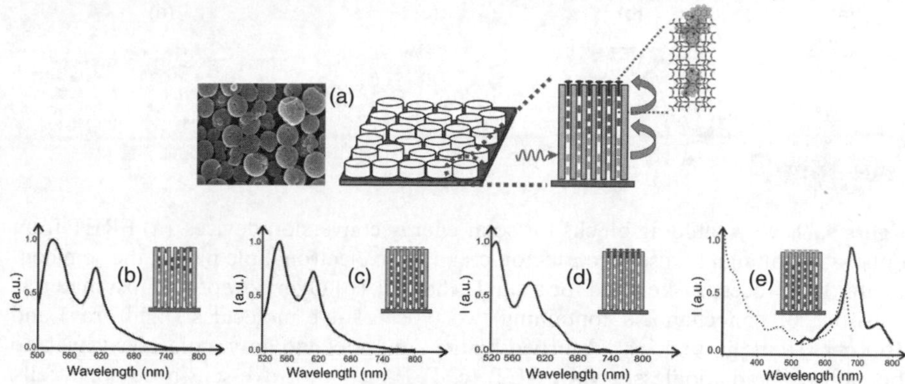

Figure 9.38. Energy transfer in c-ocML. (a) SEM images of a c-ocML along with schemes. The enlargement shows a ZL containing two dyes and modified with a stopcock, including a magnification of one channel entrance. (b–d) Emission spectra of donor (light gray) and acceptor (dark gray) loaded ZL crystals arranged as c-ocML on a glass plate. All spectra have been scaled to the same height at the maxima. (b) Ox^+,Py^+-ZL c-ocML. The emission spectrum was recorded after selective excitation of Py^+ at 460 nm. (c) (ATTO520),Ox^+-ZL c-ocML. The emission spectrum was recorded after selective excitation of ATTO520 at 460 nm. (d) (Cy02702),Py^+-ZL c-ocML. The emission spectrum was recorded after selective excitation of Py^+ at 460 nm.[34] (e) FRET in an oriented (Ru-ph4-TMS),Ox1-ZL c-ocML. Excitation (dashed) and emission (solid) spectra of a (Ru-ph4-TMS),$Ox1^+$-ZL c-ocML. The emission spectrum was observed after excitation at 450 nm, where Ru-ph4-TMS is selectively excited. The excitation spectrum was recorded at 710 nm.[35]

energy transport to the cylinder ends. We have also reported that the second stage of organization involves coupling to an external acceptor stopcock at the ends of the ZL channels, which can then trap or inject electronic excitation energy. The third stage of organization is attained by interfacing the material to an external device via a stopcock intermediate. FRET in dye–ZL materials occurs mainly along the channel axis. This implies that macroscopically organized unidirectional materials can be prepared for optimized EnT purposes as reported in Figure 9.39. These materials offer unique possibilities as building blocks for developing a new generation of antenna-sensitized solar cells. This type of dye-sensitized solar cells works by first absorbing light over a broad spectral range in the ZL–antenna material. The excitation energy migrates radiationlessly among the embedded dyes toward the stopcocks. From there, FRET to the semiconductor takes place across a very thin insulating layer. The injected electronic excitation energy can now be used to drive the charge-separation process in the active medium—see Figure 9.39c, d. We conclude that new building blocks are now ready to be tested in devices. Their size, morphology, and optical properties will need to be tailored to the specific tasks envisaged. The remaining problems to be solved require efforts at the interface of chemistry, physics, and engineering.

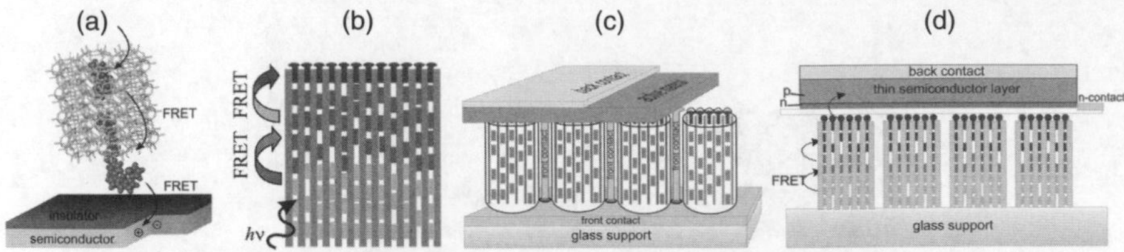

Figure 9.39. New building blocks for solar energy conversion devices. (a) FRET from a photonic antenna to a semiconductor, creating an electron–hole pair in the semiconductor. The process is sketched for a single channel. (b) Overview of the host material, consisting of nanochannels containing two types of dye molecules (light gray) and stopcocks (dark gray). Light absorbed by the light gray and gray molecules travels to the stopcock radiationlessly, via FRET. (c, d) Principle of dye-sensitized solar cells. Arranging crystals of, for example, 100-nm length with their channel axes perpendicular to the surface of a semiconductor allows for the transport of electronic excitation energy toward the ZL–semiconductor interface by FRET. The semiconductor layer can be very thin, because the electron–hole (n-p) pairs form near the surface. Transfer of electrons from the antenna to the semiconductor is prevented by introducing a thin insulating layer. The principle of operation of thin-layer silicon devices is shown in (d), while (c) represents that of organic or plastic solar cells. The white area on top of the stopcock head represents an insulating part directly integrated into it. The ZL part is enlarged with respect to the rest of the device.

THEORETICAL BACKGROUND

Electronic Transition Moment Coupling

Electronic excitation EnT from an excited molecule D^* to an acceptor A can take place if A possesses transitions, which are isoenergetic with those of D^*, as shown in Figure 9.40.

We assume as an illustrative example that the energy separation between the two vibrational states v, $v + 1$ of the donor and v', $v' + 1$ of the acceptor is the same. We further assume that the energy difference ΔE between the electronic (00′) transitions of D and A is twice this separation. ΔE reflects only fluctuations in the environment and temperature if the donor and the acceptor are of the same molecule type. Under these conditions, the following resonance energy transfer (RET) processes can take place:

$$\begin{aligned} &D^*(0') + A(0) \rightarrow D(2) + A^*(0'), \\ &D^*(0') + A(0) \rightarrow D(1) + A^*(1'), \\ &D^*(0') + A(0) \rightarrow D(0) + A^*(2'). \end{aligned} \tag{9.12}$$

We assume that two electrons are involved in a transition, one on D and one on A. The antisymmetric electronic wave functions of the initial excited state Ψ_i (D excited but not A) and of the final excited state Ψ_f (A excited but not D) can be expressed as follows, where the numbers 1 and 2 refer to the involved electrons:

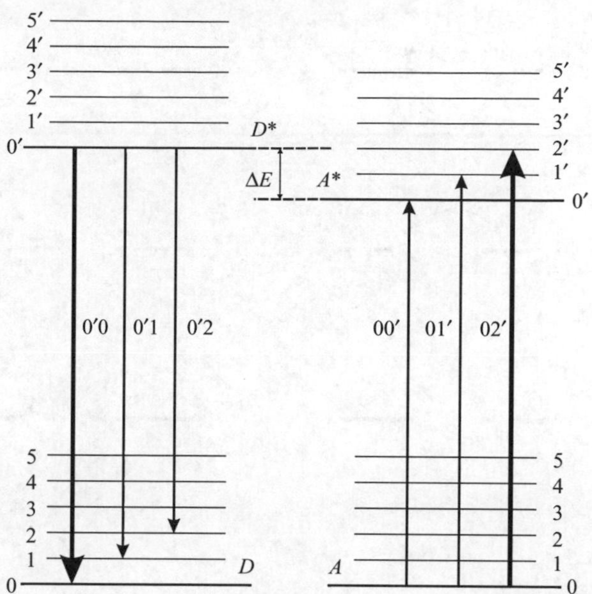

Figure 9.40. Energy levels of a donor molecule D^* and an acceptor molecule A. The vibrational states in the electronic ground state are labeled as 0, 1, 2, and so on, and those in the excited state as 0′, 1′, 2′, and so on. The arrows indicate electronic transitions for which the (D^*;A) state pairs (0′2;00′), (0′1;01′), and (0′0;02′) are in resonance.

$$\Psi_i = \frac{1}{\sqrt{2}}(\Psi_{D^*}(1)\Psi_A(2) - \Psi_{D^*}(2)\Psi_A(1)), \tag{9.13}$$

$$\Psi_f = \frac{1}{\sqrt{2}}(\Psi_D(1)\Psi_{A^*}(2) - \Psi_D(2)\Psi_{A^*}(1)). \tag{9.14}$$

The interaction term β between the initial and the final state is

$$\beta = \langle \Psi_i | H' | \Psi_f \rangle. \tag{9.15}$$

H' is the perturbation part of the Hamiltonian:

$$\hat{H} = \hat{H}_{D^*} + \hat{H}_A + H'. \tag{9.16}$$

Inserting Equations 9.13 and 9.14 in Equation 9.15 we obtain

$$\beta = \left\langle \frac{1}{\sqrt{2}}(\Psi_{D^*}(1)\Psi_A(2) - \Psi_{D^*}(2)\Psi_A(1)) \middle| H' \middle| \frac{1}{\sqrt{2}}(\Psi_D(1)\Psi_{A^*}(2) - \Psi_D(2)\Psi_{A^*}(1)) \right\rangle. \tag{9.17}$$

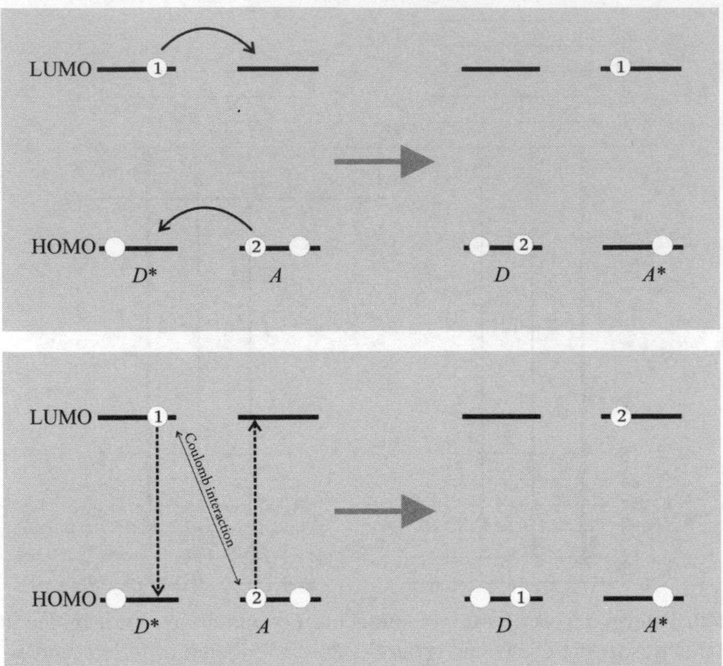

Figure 9.41. Electronic excitation energy transfer. Representation of the exchange (upper) and the Coulomb (lower) interaction mechanism. HOMO and LUMO refer to the highest occupied molecular orbital and to the lowest unoccupied molecular orbital, respectively.

This expression can be divided in two parts:

$$\langle \Psi_{D^*}(1)\Psi_A(2)|H'|\Psi_D(1)\Psi_{A^*}(2)\rangle = \langle \Psi_{D^*}(2)\Psi_A(1)|H'|\Psi_D(2)\Psi_{A^*}(1)\rangle = \beta_C, \tag{9.18}$$

$$\langle \Psi_{D^*}(1)\Psi_A(2)|H'|\Psi_D(2)\Psi_{A^*}(1)\rangle = \langle \Psi_{D^*}(2)\Psi_A(1)|H'|\Psi_D(1)\Psi_{A^*}(2)\rangle = \beta_{ex}. \tag{9.19}$$

Hence, β and also H' can be written as a sum of two terms:

$$\beta = \beta_C - \beta_{ex} \quad \text{and} \quad H' = H_C + H_{ex}. \tag{9.20}$$

The Coulomb term β_C describes a process in which the initially excited electron on D^* returns to the ground state, while an electron on A is simultaneously promoted to the excited state. The exchange term β_{ex} describes a process that can be understood as an exchange of two electrons, one on D and one on A. The Coulomb and the exchange interactions lead to two distinctly different EnT mechanisms, as illustrated in Figure 9.41. We observe that an exchange of electrons between D and A takes place in the exchange interaction. This is not the case in the Coulomb mechanism where no electrons are exchanged. The exchange term β_{ex} represents the electrostatic interaction between the charged clouds. In order to have a nonzero value for β_{ex}, overlap of the electron clouds is a prerequisite, and hence, EnT due to exchange interaction requires overlap of the wave functions of D^* and A, similar as in electron transfer reactions. This is a short range interaction. For two electrons separated by a distance r_{12} in $D^*\ldots A$, the perturbation H'_{ex} is

$$H'_{ex} = \frac{e^2}{4\pi\varepsilon_0}\frac{1}{r_{12}}, \tag{9.21}$$

where e is the elementary charge of an electron and ε_0 is the vacuum permittivity.[18]

The Coulomb term can be expanded into a sum of multiple series:

$$\beta_C = V_{dd} + V_{qd} + V_{dq} + V_{qq} + \ldots, \tag{9.22}$$

where V_{dd} is the dipole–dipole interaction, V_{qd} is the quadrupole–dipole interaction, and so on. We focus on situations where the dipole–dipole interaction term between the ETDM μ_{D^*D} and μ_{AA^*} of D and A for the transitions $D^*\rightarrow D$ and $A\rightarrow A^*$ in an environment of refractive index n is dominant. The perturbation H'_C can then be expressed by means of Equation 9.23, where l_{D^*} and l_A are the positions of the ETDM, which are separated by the distance R_{D^*A} and where κ_{D^*A} takes the relative orientation of the donor and acceptor ETDM's into account; see Figure 9.42. Equations 9.23 and 9.24 are derived in the Appendix:

$$H'_C = \frac{e^2}{4\pi\varepsilon_0 n^2}\frac{1}{R^3_{D^*A}} l_{D^*} l_A \kappa_{D^*A}, \tag{9.23}$$

$$\kappa_{D^*A} = \sin\theta_1 \sin\theta_2 \cos\phi_{12} - 2\cos\theta_1 \cos\theta_2. \tag{9.24}$$

The rate constant k_{EnT} for electronic excitation EnT for the weak coupling limit can be expressed according to Fermi's golden rule as follows[2,17,18]:

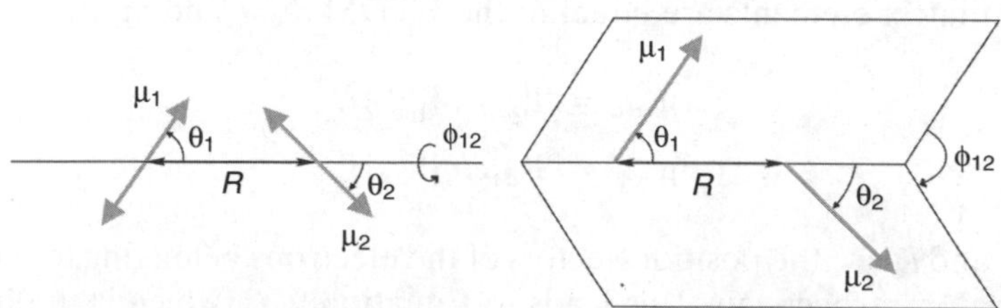

Figure 9.42. Angles describing the relative orientation of the ETDM μ_1 and μ_2 of two molecules. The numbers 1 and 2 not only refer to the corresponding electrons but are also used to identify the two molecules. Left: representation of the ETDM as oscillators. Right: vector representation of the ETDM.

$$k_{EnT} = \frac{2\pi}{\hbar}\beta_C^2\rho. \tag{9.25}$$

ρ is a measure of the density of the interacting initial $D^*...A$ and final $D...A^*$ states. It is related to the overlap between the emission spectrum of the donor and the absorption spectrum of the acceptor. Direct evaluation of Equation 9.25 by means of numerical quantum chemical calculations of different degrees of sophistication have been performed, and intermolecular electronic excitation EnT in confined space has been examined.[17,90–92]

Weak Interactions. For very weak interactions between excited and unexcited molecules, for example, couplings of less than 100 cm^{-1}, the electronic spectrum of a mixture of donors and acceptors, measured under ambient conditions, will be almost an exact superposition of the separate spectra of diluted solutions of both components. Stronger interactions lead to exciton splitting. In order to evaluate the rate constant k_{EnT} for excitation EnT from an electronically excited donor D^* and an acceptor A, we must calculate the product $\beta_C^2\rho$ (Eq. 9.25). The wave functions of the initial (i) and the final (f) states are, according to Equation 9.18:

$$\Psi_i = \Psi_{D^*}\Psi_A, \tag{9.26}$$

$$\Psi_f = \Psi_D\Psi_{A^*} \tag{9.27}$$

Inserting these along with Equation 9.23 into Equation 9.18, which we write as follows:

$$\beta_C = \langle\Psi_i|H'_C|\Psi_f\rangle, \tag{9.28}$$

we obtain the following expression:

$$\beta_C = \frac{1}{4\pi\varepsilon_0 n^2}\frac{1}{R_{D^*A}^3}|\langle\Psi_{D^*}|el_D|\Psi_D\rangle||\langle\Psi_A|el_A|\Psi_{A^*}\rangle|\kappa_{D^*A}. \tag{9.29}$$

The two matrix elements are equal to the ETDM m_{D^*D} and m_{AA^*}:

$$\mu_{D^*D} = \langle\Psi_{D^*}|el_D|\Psi_D\rangle, \tag{9.30}$$

$$\mu_{AA^*} = \langle\Psi_A|el_A|\Psi_{A^*}\rangle, \tag{9.31}$$

where l_D and l_A are the position vectors of the electrons belonging to the donor and acceptor, respectively. This leads to Equation 9.32, which is the basis for describing both exciton splitting and the rate constant k_{EnT}:

$$\beta_C = \frac{1}{4\pi\varepsilon_0 n^2}\frac{1}{R_{D^*A}^3}|\mu_{D^*D}||\mu_{AA^*}|\kappa_{D^*A}. \tag{9.32}$$

Exciton Splitting

We consider a pair of chromophores A_i and A_k at a distance R, which is such that their interaction in the electronic ground state is negligible. We further assume that the overlap of the wave functions in the electronically excited states $A_i^* \ldots A_k$ and $A_i \ldots A_k^*$ between the neighbors (A_i^* and A_k) and (A_i and A_k^*), respectively, is negligible. This means that the "optical electrons" associated with the chromophoric units preserve essentially their individual characteristic. This does, however, not necessarily mean that the interaction β_C between the electronically excited state configurations ($A_i^* \ldots A_k$) and ($A_i \ldots A_k^*$) is so weak that we can neglect it. The ETDM's $|\mu_{A*A}|$ and $|\mu_{AA*}|$ have the same value; hence, Equation 9.32 can be written as follows[19]:

$$\beta_C = \frac{1}{4\pi\varepsilon_0 n^2}\frac{|\mu_{AA^*}|^2}{R_{A^*A}{}^3}\kappa_{A^*A}, \tag{9.33}$$

where κ_{A^*A} is defined by Equation 9.24. The wave functions of the ground state and of the electronically excited states can be expressed as follows:

$$\Psi_{A_iA_k} = \Psi_{A_i}\Psi_{A_k}, \tag{9.34}$$

$$\Psi_{A_i^*A_k} = \Psi_{A_i^*}\Psi_{A_k}, \tag{9.35}$$

$$\Psi_{A_iA_k^*} = \Psi_{A_i}\Psi_{A_k^*}. \tag{9.36}$$

We denote the energy of the electronic ground state $\Psi_{A_iA_k}$ as E_0 and that of the electronically excited states $\Psi_{A_i^*A_k}$ and $\Psi_{A_iA_k^*}$ in the absence of any interaction between them as E_1. In the presence of some interaction, as expressed by the perturbation H'_C, the excited state is more accurately described by means of a linear combination of the wave functions in Equations 9.35 and 9.36:

$$\Phi(c_1,c_2) = c_1\Psi_{A_i^*A_k} + c_2\Psi_{A_iA_k^*}. \tag{9.37}$$

From this we find

$$\langle\Phi(c_1,c_2)|H|\Phi(c_1,c_2)\rangle = \varepsilon\langle\Phi(c_1,c_2)|\Phi(c_1,c_2)\rangle, \tag{9.38}$$

where H is equal to $H_i + H_k + H'_C$. Evaluating this by keeping in mind that the overlap integral $\langle\Psi_{A_i^*A_k}|\Psi_{A_iA_k^*}\rangle$ is zero, we obtain

$$\begin{vmatrix} h_{11}-\varepsilon & h_{12} \\ h_{21} & h_{22}-\varepsilon \end{vmatrix} = 0, \tag{9.39}$$

$$\Phi_+ = \frac{1}{\sqrt{2}}\left(\Psi_{A_i^*A_k} + \Psi_{A_iA_k^*}\right), \tag{9.40}$$

$$\Phi_- = \frac{1}{\sqrt{2}}\left(\Psi_{A_i^*A_k} - \Psi_{A_iA_k^*}\right). \tag{9.41}$$

The values of h_{11} and h_{22} are equal to E_1, while h_{12} and h_{21} are equal to β_C. This leads to

$$\begin{aligned} \varepsilon_+ &= E_1 + \beta_C, \\ \varepsilon_- &= E_1 - \beta_C. \end{aligned} \tag{9.42}$$

Interchanging the molecular labels i and k shows that Φ_+ is symmetric while Φ_- is antisymmetric. The excitation is on both molecules i and k in both stationary states Φ_+ and Φ_-. The excitation is collective or delocalized. The node corresponding to the minus sign in Φ_- is an excitation node. At an excitation node, the relation between the ETDM of the respective molecular centers changes phase. We explain this in Figure 9.43 (left) for an arrangement of the ETDM with $\phi_{12} = 0$ and $\theta_1 = \theta_2$, which plays an important role for dyes in nanochannels. Inserting these angle values into Equation 9.24 results in the simplified expression (Eq. 9.43) of κ_{AA^*}:

$$\kappa_{AA^*} = 1 - 3\cos^2\theta. \tag{9.43}$$

The value of β_C is the largest for in-line or collinear orientation and changes sign for parallel orientation. We illustrate this in the energy level diagram in Figure 9.43 (right). Situations with essentially collinear arrangement of the

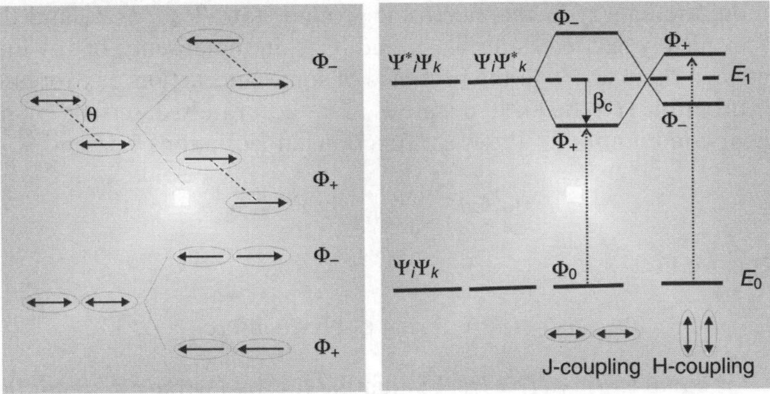

Figure 9.43. Phase relation and energy level diagram. Left: Diagram showing the phase relation of the wave functions Equations 9.40 and 9.41, which describe the interaction caused by the ETDM between the electronically excited-state configurations $A_i^* \ldots A_k$ and $A_i \ldots A_k^*$. Right: Energy level diagram showing the exciton splitting of two chromophores caused by the configuration interaction due to the ETDM. This interaction naturally causes only a splitting of electronically excited states and has no consequences on the ground state. The different splitting of the excited-state levels for ETDM-oriented collinear and for those that are parallel, as represented by means of double arrows, is due to the angle dependence of κ_{AA^*}. The allowed electronic transitions are indicated by the dotted arrows.

ETDM lead to J-coupling, while those with essentially parallel arrangement lead to H-coupling. The corresponding arrangements of the chromophores are often named as J-aggregates and H-aggregates, respectively.[20]

Crossing of the Φ_+ and Φ_- levels occurs when $\cos^2(\theta)$ is equal to 1/3, which is the case when θ is equal to the magic angle of 54.7°, according to Equation 9.43. The order of magnitude of the expected splitting $|2\beta_C|$ can be estimated using the relation between the oscillator strength f of the electronic transition and the magnitude of the ETDM,

$$f = \frac{8\pi c m_e}{3he^2}\bar{\upsilon}|\mu_{AA*}|^2, \tag{9.44}$$

where e is the elementary charge, h is Plank's constant, m_e the electron mass, c the speed of light in vacuum, and $\bar{\upsilon}$ the energy in cm^{-1} of the transition $A \to A^*$.[93–95] This equation is also useful for estimating the length of the ETDM. This is important because the validity of the dipole–dipole coupling theory depends on the condition that the distance between the ETDM of the two involved chromophores is large with respect to the length of the ETDM l_{μ^*}. Solving Equation 9.44 for μ_{A*A}, dividing it by the elementary charge and inserting the constants leads to

$$l_{\mu*} = 3.036\times10^{-6}\, cm^{0.5}\sqrt{\frac{f}{\bar{\upsilon}}}. \tag{9.45}$$

From this equation we find, as an example, that the length of the ETDM of an organic molecule with an oscillator strength of $f = 1$ absorbing light at 500 nm is 0.215 nm. This means that the distance between molecules of interest in ZL channels is in general large enough so that the dipole–dipole coupling approach can be considered as a good approximation. We illustrate this in Figure 9.44 for PR149 for which we calculate the length of the ETDM to be about 0.19 nm, while the shortest distance between two of these molecules in the channels of ZL has been estimated to be about 2.2 nm.

Inserting Equation 9.44 in the expression for β_C (Eq. 9.33) leads to

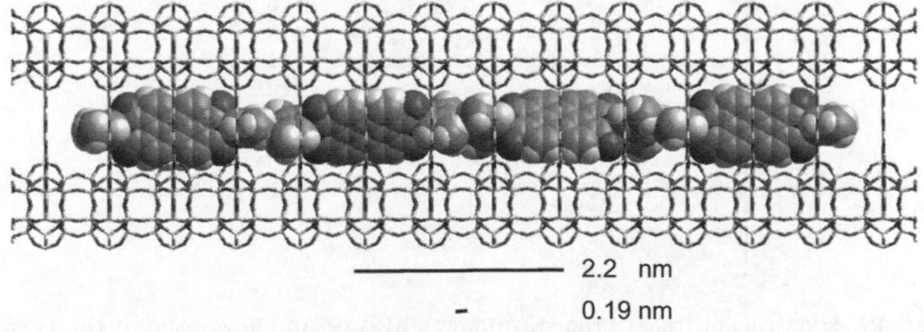

Figure 9.44. Comparison of the length of aligned PR149 (2.2 nm) with the ETDM (0.19 nm), at high packing in a channel of ZL.

$$\beta_C = \frac{3he^2}{32\pi^2 cm_e\varepsilon_0}\frac{f}{\bar{\upsilon}}\frac{\kappa_{A^*A}}{R_{A^*A}{}^3}\frac{1}{n^2}. \tag{9.46}$$

It is convenient to collect the constant term by writing

$$\beta_C = AD\frac{f}{\bar{\upsilon}}\frac{\kappa_{A^*A}}{R_{A^*A}{}^3}\frac{1}{n^2}. \tag{9.47}$$

The value of the constant AD is equal to $1.615 \times 10^{-18}\,\mathrm{m^2\,cm^{-1}}$ if we express β_C in $\mathrm{cm^{-1}}$, which is convenient.

We discuss some consequences of this by using a perylene dye (Table 9.2) as example. These dyes have been used in different experiments[21–23,49] and they are well aligned along the channel axis, meaning that the angles ϕ_{12}, θ_1, and θ_2 are approximately zero. Many derivatives of these dyes with different types of substituent R are known, where R does not affect the electronic spectra of the molecules. The substituent R, however, determines the shortest distance between two chromophores in ZL at high packing. This is useful for studying the dipole–dipole coupling strength. We show in Figure 9.45 the exciton splitting of the stationary states Φ_- and Φ_+ caused by the J-coupling as a function of separation. The shortest high packing distance that can be realized for R = CH_3 is in the order of 1.5 nm at which the level splitting of two interacting molecules is about 400 $\mathrm{cm^{-1}}$. It is about 100 $\mathrm{cm^{-1}}$ at 2.2 nm.

PC21$^+$ has an oscillator strength of about 1.4 that causes stronger J-coupling. Its ETDM is polarized parallel to the channel axis. The absorption and fluorescence spectra of PC21$^+$ reported in Figure 9.6 illustrate the consequences of the J-coupling. A nice demonstration of exciton coupling has been recently reported for Py$^+$-loaded ZL for which correlated fluorescence microscopy, fluorescence lifetime, and spectral imaging of single crystals were performed.

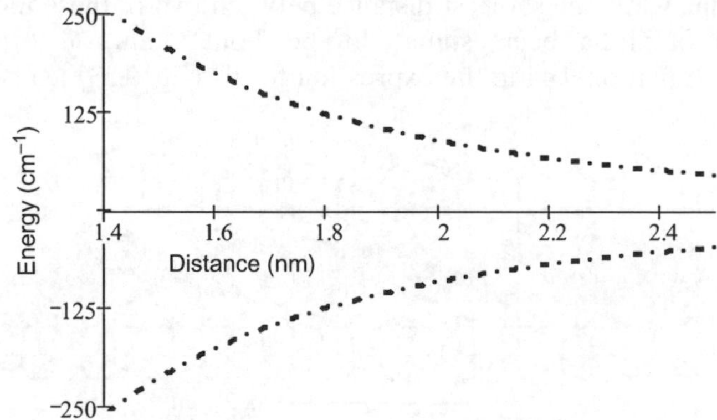

Figure 9.45. Exciton splitting of the stationary states Φ_- and Φ_+ caused by the J-coupling as a function of separation, calculated for $f = 0.8$, $\bar{\upsilon} = 20{,}000\ \mathrm{cm^{-1}}$, and $n = 1.45$ and $\theta = 0$.

At higher loading, the molecules hinder themselves when entering the channels in a process that can be understood as "traffic jam in nanochannels." This causes J-coupling in the region of the channel entrance.[23a]

The theoretical reasoning outlined so far has been limited to interactions between two neighbors. The experimentally observed phenomena in high packing situations, however, are in general expected to be due to interactions between more than two units. Generalization of the exciton splitting due to interaction of N chromophores is straightforward. It leads to Equation 9.48, where α denotes the energy in absence of any interaction and j is equal to $1, 2, \ldots, N$[19,20]:

$$\varepsilon(j,N) = \alpha + 2\beta_C \cos\left(\frac{j}{N+1}\pi\right). \tag{9.48}$$

This means that N interacting chromophores generate N levels. This causes some broadening, which affects mainly the absorption spectra. The maximum splitting $\Delta E(N)$ is equal to the difference between the levels $\varepsilon(1,N)$ and $\varepsilon(N,N)$. We see from Equation 9.49 that it converges rapidly to the value of $4|\beta_C|$:

$$\Delta E(N) = 2|\beta_C|\left(\cos\left(\frac{1}{N+1}\pi\right) - \cos\left(\frac{N}{N+1}\pi\right)\right). \tag{9.49}$$

We illustrate the contents of Equations 9.48 and 9.49 in Figure 9.46 for $N = 100$ and $\beta_C = 100\,\text{cm}^{-1}$, while the value of α has been arbitrarily chosen to be 0. The consequences of the ability to prepare densely packed dye-loaded nanochannel materials with large J-coupling coherence lengths have only very scarcely been explored.[21–23]

Influence of the J-Coupling on the Value of the ETDM. In the J-coupling case, occurring at collinear arrangement of the ETDMs, the electronic

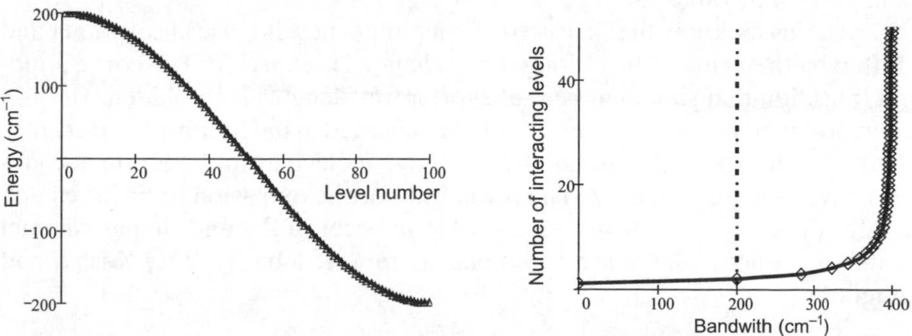

Figure 9.46. Energy levels (left) and bandwidth (right), calculated for *J*-coupling with $\beta_C = 100\,\text{cm}^{-1}$. Left: energy levels of an exciton extended over 100 chromophores. Right: bandwidth for a varying number of interacting levels. The vertical dash-dot line indicates the value for N = 2.

TABLE 9.4. ETDM and Oscillator Strength of a Molecular Dimer Exciton

In-line arrangement (J-coupling)	$M_{+}=\frac{1}{\sqrt{2}}(\mu_{A_i^*A_k}+\mu_{A_iA_k^*})=\frac{2}{\sqrt{2}}\mu_{AA^*}$	f=2f$_{monomer}$
	$M_{-}=\frac{1}{\sqrt{2}}(\mu_{A_i^*A_k}-\mu_{A_iA_k^*})=0$	f=0
Parallel arrangement (H-coupling)	$M_{-}=\frac{1}{\sqrt{2}}(\mu_{A_i^*A_k}-\mu_{A_iA_k^*})=0$	f=0
	$M_{+}=\frac{1}{\sqrt{2}}(\mu_{A_i^*A_k}+\mu_{A_iA_k^*})=\frac{2}{\sqrt{2}}\mu_{AA^*}$	f=2f$_{monomer}$

transition $\Phi_+ \leftarrow \Phi_0$ to the lower lying state is allowed, while that to the higher lying Φ_- state is forbidden (Fig. 9.43). The ETDM for a transition $(A_i \ldots A_k) \rightarrow [(A_i^* \ldots A_k) \leftrightarrow (A_i \ldots A_k^*)$ can be expressed as follows:

$$\begin{aligned} M_+ &= \langle \Psi_i \Psi_k | \hat{\mu}_i + \hat{\mu}_k | \Phi_+ \rangle, \\ M_- &= \langle \Psi_i \Psi_k | \hat{\mu}_i + \hat{\mu}_k | \Phi_- \rangle, \end{aligned} \tag{9.50}$$

where $\hat{\mu}_i$ and $\hat{\mu}_k$ are the ETDM operators of the molecules A_i and A_k. Evaluating this by taking orthogonality and normalization properties of the intramolecular state function into account we obtain

$$\begin{aligned} M_+ &= \frac{1}{\sqrt{2}}(\mu_{A_i^*A_k} + \mu_{A_iA_k^*}), \\ M_- &= \frac{1}{\sqrt{2}}(\mu_{A_i^*A_k} - \mu_{A_iA_k^*}). \end{aligned} \tag{9.51}$$

M_- is equal to zero. Thus, the ETDM of the two interacting chromophores appear as superposition of those of the individual partners. The results are summarized in Table 9.4.

From this we know that in case of J-coupling, a new intense absorption band shifted by the value of β_C to longer wavelengths is expected. The corresponding transition that should appear at shorter wavelengths is forbidden. This has been observed many times and is well documented in the literature.[20] Recently, the first examples of J-coupling between dye molecules in the ZL nanochannels have been reported.[21–23] The formalism for the extension to linear chains, as illustrated in Figure 9.44 and expected to occur in ZL and similar channel materials, follows the same reasoning as reported by McRae, Kasha, and others.[20]

FRET

We consider the EnT transfer $D^*(0') + A(0) \rightarrow D(0) + A^*(2')$, which we can also express as $D^*(0')\,A(0) \rightarrow D(0)\,A^*(2')$. The rate constant for EnT between two resonant levels can be expressed according to Equation 9.25 as follows:

$$k_{EnT(0'0;02')} = \frac{2\pi}{\hbar}\beta^2_{D*A}\rho_{(0'0;02')}, \tag{9.52}$$

where β_{D*A} is equal to β_C. In order to find all transitions that take place between D^* and A, $D^*\ldots A \rightarrow D\ldots A^*$, we must sum over all states that are in resonance. Since we are interested in condensed phase, discrete rotational states play no role. Denoting the donor states as $(d'\delta)$ and those of the acceptor as $(\alpha a')$ the transfer rate constant can be expressed as follows:

$$k_{EnT(d'\delta;\alpha a')} = \frac{2\pi}{\hbar}\beta^2_C\rho_{(d'\delta;\alpha a')}. \tag{9.53}$$

It is not sufficient to consider only the lowest vibrational state of the donor since EnT can be very fast. We must sum over all iso-energetic situations. This means that the sum involves all donor and acceptor states that are in resonance:

$$k_{EnT} = \sum k_{EnT(d'\delta;\alpha a')}. \tag{9.54}$$

There is no need to assume that the donor is in a thermally relaxed excited state. EnT can often be so fast that there is no time left for establishing thermal equilibrium. This means that k_{EnT} can be time dependent. We do not take this explicitly into account, but it is good to keep it in mind. It would in fact lead to a correction in the expression for the spectral overlap integral. Spectra in condensed phase are usually broadened due to solute solvent interactions and lattice vibrations. Hence, the initial (i) and the final (f) levels of $D^*\ldots A$ and $D\ldots A^*$, respectively, are broadened. We may therefore express the density of states ρ_E within a continuous energy range, allowing us to introduce the normalized functions $S_D(E_{D^*})$ and $S_A(E_A)$. The first expresses the probability that an excited molecule D^* emits photons of energy E_{D^*}, and the second represents the probability that A absorbs photons of energy E_A:

$$\int_{E_\gamma} S_\gamma(E_\gamma)dE_\gamma = 1. \tag{9.55}$$

$S_D(E_{D^*})$ and $S_A(E_A)$ reflect the shape of the luminescence spectrum of D^* and of the absorption spectrum of A, respectively. The resonance condition illustrated in Figure 9.47 can be expressed as follows:

$$E_{D^*} = E^D_{0'0} + \varepsilon_{D^*} - \varepsilon_D, \tag{9.56}$$

$$E_A = E^A_{0'0} + \varepsilon_{A^*} - \varepsilon_A, \tag{9.57}$$

$$E_{\text{res}} = \int E_{D^*}\delta(E_{D^*} - E_A)dE_{D^*}. \tag{9.58}$$

The rate constant for the EnT process $D^* + A \rightarrow D + A^*$ can be expressed as the integral over the resonant energy range:

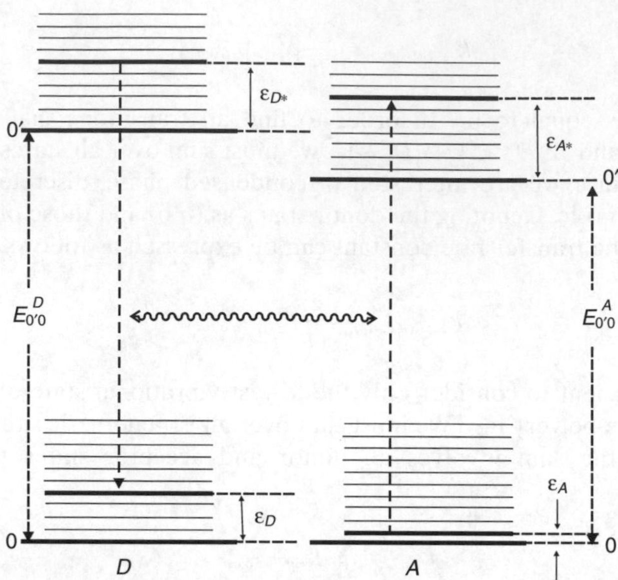

Figure 9.47. Energy level diagram of a donor molecule D and an acceptor molecule A illustrating the conditions for resonance energy transfer (RET). The wavy horizontal line symbolizes the long range interaction between the involved states.

$$k_{EnT} = \frac{2\pi}{\hbar} \int_{E_{D^*}} \int_{E_A} \beta_C^2 S_D(E_{D^*}) S_A(E_A) \delta(E_{D^*} - E_A) dE_{D^*} dE_A. \tag{9.59}$$

Inserting β_C from Equation 9.32 we get

$$k_{EnT} = \frac{2\pi}{\hbar} \left(\frac{\kappa_{D^*A}}{4\pi\varepsilon_0 n^2 R_{D^*A}^3} \right)^2 \int_{E_{D^*}} \int_{E_A} |\mu_{D^*D}|^2 S_D(E_{D^*}) |\mu_{AA^*}|^2 S_A(E_A) \delta(E_{D^*} - E_A) dE_{D^*} dE_A. \tag{9.60}$$

We evaluate this integral by using the Einstein coefficients A_{D^*D} and B_{D^*D}. The factor consisting of the ETDM multiplied by the distribution functions can be substituted by the lifetime and shape of the luminescence spectrum of the donor D^* and by the absorption spectrum of the acceptor A. We integrate over the frequency ν. Since $E = h\nu$, dE must be substituted by $hd\nu$. We also substitute the arguments E_{D^*} and E_A by ν_{D^*} and ν_A, respectively:

$$k_{EnT} = \frac{1}{\hbar^2} \left(\frac{\kappa_{D^*A}}{4\pi\varepsilon_0 n^2 R_{D^*A}^3} \right)^2 \int_{\nu_{D^*}} \int_{\nu_A} |\mu_{D^*D}|^2 S_D(\nu_{D^*}) |\mu_{AA^*}|^2 S_A(\nu_A) \delta(\nu_{D^*} - \nu_A) d\nu_{D^*} d\nu_A. \tag{9.61}$$

The connection between the ETDM and the Einstein coefficient for induced absorption or emission B_{D*D} and that for spontaneous emission, A_{D*D}, can be expressed as follows:

$$B_{D^*D} = \frac{2\pi}{3\hbar^2}\frac{1}{4\pi\varepsilon_0}\frac{1}{n^2}(\mu_{D^*D})^2, \tag{9.62}$$

$$A_{D^*D} = 8\pi\frac{h\nu_{D^*}^3}{c_0^3}n^3 B_{D^*D}, \tag{9.63}$$

where c_0 is the vacuum speed of light. These two relations are used, after some lengthy reasoning reported in the Appendix, to turn the rather complicated expression for the EnT rate constant (Eq. 9.61) into the practical form (Eq. 9.64), which was derived by Theodor Förster.[2] We distinguish between the intrinsic decay of the donor $\tau_{\text{in},D^*} = 1/k_F^{D^*}$ and the decay of the donor in the absence of FRET $\tau_{0,D^*} = \phi_{0,D^*}\tau_{\text{in},D^*}$, where ϕ_{0,D^*} is the luminescence quantum yield of the donor in the absence of FRET:

$$k_{EnT} = TF\cdot c_0^4\frac{\kappa_{D^*A}^2}{n^4 R_{D^*A}^6}\frac{1}{\tau_{in,D^*}}\int_\nu S_D(\nu)\frac{\varepsilon_A(\nu)}{\nu^4}d\nu. \tag{9.64}$$

The value of the Theodor Förster constant TF is given by

$$TF = \frac{9000\ln(10)}{128\pi^5 N_L} \quad \text{or} \quad TF = 8.785\cdot 10^{-25}\ \text{mol}. \tag{9.65}$$

We often prefer to record spectra in wave numbers (cm^{-1}), using the relation between frequency and wave number:

$$\nu = \bar{\nu}c_0. \tag{9.66}$$

The unit of $S(\nu)$ is the same as that of $1/\nu$; see Equation 9.55. Hence, expressing the spectral overlap integral in wave numbers we get

$$k_{EnT} = TF\frac{\kappa_{D^*A}^2}{n^4 R_{D^*A}^6}\frac{\phi_{0,D^*}}{\tau_{0,D^*}}\int_{\bar{\nu}} S_D(\bar{\nu})\frac{\varepsilon_A(\bar{\nu})}{\bar{\nu}^4}d\bar{\nu}. \tag{9.67}$$

This equation leads to the definition of the spectral overlap integral $J_{\bar{\nu}D^*A}$:

$$J_{\bar{\nu}D^*A} = \int_{\bar{\nu}} S_D(\bar{\nu})\frac{\varepsilon_A(\bar{\nu})}{\bar{\nu}^4}d\bar{\nu}. \tag{9.68}$$

The unit of $J_{\bar{\nu}D^*A}$ is ($\text{cm}^3\,\text{M}^{-1}$). The molar extinction coefficient $\varepsilon_A(\bar{\nu})$ is usually expressed in ($\text{M}^{-1}\,\text{cm}^{-1}$) where (M) = ($\text{mol}\,\text{L}^{-1}$). The other units are ($\tau_{0,D^*}$) = (ns), ($R_{DA}$) = (Å), and ($N_L$) = ($\text{mol}^{-1}$). With this, the unit of the EnT rate constant is (ns^{-1}):

$$[k_{EnT}] = \frac{1}{\text{mol}^{-1}} \frac{1}{\text{Å}^6} \frac{1}{\text{ns}} \frac{\text{cm}^3}{\text{mol} \cdot \text{L}^{-1}} 10^{51} = \text{ns}^{-1}. \tag{9.69}$$

The spectral overlap $J_{\bar{\nu}D^*A}$ is often expressed in ($\text{cm}^6\text{mol}^{-1}$). Using this, we write the Förster equation as follows:

$$k_{EnT} = TF \cdot 10^{-3} \frac{\kappa^2_{D^*A}}{n^4 R^6_{D^*A}} \frac{\phi_{0,D^*}}{\tau_{0,D^*}} J_{\bar{\nu}D^*A}. \tag{9.70}$$

The efficiency of FRET depends on the inverse sixth power of the intermolecular separation, making it useful over distances in the range of 1.5–10 nm. A condition for FRET is that the absorption spectrum of the acceptor overlaps with the luminescence spectrum of the donor. We address some consequences of this important equation in the following sections.

Förster EnT Radius. A special situation occurs at a donor-to-acceptor distance at which the EnT rate is equal to the luminescence decay rate. To analyze this situation for an isotropic three-dimensional system, we compare the luminescence decay rate of an excited molecule to the EnT rate to an acceptor (Eqs. 9.71 and 9.72):

$$\text{Luminescence rate of D}^*: \left(\frac{d\rho_{D^*}}{dt}\right)_{\text{luminescence of } D^*} = -\frac{1}{\tau_{0,D^*}} \rho_{D^*} \tag{9.71}$$

$$\text{EnT rate}: \left(\frac{d\rho_{D^*}}{dt}\right)_{EnT} = -k_{EnT} \rho_{D^*}. \tag{9.72}$$

At a critical distance R_0 between the excited donor D^* and the acceptor A, we observe that the rate at which D^* emits light is equal to the rate at which it transfers its excitation energy to A. This means that the rates (Eqs. 9.71 and 9.72) are equal, and we find:

$$\frac{1}{\tau_{0,D^*}} = k_{EnT}. \tag{9.73}$$

The critical donor-to-acceptor distance at which Equation 9.73 holds is the Förster radius R_0. Inserting Equation 9.70 into Equation 9.73 we find:

$$R_0^6 = TF \frac{\kappa^2_{D^*A}}{n^4} \phi_{0,D^*} J_{\bar{\nu}D^*A}. \tag{9.74}$$

From this, we obtain the Förster radius R_0 for electronic excitation EnT:

$$R_0 = \sqrt[6]{TF \frac{\kappa^2_{D^*A}}{n^4} \phi_{0,D^*} J_{\bar{\nu}D^*A}}. \tag{9.75}$$

R_0 is the donor–acceptor distance at which the probability for EnT in an isotropic three-dimensional environment is 50%. By substituting Equation 9.75 into Equation 9.70, a very useful expression (Eq. 9.76) is obtained. It allows determining the EnT rate constant as a function of distance, provided that we know the spectral overlap and the luminescence decay time of the donor in absence of FRET:

$$k_{EnT} = \frac{1}{\tau_{0,D^*}}\left(\frac{R_0}{R}\right)^6. \tag{9.76}$$

Probability for FRET. We consider the distance dependence of the probability P for EnT. For this we write

$$P = \frac{\left(\frac{d\rho}{dt}\right)_{EnT}}{\left(\frac{d\rho}{dt}\right)_{\text{luminescence}} + \left(\frac{d\rho}{dt}\right)_{EnT}}. \tag{9.77}$$

Canceling $(d\rho/dt)_{EnT}$ and using Equations 9.71 and 9.72 gives

$$P = \frac{1}{\frac{1}{\tau_{0,D^*}k_{EnT}} + 1}. \tag{9.78}$$

Using Equation 9.76 we find that the probability for FRET can also be expressed as follows:

$$P = \frac{1}{1 + (R/R_0)^6}. \tag{9.79}$$

This equation illustrates a way to measure distances in macromolecules or biological systems, by simply considering the EnT efficiency between donor and acceptor attached to the investigated object. It is only valid if the involved chromophores are embedded in a three-dimensional isotropic medium. For other conditions, it can be expressed as follows:

$$P = \frac{1}{1 + (R/R_0)^{\alpha}}, \tag{9.80}$$

where the exponent α reflects the dimensionality of the system. It is equal to six for three-dimensional systems, to four for 2D systems and to two in the 1D case. We observe in Figure 9.48 that the probability for FRET falls off more rapidly in lower dimensionality systems at the beginning, but slower at distances larger than the Förster radius. In dye–ZL materials α values of 2 have been observed, which means quasi-1D behavior.[21,33]

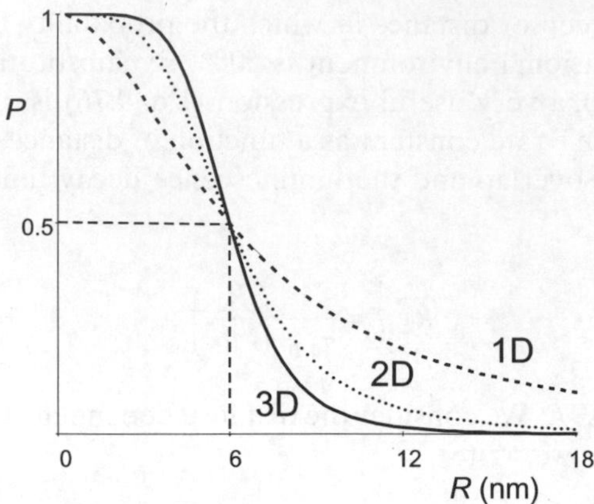

Figure 9.48. Probability for FRET versus the donor–acceptor distance in a 3D, 2D, and 1D system, calculated for a Förster radius of 6 nm.

Selection Rules. There are no strict selection rules for Förster EnT. We can, nevertheless, get a good idea by considering the following proportionality:

$$k_{EnT} \propto \frac{\phi_{D^*}}{\tau_{0,D^*}} \int_{\bar{\nu}} S_D(\bar{\nu}) \frac{\varepsilon_A(\bar{\nu})}{\bar{\nu}^4} d\bar{\nu}. \tag{9.81}$$

The EnT rate constant depends on the extinction coefficient of the acceptor. If a forbidden transition of the acceptor is involved, the FRET rate will be small. If the lifetime of the donor is large, the rate constant will scale correspondingly. Hence, the following reactions are examples of allowed FRET processes[96]:

$$\begin{aligned} &{}^1D^* + {}^1A \rightarrow {}^1D + {}^1A^* \\ &{}^1D^* + {}^3A^* \rightarrow {}^1D + {}^3A^{**} \\ &{}^3D^* + {}^1A \rightarrow {}^1D + {}^1A^* \\ &{}^3D^* + {}^3A^* \rightarrow {}^1D + {}^3A^{**}. \end{aligned} \tag{9.82}$$

We note that the rate constant decreases with increasing luminescence lifetime of the donor. The system gains time for luminescence to occur but also time for other relaxation processes. EnT from a ruthenium stopcock to an organic dye inside the channels of ZL is considered as a good example for an efficient ${}^3D^* + {}^1A \rightarrow {}^1D + {}^1A^*$ FRET process.[42]

Examples for Spectral Overlap $J_{\bar{\nu}D^*A}$ and Förster Radius R_0. The spectral overlap $J_{\bar{\nu}D^*A}$ between the emission of an electronically excited D^* and a molecule A is defined by Equation 9.68. Knowing the spectral overlap integral, we can calculate the Förster radius R_0 according to Equation 9.75. From the

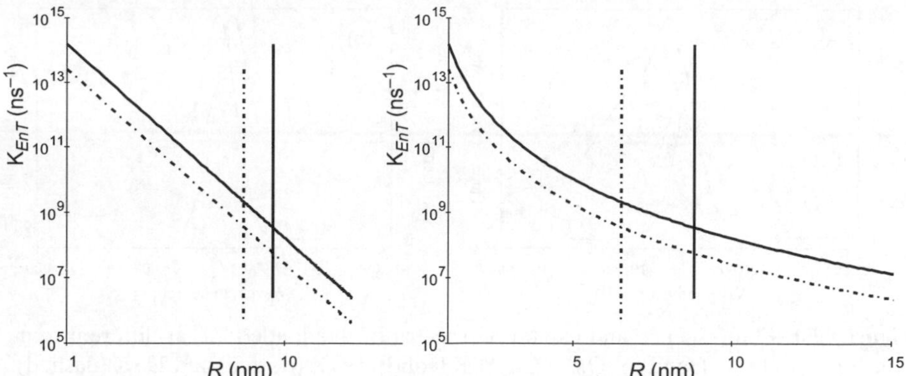

Figure 9.49. Energy transfer rate constant k_{EnT} as a function of the distance R between donor and acceptor for $\kappa^2_{D^*D} = 4$ (solid) and $\kappa^2_{D^*D} = 2/3$ (dash-dot). The vertical lines indicate the value of the Förster radius. On the left, the distance is in a logarithmic scale, while the scale is linear on the right.

Förster radius and the natural luminescence lifetime of the donor, we can calculate the FRET rate constant according to Equation 9.76. We investigate the dependence of the FRET rate constant for a typical case by considering a spectral overlap of $5 \times 10^{-10}\,\text{cm}^6\,\text{mol}$, a luminescence lifetime of 3 ns and an environment of refractive index of 1.4. These values are, for example, representative for Ox1^+ in ZL and the value of $\kappa^2_{\text{Ox1*Ox1}}$ is 4. From this we calculate a Förster radius of $R_0 = 8.8\,\text{nm}$. If we use $\kappa^2_{\text{Ox1*Ox1}} = 2/3$, which can be considered as being representative for a random orientation, we find $R_0 = 6.5\,\text{nm}$. Figure 9.49 illustrates the dependence of the EnT rate constant k_{EnT} on the distance between donor and acceptor for these two $\kappa^2_{\text{Ox1*Ox1}}$ values. The plot starts at a distance of 1 nm. At this distance, Förster's theory may not be valid since the exciton splitting is considerable. We see, however, that rate constants in the order of a few times $10^{13}\,\text{s}^{-1}$ can be derived from Förster's equation at a more realistic donor–acceptor separation of 1.5 nm.

The spectra and hence the spectral overlap $J_{\bar{\nu}D^*A}$ of the molecules depend often considerably on the environment and on the temperature. We illustrate this in Figure 9.50 for Ox^+ and Py^+ in ZL. In these two cases, $J_{\bar{\nu}D^*A}$ changes only slightly despite the increasing resolution of the vibrational structure with decreasing temperature.[12a] This, however, cannot be generalized and should always be checked for the specific systems and conditions of interest.

Luminescence Intensity Dynamics

Consider a set of noninteracting molecules A. If one of them is electronically excited to become A^*, FRET to a neighboring A can take place. This process is repeated until the excitation energy is either captured by a trap or lost by luminescence or by radiationless decay, as illustrated in Figure 9.51.

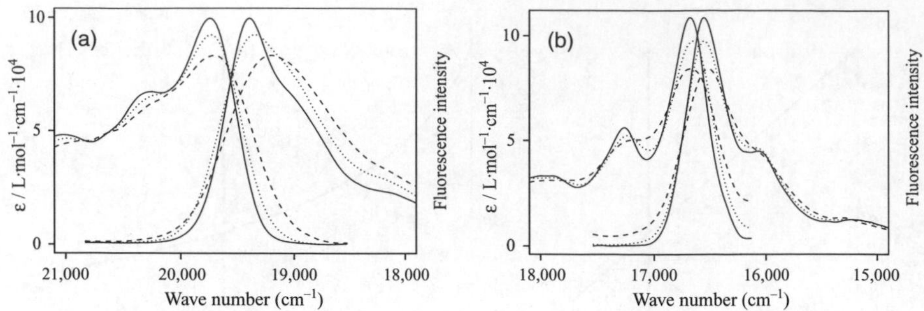

Figure 9.50. Fluorescence and excitation spectra of dye-loaded ZL at different temperatures. (a) Py^+-ZL and (b) Ox^+-ZL at 80 K (solid), 193 K (dotted) and 293 K (dashed). The fluorescence spectra have been scaled to the same height as the corresponding excitation spectra.

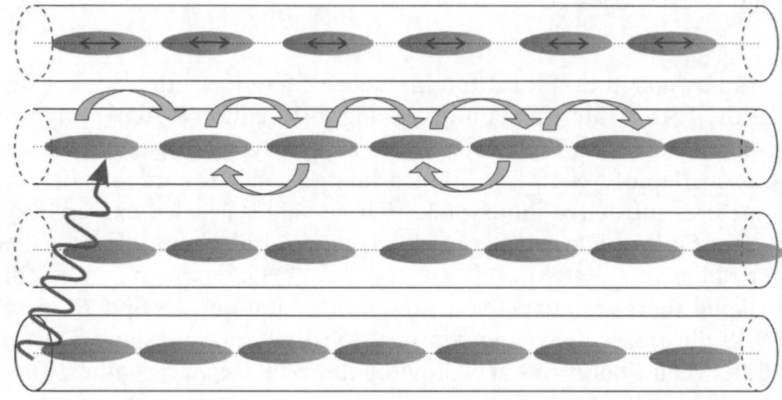

Figure 9.51. Migration of electronic excitation energy among similar molecules.

As a result, electronic excitation energy is transported in space. An important quality of this energy migration is that it cannot be observed by just measuring the luminescence decay of A^*, as we shall discuss now. We assume that a molecule A is excited electronically at time $t = 0$ by absorption of a photon. The probability that A is in the excited state at time t is $\rho_{A^*}(t)$. At $t = 0$, we have $\rho_{A^*}(0) = 1$. The decay of the excitation probability is

$$\frac{d\rho_{A^*}}{dt} = -\frac{1}{\tau_{0,A^*}}\rho_{A^*}. \tag{9.83}$$

Solving this equation gives

$$\rho_{A^*}(t) = \rho_{A^*}(0)e^{-\frac{t}{\tau_{0,A^*}}}. \tag{9.84}$$

τ_{in} is the natural (intrinsic) luminescence lifetime and corresponds to the lifetime in the absence of any other relaxation processes. We assume that all decay processes that result in deactivation of A^* are of first order with the decay constants k_i:

$$\tau_{0,A^*} = \frac{1}{\sum_i k_i} \quad \text{and} \quad \tau_{\text{in},A^*} = \frac{1}{k_F^{A^*}}. \tag{9.85}$$

The luminescence yield ϕ_L can be expressed as follows:

$$\phi_L = \frac{1}{\tau_{\text{in},A^*}} \int_0^\infty \rho_{A^*}(t)dt = \frac{\tau_{0,A^*}}{\tau_{\text{in},A^*}}. \tag{9.86}$$

We number the molecules A with the indices k and j. If there is a possibility for a radiationless transition $A_k^* \rightarrow A_j$ to occur, then the decrease of A_k^* is proportional to the probability $\rho_{A_k^*}$ that A_k is in the excited state. There is an equal increase of the excitation probability $\rho_{A_j^*}$ of the molecule A_j. Considering these processes between all molecules,

$$\frac{d\rho_{A_k^*}}{dt} = \sum_j \bar{F}_{kj}(\rho_{A_j^*} - \rho_{A_k^*}) - \frac{1}{\tau_{0,A^*}}\rho_{A_k^*}, \tag{9.87}$$

where $\bar{F}_{kj}$ is the mean rate constant for the radiationless transfer $A_k^* \rightarrow A_j$. Equation 9.87 can be written in this form, because the mean rate constants for a transfer $A_j^* \rightarrow A_k$ has the same value, $\bar{F}_{kj} = \bar{F}_{jk}$, since we are studying one kind of molecules. Emission and EnT take place in parallel (at the same time). This formulation is different from the mechanism of self-absorption and reemission where the processes are consecutive. This means that the decay of the excitation probability $(1/\tau_{0,A^*})\rho_{A^*k}(t)$ is accompanied by a balancing out of the excitation probability among the individual molecules. In order to describe not only one molecule A_k^* interacting with molecule A_j, but the behavior of an ensemble, we must sum over all cases:

$$\sum_k \frac{d\rho_{A_k^*}}{dt} = \sum_k \left(\sum_j \bar{F}_{kj}(\rho_{A_j^*} - \rho_{A_k^*}) - \frac{1}{\tau_{0,A^*}}\rho_{A_k^*} \right) \tag{9.88}$$

In this equation, the double sum cancels because $\bar{F}_{kj} = \bar{F}_{jk}$ holds:

$$\sum_k \left(\sum_j \bar{F}_{kj}(\rho_{A_j^*} - \rho_{A_k^*}) \right) = 0 \tag{9.89}$$

From this, we find

$$\sum_k \frac{d\rho_{A_k^*}}{dt} = -\frac{1}{\tau_{0,A^*}} \sum_k \rho_{A_k^*}. \tag{9.90}$$

Inserting $\rho_{A_k^*}(t)$ from Equation 9.84 leads to the following expression for the luminescence intensity, which describes the situation after excitation of an ensemble of molecules:

$$\sum_k \frac{d\rho_{A_k^*}}{dt} = -\frac{1}{\tau_{0,A^*}} \sum_k \rho_{A_k^*}(0) e^{-\frac{t}{\tau_{0,A^*}}} = -\frac{1}{\tau_{0,A^*}} e^{-\frac{t}{\tau_{0,A^*}}} \sum_k \rho_{A_k^*}(0). \tag{9.91}$$

Solving this differential equation, we find that the sum of the excitation probabilities for the individual molecules is

$$\sum_k \rho_{A_k^*}(t) = e^{-\frac{t}{\tau_{0,A^*}}} \sum_k \rho_{A_k^*}(0). \tag{9.92}$$

A more familiar way to write this equation is to express the excitation probability of A in terms of concentrations denoted as $[A^*]$:

$$[A^*](t) = [A^*]_0 e^{-\frac{t}{\tau_{0,A^*}}}. \tag{9.93}$$

This means that in a homogeneous system on average, the luminescence decay of the ensemble is not affected by energy migration. Hence, energy migration is not observed in a simple luminescence decay measurement. How can it be observed? There are several possibilities. One of them is based on the fact that under many conditions, excitation energy migration causes a change in the polarization of the emitted light. This can be observed in stationary and in time-resolved luminescence experiments.[2] Another possibility is to add luminescent traps at well-defined positions in space or to observe time- and space-resolved luminescence of an optically anisotropic material.[2,4,21,40,79]

3D Systems of Randomly Mixed Donor and Acceptor Molecules. We assume that N_D donor molecules D and N_{ac} acceptor molecules A are randomly distributed in a large volume, so that effects due to the border of the rigid system can be neglected. The molecules D and A are assumed to be at fixed positions, meaning that they cannot move. A is assumed to absorb light at lower energy than D, so that EnT can occur from D^* to A but not in the reverse direction. This situation is schematically shown in Figure 9.52.

We now discuss the decrease of the excitation probability of the donor. Any acceptor molecule A_j, at distance R_j from D^*, gives an additional channel for relaxation by means of FRET, the rate constant of which is $k_{EnT}(j)$:

$$k_{EnT}(j) = \frac{1}{\tau_{0,D^*}} \left(\frac{R_0}{R_j} \right)^6. \tag{9.94}$$

The same is true for FRET not only to A_j but also to any of the N_{ac} acceptor molecules. From this we find that the excitation probability decrease of D^* can be expressed as follows:

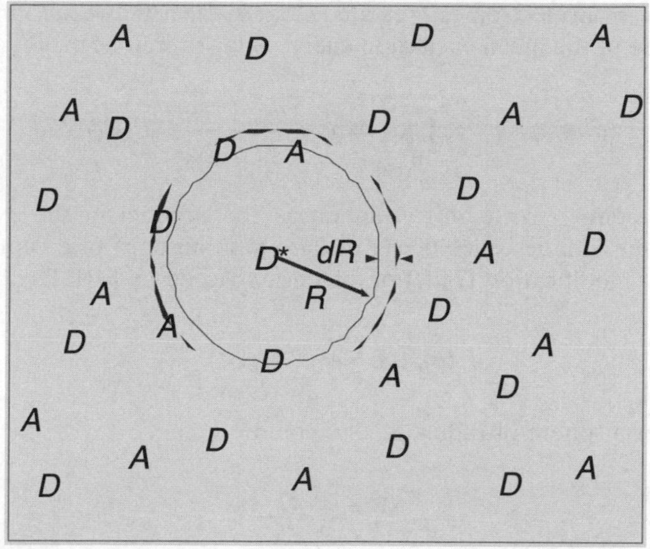

Figure 9.52. Randomly distributed donors D and acceptors A.

$$-\frac{d\rho_{D^*}}{dt} = (k_F + k_{IC})\rho_{D^*} + \left(\sum_{i=1}^{N_{ac}} k_{EnT}(i)\right)\rho_{D^*}, \tag{9.95}$$

$$(k_F + k_{IC}) = \frac{1}{\tau_{0,D^*}}. \tag{9.96}$$

This means that we have $N_{ac} + 1$ independent deactivation channels and that the excitation probability of D^* decays exponentially:

$$\rho_{D^*} = \exp\left[-\left((k_F + k_{IC}) + \sum_{i=1}^{N_{ac}} k_{EnT}(i)\right)t\right]. \tag{9.97}$$

This can also be expressed as follows:

$$(k_F + k_{IC}) + \sum_{i=1}^{N_{ac}} k_{EnT}(i) = \frac{1}{\tau_{0,D^*}} + \frac{1}{\tau_{0,D^*}} \sum_{i=1}^{N_{ac}} \left(\frac{R_0}{R_i}\right)^6. \tag{9.98}$$

We insert this in Equation 9.97 and obtain

$$\rho_{D^*} = e^{-\frac{t}{\tau_{0,D^*}}} \prod_{i=1}^{N_{ac}} \exp\left(-\left(\frac{R_0}{R_i}\right)^6 \frac{t}{\tau_{0,D^*}}\right). \tag{9.99}$$

We do not consider the decay of a single molecule D^* but that of a large number of randomly distributed molecules, which have acceptors at random distances R_i. $P(R_i)dR_i$ is the probability that a given acceptor A_i is located in the environment of an electronically excited D^* at distance between R_i and $R_i + dR_i$. The decrease of the mean excitation energy $<\rho_{D^*}(t)>$ can be expressed as

$$\langle \rho_{D^*}(t) \rangle = e^{-\frac{t}{\tau_{0,D^*}}} \prod_{i=1}^{N_{ac}} \int_0^{R_V} \exp\left(-\left(\frac{R_0}{R_i} \right)^6 \frac{t}{\tau_{0,D^*}} \right) P(R_i) dR_i. \tag{9.100}$$

It is convenient to write this equation as the product of the exponential, describing the luminescence decay of the D^* in absence of acceptors, multiplied by the modification $G_{D^*}(t)$ of the decay caused by FRET:

$$\langle \rho_{D^*}(t) \rangle = e^{-\frac{t}{\tau_{0,D^*}}} G_{D^*}(t). \tag{9.101}$$

We consider a sphere of radius R_V and volume V:

$$V = \frac{4\pi}{3} R_V^3. \tag{9.102}$$

Evaluation of the distribution function $G_{D^*}(t)$ for uniform random distribution of donors and acceptors is explained in the Appendix (Eq. 9.A61) and yields

$$G_{D^*}(t) = e^{-\sqrt{\pi} N_{ac} \left(\frac{R_0}{R_V} \right)^3 \sqrt{\frac{t}{\tau_{0,D^*}}}}. \tag{9.103}$$

The concentration of acceptor molecules c_{ac} in a spherical vessel of radius R_V in mole per liter is

$$c_{ac} = \frac{N_{ac}}{N_L} \left(\frac{4\pi}{3} R_V^3 \right)^{-1}, \tag{9.104}$$

where N_{ac} is the number of acceptors and N_L is Avogadro's number. Using the definition

$$2\gamma = \sqrt{\pi} N_{ac} \left(\frac{R_0}{R_V} \right)^3 \quad \text{or} \quad 2\gamma = \sqrt{\pi} c_{ac} N_L \frac{4\pi}{3} R_0^3, \tag{9.105}$$

we obtain

$$G_{D^*}(t) = e^{-2\gamma \left(\frac{t}{\tau_{0,D^*}} \right)^{1/2}}. \tag{9.106}$$

Excitation Probability of the Acceptor. The time evolution of the acceptor luminescence intensity is easy to describe in the simple case where the donors and acceptors can be regarded as being randomly distributed. We investigate the following processes:

$$D \longrightarrow D^* \qquad \delta\text{-pulse,} \tag{9.107}$$

$$A + D^* \xrightarrow{k_{EnT}} A^* + D \qquad \text{FRET,} \tag{9.108}$$

$$D^* \xrightarrow{k_L^D} D + h\nu \qquad \text{luminescence of the donor,} \tag{9.109}$$

$$A^* \xrightarrow{k_L^A} A + h\nu \qquad \text{luminescence of the acceptor.} \tag{9.110}$$

The FRET rate can in principle be determined from the increase in acceptor fluorescence following pulse excitation of the donor. The change in excited acceptor concentration obeys the following differential equations:

$$\frac{d[A^*]}{dt} = k_{EnT}[D^*] - \frac{1}{\tau_{0,A^*}}[A^*], \tag{9.111}$$

$$\frac{d[D^*]}{dt} = -\left(k_{EnT} + \frac{1}{\tau_{0,D^*}}\right)[D^*]. \tag{9.112}$$

Solution of these equations is, with the initial condition $[D^*](t=0) = [D^*]_0$ and the abbreviation $\tau_{D^*} = 1/(k_L^D + k_{EnT})$,

$$[A^*](t) = \left\{\frac{[D^*]_0 k_{EnT}}{1/\tau_{D^*} - 1/\tau_{0,A^*}}\right\} e^{-\frac{t}{\tau_{0,A^*}}} - \frac{[D^*]_0 k_{EnT}}{1/\tau_{D^*} - 1/\tau_{0,A^*}} e^{-\frac{t}{\tau_{D^*}}}. \tag{9.113}$$

This simplifies to Equation 9.114, if no acceptors are directly excited with the pulse:

$$[A^*](t) = \frac{[D^*]_0 k_{EnT}}{1/\tau_{D^*} - 1/\tau_{0,A}}\left(e^{-\frac{t}{\tau_{0,A^*}}} - e^{-\frac{t}{\tau_{D^*}}}\right), \tag{9.114}$$

$$[D^*](t) = [D^*] e^{-\frac{t}{\tau_{D^*}}}. \tag{9.115}$$

The luminescence intensity of the donor and acceptor are directly proportional to the time-dependent concentrations $[D^*](t)$ and $[A^*](t)$, respectively. Calculated intensities are shown in Figure 9.53, for the values indicated in the figure caption. The rise of acceptor luminescence intensity is very fast if the EnT rate constant becomes much larger than the luminescence decay rate of the acceptor.

Lower Dimensionality Systems. Lower dimensionality is expected and has also been observed in dye–ZL materials.[21,33,77] Equation 9.106 can be extended, so that it applies for any dimensionality between three and one for randomly distributed donors and acceptors, by introducing a parameter δ. δ is equal to ½ in 3D systems. It becomes smaller as the dimensionality decreases: 1/3 for 2D and 1/6 for 1D systems, respectively[97]:

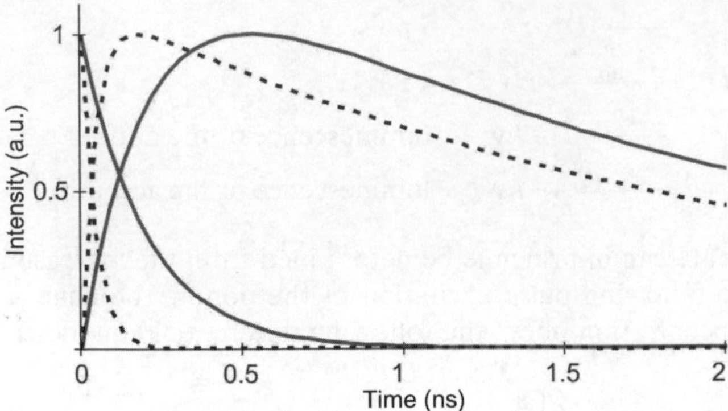

Figure 9.53. Luminescence intensity of donors and acceptors, calculated for the parameters $k_D^L = 0.5\times10^9\ \text{s}^{-1}$, $k_A^L = 0.9k_D^L$, and for the two parameter sets: $k_{EnT} = 10\times k_A^L$ (solid lines) and $k_{EnT} = 50\times k_A^L$ (dashed lines).

$$G_{D*}(t) = e^{-2\gamma\left(\frac{t}{\tau_{0,D*}}\right)^{\delta}}. \tag{9.116}$$

We define the critical concentration of acceptor molecules c_0 for a situation where γ is 1. This means that the critical concentration of acceptor molecules is represented by

$$c_0 = \frac{\alpha}{N_L}\left(\frac{4\pi}{3}R_0^3\right)^{-1}. \tag{9.117}$$

The parameter α depends on the distribution of the relative orientation of the ETDM. It is equal to $2/\sqrt{\pi}$ for random orientation but can also be smaller or larger, depending on the situation. Hence, γ can be expressed as the ratio between the actual concentration of acceptor molecules and the critical concentration c_0:

$$\gamma = \frac{c_{ac}}{c_0}. \tag{9.118}$$

We wonder about the intensity of the donor emission at constant donor but varying acceptor concentration, expressed by means of the parameter γ. This can be done by investigating the fluorescence yield as a function of γ, for otherwise constant conditions. The fluorescence yield ϕ_{D*} of the donor is proportional to the integral over the whole time range from $t = 0$ to $t = \infty$, multiplied with the proportionality constant C. We compare in a 3D system ϕ_{D*} with the luminescence yield in the absence of acceptors $\phi_{0,D*}$:

$$\phi_{D*} = C\int_0^{\infty} e^{-\frac{t}{\tau_{0,D*}}-2\gamma\sqrt{\frac{t}{\tau_{0,D*}}}}\,dt, \tag{9.119}$$

$$\phi_{0,D^*} = C\int_0^\infty e^{-\frac{t}{\tau_{0,D^*}}}dt = C\tau_{0,D^*}. \tag{9.120}$$

Dividing Equation 9.119 by Equation 9.120 we find

$$\frac{\phi_{D^*}}{\phi_{0,D^*}}(\gamma) = \frac{1}{\tau_{0,D^*}}\int_0^\infty e^{-\frac{t}{\tau_{0,D^*}}-2\gamma\sqrt{\frac{t}{\tau_{0,D^*}}}}dt. \tag{9.121}$$

Evaluation and using the abbreviation

$$g(\gamma) = \frac{\phi_{D^*}}{\phi_{0,D^*}}(\gamma) \tag{9.122}$$

gives

$$g(\gamma) = 1 - \gamma\sqrt{\pi}\exp(\gamma^2)(1 - erf(\gamma)). \tag{9.123}$$

The behavior of this donor quantum yield ratio is illustrated in Figure 9.54 where we plot $g(\gamma)$ versus γ and versus $\sqrt[3]{\gamma}$. It is not surprising, that the plot shown on the right side of Figure 9.54 resembles the behavior of the FRET probability P (Eq. 9.79 and Fig. 9.48) versus distance, because $\sqrt[3]{\gamma}$ is proportional to the donor–acceptor distance R. Similar reasoning holds for lower dimensionality systems.

We reported in Figure 9.17 that the dyes Ox^+ and Py^+ can be inserted into the channels of ZL as a random mixture and that the loading can be varied to a large extent; see Figures 9.16 and 9.17. The time-resolved luminescence behavior of this system has been analyzed in detail. The luminescence intensity

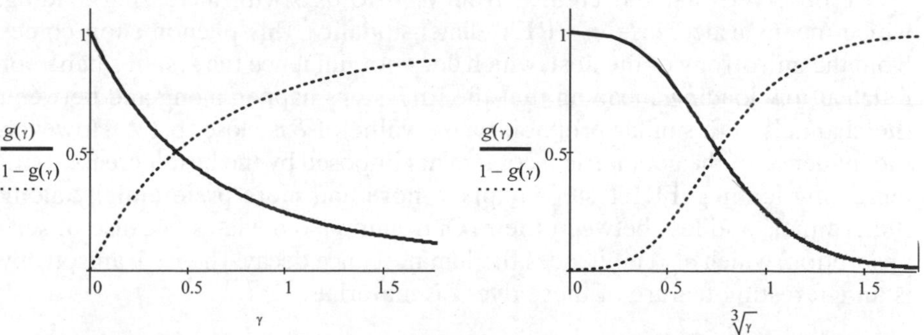

Figure 9.54. Plot of the ratio of fluorescence quantum yield $g(\gamma)$ versus γ (left) and versus $\sqrt[3]{\gamma}$ (right). We see on the left that the fluorescence quantum yield of the donor (solid line) decreases fast while that of the acceptor (dashed line) increases accordingly.

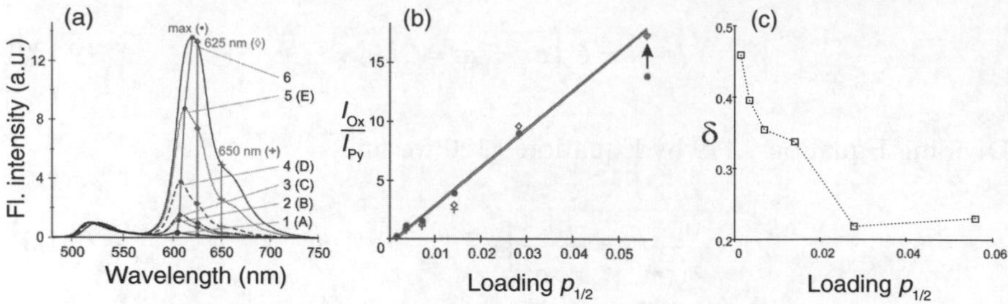

Figure 9.55. FRET between randomly mixed-dye molecules in the channels of ZL. (a) Fluorescence spectra (after excitation of Py^+ at 450 nm, scaled to the same peak height of the Py^+ emission at 520 nm) of a series of six samples containing a 1:1 mixture of Py^+ and Ox^+ with the following loadings for each dye: 1, $p_{1/2} = 0.0014$; 2, $p_{1/2} = 0.0035$; 3, $p_{1/2} = 0.007$; 4, $p_{1/2} = 0.014$; 5, $p_{1/2} = 0.028$; and 6, $p_{1/2} = 0.056$. The letters in brackets (A–E) indicate that these spectra correspond to samples A–E in Figure 9.17a. (b) Acceptor-to-donor emission ratio versus loading is linear for this system. The ratio is plotted at three acceptor wavelengths: the maximum of the emission (•), 625 nm (□), and 650 nm (+). The points represent the experimental data and the lines were fitted according to Equation 9.124. The experimental ratio for the sample with the highest loading, $p_{1/2} = 0.056$, is much lower due to self-absorption. (c) Dimensionality. The parameter δ is plotted versus the loading. In all three plots, the value at the last point ($p_{1/2} = 0.056$) is affected by self-absorption.[77]

decay of the Py^+ molecules, acting as donors, can be expressed by the following stretched exponential equation[77]:

$$I_{D^*}(t) = e^{-\frac{t}{\tau_{0,D^*}}} e^{-2\gamma\left(\frac{t}{\tau_{0,D^*}}\right)^{\delta}}. \tag{9.124}$$

Results of this study are illustrated in Figure 9.55.

We observed that δ decreases from 0.46 to 0.23 with increasing loading, which means a decrease in FRET dimensionality. This phenomenon comes from the anisotropy of the host, which does not influence the systems behavior much at low loading, meaning that the EnT steps happen along and between the channels with similar probability; the value of δ is close to 1/2. However, the influence of the geometrical constraints imposed by the host increases with increasing loading. FRET steps happen more and more preferentially along the channels and less between them. Only sample 6 deviates because of self-absorption, which also influences the luminescence decay. The EnT anisotropy is an interesting feature of these dye–ZL materials.[21,33,77,79]

Antenna Systems: Organized Chromophores. Antenna systems are supramolecular arrangements in which electronic excitation of molecules occurs in a given volume and the electronic excitation energy is then transported by a radiationless process (near-field interactions) to a well-defined location. A schematic view of an artificial dye–ZL antenna is given in Figure 9.56.

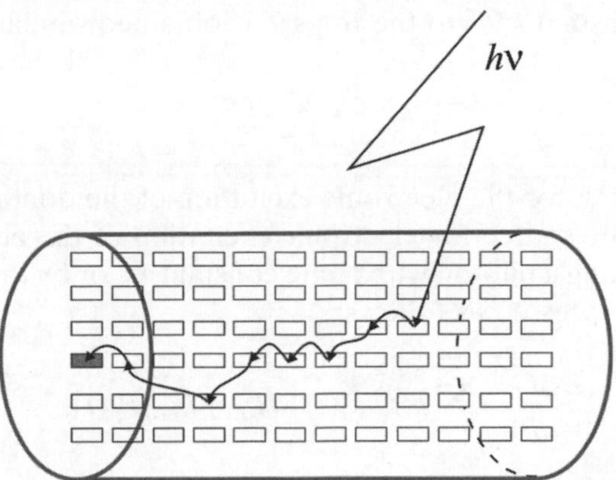

Figure 9.56. Artificial antenna with a single trap. The empty rectangles indicate donor molecules and the gray rectangle is a trap.

We consider a crystal that contains donor molecules in its body and a single acceptor or trap as illustrated in Figure 9.56, or traps on each channel end, as shown in Figures 9.1, 9.18, 9.19, and 9.20. Somewhere in the bulk, a donor is electronically excited by absorbing a photon. We are interested in knowing how fast the excitation energy migrates in the crystal depending on the occupation probability and the characteristics of the dye. We would also like to know how the front–back trapping efficiency depends on these characteristics and on the size of the crystal, and we would like to understand the luminescence decay of the donors and of the traps. Numerical methods are needed to describe the complex situations encountered. We do not know which donor D has been excited. We therefore assume that immediately after irradiation at $t = 0$, all sites i have the same excitation probability $P_i(0)$, while all traps A_T are in the ground state; hence, $P_T(0) = 0$. These probabilities change with time because of energy migration, relaxation processes, and trapping. The excitation probability $P_i(t)$ of the donors is governed by the following master equation[4]:

$$\frac{dP_i}{dt} = \sum_j P_j(t)k_{ji}^{EM} - P_i(t)k_i. \tag{9.125}$$

The sites i are populated by energy migration starting from any other site j with the corresponding rate constants k_{ji}^{EM}. They are depopulated by spontaneous emission, radiationless decay, and energy migration to sites $j \neq T$ and by FRET to trapping sites T. The energy migration rate constant k_i^{EM} is obtained by summing up the individual *rate constants* for the energy migration between an excited donor i and all surrounding acceptors j:

$$k_i^{EnT} = \sum_{i \neq j} k_{ji}^{EM}. \tag{9.126}$$

The EnT rate constant k_i^{EnT} to the traps T is obtained similarly:

$$k_i^{EnT} = \sum_T k_{Ti}^{EM}. \tag{9.127}$$

The traps are a sink for the electronic excitation of the donors. They are considered as the only source for electronic excitation of the acceptors A. They can relax by emitting a photon with a rate constant k_F^A or by thermal relaxation k_{rd}^A. This can be expressed as follows:

$$\frac{dP_T}{dt} = \sum_j k_{iT}^{EnT} P_i(t) - \left(k_F^A + k_{rd}^A\right) P_T(t). \tag{9.128}$$

Thus, the time evolution of the excitation probabilities $P_i(t)$ and $P_T(t)$ is described by Equations 9.125 and 9.128. This allows us to write the excitation probabilities for time $t + \Delta t$ as follows:

$$P_i(t+\Delta t) = P_i(t) + \frac{dP_i(t)}{dt}\Delta t. \tag{9.129}$$

Numerical solutions of these equations have been given based on the Markow chain method[4] and based on Monte Carlo simulation.[79] The general result is that quasi-1D excitation energy migration parallel to the channel axis is favored, but that it is more important for large than for nanosized crystals.

SUMMARY

Structurally organized and functionally integrated artificial systems that are capable of elaborating the energy and information input of photons to perform functions, such as processing and storing information, sensing microscopic environment on a nanoscale level, or transforming and storing solar energy, are fascinating topics of modern photochemistry. Such structures have been realized by means of host–guest chemistry where the hosts are ZL crystals of different sizes and morphologies and where the guests are dye molecules. Size, shape, and surface composition of the host play a decisive role. Its base and coat have distinctively different chemical properties. The guests, organic dye molecules, or complexes are well oriented inside the channels and can be organized into distinctive patterns. ZL crystals containing oriented fluorophores in their parallel nanochannels possess remarkable fluorescent properties. Communication between the chromophores located inside of the host and the outside world is realized via stopcock molecules of different shapes, sizes, and nature.

The preparation of different dye–ZL materials is described, along with Förster EnT experiments carried out with them. The theoretical background of FRET processes and of exciton coupling in the ID channel system is explained in detail. Further possibilities to increase the supramolecular organization into more advanced structures are discussed: the first unidirectional antenna system on a macroscopic level, organization of crystals, and communication of the crystals' interior with the environment. The robust ZL framework can be selectively functionalized at the external crystal surfaces and/or the pore entrances. This versatility can be exploited to establish communication pathways by means of EnT. Finally, the embedding of dye–ZL crystals into organic matrices opens possibilities for the development of devices such as novel LSCs or sensitized solar cells. Host–guest nano- or micro-objects can be organized into different patterns so that systems with distinct macroscopic properties result. The new materials and procedures can be applied for the development of optical and electro-optical devices, for analytical purposes, and for diagnostics. They are very attractive for investigations by means of linear and nonlinear fluorescence microscopy techniques, and they provide photochemists with fascinating possibilities for solving old problems and creating new options.

APPENDIX

Dipole–Dipole Interaction and the Orientation Factor

Dipole–dipole interactions are presented in a form that is adapted to our needs. The interaction energy V_{dd} between two dipoles μ_1 and μ_2 at distance R can be expressed as follows:

$$V_{dd} = \frac{1}{4\pi\varepsilon_0}\frac{\mu_1 \cdot \mu_2 - 3(\mathbf{n}\cdot\mu_1)(\mathbf{n}\cdot\mu_2)}{R^3}, \tag{9.A1}$$

where $\mathbf{n}$ is a unit vector in the direction of R. The dipole–dipole interaction is often the first term in the Taylor series expansion of the electrostatic interaction between two neutral molecules. We express the interaction energy V_{dd} between two dipoles in polar coordinates. We assume two fixed positive charges e_a and e_b at distance R, each of them compensated by a negative charge $-e_a$ and $-e_b$ as illustrated in Figure 9.A1:

The potential energy is[98]

$$V = \left\{ e_a e_b \left(\frac{1}{R} + \frac{1}{r_{12}} - \frac{1}{r_{1b}} - \frac{1}{r_{2a}} \right) - \frac{e_a^2}{r_{1a}} - \frac{e_b^2}{r_{2b}} \right\} \frac{1}{4\pi\varepsilon_0}. \tag{9.A2}$$

The first four terms in this expression represent the mutual interaction of two dipoles, and it is convenient to derive an approximate expression for this interaction by assuming that R is constant (since R changes only slowly in comparison to the movements of the electrons) and that the distances between e_a and $-e_a$, and also between e_b and $-e_b$, are very short with respect to the

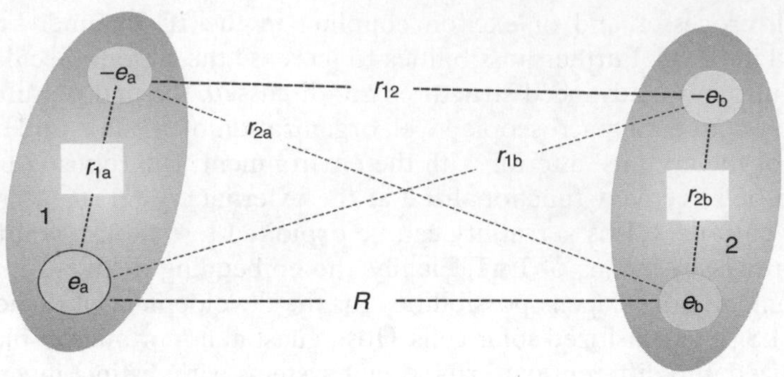

Figure 9.A1. Interaction distances between the charges in two dipoles separated by a distance R.

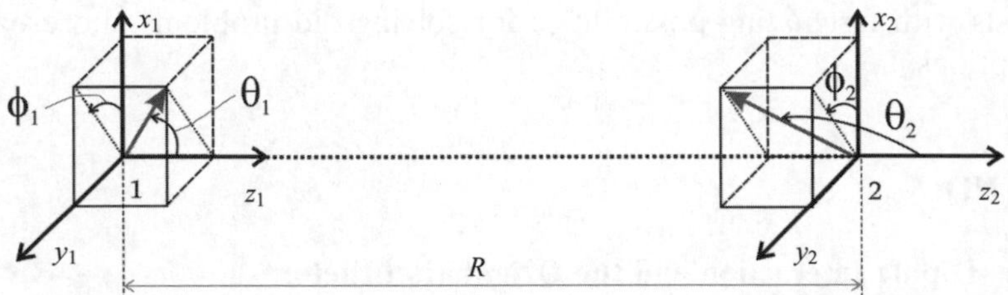

Figure 9.A2. Coordinate system used to describe the interaction between two dipoles at distance R.

distance R between the two objects 1 and 2 ($R >> r_{1a}$ and $R >> r_{2b}$). It also means, that the last two terms are of constant value and do not depend on R and, hence, also not on r_{12}, r_{2a}, and r_{1b}. This is the condition for a dipole–dipole interaction. It follows that the interaction energy V_{dd} between two dipoles can be expressed as

$$V_{dd} = \frac{e_a e_b}{4\pi\varepsilon_0}\left(\frac{1}{R} + \frac{1}{r_{12}} - \frac{1}{r_{1b}} - \frac{1}{r_{2a}}\right). \tag{9.A3}$$

This expression can be expanded in a series along the Cartesian coordinates (Fig. 9.A2):

$$V_{dd} = \frac{e_a e_b}{4\pi\varepsilon_0}\frac{1}{R^3}(x_1x_2 + y_1y_2 - 2z_1z_2) + \text{terms in } R^{-4} + \ldots \tag{9.A4}$$

Neglecting higher terms, we get

$$V_{dd} = \frac{e_a e_b}{4\pi\varepsilon_0}\frac{1}{R^3}(x_1x_2 + y_1y_2 - 2z_1z_2). \tag{9.A5}$$

This equation is equivalent to Equation 9.A1. We now express the dependence of V_{dd} on the coordinates (R, θ_1, θ_2, ϕ_{12}):

$$\begin{aligned} z_1 &= l_1 \cos\theta_1 & z_2 &= l_2 \cos\theta_2, \\ x_1 &= l_1 \sin\theta_1 \cos\phi_1 & x_2 &= l_2 \sin\theta_2 \cos\phi_2, \\ y_1 &= l_1 \sin\theta_1 \sin\phi_1 & y_2 &= l_2 \sin\theta_2 \sin\phi_2. \end{aligned} \tag{9.A6}$$

Substituting this into the term in brackets of Equation 9.A5, we obtain

$$x_1x_2 + y_1y_2 - 2z_1z_2 = \\ l_1l_2\{\sin\theta_1 \sin\theta_2 [\cos\phi_1 \cos\phi_2 + \sin\phi_1 \sin\phi_2] - 2\cos\theta_1 \cos\theta_2\}. \tag{9.A7}$$

This can be simplified by using the relations

$$\cos(\phi_1 - \phi_2) = \cos\phi_1 \cos\phi_2 + \sin\phi_1 \sin\phi_2 \text{ and } \cos\phi_{12} = \cos(\phi_1 - \phi_2), \tag{9.A8}$$

which leads to the following expression:

$$V_{dd} = \frac{e_a e_b}{4\pi\varepsilon_0} \frac{l_1 l_2}{R^3} (\sin\theta_1 \sin\theta_2 \cos\phi_{12} - 2\cos\theta_1 \cos\theta_2). \tag{9.A9}$$

By custom, the angle-dependent part V_{dd} is designated with the Greek letter κ. This leads to the final results (Eqs. 9.A10 and 9.A11)

$$\kappa_{12} = \sin\theta_1 \sin\theta_2 \cos\phi_{12} - 2\cos\theta_1 \cos\theta_2, \tag{9.A10}$$

$$V_{dd} = \frac{e_a e_b}{4\pi\varepsilon_0} \frac{l_1 l_2}{R^3} \kappa_{12}. \tag{9.A11}$$

κ_{12} is an orientation factor that describes the dependence of the dipole–dipole interaction energy V_{dd} (R, θ_1, θ_2, ϕ_{12}) on the relative orientation of the two dipoles with respect to each other. The right side of Equation 9.A11 must be divided by n^2, if the system is in an environment of refractive index n. We illustrate values for κ_{12} in Figure 9.A3. It shows that the values vary between 2 and –2. From this follows that κ_{12}^2 has values between 0 and 4.

Relation between the ETDM and the Einstein B Coefficient

Comment on the Einstein coefficients and the relation expressed in Equation 9.61.

Consider two electronic states of a molecule denoted as A and A^* (Fig. 9.A4). Associated with each electronic state is a series of nuclear motion states (vibrational, rotational, translational), which we denote as upper states u for A^* and as lower states l for A. For simplicity, we assume that each state is a singlet. It is also sufficient to consider only the vibrational states. This is not much of a restriction. If degeneracy occurs in either the electronic or

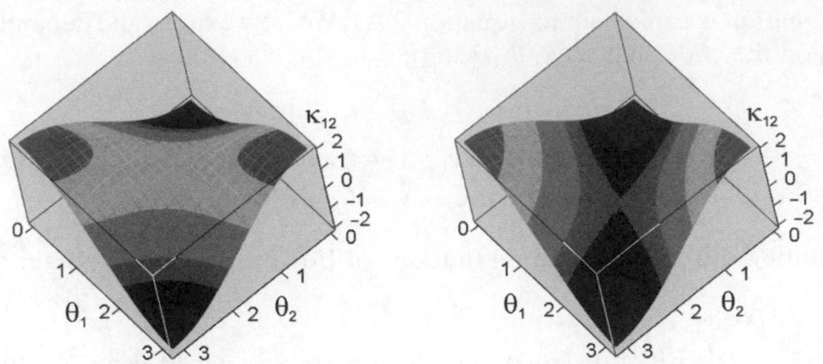

Figure 9.A3. Illustration of the values of κ_{12} for ϕ_{12} = (left) and $\phi_{12} = \pi$ (right).

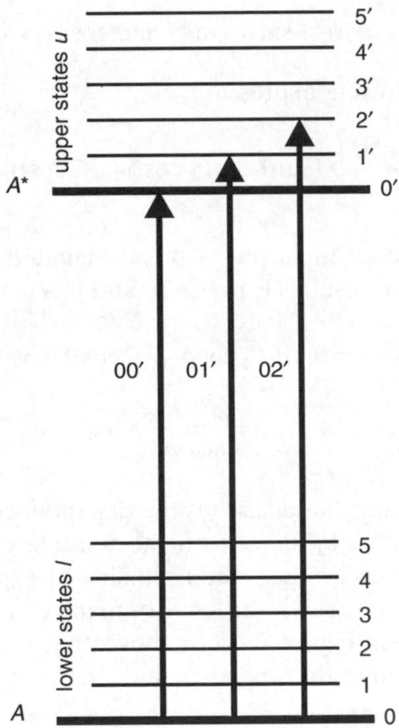

Figure 9.A4. Energy level diagram of a molecule in condensed phase. The numbers 0, 1, 2,... and 0′, 1′, 2′,... denote the vibrational levels in the ground state and in the electronically excited state.

vibrational part, they can be summed over accordingly. In order to derive the above-mentioned equation, it is necessary to shortly comment on the relationship between the Einstein coefficients *A* and *B*. We will then find the relation between the coefficient *B* and the absorption intensity. We follow closely the arguments outlined in the classical paper of Strickler and Berg.[99]

The connection between electronic transition moments and the Einstein coefficient for induced absorption or emission B_{D*D} and that for spontaneous emission, A_{D*D}, can be expressed as follows:

$$B_{D*D} = \frac{2\pi}{3\hbar^2}\frac{1}{4\pi\varepsilon_0}\frac{1}{n^2}(\mu_{D*D})^2, \tag{9.A12}$$

$$A_{D*D} = 8\pi\frac{h\nu_{D*}^3}{c_0^3}n^3 B_{D*D}, \tag{9.A13}$$

$$A_{D*D} = \frac{\nu_{D*}^3}{c_0^3}\frac{32\pi^3}{3\hbar}\frac{n}{4\pi\varepsilon_0}(\mu_{D*D})^2. \tag{9.A14}$$

The coefficient A_{D*D} is equal to the reciprocal intrinsic lifetime of D^*:

$$A_{D*D} = \frac{1}{\tau_{\mathrm{in},D*}}. \tag{9.A15}$$

Inserting this in Equation 9.A14 and rearranging leads to

$$(\mu_{D*D})^2 = \frac{1}{\tau_{\mathrm{in},D*}}\frac{c_0^3}{\nu_{D*}^3}\frac{3\hbar}{32\pi^3}\frac{4\pi\varepsilon_0}{n}. \tag{9.A16}$$

Similarly, as in Equation 9.A12, we can write

$$B_{AA*} = \frac{2\pi}{3\hbar^2}\frac{1}{4\pi\varepsilon_0}\frac{1}{n^2}(\mu_{AA*})^2. \tag{9.A17}$$

Rearranging gives

$$(\mu_{AA*})^2 = \frac{3\hbar^2}{2\pi}4\pi\varepsilon_0 n^2 B_{AA*}. \tag{9.A18}$$

We show in the section "Relation of the Einstein B coefficient and the absorption intensity" that the following relation between the molar extinction coefficient $\varepsilon_A(\nu_A)$ and the Einstein coefficient B_{AA*} holds:

$$B_{AA*}S_A(\nu_A) = \frac{10^3\ln(10)}{hN_L}\frac{c_0}{n}\frac{\varepsilon_A(\nu_A)}{\nu_A}, \tag{9.A19}$$

where N_L is Avogadro's number. From this we find

$$(\mu_{AA*})^2 S_A(\nu_A) = \frac{3\hbar}{4\pi^2}4\pi\varepsilon_0 n c_0 \frac{10^3\ln(10)}{N_L}\frac{\varepsilon_A(\nu_A)}{\nu_A}. \tag{9.A20}$$

In order to evaluate Equation 9.61, we use

$$k_{EnT} = \frac{1}{\hbar^2}\left(\frac{\kappa_{D^*A}}{4\pi\varepsilon_0 n^2 R_{DA}^3}\right)^2 \iint_{\nu_{D^*},\nu_A} Q\delta(\nu_{D^*}-\nu_A)d\nu_{D^*}d\nu_A, \tag{9.A21}$$

where Q is defined as

$$Q = \frac{1}{\hbar^2}\left(\frac{1}{R_{DA}^3}\frac{\kappa_{D^*A}}{4\pi\varepsilon_0 n^2}\right)^2 |\mu_{D^*D}|^2 S_D(\nu_{D^*})|\mu_{AA^*}|^2 S_A(\nu_A). \tag{9.A22}$$

Inserting Equations 9.A16 and 9.A20 into Equation 9.A22 leads to the following expression, because some terms cancel:

$$Q = \frac{10^3\ln(10)}{N_L}\frac{9c_0^4}{128\pi^5}\frac{1}{\tau_{\text{in},D^*}}\frac{\kappa_{D^*A}^2}{n^4 R_{DA}^6}S_D(\nu_{D^*})\frac{\varepsilon_A(\nu_A)}{\nu_{D^*}^3\nu_A}. \tag{9.A23}$$

Inserting this result in Equation 9.A21 and keeping in mind that we can set

$$\int_{\nu_{D^*}} \nu_{D^*}\delta(\nu_{D^*}-\nu_A)d\nu_{D^*} = \nu_A. \tag{9.A24}$$

Using this we find the result we have been searching for:

$$k_{EnT} = \frac{9000\ln(10)}{128\pi^5 N_{\text{L}}}c_0{}^4\frac{\kappa_{D^*A}^2}{n^4 R_{DA}^6}\frac{1}{\tau_{\text{in},D^*}}\int_{\nu} S_D(\nu)\frac{\varepsilon_A(\nu)}{\nu^4}d\nu. \tag{9.A25}$$

Relationship between the Einstein A and B Coefficients. Suppose that a large number of molecules A, immersed in a nonabsorbing medium of refractive index n, is in thermal equilibrium within a cavity in some material at temperature T. The spectral radiant density ($\text{J m}^{-3}\,\text{Hz}^{-1}$) within the medium is given by Planck's blackbody radiation law:

$$U(\nu) = \frac{8\pi h\nu^3 n^3}{c_0^3}\frac{1}{e^{h\nu/(kT)}-1}. \tag{9.A26}$$

By definition of the Einstein transition probability coefficients B_{lu}, the rate of molecules going from state l to state u by absorption of radiation is

$$N_l B_{l\to u} U(\nu_{l\to u}), \tag{9.A27}$$

where N_l is the number of molecules in state l, and $\nu_{l\to u}$ is the frequency of the transition. Molecules in state u can go to state l by spontaneous emission with probability $A_{u\to l}$, or by induced emission with probability $B_{l\to u}U(\nu_{l\to u})$. The rate at which molecules undergo this downward transition is

$$N_b[A_{u\to l} + B_{u\to l}U(\nu_{u\to l})], \tag{9.A28}$$

where $B_{u\to l} = B_{l\to u}$ and $\nu_{u\to l} = \nu_{l\to u}$. At equilibrium, the two rates must be equal:

$$N_l B_{l\to u} U(\nu_{l\to u}) = N_b[A_{u\to l} + B_{u\to l}U(\nu_{u\to l})]; \tag{9.A29}$$

hence,

$$\frac{A_{u\to l}}{B_{u\to l}} = \left(\frac{N_l}{N_b} - 1\right) U(\nu_{l\to u}). \tag{9.A30}$$

According to the Boltzmann distribution law, the ratio of molecules at equilibrium in the two states is related by

$$\frac{N_l}{N_b} = \exp\left[-\frac{h\nu_{l\to u}}{kT}\right]. \tag{9.A31}$$

Substituting this into Equation 9.A30 and using Equation 9.A26 results in the Einstein relation:

$$A_{u\to l} = \frac{8\pi h \nu_{l\to u}^3 n^3}{c_0^3} B_{u\to l}. \tag{9.A32}$$

Relation of the Einstein B Coefficient and the Absorption Intensity. In a common type of absorption measurement, a beam of essentially parallel light is passed through the sample contained in a cell having planar windows perpendicular to the light beam (Fig. 9.A5). It may be assumed, without loss of information relevant to the problem under discussion, that the reflectivity of the windows is zero, so that there is no need to consider the reflected light

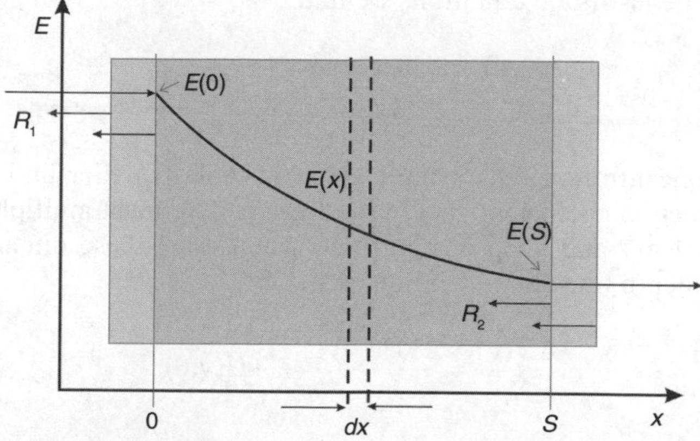

Figure 9.A5. Intensity of a light beam when passing through an absorbing medium. The reflections due to refraction index changes are indicated as R_1 and R_2.

beams R_1 and R_2. It may also be assumed that the light beam has a cross section F of uniform intensity over this area. If $U(\nu,x)$ is the spectral radiant density of the light beam after it has passed a distance x through the sample, the molar extinction coefficient $\varepsilon(x)$ can be defined by

$$\frac{U(\nu,x)}{U(\nu,0)} = 10^{-\varepsilon(\nu)Cx} = e^{-2.303\varepsilon(\nu)Cx}, \tag{9.A33}$$

where C is the concentration in mole per cubic meter and $\varepsilon(\nu)$ in square meter per mole.

The change in spectral radiant density at a short distance dx is

$$-dU(\nu,x)F = 2.303\varepsilon(\nu)U(\nu,0)CFdx. \tag{9.A34}$$

For simplicity, all molecules are assumed to be in the ground vibronic state. This may not be true at ambient temperature; however, this assumption does not affect the results. It would be possible to include many states with appropriate Boltzmann weighting factors. The number of molecules (in moles) in the volume element $dV = Fdx$ is

$$CFdx = \frac{N_l}{N_L}. \tag{9.A35}$$

The number of molecules $\Delta N(\nu)$ excited within the volume dV at position x is proportional to the number of photons, which pass the area F at x. Since they pass with speed of light, we have

$$\Delta N(\nu) = -\frac{dU(\nu,x)F}{h\nu}\frac{c_0}{n}. \tag{9.A36}$$

Combining the last three equations we find

$$\frac{\Delta N(\nu)}{N_l} = \frac{2.303\varepsilon(\nu)}{h\nu}\frac{c_0}{n}\frac{1}{N_\mathrm{L}}U(\nu,0). \tag{9.A37}$$

Since the concentration C is usually given in mole per liter and the molar decadic extinction coefficient $\varepsilon(\nu)$ in $[\mathrm{mol\ L}]^{-1}\mathrm{cm}$, we must multiply the right side by $1000\,\mathrm{L\,m^{-3}}$ and keep in mind that when using these dimensions, the penetration depth must be in centimeters:

$$\frac{\Delta N(\nu)}{N_l} = \frac{2303\varepsilon(\nu)}{h\nu}\frac{c_0}{n}\frac{1}{N_\mathrm{L}}U(\nu,0). \tag{9.A38}$$

This equation gives the probability that a molecule in state l will absorb a quantum of energy $h\nu$ and go to some excited state. To obtain the probability of going to the state u, it must be realized that this can occur with a finite range of frequencies. Hence, the equation must be integrated over this range:

$$\frac{\Delta N(\nu)}{N_l} = \frac{2303c_0}{hnN_{\mathrm{L}}}\int\frac{\varepsilon(\nu)}{\nu}U(\nu,0)d\nu. \tag{9.A39}$$

For simplicity—without loss of information—$U(\nu,0)$ can be assumed to be constant over this range and equal to $U(\nu_{l\to u})$. The value of $\varepsilon(\nu)$ must be only that for the one vibronic transition under investigation; if the spectrum is well resolved, this presents little difficulty, but it can be done in principle for any case:

$$\frac{\Delta N(\nu)}{N_l} = \left[\frac{2303c_0}{hnN_{\mathrm{L}}}\int\varepsilon(\nu)\,d\ln\nu\right]U(\nu_{l\to u}). \tag{9.A40}$$

This equation shows that for a molecule in state l in a beam of parallel light, the probability of undergoing a transition to state u is proportional to $U(\nu_{l\to u})$. The term in brackets corresponds to $B_{l\to u}$ in Equation 9.A27 where the photons might be thought of as arriving at a molecule from any direction, whereas here we consider that they all arrive from the same direction. If, however, the molecules are randomly oriented, the average probability of absorption for a large number of molecules is the same in either case for the same total radiation density.

$$B_{l\to u} = \frac{2303c_0}{hnN_{\mathrm{L}}}\int\varepsilon(\nu)d\ln\nu. \tag{9.A41}$$

The integration is over the chosen vibronic absorption band. In order to be able to easily express this for any transition, we multiply the Einstein coefficient weighted with the shape of the absorption spectrum $S_A(\nu_A)$. Hence, we can write

$$B_{AA^*}S_A(\nu_A) = \frac{2303}{hN_{\mathrm{L}}}\frac{c_0}{n}\frac{\varepsilon_A(\nu_A)}{\nu_A}, \tag{9.A42}$$

$$B_{AA^*} = \frac{2303}{hN_{\mathrm{L}}}\frac{c_0}{n}\int\varepsilon_A(\nu_A)d\ln\nu_A. \tag{9.A43}$$

Evaluation of the Distribution Function $G_{D^*}(t)$

Evaluation of the distribution function $G_{D^*}(t)$ (Eq. 9.106):

$$\langle\rho_{D^*}(t)\rangle = e^{-\frac{t}{\tau_{D^*}}}\prod_{i=1}^{N_{\mathrm{ac}}}\int_0^{R_V}\exp\left(-\left(\frac{R_0}{R_i}\right)^6\frac{t}{\tau_{0,D^*}}\right)P(R_i)dR_i = e^{-\frac{t}{\tau_{0,D^*}}}G_{D^*}(t). \tag{9.A44}$$

For a homogeneous 3D system, we assume that D^* is in the center of a sphere with radius R_V and volume V. We abbreviate the integral part in Equation 9.A44 as $J_i(t)$:

$$\langle \rho_{D^*}(t) \rangle = e^{-\frac{t}{\tau_{0,D^*}}} \prod_{i=1}^{N_{ac}} J_i(t), \tag{9.A45}$$

$$V = \frac{4\pi}{3} R_V^3. \tag{9.A46}$$

We further assume uniform statistical distribution of D and A in the three-dimensional space:

$$P(R)dR = \frac{dV}{V} = \frac{4\pi R^2}{V} dR. \tag{9.A47}$$

This distribution is the same for all acceptors A_i. Hence, we have $J(t) = J_i(t)$:

$$\langle \rho_{D^*}(t) \rangle = e^{-\frac{t}{\tau_{0,D^*}}} J(t)^{N_{ac}}. \tag{9.A48}$$

We now evaluate the integral $J(t)$:

$$J(t) = \frac{4\pi}{V} \int_0^{R_V} \left[\exp\left(-\left(\frac{R_0}{R} \right)^6 \frac{t}{\tau_{0,D^*}} \right) \right] R^2 dR. \tag{9.A49}$$

Using $X = (R_0/R)^6 \, t/\tau_{0,D^*}$, we obtain

$$J(t) = \frac{4\pi}{V} \int_0^{R_V} [\exp(-x)] R^2 dR. \tag{9.A50}$$

Comparing the Förster radius with the dimension of the total volume V,

$$x_V = \left(\frac{R_0}{R_V} \right)^6 \frac{t}{\tau_{0,D^*}} \quad \text{and} \quad V = \frac{4\pi}{3} R_V^3, \tag{9.A51}$$

we see, that x_V is much smaller than 1 since R_V is much larger than R_0 because we are considering a large volume. Substituting V and R by x_V and x, respectively,

$$J(t) = -\frac{1}{2} \sqrt{x_V} \int_{\infty}^{x_V} [\exp(-x)] \frac{dx}{\sqrt{x^3}}. \tag{9.A52}$$

Evaluation of this integral gives

$$J(t) = e^{-x_V} - \sqrt{\pi x_V} \left[1 - erf(\sqrt{x_V}) \right]. \tag{9.A53}$$

Using the approximation $x_V << 1$ as stated above, we can reduce Equation 9.A53 to

$$J(t) = 1 - \sqrt{\pi x_V}. \tag{9.A54}$$

We should keep in mind that this result may not be correct in a confined system. Inserting now Equation 9.A54 into Equation 9.A48 gives

$$\langle \rho_{D^*}(t) \rangle = e^{-\frac{t}{\tau_{0,D^*}}} \left(1 - \sqrt{\pi x_V}\right)^{N_{ac}}. \tag{9.A55}$$

Since it has been assumed that x_V is much smaller than 1, we can write

$$\left(1 - \sqrt{\pi x_V}\right)^{N_{ac}} = 1 - N_{ac}\sqrt{\pi x_V}. \tag{9.A56}$$

Equation 9.A56 can be approximated as follows for sufficiently large N_{ac}:

$$1 - N_{ac}\sqrt{\pi x_V} = e^{-N_{ac}\sqrt{\pi x_V}}. \tag{9.A57}$$

This leads to

$$\langle \rho_{D^*}(t) \rangle = e^{-\frac{t}{\tau_{0,D^*}} - \sqrt{\pi} N_{ac} \left(\frac{R_0}{R_V}\right)^3 \sqrt{\frac{t}{\tau_{0,D^*}}}}. \tag{9.A58}$$

The concentration of acceptor molecules in a spherical vessel of radius R_V in mole per liter is

$$c_{ac} = \frac{N_{ac}}{N_L} \frac{4\pi}{3R_V^3}. \tag{9.A59}$$

Using the definition

$$2\gamma = \sqrt{\pi} N_{ac} \left(\frac{R_0}{R_V}\right)^3 = \sqrt{\pi} c_{ac} N_L \frac{4\pi}{3} R_0^3, \tag{9.A60}$$

we obtain

$$\langle \rho_{D^*}(t) \rangle = e^{-\frac{t}{\tau_{0,D^*}}} e^{-2\gamma\sqrt{\frac{t}{\tau_{0,D^*}}}} = e^{-\frac{t}{\tau_{0,D^*}}} G_{D^*}(t). \tag{9.A61}$$

It is useful to define the critical concentration c_0 of acceptor molecules for a situation where γ is equal to 1. This means that the critical concentration of acceptor molecules is represented by

$$c_0 = \frac{3}{2\pi^{3/2}} \frac{1}{N_L R_0^3}. \tag{9.A62}$$

REFERENCES

1. (*a*) Pullerits, T.; Sundström, V. *Acc. Chem. Res.* **1996**, *29*, 381; (*b*) Blankenship, R.E. *Molecular Mechanisms of Photosynthesis*. Oxford: Blackwell Science, **2002**; (*c*) Balzani, V.; Credi, A.; Venturi, M. *ChemSusChem* **2008**, *1*, 26; (*d*) Benniston, A.C.; Harriman, A. *Mater. Today* **2008**, *11*, 26.
2. (*a*) Förster, T. *Ann. Phys. (Leipzig)* **1948**, *2*, 55; (*b*) Förster, T. *Fluoreszenz Organischer Verbindungen*. Göttingen: Vandenhoeck & Ruprecht, **1951**.
3. Calzaferri, G.; Gfeller, N. *J. Phys. Chem.* **1992**, *96*, 3428.
4. Gfeller, N.; Calzaferri, G. *J. Phys. Chem. B* **1997**, *101*, 1396.
5. Turro, N.J. *Acc. Chem. Res.* **2000**, *33*, 637.
6. Bein, T. In: Cejka, J.; van Bekkum, H.; Corma, A.; Schüth, F., editors. *Studies in Surface Science and Catalysis, Vol. 168: Introduction to Zeolite Science and Practice*, 3rd edition. Amsterdam: Elsevier, **2007**, pp. 611–657.
7. Ramamurthy, V. Photoprocesses of organic molecules included in zeolites. In: Ramamurthy, V., editor. *Photochemistry in Organized and Constrained Media.* New York: VCH Publishers, **1991**, pp. 429–493.
8. Megelski, S.; Calzaferri, G. *Adv. Funct. Mater.* **2001**, *11*, 277.
9. (*a*) Zabala Ruiz, A.; Brühwiler, D.; Ban, T.; Calzaferri, G. *Monatshefte Chemie.* **2005**, *136*, 77; (*b*) Zabala Ruiz, A.; Brühwiler, D.; Dieu, L.-Q.; Calzaferri, G. In: Schubert, U.; Hüsing, N.; Laine, R., editors. *Materials Syntheses, a Practical Guide.* Wien: Springer, **2008**, pp. 9–19.
10. (*a*) Ernst, S.; Weitkamp, J. *Catal. Today* **1994**, *19*, 27; (*b*) Tsapatsis, M.; Lovallo, M.; Okubo, T.; Davis, M.E.; Sadakata, M. *Chem. Mater.* **1995**, *7*, 1734; (*c*) Lee, Y.-J.; Lee, J.S.; Yoon, K.B. *Micropor. Mesopor. Mater.* **2005**, *80*, 237; (*d*) Ban, T.; Saito, H.; Naito, M.; Ohya, Y.; Takahashi, Y. *J. Por. Mater.* **2007**, *14*, 119; (*e*) Brent, R.; Anderson, M.W. *Angew. Chem. Int. Ed. Engl.* **2008**, *47*, 5327.
11. (*a*) Breck, D.W. *Zeolite Molecular Sieves*. New York: John Wiley & Sons, **1974**; (*b*) Baerlocher, Ch., and McCusker, L.B., Database of zeolite structures, **1996**, http://www.iza-structure.org/databases/; (*c*) Baerlocher, C.; Meier, W.M.; Olson, D.H. *Atlas of Zeolite Framework Types*, 5th edition. Amsterdam: Elsevier, **2001**; (*d*) Ohsuna, T.; Slater, B.; Gao, F.; Yu, J.; Sakamoto, Y.; Zhu, G.; Terasaki, O.; Vaughan, D.E.W.; Qiu, S.; Catlow, C.R.A. *Chem. Eur. J.* **2004**, *10*, 5031; (*e*) Larlus, O.; Valtchev, V.P. *Chem. Mater.* **2004**, *16*, 3381; (*f*) Lee, Y.; Kao, C.-C.; Kim, S.J.; Lee, H.-H.; Lee, D.R.; Shin, T.J.; Choi, J.Y. *Chem. Mater.* **2007**, *19*, 6252.
12. (*a*) Calzaferri, G.; Brühwiler, D.; Megelski, S.; Pfenniger, M.; Pauchard, M.; Hennessy, B.; Maas, H.; Devaux, A.; Graf, U. *Solid State Sci.* **2000**, *2*, 421; (*b*) Calzaferri, G.; Pauchard, M.; Maas, H.; Huber, S.; Khatyr, A.; Schaafsma, T. *J. Mater. Chem.* **2002**, *12*, 1; (*c*) Calzaferri, G.; Maas, H.; Pauchard, M.; Pfenniger, M.; Megelski, S.; Devaux, A. In: Neckers, D.C.; von Bünau, G.; Jenks, W.S., editors. *Advances in Photochemistry, Vol. 27.* Hoboken, NJ: John Wiley & Sons, **2002**, pp. 1–50.
13. Calzaferri, G.; Huber, S.; Maas, H.; Minkowski, C. *Angew. Chem. Int. Ed. Engl.* **2003**, *42*, 3732.
14. (*a*) Brühwiler, D.; Calzaferri, G. *Micropor. Mesopor. Mater.* **2004**, *72*, 1; (*b*) Wang, Y.; Li, H.; Gu, L.; Gan, Q.; Li, Y.; Calzaferri, G. *Micropor. Mesopor. Mater.* **2009**, *121*, 1.

15. (*a*) Schulz-Ekloff, G.; Wöhrle, D.; van Duffel, B.; Schoonheydt, R.A. *Micropor. Mesopor. Mater.* **2002**, *51*, 91; (*b*) Hashimoto, S. *J. Photochem. Photobiol. C Photochem. Rev.* **2003**, *4*, 19; (*c*) Chrétien, M.N. *Pure Appl. Chem.* **2007**, *79*, 1; (*d*) Tsotsalas, M.; Busby, M.; Gianolio, E.; Aime, S.; De Cola, L. *Chem. Mater.* **2008**, *20*, 5888; (*e*) Abeykoon, A.M.M.; Castro-Colin, M.; Anokhina, E.V.; Iliev, M.N.; Donner, W.; Jacobson, A.J.; Moss, S.C. *Phys. Rev. B* **2008**, *77*, 075333; (*f*) Hashimoto, S.; Yamaji, M. *Phys. Chem. Chem. Phys.* **2008**, *10*, 3124; (*g*) Zhu, J.; Huang, Y. *J. Phys. Chem. C* **2008**, *112*, 14241; (*h*) Busby, M.; Kerschbaumer, H.; Calzaferri, G.; De Cola, L. *Adv. Mater.* **2008**, *20*, 1614; (*i*) Brühwiler, D.; Calzaferri, G. *C. R. Chimie* **2005**, *8*, 391.
16. (*a*) Calzaferri, G. EU and US patents. EP 1335879B1, US 6932919B2, US7327012, **2008**; (*b*) Maas, H.; Calzaferri, G. *Angew. Chem. Int. Ed. Engl.* **2002**, *41*, 2284.
17. (*a*) Andrews, D.L.; Leeder, J.M. *J. Chem. Phys.* **2009**, *130*, 184504; (*b*) Beljonne, D.; Curutchet, C.; Scholes, G.D.; Silbey, R.J. *J. Phys. Chem. B* **2009**, *113*, 6583; (*c*) Scholes, G.D.; Curutchet, C.; Mennucci, B.; Cammi, R.; Tomasi, J. *J. Phys. Chem. B* **2007**, *111*, 6978; (*d*) May, V.; Kühn, O. *Charge and Energy Transfer Dynamics in Molecular Systems*, 2nd edition. Berlin: Wiley-VCH, **2004**; (*e*) Ritz, T.; Damjanović, A.; Schulten, K. *ChemPhysChem* **2002**, *3*, 243; (*f*) Andrews, D.L.; Demidov, A.A. *Resonance Energy Transfer*. New York: John Wiley & Sons, **1999**; (*g*) Wang, X.F.; Herman, B. *Fluorescence Imaging Spectroscopy and Microscopy*. New York: John Wiley & Sons, **1996**; (*h*) Tamai, N.; Yamazaki, T.; Yamazaki, I. *J. Phys. Chem.* **1987**, *91*, 841; (*i*) Anfinrud, P.; Crackel, R.L.; Struve, W.S. *J. Phys. Chem.* **1984**, *88*, 5873; (*j*) Zumofen, G.; Blumen, A. *J. Chem. Phys.* **1982**, *76*, 3713; (*k*) Kuhn, H. *J. Chem. Phys.* **1970**, *53*, 101.
18. Dexter, D.L. *J. Chem. Phys.* **1953**, *21*, 836.
19. Davydov, A.S. *Usp. Fiz. Nauk.* **1964**, *82*, 393.
20. (*a*) McRae, E.G.; Kasha, M. *Physical Progress in Radiation Biology.* New York: Academic Press, **1964**, pp. 23–42; (*b*) Kasha, M. *Radiat. Res.* **1963**, *20*, 55; (*c*) Kobayashi, T. *J-Aggregates*. London: World Scientific, **1996**; (*d*) Lenhard, J.R.; Hein, B.R. *J. Phys. Chem.* **1996**, *100*, 17287; (*e*) Bakalis, L.D.; Knoester, J. *J. Phys. Chem. B* **1999**, *103*, 6620; (*f*) Lebedenko, A.N.; Guralchuk, G.Y.; Sorokin, A.V.; Yefimova, S.L.; Malyukin, Y.V. *J. Phys. Chem. B* **2006**, *110*, 17772.
21. (*a*) Calzaferri, G.; Lutkouskaya, K. *Photochem. Photobiol. Sci.* **2008**, *7*, 879; (*b*) Calzaferri, G. *Il Nuovo Cimento* **2008**, *123B*, 1337.
22. (*a*) Calzaferri, G.; Li, H.; Brühwiler, D. *Chem. Eur. J.* **2008**, *14*, 7442; (*b*) Calzaferri, G.; Kunzmann, A.; Brühwiler, D.; Bauer, C. CH-698333, WO 2010/009560, **2010**.
23. (*a*) Busby, M.; Blum, C.; Tibben, M.; Fibikar, S.; Calzaferri, G.; Subramaniam, V.; De Cola, L. *J. Am. Chem. Soc.* **2008**, *130*, 10970; (*b*) Blum, C.; Cesa, Y.; Escalante, M.; Subramaniam, V. *J. R. Soc. Interface* **2009**, *6*, S35; (*c*) Calzaferri, G.; De Cola, L.; Busby, M.; Blum, C.; Subramaniam, V. UK0812218.6, US/PTO 12/361616. **2010**; (*d*) Calzaferri, G.; Brühwiler, D.; Meng, T.; Dieu, L.-Q.; Malinovskii, V.; Häner, R. *Chem. Eur. J.* **2010**, *16*, 11289; (*e*) Busby, M.; Devaux, A.; Blum, C.; Subramaniam, V.; Calzaferri, G.; De Cola, L. *J. Phys. Chem. C* **2011**, *115*, 5974.
24. Brühwiler, D.; Calzaferri, G.; Torres, T.; Ramm, J.H.; Gartmann, N.; Dieu, L.-Q.; López-Duarte, I.; Martínez-Díaz, M.V. *J. Mater. Chem.* **2009**, *19*, 8040.
25. (*a*) Vohra, V.; Devaux, A.; Dieu, L.-Q.; Scavia, G.; Catellani, M.; Calzaferri, G.; Botta, Ch. *Adv. Mater.* **2009**, *21*, 1146; (*b*) Vohra, V.; Calzaferri, G.; Destri, S.; Pasini, M.; Porzio, W.; Botta, Ch. *ACS Nano* **2010**, *4*, 1409–1416.

26. DeWilde, W.; Peeters, G.; Lunsford, J.H. *J. Phys. Chem.* **1980**, *84*, 2306.
27. (*a*) Corma, A.; Garcia, H. *Eur. J. Inorg. Chem.* **2004**, 1143; (*b*) Lainé, P.; Lanz, M.; Calzaferri, G. *Inorg. Chem.* **1996**, *35*, 3514; (*c*) Leiggener, M.; Calzaferri, G. *ChemPhysChem* **2004**, *5*, 1593.
28. Wang, Y.G.; Guo, Z.; Li, H.R. *J. Rare Earths Spec. Issue* **2007**, *25*, 283.
29. (*a*) Monguzzi, A.; Macchi, G.; Meinardi, F.; Tubino, R.; Burger, M.; Calzaferri, G. *Appl. Phys. Lett.* **2008**, *92*, 123301; (*b*) Li, H.; Cheng, W.; Wang, Y.; Liu, B.; Zhang, W.; Zhang, H. *Chem. Eur. J.* **2010**, *16*, 2125; (*c*) Mech, A.; Monguzzi, A.; Meinardi, F.; Mezyk, J.; Macchi, G.; Tubino, R. *J. Am. Chem. Soc.* **2010**, *132*, 4574.
30. Wang, Y.; Li, H.; Feng, Y.; Zhang, H.; Calzaferri, G.; Ren, T. *Angew. Chem. Int. Ed. Engl.* **2010**, *49*, 1434.
31. Hashimoto, S.; Hagari, M.; Matsubara, N.; Tobita, S. *Phys. Chem. Chem. Phys.* **2001**, *2*, 5043.
32. Albuquerque, R.Q.; Calzaferri, G. *Chem. Eur. J.* **2007**, *13*, 8939.
33. Minkowski, C.; Calzaferri, G. *Angew. Chem. Int. Ed. Engl.* **2005**, *44*, 5329.
34. (*a*) Zabala Ruiz, A.; Li, H.; Calzaferri, G. *Angew. Chem. Int. Ed. Engl.* **2006**, *45*, 5282; (*b*) Calzaferri, G.; Zabala Ruiz, A.; Li, H.; Huber, S. WO 2007/012216, US 60/698,480, **2007**.
35. Calzaferri, G.; Huber, S.; Devaux, A.; Zabala Ruiz, A.; Li, H.; Bossart, O.; Dieu, L.-Q. *Proc. SPIE Org. Optoelectronics Photonics II* **2006**, *6192*, 619216–619211.
36. (*a*) Suárez, S.; Devaux, A.; Bañuelos, J.; Bossart, O.; Kunzmann, A.; Calzaferri, G. *Adv. Funct. Mater.* **2007**, *17*, 2298; (*b*) Calzaferri, G.; Suarez, S.; Devaux, A.; Kunzmann, A.; Metz, H.J. EP1873202, US823975, **2007**.
37. (*a*) Megelski, S.; Lieb, A.; Pauchard, M.; Drechsler, A.; Glaus, S.; Debus, C.; Meixner, A.J.; Calzaferri, G. *J. Phys. Chem. B* **2001**, *105*, 25; (*b*) Gasecka, A.; Dieu, L.-Q.; Brühwiler, D.; Brasselet, S. *J. Phys. Chem. B* **2010**, *114*, 4192; (*c*) Fois, E.; Tabacchi, G.; Calzaferri, G. *J. Phys. Chem. C* **2010**, *114*, 10572.
38. Huber, S.; Zabala Ruiz, A.; Li, H.; Patrinoiu, G.; Botta, Ch.; Calzaferri, G. *Inorg. Chim. Acta* **2007**, *360*, 869.
39. Pauchard, M.; Devaux, A.; Calzaferri, G. *Chem. Eur. J.* **2000**, *6*, 3456.
40. Pauchard, M.; Huber, S.; Méallet-Renault, R.; Maas, H.; Pansu, R.; Calzaferri, G. *Angew. Chem. Int. Ed. Engl.* **2001**, *40*, 2839.
41. Minkowski, C.; Pansu, R.; Takano, M.; Calzaferri, G. *Adv. Funct. Mater.* **2006**, *16*, 273.
42. Bossart, O.; De Cola, L.; Welter, S.; Calzaferri, G. *Chem. Eur. J.* **2004**, *10*, 5771.
43. Coutant, M.A.; Payra, P.; Dutta, P.K. *Micropor. Mesopor. Mater.* **2003**, *60*, 79.
44. Payra, P.; Dutta, P.K. *Micropor. Mesopor. Mater.* **2003**, *64*, 109.
45. Demas, J.N.; Harris, E.W.; McBride, R.P. *J. Am. Chem. Soc.* **1977**, *99*, 3547.
46. Albuquerque, R.Q.; Popović, Z.; De Cola, L.; Calzaferri, G. *ChemPhysChem* **2006**, *7*, 1050.
47. Albuquerque, R.Q.; Zabala Ruiz, A.; Li, H.; De Cola, L.; Calzaferri, G. *Proc. SPIE Photonics Solar Energy Syst.* **2006**, *6197*, 61970B-1–61970B-5.
48. Huber, S.; Calzaferri, G. *Angew. Chem. Int. Ed. Engl.* **2004**, *43*, 6738.
49. Huber, S.; Calzaferri, G. *ChemPhysChem* **2004**, *5*, 239.

50. Marcolli, C.; Calzaferri, G. *Appl. Organomet. Chem.* **1999**, *13*, 213.

51. Yang, P.; Deng, T.; Zhao, D.; Feng, P.; Pine, D.; Chmelka, B.F.; Whitesides, G.M.; Stucky, G.D. *Science (Reports)* **1998**, *282*, 2244.

52. Walt, D.R. *Nat. Mater.* **2002**, *1*, 17.

53. Böker, A.; Lin, Y.; Chiapperini, K.; Horowitz, R.; Thompson, M.; Carreon, V.; Xu, T.; Abetz, C.; Skaff, H.; Dinsmore, A.D.; Emrick, T.; Russell, T.P. *Nat. Mater.* **2004**, *3*, 302.

54. Bashouti, M.; Salalha, W.; Brumer, M.; Zussman, E.; Lifshitz, E. *ChemPhysChem* **2006**, *7*, 102.

55. Huo, S.-J.; Xue, X.-K.; Li, Q.-X.; Xu, S.-F.; Cai, W.-B. *J. Phys. Chem. B* **2006**, *110*, 25721.

56. Dai, L.; Patil, A.; Gong, X.; Guo, Z.; Liu, L.; Liu, Y.; Zhu, D. *ChemPhysChem* **2003**, *4*, 1150.

57. Ozin, G.A.; Arsenault, A.C. *Nanochemistry: A Chemical Approach to Nanomaterials*. Cambridge, UK: RSC, **2005**.

58. Hurst, S.J.; Payne, E.K.; Qin, L.D.; Mirkin, C.A. *Angew. Chem. Int. Ed. Engl.* **2006**, *45*, 2672.

59. Bein, T. *MRS Bull.* **2005**, *30*, 713.

60. Gouzinis, A.; Tsapatsis, M. *Chem. Mater.* **1998**, *10*, 2497.

61. Yoon, K.B. *Acc. Chem. Res.* **2007**, *40*, 29.

62. Lee, J.S.; Lim, H.; Ha, K.; Cheong, H.; Yoon, K.B. *Angew. Chem. Int. Ed. Engl.* **2006**, *45*, 5288.

63. Bowden, N.; Terfort, A.; Carbeck, J.; Whitesides, G.M. *Science (Reports)* **1997**, *276*, 233.

64. (*a*) Yunus, S.; Spano, F.; Patrinoiu, G.; Bolognesi, A.; Botta, Ch.; Brühwiler, D.; Zabala Ruiz, A.; Calzaferri, G. *Adv. Funct. Mater.* **2006**, *16*, 2213; (*b*) Vohra, V.; Bolognesi, A.; Calzaferri, G.; Botta, Ch. *Langmuir* **2009**, *25*, 12019.

65. Bossart, O.; Calzaferri, G. *Micropor. Mesopor. Mater.* **2008**, *109*, 392.

66. (*a*) Li, H.; Wang, Y.; Zhang, W.; Liu, B.; Calzaferri, G. *Chem. Commun.* **2007**, 2853; (*b*) Wang, Y.; Li, H.; Liu, B.; Gan, Q.; Dong, Q.; Calzaferri, G.; Sun, Z. *J. Solid State Chem.* **2008**, *181*, 2469.

67. (*a*) Hashimoto, S.; Samata, K.; Shoji, T.; Taira, N.; Tomita, T.; Matsuo, S. *Micropor. Mesopor. Mater.* **2009**, *117*, 220; (*b*) Cucinotta, F.; Popović, Z.; Weiss, E.A.; Whitesides, G.M.; De Cola, L. *Adv. Mater.* **2009**, *21*, 1142.

68. (*a*) Cucchi, I.; Spano, F.; Giovanella, U.; Catellani, M.; Varesano, A.; Calzaferri, G.; Botta, C. *Small* **2007**, *3*, 305; (*b*) Vohra, V.; Bolognesi, A.; Calzaferri, G.; Botta, C. *Langmuir* **2010**, *26*, 1590.

69. (*a*) Huang, Z.-M.; Zhang, Y.-Z.; Kotaki, M.; Ramakrishna, S. *Compos. Sci. Technol.* **2003**, *63*, 2223; (*b*) Sun, Z.; Zussman, E.; Yarin, A.L.; Wendorff, J.H.; Greiner, A. *Adv. Mater.* **2003**, *15*, 1929; (*c*) Li, D.; Xia, Y. *Adv. Mater.* **2004**, *16*, 1151.

70. Yu, J.H.; Fridrikh, S.V.; Rutledge, G.C. *Adv. Mater.* **2004**, *16*, 1562.

71. Popović, Z.; Busby, M.; Huber, S.; Calzaferri, G.; De Cola, L. *Angew. Chem. Int. Ed. Engl.* **2007**, *46*, 8898.

72. Han, M.Y.; Gao, X.; Su, J.Z.; Nie, S. *Nat. Biotechnol.* **2001**, *19*, 631.

73. (*a*) Popović, Z.; Otter, M.; Calzaferri, G.; De Cola, L. *Angew. Chem. Int. Ed. Engl.* **2007**, *46*, 6188; (*b*) Strasser, C.A.; Otter, M.; Albuquerque, R.Q.; Höne, A.; Vida, Y.; Maier, B.; De Cola, L. *Angew. Chem. Int. Ed. Engl.* **2009**, *48*, 7928; (*c*) Tsotsalas, M.; Kopka, K.; Luppi, G.; Wagner, S.; Law, M.; Schäfers, M.; De Cola, L. *ACS Nano* **2010**, *4*, 342.
74. (*a*) Tessler, N.; Medvedev, V.; Kazes, M.; Kan, S.; Banin, U. *Science* **2002**, *295*, 1506; (*b*) Kalinina, O.; Kumacheva, E. *Chem. Mater.* **2001**, *13*, 35; (*c*) Gourevich, I.; Pham, H.; Jonkman, J.; Kumacheva, E. *Chem. Mater.* **2004**, *16*, 1472; (*d*) Avella, M.; Errico, M.; Martuscelli, E. *Nano Lett.* **2001**, *1*, 213; (*e*) Sanchez, C.; Soler-Illia, G.J.; Ribot, F.; Lalot, T.; Mayer, C.R.; Cabuil, V. *Chem. Mater.* **2001**, *13*, 3061.
75. Schneider, J.; Fanter, D.; Bauer, M.; Schomburg, C.; Wöhrle, D.; Schulz-Ekloff, G. *Micropor. Mesopor. Mater.* **2000**, *39*, 257.
76. Devaux, A.; Popović, Z.; Bossart, O.; De Cola, L.; Kunzmann, A.; Calzaferri, G. *Micropor. Mesopor. Mater.* **2006**, *90*, 69.
77. Lutkouskaya, K.; Calzaferri, G. *J. Phys. Chem. B* **2006**, *110*, 5633.
78. Pfenniger, M.; Calzaferri, G. *ChemPhysChem* **2000**, *4*, 211.
79. Yatskou, M.M.; Meyer, M.; Huber, S.; Pfenniger, M.; Calzaferri, G. *ChemPhysChem* **2003**, *4*, 567.
80. (*a*) Maas, H.; Calzaferri, G. *Spectrum* **2003**, *16*, 18; (*b*) Mass, H. Energy transfer at the frontiers of dye-loaded zeolite L. PhD thesis, University of Bern, **2003**.
81. (*a*) Dieu, L.-Q.; Devaux, A.; López-Duarte, I.; Martínez-Díaz, M.V.; Brühwiler, D.; Calzaferri, G.; Torres, T. *Chem. Commun.* **2008**, 1187; (*b*) López-Duarte, I.; Dieu, L.-Q.; Dolamic, I.; Martínez-Díaz, M.V.; Torres, T.; Calzaferri, G.; Brühwiler, D. *Chem. a Eur. J.* **2011**, *17*, 1855.
82. (*a*) Koeppe, R.; Bossart, O.; Calzaferri, G.; Sariciftci, N.S. *Sol. Energ. Mater. Sol. Cells* **2007**, *91*, 986; (*b*) Shankar, K.; Feng, X.; Grimes, C.A. *ACS Nano* **2009**, *3*, 788.
83. (*a*) Bossart, O. Zeolite L antenna material for organic light emitting diodes and organic solar cells. PhD thesis, University of Bern, **2006**; (*b*) Bossart, O.; Calzaferri, G. *Chimia* **2006**, *60*, 179.
84. Kido, J.; Hongawa, K.; Okuyama, K.; Nagai, K. *Appl. Phys. Lett.* **1993**, *63*, 2627.
85. (*a*) Ouisse, T.; Stéphan, O.; Armand, M.; Leprêtre, J.C. *J. Appl. Phys.* **2002**, *92*, 2795; (*b*) Edman, L.; Moses, D.; Heeger, A.J. *Synth. Met.* **2003**, *138*, 441.
86. Sariciftci, N.S.; Smilowitz, L.; Heeger, A.J.; Wudl, F. *Science* **1992**, *258*, 1474.
87. Nelson, J. *Curr. Opin. Solid State Mater. Sci.* **2002**, *6(1)*, 87.
88. Liu, Y.-X.; Summers, M.A.; Scully, S.R.; McGehee, M.D. *J. Appl. Phys.* **2006**, *99*, 093521.
89. (*a*) Shurcliff, W.A. *J. Opt. Soc. Am.* **1951**, *41*, 209; (*b*) Garvin, R.L. *Rev. Sci. Instrum.* **1960**, *31*, 1010; (*c*) Weber, W.; Lambe, H.J. *Appl. Opt.* **1976**, *15*, 2299; (*d*) Goetzberger, A.; Greubel, W. *Appl. Phys.* **1977**, *14*, 123; (*e*) Batchelder, J.S.; Zewail, A.H.; Cole, T. *Appl. Opt.* **1979**, *18*, 3090; (*f*) Batchelder, J.S.; Zewail, A.H.; Cole, T. *Appl. Opt.* **1981**, *20*, 3733; (*g*) Langhals, H. *Nachr. Chem. Tech. Lab.* **1980**, *28*, 716; (*h*) Kittidachachan, P.; Danos, L.; Meyer, T.J.J.; Alderman, N.; Markvart, T. *Chimia* **2007**, *61*, 780; (*i*) Brühwiler, D.; Dieu, L.-Q.; Calzaferri, G. *Chimia* **2007**, *61*, 820; (*j*) Currie, M.; Mapel, J.; Heidel, T.; Goffri, S.; Baldo, M. *Science* **2008**, *321*, 226; (*k*) Calzaferri, G.; Kunzmann, A.; Brühwiler, D.; Bauer, C.H. CH-698333, **2009**.
90. (*a*) Scholes, G.D.; Fleming, G.R. *J. Phys. Chem. B* **2000**, *104*, 1854; (*b*) Ortiz, W.; Krueger, B.P.; Kleiman, V.D.; Krause, J.L.; Roitberg, A.E. *J. Phys. Chem. B* **2005**,

109, 11512; (*c*) Closs, G.L.; Piotrowiak, P.; MacInnis, J.M.; Fleming, G.R. *J. Am. Chem. Soc.* **1988**, *110*, 2652.

91. Sancho-García, J.C.; Brédas, J.-L.; Beljonne, D.; Cornil, J.; Martínes-Álvarez, R.; Hanack, M.; Poulsen, L.; Gierschner, J.; Mack, H.-G.; Egelhaaf, H.-J.; Oelkrug, D. *J. Phys. Chem. B* **2005**, *109*, 4872.
92. Fois, E.; Gamba, A.; Medici, C.; Tabacchi, G. *Chem. Phys. Chem.* **2005**, *6*, 1917.
93. Mulliken, R.S. In: Ramsay, D.A.; Hinze, J., editors. *Selected Papers of Robert S. Mulliken*. London: University of Chicago Press, **1975**, p. 620.
94. McGlynn, S.P.; Vanquickenborne, L.G.; Kinoshita, M.; Carroll, D.G. *Introduction to Applied Quantum Chemistry*. New York: Holt, Rienehart and Winston, **1972**.
95. Atkins, P.W.; Friedman, R.S. *Molecular Quantum Mechanics*, 3rd edition. Oxford: Oxford University Press, **1997**.
96. Valeur, B. *Molecular Fluorescence*. New York: Wiley-VCH, **2002**.
97. Farinha, J.P.S.; Spiro, J.G.; Winnik, M.A. *J. Phys. Chem. B* **2001**, *105*, 4879.
98. Kauzmann, W. *Quantum Chemistry*. New York: Academic Press, **1957**.
99. Strickler, S.J.; Berg, R.A. *J. Chem. Phys.* **1962**, *37*, 814.
100. (*a*) Devaux, A.; Lutkouskaya, K.; Calzaferri, G.; Dieu, L.-Q.; Brühwiler, D.; De Cola, L.; Torres, T. *Chimia* **2007**, *61*, 626; (*b*) Popovic, Z.; Tsotsalas, M.; Busby, M.; De Cola, L.; Calzaferri, G.; Josel, H.P. PCT/EP2007/005811, China1128455, **2008**.
101. Calzaferri, G.; Méallet-Renault, R.; Brühwiler, D.; Pansu, R.; Dolamic, I.; Dienel, T; Adler, P.; Li, H; Kunzmann, A. *ChemPhysChem*, **2011**, *12*, 580.

10

CONTROLLING PHOTOREACTIONS THROUGH NONCOVALENT INTERACTIONS WITHIN ZEOLITE NANOCAGES

V. RAMAMURTHY AND JAYARAMAN SIVAGURU

Drawing inspiration from nature, chemists have employed confinement as a tool to achieve selectivity in chemical reactions.[1] Confining reactants has the power to drive chemical reactions with astonishing precision as seen in the remarkable selectivities exhibited by enzymes in chemical processes, the phenomenon of photosynthesis, and so on. Within a confined environment,[2] the familiar electronic and steric effects of solution chemistry are replaced by structural and topological factors that frequently result in products that display a level of selectivity or specificity that has hitherto been unobtainable in solution. Upon confinement, inherent reactivity of confined guests often becomes of secondary importance compared with features such as symmetry, geometric considerations, and noncovalent interactions that develop within the confined environments. This review will highlight the confinement effect offered by zeolite Y supercages,[3,4] the noncovalent interaction (viz., cation–π and cation–carbonyl interactions)[5,6] that develops upon inclusion of substrates within faujasite zeolite X/Y supercages, and the effect of such interactions in photophysical and photochemical processes).[7,8]

ZEOLITE Y SUPERCAGES: NANOCAVITY THAT OFFERS NONCOVALENT INTERACTIONS

Zeolites[3] are inorganic crystalline aluminosilicate microporous materials (particle size ranges 0.1–10 μm, 40 naturally occurring and more than 100 synthetic

Supramolecular Photochemistry: *Controlling Photochemical Processes*, First Edition. Edited by V. Ramamurthy and Yoshihisa Inoue.

forms) with three-dimensional open framework structures possessing $[SiO_4]^{4-}$ and $[AlO_4]^{5-}$ tetrahedral building blocks.[9] Zeolites used in the present work are large pore X and Y, which belong to the family of faujasite-type zeolites (Fig. 10.1). The cage structure of faujasite zeolites are constructed of openings containing four- and six-membered rings of $[SiO_4]^{4-}$ and $[AlO_4]^{5-}$ polyhedra called the sodalite cages, which are arranged tetrahedrally (Fig. 10.1) to form an even larger cage called the supercage. All the corners of these tetrahedra are linked to form channels and cages of discrete size with no two aluminum atoms sharing the same oxygen that extend periodically and regularly across the entire structure. Typically, there are eight supercages per unit cell in the case of faujasite X/Y zeolite. Large voids ca. 14 Å in diameter are interconnected by tetrahedrally disposed 12-membered ring window ca. 8-Å diameter (Fig. 10.1). The adsorbed guest molecules can diffuse through the channels and cages through the 12-membered ring window and are located in the ca. 14-Å supercage. Alkali or alkaline earth metal cation balances the overall negative charge of the zeolite framework caused by the difference in charge between the $[SiO_4]^{4-}$ and $[AlO_4]^{5-}$ tetrahedra. In X and Y zeolites, the charge compensating cations occupy three different positions (Fig. 10.1): site I (23 cations per unit cell) is located on hexagonal prism faces between the sodalite units, site II (32 cations per unit cell) is located in the open hexagonal faces, and site III (30 cations in X) is located on the walls of the larger cavity formed by the four ring oxygens.[10] Type I cations are buried deeper and are generally noninteracting with the guest molecules. Only cations of sites II and III are expected to be readily accessible to the adsorbed organic. The number of cations in faujasites is dictated by the Si:Al ratio in the framework. For example, faujasite zeolites with Si:Al ratio of 2.4 has 55 cations per unit cell, whereas faujasite zeolites with Si:Al of 40 has just 5 cations per unit cell. The cations and water molecules present are located in the cages, cavities, and channels of the zeolites. When the water is removed, the voids created within the framework can take in other molecules. This process is called "sorption," and the zeolites are said to "sorb" molecules into their void volume, that is, they act as "sorbents." The organic guests can be caged inside these voids within the zeolites. Adsorption of organic molecules on zeolites can occur both on the external and internal surfaces. Internal complexation occurs by diffusion of the guest onto the channels and cavities within the zeolites, which is size and shape selective.[7] The charge-compensating cations of the zeolite bind to the organic molecule aiding in diffusion of organics onto the zeolite. The kinetic diameter of the guest molecule must be smaller than that of the pore size of the cavity. The cations tend to come out of their original position and bind with organic molecules. The binding interactions will depend on the charge density of the cation.[10,11] The higher the charge density, the stronger the binding interaction.[6] The cations of the zeolites interact with the guest molecules and hold them tightly within the cages or channels. In the presence of water, the cations are hydrated and are shielded from the sorbed organic or inorganic guests. Zeolites are generally activated at ca. 500°C to remove most of the water

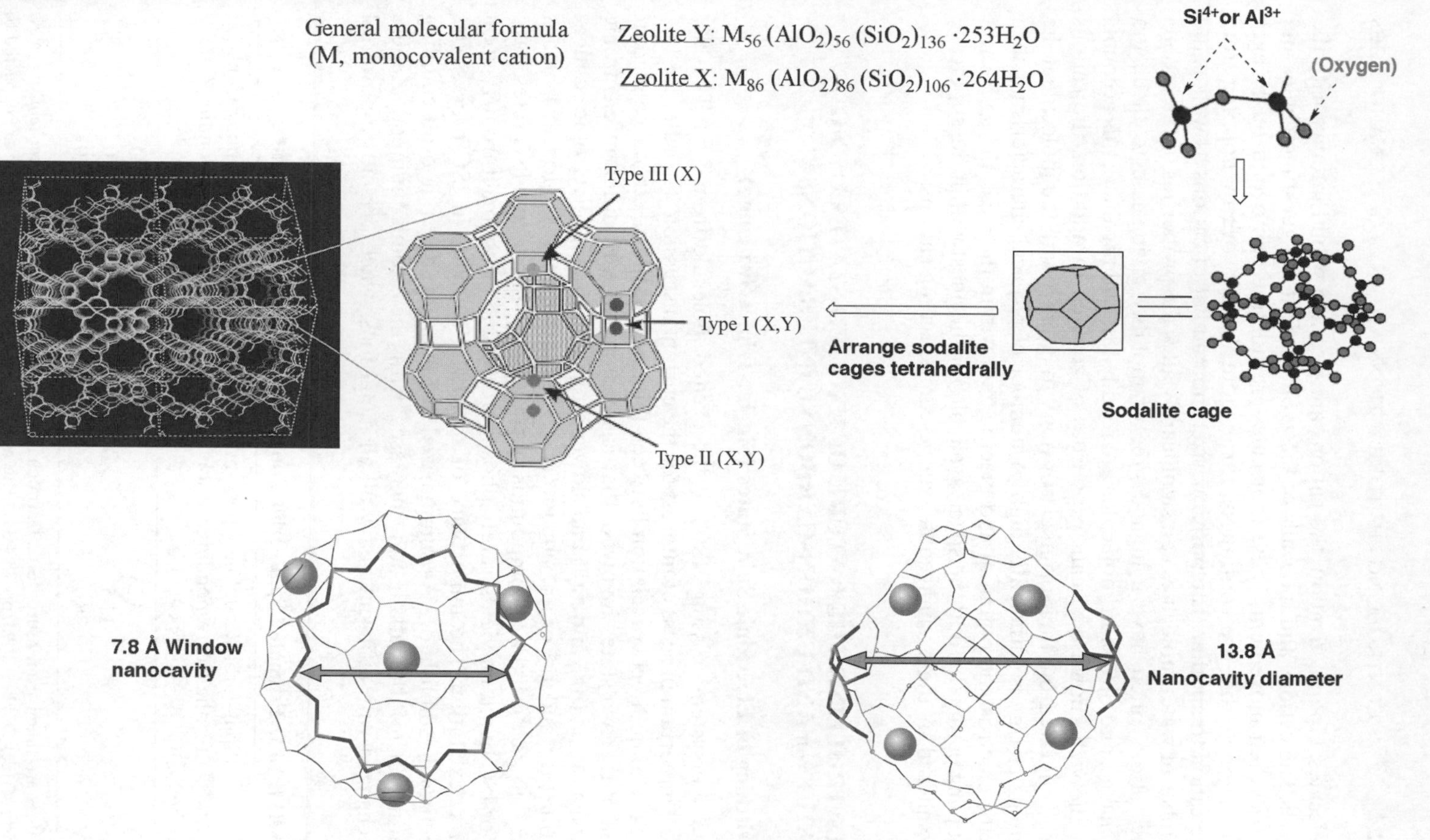

Figure 10.1. Structural features of faujasite zeolite (X and Y).

molecules. Upon activation, the cations are free and would interact with the guests. The cations and the walls of the zeolite orient the guest molecules and restrict their movement. These molecules are still free to move through cages and channels that are continuous throughout the network by hopping from one cation to another. The position, size, and number of cations as well as the number of water molecules can significantly alter the properties of the zeolite. The cations often have a high degree of mobility giving rise to facile ion exchange, and the water molecules are readily lost and regained; this accounts for the well-known desiccant properties of zeolites. In addition, the cations could migrate from their original position[11] to interact with the molecule if the binding energy is sufficiently high. A number of physical characteristics such as the electrostatic potential and electric field within the cage, the spin–orbit coupling parameter and the space available for the guest within the supercage are altered by the exchangeable charge-compensating cations.

CONTROLLING THE NATURE OF EXCITED STATES USING CATION–π AND CATION–CARBONYL INTERACTIONS

Switching of Electronic Spin States (Singlet-Triplet Switching)

The intersystem crossing (ISC) of the excited molecule from S_1 to T_1 is the most important process of interest for the current discussion. The rules for ISC suggest an indirect correlation between the rate and the electronic configuration of the two states involved, that is, a faster rate with different electronic configuration ($n\pi^*$ and $\pi\pi^*$) and slower rate with the same electronic configuration ($\pi\pi^*$ and $\pi\pi^*$ or $n\pi^*$ and $n\pi^*$).[12,13] Thus, aromatics ($\pi\pi^{*1}$ to $\pi\pi^{*3}$), olefins ($\pi\pi^{*1}$ to $\pi\pi^{*3}$), and azo compounds ($n\pi^{*1}$ and $n\pi^{*3}$) have low ISC rates, have poor ISC quantum yields, rarely phosphoresce, and rarely react from T_1 upon direct excitation. The importance of spin–orbit coupling in ISC process has been well established.[13] As summarized in Table 10.1, the spin–orbit coupling parameter of the alkali ion increases with the atomic number. Zeolite exchanged with different heavy alkali metal ions, which presents a powerful

TABLE 10.1. Estimated Spin–Orbit Coupling Constants for Metal Ions[a]

Zeolite	Spin–Orbit Coupling Constant for the Corresponding Cation (ζ cm^{-1})	Zeolite	Spin–Orbit Coupling Constant for the Corresponding Cation (ζ cm^{-1})
LiY	0.23	RbY	160
NaY	11.5	CsY	370
KY	38	TlY	3410

[a]Values are taken from Murov, S.L.; Carmichael, I.; Hug, G. *Handbook of Photochemistry*. New York: Marcel Dekker, **1993**, pp. 338–341. The numbers for the corresponding ion are expected to be different, but we expect the trend to remain the same.

matrix to optical detection of magnetic resonance (ODMR) in zero applied magnetic field, has provided information on the geometry of the alkali metal ion–aromatic (naphthalene) interaction in zeolites. The sublevel-specific dynamics for adsorbed naphthalene show a distinct increase in relative radiative character and total rate constant of the out-of-plane x-sublevel with increasing mass of the cation perturber. ODMR kinetic results suggest naphthalene to be adsorbed through its π-cloud at a cation site (cation–π interaction[6]).

The uniqueness of zeolites is the ability to observe phosphorescence from systems, which commonly fail to show this emission in organic glassy matrices even when subjected to heavy-atom effect. One could observe phosphorescence from even those organic molecules that do not phosphoresce under normal conditions. The potential of this technique is shown with three classes of molecules namely aromatics, polyenes, and azo compounds. As shown in Figure 10.2, the emission spectrum of naphthalene is profoundly affected by inclusion in faujasites.[14] For low-mass cations such as Li^+, the emission spectra show the typical naphthalene blue fluorescence. However, as the mass of the cation increases (e.g., from Li^+ to Cs^+ to Tl^+), there is a dramatic decrease in fluorescence intensity and a simultaneous appearance of the phosphorescence of naphthalene. Support for the heavy alkali metal ion enhancement of phosphorescence is evident from the dependence of the ratio of fluorescence to phosphorescence on the Cs^+ to Na^+ content in a zeolite. As seen in Figure 10.3, the phosphorescence intensity of phenanthrene increases with the Cs^+ ion content. The general nature of the above is reflected by results with other aromatic compounds. Heavy-atom-induced phosphorescence that allows the

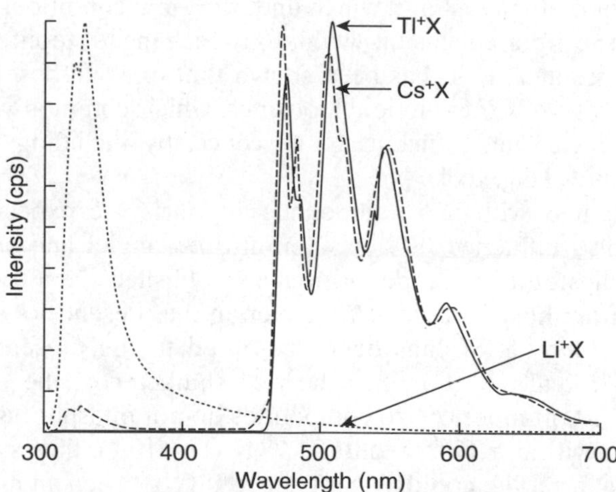

Figure 10.2. Emission spectra at 77 K of naphthalene included within LiX, CsX, and TlX. Note the difference in relative intensities of fluorescence and phosphorescence with the cation. Essentially fluorescence from LiX and phosphorescence in TlX are observed.

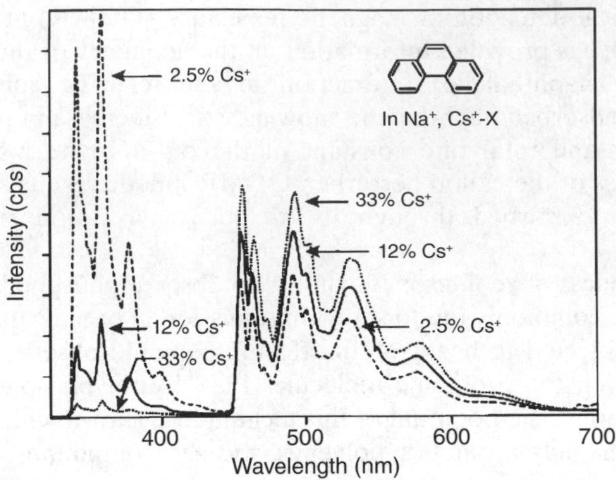

Figure 10.3. Emission spectra at 77 K of phenanthrene included within CsNaX zeolite at 77 K. Note that the intensity of phosphorescence increases with the increase in Cs^+ content.

use of ODMR in zero applied magnetic field, has provided information on the geometry of the alkali metal ion–aromatic (naphthalene) interaction in zeolites.[15] The sublevel-specific dynamics for adsorbed naphthalene show a distinct increase in relative radiative character and total rate constant of the out-of-plane x-sublevel with increasing mass of the cation perturber. Once again, ODMR kinetic results suggest naphthalene to be adsorbed through its π-cloud at a cation site (cation–π interaction[6]).

In the case of olefinic systems, which under normal conditions do not show phosphorescence, upon inclusion within Tl^+-exchanged zeolites emit from their triplet states at 77 K. It has been shown that one can observe phosphorescence from all-*trans*-α,ω-diphenylpolyenes, which commonly exhibit very low ISC efficiencies and efficient fluorescence, by including them in Tl^+-exchanged zeolites (Fig. 10.4).[14,15]

Results obtained with azo compounds supports the expectation that a zeolite could also influence ISC between an nπ* singlet and an nπ* triplet. Numerous studies on azo compounds have established them to possess very poor ISC and not phosphoresce at 77 K even in the presence of a heavy-atom perturber. The lack of ISC has been attributed to the presence of a large energy gap (>15 kcal mol^{-1}) and to the nπ* character of the excited states involved in ISC. A number of azo compounds do not phosphoresce in organic glass but do so within a Tl^+Y zeolite at 77 K (Fig. 10.5), thus supporting the expectation that a zeolite could also influence ISC between an nπ* singlet and an nπ* triplet of azo compounds.[16]

The utility of the heavy alkali ion effect in controlling product distribution in photochemical reactions is illustrated below with the three examples of acenaphthylene, dibenzobarrelene, and dibenzylketone. Irradiation of

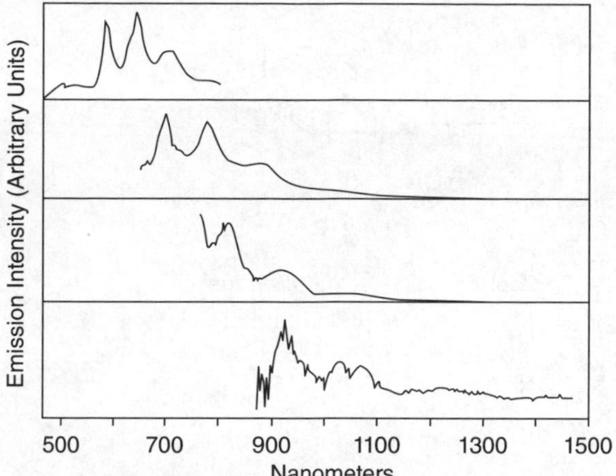

Figure 10.4. Phosphorescence spectra at 77 K of all *trans*-diphenylpolyenes (from top to bottom: stilbene; 1,4-diphenylbutadiene; 1,6-diphenylhexatriene; and 1,8-diphenyloctatetraene) included within Tl^{+}ZSM-5.

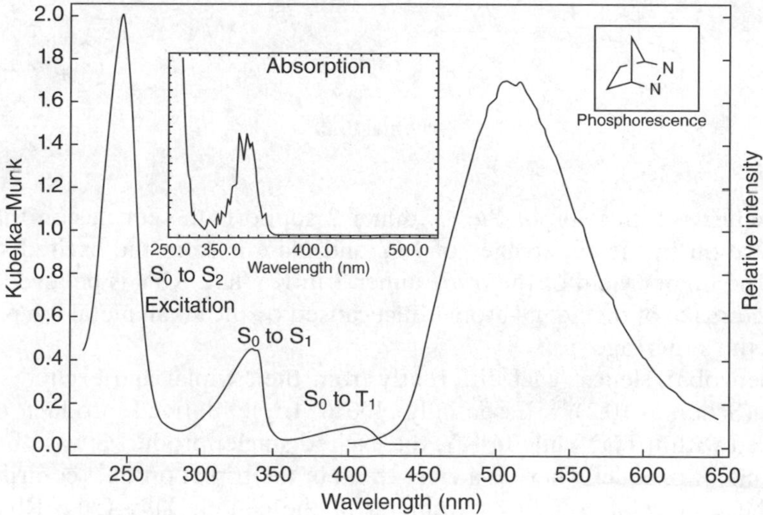

Figure 10.5. Emission and excitation spectra of diazo-(2,3)-bicyclo[2.2.1]heptane included within TlY, recorded at 77 K. Insert shows the diffuse reflectance absorption spectrum. The emission on the right is assigned to be phosphorescence. The longest wavelength band in the excitation spectrum is believed to be S_0 to T_1 transition.

acenaphthylene (**1**) in solution yields the *cis* and the *trans* dimers; the excited singlet gives predominantly the *cis* dimer **2**, whereas the triplet gives both *cis* and *trans* dimers in comparable amounts (Scheme 10.1). Photolysis of dry solid inclusion complexes of acenaphthylene in various cations (Li^+, Na^+, K^+, Rb^+)-exchanged Y zeolites gave the *cis* and *trans* dimers in varying ratios.[17]

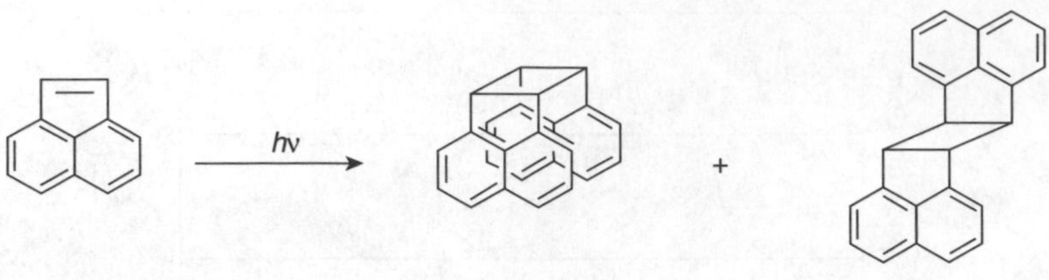

Zeolite	*cis/trans*
LiY	25
NaY	25
KY	2.3
RbY	1.5

Scheme 10.1.

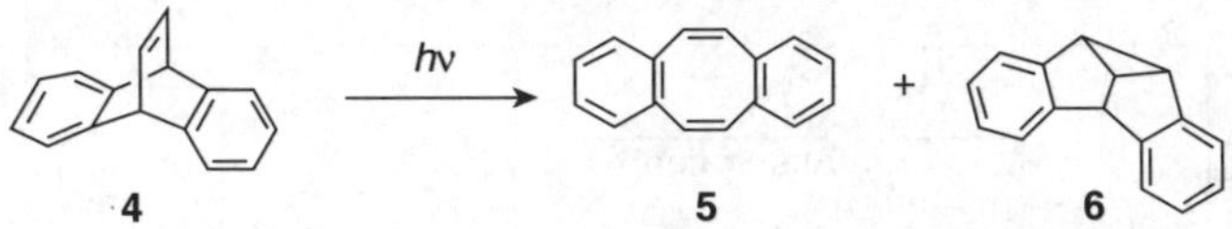

Scheme 10.2.

The exclusive formation of the *cis* dimer **2** supports the conclusion that the dimerization in the supercages of LiY and NaY is from the excited singlet state. The higher yield of the *trans* dimer **3** in KY and RbY is believed to be a consequence of the heavy-atom effect caused by the alkali metal ion present within the supercage.

Dibenzobarrelenes react differently from their triplet and excited singlet states (Scheme 10.2).[18] Essentially, 100% triplet-derived product **6** was observed within TlY, while in KY, the excited singlet product **5** was obtained as the major product. There is a clear trend in the triplet product contribution with the spin–orbit coupling parameter of the cation ($Tl^+ > Cs^+ > Rb^+ > K^+$; Table 10.1).

The last example relates to influencing product distribution via spin change of a radical pair.[19] As illustrated in Scheme 10.3, photolysis of dibenzylketone **7** leads to the formation of benzyl and phenylacyl triplet radical pair. In the absence of geminate recombination, the radical pair diffuses apart, decarbonylates, and couples to yield 1,2-diphenylethane **8**. However, with diffusional separation restraints in a medium such as zeolite, the primary triplet radical pair may undergo ISC to the singlet state from where rearrangement to *ortho*- and *para*-phenyl-substituted benzophenones **9** and **10** could occur. As shown in Scheme 10.3, the ratio of 1,2-diphenylethane to the benzophenone deriva-

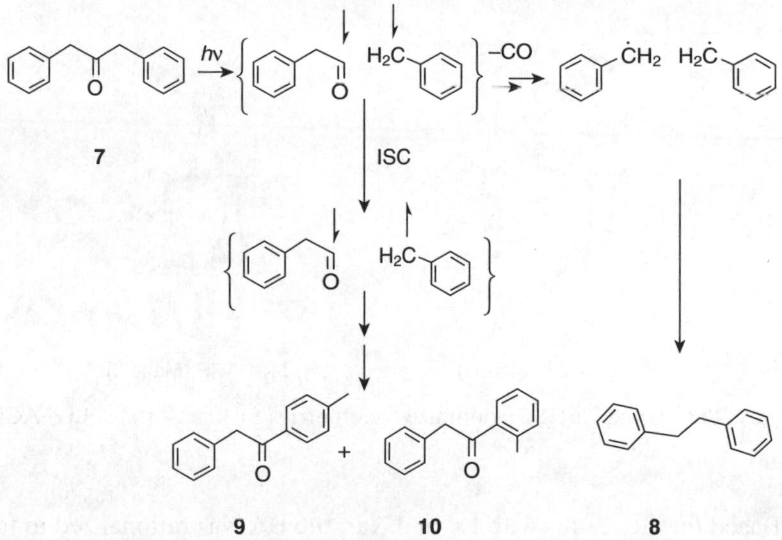

	9 + 10	8
Hexane	0	>99%
KY	20	80
RbY	30	70
CsY	63	37
T1Y	76	23

Scheme 10.3.

tives within zeolites depends on the alkali ion, with the lighter one favoring the former and the heavier one favoring the latter, respectively.

Switching of Electronic Configuration (nπ*–ππ* Switching)

Molecules possessing a carbonyl chromophore have two types of lowest excited states, nπ* and ππ*.[13] Using computational methods, solid-state nuclear magnetic resonance (NMR), steady-state/time-resolved emission experiments, and product studies, it was established that the nature of the lowest triplet state of aryl alkyl ketones and cyclic enones could be controlled with the alkali ions present in zeolites. Density functional theory calculations (B3LYP/6-31G*) suggest that Li^+ binds to acetophenone (ACP) via two modes: (1) dipolar interaction to carbonyl (mode I) and (2) cation–π interaction to phenyl ring (mode II) (Fig. 10.6).[20] Binding affinities (BAs) of Li^+ to ACP in these two modes in both the ground state (55.3 and 37.7 kcal mol^{-1}) and the triplet state (57.3 and 47.2 kcal mol^{-1}) are large. As would be expected, Li^+ binds more strongly (BA in ground state: 55.3 kcal mol^{-1}) than Na^+ ion (39.9 kcal mol^{-1}) to

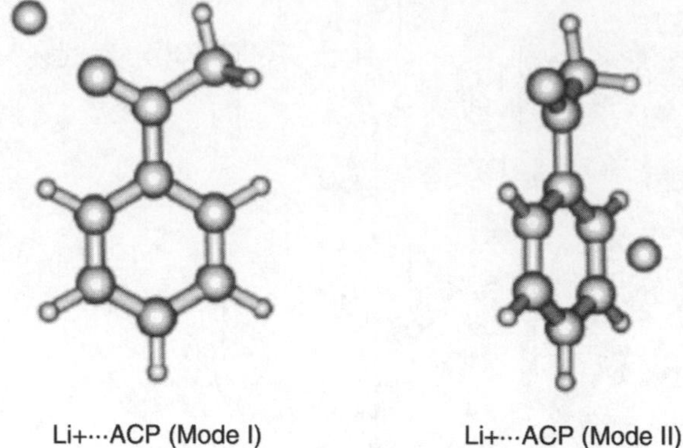

Figure 10.6. B3LYP/6-31G*-optimized geometries for the S_0 states Li^+···ACP.

ACP. Based on the results with Li^+ and Na^+, the BA was rationalized to follow the trend $Li^+ > Na^+ > K^+ > Rb^+ > Cs^+$.[20]

Solid-state NMR results supported the computational predictions.[20] The three independent measurements, static, magic angle spectra (MAS), and cross-polarized magic angle spectra (CP-MAS), of ^{13}C-enriched ACP included in MY zeolites suggest an interaction between the alkali ion and ACP molecules: (1) The line width in the case of static and MAS spectra (Figs. 10.7 and 10.8), and signal intensity in the case of CP-MAS (Fig. 10.8) spectra depend on the cation (MY), suggesting lesser mobility of ACP molecules in LiY and greater freedom of movement in CsY. (2) Greater mobility of ACP molecules on silica, which lack alkali metal ion, than within NaY zeolites, which has at least four type II alkali metal ions in a supercage, speaks for the importance of alkali metal ion···ACP interaction. (3) From the comparison of the line width in MAS spectra of ACP in NaY zeolites of varying Si:Al ratio (line width at half height: 500, 390, 350, and 300 Hz in zeolites of Si:Al ratio 2.5, 6, 15, and 40), it is obvious that ACP molecules are more restricted within NaY of high alumina content (consequently more alkali ion). For example, ACP molecules are more restricted in Y zeolites with Si:Al = 2.4 (line width in MAS spectra: 500 Hz) than in Y zeolites with Si:Al = 40 (line width in MAS spectra: 300 Hz). (4) ACP molecules are more mobile within hydrated (line width in MAS spectra: 140 Hz) than in dry NaY.

Density functional theory calculations (B3LYP/6-31G*) employed to identify the consequence of alkali metal ion binding on the ordering of excited triplet states indicated that the ordering of the frontier molecular orbitals changes both in the ground and in the excited triplet states upon Li^+ complexation (Fig. 10.9).[20] Based on the orbital energies, the lowest triplet state of free and Li^+ bound ACP is predicted to have $n\pi^*$ and $\pi\pi^*$ configurations. Time-dependent density functional theory (TDDFT) calculation besides correctly predicting the ordering in the case of unperturbed ACP and

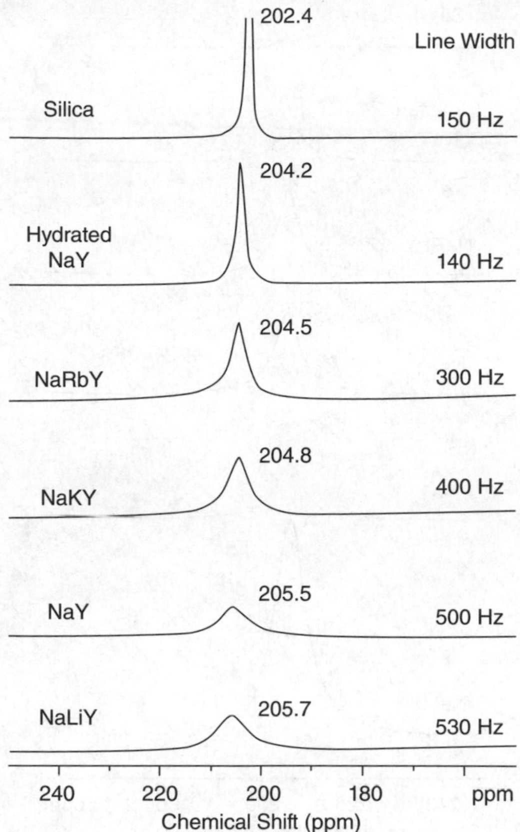

Figure 10.7. MAS NMR spectra of ^{13}C (CNO)-enriched acetophenone included in NaLiY, NaY, NaKY, NaRbY, hydrated NaY, and silica gel. Note the variation of line width with the zeolite.

4-methoxyacetophenone (MACP), identified the triplet energies within a few kilocalorie per mole ($<2\,kcal\,mol^{-1}$) of experimental values. Even more impressive was the fact that computed energy gaps between $n\pi^*$ and $\pi\pi^*$ states in ACP and MACP were consistent with the experimental observations ($<5\,kcal\,mol^{-1}$). Consistency between computed and experimental results with ACP and MACP gave us confidence in the results obtained in the case of alkali metal-ion-bound ACP. Computational prediction of interest to the studies in zeolites is the lowered energy level of $\pi\pi^*$ triplet below $n\pi^*$ triplet when Li^+ and Na^+ bind to ACP through the carbonyl oxygen.

The fingerprint features of $n\pi^*$ and $\pi\pi^*$ state emissions and their lifetimes were used to identify the nature of the lowest triplet of ACPs included within alkali ion-exchanged Y zeolites.[20] It is known that at 77 K, ACP in nonpolar as well as moderately polar organic solvents shows a structured emission (phosphorescence) with a short lifetime (millisecond range) characteristic of

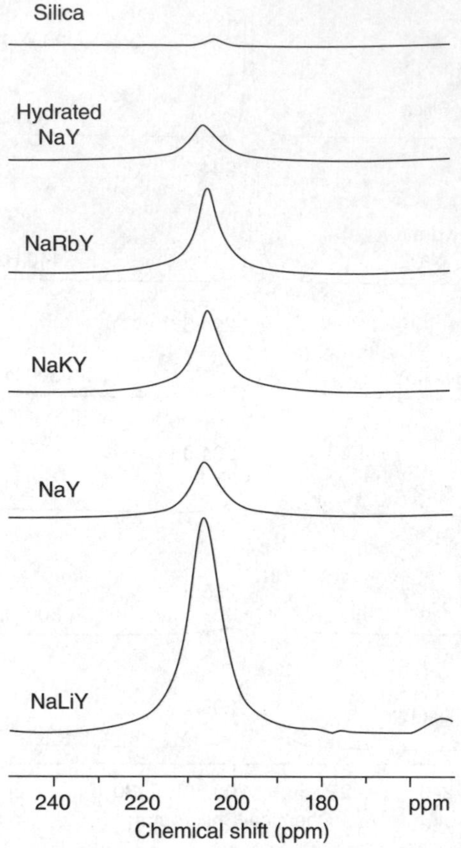

Figure 10.8. ^{1}H–^{13}C CP-MAS spectra of ^{13}C (CNO)-enriched acetophenone included in NaLiY, NaY, NaKY, NaRbY, hydrated NaY, and silica gel. Note the variation of the intensity of the signal with the zeolites.

nπ* states, while MACP shows a broad emission with a long lifetime (seconds) characteristic of ππ* triplet state. As illustrated in Figure 10.10, ACP within NaY shows, independent of the time slice, broad phosphorescence, suggesting the emitting state to be ππ* in character. Triplet lifetime measurements further confirm the ππ* character of the emitting state within NaY. The emitting state in the case of ACP-NaY has only one long component with a lifetime of 420 ms. It is important to note that the alkali metal ions that can interact strongly (Li^+, Na^+, and K^+) with the carbonyl chromophore show only ππ*-like phosphorescence. The change in the nature of the lowest triplet state from ππ* to nπ* state on co-inclusion of methanol accurately identifies the source of state switch within NaY to be the binding of alkali metal ion to carbonyl (Fig. 10.10). Were the polarity of zeolites the causative factor for the state switch, inclusion of methanol in the zeolites, should have stabilized the ππ* state. The observed

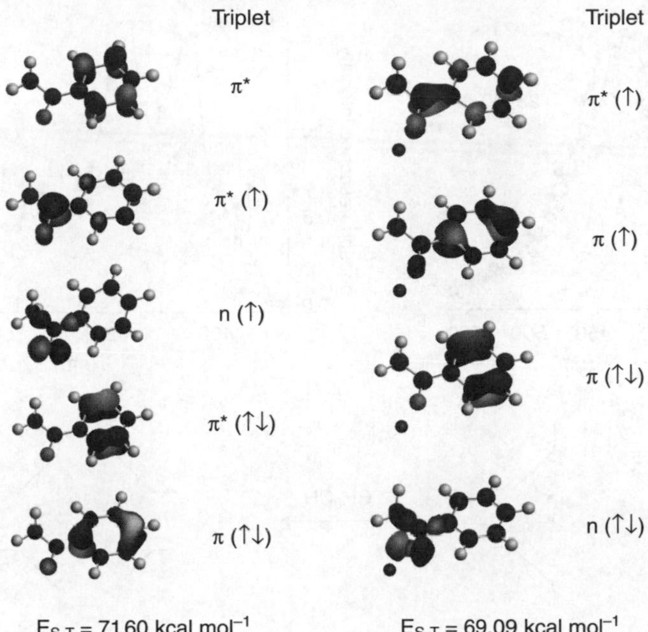

Figure 10.9. Orbital diagram showing the nature of triplet excitation in the triplet state of ACP and Li^+···ACP (complexed to CO) computed at B3LYP/6-31G* level. Computed triplet energies of T_1 are included at the bottom.

effect is consistent with the model where the binding of methanol to the alkali metal ions would release ACP to exhibit its inherent triplet nπ* character.

When Li^+ binds to ACP via the phenyl ring (mode II in Fig. 10.6), the lowest triplet is predicted (B3LYP/6-31G*) to have nπ* character. The observed phosphorescence emission within LiY and NaY is characteristic of ππ* state, suggesting alternate binding of the cations to ACP. In fact, B3LYP/6-31G* computation shows energetic preference of binding to carbonyl oxygen over the phenyl ring by 17.6 kcal mol^{-1}. Thus, a comprehensive view of the results of computation, solid-state NMR, and photophysical studies clearly establish the source of the triplet state switching observed with ACP in MY zeolites to be due to alkali metal ion···ACP interaction via the carbonyl oxygen.

According to the CIS(D)/6-31+G* calculations, similar to ACP, the character of the lowest triplet states of cyclopentenone and cyclohexenone are also expected to switch from nπ* to ππ* due to interaction with alkali metal ions.[21] Upon irradiation of alicyclic enones within zeolites, a most remarkable consequence of alkali metal ion-controlled state switching can be seen in the photoproduct distributions.[22,23] Steroidal enones such as cholesteryl acetate **(11)** yield product **13** from nπ* triplet state in solution (Scheme 10.4). However, within NaY zeolite, the major product **12** is derived from the ππ* triplet state. Formation of the β-hydrogen insertion product **12** within zeolite is consistent

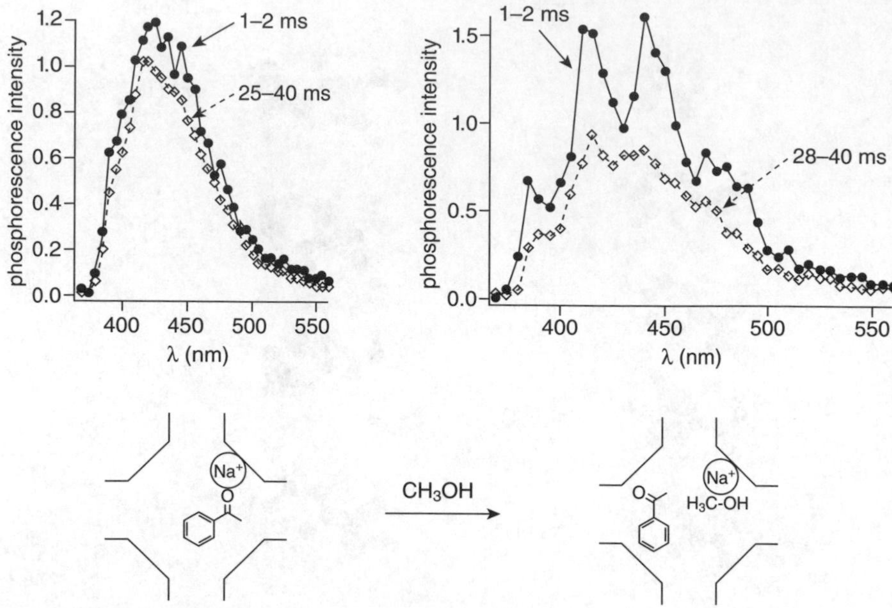

Figure 10.10. (Left) Time-resolved phosphorescence at 77 K. (Right) Time-resolved phosphorescence pulse (λ_{ex} = 320 nm) at 77 K.

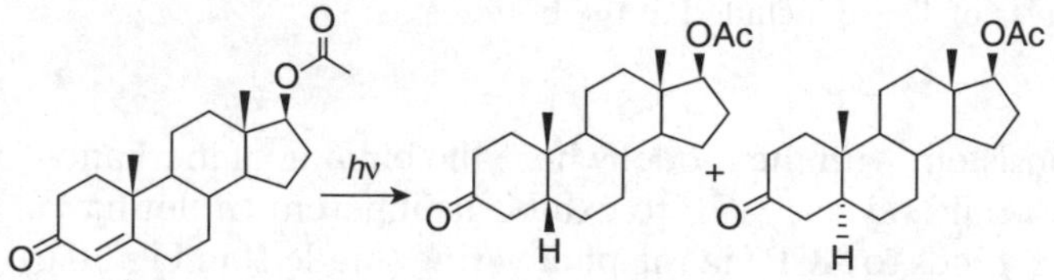

hν

5-β (ππ*) 5-α (nπ*)

11	12	13
Hexane	No reaction	
2-Propanol	27%	73%
NaY (Si:Al: 2.5	85	15
NaY (Si:Al: 300)	No reaction	

Scheme 10.4.

with its origin from the ππ* state.[24] Androstenedione (**14**) has been established to react in solution mainly from the cyclopentanone D-ring (Scheme 10.5). No products due to reactions from the D-ring are seen upon irradiation of androstenedione included within NaY. More importantly within NaY, similar to cholesteryl acetate, the photoreduction product **18** via β-hydrogen insertion is obtained as the major product. The changes in the reactivity of cholesteryl acetate and androstenedione when included within NaY is believed to be due

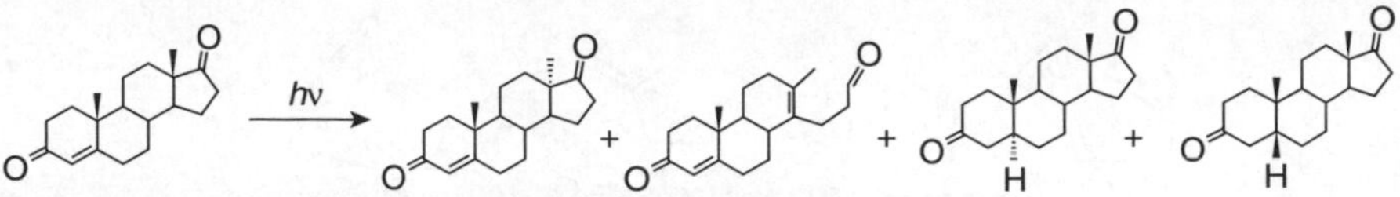

14	15 (nπ*)	16 (nπ*)	17 (nπ*)	18 (nπ*)
Hexane	40%	4%	–	–
2-Propanol	18	2	14	–
Na Y (Si:Al: 2.5)	–	–	15	85
High Silica Y (Si:Al: >285)	30%	–	–	–

Scheme 10.5.

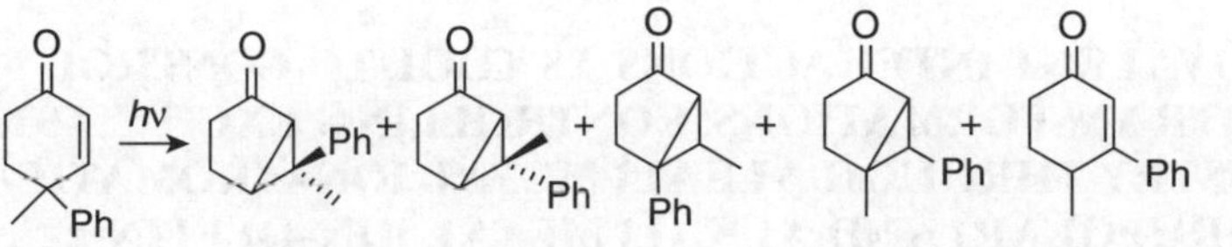

	$^3\pi\pi^*$ (20, 21, 22)	$^3n\pi^*$ (23, 24)
Benzene	0%	100%
Ethanol	33	67
NaY/hexane	70	30

Scheme 10.6.

to the lowering of the ππ* excited state of the cyclohexenone A ring. Enone **11** does not react while enone **14** gives only nπ*-derived product **15** in Y–Sil, the zeolite with no cation. These observations emphasize the role of cations in altering the nature of the lowest triplet.

As shown in Scheme 10.6, of the several products on excitation of 4-methyl-4-phenyl-2-cyclohexenone (**19**), products **23** and **24** have been established to arise from the nπ* triplet and products **20**–**22** from the ππ* triplet.[23] Consistent with the behavior of cholesteryl acetate and androstenedione, direct irradiation of enone **19** included within NaY gave higher yields of products **20**–**22** (70%) derived from the ππ* triplet than in nonpolar benzene (0%) or moderately polar ethanol (33%).

Triplet sensitization of 3-methyl-3-(1-cyclopentenyl)butan-2-one (**25**) yields the 1,3-acyl migration product **26** from the nπ* triplet (and nπ* singlet) and the oxa-di-π-methane product **27** from the ππ* triplet (Scheme 10.7).[23] The oxa-di-π-methane product **27** was obtained in higher yield within zeolite than in nonpolar or polar solvents. The selectivity in favor of the ππ* triplet product observed in zeolites once again is consistent with the model that alkali ions lower the ππ* triplet below the nπ* triplet state upon interaction with enones.

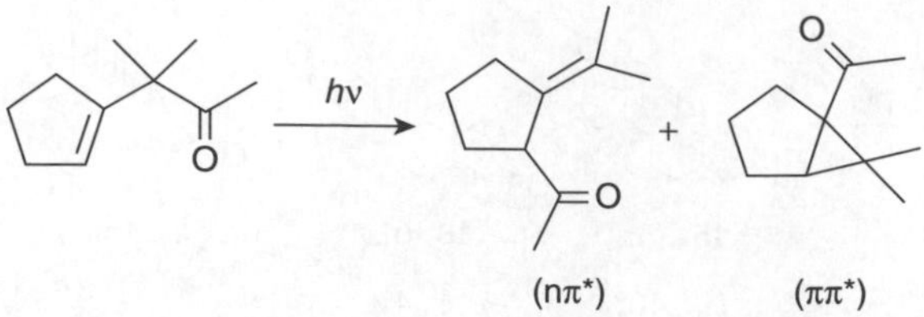

25		26	27
		($n\pi^*$)	($\pi\pi^*$)
	Hexane	100%	0%
	Ethanol	78	22
	KY/hexnae	26	74

Scheme 10.7.

NONCOVALENT INTERACTIONS AS TOOL TO CONTROL PHOTOTRANSFORMATIONS: CONTROLLING EXCITED-STATE CHEMISTRY THROUGH ALKALI METAL ION–AROMATIC-π (QUADRUPOLAR) AND ALKALI METAL ION–OLEFIN-π INTERACTIONS

Direct evidence in favor of alkali metal ion site for the location of benzene, xylenes, and mesitylene within NaY based on powder neutron diffraction studies at low temperatures (<10 K) exists.[25,26] The stability of the alkali metal ion site is believed to be due to cation–π-type electrostatic interaction (quadrupolar).[6] Such an interaction most likely plays a role during the adsorption of larger aromatic molecules such as naphthalene, phenanthrene, and pyrene within zeolites. For example, ^{2}H NMR spectra of larger aromatic molecules (e.g., phenanthrene-D_{10}) in MX zeolites consist of the two components, narrow and static, where the relative amounts vary with temperature.[27] The heat of dissociation of the ion–phenanthrene complex has been estimated on the assumption that the static component is due to alkali metal ion–phenanthrene-bound state and the narrow component to a free state (unbound to alkali metal ions). Supporting the general notion that aromatic molecules interact with the alkali metal ions present in a supercage, a linear relationship between the heat of dissociation and charge density of the alkali metal ion has been observed (enthalpies of interaction: Na^+ −14.9; K^+ −11.0; Cs^+ −7.9 kcal mol^{-1}).

This section will feature the effects on the excited-state behavior of guest molecules resulting from the electrostatic interaction between alkali metal ions and aromatic molecules within zeolites. The role of alkali metal ions during adsorption of aromatic molecules within a zeolite is revealed by the fluorescence spectra of phenanthrene (Fig. 10.11).[28] Only the monomer fluorescence is observed upon excitation of phenanthrene included within anhydrous zeolite (one molecule per 10 supercages). However, on adsorption of water by the zeolite, in addition to the monomer emission, fluorescence from microcrystal is also detected (Fig. 10.11). The association of phenanthrene to

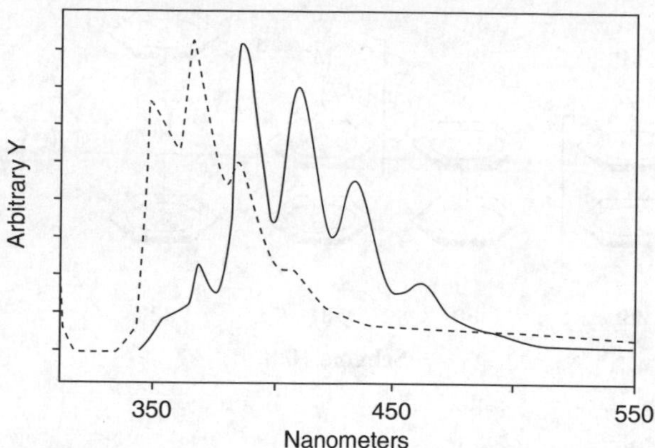

Figure 10.11. Fluorescence spectra of phenanthrene. <S>: 0.4, included within "dry" and "wet"; NaY: ---"dry" NaY, λ_{ex}: 293 nm; — "wet" NaY, λ_{ex}: 320 nm.

TABLE 10.2. Binding Affinities (kcal mol^{-1}) Computed at MP2 Level for M$^+$···Benzene···Benzene System[a]

Metal Ion	Structure **28**	Structure **29**	Structure **30**	Structure **31**	Structure **32**	Structure **33**
Li	43.78	48.62	2.08	6.93	81.1	75.8
Na	29.66	33.34	2.08	5.77	56.2	53.9
K	16.74	19.68	2.08	5.03	32.3	32.6
Rb	14.63	17.04	2.08	4.50	27.7	28.2
Cs	11.94	14.04	2.08	4.19	22.6	22.7

[a]See Scheme 10.8 for description of the structures.

form phenanthrene microcrystals is prompted by the displacement of phenanthrene, by water, from the internal to external surface. It is believed that under anhydrous conditions, phenanthrene is adsorbed onto the internal surfaces of zeolite through alkali metal ion–arene interaction, whereas in a hydrated zeolite, the stronger binding of water molecules to the alkali metal ion displaces the phenanthrene molecules to the outer surface. Consistent with this view, dehydration of the hydrated sample led to the complete disappearance of emission due to microcrystals, and detection of emission was from the monomer only.

One of the consequences of alkali metal ion coordination to the arene is its polarization such that the open face would attract an electron-rich molecule, which in this case is another arene molecule. Computational results strongly support this intuition.[29] The computed alkali metal ion–benzene BAs presented in Table 10.2 (Scheme 10.8) compare quite well with the available experimental data.[30] More important in the context of the present study is the prediction that an alkali-ion-coordinated benzene ring would interact more

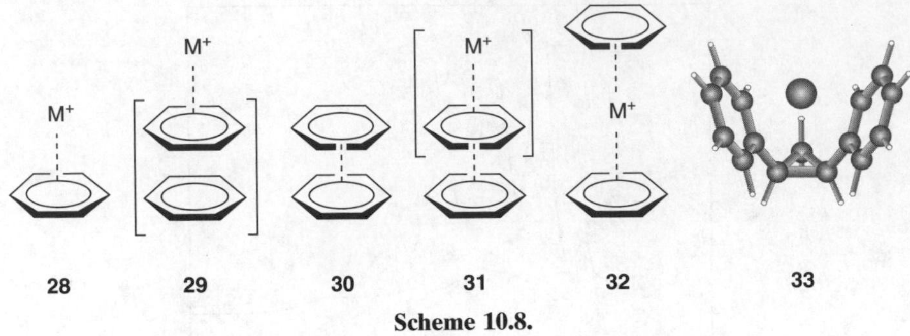

Scheme 10.8.

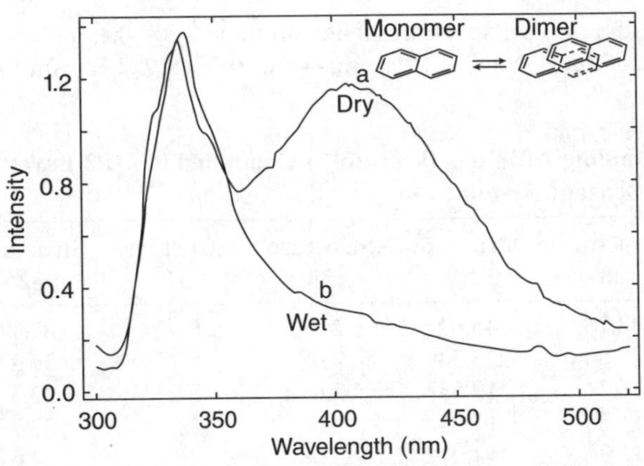

Figure 10.12. Fluorescence spectra of naphthalene included within NaY. Dry sample (a) shows both monomer and excimer emissions, and the wet sample (b) shows essentially monomer emission.

strongly with a second benzene molecule. The BA of a benzene molecule to a Li^+-complexed benzene ring is as large as 6.9 kcal mol^{-1} at the MP2/6-31G* level. Even with the weakly coordinating Cs^+ ion, the corresponding BA is double that of benzene–benzene BA (2.08 kcal mol^{-1}) at the same theoretical level. Alkali-ion-prompted ground-state aggregation of arenes is readily detected by excimeric emission of naphthalene, anthracene, and pyrene.[29,31] This phenomenon is illustrated below with the photophysical behavior of naphthalene within MY zeolite.[29]

The recorded emission spectrum at room temperature of naphthalene included in dry NaY consists of two components, from monomer and the excimer fluorescence (Fig. 10.12). The following observations suggest that the latter results from pre-aggregated dimers (static excimers) present within the supercages of NaY. The two emissions have slightly different excitation spectra. No growing in of the excimer on a nanosecond timescale was noticed when monomer and excimer decays were monitored by time-resolved single photon counting. The absence of any negative pre-exponential term for excimer decay also implies that the long wavelength emission is not due to dynamic excimers. Furthermore, the ratio of the excimer to monomer emission increased slightly upon lowering the temperature. Had the broad emission in the region 380–480 nm been due to a dynamic excimer, lowering of the temperature would have resulted in a decrease of the excimer emission. Dynamic excimer formation would require a decrease in lifetime with an increase in loading level, while the static excimer formation would not affect the lifetime at all. Varying the loading level between 0.002 and 0.2 (average number of molecules per supercage) did not alter the naphthalene monomer lifetime significantly. Thus, consistent with the model that the alkali metal ions coordinated to water molecules would have lesser polarizing power, the excimer emission can be turned on or off by co-adsorbed water (Fig. 10.12).

Manipulating Photoisomerization of Diphenylcyclopropanes within Zeolites

Based on the optimized structures, the most energetically stable arrangement of a single alkali ion and two aromatic molecules is the sandwich structure in which the alkali ion is present between the two aromatic molecules (structure **32** in Scheme 10.8; Table 10.2). Although the sandwich type of alkali ion binding to two benzene rings is less likely within a zeolite, a predesigned molecule with a cavity such as the bowl-shaped *cis*-1,2-diphenylcyclopropane (DPC) with optimally poised phenyl rings may be expected to bind to an alkali ion within a zeolite (structure **33** in Scheme 10.8). Ab initio calculations on alkali metal ion complexes with *cis*-DPC reveal that the BAs are comparable to those of sandwich structures with free benzene ligands and are also alkali metal ion dependent (Table 10.2).[32] In contrast to *cis*-DPC, in the *trans* isomer, the alkali metal ion would bind only to a single benzene ring. This insight allowed us to selectively convert *trans*-DPC **35a** to the *cis* isomer **34a** (Scheme 10.9). Upon triplet sensitization, DPC in solution gives a photostationary state consisting of *cis* and *trans* isomers in the ratio 45:55. However, within NaY upon triplet sensitization, the photostationary state was enriched with the *cis* isomer up to 92%.

In the bowl-type structure **33**, any motion disrupting the cooperative interaction of the alkali metal ion to the two phenyl groups would be resisted both in the ground and excited states. Since conversion from the *cis* geometry to the *trans* via 1,3-diradical intermediate involves such a disruption, this process is expected to have a barrier in the excited state within a zeolite. Consistent with the postulate that the alkali metal ion is most important for the *cis* enrichment within zeolites, inclusion of water (a better coordinator to the alkali ion) within LiY resulted in a photostationary state similar to that in solution (*cis*:*trans*, 49:51). Furthermore, the *cis*:*trans* ratio in Cs^+, which has small interaction energy (Table 10.2), gives much less *cis* (CsY: 65%) in the photostationary state than Li^+ (LiY: 95%).

34 *cis* $\xrightarrow{h\nu \text{ (or) Sens}^3}$ **35** *trans*

cis to *trans* at the photostationary state

	Solution	LiY	NaY	KY
a: R = –H	45:55	95:5	92:8	88:12
b: = $-CONHCH_2CH_2CH_2CH_3$	48:52	12:88	40:60	35:65
c: = $-CONHCH_2Ph$	50:50	11:89	35:65	58:42

Scheme 10.9.

If the above model is correct, isomerization from the *cis* to the *trans* geometry should be possible in systems with preferential binding of the alkali metal ion to a site other than the *cis*-diphenyl groups. The photobehavior of the amides of 2,3-diphenylcyclopropane-1-carboxylic acid (**34/35b** and **34/35c**; Scheme 10.9) is consistent with this expectation.[33] For example, the alkyl amides of 2,3-diphenylcyclopropane-1-carboxylic acid upon triplet sensitization within NaY gave both *trans* and *cis* isomers similar to that in solution (Scheme 10.9). Computational (B3LYP/6-31G*) results on methyl amide of 2,3-diphenylcyclopropane-1-carboxylic acid provide an insight into the excited-state behavior of the above molecules. Computation suggested the possibility of three structures of nearly equal BAs in which the alkali metal ion bound either solely to the amide or cooperatively to the amide and the phenyl groups or to the two phenyl groups of methyl amide of 2,3-diphenylcyclopropane-1-carboxylic acid (Fig. 10.13). No barrier for rotation of *cis* to the *trans* geometry is expected in structure 1 and structure 3 shown in Figure 10.13.

Exploitation of alkali metal ion–diphenyl cooperative binding to control photoproduct distributions is further exemplified by the excited-state chemistry of benzyl aryl ethers. For example, photolysis of benzyl phenyl ether **36** in hexane results in the formation of four products (**37–40**) in nearly equal amounts via cleavage of the benzyl–ether oxygen bond (Scheme 10.10).[29,34] On the other hand, irradiation of benzyl phenyl ether included in NaY gave *ortho*-benzyl phenol **39** as the major product. According to computation (B3LYP/6-31G*), a sandwich-type structure shown in Figure 10.14 is preferred by the Li^+–benzyl phenyl ether complex. Upon irradiation of the above sandwich-type Li^+ complex, the two fragments of the molecule resulting from benzyl–oxygen bond cleavage would be subjected to translational and rotational restrictions by the alkali metal ion. While in solution the two fragments would be free to diffuse to yield *ortho* and *para* rearranged products, within zeolites the metal ion would hold and direct them to a product requiring least motion,

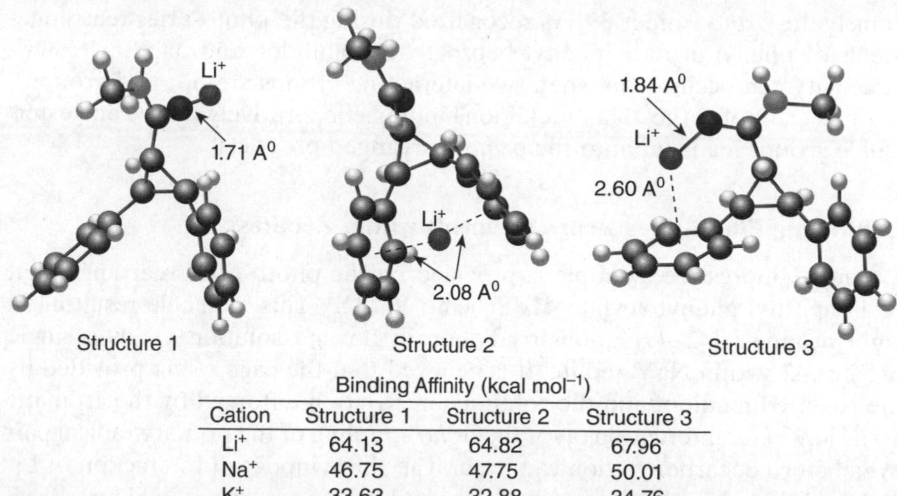

Binding Affinity (kcal mol^{-1})

Cation	Structure 1	Structure 2	Structure 3
Li^+	64.13	64.82	67.96
Na^+	46.75	47.75	50.01
K^+	33.63	32.88	34.76

Figure 10.13. B3LYP/6-31G* optimized geometries of complexes of Li^+ ion with methyl amide of 2β,3β-diphenylcyclopropane-1α-carboxylic acid. Binding affinities with Li^+, Na^+, and K^+ are also included. Note the difference in the location of cation in the three structures. The three structures have almost the same energy.

36 $\xrightarrow{h\nu}$ **37** + **38** + **39** + **40**

	37	38	39	40
Hexane	28	16	30	26
NaY	7	–	85	10

Scheme 10.10.

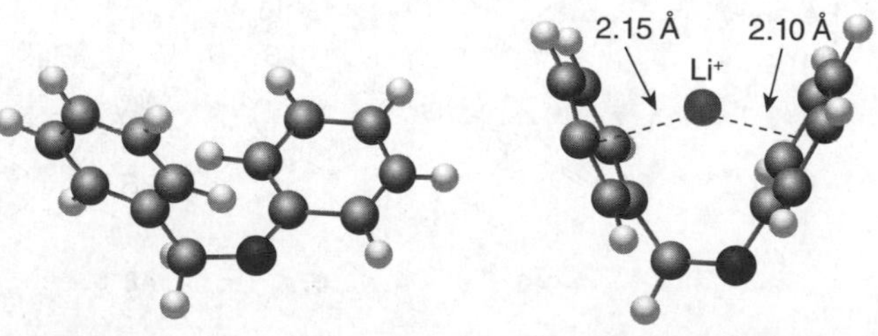

Figure 10.14. B3LYP/6-31G* optimized geometries of benzyl phenyl ether and its complex with Li^+. Binding affinity (BA) for the later complex is indicated at the bottom.

namely the *ortho* isomer **39**. As recognized during the photo-Fries rearrangements of phenyl acetate, phenyl benzoate, acetanilides, and naphthyl esters, selectivity can occur even when two interacting groups are not aryl groups.[34] In these examples, the alkali metal ion binding cooperatively to aryl and ester/amide groups leads them to the *ortho* rearranged products.

Controlling Photo-Fries Rearrangements within Zeolites

The most impressive example is provided by the photo-Fries rearrangement of 1-naphthyl phenyl acylate **41** (Scheme 10.11).[35] This molecule resulting in eight products **(42–49)** upon irradiation in hexane solution yields a single product **42** within NaY zeolite. It is believed that the cage effect provided by the confined medium and the rotational restriction enforced by the aromatic alkali ion–ester interaction favor the *ortho* coupling of the primary radical pair even before decarbonylation can occur. The three modes of interaction of Li^+ to 1-naphthyl phenyl acylate possessing similar BAs are shown in Figure 10.15. A feature common to all three structures is the cooperative binding of the alkali metal ion to two parts of the molecule that would enforce restrictions on the two fragments resulting from photocleavage of the naphthoxy and carbonyl bond. Under such conditions, *ortho* rearrangement would be favored.

The final example concerns the photolysis of α-alkyldeoxybenzoin **50** (Scheme 10.12).[36] This molecule upon irradiation in solution gives products of Norrish type I and type II processes with the latter being the major one. On the other hand, within MY zeolites, Norrish type I products (**51–53**) are favored over the Norrish type II products **54** and **55** (Scheme 10.12). The Norrish type

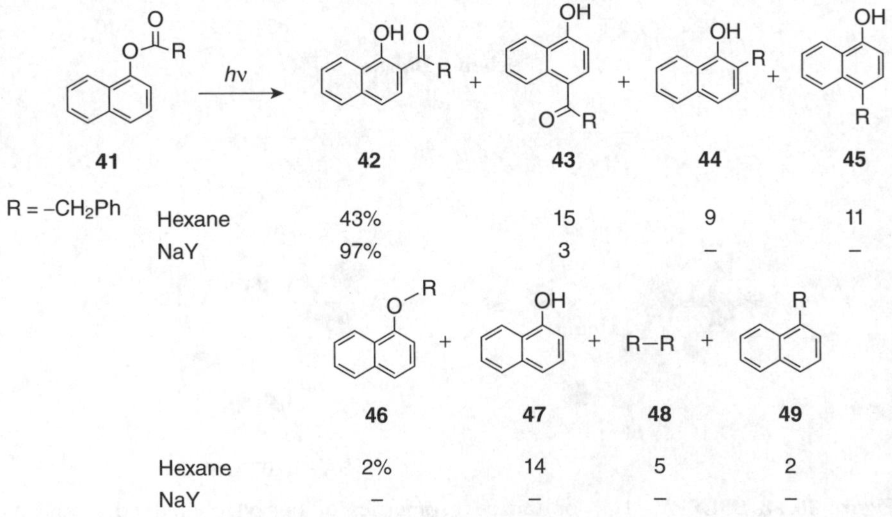

41 —$h\nu$→ **42** + **43** + **44** + **45**

R = –CH_2Ph

	42	43	44	45
Hexane	43%	15	9	11
NaY	97%	3	–	–

+ **46** + **47** + R–R (**48**) + **49**

	46	47	48	49
Hexane	2%	14	5	2
NaY	–	–	–	–

Scheme 10.11.

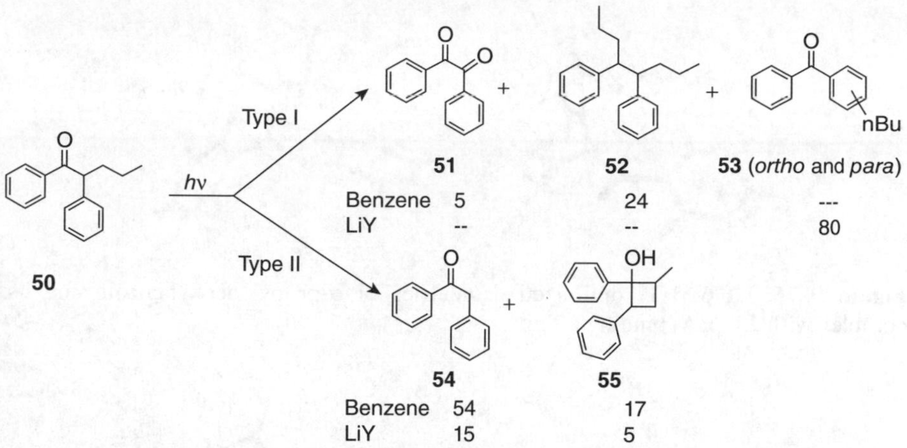

	51	52	53 (*ortho* and *para*)
Benzene	5	24	---
LiY	--	--	80

	54	55
Benzene	54	17
LiY	15	5

Scheme 10.12.

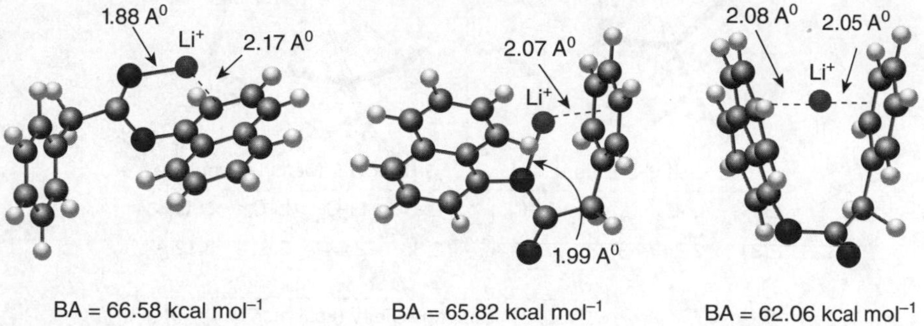

BA = 66.58 kcal mol^{-1} BA = 65.82 kcal mol^{-1} BA = 62.06 kcal mol^{-1}

Figure 10.15. B3LYP/6-31G* optimized geometries and binding affinities (BAs) of Li^+ complex with 1-naphthyl phenyl acylate.

II reaction requiring a conformation in which the γ-hydrogen approaches the carbonyl oxygen along the *n*-orbital is preferred in solution (Fig. 10.16). As seen in Figure 10.16, Li^+ through cooperative interaction with the phenyl and carbonyl groups of α-alkyldeoxybenzoins holds the molecule in a conformation where the γ-hydrogen abstraction could reach the carbonyl only along the π-face that would not result in hydrogen abstraction. Thus, the switch in the conformational preference enforced by the alkali ion turns off the generally favored Norrish type II process.

Regiocontrol during Photooxidation with Singlet Oxygen within Zeolites

Computed BAs of alkali metal ions to two representative olefins listed in Figure 10.17 show a cation–π interaction.[6,37] As seen with 5-phenyl-2-methyl-2-pentene, the presence of an aryl substituent strengthens binding through

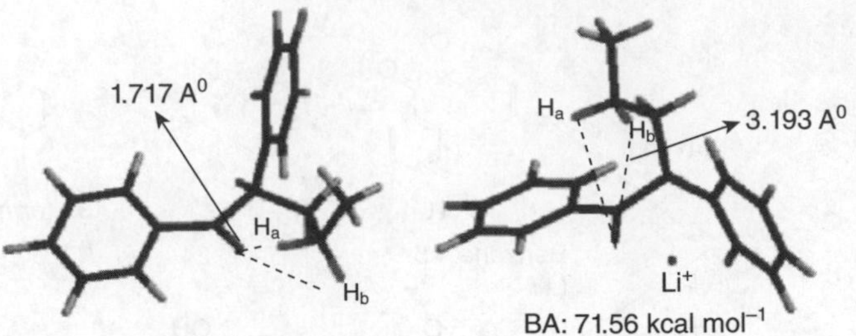

Figure 10.16. HF/6-31G* optimized geometries of α-propyl deoxybenzoin and its complex with Li^+. BA, binding affinity.

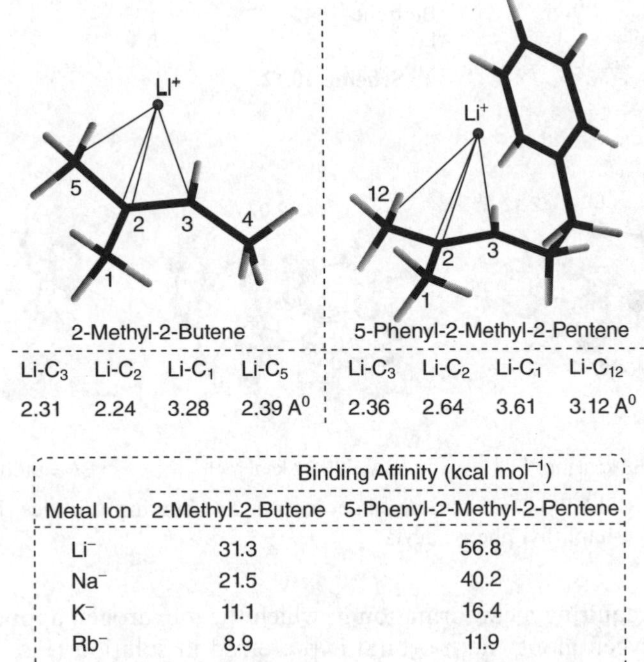

	Binding Affinity (kcal mol−1)	
Metal Ion	2-Methyl-2-Butene	5-Phenyl-2-Methyl-2-Pentene
Li−	31.3	56.8
Na−	21.5	40.2
K−	11.1	16.4
Rb−	8.9	11.9
Cs−	6.4	7.5

Figure 10.17. Calculated (B3LYP/6-31G*) key Li^+–carbon distances of Li^+ complexes (left) of 2-methyl-2-butene and (right) 5-phenyl-2-methyl-2-pentene. Binding affinities are provided below the structures.

cooperative interaction between the metal ion and olefin and aryl groups. In both the structures, Li^+ is shifted toward the less substituted side of the olefin (compare the distance between Li^+ and C_1 and C_5 in the case of 2-methyl-2-butene and between Li^+ and C_1 and C_{12} in the case of 5-phenyl-2-methyl-2-pentene). Furthermore, the ion is much closer to the less substituted olefinic carbon (compare the distance between Li^+ and C_1 and C_2). At the HF/6-31G**

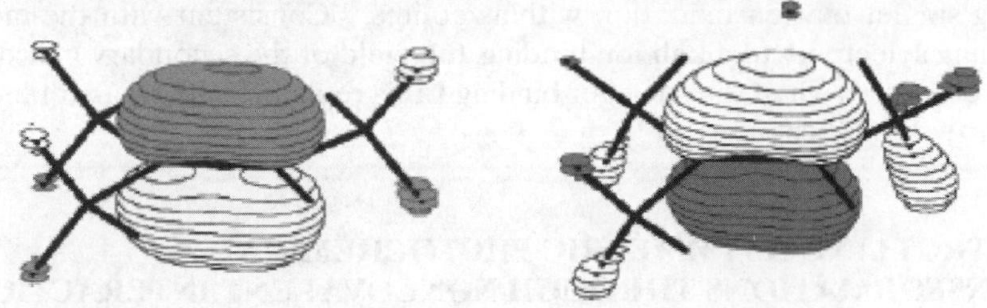

Figure 10.18. The π-HOMO of 2-methyl-2-butene (left) and its Li^+ complex (right) calculated at HF/6-31G* level.

Zeolite Y/Thionin, Hexane, *hv*, O_2: **56** → **57** (OOH) + **58** (OOH)

	57	58
Acetonitrile/RB	49	51
LiY/Thionin	74	26
NaY/Thionin	74	26
RbY/Thionin	59	41
CsY/Thionin	50	50

Zeolite Y/Thionin, Hexane, *hv*, O_2: **59** → **60** (OOH) + **61** (HOO)

	60	61
Acetonitrile/RB	53	47
LiY/Thionin	95	5
NaY/Thionin	88	12
RbY/Thionin	70	30
CsY/Thionin	52	48

Scheme 10.13.

level, an interesting orbital polarization is seen for the Li^+–2-methyl-2-butene complex (Fig. 10.18). The highest occupied molecular orbital (HOMO) lobes of the olefin are relatively larger on the less substituted olefin center. The above differences in the electronic and structural characteristics between free and alkali-ion-bound olefins suggested that the nature of the electrophilic addition of singlet oxygen to trisubstituted olefins would be different within zeolites.

It has been established that singlet oxygen adds to a trisubstituted olefin (with allylic hydrogens) from the more substituted side and yields equal amounts of both secondary and tertiary hydroperoxides (Scheme 10.13).[38] Based on the structural data presented in Figure 10.17, the alkali-ion-complexed olefin may be expected to be attacked by the singlet oxygen from the less substituted side (the electrophilic addition would be steered by M^+, which is present on the less substituted side; Fig. 10.17) to form an initial interaction with the carbon (C_2) with less substitution and higher electron density (Fig. 10.18). Such an addition would lead to hydrogen abstraction from the terminal methyl groups leading to the secondary hydroperoxide at the expense of the tertiary hydroperoxide, that is, the nonselective reaction would become selective within zeolites. Results presented for the two olefins **56** and **59** in Scheme 10.13 indicate that the expectations have been realized.[37,39] Although the details of the mechanism are still being debated, following our initial suggestion, several groups agree on the importance of alkali ion binding to olefins during singlet oxygen oxidation within zeolites.[40] Consistent with the model assigning selectivity to alkali ion binding, the yield of the secondary hydroperoxide decreases when the stronger binding Li^+ is replaced with weaker binding Cs^+ in zeolites (Scheme 10.13).

CONTROLLING ASYMMETRIC PHOTOCHEMICAL TRANSFORMATIONS THROUGH NONCOVALENT INTERACTIONS

Following the communication by Hammond and Cole in 1965 on the use of optically active sensitizers in the photosensitized isomerization of *cis*-diphenylcyclopropane,[41] several groups have performed enantio- and diastereoselective phototransformations, both in solution and the solid state.[42] Despite considerable efforts, the enantiomeric excesses (ee's) obtained under ambient conditions in solution continue to be less than 50%. The best results in solution have been obtained via the chiral auxiliary methodology, yielding in select examples' diastereomeric excesses (de's) close to 100%.[43] Asymmetric photochemistry in the crystalline state is based on the "chance" crystallization of achiral molecules in chiral space groups.[44] Because of the limited probability of this happening, there are relatively few examples of asymmetric induction during photolysis of achiral molecules in the crystalline state.[45] A more general methodology known as the "ionic chiral auxiliary approach" introduced by Scheffer facilitates the crystallization of achiral molecules in chiral space groups.[46–49] On the basis of this strategy, Scheffer and coworkers provided a number of examples that yield photoproducts in very high ee (or de) in the crystalline state. Recognizing the problem of crystallizing achiral molecules in chiral space groups, several researchers have explored chiral hosts as the reaction medium. The earliest such report is that of Natta on the photopolymerization of 1,3-dienes included in the channels of optically active perhydrotriphenylene,[50] and the most successful ones are those of Tanaka and Toda using chiral diol hosts.[51] While crystalline and host–guest assemblies have been extremely useful in conducting enantioselective photoreactions, their general applicability has been limited. Not all molecules crystallize either alone or in the presence of organic host systems. The reactivity of molecules in the crystalline state and in solid host–guest assemblies is controlled by the details of molecular packing. Currently, molecular packing, and consequently, the chemical reactivity in the crystalline state cannot be reliably predicted. Therefore, even after successfully crystallizing a molecule in a chiral space group or complexing a molecule with a chiral host or a chiral auxiliary, there is no guarantee that the guest will react in the crystalline state. One approach has been to employ readily available and inexpensive zeolites as media to bring about asymmetric induction in photochemical reactions. In this section, the different approaches that could be employed to bring about chiral induction during photoreaction within zeolites will be discussed.

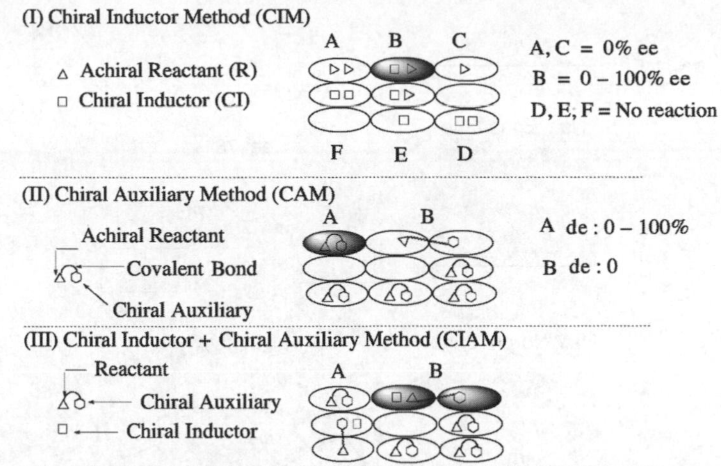

Figure 10.19. Probable occupancy of reactant and chiral inductors within zeolites. The three models represent the three approaches used in this study.

Chiral Inductor Method (CIM)

To provide the asymmetric environment lacking in zeolites during the reaction, a chiral source must be introduced. For this purpose, the approach referred to as CIM, the nonchiral interior surface of the zeolite is rendered "locally chiral" by adsorption of a chiral inductor such as ephedrine.[49,52] This simple method affords easy isolation of the product, as the chiral inductor and the reactant are not connected through either a covalent or an ionic bond. The chiral inductor that is used to modify the zeolite interior will determine the magnitude of the enantioselectivity of the photoproduct.

The strategy of employing chirally modified zeolites as reaction media requires the inclusion of two different molecules, a chiral inductor (CI) and a reactant (R), within the interior space of an achiral zeolite and by its very nature does not allow quantitative asymmetric induction. The six possible occupancies of two different molecules CI and R included within a given zeolite are shown in Figure 10.19I: cages containing two R molecules (type A), one R and one CI (type B), single R (type C), two CIs (type D), a single CI (type E), and no CI or R molecules (type F). The products obtained from the photoreaction of R represent the sum of reactions that occur in cages of types A, B, and C, of which B alone leads to asymmetric induction depending on the interaction between the chiral inductor and the reactant. Obtaining high asymmetric induction therefore requires the placement of each reactant molecule next to a chiral inductor (type B situation), that is, enhancement of the ratio of type B cages to the sum of types A and C.

To examine the viability of CIM, a number of photoreactions were explored viz., electrocyclic reactions,[53,54] the Zimmerman di-π-methane reaction,[55] the

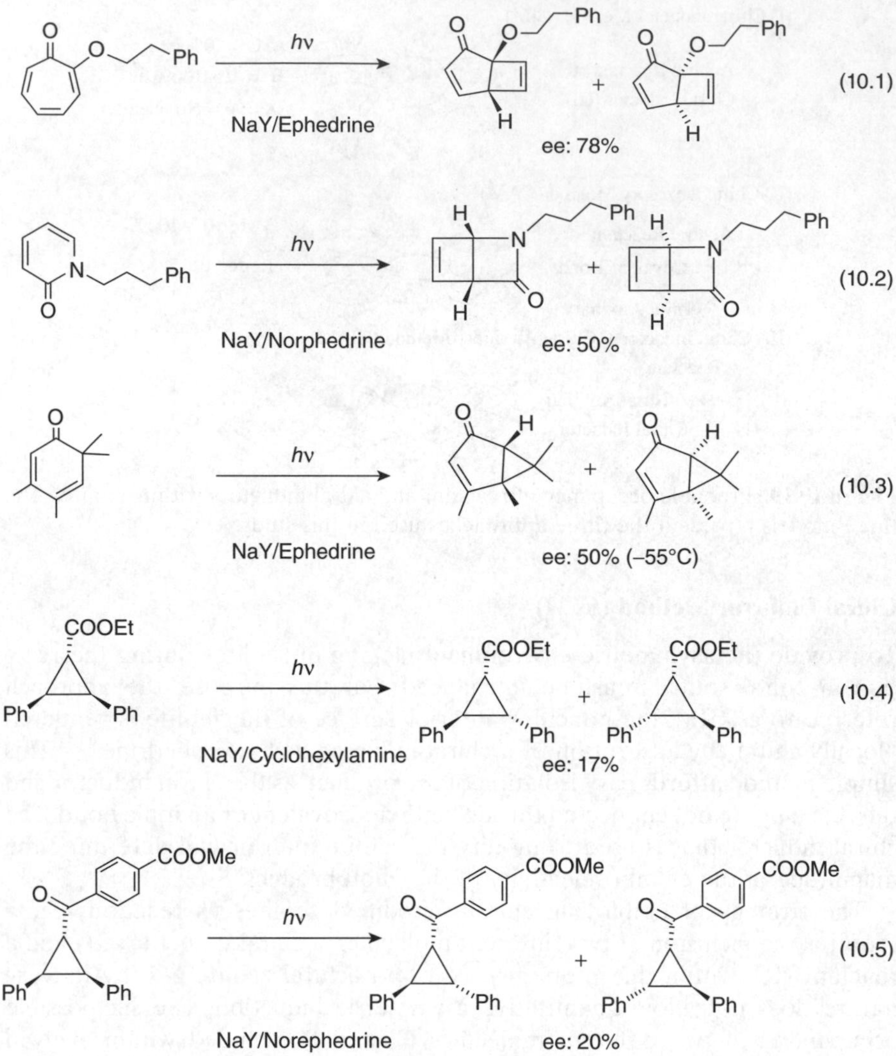

hν, NaY/Ephedrine, ee: 78% (10.1)

hν, NaY/Norphedrine, ee: 50% (10.2)

hν, NaY/Ephedrine, ee: 50% (−55°C) (10.3)

hν, NaY/Cyclohexylamine, ee: 17% (10.4)

hν, NaY/Norephedrine, ee: 20% (10.5)

Scheme 10.14.

oxa-di-π-methane rearrangement,[56] Yang cyclization,[57,58] geometric isomerization of DPC derivatives,[59–65] and the Schenk-ene reaction.[55] All these reactions gave racemic products even in the presence of chiral inductors in solution (see Schemes 10.14 and 10.15 for selected examples). Although most systems gave moderate ee's (15–50%), two reactions gave respectable ee's within NaY, namely, the photocyclization of tropolone phenylethyl ether (78%, Eq. 10.1, Scheme 10.14)[53,54] and photocyclization of 1-(3-phenylpropyl)-2-pyridone (50%, Eq. 10.2, Scheme 10.14).[66] Considering that in solution all systems gave

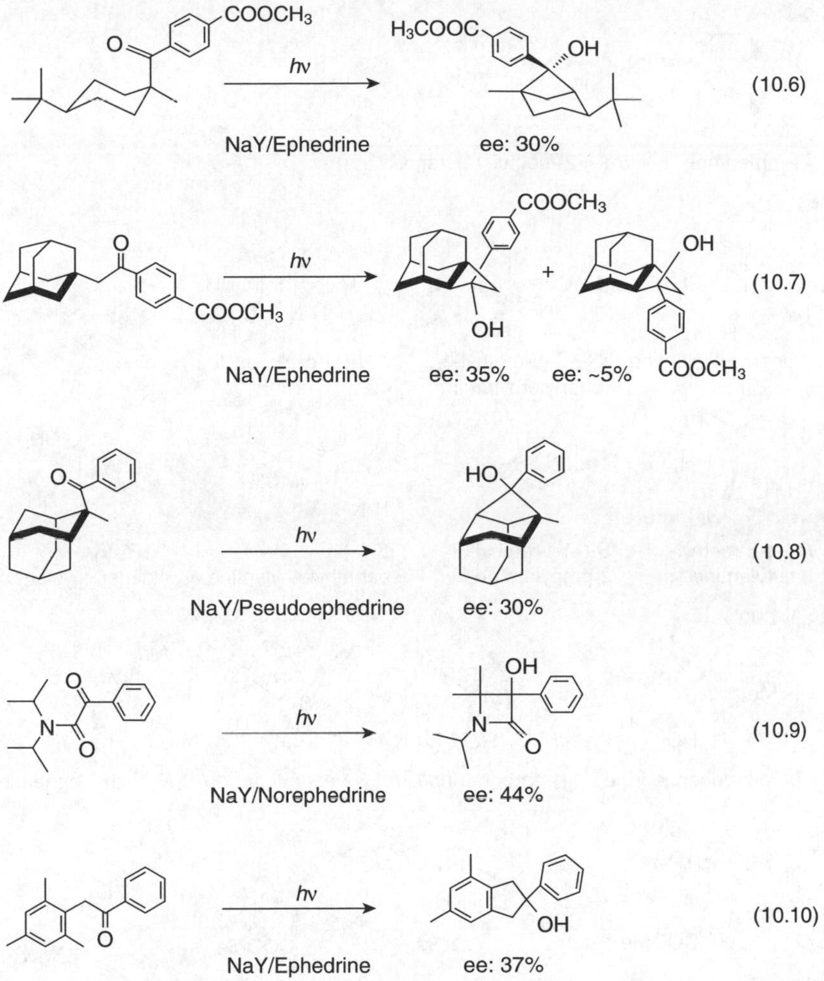

$COOCH_3$ — O — $h\nu$ → H_3COOC — OH (10.6)

NaY/Ephedrine ee: 30%

$COOCH_3$ — O — $h\nu$ → + OH (10.7)

$COOCH_3$ OH

NaY/Ephedrine ee: 35% ee: ~5% $COOCH_3$

O — $h\nu$ → HO (10.8)

NaY/Pseudoephedrine ee: 30%

O — N — O — $h\nu$ → OH — N — O (10.9)

NaY/Norephedrine ee: 44%

O — $h\nu$ → OH (10.10)

NaY/Ephedrine ee: 37%

Scheme 10.15.

racemic products even in the presence of a chiral inductor, the CIM results within zeolites are both significant and mechanistically interesting. In general, best results were obtained with ephedrine, norephedrine, or pseudoephedrine as the chiral inductor (for structures of chiral inductors and chiral auxiliaries, see Scheme 10.16).[59]

Chiral Auxiliary Method (CAM)

Unless a strategy that places each reactant molecule next to a chiral inductor within a zeolite is developed, high stereoselectivity is unlikely by CIM. This

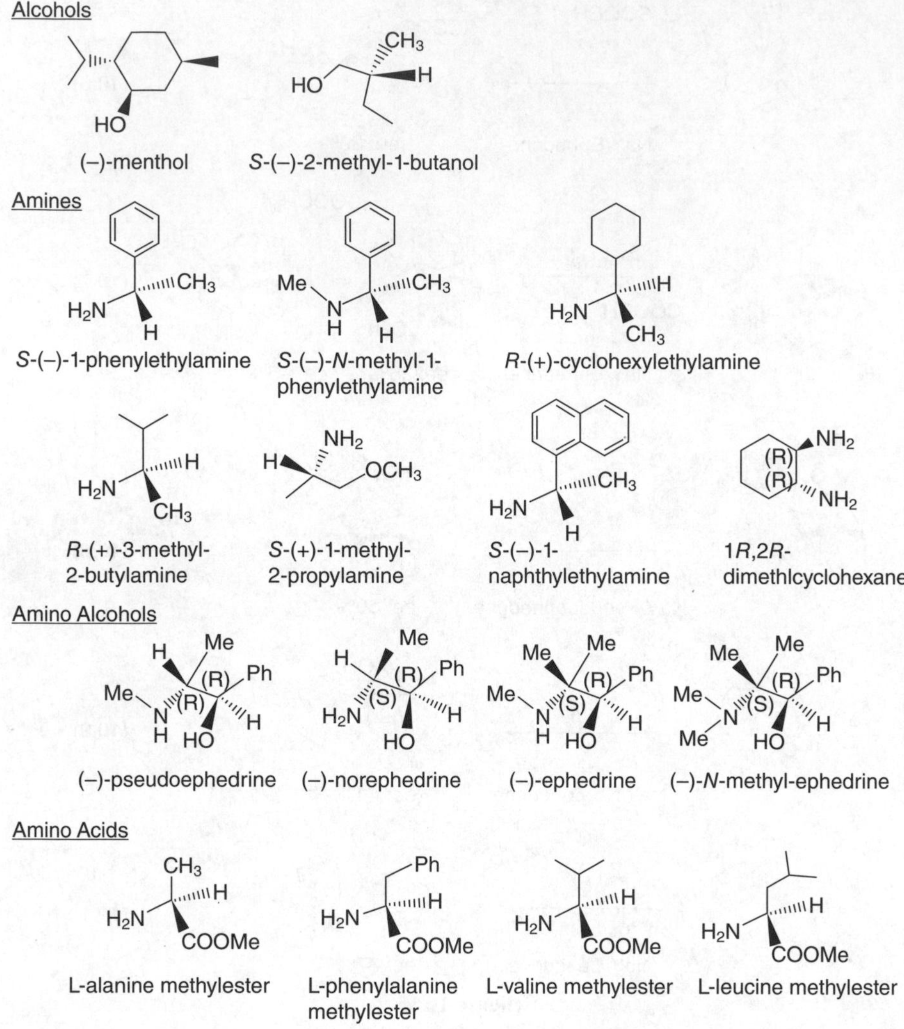

Scheme 10.16. Selected chiral inductors.

led to the CAM in which the chiral perturber is connected to the reactant via a covalent bond.[46,47,54,60,63,64,67] In this approach, most cages are expected to contain both the reactant as well as the chiral inductor components within the same cage (Fig. 10.19II). Asymmetric induction using CAM was explored for several systems (for selected examples, see Scheme 10.17; for the sake of variety, examples with different chiral auxiliaries are provided) and was found that the de's obtained within zeolites are far superior to those in solution. More than 75% de (diastereomeric excess) has routinely been obtained within MY zeolites for systems that yield photoproducts in 1:1 diastereomeric ratio in solution.

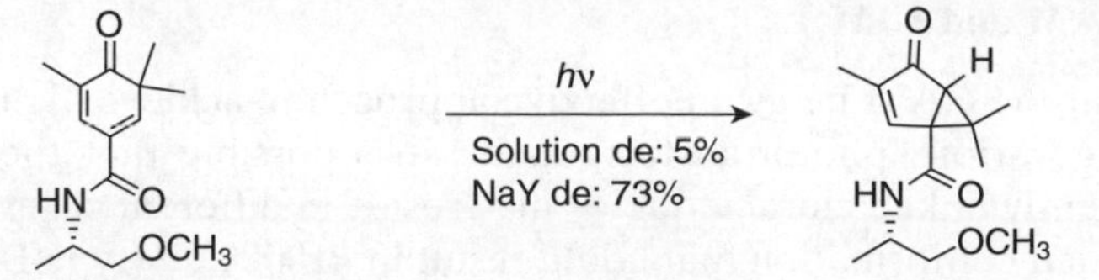

$h\nu$ — Solution de: 5%; NaY de: 73% (10.11)

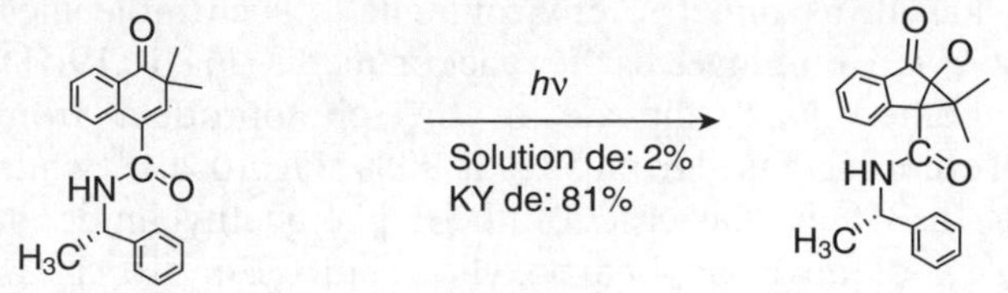

$h\nu$ — Solution de: 2%; KY de: 81% (10.12)

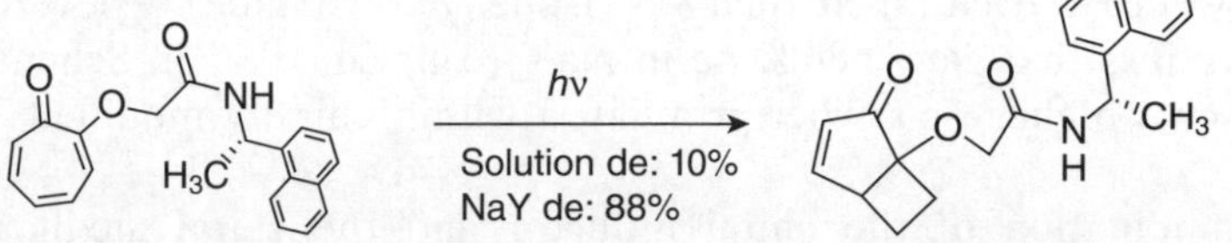

$h\nu$ — Solution de: 10%; NaY de: 88% (10.13)

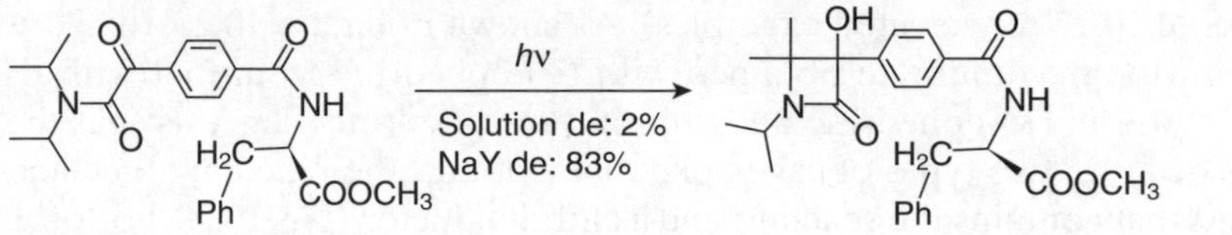

$h\nu$ — Solution de: 2%; NaY de: 83% (10.14)

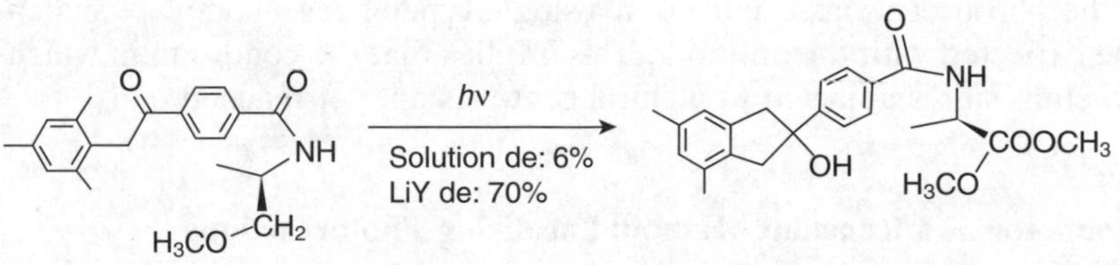

$h\nu$ — Solution de: 6%; LiY de: 70% (10.15)

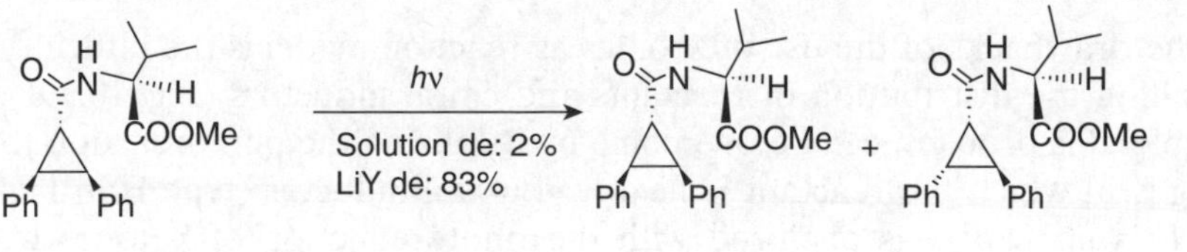

$h\nu$ — Solution de: 2%; LiY de: 83% (10.16)

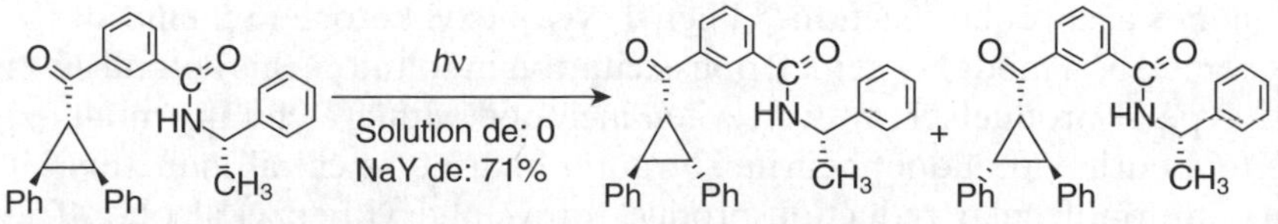

$h\nu$ — Solution de: 0; NaY de: 71% (10.17)

Scheme 10.17. Selected examples of asymmetric photoreaction explored using CAM. Chiral alcohols and amine were connected to the reactants as esters and amides, respectively.

A Combination of CAM and CIM

In spite of the fact that the CAM being an effective approach to achieve high stereoselection during various phototransformations, it is possible that the reactant and its covalently linked chiral inductor may reside in different cages by adopting an extended conformation that could result in <100% de (type B in Fig. 10.19II). To provide an asymmetric environment to such molecules, chirally modified Y zeolites were employed as the reaction media (Fig. 10.19III). Within (–)-ephedrine-modified NaY, the de in the photoproduct from tropolone-2-methylbutyl ether increased from 53% to 90% (Fig. 10.20),[46] while it decreased from 59% to 3% in the case of the 1-phenylethylamide of 2,6,6-trimethylcyclohexa-2,4-diene-1-one-4-carboxylic acid (Eq. 10.11 in Scheme 10.17 where the chiral auxiliary is 1-phenylethylamide).[56] Several tropolone derivatives that gave >80% de in NaY (e.g., Eq. 10.13 in Scheme 10.17) did not give higher de's when irradiated within chirally modified Y zeolites.

Thus, the combination of the chiral inductor and the chiral auxiliary approach had led to a limited success, and it is still somewhat unpredictable (see Scheme 10.17 for selected examples). As shown in Figure 10.20, the 20% decrease in the maximum de obtained with (+)-ephedrine from that with its antipode (90% in (–)-ephedrine and 70% in (+)-ephedrine) suggests that the reactions occur in two types of cages, one that contains the reactant alone and the second that contains the reactant and a chiral inductor (Fig. 10.20, bottom). Had all the photoreactions occurred in a single type of cage, complete switch would be expected with the antipode. This implies that the condition in which every reactant site is adjacent to a chiral center is not yet reached.

Chiral Inductor as a Reactant Method: Localizing Photoreactions to Specific Cages

One of the drawbacks of the use of zeolites as reaction media is the difficulty in controlling the distribution of reactants and chiral inductors (Fig. 10.19I). In principle, this problem can be overcome by localizing the photoreaction to those cages in which the reactant is next to a chiral inductor (type B in Fig. 10.19I). This concept was explored with the photoreduction of ketones by chiral amines as a probe reaction.[68] Phenyl cyclohexyl ketone (Scheme 10.18) was examined as a model system. Upon excitation in solution, this ketone gives Norrish type II products. However, when included within a chirally modified zeolite (ephedrine, pseudoephedrine, or norephedrine as chiral inductors), it gave the intermolecular reduction product, α-cyclohexyl benzyl alcohol. The ratio of photoreduction to Norrish type II product was dependent on the nature (primary, secondary, or tertiary) and amount of the chiral amine that acts as an electron donor. These observations indicate that photoreduction occurs only in cages that contain a chiral inductor. Using norephedrine as the chiral inductor, the ee obtained for the reduction product was 68%. As

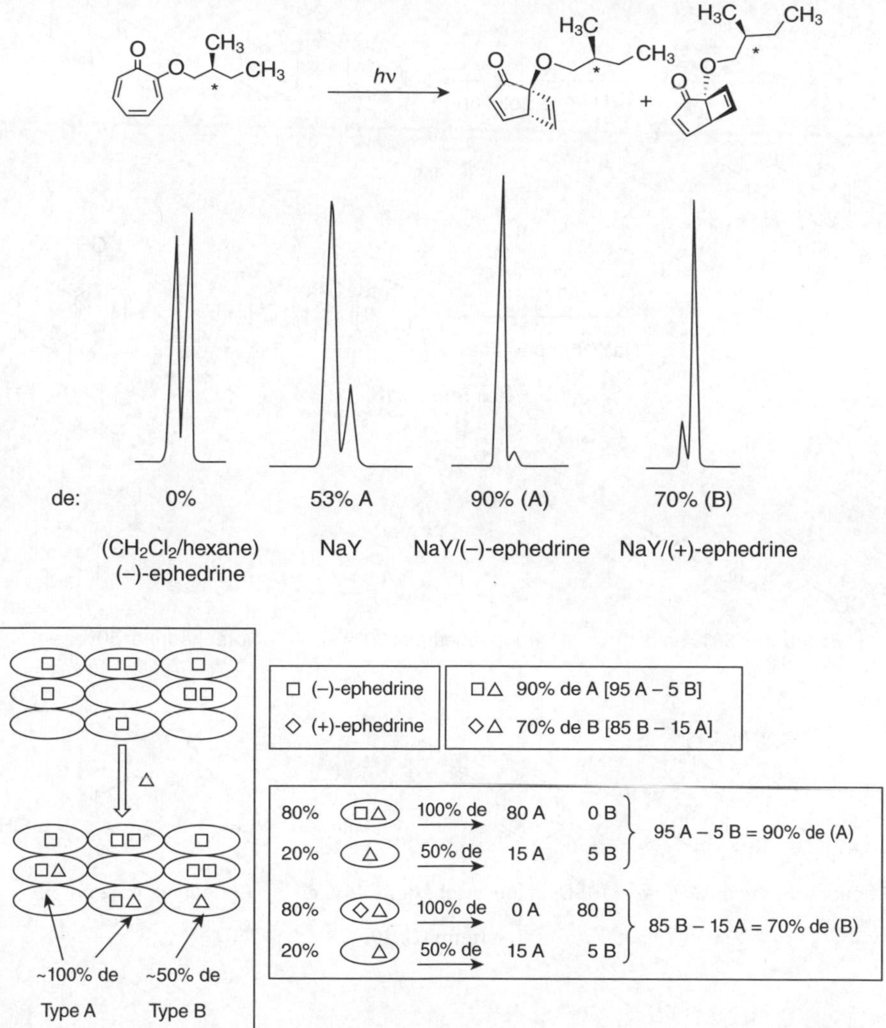

Figure 10.20. Results on the photocyclization of (*S*)-tropolone 2-methylbutyl ether within chirally modified NaY zeolite. The % diastereomeric excess (de) and the isomer enhanced are shown on the gas chromatograph (GC) traces. Note that the extent of chiral induction brought forth by the antipodes is not the same with achiral and chiral tropolone ether. The lower de obtained with (+)-ephedrine could be understood on the basis of the model presented in the bottom part of the figure.

expected, the configuration of the photoproduct was reversed by using the antipode of the chiral inductor. It is important to note that under similar conditions in solution, no ee was obtained in the reduction product that is formed in minor amounts. It was shown that the strategy presented above with phenyl cyclohexyl ketone is general by investigating a number of aryl alkyl and diaryl ketones (for selected examples, see Scheme 10.19). Despite the fact that the

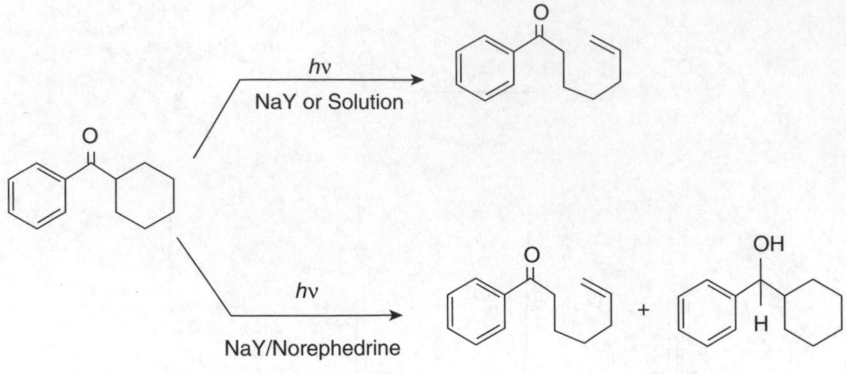

$h\nu$
NaY or Solution

$h\nu$
NaY/Norephedrine

Scheme 10.18.

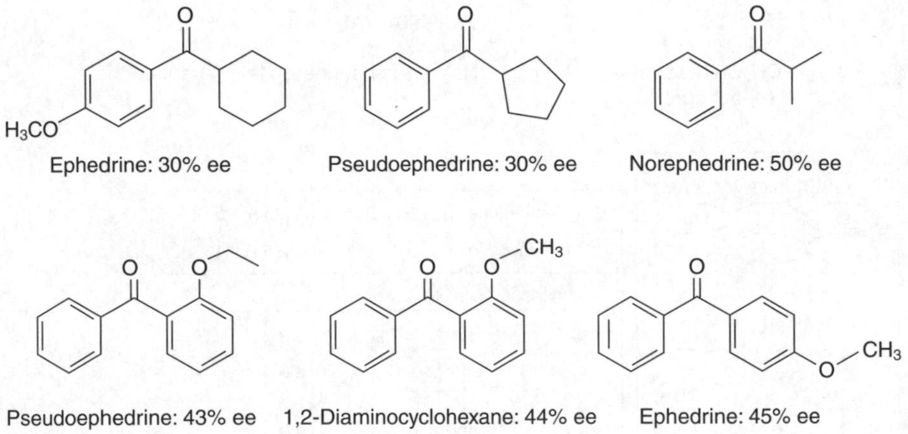

Ephedrine: 30% ee
Pseudoephedrine: 30% ee
Norephedrine: 50% ee

Pseudoephedrine: 43% ee
1,2-Diaminocyclohexane: 44% ee
Ephedrine: 45% ee

Scheme 10.19.

entire reaction occurs within chirally modified cages, the ee is still not quantitative, suggesting that there are yet unknown factors that needs to be optimized.

Role of Metal Ions in Stereoselective Phototransformations within Zeolites

Having established the superiority of zeolites over solution media for asymmetric induction during photoreactions, one needs to gain an insight into how zeolites enhance the power of chiral inductors and chiral auxiliaries. Of the four methods discussed above, the CAM is more general, normally gives high de (>70%), and is more amenable to mechanistic understanding. In this context, detailed investigations were carried out with seven independent systems within alkali ion-exchanged Y zeolites (photocyclization of tropolone

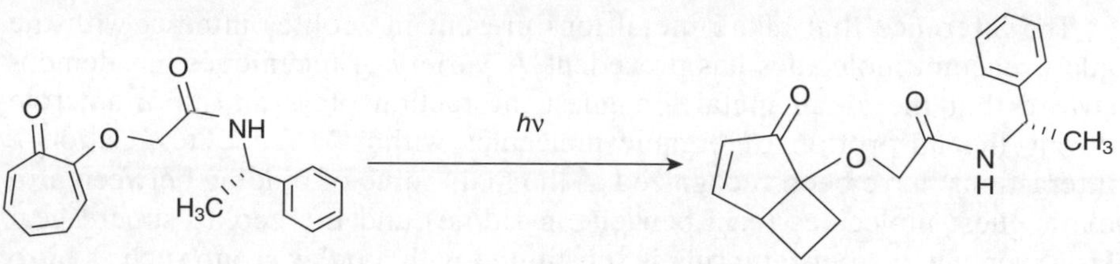

Medium	% de	NaY (si:Al ratio)	% de
Solution	2	2.4	85
LiY	64	6	22
NaY	85	15	16
KY	26	40	4
RbY	5	NaY, 2.4, hydrated	3
CsY	5		

Scheme 10.20.

alkyl ethers, γ-hydrogen abstraction reaction of 2-benzoyladamantane-2-carboxylic acid derivatives[58] and *N*,*N*-bis(1-methylethyl)-α-oxo-benzeneacetamide-4-carboxylic acid derivatives,[57] oxa-di-π-methane rearrangements of 2,2-dimethyl-1-(2H)-naphthalenone derivatives and 2,6,6-trimethyl-2.4-cyclohexadieneone-4-carboxylic acid derivatives,[56,66] and photoisomerization of 2β,3β-diphenylcyclopropane-1α-carboxylic acid derivatives and 2β,3β-diphenyl-1α-benzoylcyclopropane derivatives;[59–65] Scheme 10.17) with a variety of appended chiral auxiliaries. A few generalizations that are presented below have been reached based on the studies with these systems. However, a model with predictive powers is still not available.

For all the systems investigated, the effectiveness of a chiral auxiliary is enhanced within a zeolite compared with that in solution. As representative of the behavior of all the molecules investigated, results from one system (photocyclization of the amide derived from tropolone acetic acid and 1-phenylethylamine; Scheme 10.20) are highlighted below.[69] (1) The de was dependent on the nature of the alkali metal ion. (2) The de varied with the water content of the NaY used. (3) The de was dependent on the number of alkali ions present within a zeolite. (4) The de upon irradiation of the tropolone derivative adsorbed on silica gel, a surface that does not contain alkali metal cations, was negligible (<5%). These observations lead us to conclude that alkali metal ions present in zeolites play an important role in the asymmetric induction process. Water molecules that hydrate the alkali metal ions probably disrupt the close interaction between the alkali metal ion and the reactant molecule and thus decrease the observed de.

The inference that alkali metal ions present in zeolites interact with the guest reactant molecules has precedent. A variety of techniques has demonstrated that the alkali metal ion–guest interaction plays an important role during the adsorption of organic molecules within MY zeolites. Cation–π interactions[6] have been recognized as the main force of binding between aromatic guest molecules (e.g., benzene, pyridine) and the zeolite supercage.[25] However, when a benzene ring is substituted with a polar group such as nitro or carbonyl (dinitrobenzene, ACP), the primary interaction is between the alkali ions and the oxygen atoms of the nitro or carbonyl groups (due to cation–dipolar interaction).[4,20] How does the alkali ion–reactant interaction influence the power of a chiral auxiliary? Computational studies indicate that the conformations of free and Na^+-bound phenylalanine are different.[70] Furthermore, in this case, alkali ion···π, alkali ion···OC dipolar, and alkali ion···nitrogen lone pair interactions restrict the conformational mobility of Na^+-bound phenylalanine and other organic molecules.[5,71,72] It is believed that the enhanced asymmetric induction within a zeolite (with respect to that in solution) could be understood on a similar basis. Two examples discussed below provide a glimpse of how an alkali ion might influence asymmetric induction within a zeolite.

The enhanced de in the products derived via γ-hydrogen abstraction from several chiral esters of 2-benzoyladamantane-2-carboxylic acid within a zeolite compared with that in solution (Scheme 10.21) may be understood on the basis of conformational restriction brought forth by the alkali ion–reactant interac-

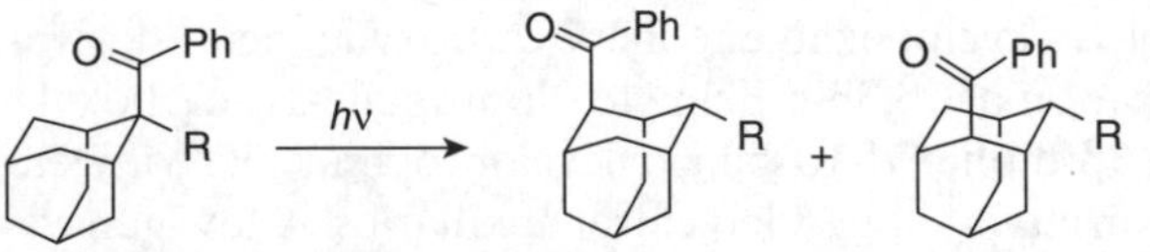

Diastereomeric excess on the product δ-ketoesters upon irradiation of the chiral esters of 2-benzoyladamantane-2-carboxylic acids

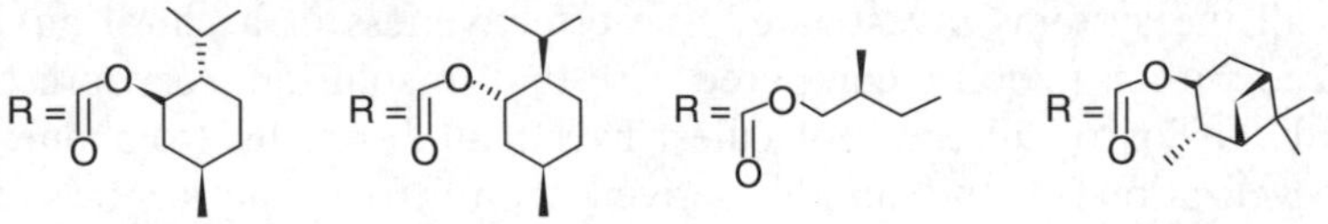

Medium	(–)-Menthylester	(–)-Isomenthylester	*S*(–)-2-methylbutylester	(+)-Isopinocampheolester
CH_3CN	22(A)	14(B)	5(A)	5(B)
LiY	79(B)	65(A)	26(B)	52(B)
NaY	60(B)	79(A)	54(B)	62(B)
KY	31(B)	3(A)	30(B)	4(A)

Scheme 10.21.

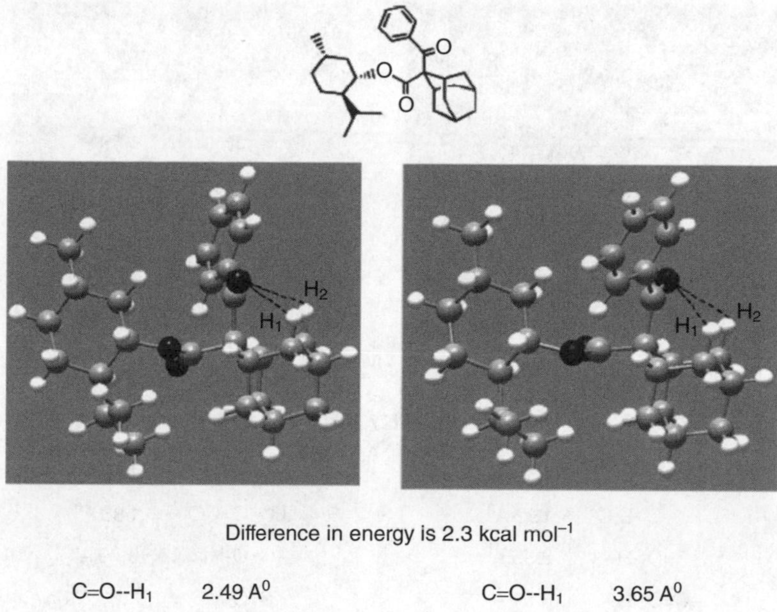

Figure 10.21. Two conformations of the menthyl ester of 2-benzoyladamantane-2-carboxylic acid as computed at the RB3LYP/6-31G(d) level. Note that the two conformations in which the carbonyl group is tilted toward the two prochiral hydrogens have different energies. The distances between the carbonyl oxygen and the prochiral hydrogens are included.

tion.[58] For example, density functional calculations (B3LYP/6-31G*)[73] suggest that the menthyl ester of 2-benzoyladamantane-2-carboxylic acid has two conformations, differing by 2.3 kcal mol^{-1}, in which the carbonyl chromophore is tilted toward one or the other diastereotopic γ-hydrogens at C4 and C6 (Fig. 10.21). This would allow equilibrium between the two conformers leading to negligible de in solution. On the other hand, in the presence of Li^+, there is a larger difference in energy between the two conformers (Fig. 10.22; 10.4 kcal mol^{-1}). Closer examination of the two structures indicates that the nature of the interaction between Li^+ and the menthyl ester of 2-benzoyladamatane-2-carboxylic acid is different in the two conformers. In structure "a" the interaction is between Li^+ and the oxygen atoms of the ketone and the ether oxygens of the ester group (CO–**O**–C), and in structure "b" the interaction is between Li^+ and the carbonyl oxygens of the keto and the ester (C**O**–O–C) groups. In these structures, the cation acts as "glue" to restrict the relative motions of the reactive and chiral auxiliary portions of the molecule (Fig. 10.22). By this process, the chiral auxiliary is able to exert a strong influence on the γ-hydrogen abstraction reaction. Once interconversion between the conformers is restricted, the reaction will take place from the

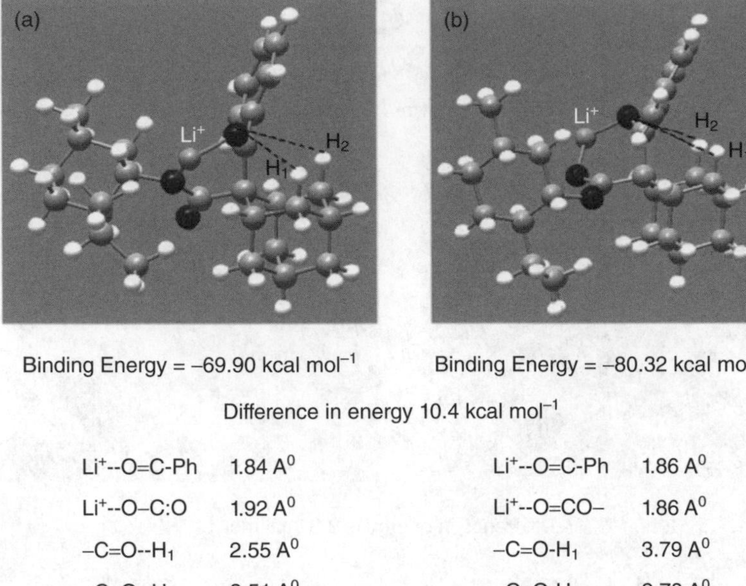

Figure 10.22. The most stable structure computed (RB3LYP/6-31G*) for the Li^+ complex of the menthyl ester of 2-benzoyladamantane-2-carboxylic acid. In the two structures, the sites of interaction of Li^+ are different. The carbonyl group is tilted toward different prochiral hydrogens in the two structures. Note that the binding energies are considerably different for the two structures.

more stable of the two conformers. Thus, while the chiral auxiliary is essential to differentiate the two diastereotopic hydrogens, the cation is important to restrict bond rotations and freeze the molecule in a conformation that leads primarily to a single cyclobutanol diastereomer.

A comparison of the effectiveness of aryl and alkyl amides as chiral auxiliaries further reveals the importance of alkali ion–chiral auxiliary interaction within zeolites. Results on three systems are highlighted in Scheme 10.22.[59,69] Generally, higher de is obtained with aryl amides than with alkyl amides as chiral auxiliaries. The important question concerning the reason for the contrasting behavior of the aryl and the alkyl chiral auxiliaries was answered with the help of ab initio computations at the Hartree–Fock level (HF/3-21G).[73] Results on one system discussed below illustrate the point. Computations carried out with 2β,3β-diphenylcyclopropane-1α-carboxamides of 1-phenylethylamine and 1-cyclohexylethylamine revealed that Li^+ binds to the substrate in several different geometries,[64] and of these, the one shown in Figure 10.23 is the most stable. In the structure shown for the 1-phenylethylamide, Li^+ interacts simultaneously with both the phenyl group and the amide carbonyl oxygen. Such an interaction is expected to reduce the rotational freedom of the chiral auxiliary and thus make it "rigid." On the

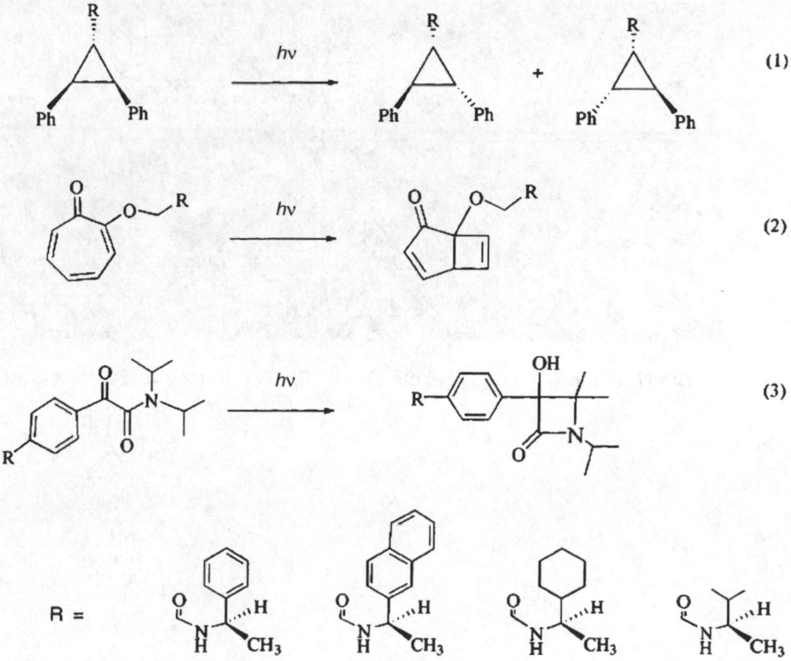

(1)

(2)

(3)

R =					
Reaction 10.18	de in LiY	80%	70%	29%	7%
Reaction 10.19	de in NaY	85%	88%	45%	45%
Reaction 10.20	de in NaY	62%	54%	22%	30%

Scheme 10.22.

other hand, in the case of the 1-cyclohexylethylamide, the cation interacts primarily with the amide carbonyl oxygen via a dipolar-type interaction and does not interact with the chiral auxiliary part (cyclohexyl group). Such a type of interaction would have no effect on the rotational mobility of the chiral auxiliary. Thus, a model based on the difference in flexibility of the chiral auxiliary due to differences in cation binding between aryl and alkyl chiral auxiliaries accounts for the observed variation in de between the two classes of chiral auxiliary (Scheme 10.22).[62–64] The prominent interaction between the alkali metal ion and the aryl group can be characterized as a cation–π interaction.[6]

While one might question the relevancy of the computed "free-space" cation-organic structures to those within a "confined space," the fact that the photochemical behavior of a large number of molecules can be understood on the basis of a model developed with computed structures suggests that simple ab initio computations are useful in understanding reactions within zeolites. The proposed model is based on the concept that alkali ion–reactant

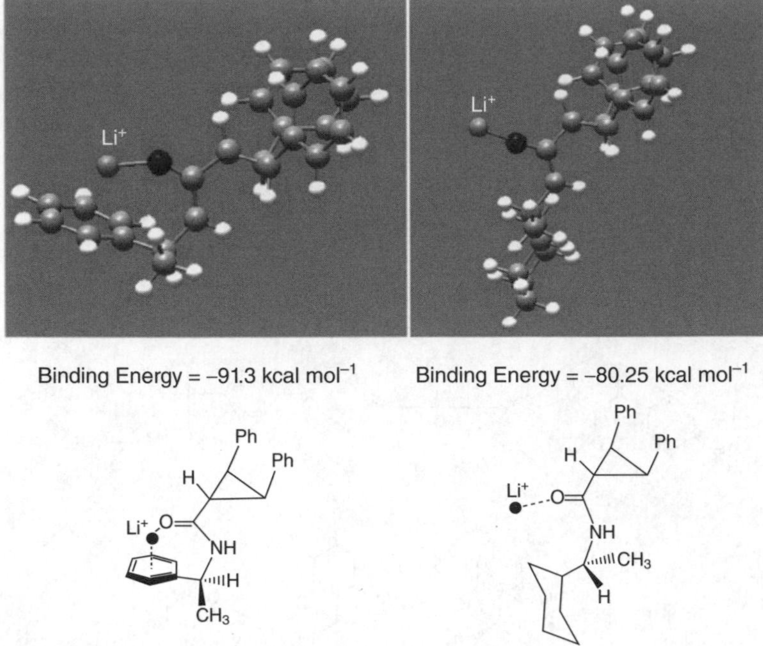

Figure 10.23. RHF/3-21G optimized structures of Li^+ bound 2β, 3β-diphenylcyclopropane-1α-carboxamides of 1-phenylethylamine (a) and 1-cyclohexylethylamine (b). Binding affinities are included at the bottom of each structure.

interactions are responsible (at least partially) for the asymmetric induction within zeolites. The alkali metal ion enhances the power of a chiral auxiliary by restricting the conformational flexibility of the chiral auxiliary and/or the reacting parts of the molecule.

Alkali Metal Ion and Reactive State-Dependent Diastereomer Switch

The alkali ions control not only the extent of de but also the diastereomer that is enhanced. Eight of the 11 amides of 2β,3β-diphenylcyclopropane-1α-carboxylic acid examined displayed alkali ion-dependent diastereomer switch (for selected examples, see Scheme 10.23).[60,64] The most remarkable example is provided by the amide derived from L-valine methyl ester (**62b**; Fig. 10.24). In this case, the de in LiY is 83% favoring the B isomer (*SS* isomer in **63b**), whereas in KY, it is 80% in favor of the A isomer (*RR* isomer in **63b**). Additionally, the alkali ion-dependent diastereomer switch can be turned off by methylating the amide nitrogen (**62d** and **62e**, Scheme 10.23).[64] For example, while the 2β,3β-diphenylcyclopropane-1-α-carboxamide derived from *S*(–)-1-phenylethylamine (**62a**) and (–)-norephedrine (**62c**) showed an alkali ion-dependent diastereomer switch, the corresponding *N*-methyl amides derived

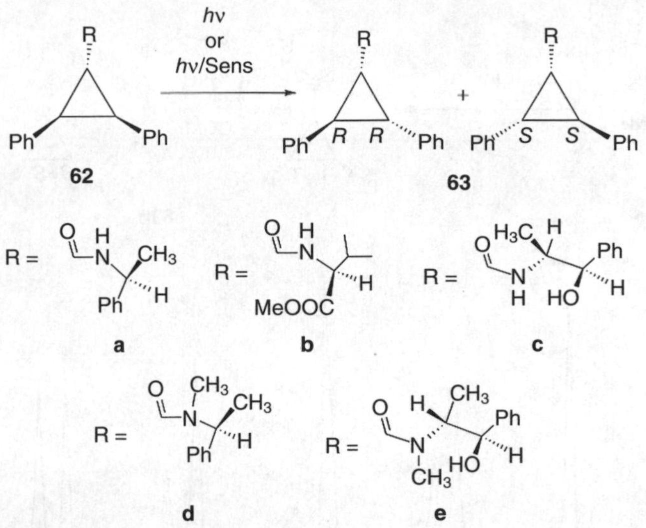

Compound	Mode	LiY	NaY	KY
62a	Direct	80-B	28-A	14-A
62a	Sens	33-B	40-A	61-A
62b	Direct	83-B	21-A	80-A
62b	Sens	52-B	18-A	81-A
62c	Direct	25-B	60-A	13-A
62c	Sens	23-B	59-B	43-B
62e	Direct	17-A	30-A	20-A
62e	Direct	58-A	89-A	15-A

A and B refer to the first and the second diastereomer peaks in the GC trace. In the case of photoproducts from **62a** and **62b**, the *trans* isomer absolute configuration for GC Peak A is *RR* isomer and GC Peak B is *SS* isomer. Sens, 4'-methoxyacetophenone.

Scheme 10.23.

from *S*(–)-*N*-methyl-1-phenylethylamine (**62d**) and (–)-pseudoephedrine (**62e**) did not show a cation-dependent diastereomer switch (Scheme 10.23).

The geometric isomerization of the amides of 2β,3β-diphenylcyclopropane-1α-carboxylic acid within zeolites could also be brought about by triplet sensitization with *para*-methoxyacetophenone as the triplet sensitizer. Three

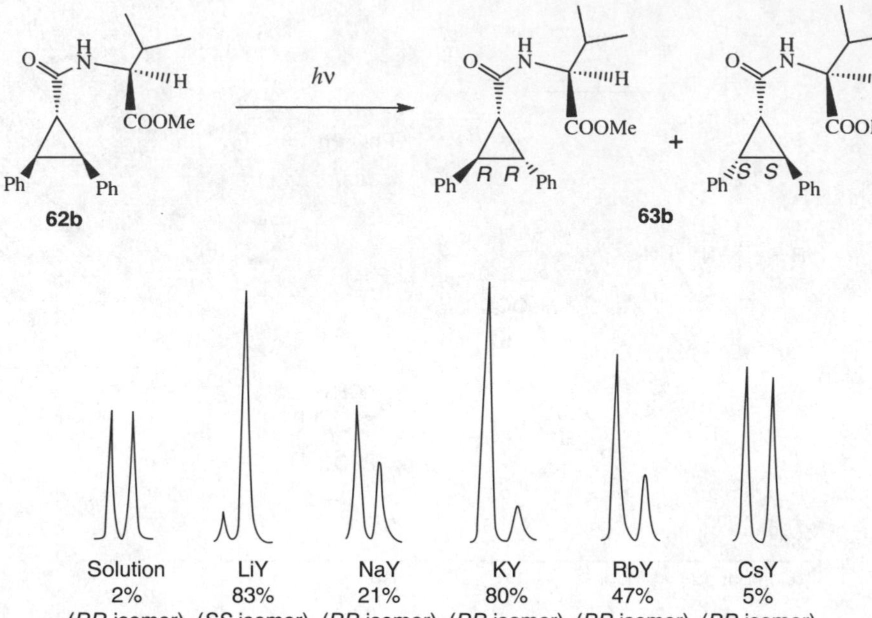

Figure 10.24. GC traces of the *trans* diastereomers upon irradiation of **62b**. Note the difference in the peaks being enhanced within LiY and KY.

examples (**62a–c**; Scheme 10.23) were investigated in this context.[62] As seen in Scheme 10.23, the de obtained under direct excitation and *para*-methoxyacetophenone sensitization were different; however, among the three examples, no clear pattern emerges. Remarkably, within NaY, direct excitation of **62c** resulted in a 60% excess of diastereomer A, whereas upon triplet sensitization, diastereomer B was favored in 59% excess. Since the mechanism of isomerization need not be the same from S_1 and T_1 states, the reactive state-dependent diastereomer switch is not totally unexpected.[65] By investigating the time-dependent photochemistry of the pure diastereomers of 2α,3β-diphenylcyclopropane-1α-carboxamide of 1-phenyl ethylamine within zeolites, it was established that the isomerization from the S_1 state proceeds through a zwitterionic intermediate, while that from the T_1 state proceeds via a triplet diradical (Scheme 10.24).[74] The observed switch in the favored diastereomer in the case of **62c** (Scheme 10.23) under the two conditions is most likely the result of variations in the ability of the chiral auxiliary to stabilize the intermediates with different characters, that is, diradical and zwitterionic (Scheme 10.24).

As shown in Scheme 10.23 (Fig. 10.24), the cation not only controls the extent of diastereoselectivity but also the isomer being enhanced. For the sake of brevity, only amide derivatives **62a** and **63b** will be discussed. The most

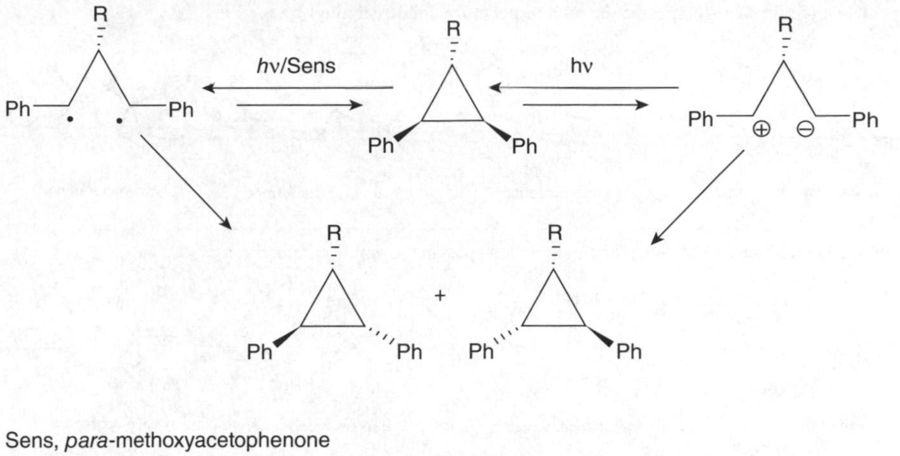

Scheme 10.24.

remarkable example is provided by the amide derived from L-valine methyl ester **62b** (Fig. 10.24). Within LiY, the *trans*-**63b**-*SS*-isomer was favored to 83% in excess, whereas in KY, the *trans*-**63b**-*RR*-isomer was favored to 80% in excess. In the case of **62a** (Scheme 10.23), LiY favored *trans*-**63a**-*SS* (80% excess) and NaY favored *trans*-**63a**-*RR* (28% excess). Similar striking results were also observed during triplet sensitization of **62a** and **62b** (Scheme 10.23). Triplet sensitization of **62b** within LiY *trans*-**63b**-*SS* was favored (de 52%), whereas in KY, *trans*-**63b**-*RR* was favored (81% de). In the case of **62a**, LiY favored the formation of *trans*-**63a**-*SS* (33% de) and KY favored *trans*-**63a**-*RR* (61% de). One possibility is that different cations bind to different sites on the reactant molecule, leading to cation-dependent diastereomer switch. This suggestion, although in this case has no experimental support, has literature precedence. On the basis of density functional calculations and low-energy, collisionally activated, and thermal radiative dissociation experiments, a difference in binding pattern between Li^+ and K^+ ions with glycine, valine, and arginine has been proposed.[72] For example, Li^+ binds to these molecules through N, O coordination (oxygen on the carboxyl group and nitrogen on the amino group), whereas K^+ binds through O, O coordination (oxygens on the carboxyl moiety). A similar switch in the binding site has also been reported during the interaction of Li^+ and Cs^+ to arginine. Such a phenomenon could be involved within a zeolite and responsible for the observed cation-controlled diastereomer switch within a zeolite. A similar alkali ion-dependent diastereomer switch was observed in the case of the amides derived from *N*,*N*-bis-(1-methylethyl)-α-oxo-benzeneacetamide-4-carboxylic acid (Eq. 10.14, Scheme 10.17).[75] To rationalize the observed cation-dependent diastereomer switch, the geometric isomerization of the amides of 2β,3β-diphenylcyclopropane-1α-carboxylic acid derivative **62** (Scheme 10.23) will be discussed in detail.

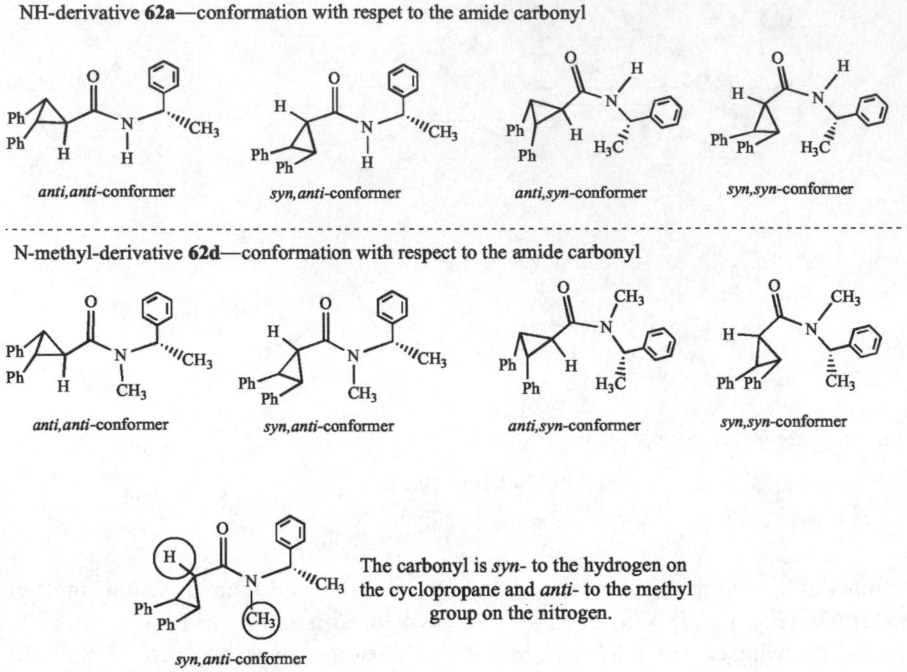

NH-derivative **62a**—conformation with respet to the amide carbonyl

anti,anti-conformer *syn,anti*-conformer *anti,syn*-conformer *syn,syn*-conformer

N-methyl-derivative **62d**—conformation with respect to the amide carbonyl

anti,anti-conformer *syn,anti*-conformer *anti,syn*-conformer *syn,syn*-conformer

The carbonyl is *syn*- to the hydrogen on the cyclopropane and *anti*- to the methyl group on the nitrogen.

syn,anti-conformer

Scheme 10.25.

To our knowledge, no alkali metal ion-dependent diastereomer switch has been reported previously, and it is clear that this unusual phenomenon needs further study.[64] Cation-bound computed structures once again provided us a rationale to understand the cation-dependent diastereomer switch. As mentioned above, diastereomer switch during both direct excitation and triplet sensitization depends on whether the amide nitrogen is substituted with a hydrogen atom or methyl group. As illustrated in Scheme 10.25, the amides investigated (examples of **62a** and **62d** are provided) could adopt four conformations, based on the rotation around cyclopropyl carbonyl and carbonyl-N of the amide linkage. For simplicity, the nomenclatures shown in Scheme 10.25 was employed to identify the four conformers. Restricted rotation around the carbonyl-N bond in amides is well documented in the literature.[76] For example, in the case of *N*-methylformamide, the energy for rotation (E_{rot}) was calculated to be around 0.4 kcal mol^{-1} and in the case of *N,N*-dimethylformamide, the energy for rotation (E_{rot}) was estimated to be around 3.82 kcal mol^{-1}.[76] Based on this, one would expect the rotational barrier for interconversion between *syn* and *anti* conformations (with respect to the carbonyl-N bond) as in **62a** to be much smaller than in **62d**. ^{1}H NMR and ^{1}H–^{1}H COSY spectra confirmed that **62d** exists as two distinct conformers, while **62a** exists as a single averaged conformer in solution. Additionally, based on NMR, it was confirmed that the two conformers (present in equal amounts) of **62d** are thermodynamically stable and do not interconvert at room temperature in solution. The NMR results suggest, as expected, a higher barrier for rotation around the amide bond (OC–NCH_3) in **62d** than (OC–NH) in **62a**. It is quite likely that the rotation around the cyclopropyl–carbonyl bond has a small barrier; hence, *syn* and *anti* conformations around this bond equilibrate fairly rapidly in both N–H and N–CH_3 systems in solution. It was visualized that within a zeolite, depending on the cation, one of the four conformers of **62a** would be preferred leading to cation-dependent diastereomer switch. In other words, the cation-dependent diastereomer switch has its origin on the cation-dependent conformer preference within a zeolite.

To gain a better insight on the role of cation binding to different conformers of **62a** and **62d** and its effect on the observed cation-dependent diastereomer switch, density functional theory and ab initio calculations were performed at the RB3LYP/6-31G* level.[73] BAs for Li^+-bound structures for **62a** and **62d** (RB3LYP/6-31G* level) are given in Figures 10.25 and 10.26. In the case of Li^+-bound **62a**, of the four conformers examined (*anti,anti-*; *syn,anti-*; *anti,syn-*; and *syn,syn*; Fig. 10.25), the first two were more stable than the latter two; the Li^+-bound *anti,anti-* and *syn,anti-*conformers have almost the same energy (Fig. 10.25). On the other hand, in the case of **62d**, cation binding leads to only one most stable conformer, that is, *anti,anti-*conformer (Fig. 10.26). During geometry optimization, surprisingly, the Li^+-bound *syn,anti-*conformer of **62d** transformed itself to the Li^+-bound *anti,anti-*conformer (Fig. 10.26). This is most likely due to steric interaction between the N–CH_3 and cyclopropyl C–H hydrogens in the cation-bound *syn,anti-*conformation. If this trend is maintained within a zeolite, cation-bound **62d**, irrespective of the cation, will adopt only the *anti,anti-*conformation. This could translate into the cation-independent diastereomer enhancement. In the case of **62a**, according to NMR, all four conformers (*anti,anti-*; *syn,anti-*; *anti,syn-*; and *syn,syn-*) having nearly the same energies equilibrate in solution. According to calculated data, upon Li^+ binding, *anti,anti-* and *syn,anti-*conformers are distinctly more stable than *anti,syn-* and *syn,syn-*conformers. Thus, within a zeolite, only these two conformers are expected to be present. It is quite possible that within a zeolite, depending on the alkali ion, one of the two conformers is preferred, leading to cation-dependent diastereomer switch. Interestingly, computational results discussed below suggest that Li^+-bound *anti,anti-* and *syn,anti-*conformers of **62a** would give different amounts of *trans*-**63a**-*RR* and *trans*-**62a**-*SS*. This provides a hope that the model based on cation-controlled conformational preference may have some validity. As shown in Figures 10.27 and 10.28, the *trans*-**63a**-*SS* is expected to be formed from **62a** via phenyl rotation at the C_3-carbon and *trans*-**63a**-*RR* via phenyl rotation at the C_2-carbon. The question is whether both *anti,anti-* and *syn,anti-*conformers of **62a** would give the same amounts *trans*-**63a**-*RR* and *trans*-**63a**-*SS*. In the absence of cations, both rotations are equally possible, and therefore, diastereoselectivity is expected to be small. However, Li^+ binding to the two conformers makes a difference. As seen in Figure 10.27, *trans*-**63a**-*SS* is more easily formed from the Li^+-bound *anti,*

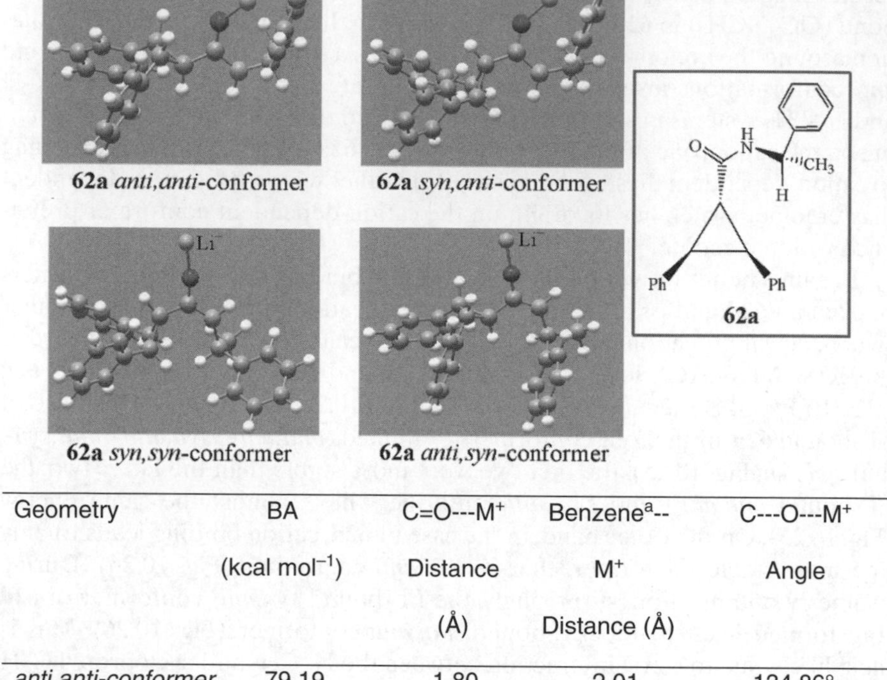

Geometry	BA (kcal mol^{-1})	C=O---M$^+$ Distance (Å)	Benzene[a]--M$^+$ Distance (Å)	C---O---M$^+$ Angle
anti,anti-conformer	79.19	1.80	2.01	134.86°
syn,anti-conformer	80.51	1.80	2.01	136.43°
anti,syn-conformer	64.39	1.70	–	175.80°
syn,syn-conformer	68.64	1.71	–	168.99°

a The benzene ring of the chiral auxiliary.

Figure 10.25. Binding affinities and structural details of Li$^+$ bound to various conformers of **62a** optimized at RB3LYP/6-31G*.

anti-conformer of **62a**. Rotation around the C_3-carbon is helped by the developing cation–π interaction[6] with the C_3-phenyl. Rotation around the C_2-carbon does not result in any such stabilization. Based on this, one would predict that the *trans*-**63a**-*SS* isomer would be formed in excess upon excitation of Li$^+$-bound *anti,anti*-conformer of **62a**. On the other hand, as seen in Figure 10.28, computational results suggest that *trans*-**63a**-*SS* and *trans*-**63a**-*RR* isomers would be formed in near equal amounts from *syn,anti*-conformer of **62a**. As illustrated in Figure 10.28, rotation of C_2-phenyl or C_3-phenyl does not result in any significant extra stabilization by Li$^+$ ion. Thus, it is clear that Li$^+$-bound *anti,anti*- and *syn,anti*-conformers of **63a** would not give *trans*-**63a** with the

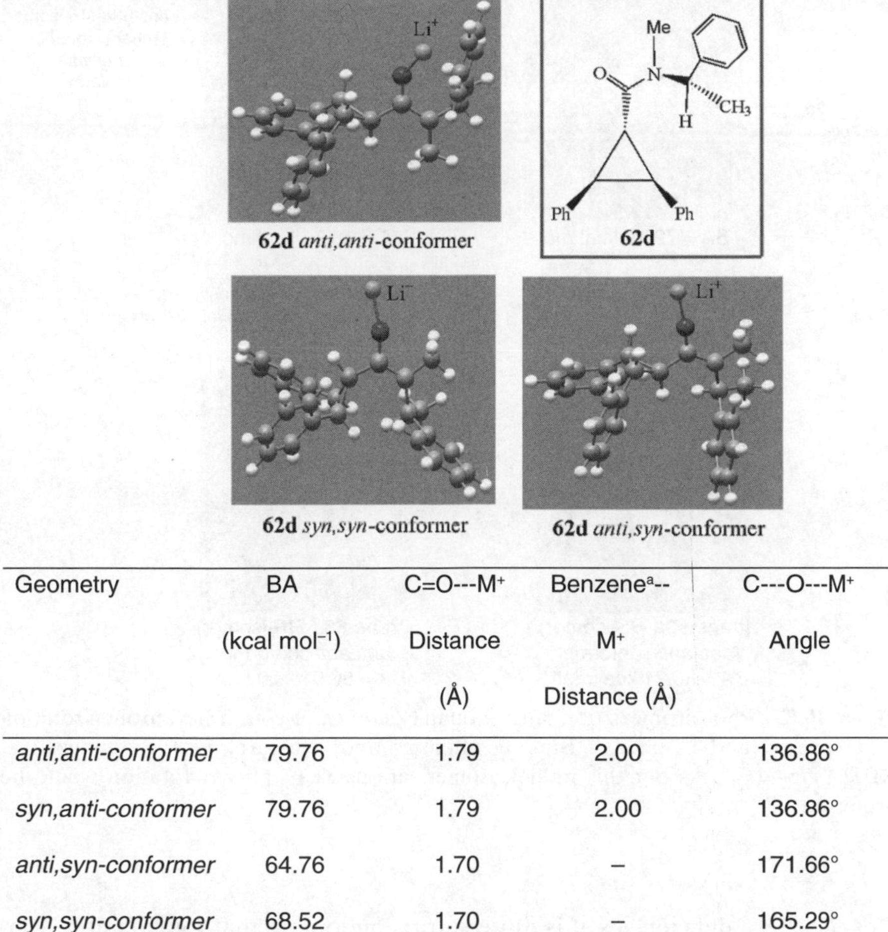

Geometry	BA (kcal mol^{-1})	C=O---M$^+$ Distance (Å)	Benzene[a]--M$^+$ Distance (Å)	C---O---M$^+$ Angle
anti,anti-conformer	79.76	1.79	2.00	136.86°
syn,anti-conformer	79.76	1.79	2.00	136.86°
anti,syn-conformer	64.76	1.70	–	171.66°
syn,syn-conformer	68.52	1.70	–	165.29°

a syn, *anti*-Conformer **62d** optimization resulted in anti, *anti*-conformer **62d,** and hence, the binding affinity is the same.

Figure 10.26. Binding affinities and structural details of Li^+ bound to various conformers of **62d** optimized at RB3LYP/6-31G*.

same diastereoselectivity. How much de is obtained and which isomer is obtained in excess will depend on which one of the two conformers is preferred within a zeolite. Selective formation of *trans*-**63a**-*SS* within LiY is an indication of *anti,anti*-conformer being preferred when **62a** is included within the zeolite. Undoubtedly, there are problems in the above approach. Computations refer to ground-state structures, while reactions originate from excited singlet and triplet states. Computations deal with closed shell molecules, while the reaction involves diradical and zwitterionic intermediates.

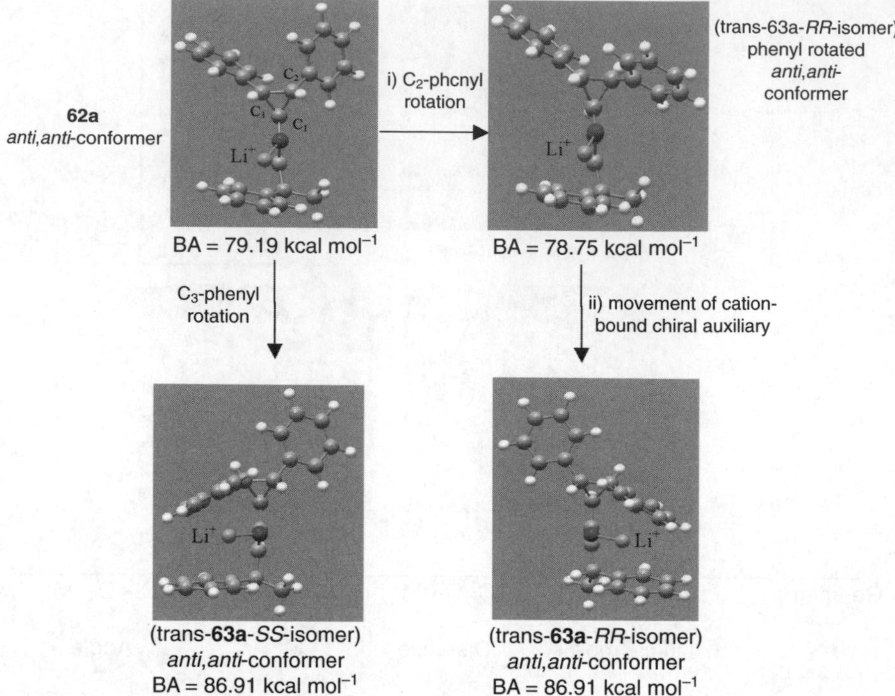

Figure 10.27. Photoisomeriztion of Li^+-bound **62a** *anti,anti*-conformer through rotation of C_2-phenyl and C_3-phenyl. Binding affinities and structural details optimized at RB3LYP/6-31G*. As per this model, isomerization via C_3-phenyl rotation would be preferred.

Despite these deficiencies, it is quite surprising to note that the ab initio computational data of the ground-state structures of Li^+-**62a** complexes provided an insight into the cause of diastereoselectivty within alkali ion-exchanged Y zeolites. The fact that different amounts of diastereoselectivities were obtained for the same systems during direct excitation and triplet sensitization suggests that nature of excited state and the intermediates involved should be taken into consideration while predicting asymmetric induction within zeolites.

A Few Experimental Guidelines for Employing Zeolite for Asymmetric Photoreaction

It would not be surprising if one had the feeling at this stage that asymmetric photochemistry within zeolites lacks predictability. While this may be true, there is enough information available for this strategy to be used generally. While quantitative ee's or de's are unlikely, modest enantio- and diastereoselectivities are a distinct possibility. For best results, one should use "dry" condi-

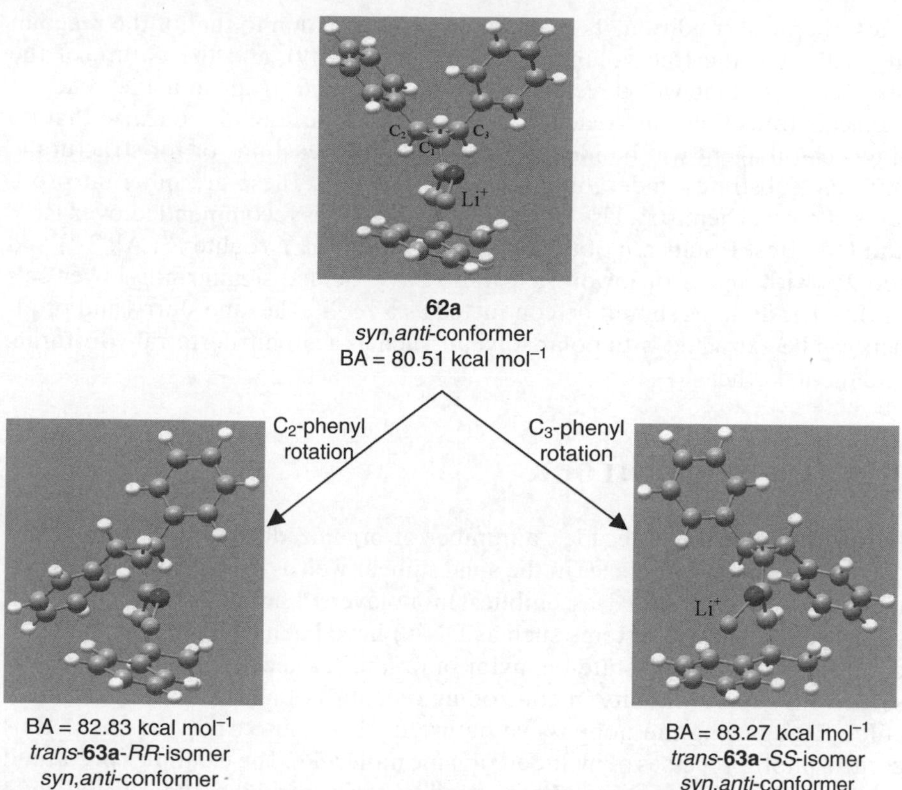

Figure 10.28. Photoisomeriztion of Li^+-bound **62a** *syn,anti*-conformer through rotation of C_2-phenyl and C_3-phenyl. Binding affinities and structural details optimized at RB3LYP/6-31G*. As per this model, isomerization via both C_3-phenyl and C_2-phenyl rotation would be equally preferred.

tions. That is, zeolites must be activated at 500°C prior to use, sample loading and washing should be carried out with dry hexane, and the reactant-loaded zeolite should be dried using a vacuum line ($>10^{-3}$Torr) prior to irradiation. During the chiral inductor strategy, one should try to achieve a loading level of one chiral inductor molecule per supercage. Chiral inductors that generally work well are ephedrine, pseudoephedrine, and norephedrine (Scheme 10.16). The chiral auxiliary strategy generally works better than the chiral inductor approach. With some manipulation, de's in the range of 70–90% could be achieved for most systems. The best chiral auxiliaries are derived from alcohols such as menthol; amines such as 1-phenylethylamine; amino alcohols such as ephedrine, pseudoephedrine, norephedrine; and amino acid derivatives such as valine methyl ester, alanine methyl ester, phenylalanine methyl ester, and leucine methyl ester (Scheme 10.16). The suitability of a chiral inductor or chiral auxiliary for a particular study depends on its inertness under the given photochemical condition, its shape and size (in relation to that of the reactant molecule and the free volume of the zeolite cavity), and the nature of the interaction(s) that will develop between the chiral agent and the reactant molecule/transition state/reactive intermediate. One should recognize that no single chiral agent will be ideal for two different reactions or for structurally differing substrates undergoing the same reaction. These are inherent problems of chiral chemistry. Use of LiY, NaY, and KY is recommended over RbY and CsY. Best results are obtained with commercial Y zeolite (Si:Al: 2.4) and zeolites with higher or lower Si:Al ratio have yielded significantly lower ee's or de's. Irradiations should be conducted as a zeolite–hexane slurry, and products can be extracted with polar solvents such as acetonitrile, tetrahydrofuran, and diethyl ether.

SUMMARY AND OUTLOOK

During the past three decades, a number of organized assemblies (molecular crystals, inclusion complexes in the solid state as well as in solution, liquid crystals, micelles and related assemblies, monolayers, Langmuir–Blodgett films, surfaces, and natural systems such as DNA) have been examined as media in controlling the excited-state behavior of organic molecules.[1] Zeolites are versatile with superior ability in controlling reactions of a large variety of molecules. Recognizing the nonpassive nature of the zeolite cavity has helped us control photoprocesses of included organic molecules. The examples presented in this article demonstrate the possibility for controlling the excited-state behavior of organic molecules through exploitation of the alkali ions present in zeolites. Examination of the zeolite interior, in which the reactant molecule is held, suggests that the most likely factor responsible for altering the reactivity compared with solution is the difference in conformational preference for the reactant molecules (and chiral inductors) and the nonbonding interaction that develop within the zeolite nanocavitiy. Nonbonding interactions can be effectively employed to alter the photophysics of organic molecules.

Confinement and alkali metal interaction with the encapsulated guests could be employed to control photochemical reactivity. Most notably, controlling variety of asymmetric phototransformations within zeolites using confinement and nonbonding interaction has opened up opportunities to employ zeolites as an organized nanocavity to achieve high stereoselection during chemical transformations. Additionally, the cost and environmentally benign nature of the medium and the "greenness" of the reagent (light) used to register the selectivity/chirality of the product justify further exploration of zeolite-mediated photochemical photophysical processes. Examples provided above bring out the remarkable similarity between enzymes and zeolites. Zeolites similar to enzymes provide a confined space and an active site (alkali metal ion) where a molecule could be manipulated. Zeolites are far more versatile and unique to be used only as shape-selective catalysts.

REFERENCES

1. Ramamurthy, V., ed. *Photochemistry in Organized and Constrained Media*. New York: Wiley-VCH, **1991**.
2. Dodziuk, H. *Introduction to Supramolecular Chemistry*. Dordecht: Kluwer Academic Publishers, **2002**.
3. (*a*) Breck, D.W. *Zeolite Molecular Sieves, Chemistry and Use*. New York: John Wiley & Sons, **1991**; (*b*) Dyer, A. *An Introduction to Zeolite Molecular Sieves*. New York: John Wiley & Sons, **1988**.
4. Kirschhock, C.; Fuess, H. *Zeolites.* **1996**, *17*, 381.
5. (*a*) Dunbar, R.C. *J. Phys. Chem. A* **2000**, *104*, 8067; (*b*) Raber, D.J.; Raber, N.K.; Chandrasekhar, J.; Schleyer, P.V.R. *Inorg. Chem.* **1984**, *23*, 4076.
6. Ma, J.C.; Dougherty, D.A. *Chem. Rev.* **1997**, *97*, 1303.
7. Turro, N.J.; Cheng, C.C.; Abrams, L.; Corbin, D.R. *J. Am. Chem. Soc.* **1987**, *109*, 2449.
8. Turro, N.J. *Acc. Chem. Res.* **2000**, *33*, 637.
9. Szostak, R. *Handbook of Molecular Sieves*. New York: Van Nostrand Reinhold, **1992**.
10. (*a*) Lim, K.H.; Grey, C.P. *J. Am. Chem. Soc.* **2000**, *122*, 9768; (*b*) David, H.O. *Zeolites* **1995**, *15*, 439.
11. Grey, C.P.; Poshni, F.I.; Gualtieri, A.F.; Norby, P.; Hanson, J.C.; Corbin, D.R.J. *Am. Chem. Soc.* **1997**, *119*, 1981.
12. (*a*) Cohen, M.D. *Angew. Chem. Int. Ed. Engl.* **1975**, *14*, 386; (*b*) Weiss, R.G.; Ramamurthy, V.; Hammond, G.S. *Acc. Chem. Res.* **1993**, *26*, 530.
13. Turro, N.J. *Modern Molecular Photochemistry*. New York: University Science Books, **1991**.
14. (*a*) Ramamurthy, V.; Caspar, J.V.; Kuo, E.W.; Corbin, D.R.; Eaton, D.F. *J. Am. Chem. Soc.* **1992**, *114*, 3882; (b) Caspar, J.V.; Ramamurthy, V.; Corbin, D.R. *Coord. Chem. Rev.* **1990**, *97*, 225.
15. Ramamurthy, V.; Caspar, J.V.; Corbin, D.R.; Schlyer, B.D.; Maki, A.H. *J. Phys. Chem.* **1990**, *94*, 3391.
16. Uppili, S.; Marti, V.; Nikolaus, A.; Jockusch, S.; Adam, W.; Engel, P.S.; Turro, N.J.; Ramamurthy, V. *J. Am. Chem. Soc.* **2000**, *122*, 11025.
17. Ramamurthy, V.; Corbin, D.R.; Kumar, C.V.; Turro, N.J. *Tetrahedron Lett.* **1990**, *31*, 47.
18. Pitchumani, K.; Warrier, M.; Scheffer, J.R.; Ramamurthy, V. *Chem. Commun.* **1998**, 1197.
19. Warrier, M.; Turro, N.J.; Ramamurthy, V. *Tetrahedron Lett.* **2000**, *41*, 7163.
20. Shailaja, J.; Lakshminarasimhan, P.H.; Pradhan, A.R.; Sunoj, R.B.; Jockusch, S.; Karthikeyan, S.; Uppili, S.; Chandrasekhar, J.; Turro, N.J.; Ramamurthy, V. *J. Phys. Chem. A* **2003**, *107*, 3187.
21. Sunoj, R.B.; Lakshminarasimhan, P.; Ramamurthy, V.; Chandrasekhar, J. *J. Comput. Chem.* **2001**, *22*, 1598.
22. Jayathirtha Rao, V.; Uppili, S.R.; Corbin, D.R.; Schwarz, S.; Lustig, S.R.; Ramamurthy, V. *J. Am. Chem. Soc.* **1998**, *120*, 2480.
23. Uppili, S.; Takagi, S.; Sunoj, R.B.; Lakshminarasimhan, P.; Chandrasekhar, J.; Ramamurthy, V. *Tetrahedron Lett.* **2001**, *42*, 2079.

24. (*a*) Chan, A.C.; Schuster, D.I. *J. Am. Chem. Soc.* **1986**, *108*, 4561; (*b*) Schuster, D.I.; Woning, J.; Kaprinidis, N.A.; Pan, Y.; Cai, B.; Barra, M.; Rhodes, C.A. *J. Am. Chem. Soc.* **1992**, *114*, 7029.
25. Fitch, A.N.; Jobic, H.; Renouprez, A.J. *Chem. Soc. Chem. Commun.* **1985**, 284.
26. Goyal, R.; Fitch, A.N.; Jobic, H. *J. Chem. Soc. Chem. Commun.* **1990**, 1152.
27. Hepp, M.A.; Ramamurthy, V.; Corbin, D.R.; Dybowski, C. *J. Phys. Chem.* **1992**, *96*, 2629.
28. Ramamurthy, V. *Mol. Cryst. Liq. Cryst. Sci. Technol. Sect. A* **1994**, *240*, 53.
29. Thomas, K.J.; Sunoj, R.B.; Chandrasekhar, J.; Ramamurthy, V. *Langmuir* **2000**, *16*, 4912.
30. Nicholas, J.B.; Hay, B.P.; Dixon, D.A. *J. Phys. Chem. A* **1999**, *103*, 1394.
31. (*a*) Hashimoto, S.; Ikuta, S.; Asahi, T.; Masuhara, H. *Langmuir* **1998**, *14*, 4284; (*b*) Iu, K.K.; Thomas, J.K. *Langmuir* **1990**, *6*, 471; (*c*) Ramamurthy, V.; Sanderson, D.R.; Eaton, D.F. *J. Phys. Chem.* **1993**, *97*, 13380.
32. Lakshminarasimhan, P.; Sunoj, R.B.; Chandrasekhar, J.; Ramamurthy, V. *J. Am. Chem. Soc.* **2000**, *122*, 4815.
33. Kaanumalle, L.S.; Sivaguru, J.; Sunoj, R.B.; Lakshminarasimhan, P.H.; Chandrasekhar, J.; Ramamurthy, V. *J. Org. Chem.* **2002**, *67*, 8711.
34. (*a*) Gu, W.; Warrier, M.; Schoon, B.; Ramamurthy, V.; Weiss, R.G. *Langmuir* **2000**, *16*, 6977; (*b*) Pitchumani, K.; Warrier, M.; Cui, C.; Weiss, R.G.; Ramamurthy, V. *Tetrahedron Lett.* **1996**, *37*, 6251; (*c*) Pitchumani, K.; Warrier, M.; Ramamurthy, V. *J. Am. Chem. Soc.* **1996**, *118*, 9428; (*d*) Pitchumani, K.; Warrier, M.; Ramamurthy, V. *Res. Chem. Intermed.* **1999**, *25*, 623.
35. Gu, W.; Warrier, M.; Ramamurthy, V.; Weiss, R.G. *J. Am. Chem. Soc.* **1999**, *121*, 9467.
36. (*a*) Corbin, D.R.; Eaton, D.F.; Ramamurthy, V. *J. Org. Chem.* **1988**, *53*, 5384; (*b*) Corbin, D.R.; Eaton, D.F.; Ramamurthy, V. *J. Am. Chem. Soc.* **1988**, *110*, 4848.
37. Shailaja, J.; Sivaguru, J.; Robbins, R.J.; Ramamurthy, V.; Sunoj, R.B.; Chandrasekhar, J. *Tetrahedron* **2000**, *56*, 6927.
38. (*a*) Frimer, A.A., editor. *Singlet Oxygen*, Vol. 1–4. Boca Raton, FL: CRC Press, **1985**; (*b*) Wasserman, H.H.; Murray, R.W., editors. *Singlet Oxygen*. New York: Academic, **1979**.
39. (*a*) Kaanumalle, L.S.; Shailaja, J.; Robbins, R.J.; Ramamurthy, V. *J. Photochem. Photobiol. A Chem.* **2002**, *153*, 55; (*b*) Li, X.; Ramamurthy, V. *J. Am. Chem. Soc.* **1996**, *118*, 10666; (*c*) Robbins, R.J.; Ramamurthy, V. *Chem. Commun.* **1997**, 1071.
40. (*a*) Pace, A.; Clennan, E.L. *J. Am. Chem. Soc.* **2002**, *124*, 11236; (*b*) Poon, T.; Sivaguru, J.; Franz, R.; Jockusch, S.; Martinez, C.; Washington, I.; Adam, W.; Inoue, Y.; Turro, N.J. *J. Am. Chem. Soc.* **2004**, *126*, 10498; (*c*) Sivaguru, J.; Poon, T.; Franz, R.; Jockusch, S.; Adam, W.; Turro, N.J. *J. Am. Chem. Soc.* **2004**, *126*, 10816; (*d*) Sivaguru, J.; Poon, T.; Hooper, C.; Saito, H.; Solomon, M.; Jockusch, S.; Adam, W.; Inoue, Y.; Turro, N.J. *Tetrahedron* **2006**, *62*, 10647; (*e*) Sivaguru, J.; Saito, H.; Solomon M.R.; Kaanumalle, L.S.; Poon, T.; Jockusch, S.; Adam, W.; Ramamurthy, V.; Inoue, Y.; Turro, N.J. *Photochem. Photobiol.* **2006**, *82*, 123; (*f*) Stratakis, M.; Kosmas, G. *Tetrahedron Lett.* **2001**, *42*, 6007.
41. Hammond, G.S.; Cole, R.S. *J. Am. Chem. Soc.* **1965**, *87*, 3256.

42. (*a*) Inoue, Y. *Chem. Rev.* **1992**, *92*, 741; (*b*) Inoue, Y.; Ramamurthy, V., editors. *Chiral Photochemistry*. New York: Marcel Dekker, **2004**; (*c*) Rau, H. *Chem. Rev.* **1983**, *83*, 535.
43. (*a*) Buschman, H.; Scharf, H.-D.; Hoffmann, N.; Esser, P. *Angew. Chem. Int. Ed. Engl.* **1991**, *30*, 477; (*b*) Faure, S.; Piva-le-Blanc, S.; Bertrand, C.; Pete, J.-P.; Faure, R.; Piva, O. *J. Org. Chem.* **2002**, *67*, 1061.
44. Elgavi, A.; Green, S.B.; Schmidt, G.M.J. *J. Am. Chem. Soc.* **2058**, *1973*, 95.
45. Sakamoto, M. *Chem. Eur. J.* **1997**, *3*, 684.
46. Cheung, E.; Chong, K.C.W.; Jayaraman, S.; Ramamurthy, V.; Scheffer, J.R.; Trotter, J. *Org. Lett.* **2000**, *2*, 2801.
47. Chong, K.C.W.; Sivaguru, J.; Shichi, T.; Yoshimi, Y.; Ramamurthy, V.; Scheffer, J.R. *J. Am. Chem. Soc.* **2002**, *124*, 2858.
48. (*a*) Gamlin, J.N.; Jones, R.; Leibovitch, M.; Patrick, B.; Scheffer, J.R.; Trotter, J. *Acc. Chem. Res.* **1996**, *29*, 203; (*b*) Garcia-Garibay, M.; Scheffer, J.R.; Trotter, J.; Wireko, F. *J. Am. Chem. Soc.* **1989**, *111*, 4985; (*c*) Scheffer, J.R.; Dzakpasu, A.A. *J. Am. Chem. Soc.* **1978**, *100*, 2163.
49. Leibovitch, M.; Olovsson, G.; Sundarababu, G.; Ramamurthy, V.; Scheffer, J.R.; Trotter, J. *J. Am. Chem. Soc.* **1996**, *118*, 1219.
50. Farina, M.; Audisio, G.; Natta, G. *J. Am. Chem. Soc.* **1967**, *89*, 5071.
51. (*a*) Tanaka, K.; Toda, F. *Chem. Rev.* **2000**, *100*, 1025; (*b*) Toda, F. *Acc. Chem. Res.* **1995**, *28*, 480.
52. Sundarababu, G.; Leibovitch, M.; Corbin, D.R.; Scheffer, J.R.; Ramamurthy, V. *Chem. Commun.* **1996**, 2159.
53. (*a*) Joy, A.; Uppili, S.; Netherton, M.R.; Scheffer, J.R.; Ramamurthy, V. *J. Am. Chem. Soc.* **2000**, *122*, 728; (*b*) Joy, A.; Ramamurthy, V.; Scheffer, J.R.; Corbin, D.R. *Chem. Commun.* **1998**, 1379.
54. Joy, A.; Scheffer, J.R.; Ramamurthy, V. *Org. Lett.* **2000**, *2*, 119.
55. Joy, A.; Robbins, R.; Pitchumani, K.; Ramamurthy, V. *Tetrahedron Lett.* **1997**, *38*, 8825.
56. Uppili, S.; Ramamurthy, V. *Org. Lett.* **2002**, *4*, 87.
57. Natarajan, A.; Wang, K.; Ramamurthy, V.; Scheffer, J.R.; Patrick, B. *Org. Lett.* **2002**, *4*, 1443.
58. Natarajan, A.; Joy, A.; Kaanumalle, L.S.; Scheffer, J.R.; Ramamurthy, V. *J. Org. Chem.* **2002**, *67*, 8339.
59. Sivaguru, J.; Natarajan, A.; Kaanumalle, L.S.; Shailaja, J.; Uppili, S.; Joy, A.; Ramamurthy, V. *Acc. Chem. Res.* **2003**, *36*, 509.
60. Sivaguru, J.; Scheffer, J.R.; Chandarasekhar, J.; Ramamurthy, V. *Chem. Commun.* **2002**, 830.
61. (*a*) Sivaguru, J.; Shailaja, J.; Ramamurthy, V. In: Auerbach, S.M.; Carrado, K.A.; Dutta, P.K., editors. *Handbook of Zeolite Science and Technology*. New York: Marcel Dekker, **2003**, p. 515; (*b*) Sivaguru, J.; Shailaja, J.; Uppili, S.; Ponchot, K.; Joy, A.; Arunkumar, N.; Ramamurthy, V. Achieving enantio and diastereoselectivities in photoreactions through the use of a confined space. In: Toda, F., editor. *Organic Solid-State Reactions*. Boston: Kluwer Academic, **2002**, pp. 159–188.
62. Sivaguru, J.; Shichi, T.; Ramamurthy, V. *Org. Lett.* **2002**, *4*, 4221.

63. Sivaguru, J.; Sunoj, R.B.; Wada, T.; Origane, Y.; Inoue, Y.; Ramamurthy, V. *J. Org. Chem.* **2004**, *69*, 5528.
64. Sivaguru, J.; Sunoj, R.B.; Wada, T.; Origane, Y.; Inoue, Y.; Ramamurthy, V. *J. Org. Chem.* **2004**, *69*, 6533.
65. Sivaguru, J.; Wada, T.; Origane, Y.; Inoue, Y.; Ramamurthy, V. *Photochem. Photobiol. Sci.* **2005**, *4*, 119.
66. Sivasubramanian, K.; Kaanumalle, L.S.; Uppili, S.; Ramamurthy, V. *Org. Biomol. Chem.* **2007**, *5*, 1569.
67. (*a*) Koodanjeri, S.; Joy, A.; Ramamurthy, V. *Tetrahedron* **2000**, *56*, 7003; (*b*) Jayaraman, S.; Uppili, S.; Natarajan, A.; Joy, A.; Chong, K.C.W.; Netherton, M.R.; Zenova, A.; Scheffer, J.R.; Ramamurthy, V. *Tetrahedron Lett.* **2000**, *41*, 8231.
68. Shailaja, J.; Ponchot, K.J.; Ramamurthy, V. *Org. Lett.* **2000**, *2*, 937.
69. Kaanumalle, L.S.; Sivaguru, J.; Arunkumar, N.; Karthikeyan, S.; Ramamurthy, V. *Chem. Commun.* **2003**, 116.
70. Gapeev, A.; Dunbar, R.C. *J. Am. Chem. Soc.* **2001**, *123*, 8360.
71. (*a*) Wyttenbach, T.; Witt, M.; Bowers, M.T. *J. Am. Chem. Soc.* **2000**, *122*, 3458; (*b*) Siu, F.M.; Ma, N.L.; Tsang, C.W. *J. Am. Chem. Soc.* **2001**, *123*, 3397; (*c*) Hoyau, S.; Ohanessian, G. *Chem. Eur. J.* **1998**, *4*, 1561; (*d*) Hoyau, S.; Norrman, K.; McMahon, T.B.; Ohanessian, G. *J. Am. Chem. Soc.* **1999**, *121*, 8864.
72. (*a*) Jockusch, R.A.; Price, W.D.; Williams, E.R. *J. Phys. Chem. A* **1999**, *103*, 9266; (*b*) Jockusch, R.A.; Lemoff, A.S.; Williams, E.R. *J. Am. Chem. Soc.* **2001**, *123*, 12255.
73. Gaussian, 98, Revision A.9. Pittsburgh, PA: Gaussian, **1998**.
74. Sivaguru, J.; Jockusch, S.; Turro, N.J.; Ramamurthy, V. *Photochem. Photobiol. Sci.* **2003**, *2*, 1101.
75. Natarajan, A.; Kaanumalle, L.S.; Ramamurthy, V. In: Horspool, W.; Lenci, F., editors. *CRC Handbook of Organic Photochemistry and Photobiology*, 2nd edition. Boca Raton, FL: CRC Press, LLC, **2004**, p. 107/1.
76. (*a*) Alemán, C. *J. Phys. Chem. A* **2002**, *106*, 1441; (*b*) LeMaster, C.B.; True, N.S. *J. Phys. Chem.* **1989**, *93*, 1307; (*c*) Ross, B.D.; True, N.S. *J. Am. Chem. Soc.* **1984**, *106*, 2451; (*d*) Wiberg, K.B.; Rablen, P.R.; Rush, D.J.; Keith, T.A. *J. Am. Chem. Soc.* **1995**, *117*, 4261; (*e*) Wiberg, K.B.; Rush, D.J. *J. Am. Chem. Soc.* **2001**, *123*, 2038; (*f*) Wiberg, K.B.; Rush, D.J. *J. Org. Chem.* **2002**, *67*, 826.

11

PHOTOCHEMICAL AND PHOTOPHYSICAL STUDIES OF AND IN BULK POLYMERS

SHIBU ABRAHAM AND RICHARD G. WEISS

INTRODUCTION

This chapter deals with the use of bulk polymers for photophysical and photochemical applications,[1–3] and it excludes, by design, the large volume of literature dealing with solution state investigations of polymers. Host–guest photochemistry, to which this chapter pertains in some respects, has been utilized extensively to study the dynamics of photoreactions and to probe the microstructural features of the host. The selectivity and efficiency of organic reactions in these systems depend on the strength of noncovalent interactions.[4] A large number of host materials whose guest sites have "soft" and "hard" walls have been employed to perform these reactions. Commonly used soft-matter hosts include micelles, liquid crystals, and polymers. In this chapter, the properties that should be considered when selecting a bulk polymer for conducting a photochemical or photophysical process are discussed. In addition, several examples of photoprocesses are described to demonstrate the relationship between the host structure and guest response when one or the other is excited electronically.

The reactivities of guest molecules may be influenced significantly when they are embedded in a polymer matrix above or below its glass transition temperature. The voids (i.e., "free" spaces) in polymeric media are "soft" above the glass transition temperature. They provide a "cage" or reactor space that may be able to change its shape and volume either slower or faster than

Supramolecular Photochemistry: Controlling Photochemical Processes, First Edition. Edited by V. Ramamurthy and Yoshihisa Inoue.

the intrinsic molecular shape changes of the guest, depending on the nature of the processes. In isotropic solutions, such considerations can be treated by the Kramers theory.[5] This type of host relaxation and specific polymer–guest interactions are the major factors influencing the selectivity of product formation. They do so by affecting the conformational lability or equilibrium and translational mobility of guests in their ground state and the corresponding excited-state properties, including lifetimes. As a corollary, appropriately selected, environmentally sensitive guests can probe the physical properties of the reaction cages of polymers.

SELECTING A BULK POLYMER FOR PHOTOCHEMICAL/ PHOTOPHYSICAL APPLICATIONS

A strong correlation can be derived between the bulk properties and physical parameters of polymers, such as density, molecular weight dispersity, branching frequency, and tacticity.[6,7] The following sections describe the chemical and physical properties of bulk polymers, with an emphasis on understanding how the major structural factors affect material properties, especially as they may pertain to photochemical and photophysical applications.

"Polarity"[6]

The nature of the interactions between a polymer and a solvent is an important part of selecting a polymer because the same considerations apply to polymer–guest molecule interactions in many aspects. The free energy of mixing ($\Delta G_m = \Delta H_m - T\Delta S_m$) determines the solubility and miscibility of an amorphous polymer in a solvent. The fundamental parameter required to define these interactions is the specific cohesion energy, which is directly related to the solubility parameter of Hildebrand, δ.[8] This parameter describes the enthalpy change upon mixing. Complete miscibility occurs when solubility parameters are similar. The solubility parameters of solvents (δ_s) can be measured directly from the energy of vaporization, while those of polymers (δ_p) are measured indirectly.[8] A polymer can be dissolved by a solvent if $\delta_p \cong \delta_s$. Thus, high solubility and good compatibility are predicted for two species with similar values of δ and with approximately equal molar volumes and surface tension coefficients. These parameters provide information about the cohesive and adhesive properties of polymers, and also help to identify compatible solvents for their swelling. For a binary mixture, the polymer–solvent interaction parameter (χ) at temperature (T) is related to the solubility parameters of the solvent (δ_1) and polymer (δ_2) by Equation 11.1,[8] where R is the ideal gas constant and V_1 is the molar volume of the solvent:

$$\chi = \frac{V_1}{RT}(\delta_1 - \delta_2)^2 + 0.34. \tag{11.1}$$

This equation contains enthalpic and entropic components, and it takes the entropic component to be a constant (0.34). Miscibility occurs when the "χ" is positive and close to zero.

Density and Bulk Free Volume

Polymers, as do other molecules in condensed phases, contain "void" spaces as a result of packing constraints. In polyatomic molecules, vacant spaces exist between nonbonded atoms at distances greater than the sum of their van der Waals radii even when the individual components are closely packed. In amorphous polymers, the term "free volume" (V_f) refers to the difference between the total volume of the material (V) and the van der Waals volume occupied by the constituent polymer chains (V_{occ}): $V_f = V - V_{occ}$, and V_f is usually $\approx(0.3\text{–}0.4)V$.[9] This free volume can be correlated with the physical properties of polymers. It reduces the density (specific volume is the inverse of density) and crystallinity of polymers.* For example, the density of completely crystalline polyethylene (**PE**) (ca. $1\,\text{g cm}^{-3}$) is much larger than that of amorphous **PE** ($\leq 0.9\,\text{g cm}^{-3}$) as a result of chain packing differences.[10] Among the various free volume theories, the one proposed by Vrentas et al.[11] is used frequently because it has high predictive capabilities for physical properties such as diffusivity and viscosity.[12] This theory is based on the concept of mean "hole" free volume, which is distributed throughout the bulk of a polymer and separates the specific volume into three different types: occupied volume (i.e., the volume at 0 K), interstitial volume, and hole free volume. An increase in temperature creates holes or vacancies whose size (and location) change with time due to the motions of polymer chains.[12]

Hole Free Volumes

An important spectroscopic method for measuring the mean size of the holes in a material is positron annihilation spectroscopy (PALS).[13] The relatively long-lived species *ortho*-positronium (*o*-Ps) ($\tau_3 = 1.1\text{–}2.6\,\text{ns}$) are trapped in holes due to repulsion between molecules and the *o*-Ps. The probability that an *o*-Ps will be annihilated in the polymer matrix is proportional to the density of nearby electrons as well as the size of the holes (pick-off annihilation). Variations of *o*-Ps lifetime and intensity with temperature measure changes in free volume and its distribution. The temporal decay curve for the *o*-Ps can be fitted to yield the salient component τ_3 (Eq. 11.2):

$$\tau_3 = \frac{1}{2}\left[1 - \frac{R}{R+\Delta R} + \frac{1}{2\pi}\sin\frac{2\pi R}{R+\Delta R}\right]^{-1}. \tag{11.2}$$

*Throughout the chapter, different polyethylenes will be identified as **PEXX**, where **XX** is the percent of crystallinity; s and u after **PEXX** indicate stretched and unstretched films, respectively.

Here, R is the radius of a presumed spherical hole with a surface electron layer of thickness ΔR. The mean hole radius $\langle R \rangle$ can be used to calculate the mean hole volume ($\langle v_f \rangle = (4/3)\pi \langle R \rangle^3$).

An important method for reducing hole free volume in polymer films is mechanical stretching (or "drawing"). This reduction lowers the rate of diffusion of guest molecules and reactive intermediates.[2] In addition, there is a net translocation of guest molecules from amorphous to interfacial regions (i.e., at the interfaces between crystallites and amorphous chains) when a **PE** film is stretched, and guest molecules with aspect ratio not equal to one are oriented with respect to the stretching direction.[14] This form of orientation has been exploited to assign the direction of transition dipoles in electronic transitions.[15]

Phase Transitions and Free Volume: Glass Transition Temperature (T_g) and Relaxations

Below T_g, the small decrease in free volume with decreasing temperature has been attributed to "crystal-like contractions of the lattices" due to decreased oscillations of the chain segments.[16] At the glass transition temperature, no discontinuity is observed in first-order properties such as volume and internal energy. On the other hand, second-order properties such as the thermal expansion coefficient and heat capacity do exhibit abrupt changes.[16] For a true second-order transition, the glass transition can be described as illustrated in Equation 11.3.[6] In the glass, polymer chains are trapped in a thermodynamically nonequilibrated state characterized by high energy and excess volume. Glass transitions arise from the kinetic limitations on the rate of the internal movements with temperature. The amorphous regions are stiff and lack significant motions of chain segments over short periods of time; only short range rotational and translational motions occur below T_g:

$$\frac{dT_g}{dp} = \frac{T_g V(T_g)\Delta\alpha}{\Delta C_p} = \frac{\Delta\kappa}{\Delta\alpha}. \tag{11.3}$$

In Equation 11.3, p is the pressure, $V(T_g)$ is the molar volume at T_g, and ΔC_p is the molar heat capacity difference between the rubbery and glassy states. $\Delta\alpha$ is the thermal expansion coefficient difference, and $\Delta\kappa$ is the compressibility difference. As a result of the mobility of the polymer chains, molecular rearrangement or slow movement of matter (creep) occurs on application of a mechanical stress.[17] The chains slowly relax toward their initial spatial arrangement over time when the stress is released. The dynamics of relaxations and the thermodynamic properties of a polymer are related.[17]

For example, **PE** is known to exhibit three distinct mechanical relaxations: α (303–393 K), β (243–283 K), and γ (123–153 K). The temperature ranges reflect the large structural differences that can exist among the diverse family of **PE** types.[18] The origins of these relaxation processes in **PE** pertain to the movement of guest molecules in a polymer cavity at a specific temperature. The α-relaxation is attributed to reorientational motions within the crystalline regions; it does not occur in completely amorphous **PE**. The temperature of the α-relaxation depends on the frequency and length of the chain branches.[17] β-Relaxations originate in the crystalline and amorphous regions.[18] In the latter, these transitions are associated with the movement of large chain segments. In "low-density" **PE** (**LDPE**), β-relaxation is attributed to the mobility of chain branches. γ-Relaxation, the lowest temperature transition, arises from relatively short segmental or localized motions of chains either in the amorphous part or folded chains within the interfacial regions.[18]

The movement of chain segments in polymers and their rates can be investigated by optical methods as discussed in the sections on rates of relaxation of chain segments and monitors of local relaxation phenomenon.

Dependence of Free Volume on Temperature

Above the glass transition temperature, the rates and magnitudes of segmental motions of polymer chains increase significantly, leading to an increase in the hole free volume.[19] The free volume fraction can be represented by an Arrhenius-like equation, $h = V_f/V = A\bullet\exp(-H_h/(k_BT))$; H_h is the vacancy formation enthalpy.[9] The free volume fraction at T_g, h_g, is not the same for all polymers; it increases from ≈0.02 for polymers with T_g at 200 K to 0.08 for polymers with T_g at 400 K.[9] At temperatures higher than T_g, for example, at $T = T_g + 200\,\text{K}$, values of h (h_T) as high as 0.2–0.3 have been found.[9]

The volumetric behavior of a polymer with temperature is shown schematically in Figure 11.1. Cooling an amorphous material to a temperature below its melting temperature, T_m, results in a continuous decrease in the specific volume. Additional cooling further reduces the specific volume in the amorphous regions until T_g is reached, where the extra hole free volume is trapped. Over time at isothermal conditions, the polymer chains below T_g can relax slowly, and the nonequilibrium liquid volume curve approaches the equilibrium volume curve.

Dependence of Properties of Polymers on Their Average Molecular Weight (<M_w>)

Most polymers are made by procedures that result in a distribution of chain lengths. Number average (M_n) and weight average (M_w) molecular weights (Eqs. 11.4 and 11.5) are two commonly used expressions that describe the distributions.[20] Photoinduced cleavage and cross-linking reactions of polymers cause changes in the molecular weight distribution. Many physical properties of polymers, such as their second-order transition temperatures, viscosity–temperature coefficients, and specific volumes change with increasing molecular weight up to a limit.[16] Increases in molar mass and cross-linking also

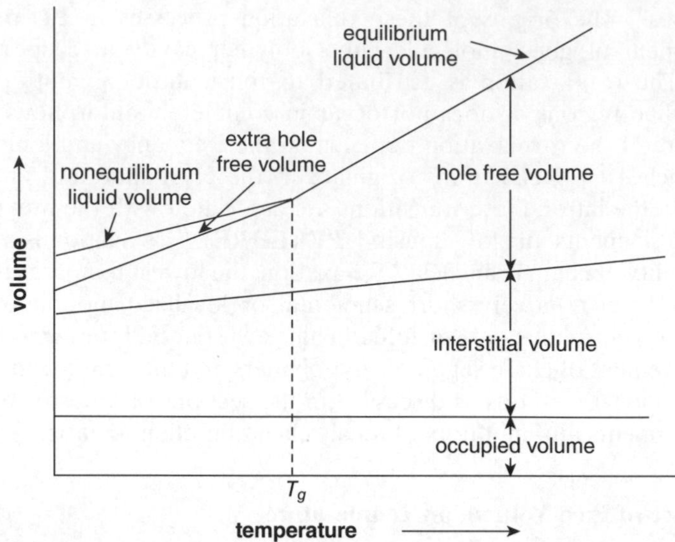

Figure 11.1. Relationship between the free volume and temperature for a polymer above and below its glass transition temperature.[12]

increase the gel content, which is defined as the insoluble component of a polymer in a good solvent:

$$M_n = \sum_i n_i m_i \Big/ \sum_i n_i = P_w M_o, \tag{11.4}$$

$$M_w = \sum_i n_i m_i^2 \Big/ \sum_i n_i m_i = P_n M_o. \tag{11.5}$$

In Equations 11.4 and 11.5, n_i is the number of units with weight m_i and M_o is the molecular weight of the monomer. P_n and P_w are the number and weight degrees of polymerization.

Polydispersity is defined as the ratio of M_n and M_w. In some applications, monodisperse polymers are sought because they permit a more precise characterization of the medium. In others, the distributions of chain lengths or chain branching (vide infra) imbue needed properties in the bulk that are not found in the monodisperse polymers.

DOPING METHODS

A wide variety of methods are available for incorporating (doping) guest molecules into polymers. These include swelling/deswelling, spin casting, sublimation, coprecipitation, and film casting via slow evaporation.

Swelling and Deswelling

Solvents with a polymer–solvent interaction parameter (χ) marginally greater than 0.5 can swell polymers.[21] The degree of swelling depends on the solubility of the solvent molecules in the polymer matrices. The solvent molecules penetrate the entangled networks of polymer chains, thereby "softening" the walls of potential guest sites. If guest molecules are entrained in the solvent, they as well can be introduced into the polymer. Typically, a polymer film is immersed in a solution of swelling solvent containing dopants. If the rate of diffusion of solvent molecules out of the polymer is faster than that of the guest molecules, the latter are trapped in polymer cavities whose size and shape they define. This constitutes a templating effect.[2]

Film Casting via Slow Evaporation

Dissolving a host polymer and its guest molecules in a "good" solvent and casting a film from them is a standard, one-step procedure for preparing doped films. The concentrations of dopants can be varied easily in this way. Slow evaporation of solvent can result in good optical quality films with smooth surfaces. However, slow evaporation of solvent may allow aggregation of the guest molecules, should they have a tendency to do so, as is the case of perylenediamide in polyamides.[22] The type of solvent, evaporation rate, thickness, and temperature can be varied to produce films with different sites for the guest molecules. For example, depending on the solvent used for casting, conductivity differs by orders of magnitude and the polaron absorption peaks shift significantly for doped (protonated) polyaniline films.[23,24]

Coprecipitation

Coprecipitation is used for the preparation of multiphase, immiscible polymer blends. It is simpler than other methods in which phase separation is achieved chemically by copolymerization.[25] For example, a polymer blend of diacetylene-containing polyester and poly[ethylene-*co*-(vinyl acetate)] can be prepared by dissolving them in 50/50 (v/v) tetrahydrofuran (THF)/toluene followed by coprecipitation using 20/80 (v/v) methanol/water.[26] Another application of this method is in the preparation of dye–polymer mixtures in which slow solvent evaporation can lead to instability, particularly in highly doped polymers. Additionally, optical quality films of dye-dispersed polymers have been obtained by coprecipitation methods with good control over the size and shape of the associated molecules.[27]

Spin Casting

The processing of polymer films by casting their solutions on a spin plate is a simple conventional method. In it, a solution of the polymer and guest molecule at a desired concentration is placed usually dropwise onto a spinning disk. The rate of spinning and position of the sample on the spin plate determine the thickness and homogeneity of the film.[28,29] As an example, uniform films of a host poly(*N*-vinylcarbazole) and guest coumarin 6, with electroluminescent properties, have been obtained by spin casting.[30]

The conformational change of polymer chains from coil-like to an extended form can be transferred to the films by the spin-casting process. Such morphological changes can affect the photophysical properties of polymers because of the alignment of polymer chains. For example, the forces exerted during spin casting of doped polyaniline influence the absorption maximum of the polaron band as a result of the incorporation of "π-defects" caused by ring twisting.[31]

Sublimation

Guest molecules can be sublimated onto and then diffused into polymer films. This process has an advantage over several of the other methods in that guest molecules are dispersed in a polymer film with no deformation of its original shape. For example, poly(methyl methacrylate) (**PMMA**) films with densely dispersed organofluoro compounds have been prepared by sublimation in a vacuum apparatus followed by diffusion into the polymer.[32] *cis*-1,2-Dicyano-1,2-bis(2,4,5-trimethyl-3-thienyl)ethane, a photochromic diarylethene, has been found to disperse more efficiently into poly(bisphenol A carbonate) than into methacrylate polymers using this doping method.[33] Also, multilayered electroluminescent devices containing two or more vacuum-sublimed dyes have been found to exhibit higher performances than those prepared by other methods.[34]

PROPERTIES OF POLYMERS FOR SPECTROSCOPIC INVESTIGATIONS

Transparency in the Near UV-Vis Wavelength Region

The selection of an appropriate polymer as a medium for a photoinduced process of a guest molecule depends on a number of criteria, such as the inertness of the polymer under the irradiation conditions and its transparency or reflectivity in the near UV-Vis range (i.e., where the guest molecules absorb). The inertness of a host polymer to the incident radiation is very important in order to avoid its interference with the course and efficiency of the guest reactions. Polymers such as partially crystalline **PE** are transparent in the near-UV-Vis regions, but they do scatter radiation from crystallite surfaces (as a result of differences between the indices of refraction of amorphous and crystalline domains), and they do contain some centers of unsaturation and allylic H atoms that can react with intermediates and excited states of many guest molecules.[35] Regardless of what is known about the absorption properties of a type of chromophore in solution, it is imperative to determine the absorption spectra of a polymer containing it before adopting it for a particular application—the aggregate absorption of the chromophores within the polymer may result in unwanted bathochromic shifts.

Chromophores or Other Reactive Groups in a Polymer

Many commercial polymers contain small concentrations of impurities or chain initiating/terminating groups that can interact with guest molecules to alter their expected behavior based on the bulk formula of a host polymer matrix (see the section on "How Does the Reactivity of a Guest Molecule Depend on Polymer Groups?"). For example, the decay of benzophenone phosphorescence is quenched in **PMMA** glasses, even though they are considered inert matrices for many photophysical and photochemical processes.[36] Irradiation of guest molecules, may lead to the sensitized excitation of groups in the polymer host as well or groups in the polymer may sensitize the excited-state formation of guest molecule. Both scenarios have been reported in the literature.[37,38]

Introduction of UV absorbing chromophores is a common method to make environmentally degradable polymers. The $n \rightarrow \pi^*$ transition of carbonyl groups, for example, facilitates the absorption of $\lambda > 300$ nm radiation and subsequent H abstraction or Norrish type I reactions.[39] Although ethylene-based polymers are transparent in the near-UV region, the absorption spectra of ethylene-carbon monoxide copolymers extend to >300 nm, and they are degraded rapidly in sunlight.[39] Also, reversible cycloaddition reactions of the anthryl chromophores in polymers have been used for photoswitching applications.[40] The cleavage reaction of the anthryl cycloadduct requires radiation with $\lambda < 300$ nm. Heterocyclic functional groups such as esters present in anthryl-based polymers interfere with cleavage reactions by absorbing this radiation and limit the efficiency of these polymers in photoswitching applications.[41]

SPECIFIC PROPERTIES OF POLYMERS

Chain Branching Frequency and Length of Branches

Chain branching in polymers influences their viscosity, solubility, and density.[42,43] For example, many branched polysaccharides are water soluble, while linear (unbranched) cellulose is not. LDPEs are categorized on the basis of their lower densities and, structurally, on the basis of having high branching frequencies;[44] high-density **PEs (HDPEs)** have densities ca. 0.94 g cm^{-3} [44] and fewer and longer branches. Although, these definitions are somewhat arbitrary, **LDPEs** usually have ≤50% crystallinity and **HDPEs** have ≥50% crystallinity. Decreasing the length of branches increases the T_g of **PEs** with branches in each repeat unit of the chain, whereas T_g is almost unaffected by branch length

A-A-A-A-A-A (HOMO) **A-B-A-B-A-B-A** (ALTERNATING)

A-A-A-B-B-B (BLOCK) **A-A-B-A-B-B-B-A-B** (RANDOM)

Figure 11.2. Examples of some possible arrangements of monomer units in copolymers.

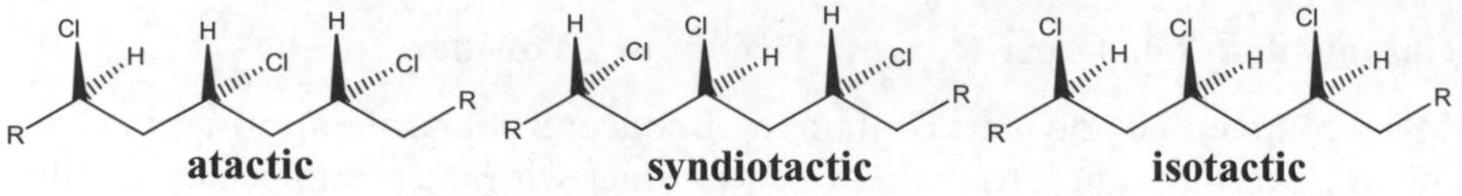

atactic **syndiotactic** **isotactic**

Figure 11.3. Arrangement of Cl and H atoms in triad segments of atactic, syndiotactic and isotactic poly(vinyl chloride) (**PVC**).[183]

when there are few along a chain. The Hildebrand parameter also decreases with increasing branch length.

In copolymers, the specific nature and distribution of the monomers affect the physical properties of the bulk (such as solubility and melting temperature).[17] Some possible arrangements of monomer units within a chain are shown in Figure 11.2. The noncrystalline portions of a bulk copolymer usually have randomly distributed monomer units.

Regio- and stereo-relationships among groups on a polymer chain affect polymer properties as well.[45,46] When monomer units of olefinic polymers contain an asymmetric center along the backbone of the chain, the different arrangements of the substitutents result in stereoisomers—atactic (random), syndiotatic (opposite), and isotactic (same)—as shown in Figure 11.3. Isotactic or syndiotatic polymers are more crystalline than their atactic counterparts due to packing constraints imposed by the branches.[20]

End Groups and/or Residual Catalysts as Quenchers or Chromophores

The presence of residual catalysts or functional groups of branches or on chain ends (placed there as chain initiating or chain terminating units) may also affect the ability of a polymer to serve as an "inert" host for photochemical or photophysical processes of guest molecules. Modification of optical properties of lumophores by polyacids such as poly(methacrylic acid) and poly(acrylic acid) is one such example.[47] The presence of free acid groups in these polymers strongly influences the conformations of their chains and can restrict the mobility of guest lumophores. Also, the excited states of groups, such as 1-pyrenyl (**1-Py**), can aggregate and lose luminescence intensity when attached to the ends of polymer chains, as has been found for pyrenyl groups attached to the ends of poly(ethylene oxide) (PEO) chains.[48]

End groups may also quench the excited states of guest molecules even if the monomer units along the intervening chain do not. In general, the segments at the end of each homo-polymer chain possess greater free volume

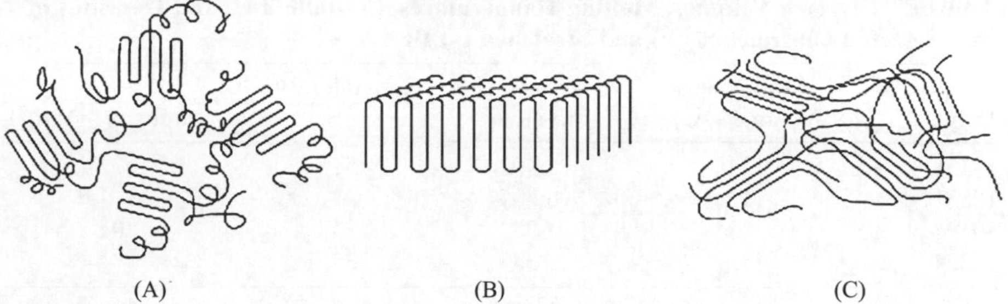

Figure 11.4. Models of the microstructure of a partially crystalline polymer: (A) chains embedded in crystalline and amorphous regions, (B) a tightly folded lamellar crystal, and (C) a fringed micellar organization. (Reprinted with permission from Reference 51. Copyright 2009 American Chemical Society.)

than interior sections. The volume of end groups, especially bulky ones, can change the van der Waals volume and molecular refraction of the polymers. When the end groups are able to form hydrogen bonds, they can promote strong interchain interactions that affect the movement of guest molecules within the polymer matrix.[49]

Degree of Crystallinity

PE crystals possess chain-folded lamellar fibers with thicknesses of 100–200 Å and widths in the micrometer range.[50] The crystals are embedded in the amorphous phase, forming interconnected networks (Fig. 11.4).[51] Bundles of different small crystalline regions can also be randomly connected by amorphous regions (fringed-micelle model, Fig. 11.4C). Guest molecules can enter the noncrystalline regions, but not the crystalline parts, of polymers.[2,52] The chains in the interfacial region are packed in an amorphous fashion, and their conformations are influenced by the crystallites that constitute their surface boundary.

Electron–phonon coupling between polymer chains and the excited state of a guest molecule can modulate luminescence properties. For example, pyrene emission contains a band at 365 nm in partially crystalline **PE** that is, in addition to the normal vibronic bands, observed in common solvents.[53] This fluorescence band is attributed to pyrene molecules located at the interfacial regions.[53] The rigid interfacial cavities in which pyrene molecules are located result in emission from high-energy, unrelaxed excited-singlet states.

Although partially crystalline polymers do not have well-defined melting points or heats of fusion, their percentages of crystallinity can be calculated by differential scanning calorimetry (DSC) from the heat of melting (ΔH) or from the ratio between areas under crystal diffraction peaks and the total area in X-ray diffractograms (XRDs) when the heats and diffractograms of single

TABLE 11.1. Free Volumes, Melting Temperatures, Crystallinities, and Densities of Two Selected Unstretched (u) and Stretched (s) PEs[2]

Polymer	Hole Free Volume (Å^3)	mp (K)[a]	Crystallinity (±2%)	Density (g cm^{-3})
$PE51_{(u)}$	124	402	51	0.94
$PE51_{(s)}$	113	398,405	52	—
$PE24_{(u)}$	142	382	24	0.92
$PE24_{(s)}$	122	383	33	—

[a]From endotherms in the first heating curves in DSC thermograms.

crystals of model compounds or polymers are known. Thus, % crystallinity of a partially crystalline **PE** equals $100 \times (\Delta H/\Delta H^*)$, where ΔH^* is the heat for melting an equal weight of a **PE** single crystal. Some representative values, along with other properties of **PE**s, are presented in Table 11.1.

Plastic and Rubbery Polymers

Plastics and rubbers are elastomeric materials. Plastics cannot be distorted significantly without fracturing, whereas rubbers can be stretched. Plastics can be molded or extruded (under some conditions) into an object that retains its shape, while rubbers return almost to their original shape when the stretching force is removed. At temperatures lower than T_g, the molecular mobility of polymers (and guest molecules) is slower than that in the rubbery state. For example, the quantum yield of Norrish type II photolysis of vinyl-ketone copolymers decreases with decreasing temperature below T_g and is negligible at temperatures below the freezing of local mode relaxations of the main chains.[36] Reactivity of these polymers increases with increasing temperature, and, in the rubbery state above T_g, the quantum yield is the same as that in solution.

Microscopic Properties of Polymers

Site Types for Guest Molecules. The nature and extent of interaction of guest molecules with a host depends on the type of reaction sites available. Generally, reactions requiring minimum distortions of surfaces of an occupied site are favored. A photoreactive matrix can be considered to be an ensemble of chromophore sites. The concept of such sites is related to that of cavities introduced by Schmidt and Cohen in the photochemistry of organic crystals.[54] They viewed reaction cavities as spaces of specific size and shape whose surfaces surround the guest molecules. It is useful to consider guest sites in bulk polymers as reaction cavities as well, even though they are not as homogeneous as in a crystal.

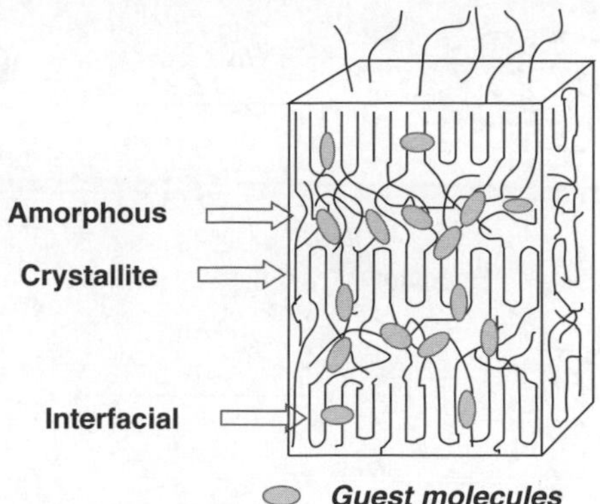

Figure 11.5. Cartoon representation of site types available to guest molecules in a **PE** matrix. (Adapted from Reference 52b. Copyright 2009, with permission from Elsevier.)

The distribution of site types can be altered physically by temperature or by stretching polymer films.[2] In partially crystalline **PEs**, sites available to a guest molecule are in the amorphous part and in interfacial regions (along the lateral surfaces of crystallites), as shown in Figure 11.5.

Differences in the conformation of the chains of a polymer host can change the concentration of guest sites. For example, the ability of intramolecular excimers to form in poly(2-vinyl naphthalene) that had been doped into a polystyrene (PS) matrix changed with time due to relaxations near the glass transition temperature like those mentioned in the section on "Phase Transitions and Free Volume."[55] This observation is a specific example of how history of a polymer matrix (e.g., the temperature or manner in which a sample is prepared) can affect the photochemical and photophysical properties of the guests.[55]

Dependence of Photoreactions on Site Type. Lattice restraints on the motion of the atoms of reactants in reaction cavities provide a topochemical control over reactivity.[54] The site characteristics can be as important as the molecular properties in defining the efficiency or quantum yield of reactions of guest species. Reaction can occur in a medium only if sufficient space along the needed coordinates is available for the excited-state molecules to transform into photoproducts; the cavity walls in a constrained medium permit only limited movement of guests. Generally, reactions requiring less distortion of occupied sites occur more readily than those requiring significant changes in a constraining media. When the relaxation times of cavity walls are much longer than the time required for transformation of a reactant to its products, the shape changes of the guest molecule must conform to the space available initially (Fig. 11.6A). Examples include sites provided by the crystalline surfaces of silica, clay, or zeolites. In media with softer walls (such as micelles, microemulsions, liquid crystals, and polymers), reaction cavities can respond more rapidly to shape changes of guests as the reaction

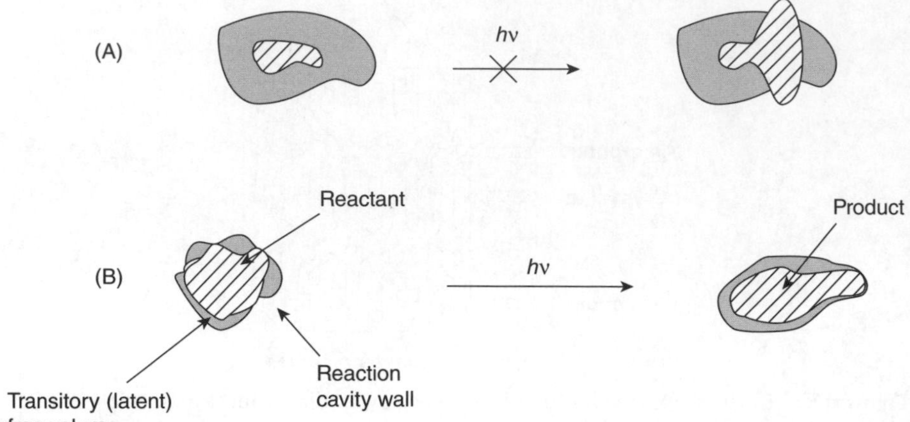

Figure 11.6. Cartoon representation of hard (A) and soft (B) reaction cavities. (Reprinted with permission from Reference 4. Copyright 2009 American Chemical Society.)

coordinate progresses; the guests are more mobile in cavities with "soft" walls (Fig. 11.6B).

The size and shape of occupied sites, as well as strain imposed by cavity walls, can alter the reactivity of guest molecules when they are introduced into a polymer. Such host–guest interactions facilitate or inhibit guest reactions by very different dynamic interactions than those of isotropic media.[4] Photoreactions of guest molecules can, therefore, be used to probe the micromorphology of a bulk polymer if the product distribution is directly related to the distribution of site configurations.

Guest Molecules as Reactants or Probes: Are Guest Molecules Aggregated or Dispersed?

Available space between polymer chains defines the efficiency of inter- or intramolecular interactions in polymers.[50] Shifts in excimer emission and excimer-to-monomer emission intensity ratios are qualitative measures of the efficiency of aggregation. The excimer band of a guest molecule, methyl 4-(1-pyrenyl)butyrate (**P-1**, Chart 11.1) in PS films, is blue shifted by 40 nm compared with that in fluid solutions.[56] The dearth of emission from excimers of **P-1** in PS has been attributed to the difficulty of molecular pairs to attain the necessary complex geometry. With increasing temperature, this excimer band becomes red shifted and reaches the wavelength maximum at 348 K ($T_g = 327$ K) found in fluid media.[56] These spectral changes have been attributed to increased mobility of polymer chains at higher temperatures.

In **PE** films, some α,ω-bis(1-pyrenyl)alkanes (**P*n*P**, where *n*, the number of carbons atoms in the alkane chain) (Chart 11.1) can form intramolecular

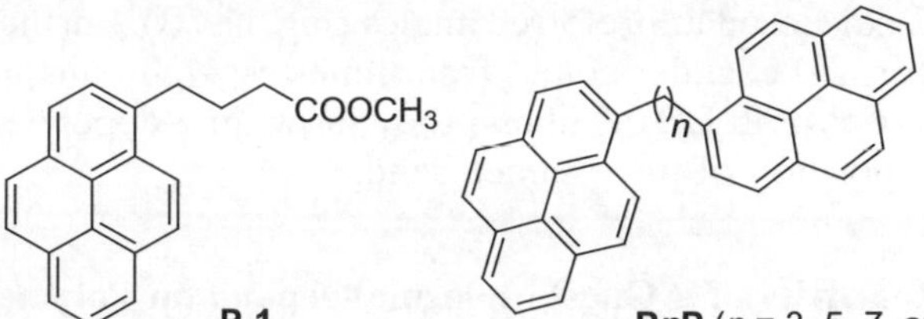

COOCH$_3$

P-1 **P*n*P** (n = 3, 5, 7, and 12)

Chart 11.1.

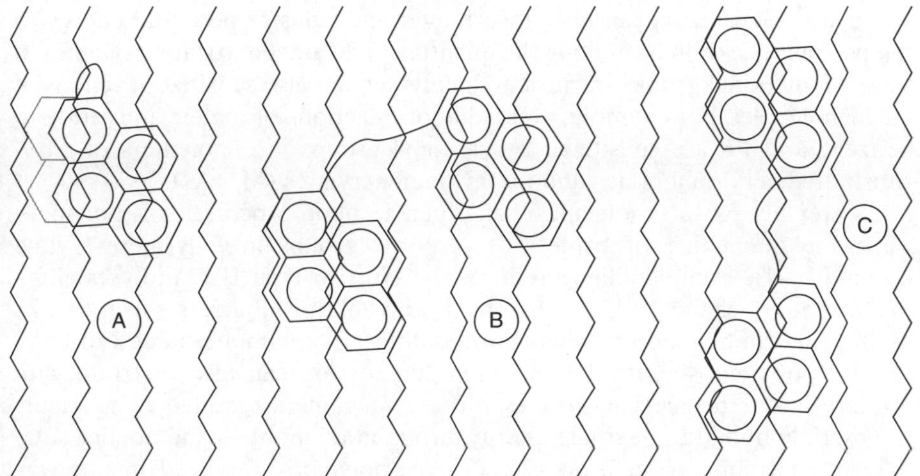

Figure 11.7. Aggregate and monomer conformations of α,ω-bis(1-pyrenyl)alkanes (**P*n*P**) in **PE** films: (A) sandwich conformation, (B) long pyrenyl axes are parallel and not overlapping, and (C) long pyrenyl axes make an angle of 70° and are not overlapping. (Reprinted with permission from Reference 50. Copyright 2009 American Chemical Society.)

excimers.[50] Some possible conformations for these **P*n*P** molecules on a lateral lamellar surface of a **PE** microcrystallite are shown in Figure 11.7. Face-to-face stacking of the pyrenyl groups is involved in the formation of classical excimers (Fig. 11.7A), whereas the extended conformation in Figure 11.7C can provide monomer-like luminescence only. The conformations of **P*n*Ps** in **PE** films are not optimal for static (ground-state controlled) interactions and fluorescence spectra of higher homologs (n > 3) lacks or exhibits very weak intensities for the broad, red-shifted band characteristic of classical pyrenyl excimers. Fluorescence decay histograms of the **P*n*P** in **PE** films possess two decay constants, and the short decay component (~50 ns) is assigned to interacting pyrenyl groups at the proximity and orientations for excimer-like emission.[50] This "nonclassical" excimer emission arises from conformations in which pyrenyl groups are disposed at preferred angles (Fig. 11.7B). Further evidence for dynamically formed excimer comes from time-resolved emission spectral (TRES) analyses of **P3P** in **LDPE** films—they show time-dependent changes in the nature and position of the excimer band.[50]

How Does the Reactivity of a Guest Molecule Depend on Polymer Groups That Can Be Sensitizers, Quenchers, or Reagents?

As mentioned in the section on "Chromophores or Other Reactive Groups in a Polymer," residual amounts of sensitizers in a polymer can absorb light in competition with guest molecules, causing changes in guest reactions and reactivity.[57] Sensitizers can also absorb light and transfer potential energy to the polymer backbone, affecting the quantum yield and polarization of emitted light.[58] Functional groups intrinsic to a polymer can also sensitize reactions of guest molecules. For example, cycloaddition reactions of maleic anhydride to benzene and PS are sensitized by carbonyl groups in copolymers of *para*-substituted acrylophenones and methyl methacrylates (MMAs).[59]

Conversely, guest molecules can quench photoprocesses of polymers. Thus, the quenching of triplets of carbonyl groups in poly(methyl vinyl ketone) by 1,3-cyclooctadiene reduces the Norrish type II photolyses along the polymer chains.[60] UV absorbers and photostabilizers, such as *o*-hydroxybenzophenone or benzotriazoles, that are commonly added to commercial polymers to retard their photodegradation can also interfere with photoreactions of guest molecules.[61] These reagents are added to polymers to absorb light and to dissipate energy through radiationless and nondegradative internal conversion processes. Clearly, polymers stabilized in this way should be extracted appropriately beforehand if they are to be used as media to examine specific photoreactions or photophysical properties of guest molecules.

Surface and Bulk Properties

When reporter groups are placed on a polymer surface by physical or chemical methods, their stability depends on what is at the interface. For example, a monolayer of azobenzene molecules introduced onto an aminosilylated silica surface exhibits photoresponsive properties.[62] Irradiation of these surfaces results in a "surface energy gradient" due to photoisomerization, leading to directional motion of a liquid droplet placed on the substrate. In other examples, where the bulk material is modified using photoresponsive groups,[63] irradiation can cause bending of the entire film (see the section on "Other Photon-Based Applications"). Of course, groups attached at interior sites are more resistant to degradation by the exterior environment than ones at or very near the surfaces. Thus, fluorescence of lumophores attached to the interior chains of **PE** films are not quenched significantly when they are placed in liquids that cannot dissolve or swell the polymer, but contain quenchers.[64]

Diffusion Studies

In isotropic solutions of low viscosity, translational and rotational diffusion are usually faster than the rates of deactivation of excited states so that small differences in chromophore environments in low-viscosity solutions are difficult to detect by steady-state measurements.[65] In bulk polymers, diffusion of guest molecules depends on available free space and the mobility of the polymer chains that constitute the walls of the reaction cavities. As expected on the basis of viscosity considerations, the rates of translational diffusion of small molecules in glassy polymers are slower than in rubbery polymers and the latter are lower than in low-viscosity liquids.[56]

For example, the mobility of polymer chains and side groups of polydimethylsiloxane (**PDMS**) has been related to quenching rate constant (k_q) of pyrene fluorescence by phthalic anhydride: $k_q \propto (T/\eta)^{E_a/(2.3+E_\eta)}$, where η is the viscosity at temperature (T) and, E_a and E are the activation energy of quenching and viscous flow, respectively.[66] Rotations of **PDMS** side groups create spaces for guest diffusion. Because the activation energy ratio, (E_a/E), is <1 in viscous **PDMS**, it is reasonable to conclude that diffusion is assisted primarily by movement of the CH_3 groups rather than long segments of the polymer chain.[66]

Rates of rotational or translational diffusion and conformational changes of chromophores can also yield information about their average environments in polymers.[56,67] Exciplex formation is attenuated when the constituent molecules have difficulties to adopt their preferred orientations or cannot diffuse rapidly in a polymer. Thus, amount of intermolecular exciplex formation between anthracene or 4-(1-pyrenyl) butyrate and dimethyl toluidine is lower in PS films compared with that in fluid media, but increases as temperature approaches T_g or T_m from lower values (and chain mobility increases).[56] Absence of intramolecular exciplex formation in these polymer matrices has been attributed to slower folding of alkyl groups linking the exciplex constituents (e.g., rate of folding process $<10^6\,s^{-1}$, $T < 348\,K$ for *N,N*-dimethyl-4-[3-(pyren-1-yl)propyl]aniline) (Chart 11.2, **Py3Anil**).[56] Also, from their diffusion coefficients, $\sim10^{-8}\,cm^2\,s^{-1}$ in **PE31** at 385 K, anthracene guests are calculated to be able to move a distance of about one molecular diameter (4–5 Å) during their excited-singlet state lifetime (~5 ns).[68]

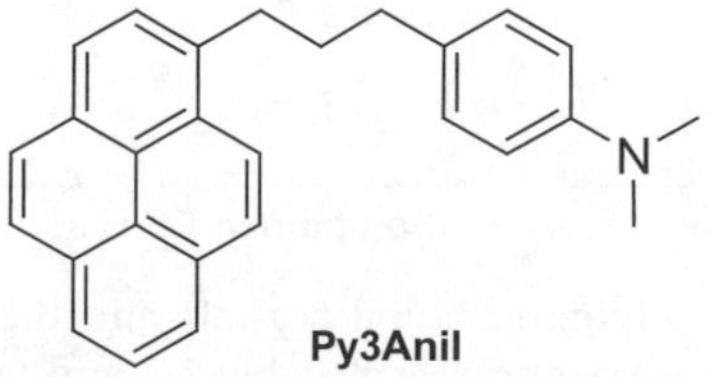

Py3Anil

Chart 11.2.

SPECTROSCOPIC APPLICATIONS

Neat Polymers

Liquid crystalline (LC) polymers have been widely investigated for optoelectronic and data storage applications.[69,70] Two important advantages of polymeric LC films over low-molecular-weight LCs for such applications include broader mesomorphic temperature ranges and higher image stabilities below the glass transition temperature.[63,71] In another application, photoisomerizations of groups in polymeric LCs have been used to initiate isothermal phase transitions or photoswitching when the isomers have significantly different shapes.[72] For example, the more globular *cis* isomers of azobenzene groups are more destabilizing to LC phases than their thermodynamically more stable and more rodlike *trans* isomers.[73] When thermal reversion of the *cis* isomers is rapid, data embedded by selective irradiation of the *trans* isomers cannot be stored for long periods in films. Using the azobenzene-containing mesogenic polymeric LC, **poly-azo** (Fig. 11.8), images can be "frozen" in place for periods as long as desired and then erased by warming the polymer matrix to allow the *cis* to return to its *trans* form.[71] A

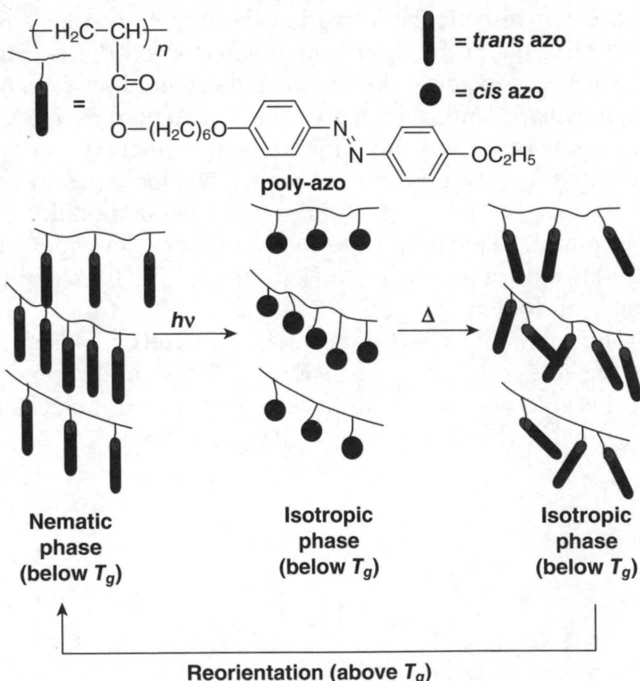

Figure 11.8. Azobenzene polymeric liquid crystals and the orientation of mesogens after *trans*→*cis*→*trans* isomerization cycles below and above T_g (redrawn from Reference 71).

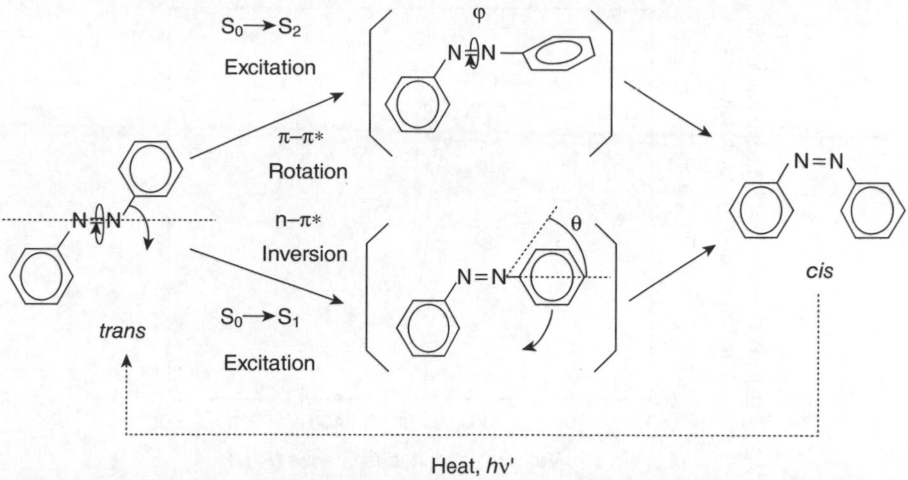

Figure 11.9. Proposed mechanisms for photoisomerization of azobenzene. (Reprinted with permission from Reference 184. Copyright 2009 American Chemical Society.)

nematic–isotropic phase transition has been achieved below T_g (318K) on exposing **poly-azo** to light, and this transition has been utilized to record images that take advantage of the large difference between the birefringence of the nematic and isotropic phases. Again, raising the temperature above T_g enables the *cis*-azobenzene groups to isomerize thermally to their *trans* forms and erase the image (Fig. 11.8).

Photochromic Guests in Polymer Hosts

Some commonly used dopants in bulk polymers contain azobenzene, stilbene, spiropyran, or thioindigo groups.[69] Each of these undergoes a photoinduced interconversion whose mechanism depends on the substituents and local steric factors imposed by the polymer matrix. Azo groups can undergo photoisomerization via two distinct pathways: inversion from $^1(n,\pi^*)$ states and rotation from $^1(\pi,\pi^*)$ states (Fig. 11.9). The ability of azobenzenes to isomerize in restricted environments is usually attributed to the less spatially demanding inversion mechanism with its semilinear transition state;[74] the rotation mechanism is not permitted sterically if the cavity does not have sufficient free volume in a direction perpendicular to the plane defined by the C–N–N–C atoms of the ground state.[75] A minimum critical free volume of $0.12\,nm^3$ for the inversion and $0.38\,nm^3$ for the rotation mechanisms has been calculated.[76]

Victor and Torkelson reported a general technique for measuring the distribution of local free volume in glassy polymers, such as PS, based on photoisomerization reactions.[77] They measured the amount of probe photoisomerization in a glassy polymer relative to that in a dilute solution of toluene as a

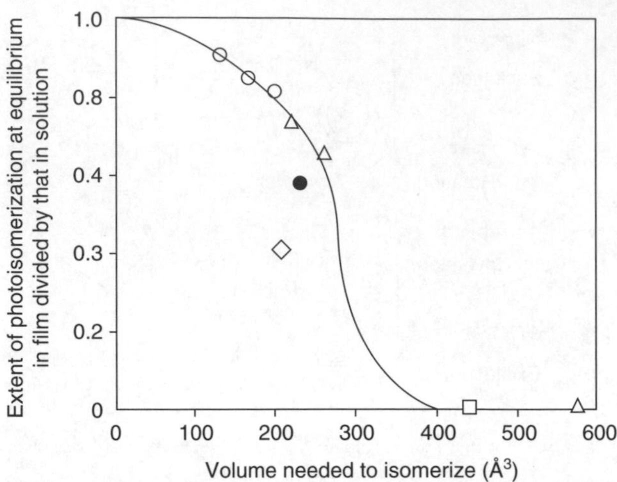

Figure 11.10. Cumulative distribution of local free volume in an unannealed polystyrene glass at 298 K measured by photochromic and fluorescent probes (the van der Waals volume of each molecule in its *trans* state is in parentheses): (○) azo (<170 Å^3), (Δ) stilbene (151 Å^3), (□) 1,3-di(l-pyrenyl)propane, (◇) thioindigo (197 Å^3), and (•) 2,2′-azonaphthalene (216 Å^3). (Reprinted with permission from Reference 77. Copyright 2009 American Chemical Society.)

function of the volume required for photoisomerization of the probes (Fig. 11.10). The excess space (or rearrangement of space) necessary to effect isomerization, beyond the van der Waals volumes occupied by their ground states, is either intrinsic to the guest sites or must become available to them during the isomerization process. The data in Figure 11.10 are interpreted to mean that more than 90% of the sites in the PS have between 120 and 130 Å^3 of free volume and none has >400 Å^3; ca. 50% of the sites have 260–280 Å^3. Similar values are obtained from positron annihilation spectroscopic studies.[78]

A cross-linked, nematic, liquid crystalline elastomer (**LCE**) film doped with an azo-dye, Disperse Orange I (Chart 11.3), has been shown to exhibit a mechanical response to nonuniform illumination.[79] The **LCE** consists of a poly(methylsiloxane) (**PMS**) backbone with mesogenic aromatic side-chain units. The samples are lightly cross-linked first, using the trifunctional cross-linker, **C9Ph**. Then, they are stretched unidirectionally to align the mesogens and to establish a nematic order. Finally, they are cross-linked further to form a transparent birefringent monodomain. Once irradiated, the **LCE** undergoes a large and rapid light-induced bending deformation.

Mechanical strain acts as an external field to align the liquid crystal, and the orientational order acts as an external body stress leading to elastic strain. Upon photoisomerization of the azo dye, the degree of orientational order is reduced and the liquid crystal can reorient. In the **LCE**, the change in the orientational order results in internal stress due to the strong coupling between

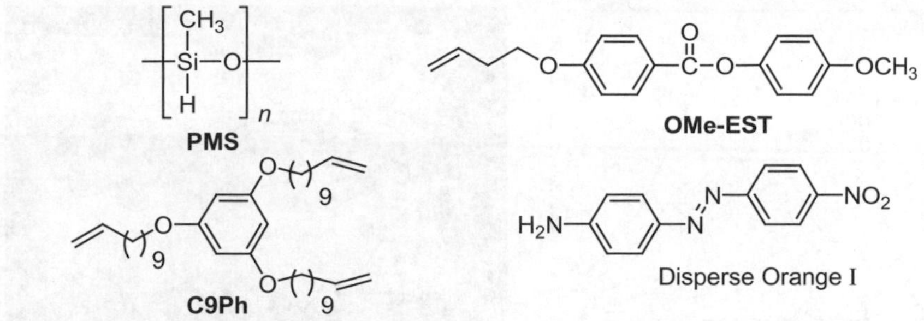

PMS

OMe-EST

C9Ph

Disperse Orange I

Chart 11.3. Structures of the components and dye in a light-responsive **LCE**: the poly(methylsiloxane) (**PMS**) backbone, the mesogenic aromatic side chain (**OMe-EST**), the trifunctional cross-linker (**C9Ph**), and the azo dye, Disperse Orange I.

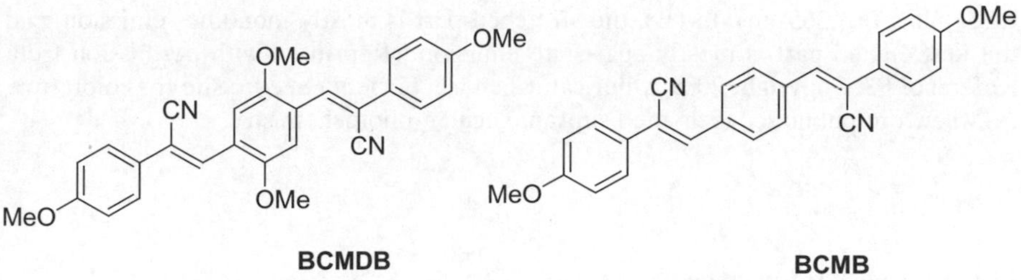

BCMDB

BCMB

Chart 11.4. Structures of photoluminescent dyes.[83]

the orientational order and mechanical strain. This combination ultimately leads to shape changes of the film. In this way, the dye-doped **LCE** was bent away from the irradiation source by more than 60°.

Another application involves the formation of luminescent aggregates (excimers) in polymeric hosts to extract information about the molecular and macroscopic properties of the polymer.[80] Elegant examples from the group of Weder have shown that binary blends of cyano-substituted oligo-phenylene vinylenes (**BCMDB** and **BCMB**, Chart 11.4) in linear **LDPE** (**PE51**) or polypropylene can be tuned to produce almost any ratio of monomer-to-excimer emission. The oligomeric additives reside in the amorphous fraction of the polymer. The experimental variables are composition, processing conditions, and temperature.[81,82] The binary blends of **PE51** and photoluminescent dyes have been used as internal molecular strain sensors.[83]

At low concentrations, the photoluminescent dyes are incorporated in the amorphous fraction of **PE51** in a molecularly dissolved state; at higher concentrations, excimers are formed. Solid-state deformation had a pronounced effect on the emission characteristics of the blend films. When films were stretched to a draw ratio of 500%, monomer-to-excimer emission ratios increased by a factor of 10 and significant color changes were noted (Fig. 11.11).

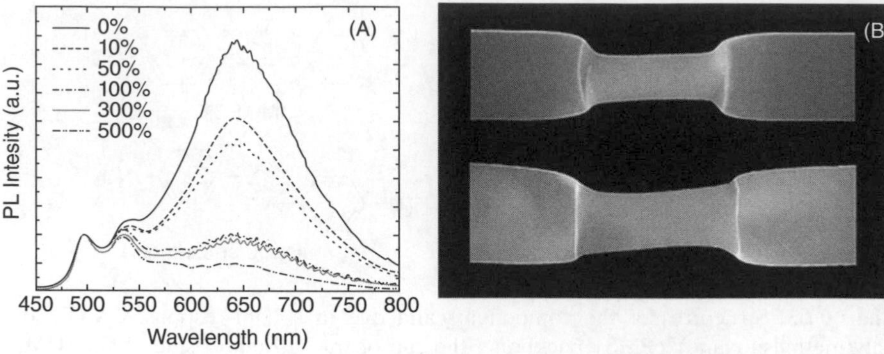

Figure 11.11. (A) Photoluminescence (PL) spectra of **PE51** films and 0.20% (w/w) **BCMDB** as a function of draw ratio and (B) visual appearance of luminescence from 5× stretched and unstretched portions of **BCMDB** (top) and **BCMB** (bottom) doped **PE51** films (λ_{ex} 365 nm). In (B), the stretched part is mostly monomer emission, and the unstretched part is mostly aggregate emission. (Reprinted with permission from Reference 83. Copyright 2009 American Chemical Society.) See ftp site for color: ftp://ftp.wiley.com/public/sci_tech_med/supramolecular_photochemistry.

Polymers as Photosensors[84]

Although a large number of sensors is known for detecting very small amounts of analytes in solution, fewer solid-state sensors have been reported[85] due to problems associated with quantum yields and optical stability.[86] Polymer films with luminescent groups or doped molecules, such as **1-Py** and anthracene, are used as oxygen sensors because their excited states are efficiently quenched by triplet (ground-state) oxygen. Polymer films with luminescent dyes containing transition metals, especially Pd and Pt, are particularly useful for sensing applications due to their stability, long-lived triplet excited states, and high luminescence quantum yields.[87]

Amplification of Fluorescence Quenching

Solid-state sensors of nitroaromatic and quinone compounds have been reported by Swager and coworkers.[88,89] These sensors utilize the fluorescence quenching of conjugated poly(phenelyene ethylene) films (**P-2**, Chart 11.5). The mechanism of the fluorescence response in these systems involves photoinduced electron transfer from the polymer donor to the electron-deficient analytes, which bind to the polymers via tight π-complexes.

Whitten and coworkers studied the rapid quenching of the fluorescence from a monolayer of the sulfonated polymer, poly(2-methoxy-5-propyloxy sulfonate phenylene vinylene) (**MPS-PPV**, Chart 11.5) on a glass substrate by

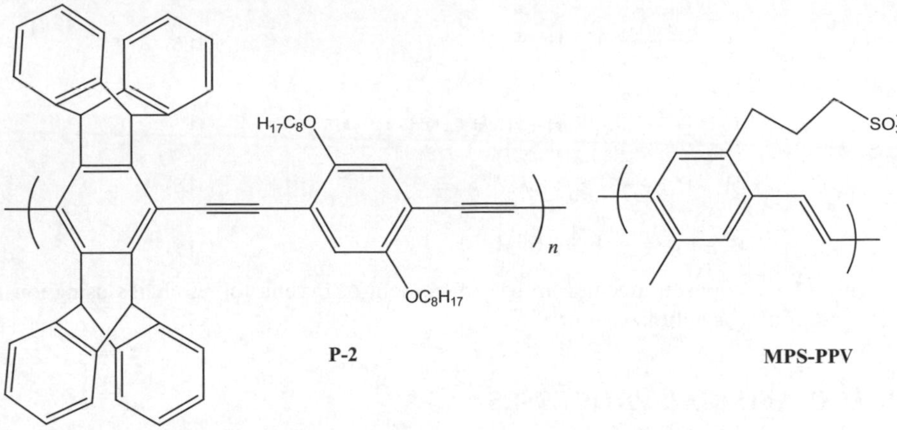

P-2 MPS-PPV

Chart 11.5. Structures of conjugated polymers used as sensors.

exposing it to vapors of nitroaromatics. The system can detect as low as $<8 \times 10^{-9}$ M of analyte.[90]

Polymers to Determine the Kinetic Energy (KE) of High-Energy (Ionizing Radiation) Particles

Conventional dosimeters measure the total energy deposited, but do not measure the KE of the impinging high-energy particles. A relatively simple method to measure the KE has been developed using stacks of pyrene-doped **PE** films of known thickness that allow a profile of the depth of penetration of the impinging particles to be determined.[91] In this experiment, the KE of the particles is transformed into potential energy and then into chemical reactions—covalent attachment of some pyrene molecules to **PE** chains—whose concentration depends on the depth of penetration. The profile of concentration of attached pyrene versus the depth of the films within a stack (from optical density changes and shapes of the emission spectra) can be correlated with the particle energy and, in some cases, its type. Principal steps involved in the mechanism of attachment of pyrene to **PE** are shown in Equations 11.6–11.9 (Scheme 11.1). Polymer-based radicals are formed, along with a small amount of pyrene excited states and radicals, upon bombardment of the **PE** films. The polymer radicals add to the pyrene molecules (Eqs. 11.8 and 11.9) to yield covalently attached pyrenyl groups in the **PEs**. The films are extracted to remove unreacted pyrene or photoproducts from pyrene that are not linked covalently to the polymer chains, leaving for analysis only those pyrenyl groups that have become covalently attached.

$$\mathrm{PE} \xrightarrow[\text{Radiation}]{\text{Ionizing}} \mathrm{R}^{\#} + \mathrm{H}^{\#} \tag{11.6}$$

$$\mathrm{PyH} \xrightarrow[\text{Radiation}]{\text{Ionizing}} {}^{1}\mathrm{PyH} + {}^{3}\mathrm{PyH} + \mathrm{PyH}^{-} + \dot{\mathrm{P}}\mathrm{y}\dot{\mathrm{H}}^{+} \tag{11.7}$$

$$\mathrm{R}^{\#} + \mathrm{PyH} \longrightarrow \mathrm{R\text{-}Py\text{-}H}^{\#} \tag{11.8}$$

$$\mathrm{R}^{\#} + \mathrm{PyH} \longrightarrow \mathrm{PyH} + \mathrm{H}\cdot \tag{11.9}$$

Scheme 11.1. Proposed mechanism for attachment of pyrene to **PE** chains using ionizing radiation ($^{\#}$ symbolizes + or •).[91]

PHOTOPHYSICAL PROCESSES

Measurement of Translational Diffusion of Small Molecules in Polymer Films

Lumophores such as **1-Py** and 9-anthrylmethyl (**9-AnCH$_2$**) that are covalently attached to polymer chains of **PE** films have been used to determine the accessibility of a series of *N*,*N*-dialkylanilines (**DAA**). The **DAA** are efficient quenchers of the **1-Py** lumophore excited-singlet states via Dexter (contact) mechanisms.[92–94] As the quenchers diffuse into the films, the fluorescence intensity decreases. These temporal changes can then be related to the diffusion coefficients of the **DAA**, as well as their ability to occupy different types of sites within a polymer. The reverse process, diffusion of **DAA** molecules from a film into a neat solvent bath, leads to an increase in the fluorescence intensity of the appended **1-Py** or **9-AnCH$_2$** groups. The relative fluorescence intensity I_{rel}(out) for out-of-diffusion at time = t is given in Equation 11.10:

$$I_{\text{ref}}(\text{out}) = \frac{I_t - I_0}{I_\infty - I_0}, \tag{11.10}$$

where I_t, I_∞, and I_0 are the fluorescence intensities measured at time = t, at "infinite" time, and the moment when a doped film is placed in a liquid into which the **DAA** diffuse, respectively. An analogous equation yields I_{rel}(in). I_0 is calculated from the intercept of the extrapolated slope in a plot of I_t versus $t^{1/2}$ using the "early-time" expression of Fick's second law (Eq. 11.11):

$$I_t = I_0 - (I_0 - I_\infty)\left(\frac{4}{l}\right)\left(\frac{D}{\pi}\right)^{1/2} t^{1/2}. \tag{11.11}$$

In this equation, l is the film thickness and D is the diffusion coefficient. The value of D was calculated by fitting I_0(out) at each t to a series expansion of Fick's second law truncated after the first 16 terms (Eq. 11.12). Analogous

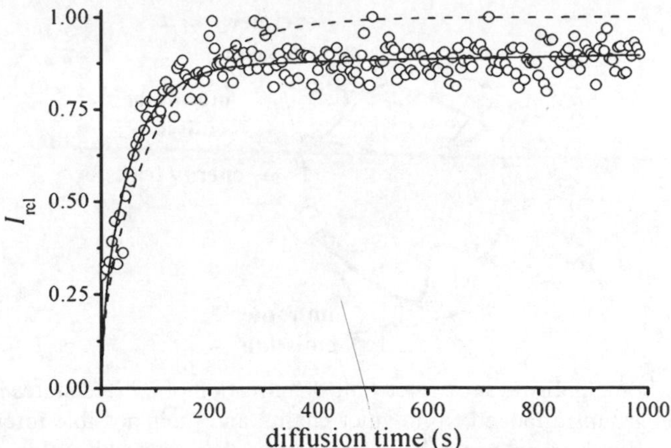

Figure 11.12. Fluorescence out-diffusion data for **DODA** doped **PE42**-anthracene film at 313 K, and best-fit curves according to Equation 11.12 (dashed-line) and a modified expression (solid line) mentioned in the text. (Reprinted from Reference 94. Copyright 2009, with permission from Elsevier.)

expressions have been derived to follow the diffusion of the **DAA** into a film from a liquid reservoir:

$$I_{\text{ref}}(\text{out}) = 1 - \sum_{n=0}^{15} \frac{8}{(2n+1)^2 \pi^2} \exp - \left[\frac{(2n+1)^2 \pi^2 t}{l^2} \right]. \quad (11.12)$$

Although good fits to the diffusion data for *N,N*-dimethylaniline (**DMA**) were obtained from Equation 11.12, the histogram for *N,N*-dioctadecylaniline (**DODA**) required a modified expression with two diffusion coefficients, *D*1 and *D*2, and their fractional contributions, *C* and (1-*C*), respectively,[94] indicating that two parallel pathways for diffusion are occurring (Fig. 11.12). These fluorescence measurements reveal information about micro- and macrostructures of the host by correlating the quenching rate constants with diffusional rate constants. The effective accessibilities of the **DAA**s to sites in which the probes are located and their rates of diffusion within the films provide additional insights into the film properties at the molecular chain/segment distance scale. Also, the longer excited-singlet state lifetime of **1-Py** allows it to assess events occurring in a larger surrounding volume than those reported by the much short-lived excited-singlet state of **9-AnCH$_2$**. In this way, excited-state lifetimes can be tuned to interrogate different volumes of the host space.

Rates of Electron or Exciton Migration

Photophysical properties of the chromophores in polymers are a signature of the interactions of the polymer chains. Some of the association/dissociation

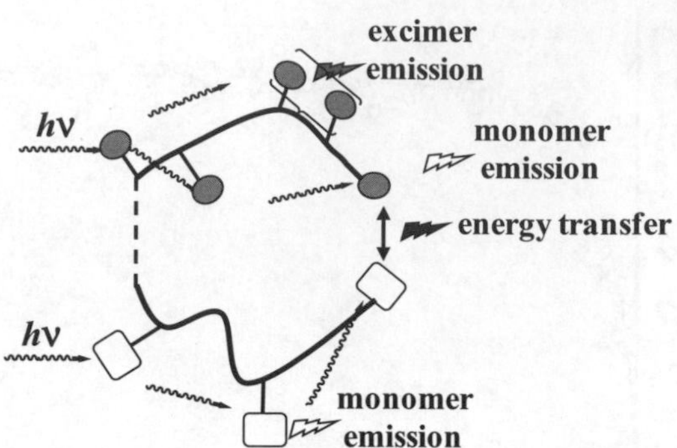

Scheme 11.2. Some pathways for electronic deactivation of excited states in polymers; chromophoric groups attached to polymer chains, and their possible interactions are shown. Lines represent polymer chain backbones; gray and white objects represent chromophoric donor–acceptor groups.[61]

pathways of the chromophores with a concomitant modification in the optical properties in a polymer are represented in Scheme 11.2.

Exciton energy transfer can proceed from an excited-state guest molecule to a ground-state guest or to a polymer under conditions of appropriate energetics. In the exchange (Dexter) mechanism, "hopping" from one polymer group to another along or between chains can occur if the monomer units contain functional groups amenable to such transfer. In polyaniline, an important conducting polymer, formation of charge-separated species or polarons occurs first. These polarons then migrate from one nitrogen center to another through the conjugated chain.[95,96]

Radiative energy transfer between anthracene molecules doped into **PE** films has also been reported.[68] This type of energy transfer is indicated by a progressive decrease in the 0,0 emission band of anthracene with increases in its concentration. As noted in the section on "Chromophores or Other Reactive Groups in a Polymer," energy transfer can initiate or prevent photodegradation of polymers[97,98] (e.g., Norrish type I and type II reactions discussed in the section on "Photo-Cleavage Reactions").

Winnik and coworkers reported direct, nonradiative energy transfer in films of poly(isoprene-*b*-methyl methacrylate) diblock copolymers containing phenanthrene as a donor group in one polymer (**PI-Ph-PMMA**) and anthracene as an acceptor in the other polymer (**PI-An-PMMA**).[99] The polymers were labeled at the block junctions as shown in Chart 11.6. Films were prepared by mixing individual polymers in a solvent and casting the solution. The proximity of donor and acceptor groups within R_o, the critical Förster distance, resulted in efficient energy transfer.

This type of nonradiative energy transfer is a convenient method for studying the interactions of polymer chains.[99–102] The distance between interacting groups is obtained from analyses of temporal luminescence decay curves. A

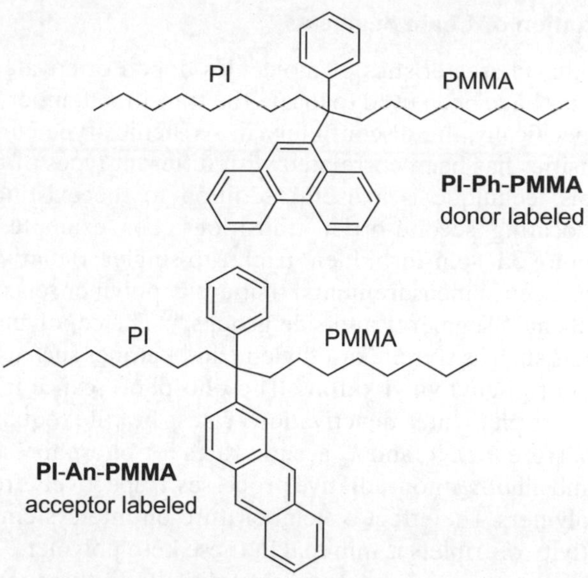

PI PMMA

PI-Ph-PMMA
donor labeled

PI PMMA

PI-An-PMMA
acceptor labeled

Chart 11.6. Structures of polymers labeled with phenanthrene donors and anthracene acceptors.[99]

single exponential decay was observed for phenanthrene polymer in the absence of any acceptor.[101] When the anthracene labeled was introduced, a rapid decay component was observed as a result of energy transfer. For a random distribution of donors and acceptors, the donor fluorescence intensity decay $I_D(t)$ can be described by a modified Förster equation (Eq. 11.14) [99]:

$$I_D(t) = \exp\left[-\frac{t'}{\tau_D}\right] \exp\left[-P\left(\frac{t'}{\tau_D}\right)^{0.5}\right], \tag{11.13}$$

$$P = \frac{4}{3}\pi^{3/2}\left(\frac{3}{2}\langle\kappa^2\rangle\right)^{1/2} C_A R_o^3. \tag{11.14}$$

t' is the fluorescence decay time, τ_D is the donor lifetime in the absence of quenchers, $\langle\kappa^2\rangle$ is the orientation factor, and C_A describes the acceptor concentration. Using this method, the characteristic distance for energy transfer between phenanthrene donor and anthracene acceptor (the Förster radius R_o) can be calculated.[102] The local concentration of the dyes within the interface is also obtained from the fluorescence decay analysis. The thickness of the interface obtained for mixtures of **PI-Ph-PMMA** and **PI-An-PMMA**, 1.6 ± 0.1 nm, is in agreement with values obtained from other experiments,[99,101] and it does not change when different amounts of **PI-Ph-PMMA** and **PI-An-PMMA** are mixed.

Rates of Relaxation of Chain Segments

The luminescence characteristics of molecules doped or covalently attached to polymer chains have been used to locate the transition temperatures of bulk polymers.[103] Specifically, the discontinuity in Arrhenius-type curves for luminescence intensities has been correlated with different types of relaxations in polymers.[68] This technique is a useful addition to the existing mechanical methods for locating second-order transitions. For example, Guillet and coworkers employed spin-forbidden triplet-to-singlet radiative transitions (N.B., phosphorescence measurements) to identify polymer relaxations associated with small chain segments and side groups.[103] The copolymers employed contained olefins such as styrene or ethylene and ketones such as methyl vinyl ketone (MVK) or phenyl vinyl ketone. The phosphorescence intensity (I_p) is related to the triplet-state deactivation rates by the equation, $I_p \propto k_p/(k_p + k_{rxn} + k_n)$. Here k_p, k_{rxn}, and k_n are the rates for phosphorescence, chemical reactions, and all other nonradiative processes, respectively, from the triplet state of the polymers. The effect of temperature on intersystem crossing and chemical reactivity of triplets is minimal in these keto polymers. The phosphorescence decay, therefore, can be correlated directly to the nonradiative processes, which, in turn, are directly related to the movements of polymer segments and side groups. With increasing temperature, quenching of the triplet state increased as a result of increased oxygen diffusion that is assisted by the facilitated motions of small segments of the polymer chains. A plot of the logarithm of the I_p versus $1/T$ could be fitted well to a straight line with a positive slope $\Delta E_Q/R$, where E_Q is the activation energy for quenching. The onset temperatures between linear regions in these Arrhenius plots of phosphorescence intensity correlate with the onset of the appearance or freezing of segmental motions in these polymers.

Monitors of Local Relaxation Phenomena

When fluorescent groups attached to polymer chains are directly involved in relaxation processes, polymer mobility and fluorophore deactivation are strongly correlated. 1-Pyrenylmethyl (**PyCH$_2$**) and anthryl groups covalently attached to the chains of several types of **PE** films have been shown to be able to detect the onset temperatures of relaxation processes.[104] The sensitivity of the lumophores to a particular relaxation process depends on how strongly the lumophores interact with the local chain motions of the polymer and how similar are the time constants for fluorescence decay and chain relaxation. Thus, fluorescence efficiency from covalently attached anthryl groups (with shorter-lived excited singlet states) is more sensitive to very rapid motions,[105] while covalently attached pyrenyl groups (with longer-lived excited-singlet states) are more sensitive to slower motions from larger segments of **PE** chains. The onset of relaxation processes in **PE** films has been determined from discontinuities in the emission intensities (as total integrated spectral areas), intensities of wavelength maxima, and the full widths at half maxima (FWHM) of 0,0 emission bands versus temperature (e.g., Fig. 11.13). Typically, the

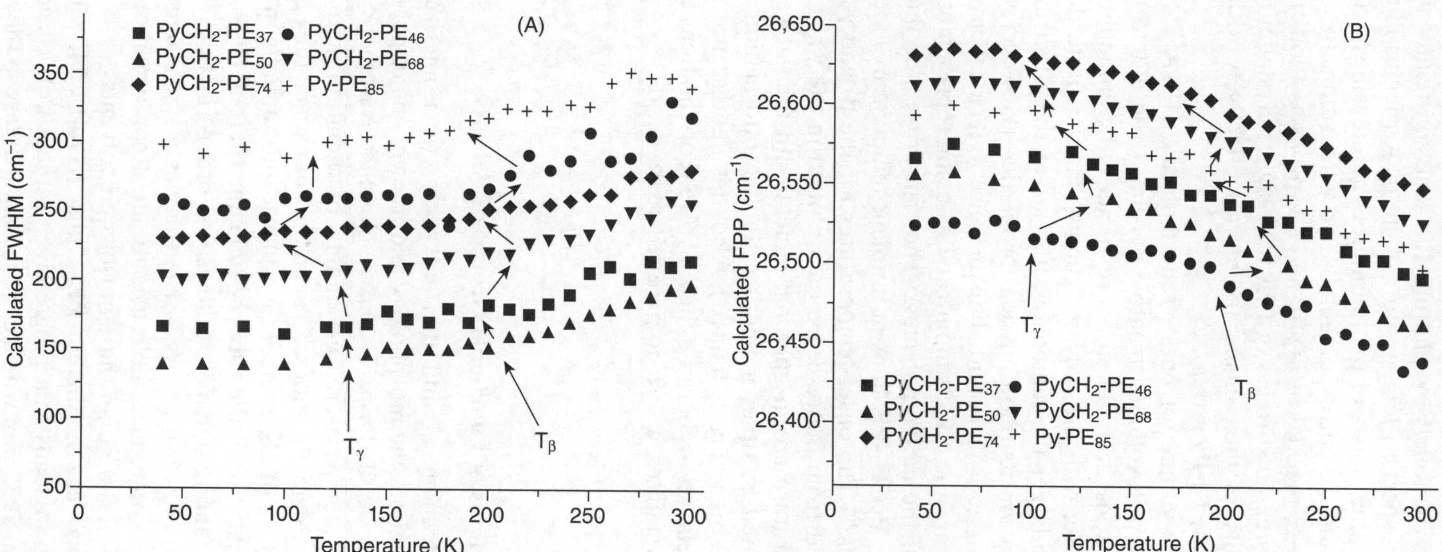

Figure 11.13. Changes of FWHM (A) and fluorescence peak maximum (FPP) (B) of the 0,0 emission bands of **$PyCH_2$-PE** films as a function of temperature. The number after **PE** indicates the crystallinity of the film. (Reprinted with permission from Reference 104. Copyright 2009 Koninklijke Brill NV.)

transition onsets are identified by plotting the emission-related quantities either directly with respect to temperature or in an Arrhenius-like fashion. The discontinuities in the slopes of the luminescence intensities over specific temperature ranges and the corresponding β- and γ-transitions to which they have been ascribed are marked in Figure 11.13. An observation from these studies is that the increased stiffness of **PEs** with higher crystallinity reduces significantly the ability of their chains to interact effectively with excited-singlet states of attached **1-Py** groups.[106]

The van der Waals volume of one pyrene molecule (184 $Å^3$ [104]) is much larger than mean free void volumes in native **PE** films (from 120 to 153 $Å^3$ [104]). The lumophore motions are, therefore, restricted—the lumophores must remain in close contact with the polymer chains that constitute their cavity walls. In general, the larger the lumophore, the larger the effect induced by the local environment on its photophysical properties. (However, the larger the lumophore, the larger is its potential disturbance to its local environment, and the less reliable are the effects monitored as being representative of the microenvironment of the native polymer!) The time-correlated fluorescence decay profiles of the **PyCH$_2$-PE** films at ambient temperatures yield singlet lifetimes of ca. 200 ns.[105] The onset temperatures for the β- and γ-relaxation processes as obtained from fluorescence data plotted in the ways described above versus temperature are in good agreement with the values reported from the data obtained by other methods (although information about the γ-relaxations from fluorescence is limited). Also, this fluorescence method is unable to detect α-relaxations because they involve reorientations of ordered chains within the crystallites, where no pyrenyl groups reside.

REACTIONS

Photoinitiated Reactions of Functional Groups of Polymers[107]

Isomerizations. Photochemical and thermal geometric isomerizations can provide information about the morphology and free volume of polymers.[108] A change from a more extended *trans* to a more globular *cis* isomer can be controlled externally by temperature, wavelength of radiation, and the space available within the polymer cavities.[109] As one might expect, the efficiency of *trans*→*cis* isomerization of azo moieties is reduced substantially in the glassy state of polymer hosts.[110,111] In the rubbery state, the thermal *cis*→*trans* isomerization of azo-containing polymers follows first-order kinetics.[109] However, a portion of the isomerization is very fast in the glassy state, suggesting that polymer cavities force the guest molecules to adopt conformations that are closer to the transition state than those in isotropic solvents.

Irradiation of azobenzene-linked **PE** films (**Azo-PE24**, Chart 11.7) using 367-nm radiation, leads to a progressive decrease in the *trans* (π→π*) absorption maximum at 326 nm with a concomitant increase in the *cis* (n→π*)

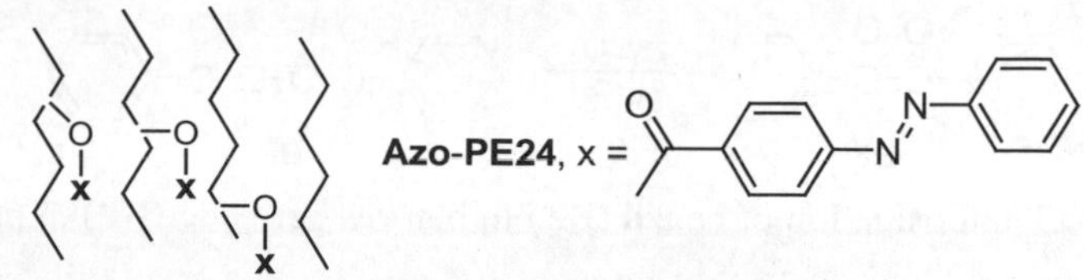

Chart 11.7. Cartoon representation of the structure of an azobenzene-linked **PE24** film.[109]

absorption maximum at 440 nm.[109] The kinetics of the thermally induced *cis*→*trans* isomerization of the azobenzenyl groups has been investigated at different temperatures and stretching ratios. Comparison between the isomerization rates of the stretched and unstretched films suggests that probe translocations takes place upon stretching.[2] Isomerization reactions follow first-order kinetics in unstretched films at all temperatures and in a stretched film at 343 K. The rate constants, k, can be calculated from the slopes of plots of $\ln([cis]_t/[cis]_0)$ versus time, t (Eq. 11.15):

$$\ln([cis]_t/[cis]_0) = \ln[(OD_\infty - OD_t)/(OD_\infty - OD_0)] = -kt. \qquad (11.15)$$

In Equation 11.15, OD_0, OD_∞, and OD_t are the optical densities at 326 nm at time 0, ∞, and t, respectively, and $[cis]_t$ and $[cis]_0$ are the concentrations of the *cis* isomer at time t and 0, respectively. The calculated activation energy (E_a) and pre-exponential factor (A) from Arrhenius plots of the values of k at different temperatures are similar to values found for *cis–trans* isomerizations for azobenzenes in liquid solutions.

The isomerization was found to be more complex in stretched **PE24** films. Reasonable fits of the kinetic data from stretched **Azo-PE24** below 343 K were obtained only after addition of a second first-order component to Equation 11.15 (i.e., Eq. 11.16): $k_1 = 6.35 \times 10^{-5}\,s^{-1}$ and $k_2 = 1.87 \times 10^{-5}\,s^{-1}$ 303 K. The dual isomerization processes were ascribed to differences in free volume and wall flexibility of sites in the interfacial and amorphous regions of the **PE** matrix:

$$[cis]_t/[cis]_0 = \alpha \exp(-k_1 t) + (1-\alpha)\exp(-k_2 t). \qquad (11.16)$$

In Equation 11.16, α is the fraction of the faster component. The calculated activation energies for both isomerization routes in the stretched film (15.7 ± 2.2 and 19.4 ± 1.4 kcal mol^{-1}) are lower than in the unstretched one (23.1 ± 0.7 kcal mol^{-1}). The corresponding pre-exponential factors are also smaller (7.3 ± 1.2 and 9.2 ± 0.7 s^{-1} for the stretched film and 11.8 ± 0.5 s^{-1} for the unstretched film). Even though the activation energy parameters indicate that stretching facilitates isomerization, the lower pre-exponential factors indicate that isomerization is more difficult from an entropic standpoint.

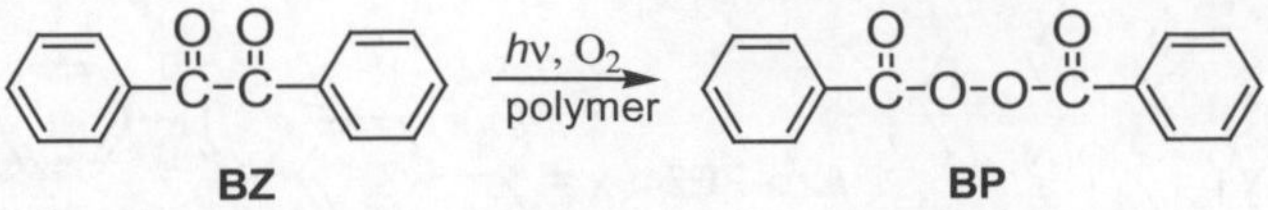

$$C_6H_5\text{-}C(=O)\text{-}C(=O)\text{-}C_6H_5 \xrightarrow[\text{polymer}]{h\nu,\ O_2} C_6H_5\text{-}C(=O)\text{-}O\text{-}O\text{-}C(=O)\text{-}C_6H_5$$

BZ **BP**

Scheme 11.3. Photooxidation of benzil (**BZ**) to benzoyl peroxide (**BP**) in polystyrene films.[112]

As films are stretched, the *cis* isomers in the interfacial sites (where the rates of chain relaxations are slower) are predicted to experience more strain due to a larger decrease in free volume. These *cis* groups, therefore, revert back to their *trans* forms more rapidly than in isotropic solutions because they are in conformations that are closer to the transition state.

Bimolecular Reactions. The excited states of benzil (**BZ**), an important industrial dicarbonyl compound, can either abstract a hydrogen atom leading to ketyl radicals in the absence of oxygen or, upon photooxidation, it can yield a mixture of products that includes phenyl benzoate, benzoic acid, biphenyl, and benzoyl peroxide (**BP**).[112] Selective conversion of **BZ** to **BP** (Scheme 11.3) can be achieved by irradiating **BZ** within polymer films.[112]

The **BZ**→**BP** photoconversion can be followed by Fourier-transform infrared (FT-IR) spectroscopy by monitoring the decrease in absorbance from the 1,2-dicarbonyl group of **BZ** (1650–1700 cm^{-1}) and the increase in absorbance of the **BP** (1750–1800 cm^{-1}) as shown in Figure 11.14A. Plots of ln A_0/A (where A and A_0 are the **BZ** absorbances at 1684 cm^{-1} at times t and 0, respectively) versus time for **BZ** consumption are linear when the irradiation flux is constant and the optical densities of **BZ** at the irradiating wavelengths remain <2. The relative rates of reaction are sensitive to the nature of the glassy polymer matrix and they decrease in the order: **PS** > bisphenol A polycarbonate (**PC**) > poly(vinyl chloride) (**PVC**) > bisphenol A polysulfone (**PSF**) > **PMMA** (Fig. 11.14B).

The photooxidation reaction proceeds from the excited triplet state of **BZ** (^{3}**BZ**). Competing deactivation pathways, shown in Scheme 11.4, include quenching of ^{3}**BZ** with ground-state oxygen ($^3O_2$) (Eq. 11.19), radiationless deactivation (Eq. 11.20), phosphorescence (Eq. 11.21), and reaction with the polymer matrix (Eq. 11.22).

Photooxidation of **BZ** to **BP** can also be observed in **BZ**-linked aromatic molecules or when a copolymer containing a monomer with a **BZ** group is irradiated.[113] Thus, when films of the copolymer, 1-{4-(2-methacryloxyethoxyphenyl)}-2-phenyl-1,2-ethanedione (**BZMA**) and styrene (**S**) (**BZMA/S**) were irradiated, **BZMA** photooxidation was found to be 10 times faster than that of **BZ** in **PS**.[113]

Cleavage Reactions. Photodegradable plastics contain groups such as carbonyls, carboxyls, or peroxides. An example of a commercial photodegradable polymer is Ecolyte, which has large amounts of carbonyl groups

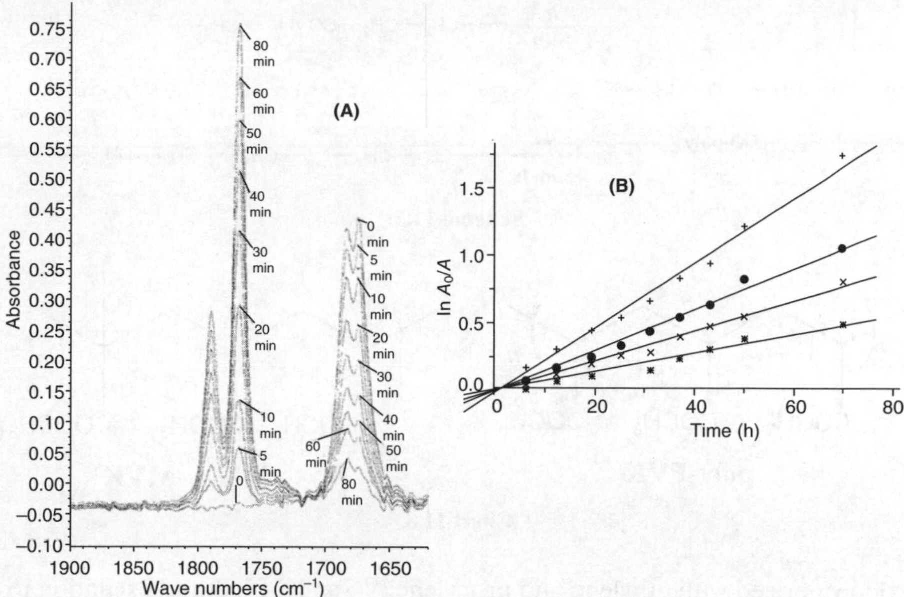

Figure 11.14. (A) FT-IR spectra of **BZ** in a **PS** matrix irradiated for different periods in air in a Spectromat apparatus and (B) plots of **BZ** absorbance at 1684 cm^{-1} in **PS** (+), **PC** (•), **PSF** (x), and **PMMA** (✳) films. (Reprinted with permission from Reference 112. Copyright 2009 Wiley-VCH Verlag GmbH & Co. KGaA.)

$$\mathrm{BZ} \xrightarrow{I_{abs}} {}^{1}\mathrm{BZ} \xrightarrow{\Phi_{ISC}} {}^{3}\mathrm{BZ} \tag{11.17}$$

$${}^{3}\mathrm{BZ} + {}^{3}\mathrm{O_2} \xrightarrow{k_{ox}} \mathrm{BP} \tag{11.18}$$

$${}^{3}\mathrm{BZ} + {}^{3}\mathrm{O_2} \xrightarrow{k_q} \mathrm{BZ} + {}^{1}\mathrm{O_2}\ (+\ {}^{3}\mathrm{O_2}) \tag{11.19}$$

$${}^{3}\mathrm{BZ} \xrightarrow{k_d} \mathrm{BZ} + \Delta \tag{11.20}$$

$${}^{3}\mathrm{BZ} \xrightarrow{k_{ph}} \mathrm{BZ} + h\nu \tag{11.21}$$

$${}^{3}\mathrm{BZ} + \text{polymer} \xrightarrow{k_p} \text{adducts} \tag{11.22}$$

Scheme 11.4. Proposed mechanism of photochemical reaction of benzil with molecular oxygen and deactivation pathways in polymer matrices.[112]

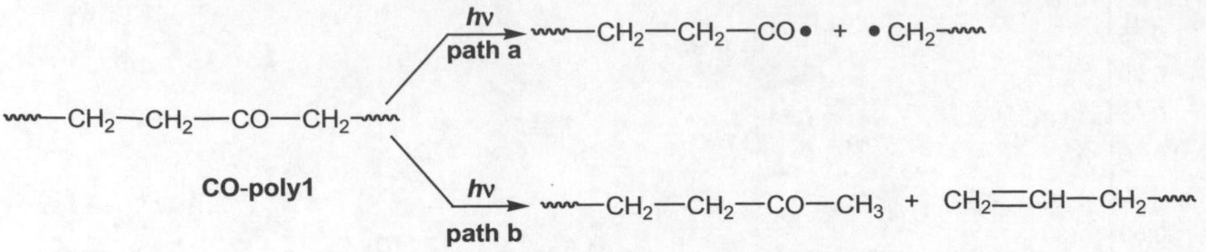

$\sim\!\sim CH_2-CH_2-CO-CH_2\sim\!\sim$ **CO-poly1**

$\xrightarrow[\text{path a}]{h\nu}$ $\sim\!\sim CH_2-CH_2-CO\bullet \; + \; \bullet CH_2\sim\!\sim$

$\xrightarrow[\text{path b}]{h\nu}$ $\sim\!\sim CH_2-CH_2-CO-CH_3 \; + \; CH_2=CH-CH_2\sim\!\sim$

Scheme 11.5.

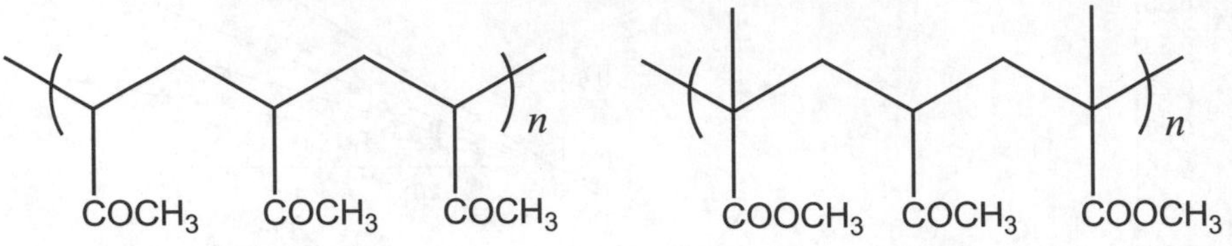

$COCH_3$ $COCH_3$ $COCH_3$ (n) **poly-PVK**

$COOCH_3$ $COCH_3$ $COOCH_3$ (n) **MMA–MVK**

Chart 11.8.

copolymerized with ethylene and propylene.[114] Sunlight-induced excitation to n,π* electronic states initiates Norrish type I and Norrish type II reactions that cleave the polymer chains in these polymers and facilitate their ingestion by microorganisms.[114]

Aliphatic ketones undergo some dissociation by Norrish type I cleavage processes, and ketones containing γ-hydrogen atoms undergo type II cleavage processes, also, via formation of truncated ketones and alkenes.[39] In an ethylene-carbon monoxide copolymer, **CO-poly1**, at near-ambient temperatures, the quantum yield of the type II reaction (Φ_{II} ~0.02; path b) is higher than the quantum yield of the type I reaction (Φ_I ~0.001; path a) (Scheme 11.5).[114] However, Φ_{II} is nearly zero below the glass transition temperature because the restrictions to molecular movements do not allow the requisite six-membered transition state for the initial type II reaction step to be attained.

Norrish type II is a major reaction pathway in poly(methyl vinyl ketone) (**poly-MVK**, Chart 11.8) also.[115] The quantum yield for Norrish type II reactions is found to be nearly an order magnitude higher in the copolymer of **MMA** and a small amount of **MVK** (**MMA–MVK**, Chart 11.8) than in **poly-MVK**.[115]

Energy transfer and self-quenching of excited states occur between adjacent carbonyl groups located at regular intervals along the polymer chain in **poly-MVK**. The excited triplet-state lifetime of the carbonyl groups (10^{-8}–10^{-9}s)[98] is sufficient for polymer chains above T_g to undergo a number of rotations, allowing the excitation energy to move along the polymer chains. As shown in Scheme 11.6, the type II process in **MMA–MVK** involves the formation of a seven-membered transition state.[115] Surprisingly, although the glass transition temperature for **MMA–MVK** is 348 K, it undergoes some Norrish type II reaction even at 298 K ($\Phi_{II} = 0.0065$). The efficiency increases with

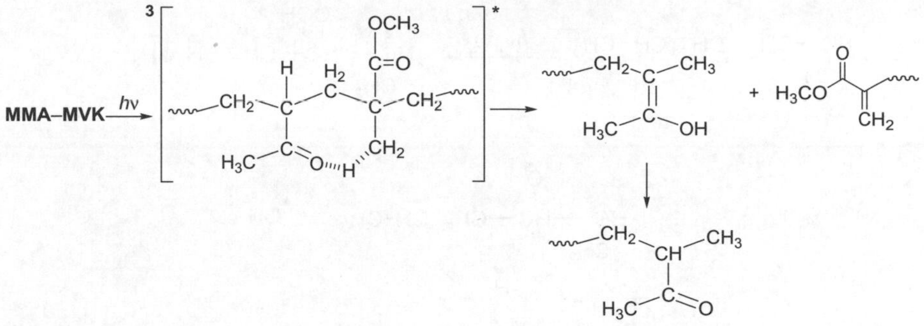

Scheme 11.6. Norrish type II reaction in **MMA–MVK** copolymer through a seven-membered ring transition state.[115]

increasing temperature and is 0.12 at 359 K.[115] The larger values of Φ_{II} in **MMA–MVK** compared with those in **poly-MVK** have been attributed to the absence of easy quenching by neighboring carbonyl groups in the former polymer. Energy transfer from an excited carbonyl to a carboxy (ester) group is endothermic.

Photooxidation of polyolefin-based plastics leads to cleavage reactions at unsaturated positions and formation of hydroperoxides at allylic positions.[114] The mechanism of the cleavage reactions in hexadiene copolymers has been studied by analyzing photolabile intermediates and photoproducts. Upon photooxidation of poly(propylene-co-hexadiene) copolymers, FT-IR spectra show a broad band at 3450 cm^{-1} (from alcohols and hydroperoxides) and concomitant disappearance of a band at 966 cm^{-1} (attributed to double bonds).[114] After protracted irradiation, bands at 1710, 1720, 1735, and 1780 cm^{-1} (attributable to aliphatic acids, carbonyl groups, esters, and γ-lactones, respectively) are observed. Shorter irradiation times result in the appearance of additional bands at 1701, 1680, and 1625 cm^{-1} (assigned to α,β-unsaturated acids, carbonyl groups, and conjugated double bonds, respectively).[114] The amounts of carbonyl and hydroxyl groups formed correlate with the loss of C=C groups. The mechanism in Scheme 11.7 has been proposed to explain the photoinduced reactions of these unsaturated polyolefins.[114]

Irradiation of 1-phenyl-2-(4-propenoylphenyl)-1,2-ethanedione/styrene (**PCOCO/S**) copolymer films in air, first at >400 nm and then at 366 nm, results in cleavage reactions also.[116] Upon irradiation, the benzil carbonyl groups in **PCOCO/S** are first converted almost quantitatively to benzoyl peroxides (**BPG**). They undergo thermally and photochemically induced cleavage reactions (Scheme 11.8). Intrachromophoric sensitization causes photodecomposition of these peroxides in which the potential energy of the excited states of peroxide groups in **BPG** (>70 kcal mol^{-1}) are converted to KE for homolytic cleavage of –O–O– bonds (E_D ~33 kcal mol^{-1}).[116] The resultant polymeric radicals can be used to cross-link polymer chains (see the section on "Cross-Linking Reactions").

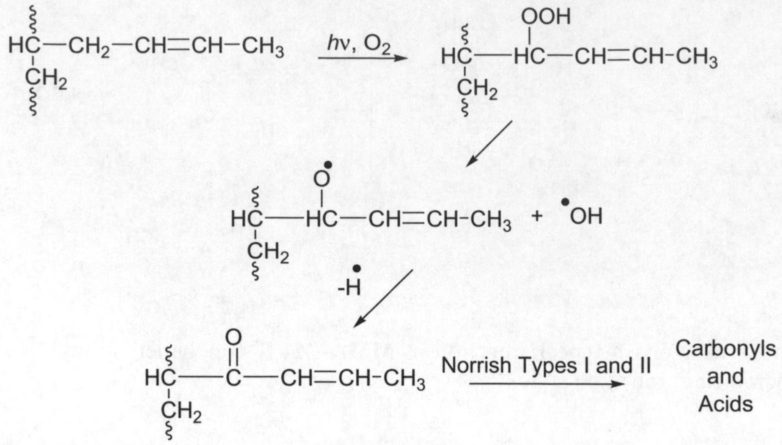

HC—CH2—CH═CH—CH3
CH2
hν, O2
OOH
HC—HC—CH═CH-CH3
CH2
O•
HC—HC—CH═CH-CH3 + •OH
CH2
-H•
O
HC—C—CH═CH-CH3
CH2
Norrish Types I and II
Carbonyls and Acids

Scheme 11.7. Proposed photooxidation mechanism of hexadiene-based copolymers.[114]

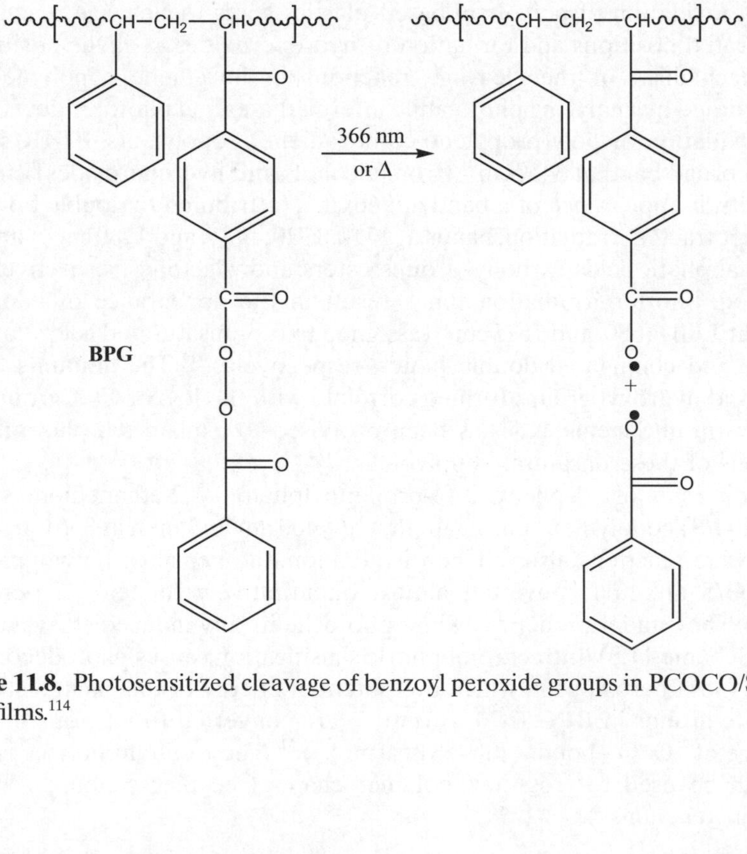

CH-CH2—CH
C═O
C═O
O
O
C═O
BPG
366 nm or Δ
CH-CH2—CH
C═O
C═O
O•
+
•O
C═O

Scheme 11.8. Photosensitized cleavage of benzoyl peroxide groups in PCOCO/S copolymer films.[114]

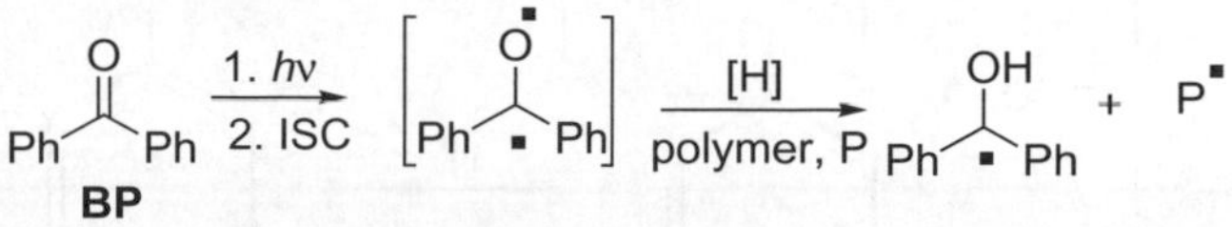

Ph–C(=O)–Ph (BP) —1. $h\nu$, 2. ISC→ [Ph–C(O•)•–Ph] —[H], polymer, P→ Ph–C(OH)•–Ph + P•

Scheme 11.9. Benzophenone (**BP**) photoinitiated formation polymer radicals.[117]

Metal complexes that can act as photosenzitizers or photointiators can enhance cleavage reactions of plastics also. For example, **PEO** forms coordination complexes with iron (III) chloride (a Fenton reagent), which accelerate photodecomposition via Cl• radicals.[114]

Cross-Linking Reactions. Cross-linking of commercial polymers as films and filaments is widely used in many industrial polymer-processing applications, especially for developing photoresists, electronic chips, and radiation curing of printing inks or adhesives.[117] Irradiation of **PEs** with high-energy radiation and in the presence of sensitizers can induce cross-linking whose efficiency depends on the temperature and type of cross-linker. Photoinitiators (sensitizers) such as benzophenone (**BP**) and quinones are commonly used for cross-linking **PEs**. The relatively unreactive ketyl radicals formed as intermediates either dimerize or combine with a polymer radical, P•, to initiate the cross-linking (Scheme 11.9).

As discussed in the section on "Cleavage Reactions," radicals are formed upon cleavage reactions of benzo-peroxides (**BPG** from **PCOCO/S** copolymers, Scheme 11.8). These radicals either abstract hydrogen atoms from the polymer chains of PS or attach themselves to a phenyl ring that is part of another side chain (Scheme 11.10).[116]

Another example is the use of oxetane-functionalized polymers in organic light-emitting diodes (OLEDs), in which the conversion of soluble polymers to insoluble ones by photo cross-linking is used for fabrication of muticolor (RGB), electroluminescent, high-resolution displays.[118] Upon photochemical curing, insoluble networks are formed by cross-linking of oxetane functionalized polymer **ox-poly** (Chart 11.9). The oxetanes are polymerized by a photoacid that opens the oxetane rings and starts the polymerization.

Polymers can be also cross-linked by cycloaddition reactions as well. Poly(vinyl cinnamate)s are known to react by three principal routes: cycloadditions, photo-Fries rearrangements, and isomerizations.[119] *Cis*⇌*trans* isomerization, the predominant photoreaction in solution,[120] is inhibited in solid polymer matrices for the reasons discussed in the section on "Photochromic Guests in Polymer Hosts." Cycloadditions and photo-Fries rearrangements are the major reaction routes in solid and LC films of poly(cinnamates).[119] Solid poly(vinyl cinnamate) undergoes [2 + 2] cycloaddition reactions, yielding truxinate and truxillate stereoisomers principally.[121] The preponderance of truxinates (head-to-head dimers) over truxillates (head-to-tail dimers) is attributed

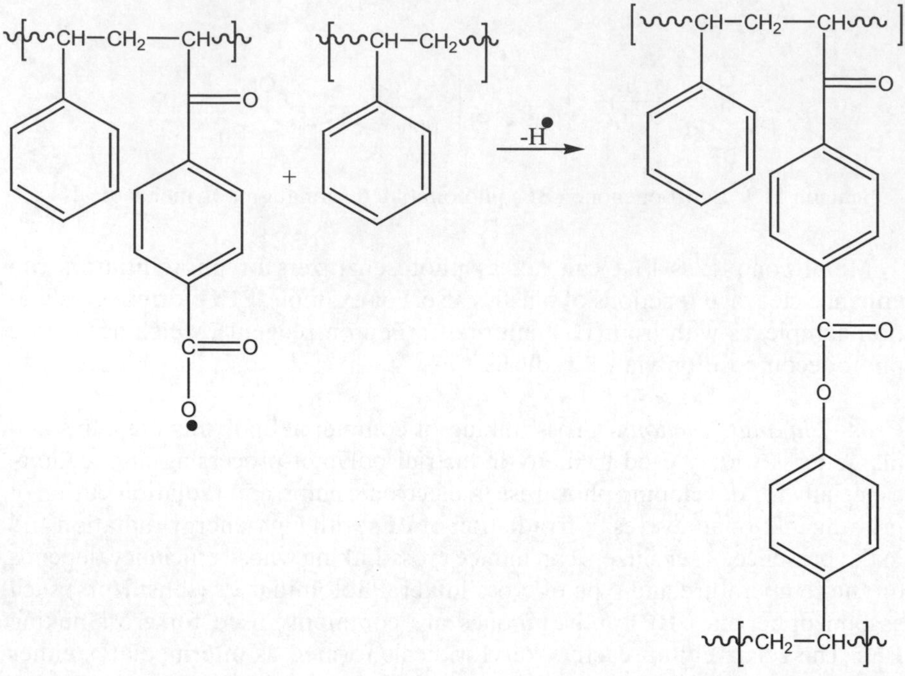

Scheme 11.10. Cross-linking reactions of ester groups with polystyrene chains.[116]

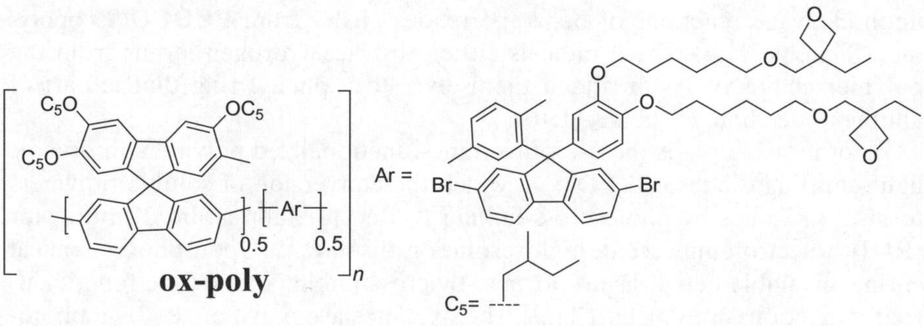

Chart 11.9. Structure of oxetane-functionalized spirobifluorene-*co*-fluorene polymer.[118]

to a sheet-like arrangement of the principal polymer chains and a parallel orientation of their cinnamate side chains.[121] A driving force for this arrangements is the preferential interactions of the polar side chains and low polarity backbone, as well as strong dipolar interactions between cinnamoyl groups.[121] As expected, the overall quantum yield of cycloaddition in the solid decreases as the reaction progresses, and becomes almost zero when about 50% of the cinnamoyl groups remain. The lack of additional reaction is attributed to a "stiffened" environment that makes attainment of the optimal geometry for

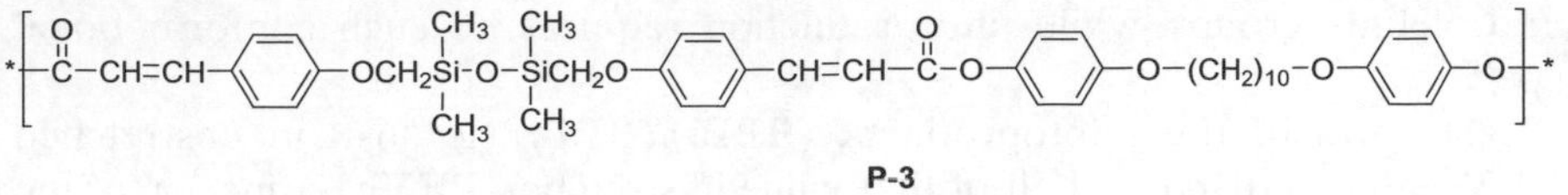

Chart 11.10. Structure of a siloxane-linked polycinnamate.[123]

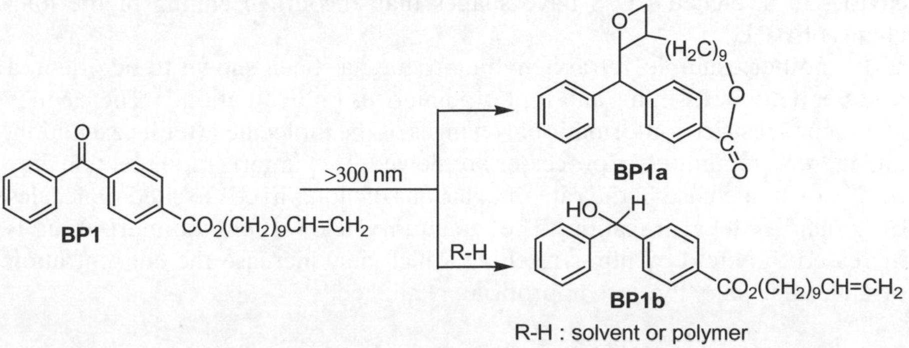

Scheme 11.11. Phototransformation of ω-undecylenyl benzophenone-4-carboxylate (**BP1**).[124]

cycloaddition very difficult and to a larger mole fraction of sites whose cinnamoyl-cinnamoyl orientations are intrinsically unable to support cycloadditions. To a certain extent, the distribution of cyclodimers can be correlated with the distribution of cinnamoyl-cinnamoyl orientations before irradiation.

A main-chain polycinnamate with siloxane linkages, **P-3** (Chart 11.10), also undergoes [2 + 2] cycloadditions (and cross-linking) and photo-Fries rearrangements.[122] A reasonable positive contrast image has been formed from films of these polymers because the photo-Fries products are yellow. The aliphatic LC ester obtained by replacing the aryl group of the cinnamate ester by an alkyl group undergoes rapid cyloaddition reactions as well, but no photo-Fries reaction products were detected.[123]

Processes of Guest Molecules Doped in Polymers

Mediation of Conformational Preferences. As discussed earlier,[54] the shapes and sizes of cavities can affect the nature of products from guest molecules by altering their conformations. This control is exemplified by the photoreactions of the conformationally labile dopant, ω-undecylenyl benzophenone-4-carboxylate (**BP1**), in unstretched and stretched **PE35** films.[124] Irradiation leads to either an oxetane (**BP1a**) via an intramolecular Paternò–Büchi reaction or a photoreduced product (**BP1b**) via H-abstraction from the medium (Scheme 11.11). Both processes occur from the $^3(n,\pi^*)$ state of the ketone. Formation of the oxetane requires a specific orientation between the carbonyl and olefinic groups, while photoreduction requires no such conformational specificity.

The ratio of the photoproducts, **(BP1b)**/**(BP1a)**, is ~6–8 in unstretched **PE35** and less than one-half of that value in stretched **PE35**. Formation of the cyclized product (and conformations that bring the vinyl and carbonyl groups closer) is clearly aided by film stretching. These data indicate that the smaller cavities in stretched **PE35** have shapes that encourage coiling of the long chains of **BP1**.

In another example, tetraphenylbutatriene has been shown to be oriented preferentially in **PE** films and it photodimerizes on irradiation.[125] The cavities in **PE** films restrict conformational changes of the molecule after its excitation, and thereby, radiationless processes are slowed. The dimerization reaction may be due to an increase in the rate of collisions of "long-lived" excited molecules or ground-state aggregation. The quantum yield of photodimerization is increased threefold by film stretching, which may increase the concentration of molecular pairs that are appropriately aligned.[125]

Creating a Templated Cage in which Reaction Occurs. The bichromophoric naphthyl derivatives (**N-Pn-N**) undergo only intramolecular photocycloaddition reactions in **PE42** films; in isotropic liquids such as hexane, both intra- and intermolecular photocycloadditions are observed (Scheme 11.12). Compartmentalization of the **N-Pn-N** molecules in **PE42** does not allow them during their excited-state lifetimes to encounter another ground-state **N-Pn-N** molecule (and produce an intermolecular adduct). The results also indicate that only one guest molecule is in each site in these films when the **N-Pn-N** concentrations are $<10^{-3}$ mol (g-film)$^{-1}$. Thus, the excited molecules either react intramolecularly or return to their ground states. Eventually, all of the **N-Pn-N** are excited in a conformation suitable for intermolecular addition.

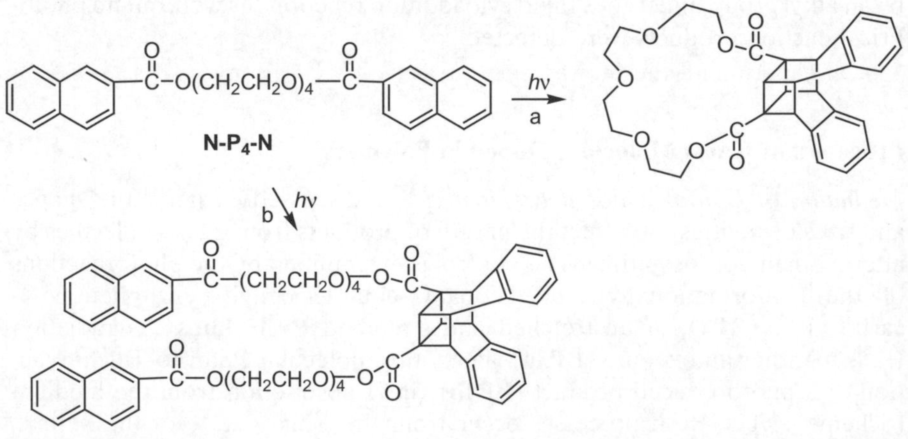

$C-O(CH_2CH_2O)_4-C$

N-P$_4$-N

$h\nu$ a

$h\nu$ b

$C-(CH_2CH_2O)_4-O-C$

$C-O(CH_2CH_2O)_4-O-C$

Scheme 11.12. Photoproducts from (a) intramolecular and (b) intermolecular cycloaddition reactions of **N-P$_4$-N**.[186]

Because the sizes and shapes of the guest sites are altered by stretching **PE42** films, so are the dimerization efficiencies: intramolecular dimerization is ca. 1.4 times faster in stretched films than in unstretched ones. The free volume of guest sites in **PE42** films decreases as they are stretched and the probability that two naphthyl rings will be vicinal increases. These results demonstrate that **PE42** films can be used to synthesize macrocycles at concentrations that would lead to intermolecular or polymerization reactions in fluid solutions.

Rates of Conformational Changes. The kinetics of inter- or intramolecular aggregation in polymers containing luminescent end groups, such as pyrenyl, can be used to monitor the probability of reactions between different polymer chains.[126] In the case of pyrenyl-labeled polymers, the relative contributions of the monomer and excimer (as well as their transients) to the total fluorescence defines the efficiency of their diffusion-controlled reactions at various temperatures.[127] When one of the pyrenyl groups of a **P*n*P** is electronically excited in a polymer host, intramolecular chain dynamics are controlled by the mobility of the polymer chains (section on "Guest Molecules as Reactants or Probes"). These measurements explore what transpires within one site because the processes involved are unimolecular.[128]

By contrast, activation energies of translational diffusion or diffusion coefficients of small molecules in polymer matrices give information about processes in which molecules move from one site to another, and they also depend on the conformational dynamics of polymer chains.[129,130] The activation energy of diffusion depends on the energy required to move intertwined (reptating) chain segments within the host matrix. Migration of guest molecules within a bulk polymer requires the synchronous relaxation of at least two vicinal polymethylene chain segments, one opening a space for the guest to enter and the other closing the space in which the guest resided. In polymer films, the out-diffusion of guest molecules from occupied sites represents the slow separation of a fluorophore-quencher contact pair by at least one polymethylene chain or the escape of a quencher molecule from a donor-occupied site.[92] In a typical experiment, the fluorescence of **DAAs** as diffusants in **PE35** films, is monitored as it diffuses into an inorganic acid solution (out-diffusion).[131] Excitation of **DAA** molecules within the film leads to fluorescence; fluorescence of **DAA** in the aqueous portion is attenuated severely due to quenching of the excited-singlet states by the acid.

Diffusion coefficients (D) for **DAA** in **PE** films can be calculated from plots of fluorescence intensity changes versus time according to Equation 11.11 and as discussed in the section on "Measurement of Translational Diffusion of Small Molecules." Using this method, the out-diffusion coefficient for **DMA** in **PE35** films is $4 \times 10^{-9}\,cm^2\,s^{-1}$ at 288 K and $2 \times 10^{-8}\,cm^2\,s^{-1}$ at 307 K.[131] An "all-time" solution to a Fickian diffusion process (Eq. 11.12) can also be used to calculate the diffusion coefficients in a polymer matrix with an initial even distribution of dopants. The diffusion coefficient, $6.5 \times 10^{-8}\,cm^2\,s^{-1}$ (298 K), based on Equation 11.12 is comparable to the values from Equation 11.11.

Arrhenius treatments ($D = D_0 e^{-(E_D/RT)}$) provide activation energies of diffusion (E_D). The magnitude of the E_D for **DMA**, ~15 kcal mol^{-1}, in **PE35** films is rather large compared with the activation energies for relaxation processes in **PE35**; α-relaxation, with the highest energy value, is less than ca. 10 kcal mol^{-1}.[131] The very high activation energy for diffusion suggests that movement of guest molecules within the **PE35** matrix requires motions of polymethylene chains in addition to those experienced by the polymer normally.

Types of Photoreactions of Guest Molecules

Unimolecular Rearrangements. Enone Isomerizations. Zimmerman and coworkers correlated molecular motion of the component atoms with their free volume for the photochemical rearrangement of three different photoactive molecules: 4,5,5-triphenylcyclohex-2-en-1-one (**3PhCO**); 1,1-dicyano-3,3,5,5-tetraphenyl-1,4-pentadiene; and 2,2-dimethyl-1,1-diphenyl-3-(2,2-diphenylvinyl)cyclopropane.[132] The behavior of the excited states of these molecules has been studied in solution, in polymer glasses, and in their crystal lattices. The nature of the photoproducts depends on the medium employed. Whereas irradiation of **3PhCO** in benzene solution leads to triphenylcyclohexanone (**3PhCOa**), the *exo* stereoisomer of 4,4,6-triphenylbicyclo[3.1.0] hexanone (**3PhCOb**), and the cyclobutanone (**3PhCOc**), irradiation in the neat crystal results in the formation of the *endo* stereoisomer of 4,4,6-triphenylbicyclic ketone (**3PhCOd**) and the benzobicyclic ketone (**3PhCOe**). However, only **3PhCOd** was found from irradiations of **3PhCO** in glassy **PMMA** films (Scheme 11.13).

The proposed reaction mechanisms involve the formation of biradical intermediates (Scheme 11.13). When multiple reaction pathways are available, the one with the least motion and minimum molecular volume displacement (ΔDV) is preferred. ΔDV is defined as "the new volume in space occupied by the target molecule." The absence of the *endo* stereoisomer (**3PhCOd**) in the solution state irradiation is attributed to intramolecular van der Waals repulsions between the *endo*-phenyl groups. The biradical intermediate **3PhCOd** leads to the formation of the *endo* isomer. Changes in the sum of atomic coordinate and volumes (i.e., ΔDV) are smallest for the formation of **3PhCOd** . The compact nature of this transition state is cited as the reason for its exclusive formation in the **PMMA** matrices in spite of the intramolecular van der Waals repulsions; transition states that require minimum volume and shape changes are preferred in the glassy states of **PMMA** and in the neat crystal. Thus, intermolecular forces drive the reaction rather than intramolecular ones in the media in which the cages have "hard" walls.

Photodecarboxylation Reactions. The most common photoreaction of aromatic esters in solution is photo-Fries rearrangements. However, decarboxylation products can be obtained from esters with bulky substituents *ortho* to

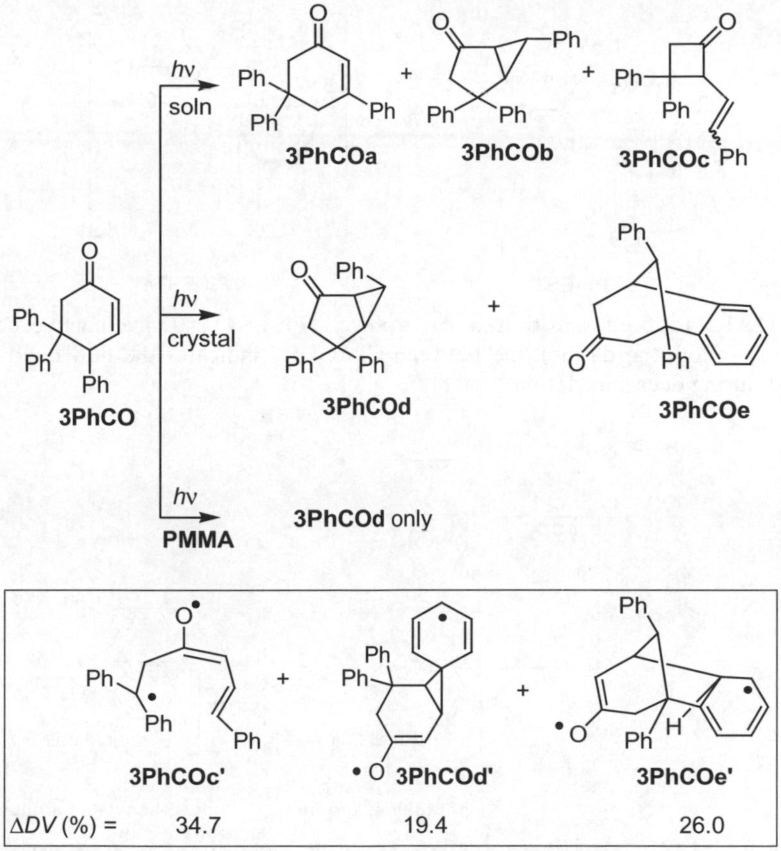

$h\nu$ soln → 3PhCOa + 3PhCOb + 3PhCOc

$h\nu$ crystal → 3PhCOd + 3PhCOe

$h\nu$ PMMA → **3PhCOd** only

3PhCO

	3PhCOc'	3PhCOd'	3PhCOe'
ΔDV (%) =	34.7	19.4	26.0

Scheme 11.13. Photochemical reactions of 4,5,5-triphenylcyclohex-2-en-1-one (**3PhCO**) in different media. Selected biradical intermediates proposed in the mechanism are shown with their displacement volumes for the formation of ΔDV (%).[132]

the ester group.[44] Some polymer matrices also enhance the yields of decarboxylation products. These reactions offer an alternative method for the preparation of biaryls and alkyl aromatics, especially those in which the carbon center attached to the aromatic ring is chiral.[133] The decarboxylation products are formed by a concerted reaction of the excited-singlet state of the aryl esters in which the stereochemistry of the carbon atom adjacent to the carboxy carbon is retained in the decarboxylated photoproduct (Chart 11.11); photo-Fries or decarbonylated photoproducts are formed from stepwise reactions involving radical pairs (Scheme 11.14).

The number of examples of dominant photodecarboxylation reactions of aromatic esters is much smaller than the citations to photo-Fries rearrangements. Even fewer examples of photodecarboxylations of aryl esters without *ortho* substituents are available.[44]

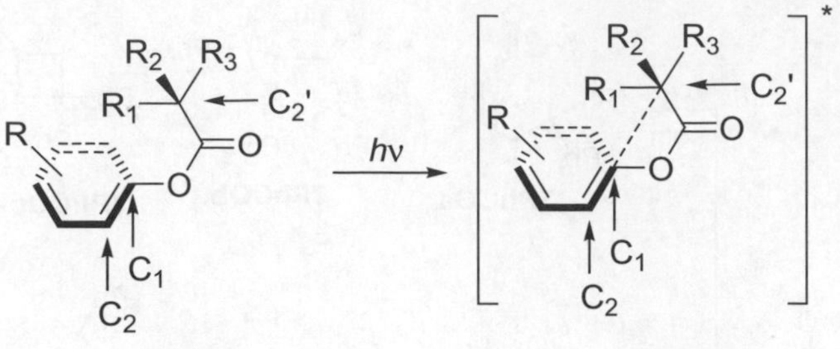

Chart 11.11. Proposed transition state of a phenyl (**Ph-EST**) ester leading to concerted decarboxylation; the dashed line between C_1 and C_2' indicates the new bond being formed during decarboxylation.[44]

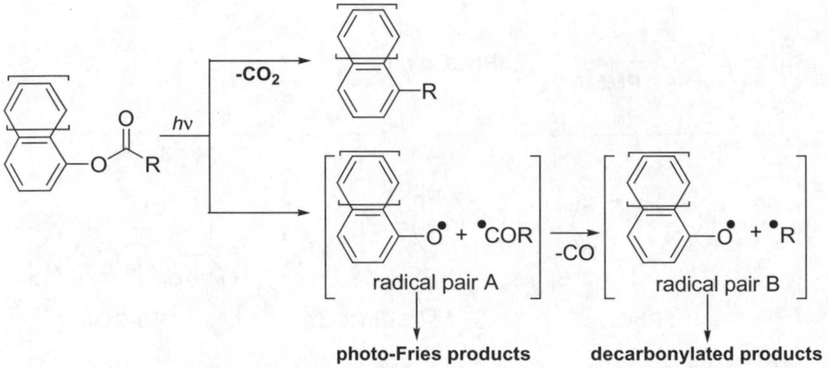

Scheme 11.14. General photochemical reaction pathways of aromatic esters.[44]

A series of unsubstituted phenyl and naphthyl esters have been investigated in solutions and in **PE** films to understand the mechanisms of their phototransformations and to find conditions where photodecarboxylation is an important component of the reactions.[44] As expected, irradiation of these esters in solution leads to the formation of very small amounts of decarboxylation products and very large amounts of photo-Fries and decarbonylation products. However, the relative yields of photodecarboxylation are increased markedly in **PE** films, and they increase further as temperature is lowered or films are stretched (and free volume is decreased) (Fig. 11.15).[44] The conformation shown in Chart 11.11 is achieved either intramolecularly by steric constraints imposed by *ortho* substituents or intermolecularly by interactions with the walls of the cavities in the **PE** media.

The increases in decarboxylation yields at lower temperatures and in stretched films arise from less efficient formation of radical pair A (Scheme 11.14) and/or its more efficient return to starting ester. Stretching or cooling stiffens the **PE** chains that constitute the walls of the reaction cavities.

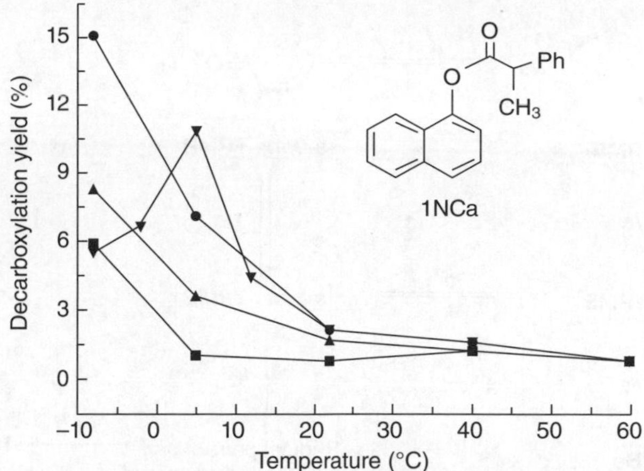

Figure 11.15. Temperature dependence of the relative decarboxylation product yields from irradiations of naphth-1-yl 2-phenylpropanoate (**1NCa**) in stretched (s) and unstretched (u) **PE** films: ■ **PE26**$_{(u)}$; • **PE26**$_{(s)}$; ▲**PE51**$_{(u)}$; ▼**PE51**$_{(s)}$; numbers after **PE** represents % crystallinity. (Adapted from Reference 44. Copyright 2009, with permission from Elsevier.)

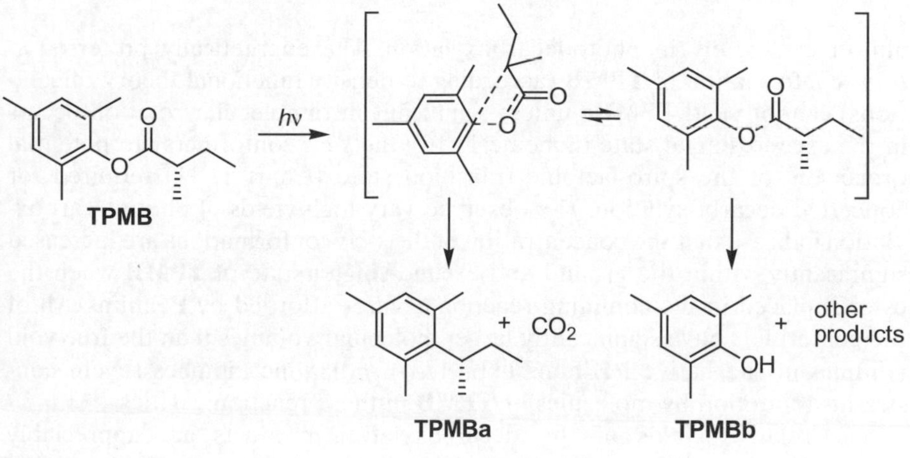

Scheme 11.15. Photoreactions of 2,4,6-trimethylphenyl (*S*)-2-methylbutanoate (**TPMB**).[134]

Therefore, motions of singlet radical pairs from their initial positions after lysis of their parent esters are inhibited.

Photoreactions of 2,4,6-trimethylphenyl (*S*)-2-methylbutanoate (**TPMB**, Scheme 11.15) in a series of solvents at different temperatures always resulted in considerable amounts of lysis (e.g., **TPMBb** and other products).[134,135] However, under optimal conditions, irradiation of **TPMB** in **PE** films resulted

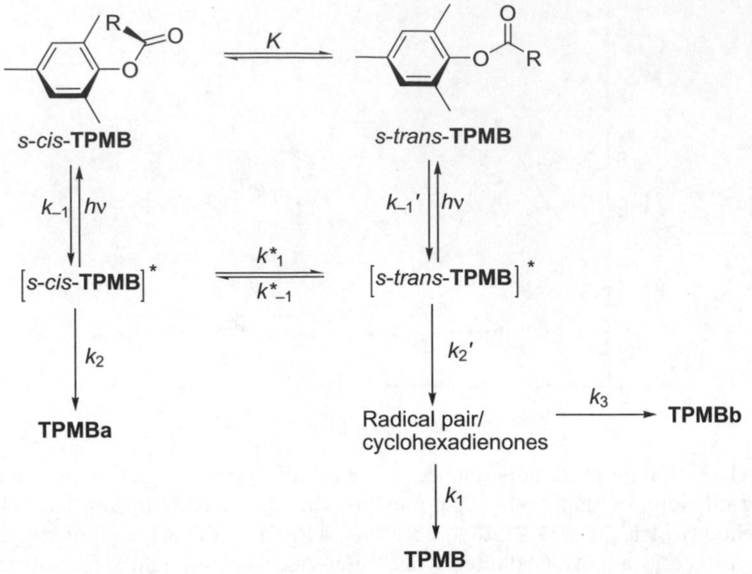

R
O
O
K
O
O
R
s-cis-**TPMB**
s-trans-**TPMB**
k_{-1} $h\nu$
k_{-1}' $h\nu$
k^*_1
[*s-cis*-**TPMB**]*
k^*_{-1}
[*s-trans*-**TPMB**]*
k_2
k_2'
k_3
TPMBa
Radical pair/ cyclohexadienones
TPMBb
k_1
TPMB

Scheme 11.16. Possible kinetic steps involved in *s-cis* and *s-trans* equilibration and photoreactions of 2,4,6-trimethylphenyl (*S*)-2-methylbutanoate (**TPMB**).[134]

almost exclusively in photodecarboxylation. The energetically preferred *s-trans* conformation of **TPMB** (according to density functional theory calculations) cannot yield **TPMBa** unless significant intramolecular rotation occurs in the excited-singlet state (Scheme 11.16); the *s-cis* conformers are potential precursors of the spiro-lactonic transition state (Chart 11.11) required for concerted decarboxylation. The observed very high yields of photodecarboxylation indicate that the concentration of the *s-cis* conformations are increased significantly within the ground and excited-singlet state of **TPMB** when the ester is placed in the confining reaction cavities afforded by **PE** films. All of the conformers have significantly larger molecular volumes than the free void volumes of the native **PE** films (113–177 Å^3 [134]), which implies severe constraints to motion by molecules of **TPMB** in these reaction cavities.

The relative yields of the decarboxylation products are appreciably decreased when the films are stretched. However, no clear trend is discernible between crystallinity of the **PE** films and relative yields of the decarboxylated product, **TPMBa** (Table 11.2). The equilibrium between *s-cis* and *s-trans* conformers is an important factor in determining the **TPMBa/TPMBb** product ratios. High yields of photodecarboxylation suggest that $k_1^* \ll k_2$ and $k_{-1}^* \ll k_2'$ (Scheme 11.16). The two unstretched **HDPE (PE68** and **PE74)** matrices yield only trace amounts of **TPMBb**, the major product being **TPMBa**. These results suggest that the reaction cavities of **PE68** and **PE74** either force all the molecules of **16** to adopt *s-cis* conformations or they prevent radical pairs from forming or from reacting in pathways other than recombinations to reform

TABLE 11.2. Photodecarboxylation of 2,4,6-Trimethylphenyl (*S*)-2-Methylbutanoate (TPMB) in PE Films at 296 K

	Yields (%)[a]		
Film	TPMBa[b]	TPMBb[b]	[TPMBa]/[TPMBb][c]
$PE0_{(u)}$	59 ± 7	13 ± 2	5
$PE46_{(u)}$	18 ± 3	7 ± 1	3
$PE46_{(s)}$	42 ± 23	31 ± 17	1
$PE50_{(u)}$	3 ± 1	5 ± 2	0.6
$PE50_{(s)}$	2 ± 1	4 ± 1	0.5
$PE68_{(u)}$	44 ± 5	1 ± 1	[d]
$PE68_{(s)}$	51 ± 8	11 ± 2	5
$PE74_{(u)}$	32 ± 4	1 ± 1	[d]
$PE74_{(s)}$	38 ± 13	10 ± 3	4

Enantiomeric excess (ee) values of TPMBa are >98% in PE films except in PE68(s) for which ee > 95%. The lower product yields are attributed to the concomitant formation of photo-Fries and cage-escape products.[135]

[a]Based on consumed **TPMB**.

[b]Limits of detection were 1%; no **TPMBb** was detected for yields listed as 1 ± 1%.

[c]Approximate values only due to high error limits.

[d]Amount of **TPMBb** was very small, leading to a very high **TPMBa/TPMBb** ratio.

TPMB. A linear relationship exists between ln(**TPMBa/TPMBb**) and the reciprocal of temperature within the range 275–338 K in **PE46** films, but is nonlinear in methylcyclohexane (Fig. 11.16). The linearity of the plot in **PE** films suggests very small conformational changes for **TPMB** during its short excited-singlet lifetime, and maintenance of a large *s-cis*/*s-trans* ratio in its ground and excited-singlet states.[135] The absence of a linear correlation in methylcyclohexane suggests that different steps in Scheme 11.16 could be rate determining at different temperatures. More specifically, the conformational interchange rate constants, k^*_1 and k^*_{-1}, may be more temperature dependent in the fluid medium than in the **PE** films.

Photo-Fries Reactions in Polymer Matrices. The photo-Fries rearrangements of aryl esters have been investigated extensively in isotropic solutions.[136–138] Photolysis proceeds from an excited-singlet (S_1) state in most cases (e.g., the $^1(\pi,\pi^*)$ ^1L_b state for phenyl acetate), although the contribution to product formation from triplet pathways has also been reported in some cases.[139,140] The homolytic dissociation of the (O=)C–O bond gives rise to a primary geminate radical pair (radical pair A in Scheme 11.15) that can either regenerate the starting ester, recombine to yield rearrangement products, or escape from its initial cage and react. For example, the acetyl radical from phenyl acetate can add preferably to the *ortho* and *para* positions of the phenoxy ring in the cage to yield keto intermediates that enolize (a tautomerization), providing the isolated *o*- and *p*-acetyl phenol products. The ratio of the *ortho*- and

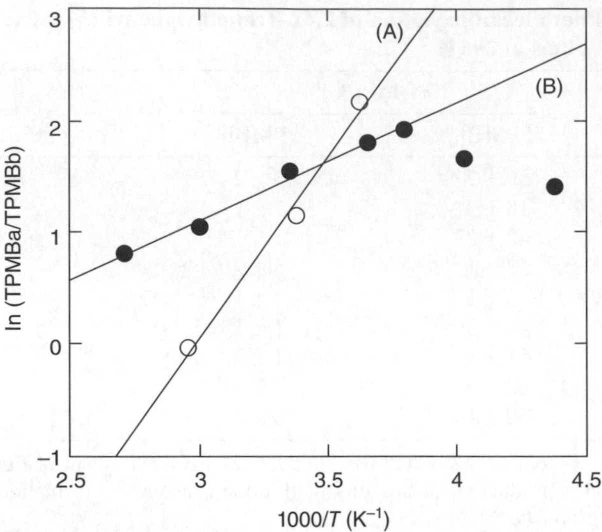

Figure 11.16. Temperature dependence of photodecarboxylation (**TPMBa**) and cage-escape product (**TPMBb**) from irradiation of **TPMB** in $\mathbf{PE46}_{(u)}$ (A) and in methylcyclohexane solutions (B). (Reprinted with permission from Reference 134. Copyright 2009 American Chemical Society.)

para-substituted products depends on the temperature of the medium and the structure of the aryl acylate.[141] The loss of carbon monoxide from radical pair A results in the formation of a secondary radical pair B (Scheme 11.15). Radical pair B reacts in an analogous fashion to radical pair A, leading to *ortho*- and *para*-substituted products. The amounts of in-cage- and out-of-cage-derived photoproducts provide an indirect method to measure "cage reactivity."

The photo-Fries reactions of acetyl and myristoyl esters of 2-naphthol have been investigated in **PE** films (Scheme 11.17).[142] Irradiation of 2-naphthyl esters in solution leads mostly to 1-acyl-2-naphthols. For example, the relative yield of **2NC-1** from irradiation of **2NCb** in *tert*-butyl alcohol is 86% (Table 11.3). Both kinetic and thermodynamic factors favor combination of the myristoyl radical at C-1 of 2-naphthoxy. In solution, cage escape can be competitive with in-cage recombination. The 2-naphthoxy radicals that escape yield 2-naphthol eventually. The length of the alkyl chain of the acyl radical and the shape of both radicals play important roles in determining product selectivity. In a clear demonstration of the manner in which polymer matrices can control photoreactivity, it has been found that the major products from liquid-phase irradiations of long alkyl myristoyl derivative (**2NCb**), **2NC-1**, and 2-naphthol, are totally absent upon irradiation of the same ester in **PE35** films; the more rodlike photoproduct, **2NC-3**, was found instead (Table 11.3). A small amount of **2NC-6** was also detected. The shape of the preferred solution product (**2NC-1**) is very different from that of **2NCb**. The product selectivity in **PE** films can be explained by assuming "cylindrical" shapes for the cavity sites. The absence

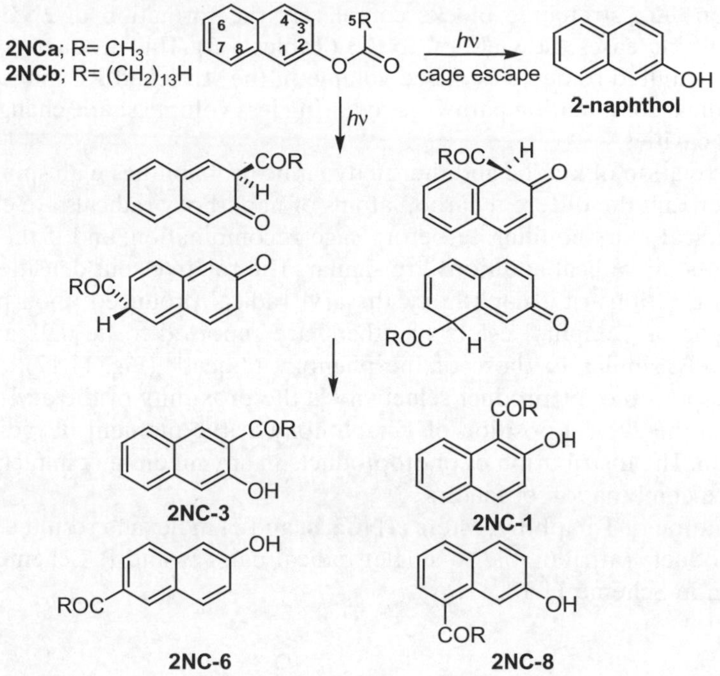

Scheme 11.17. Phototransformations of 2-naphthyl esters showing keto intermediates and their tautomerization to the isolated products.[142]

TABLE 11.3. Photoproduct Distribution from Irradiations of 10^{-3}–10^{-4} M 2-Naphthyl Acylates (2NC) in N_2-Saturated Media at 298 K[142]

		Relative Product Yields (%)			
Compound	Medium	2-Naphthol	2NC-1	2NC-3	2NC-6
2NCa	*tert*-Butyl alcohol	32 ± 1	31 ± 1	11 ± 1	17 ± 1
2NCb	*tert*-Butyl alcohol	[a]	86 ± 4	[a]	14 ± 1
2NCa	$PE35_u$	21 ± 1	31 ± 1	20 ± 1	28 ± 1
2NCa	$PE35_s$	32 ± 2	[a]	31 ± 1	37 ± 1
2NCb	$PE35_u$	[a]	[a]	75 ± 2	25 ± 2
2NCb	$PE35_s$	[a]	[a]	92 ± 2	8 ± 2

[a]Not detected.

of cage-escape products after irradiation of **2NCb** in **PE** films arises from the much slower diffusion of myristoyl and 2-naphthoxy radicals from the cavities in which they are formed than the rates at which they combine.

As noted, the relative yields of photoproducts can differ in unstretched and stretched polymer films. As discussed in the section on "Hole Free Volumes," the application of a macroscopic force (i.e., film stretching) causes large microscopic changes to the polymer morphology. In the experiments under discussion, film stretching blocks completely the formation of **2NC-1** from **2NCa**, but increases the yield of **2NC-3** (Table 11.3). This product selectivity can be attributed to decreased free volume in the stretched **PE** matrices that favor radical combination pathways requiring less volume/shape change to the reaction cavities.

A correlation of kinetic and thermodynamic probabilities with spin density on oxygen and the different carbon atoms of naphthoxy radicals are expected if the radical pairs equilibrate before cage recombination, and if their steric constraints for radical additions are similar. The relative spin densities at the 2- and 4-positions of 1-naphthoxy, the aryl radical produced upon photoinduced lysis of 1-naphtyl esters or ethers, are reported to be 0.91 and 0.94, respectively, similar to those of the phenoxy radical[10] (Fig. 11.17). Another factor responsible for product selectivity is the proximity of the acyl or alkyl radical to the 2- or 4-position of 1-naphthoxy at the moment of radical pair formation. The distribution of photoproducts in any medium is, in fact, dependent on a combination of factors.[143]

Irradiation of 1-naphthyl esters (**1NCa**, **b**, and **c**) in hexane resulted in eight photoproducts (attributable to singlet radical pairs A and B, Scheme 11.14), as shown in Scheme 11.18.[144]

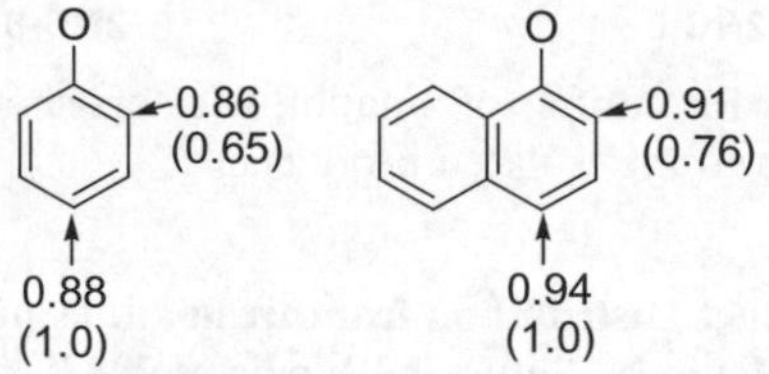

O
0.86 (0.65)
0.88 (1.0)
O
0.91 (0.76)
0.94 (1.0)

Figure 11.17. Relative electron spin densities at selected positions of phenoxy and 1-naphthoxy radicals from ab initio calculations[2,10] and from electron paramagnetic resonance experiments (in parentheses).[185]

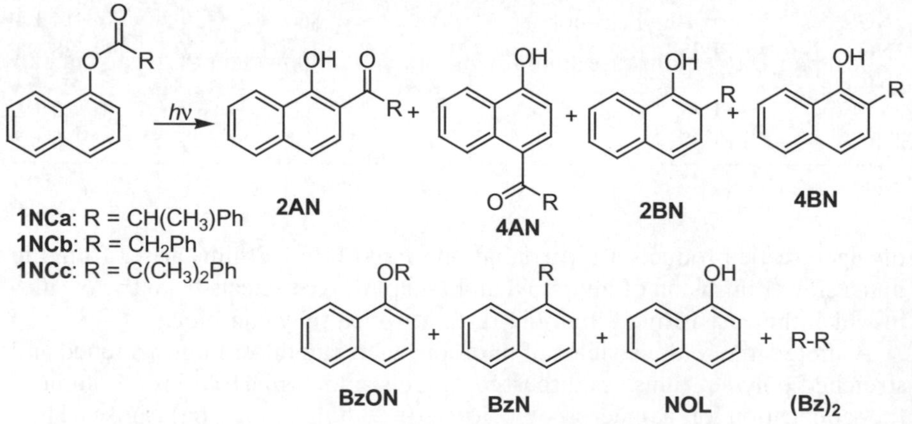

O O R
hν
OH O R
OH
O R
OH R
OH R
1NCa: R = $CH(CH_3)Ph$
1NCb: R = CH_2Ph
1NCc: R = $C(CH_3)_2Ph$
2AN **4AN** **2BN** **4BN**
OR R OH R-R
BzON **BzN** **NOL** $(Bz)_2$

Scheme 11.18. Photo-Fries reactions of 1-naphthyl esters.[144]

TABLE 11.4. Relative Yields (%) of Selected Photoproducts from 3–7 mmol kg^{-1} 1NCa in PE Films at 295 K

	Relative Photoproduct Yields				
Film	2AN	4AN	2BN	4BN	BzON
PE21$_{(u)}$	80.0 ± 0.2	9.3 ± 0.2	0.6 ± 0.3	3.0 ± 0.2	2.1 ± 0.1
PE21$_{(s)}$	65.1 ± 0.8	8.5 ± 0.5	3.7 ± 0.2	8.3 ± 0.8	4.4 ± 0.2
PE26$_{(u)}$	82.7 ± 0.8	6.3 ± 1.1	1.2 ± 0.7	3.5 ± 0.9	1.7 ± 0.3
PE26$_{(s)}$	74.7 ± 1.4	6.0 ± 1.5	2.7 ± 0.2	8.0 ± 0.5	3.0 ± 0.5
PE51$_{(u)}$	79.1 ± 1.0	6.9 ± 0.5	1.1 ± 0.7	5.0 ± 0.6	2.9 ± 0.7
PE51$_{(s)}$	66.2 ± 1.0	5.5 ± 0.1	4.3 ± 1.3	10.9 ± 1.8	4.6 ± 0.3

PE21 films were irradiated at 278 K.[2]

In solution, out-of-cage recombinations of the initially formed radical pairs result in the formation of significant amounts of **(Bz)$_2$**, **BN**, and **BzON**. The other recombination photoproducts, 2- and 4-phenylacyl-1-naphthols (**2AN** and **4AN**), are formed almost exclusively by in-cage processes. No **(Bz)$_2$** is detected from irradiation of **1NCa** in the presence of 2×10^{-2} M benzenethiol, a scavenger of the benzylic radicals that escape from their initial cages. Based on the observation that the [**2AN**]/[**4AN**] ratio is constant but the [**BN**]/[**AN**] ratio approaches zero as the thiol concentration is increased, it was concluded that the **AN** products are formed in-cage even in hexane. The presence of **(Bz)$_2$** (in the absence of benzenethiol) suggests that some of the **BzN** and **BzON** are derived from out-of-cage combination of radicals, while the absence of **(Bz)$_2$** indicates complete in-cage combination.

When **1NCa**, **b**, and **c** were irradiated in **PE71** films,[144] no **(Bz)$_2$** was detected (Table 11.4). However, the soft cavity walls in the **PE** films allow the formation of four in-cage products. At room temperature, the **2AN/4AN** ratios (>10) and **2BN/4BN** ratios (<1) emanating from the radical pairs A and B, respectively, show that the soft walls in **PE** cages and longer lifetimes of radical pair B (N.B., radical pair B must be derived from decarbonylation of radical pair A) allow radical pair B to equilibrate spatially before yielding products, but radical pair A does not. The small amounts of the **4AN** product suggest that the mobility of radical pair A during its lifetime ($\leq 10^{-6}$ s in **PE** matrices, *vide ante*) is low.

The rotational correlation time for molecules like **1NC** in **PE** is ≤50 ns.[144] During this period, most of the radical pair A reacts, limiting the formation of radical pair B and the decarbonylated products (<13%). The free volumes of holes in the **PE** films employed (<142 Å^3)[2] are smaller than van der Waals volume of the naphthyl esters irradiated (e.g., the volume of **1NCb** is 237 Å^3).[2] As such, their cavity walls are expected to restrict the mobility of the radical pairs, leading to preferential bond formation at the nearer 2-position. Consistent with this hypothesis, **2AN** is the major product.

Combinations of radical pairs derived from **1NCa–c** have been investigated in **PE** films of different crystallinity: **PE21**, **PE26**, and **PE51**.[2] In all of these polymers, the relative yield of major photoproduct, **2AN**, was at least six times larger than that of the corresponding **4AN** (see e.g., results for irradiations of **1NCa**; Table 11.4). Additionally, the photoproduct selectivity was affected by film stretching. When **1NCb** was the substrate in the **PE** films, film stretching increased the rate constants for the formation of **2AN** from radical pair A six times more than the rate constants for the formation of the corresponding **4AN**. Film stretching also increased the sum of relative yields of the three in-cage decarbonylation products: **2BN**, **BzON**, and **4BN**. The reason for this increase, indicating *longer* lifetimes of some radical pairs A in the stretched films, may be attributable to a fraction of naphthyl esters residing in cavities in interfacial or other sites where radical movements are more difficult than in the "average" sites.

Consistent with this hypothesis, fluorescence lifetime measurements of the naphthyl chromophores of **1NCb** suggest that they reside in more than one (distinguishable) cage type in **PE** films. A shorter decay component, $\tau_s \approx 8\,\text{ns}$, accounts for ca. 90% of the total fluorescence in unstretched **PE26** and ca. 75% in the stretched film. In **PE51**, the decay constants are slightly shorter than in **PE26**. The longer decay component, $\tau_l \approx 18\,\text{ns}$, was assigned to fluorescence from molecules in the more rigid interfacial sites and those with τ_s to molecules in the less rigid amorphous cavities. These data suggest that the majority of molecules reside (and react) in cages within the amorphous parts of **PE26** and **PE51**, even after the films are stretched.

No general methods are available for measuring the absolute combination rates of short-lived germinal radical pairs.[145] Commonly used techniques involve electron spin resonance (ESR) and laser flash photolysis.[146] However, knowledge of the rates of decarbonylation from acyl radicals in radical pair A and the relative photoproduct yields from irradiations of aryl esters when all of the radical pair combinations are in-cage (as is the case in many polymer films) can be used to calculate absolute rates.[147] For example, as noted, homolysis of the (O=)C–O bond of **1NCb** excited-singlet states generates a phenylacyl and a 1-naphthoxy radical-pair A; this is the start of a "radical clock" for radical pair A, and the known rate constants (i.e., inverse lifetimes) for loss of CO from the acyl radicals provides an absolute scale for determining the rate constants for combinations of radical pair A by the pathways leading to the keto tautomers of **2-AN** and **4AN**, k_{2A} and k_{4A}, respectively (Eqs. 11.23 and 11.24).

The decarbonylation rate constants ($k_{-\text{CO}}$) are $4.8 \times 10^6\,\text{s}^{-1}$ and $4.0 \times 10^7\,\text{s}^{-1}$ for phenylacetyl (from **1NCb**) and 2-phenylpropanoyl (from **1NCa**), respectively, at 295 K in isooctane.[145,148] They are somewhat sensitive to solvent polarity but not to medium viscosity, even in polymers such as **PE**[149]:

$$k_{2A} \cong \frac{k_{-\text{CO}}[2\text{AN}]}{[2\text{BN}]+[4\text{BN}]+[\text{BzON}]}, \tag{11.23}$$

$$k_{4A} \cong \frac{k_{-\text{CO}}[4\text{AN}]}{[2\text{BN}]+[4\text{BN}]+[\text{BzON}]}. \tag{11.24}$$

The singlet radical pairs generated during photo-Fries rearrangements of 1-naphthyl acetate recombine in less than 1 ns in acetonitrile solutions at room temperature.[140,150] In anisotropic, viscous media, recombinations are slowed by the cages in which the reactant molecules reside. Absolute values of k_{2A} and k_{4A} for naphthyl esters in the polyolefinic films are one to two orders of magnitude slower (at 295 K, k_{2A} ~$10^8\,s^{-1}$ from Eq. 11.23 and k_{4A} ~$10^7\,s^{-1}$ from Eq. 11.24) than in low-viscosity, isotropic media.[145] Much of the difference in the relative yields of the decarbonylation products from **1NCa** and **1NCb** in the **PE** films can be attributed to differences in the rates of decarbonylation: The phenylacetyl radical loses CO more slowly than the 2-phenylpropanoyl radical.

The change in stereochemistry of photoproducts from the irradiation of chiral aryl acylates is another tool to monitor the motions leading to radical pair recombinations. Although the rates of the two processes can be mediated by polarity and viscosity of the solvent[151,152] or by changing the spin multiplicity of the radical pairs,[153] stereochemistry is not retained in the photoproducts in many coupling reactions of prochiral radical pairs in fluid media.[153] However, cavities in constrained polymeric media have been shown to provide substantial enantioselectivity in some radical pair reactions. Significant enantioselectivity was found during the recombination of decarbonylated radical pairs from the irradiation of 1-naphthyl (*R*)-2-phenylpropanoate ((*R*)-**1NCc**) in **PE** films (Scheme 11.19).[154] The carbonylated pair A and decarbonylated pair B yield the in-cage recombination products **1NCc-2** and **1NCc-3**, as well as the decarbonylated products **1NCc-4**, **1NCc-5**, and **1NCc-6** (isolated after tautomerization of the initial keto products); cage escape leads to the formation of 1-naphthol, styrene, 2-phenylpropane, and 2,3-diphenylbutane.

The photo-Fries rearrangement products from irradiations of (*R*)-**1NCc** in hexane, **1NCc-2** and **1NCc-3**, were found to retain ~99% of their enantiomeric purity while the decarbonylated products, **1NCc-4**, **1NCc-5**, and **1NCc-6**, were

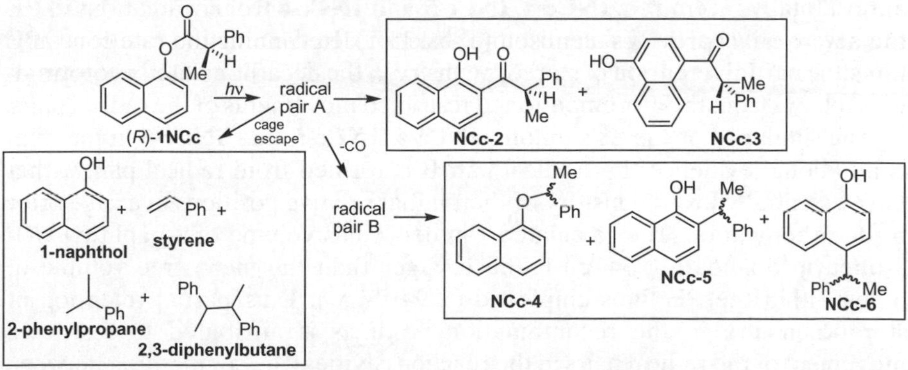

Scheme 11.19. Proposed reaction pathways for photoreactions of 1-naphthyl (*R*)-2-phenylpropanoate ((*R*)-**1NCc**).[154]

TABLE 11.5. Enantiomeric Excesses (%)[a] of Photoproducts from Irradiations of (*R*)-1NCc in PE Films and in Hexane at Room Temperature[154]

Medium	1NCc-2	1NCc-3	1NCc-4	1NCc-5	1NCc-6
Hexane	99	99	0.8	1.0	0.2
PE0$_{(u)}$	99	99	19.3	21.3	21.6[b]
PE0$_{(s)}$	99	99	18.9	20.9	23.9
PE46$_{(u)}$	97	99	12.3	15.6[b]	16.5
PE46$_{(s)}$	76	99	7.1	7.2[b]	9.6
PE74$_{(u)}$	97	99	16.8	13.4	10.0
PE74$_{(s)}$	81	99	6.6	7.6	9.0

[a]±1% except as indicated.
[b]±2%.

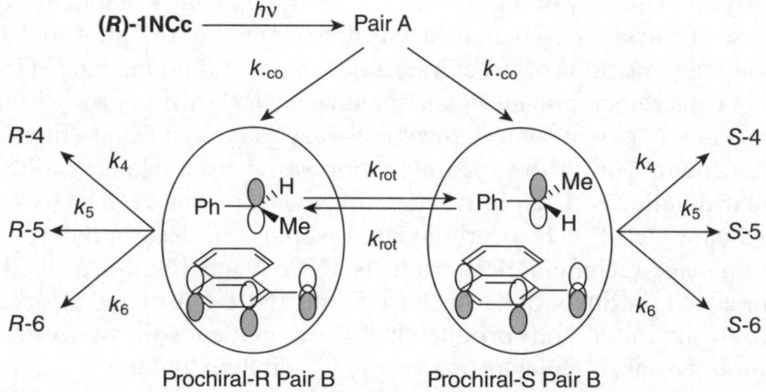

Scheme 11.20. Possible mechanistic steps for combinations of radical pair B generated from irradiations of 1-naphthyl (*R*)-2-phenylpropanoate ((*R*)-**1NCc**). (Reprinted with permission from Reference 154. Copyright 2009 American Chemical Society.)

almost totally racemized; **1NCc-4**, **1NCc-5**, and **1NCc-6** from irradiations in **PE** films were only partially racemized (Table 11.5). In the reaction cavities of **PE** films, the partial retention of stereochemistry in the decarbonylated photoproducts follows from the aforementioned restricted movements of the radical pairs.

The similarity of the ee's among **1NCc-4**, **1NCc-5**, and **1NCc-6** in one film is additional evidence that radical pair B is formed from radical pair A that have essentially lost the history of their initial relative positions in a cage prior to decarbonylation. The calculated van der Waals volume of 1-naphthyl (*R*)-2-phenylpropanoate (254 Å^3) is much larger than the mean free volume of holes within the **PE** films employed (129–177 Å^3). Thus, partial retention of stereochemistry in the recombination products is attributed to restricted movement of the radical pairs in the reaction cavities. The competition between the rates of racemization and product formation determines the enantiopurity of the products, as shown in Scheme 11.20.

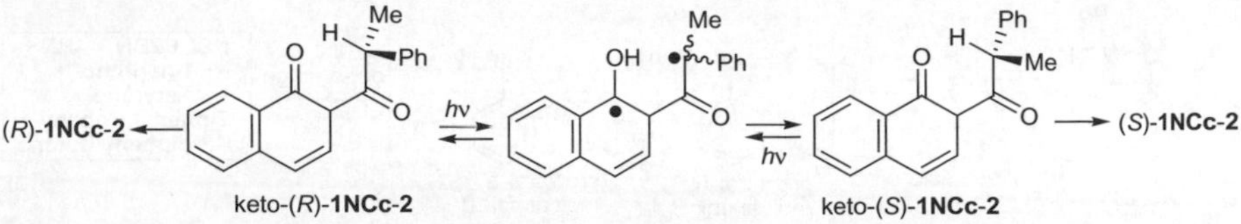

Scheme 11.21. Proposed mechanism for racemization of **1NCc-2** via reversible H-atom abstractions of the keto intermediate.[154]

Surprisingly, the ee values of the decarbonylated products tend to increase with decreasing crystallinity and increasing mean hole free volume. One of the basic requirement for a nonzero ee value in products **1NCc-4**, **1NCc-5**, and **1NCc-6** is that $k_{rot} < k_4 + k_5 + k_6$; these rate constants are weighted averages for reactions of radical pairs in the amorphous and interfacial regions. The softer walls in amorphous cavities facilitate the translational motions needed to orient radical pair B for the formation of the decarbonylated photoproducts: the translational and tumbling rates by the planar 1-phenylethyl radicals must be at least comparable if some of the chirality in the initially formed prochiral radical pair B is to be retained in the decarbonylated products. The translational motions associated with k_4, k_5, and k_6 are assumed to be more difficult in interfacial cavities than in amorphous ones due to the relatively lower mobility of the stiffer chains in the interfacial regions: $[k_{rot}/(k_4 + k_5 + k_6)]_{interfacial} > [k_{rot}/(k_4 + k_5 + k_6)]_{amorphous}$. According to this hypothesis, the observed increase in the ee value with decreasing crystallinity arises from differences in the locations of radical pairs in **PE** films.

Radical pair A retains nearly 100% ee in its photo-Fries products in most of the **PE** films (Table 11.5). However, significant loss of enantiopurity is observed in **1NCc-2** from irradiation in films having significant crystallinity (i.e., interfacial content). The loss of enantiopurity can be attributed to reversible γ-H abstraction reactions by the long-lived keto tautomers of **1NCc-2**, especially those in interfacial cavities (Scheme 11.21) that undergo.

Photo-Claisen Reactions and Their Space Requirements Compared with Those of Photo-Fries Reactions. In **PE** films, photo-Claisen rearrangements of benzyl phenyl ether and benzyl 1-naphthyl ether are less regioselective than those of the photo-Fries reactions of the corresponding esters.[155] Both excited-singlet and triplet states are reactive in photo-Claisen reactions, even though the triplet component is very small.[156] The regio- and stereoselectivities of the combination of a prochiral radical pair obtained directly from 1-naphthyl (*R*)-1-phenyl ether ((*R*)-**NC**) demonstrate the role of the templating effect of polymer cages on the translational and tumbling motions of a *directly formed* radical pair B (Scheme 11.22).[157]

As mentioned in the discussion of photo-Fries reactions, the translational motion of radicals is faster in amorphous sites than in interfacial regions. The overall stereo- and regioselectivities are weighted averages of the fraction of

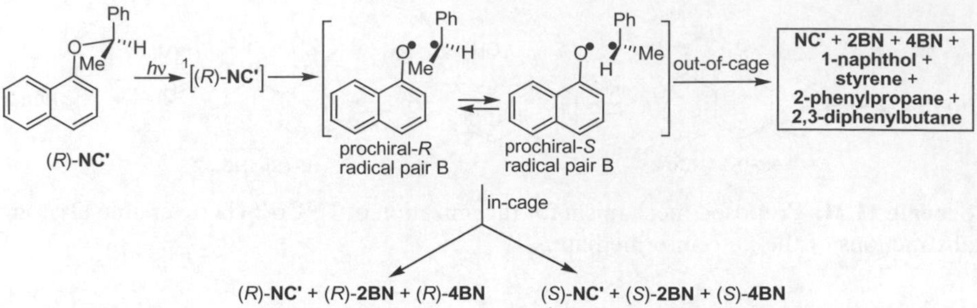

Scheme 11.22. Proposed mechanism of photo-Claisen reactions of (***R***)-**NC′**.[157]

reactions occurring in these two site types. As expected, 2**BN**/4**BN** ratios and ee values of photoproducts from (R)-**NC** are higher in the **PE** films than in isotropic solvents of comparable polarity, and, consistent with the results from photo-Fries reactions, the ee values of the photoproducts from (R)-**NC** are higher in completely amorphous **PE0** films (lacking interfacial sites) than in partially crystalline **PE46** or **PE74** films. The efficiency of racemization during recombinations of radical pair B leading to **NC** , S, is given in Equation 11.25[158]:

$$S = \ln\left(\mathrm{ee}_{(R)\text{-NC}'} - \mathrm{ee}_{\mathrm{NC}'}\right)/\ln\left(1 - \text{conversion fraction}\right). \qquad (11.25)$$

Here, $\mathrm{ee}_{\mathbf{NC}}$ is the ee content of (R)-**NC** at different photoconversions and $\mathrm{ee}_{(\boldsymbol{R})\text{-}\mathbf{NC}}$ is the ee prior to irradiation. Racemization efficiencies during recombination of radical pair B are 0.23 and 0.24 in unstretched and stretched **PE0**, while those in **PE46** are 0.27 and 0.44. This efficiency is a measure of the relative rates of tumbling and translational motions of the radicals in radical pair B. The S values are unaffected by film stretching in **PE0** films, but increase in stretched **PE46** films due to translocations of guest molecules to more rigid interfacial sites and an overall decrease in hole volume. In addition, photoproduct distributions from irradiations of several aryl ethers indicate that photo-Claisen reactions are less sensitive than photo-Fries reactions to changes in available hole free volume.[155]

Cages in **PE** resist the creation of additional space in the dimension orthogonal to the aryloxy plane (z-axis) of a radical in a pair. Since the formation of **AN** keto precursors (during photo-Fries reactions) requires the creation of more space along the z-axis than do BN keto tautomers (during photo-Claisen reactions) (Fig. 11.18), **PE** cages affect the relative rates of combination of the former more than the latter. Thus, the observed greater photoproduct selectivity and rate diversity for formation of the **AN** isomers may be due to spatial constraints imposed anisotropically by the media rather than steric or electronic factors intrinsic to the reacting species.

Photo-Cleavage Reactions. Norrish type I and type II cleavage reactions of ketones[1,159,160] offer another opportunity to study cage effects in constrained

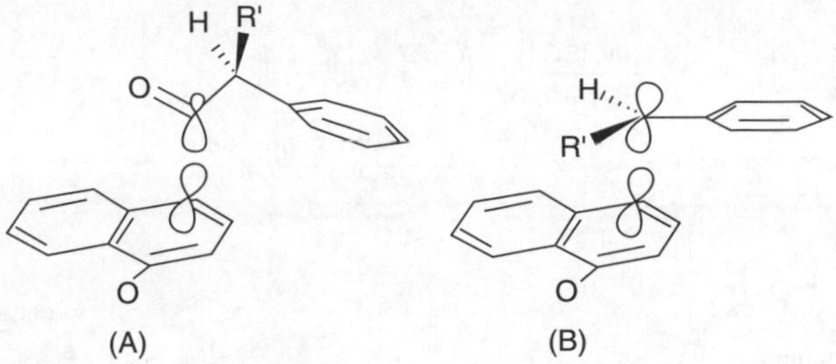

Figure 11.18. Approximate orientations for addition to a 1-naphthoxy radical by (A) an acyl and (B) a benzylic radical. Addition to C-4 is shown. The *z*-axis is orthogonal to the 1-naphthoxy plane. (Reproduced from Reference 2. Copyright 2009, with permission from Elsevier.)

polymeric media. Dibenzyl ketones (**ACOB$_n$**) undergo efficient Norrish type I cleavage reactions from their triplet excited states.[161] The facility of the cleavage is a result of the influence of the two aromatic rings on the (O=)C–C bond cleavage (α-cleavage). The triplet energies of dibenzyl ketone (**ACOB$_0$**) and acetone are almost the same (E_T ~80 kcal),[162] but the phenyl groups in **ACOB$_0$** lower the activation energy for α-cleavage from the triplet state and increases the quantum yield of radical formation to nearly one; α-cleavage from triplet states of acetone at room temperature in solution is very inefficient. The rapid photochemical α-cleavage from the triplet state of an **ACOB$_n$** yields a spin- and composition-correlated triplet geminate radical pair. Decarbonylation from the triplet radical pair leads to the formation of a pair of two benzylic radicals ($A^•/B^•$), which, in the absence of radical traps, either combine within their cages to yield AB or escape and then recombine leading to three different diaryl ethanes: AA, AB, and BB in a 1:2:1 ratio (Scheme 11.23).

Information about the dynamics of radicals in microheterogenous systems can be obtained from product analyses and/or kinetic studies. The combination of benzylic radical pairs requires that they be within van der Waals contact distances, be oriented appropriately, and have singlet character.[163] The triplet benzylic radical pair, therefore, must undergo intersystem crossing before or during C–C bond formation. If the size of the sites occupied by a pair of benzylic radicals is very small, the triplet-singlet energy gap will be very high, and intersystem crossing (ISC) will be very slow. Although the efficiency of ISC is higher in larger guest sites, so is the probability of cage escape. Thus, the reactivity in cages and the rate constants of those reactions, k_{cage}, are based on a convolution of spin and radical diffusion considerations, both of which depend on the space available in the cage where a radical pair resides.

The relative photoproduct yields from **ACOB$_1$** (whose derived benzylic radicals have very similar rates of diffusion), can be used to determine the fraction of products, which are formed in-cage and out-of-cage (i.e., the cage recombination factor, F_c; Eq. 11.26). The value of F_c is very sensitive to the local host environment.[164] In the case of **ACOB$_1$** the excited-singlet state

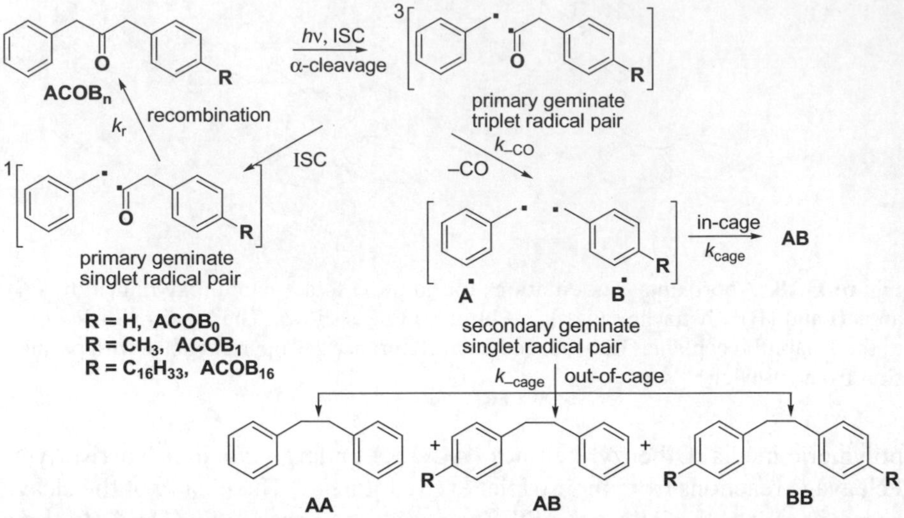

Scheme 11.23. Possible pathways of Norrish type I reactions of dibenzyl ketones and the formation of diaryl ethanes.[149]

($\tau = 3.6 \times 10^{-9}$ s at 295 K in benzene) undergoes ISC to its triplet state quantitatively (ϕ_{ST} ~1; $k_{ISC} > 10^8\,s^{-1}$ at 295 K in benzene). The triplet state then undergoes α-cleavage very rapidly (~10^{-10} s), followed by decarbonylation of the arylacetyl group ($k_{-CO} > 10^6\,s^{-1}$ at 295 K in hexane[163,165]). The residence time of benzylic radicals (or other species of similar size) in solvent cages in a nonviscous isotropic medium is ~10^{-10} s, allowing cage escape because the time required for the recombination of triplet radical pairs is ~10^{-6} s: F_c is ~0 at ambient temperatures and increases only slightly with decreasing temperature:

$$F_c = \frac{[AB]-[AA]-[BB]}{[AB]+[AA]+[BB]}. \tag{11.26}$$

The reactivity and photoproduct distribution of dibenzyl ketones in **PE** films depend mainly on the mean free volume. The free volume available in partially crystalline **PE** (113 Å^3 for **PE** with 68% crystallinity[163]) is comparable to the calculated molecular volume of **ACOB₁**. As a result, the "residence time" of the triplet radical pairs in their reaction cages is comparable to the spin evolution times, and high values of F_c are found. Thus, the rate constant for the formation of the in-cage radical pair combination products, k_{cage}, is dependent on the size of the reaction cage, with ISC as the rate-determining step.[166] The relationship between F_c and k_{cage} in **ACOB$_1$** and other **ACOB$_n$** in which the rates of radical diffusion are comparable is given by the simple expression in Equation 11.27. See Scheme 11.23 for these rate constants:

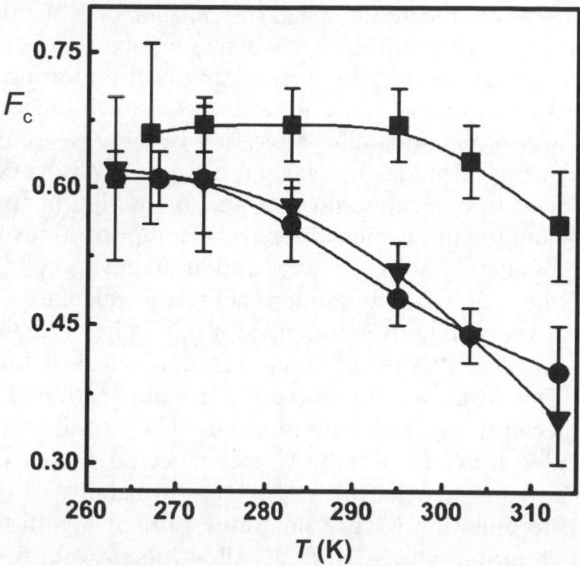

Figure 11.19. Dependence of F_c on temperature during irradiation of **ACOB$_1$** in unstretched films of **PE46** (triangles), **PE47** (circles), and **PE68** (squares). (Reprinted from Reference 163 with permission of The Royal Society of Chemistry [RSC] for the European Society for Photobiology, the European Photochemistry Association, and the RSC.)

$$F_c = \left(\frac{k_{cage}}{k_{cage} + k_{-cage}}\right)\left(\frac{k_{-CO}}{k_{-CO} + k_{-cage}}\right). \tag{11.27}$$

They are dependent on temperature and the structural properties of the **PE** films. The F_c values in **PE** films decrease with decreasing crystallinity: $F_c = 0.67$ in **PE68** and 0.51 in **PE46**.[162] They also vary with temperature because the temperature dependencies of the rate constants in Eq. 11.21 are not the same (Fig. 11.19).[163]

When **ACOB$_1$** was irradiated near the glass transition temperature, F_c was found to depend on the degree of conversion of **ACOB$_1$**.[149] At these temperatures, the rotational and translational mobility of **ACOB$_1$** is reduced due to the near cessation of the long segmental chain motions of the matrix. As a result, the molecules of **ACOB$_1$** are unable to change their conformations rapidly, and different conformations and sites can allow differing amounts of cage escape, return to **ACOB$_1$**, and in-cage reactivity.

The calculation of the cage recombination factor by steady-state methods may lead to wrong values within a polymer matrix (or a microphase separated medium[167]) when the rates of diffusion of radicals within a pair are very different.[149] Thus, higher than statistical amounts of cross-termination (AA and BB type) products are observed during the photolysis of dibenzyl ketones in sodium dodecyl sulfate micelles.[167] High mobility of one of the radicals can also enhance the AB-type product even when reaction occurs out-of-cage due to the persistent radical effect.[160] An increase in the cross-termination product may be observed when one of the transients has a much longer life than the other, and they are formed at similar rates.[167,168] In these cases, Eq. 11.20 is no longer valid and a kinetic model, involving the transient absorption of radicals, has been developed to determine the true recombination factor, F_{cAB}.[149] This model uses k_{-CO} and the intensities of the transient absorptions from the benzylic radicals immediately after the flash and after the completion of the in-cage portion of the radical combination reactions to calculate F_{cAB}.

When 1-(4-hexadecylphenyl)-3-phenyl-2-propanone (**ACOB$_{16}$**, Scheme 11.23) was irradiated in **PEs** of different crystallinities, only the AB product was detected.[149] This result was interpreted as a manifestation of the persistent radical effect because the rates of translational diffusion of benzyl ($A^•$) and the 4-hexadecyltolyl ($B^•$) radicals are expected to be very different, especially in a polymer matrix. In essence, the probability of one $B^•$ finding another for self-termination is very small due to their low diffusivity in **PE**; however, the high mobility of $A^•$ radicals allows them to diffuse to the sites occupied by $B^•$ radicals to form the AB product. As a result, the *efficiency* of out-of-cage formation of AB is greatly enhanced. The actual scenarios are somewhat more complicated, and the original references should be consulted for details.[149,169] Regardless, the kinetic model can separate the in-cage and out-of-cage (persistent radical effect) components of the reaction even when both yield the AB product. The concept of a persistent radical effect is well explored in organic reactions,[161] and, in the present case, the steady-state population inequality of the $A^•$ and $B^•$ radicals is a major factor in product selectivity.

For purposes of analyses, the transient absorption spectra of the radicals are separated into three temporal regions: an "instantaneous" rise, a time-resolved rise, and a decay (Fig. 11.20A). The sharp initial rise is attributed to fast homolysis of the **ACOB$_n$** triplets to form benzylic and aryl acetyl radicals. The protracted rise at $<10^{-6}$ s is caused by the decarbonylation of the arylacetyl radicals. Triplet radical-pair combinations (k_{cage}) and radical escape from geminate reaction cages occurs in the $<2 \times 10^{-5}$ s time period, during which the transient intensity decays as in-cage, first-order combinations of the benzylic radicals occur. Because the transient absorption traces maintain some intensity at least until a millisecond after the initial lysing laser pulse, a significant fraction of radicals must have escaped from their original cages. That $F_c = 1$ in the case of **ACOB$_{16}$** could have been explained simply by cage recombination were it not for the fact that benzylic radicals remained much longer than their expected lifetime, a few microseconds.

When the intensity of the radiation absorbed by **ACOB$_n$** is invariant with time and conversion (steady-state photolysis), the concentrations of the transient species remain nearly constant. Since intermediates are formed very rapidly compared with the total photolytic process, the time required for the

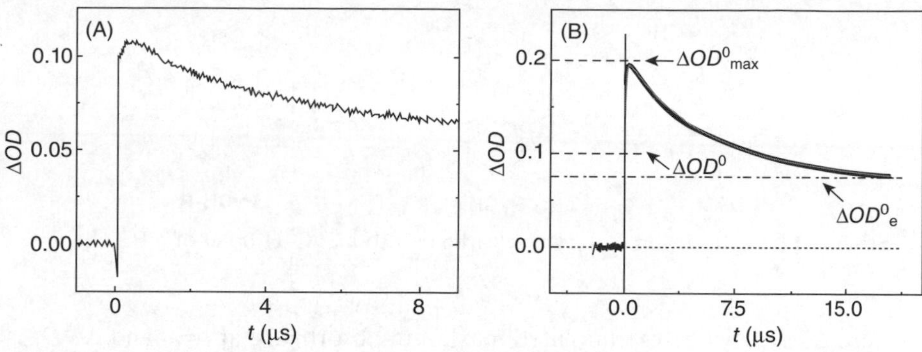

Figure 11.20. Transient absorption profiles in **PE68** monitored at 320 nm: (A) up to ca. 8 μs after laser-pulsed photolysis of **ACOB$_{16}$**; note the "instantaneous" and protracted rise portions; and (B) for **ACOB$_1$** (black) and the best fit according to the model (red), as well as the estimated values of ΔOD^0, $\Delta OD^0_{\max}$, and ΔOD^0_e (Reprinted with permission from Reference 149. Copyright 2009, American Chemical Society.) See ftp site for color: ftp://ftp.wiley.com/public/sci_tech_med/supramolecular_photochemistry.

intermediates to reach the steady-state condition is negligibly short. The usual static form of steady-state equation can then be rewritten in a dynamic final form that relates F_c and the true cage effect, F_{cAB} (Eq. 11.28):

$$F_c = \frac{F_{cAB} + R_-(1 - F_{cAB})}{F_{cAB} + R_+(1 - R_{cAB})}, \tag{11.28}$$

where R_- and R_+ are two complex expressions that can be evaluated under limiting conditions. As mentioned, F_{cAB} is a function of the rate constants, k_{cage}, k_{-cage}, and k_{-CO} (Eq. 11.27). The values of these rate constants are either known independently (k_{-CO}) or are obtained from fitting of the transient absorption traces. F_c includes all reaction routes to AB products as though they arise from in-cage radical pair combinations, while F_{cAB} includes only true in-cage recombinations. There are two practical limiting conditions for F_c (Eq. 11.28): (1) diffusion coefficients of the two benzylic radicals, $A^\bullet$ and $B^\bullet$, are the same; and (2) diffusion of one of the radicals is much faster than the other. In the first case, F_{cAB} values from laser-flash experiments (*vide infra*) are the same as those for F_c from steady-state irradiations; in the second case, the value of F_c always approaches unity, regardless of the mechanistic origin of the AB product.

More simply, transient absorption measurements such as those in Figure 11.20B can be used to approximate the values of F_{cAB} (Eq. 11.29):

$$F_{cAB} = 1 - \frac{\Delta OD^0_e}{\Delta OD^0_{\max}}, \tag{11.29}$$

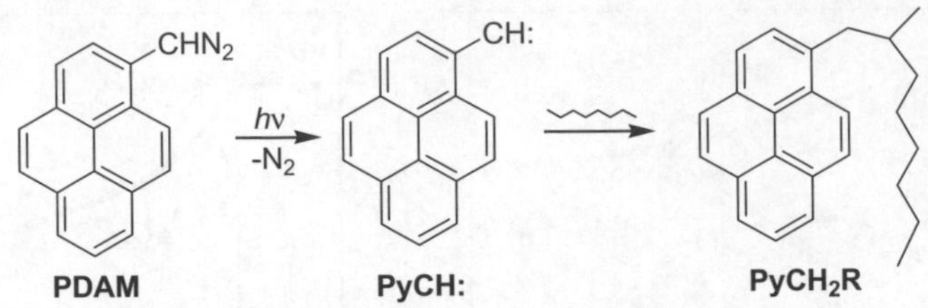

Scheme 11.24. Insertion of 1-pyrenylcarbene into a 2^0 C–H bond of a PE chain.[171]

where ΔOD^0_{max} is the extrapolated maximum absorbance at $t = 0$ and ΔOD^0_e is the directly proportional to the total concentrations of benzylic radicals, which escape from their initial cages.

REACTIONS OF GUEST MOLECULES TO MODIFY POLYMERS AND APPLICATIONS OF MODIFIED POLYMERS

Polymer surfaces can be modified by a variety of methods to make them amenable to specific applications. These methods include grafting, surface blending, copolymerization with functionalized monomers, and chemical modification of preformed polymers.[64] Whereas doped host–guest systems report only an average environment due to the mobility of the guest molecules and their translocation among sites, covalent attachment of reporter groups ensures that they sense a specific environment over long periods. As mentioned in the section on "Polymers to Determine the Kinetic Energy of High-Energy Particles," polycyclic aromatic hydrocarbons, such as pyrene, can be covalently attached to the chains of **PE** films when exposed to UV or ionizing radiation. For example, when ultrahigh molecular weight polyethylene (**UHMPE**) films are doped with pyrene and bombarded with 4.5 MeV protons or irradiated with >300-nm photons under oxygen-free conditions, 1-Py groups are covalently attached to the polymer chains.[170] The sequential absorption of two UV (>250 nm) photons has been proposed to be necessary for the attachment.[170] Bombardment by protons, particles, and electrons results in attachment of pyrene molecules at one or two positions, while irradiation by comparable doses of >300-nm photons resulted in only monosubstituted pyrenes.[91] Also, **PE** crystallinity does not influence the selectivity of the attachment reaction. However, higher doses of radiation result in cross-linking and scission of **PE** chains. Pyrenyl groups can also be attached selectively to polymer chains by UV irradiation of 1-pyrenyldiazomethane (**PDAM**) in films.[171] Irradiation of ethylene-co-vinyl acetate copolymers doped with **PDAM** produced a reactive carbene that added to the polymer chains (Scheme 11.24). These additions occur at terminal methyl groups (1^0), methylene chains (2^0), tertiary (3^0) positions, and across residual vinylic groups of the PE chains. The intrinsic selectivity of the carbene to add to specific locations of a **PE** is

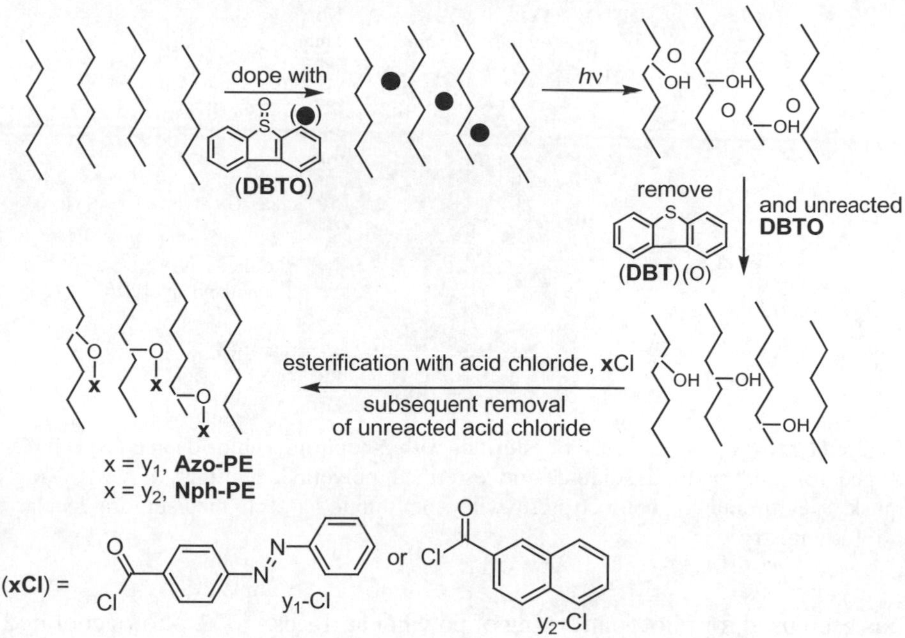

Scheme 11.25. Protocol for the interior modification and derivatization of amorphous regions to prepare **PE** films with covalently linked chromophores and lumophores.[64]

not manifested because the pyrenyl carbene cannot diffuse as rapidly to the most reactive sites as it reacts with less attractive ones.

Azobenzene moieties can be attached to chains of **PE24** films by esterification of 4-(phenylazo)benzoyl chloride at interior sites that have been modified to contain a hydroxyl group (Scheme 11.25).[64] In this method, **PE** chains are derivatized by irradiating ($\lambda > 300\,nm$) films doped with dibenzothiophene-5-oxide (**DBTO**). Ground-state oxygen atoms (^3O_P) are formed, and they insert into nearby C–H bonds of the polymer. Reaction of the hydroxyl groups with acid chlorides (containing chromophoric/lumophoric groups) result in the esters. Other applications of modified films of this type have been discussed in the sections on "Isomerizations" and "Other Photon-Based Applications."

Another application of covalent attachment is the reaction of the hydroxylated chains of **PE42** with fluorescent reporter groups to make permanent fluorescence images. By irradiating through a mask, hydroxylation occurred only in the exposed regions of the **DBTO**-doped film. The hydroxyl groups of **PE** chains were esterified then with 2-naphthoyl chloride (Scheme 11.25). The fluorescence from the naphthyl group distinctly separates the modified and unmodified regions of the film, illustrating their image recording applications (Fig. 11.21).

The photo-Fries rearrangement of a photosensitive polymer, the poly(bicyclo[2.2.1]hept-5-ene-2,3-dicarboxylic acid, diphenyl ester) (**poly-1**),

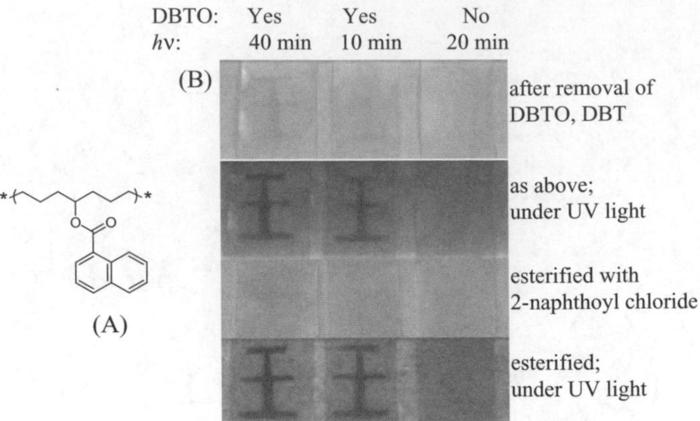

Figure 11.21. (A) Chains of **PE42** esterified with 2-naphthoyl chloride and (B) **DBTO**-doped (or undoped), irradiated, and esterified polyethylene films with a "Wang" mask. See ftp site for color: ftp://ftp.wiley.com/public/sci_tech_med/supramolecular_photochemistry.

has been used for photopatterning of polymer surfaces.[172] The photogenerated hydroxyl groups were selectively functionalized with a variety of reagents to obtain patterned surfaces. Reaction of hydroxyl groups with acid chlorides formed esters and reaction of carbonyl groups with 2,4-dinitrophenylhydrazine yielded hydrazones. Fluorescent patterns were formed upon reaction of the hydroxyl groups of the photo-Fries products with the sulfonic acid group of dansyl chloride, a fluorescent dye. Fe^{3+} ions were selectively deposited onto the illuminated areas because the hydroxyl ketones act as ligands for metal ions. Intensely colored thiocyanate complexes were generated by subsequent treatment of immobilized Fe^{3+}-containing polymers with ammonium rhodanide. The various pathways are illustrated in Scheme 11.26.

OTHER PHOTON-BASED APPLICATIONS

Plastics have been used in an enormous number of applications in organic electronics since the discovery of conducting polymers.[173] Conjugated polymers (CPs) are a promising class of semiconductors for a wide variety of optical applications, such as optical fibers, lasers, organic solar cells, and optical data storage devices.[174,175] Advantages of CPs include good performance, easy processability, tunability, and low cost.[176] Another significant advantage of CP over conventional dyes (many of which undergo self-quenching at higher concentrations) is that the solid films can have quantum efficiencies as high as 75%.[175]

A recent development in the field of polymer electronics involves surface patterning through the self-assembly of polymers. By using polymerizable

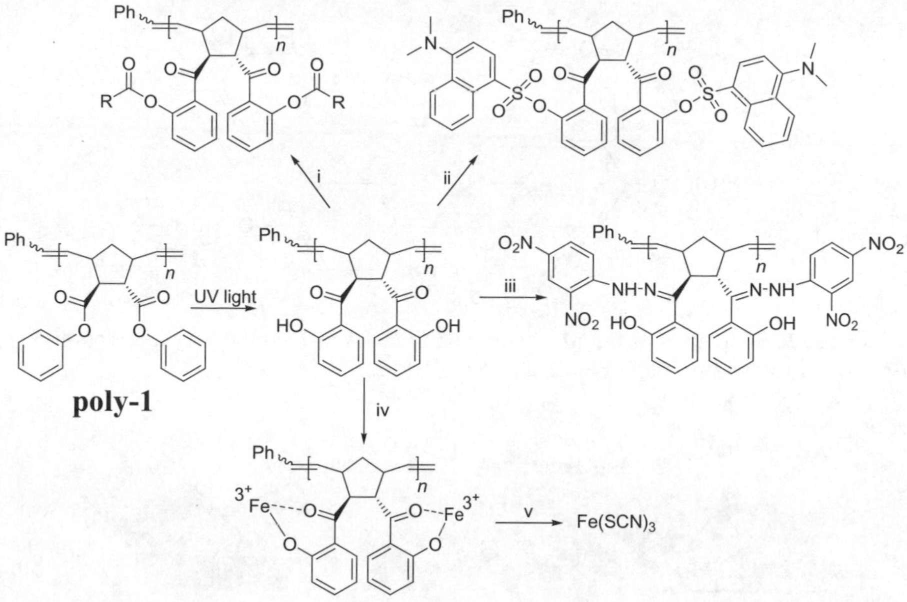

Scheme 11.26. Photo-Fries reaction of **poly-1** and subsequent postexposure reactions.[172] (i) Vapors of $RCOCl/CH_2Cl_2$; R=CH_3, C_6H_5, C_3F_7; (ii) dansyl chloride, triethylamine, acetonitrile; (iii) 2,4-dinitrophenylhydrazine, ethanol, HCl (aq); (iv) $FeCl_3$, ethanol; (v) NH_4SCN, ethanol.

reactive precursors, high-resolution patterns can be obtained on surfaces. In photolithographic techniques, the selective irradiation of polymers was used to generate specific patterns in the exposed regions.[177] Holographic gratings have been developed using the photochemical phase transitions in polymer LCs containing azobenzene moieties.[178]

Polymeric materials containing azobenzene groups have been made to bend along selected directions when irradiated with linearly polarized light.[179] The *trans→cis* isomerization of azobenzene groups preferentially along one face of a film results in a volume contraction that is sufficiently large to induce the shape changes of the material. Because the films are optically dense at 366 nm—the extinction coefficient of azobenzene moieties is about $2.0 \times 10^4\,L\,mol^{-1}\,cm^{-1}$ at 360 nm[180]—the vast majority of the isomerization occurs near the surface exposed to light. In this way, light energy is converted to mechanical energy with a macroscopically observable shape change that classifies the films as "artificial muscles!"[63] One film that exhibits the ability to bend has been prepared by thermal polymerization of the azobenzene-containing liquid crystal monomer (**Azo-1**) and the diacrylate cross-linker (**Azo-2**) (Chart 11.12). The films bend toward the direction of irradiation (366 nm) and parallel to the direction of the electric vector of the polarized light.[179]

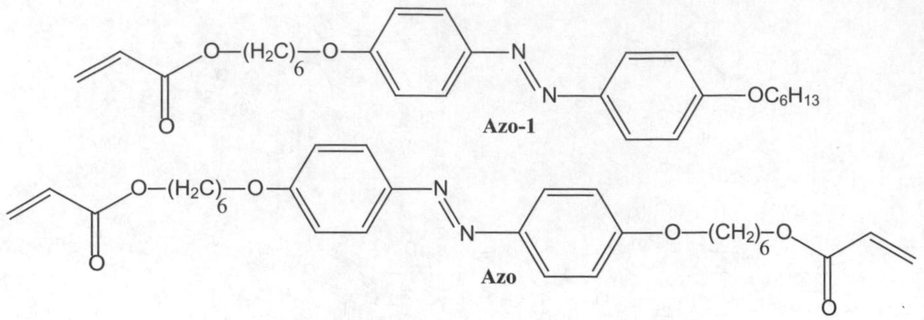

Chart 11.12. Structures of liquid crystal monomer (Azo-1) and diacrylate cross-linker (Azo-2) employed to make bendable polymer film.

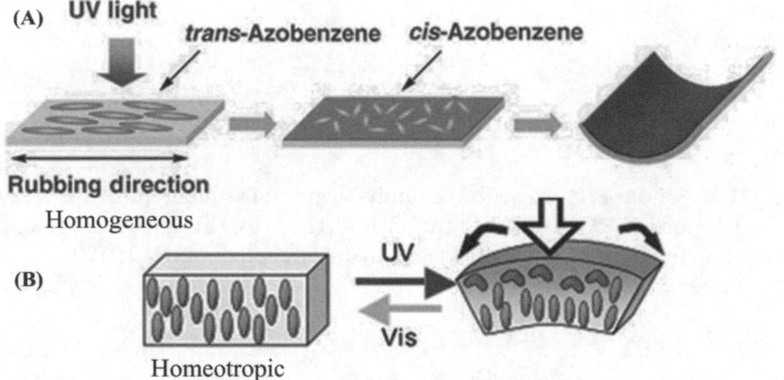

Scheme 11.27. Bending mechanism in films with azobenzene groups aligned in homogeneous and homeotropic arrangements. (Reprinted with permission from Reference[63]. Copyright 2009, Wiley-VCH Verlag GmbH & Co. KGaA.) See ftp site for color: ftp://ftp.wiley.com/public/sci_tech_med/supramolecular_photochemistry.

As mentioned, changes in the alignment of the azobenzene moieties occur only near a film surface. This results in an uneven distribution of anisotropic deformations perpendicular to the film surface, and bending of the film: the bending behavior of polymer films depends on the initial alignment of the photoactive mesogens; films in which azobenzene groups are initially aligned, preferentially, homeotropically (i.e., with their long axes perpendicular to the film surface) and homogeneously (i.e., with the long axes parallel to the surface of the film) have been prepared.[180] The surface exposed to radiation expands in homeotropic films, while that in the homogeneous film contracts along the alignment direction (Scheme 11.27).

This concept has been extended to azobenzene-containing polymers that are laminated onto one surface of **PE** films. Photochemically induced *trans→cis* and *cis→trans* isomerizations of the azobenzene moieties allows the films to walk or behave as robotic arms![181]

In a remarkable reversal of the concept to convert electromagnetic energy into mechanical force, films have been developed that convert mechanical force (film stretching) to effect what is usually considered a photochemical reaction, the ring opening of a spiropyran to its merocyanine form.[182] This "mechanochromic" reaction takes place when the two halves of a spiropyran are attached to different **PMMA** chains. The slippage of one of the chains with respect to the other as a film is extended induces the spiro ring to open because the merocyanine form is longer than the spiro form. These changes are shown conceptually in Figure 11.22.

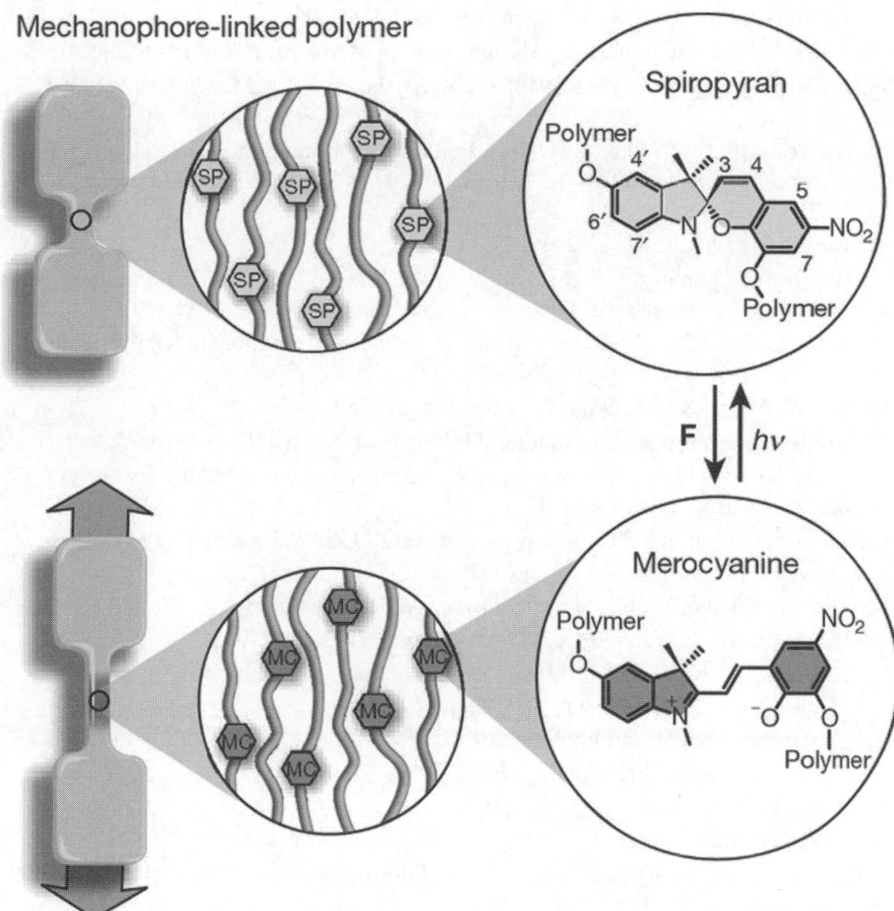

Figure 11.22. Conceptual representation of the conversion of mechanical force into kinetic energy to open a spiropyran ring and the use of light to close the merocyanine subsequently. (Reprinted with permission from Macmillan Publishers Ltd:[Nature] [Reference 182], Copyright 2009.) See ftp site for color: ftp://ftp.wiley.com/public/sci_tech_med/supramolecular_photochemistry.

ACKNOWLEDGMENTS

The authors thank the U.S. National Science Foundation and the Petroleum Research Fund of the American Chemical Society for their support of the part of the research in this chapter that was performed at Georgetown.

REFERENCES

1. Ramamurthy, V. *Photochemistry in Organized and Constrained Media*. New York: VCH Publishers, **1991**.
2. Gu, W.; Weiss, R.G. *J. Photochem. Photobiol. C* **2001**, *2*, 117.
3. Turro, N.J. *Proc. Natl. Acad. Sci. U.S.A.* **2002**, *99*, 4805.
4. Weiss, R.G.; Ramamurthy, V.; Hammond, G.S. *Acc. Chem. Res.* **1993**, *26*, 530.
5. (*a*) Kramers, H.A. *Physica* **1940**, *7*, 284; (*b*) Janssen, J.A.M. *Physica A* **1988**, *152*, 145.
6. Andrews, R.J.; Grulke, E.A. In: Brandup, J.; Immergut, E.H.; Grulke, E.A., editors. *Polymers Handbook*, 4th edition. New York: Wiley-Interscience, **1999**, VI/193–277.
7. Aleshin, A.N. *Adv. Mater.* **2006**, *18*, 17.
8. Brandup, J.; Immergut, E.H.; Grulke, E.A. *Polymers Handbook*, 4th edition. New York: Wiley-Interscience, **1999**.
9. Dlubek, G.; De, U.; Pionteck, J.; Arutyunov, N.Y.; Edelmann, M.; Krause-Rehberg, R. *Macromol. Chem. Phys.* **2005**, *206*, 827.
10. Gu, W.; Weiss, R.G. *J. Org. Chem.* **2001**, *66*, 1775.
11. Vrentas, J.S.; Vrentas, C.M.; Duda, J.L. *Polym. J. (Tokyo, Jpn.)* **1993**, *25*, 99.
12. Duda, J.L.; Zielinski, J.M. In: Neogi, P., editor. *Diffusion in Polymers*. New York: Marcel Decker, **1996**, 143–171.
13. Zimerman, O.E.; Cui, C.; Wang, X.; Atvars, T.D.Z.; Weiss, R.G. *Polymer* **1998**, *39*, 1177.
14. Thulstrup, E.W.; Michl, J. *J. Am. Chem. Soc.* **1982**, *104*, 5594.
15. Wang, C.; Xu, J.; Weiss, R.G. *J. Phys. Chem. B* **2003**, *107*, 7015.
16. Fox, T.G.; Flory, P.J. *J. Appl. Phys.* **1950**, *21*, 581.
17. McCrum, N.G.; Read, B.E.; Williams, G. *Anelastic and Dielectric Effects in Polymeric Solids*. London: Wiley, **1976**.
18. Zhou, H.; Wilkes, G.L. *Macromolecules* **1997**, *30*, 2412.
19. Wang, C.L.; Hirade, T.; Maurer, F.H.J.; Eldrup, M.; Pedersen, N.J. *J. Chem. Phys.* **1998**, *108*, 4654.
20. Guillet, J. *Polymer Photophysics and Photochemistry*. New York: Cambridge University Press, **1985**.
21. Barton, A.F.M. *CRC Handbook of Polymer-Liquid Interaction Parameters and Solubility Parameters*. Boca Raton, FL: CRC Press, **1990**.
22. Rumyantsev, B.M.; Berendyaev, V.I.; Tsegel'skaya, A.Y.; Kotov, B.V. *Mol. Cryst. Liq. Cryst.* **2002**, *384*, 61.
23. Avlyanov, J.K.; Min, Y.; MacDiarmid, A.G.; Epstein, A.J. *Synth. Met.* **1995**, *72*, 65.

24. MacDiarmid, A.G.; Epstein, A.J. *Synth. Met.* **1994**, *65*, 103.
25. Leroux, F.; Besse, J.-P. *Chem. Mater.* **2001**, *13*, 3507.
26. Lovell, P.A.; Stanford, J.L.; Wang, Y.-F.; Young, R.J. *Macromolecules* **1998**, *31*, 834.
27. Hiraga, T.; Takarada, S.; Tanaka, N.; Hayamizu, K.; Moriya, T. *Jpn. J. Appl. Phys.* **1994**, *33*, 5051.
28. Braun, D.; Heeger, A.J. *Appl. Phys. Lett.* **1991**, *58*, 1982.
29. Shi, Y.; Liu, J.; Yang, Y. *J. Appl. Phys.* **2000**, *87*, 4254.
30. Liu, C.; Zou, X.; Yin, S.; Zhang, W. *Thin Solid Films* **2004**, *466*, 279.
31. Feng, J.; MacDiarmid, A.G.; Epstein, A.J. *Synth. Met. Int. Conf. Sci. Technol. Synth. Metals* **1997**, *84*, 131.
32. Mochizuki, H.; Mizokuro, T.; Yamamoto, N.; Tanigaki, N.; Hiraga, T. *J. Photopolym. Sci. Tech.* **2003**, *16*, 199.
33. Mizokuro, T.; Mochizuki, H.; Yamamoto, N.; Horiuchi, S.; Tanigaki, N.; Hiraga, T.; Tanaka, N. *J. Photopolym. Sci. Tech.* **2003**, *16*, 195.
34. Tsutsui, T.; Aminaka, E.; Lin, C.P.; Kim, D.-U. *Philos. Trans. R. Soc. Lond. A* **1997**, *355*, 801.
35. Feng, J.; Winnik, M.A.; Shivers, R.R.; Clubb, B. *Macromolecules* **1995**, *28*, 7671.
36. Horie, K.; Mita, I. *Adv. Polym. Sci.* **1989**, *88*, 77.
37. Montali, A.; Bastiaansen, C.; Smith, P.; Weder, C. *Nature* **1998**, *392*, 261.
38. Nishikubo, T.; Kawashima, T.; Inomata, K.; Kameyama, A. *Macromolecules* **1992**, *25*, 2312.
39. Hartley, G.E.; Guillet, J.E. *Macromolecules* **1968**, *1*, 165.
40. Goldbach, J.T.; Russell, T.P.; Penelle, J. *Macromolecules* **2002**, *35*, 4271.
41. Coursan, M.; Desvergne, J.P.; Deffieux, A. *Macromol. Chem. Phys.* **1996**, *197*, 1599.
42. Kim, Y.H.; Webster, O.W. *Macromolecules* **1992**, *25*, 5561.
43. Hawker, C.J.; Lee, R.; Frechet, J.M.J. *J. Am. Chem. Soc.* **1991**, *113*, 4583.
44. Gu, W.; Abdallah, D.J.; Weiss, R.G. *J. Photochem. Photobiol. A* **2001**, *139*, 79.
45. Xu, B.; Holdcroft, S. *Macromolecules* **1993**, *26*, 4457.
46. Chen, F.; Mehta, P.G.; Takiff, L.; McCullough, R.D. *J. Mater. Chem.* **1996**, *6*, 1673.
47. Chu, D.Y.; Thomas, J.K. *J. Phys. Chem.* **1985**, *89*, 4065.
48. Cuniberti, C.; Perico, A. *Eur. Polym. J.* **1977**, *13*, 369.
49. Askadskii, A.A. *Physical Properties of Polymers*. Amsterdam: Gordon and Breach Science Publishers, **1996**.
50. Zimerman, O.E.; Weiss, R.G. *J. Phys. Chem. A.* **1998**, *102*, 5364.
51. Corradini, P.; Auriemma, F.; DeRosa, C. *Acc. Chem. Res.* **2006**, *39*, 314.
52. (*a*) Phillips, P.J. *Chem. Rev.* **1990**, *90*, 425; (*b*) Jang, Y.T.; Phillips, P.J.; Thulstrup, E.W. *Chem. Phys. Lett.* **1982**, *93*, 66.
53. Vigil, M.R.; Bravo, J.; Atvars, T.D.Z.; Baselga, J. *Macromolecules* **1997**, *30*, 4871.
54. Cohen, M.D. *Angew. Chem. Int. Ed. Engl.* **1975**, *14*, 386.
55. Frank, C.W. *Macromolecules* **1975**, *8*, 305.
56. Farid, S.; Martic, P.A.; Daly, R.C.; Thompson, D.R.; Specht, D.P.; Hartman, S.E.; Williams, J.L.R. *Pure Appl. Chem.* **1979**, *51*, 241.

57. Ono, Y.; Kawashima, N.; Kudo, H.; Nagai, T.; Nishikubo, T. *Polym. J. (Tokyo, Jpn.)* **2005**, *37*, 246.
58. Palmans, A.R.A.; Smith, P.; Weder, C. *Macromolecules* **1999**, *32*, 4677.
59. Hrdlovič, P.; Zahumenský, L.; Lukáč, I.; Sláma, P. *J. Polym. Sci.* **1978**, *16*, 877.
60. Dan, E.; Somersall, A.C.; Guillet, J.E. *Macromolecules* **1973**, *6*, 228.
61. Scully, A.D.; Ghiggino, K.P. In: Allen, N.S.; Edge, M.; Bellobono, I.R.; Selli, E., editors. *Current Trends in Polymer Photochemistry*. Hertfordshire: Ellis Horwood Ltd, **1995**.
62. Ichimura, K.; Oh, S.-K.; Nakagawa, M. *Science* **2000**, *288*, 1624.
63. Ikeda, T.; Mamiya, J.-I.; Yu, Y. *Angew. Chem. Int. Ed. Engl.* **2007**, *46*, 506.
64. Wang, C.; Weiss, R.G. *Macromolecules* **1999**, *32*, 7032.
65. Diepens, M.; Gijsman, P. *Polym. Degrad. Stab.* **2007**, *92*, 397.
66. Chu, D.Y.; Thomas, J.K. *J. Phys. Chem.* **1989**, *93*, 6250.
67. Benten, H.; Ohkita, H.; Ito, S.; Yamamoto, M.; Tohda, Y.; Tani, K. *J. Phys. Chem. B* **2004**, *108*, 16457.
68. Talhavini, M.; Atvars, T.D.Z.; Cui, C.; Weiss, R.G. *Polymer* **1996**, *37*, 4365.
69. Ichimura, K. *Chem. Rev.* **2000**, *100*, 1847.
70. Percec, V.; Tomazos, D. *Adv. Mater.* **1992**, *4*, 548.
71. Ikeda, T.; Tsutsumi, O. *Science* **1995**, *268*, 1873.
72. Shibaev, V.; Bobrovsky, A.; Boiko, N. *Prog. Polym. Sci.* **2003**, *28*, 729.
73. Das, S.; Davis, R. *Proc. Indian Natl. Sci. Acad.* **2003**, *69 A*, 109.
74. Naito, T.; Horie, K.; Mita, I. *Macromolecules* **1991**, *24*, 2907.
75. Dugave, C.; Demange, L. *Chem. Rev.* **2003**, *103*, 2475.
76. Xie, S.; Natansohn, A.; Rochon, P. *Chem. Mater.* **1993**, *5*, 403.
77. Victor, J.G.; Torkelson, J.M. *Macromolecules* **1987**, *20*, 2241.
78. Arifov, P.U.; Vasserman, S.N.; Dontsov, A.A.; Tishin, S.A. *Dokl. Phys. Chem.* **1984**, *8*, 661.
79. Lopez, M.C.; Finkelmann, H.; Muhoray, P.P.; Shelley, M. *Nat. Mater.* **2004**, *3*, 307.
80. Semerak, S.N.; Frank, C.W. *Macromolecules* **1984**, *17*, 1148.
81. Löwe, C.; Weder, C. *Adv. Mater.* **2002**, *14*, 1625.
82. Crenshaw, B.R.; Weder, C. *Adv. Mater.* **2005**, *17*, 1471.
83. Crenshaw, B.R.; Weder, C. *Chem. Mater.* **2003**, *15*, 4717.
84. de Silva, A.P.; Gunaratne, H.Q.N.; Gunnlaugsson, T.; Huxley, A.J.M.; McCoy, C.P.; Rademacher, J.T.; Rice, T.E. *Chem. Rev.* **1997**, *97*, 1515.
85. Winnik, F.M. *Chem. Rev.* **1993**, *93*, 587.
86. Jenekhe, S.A.; Osaheni, J.A. *Science* **1994**, *265*, 765.
87. Lu, X.; Winnik, M.A. *Chem. Mater.* **2001**, *13*, 3449.
88. Thomas, S.W. III; Amara, J.P.; Bjork, R.E.; Swager, T.M. *Chem. Commun.* **2005**, 4572.
89. Yang, J.-S.; Swager, T.M. *J. Am. Chem. Soc.* **1998**, *120*, 5321.
90. Chen, L.; McBranch, D.W.; Wang, H.-L.; Helgeson, R.; Wudl, F.; Whitten, D.G. *Proc. Natl. Acad. Sci. U.S.A.* **1999**, *96*, 12287.

91. Brown, G.O.; Guardala, N.A.; Price, J.L.; Weiss, R.G. *J. Phys. Chem. A* **2003**, *107*, 3543.

92. Jenkins, R.M.; Hammond, G.S.; Weiss, R.G. *J. Phys. Chem.* **1992**, *96*, 496.

93. Taraszka, J.A.; Weiss, R.G. *Macromolecules* **1997**, *30*, 2467.

94. Schurr, O.; Weiss, R.G. *Polymer* **2004**, *45*, 5713.

95. Shimano, J.Y.; MacDiarmid, A.G. *Synth. Met.* **2001**, *123*, 251.

96. Heeger, A.J. *J. Phys. Chem. B* **2001**, *105*, 8475.

97. Heskins, M.; Guillet, J.E. *Macromolecules* **1968**, *1*, 97.

98. Heskins, M.; Guillet, J.E. *Macromolecules* **1970**, *3*, 224.

99. Yang, J.; Lu, J.; Rharbi, Y.; Cao, L.; Winnik, M.A.; Zhang, Y.; Wiesner, U.B. *Macromolecules* **2003**, *36*, 4485.

100. Oh, J.K.; Tomba, P.; Ye, X.; Eley, R.; Rademacher, J.; Farwaha, R.; Winnik, M.A. *Macromolecules* **2003**, *36*, 5804.

101. Ni, S.; Zhang, P.; Wang, Y.; Winnik, M.A. *Macromolecules* **1994**, *27*, 5742.

102. Roller, R.S.; Winnik, M.A. *J. Phys. Chem. B* **2005**, *109*, 12261.

103. Somersall, A.C.; Dan, E.; Guillet, J.E. *Macromolecules* **1974**, *7*, 233.

104. Xu, J.; Luo, C.; Atvars, T.D.Z.; Weiss, R.G. *Res. Chem. Intermed.* **2004**, *30*, 509.

105. Schurr, O.; Yamaki, S.B.; Wang, C.; Atvars, T.D.Z.; Weiss, R.G. *Macromolecules* **2003**, *36*, 3485.

106. Luo, C.; Atvars, T.D.Z.; Meakin, P.; Hill, A.J.; Weiss, R.G. *J. Am. Chem. Soc.* **2003**, *125*, 11879.

107. Kumar, G.S.; Neckers, D.C. *Chem. Rev.* **1989**, *89*, 1915.

108. Sung, C.S.P.; Gould, I.R.; Turro, N.J. *Macromolecules* **1984**, *17*, 1447.

109. Wang, C.; Weiss, R.G. *Macromolecules* **2003**, *36*, 3833.

110. Paik, C.S.; Morawetz, H. *Macromolecules* **1972**, *5*, 171.

111. Sarkar, N.; Sarkar, A.; Sivaram, S. *J. Appl. Polym. Sci.* **2001**, *81*, 2923.

112. Kósa, C.; Lukáč, I.; Weiss, R.G. *Macromol. Chem. Phys.* **1999**, *200*, 1080.

113. Kósa, C.; Lukáč, I.; Weiss, R.G. *Macromolecules* **2000**, *33*, 4015.

114. Scoponi, M.; Pradela, F.; Carassiti, V. *Coord. Chem. Rev.* **1993**, *125*, 219.

115. Amerik, Y.; Guillet, J.E. *Macromolecules* **1971**, *4*, 375.

116. Mosnáček, J.; Weiss, R.G.; Lukáč, I. *Macromolecules* **2002**, *35*, 3870.

117. Rånby, B. In: Allen, N.S.; Edge, M.; Bellobono, I.R.; Selli, E., editors. *Current Trends in Polymer Photochemistry*. Hertfordshire: Ellis Horwood Ltd, **1995**.

118. Müller, C.D.; Falcou, A.; Reckefuss, N.; Rojahn, M.; Wiederhirn, V.; Rudati, P.; Frohne, H.; Nuyken, O.; Becker, H.; Meerholz, K. *Nature* **2003**, *421*, 829.

119. Subramanian, P.; Creed, D.; Griffin, A.C.; Hoyle, C.E.; Venkataram, K. *J. Photochem. Photobiol. A* **1991**, *61*, 317.

120. Rennert, J.; Soloway, S.; Waltcher, I.; Leong, B. *J. Am. Chem. Soc.* **1972**, *94*, 7242.

121. Egerton, P.L.; Pitts, E.; Reiser, A. *Macromolecules* **1981**, *14*, 95.

122. Griffin, A.C.; Hoyle, C.E.; Gross, J.R.D.; Venkataram, K.; Creed, D. *Makromol. Chem. Rapid Commun.* **1988**, *9*, 463.

123. Creed, D.; Griffin, A.C.; Gross, J.R.D.; Hoyle, C.E.; Venkataram, K. *Mol. Cryst. Liq. Cryst.* **1988**, *155*, 57.

124. Ramesh, V.; Weiss, R.G. *Macromolecules* **1986**, *19*, 1486.
125. Aviv, G.; Sagiv, J.; Yogev, A. *Mol. Cryst. Liq. Cryst.* **1976**, *36*, 349.
126. Strukelj, M.; Martinho, J.M.G.; Winnik, M.A.; Quirk, R.P. *Macromolecules* **1991**, *24*, 2488.
127. Martinho, J.M.G.; Castanheira, E.M.S.; Reis e Sousa, A.T.; Saghbini, S.; André, J.C.; Winnik, M.A. *Macromolecules* **1995**, *28*, 1167.
128. Freeman, B.D.; Bokobza, L.; Sergot, P.; Monnerie, L.; Schryver, F.C.D. *Macromolecules* **1990**, *23*, 2566.
129. Naciri, J.; Weiss, R.G. *Macromolecules* **1989**, *22*, 3928.
130. He, Z.; Hammond, G.S.; Weiss, R.G. *Macromolecules* **1992**, *25*, 501.
131. Lu, L.; Weiss, R.G. *Macromolecules* **1994**, *27*, 219.
132. Zimmerman, H.E.; O'Brien, M.E. *J. Org. Chem.* **1994**, *59*, 1809.
133. Finnegan, R.A.; Mattice, J.J. *Tetrahedron* **1965**, *21*, 1015.
134. Mori, T.; Inoue, Y.; Weiss, R.G. *Org. Lett.* **2003**, *5*, 4661.
135. Mori, T.; Weiss, R.G.; Inoue, Y. *J. Am. Chem. Soc.* **2004**, *126*, 8961.
136. Belluš, D.; Hrdlovič, P. *Chem. Rev.* **1967**, *67*, 599.
137. Belluš, D. *Adv. Photochem.* **1971**, *8*, 109.
138. Kobsa, H. *J. Org. Chem.* **1962**, *27*, 2293.
139. Lally, J.M.; Spillane, W.J. *J. Chem. Soc. Chem. Commun.* **1987**, 8.
140. Gritsan, N.P.; Tsentalovich, Y.P.; Yurkovskaya, A.V.; Sagdeev, R.Z. *J. Phys. Chem.* **1996**, *100*, 4448.
141. Chatgilialoglu, C.; Crich, D.; Komatsu, M.; Ryu, I. *Chem. Rev.* **1999**, *99*, 1991.
142. Cui, C.; Weiss, R.G. *J. Am. Chem. Soc.* **1993**, *115*, 9820.
143. Höfler, T.; Grießer, T.; Gruber, M.; Jakopic, G.; Trimmel, G.; Kern, W. *Macromol. Chem. Phys.* **2008**, *209*, 488.
144. Gu, W.; Warrier, M.; Ramamurthy, V.; Weiss, R.G. *J. Am. Chem. Soc.* **1999**, *121*, 9467.
145. Gu, W.; Weiss, R.G. *Tetrahedron* **2000**, *56*, 6913.
146. Griller, D.; Ingold, K.U. *Acc. Chem. Res.* **1980**, *13*, 317.
147. Horner, J.H.; Tanaka, N.; Newcomb, M. *J. Am. Chem. Soc.* **1998**, *120*, 10379.
148. (*a*) Turro, N.J.; Gould, I.R.; Baretz, B.H. *J. Phys. Chem.* **1983**, *87*, 531; (*b*) Zhang, X.; Nau, W.M. *J. Phys. Org. Chem.* **2000**, *13*, 634; (*c*) Lunazzi, L.; Ingold, K.U.; Scaiano, J.C. *J. Phys. Chem.* **2002**, *87*, 529.
149. Chesta, C.A.; Mohanty, J.; Nau, W.M.; Bhattacharjee, U.; Weiss, R.G. *J. Am. Chem. Soc.* **2007**, *129*, 5012.
150. Nakagaki, R.; Hiramatsu, M.; Watanabe, T.; Tanimoto, Y.; Nagakura, S. *J. Phys. Chem.* **1985**, *89*, 3222.
151. Koenig, T.; Owens, J.M. *J. Am. Chem. Soc.* **1973**, *95*, 8484.
152. Koenig, T.; Owens, J.M. *J. Am. Chem. Soc.* **1974**, *96*, 4052.
153. Bhanthumnavin, W.; Bentrude, W.G. *J. Org. Chem.* **2001**, *66*, 980.
154. Xu, J.; Weiss, R.G. *Org. Lett.* **2003**, *5*, 3077.
155. Gu, W.; Warrier, M.; Schoon, B.; Ramamurthy, V.; Weiss, R.G. *Langmuir* **2000**, *16*, 6977.

156. Adam, W.; Fischer, H.; Hansen, H.J.; Heimgartner, H.; Schmid, H.; Waespe, H.R. *Angew. Chem. Int. Ed. Engl.* **1973**, *12*, 662.

157. Xu, J.; Weiss, R.G. *Photochem. Photobiol. Sci.* **2005**, *4*, 348.

158. Tarasov, V.; Ghatlia, N.D.; Buchachenko, A.; Turro, N.J. *J. Phys. Chem.* **1991**, *95*, 10220.

159. Doubleday, C.; Turro, N.J.; Wang, J.F. *Acc. Chem. Res.* **1989**, *22*, 199.

160. Turro, N.J.; Lei, X.-G.; Li, W.; Liu, Z.; McDermott, A.; Ottaviani, M.F.; Abrams, L. *J. Am. Chem. Soc.* **2000**, *122*, 11649.

161. Noh, T.; Step, E.; Turro, N.J. *J. Photochem. Photobiol. A* **1993**, *72*, 133.

162. Engel, P.S. *J. Am. Chem. Soc.* **1970**, *92*, 6074.

163. Bhattacharjee, U.; Chesta, C.A.; Weiss, R.G. *Photochem. Photobiol. Sci.* **2004**, *3*, 287.

164. Hrovat, D.A.; Liu, J.H.; Turro, N.J.; Weiss, R.G. *J. Am. Chem. Soc.* **1984**, *106*, 5291.

165. Tsentalovich, Y.P.; Fischer, H. *J. Chem. Soc. Perkin 2* **1994**, 729.

166. Tarasov, V.F.; Ghatlia, N.D.; Buchachenko, A.L.; Turro, N.J. *J. Am. Chem. Soc.* **1992**, *114*, 9517.

167. Kleinman, M.H.; Shevchenko, T.; Bohne, C. *Photochem. Photobiol.* **1998**, *67*, 198.

168. Griller, D.; Ingold, K.U. *Acc. Chem. Res.* **1976**, *9*, 13.

169. Fischer, H. *Chem. Rev.* **2001**, *101*, 3581.

170. Luo, C.; Guardala, N.A.; Price, J.L.; Chodak, I.; Zimerman, O.; Weiss, R.G. *Macromolecules* **2002**, *35*, 4690.

171. Yamaki, S.B.; Atvars, T.D.Z.; Weiss, R.G. *Photochem. Photobiol. Sci.* **2002**, *1*, 649.

172. Griesser, T.; Höfler, T.; Temmel, S.; Kern, W.; Trimmel, G. *Chem. Mater.* **2007**, *19*, 3011.

173. Burroughes, J.H.; Bradley, D.D.C.; Brown, A.R.; Marks, R.N.; Mackay, K.; Friend, R.H.; Burns, P.L.; Holmes, A.B. *Nature* **1990**, *347*, 539.

174. Tessler, N.; Dento, G.J.; Friend, R.H. *Nature* **1996**, *382*, 695.

175. Hide, F.; Díaz-García, M.A.; Schwartz, B.J.; Heeger, A.J. *Acc. Chem. Res.* **1997**, *30*, 430.

176. Kraft, A.; Grimsdale, A.C.; Holmes, A.B. *Angew. Chem. Int. Ed. Engl.* **1998**, *37*, 402.

177. Nie, Z.; Kumacheva, E. *Nature* **2008**, *7*, 277.

178. Ikeda, T. *J. Mater. Chem.* **2003**, *13*, 2037.

179. Yu, Y.; Nakano, M.; Ikeda, T. *Nature* **2003**, *425*, 145.

180. Kondo, M.; Yu, Y.; Ikeda, T. *Angew. Chem. Int. Ed. Engl.* **2006**, *45*, 1378.

181. Yamada, M.; Kondo, M.; Miyasato, R.; Naka, Y.; Mamiya, J.; Kinoshita, M.; Shishido, A.; Yu, Y.; Barrett, C.J.; Ikeda, T. *J. Mater. Chem.* **2009**, *19*, 60.

182. Davis, D.A.; Hamilton, A.; Yang, J.; Cremar, L.D.; Gough, D.V.; Potisek, S.L.; Ong, M.T.; Braun, P.V.; Martínez, T.J.; White, S.R.; Moore, J.S.; Sottos, N.R. *Nature* **2009**, *459*, 68.

183. Carraher, C.E. *Polymer Chemistry*, 7th edition. Boca Raton, FL: CRC Press, Taylor and Francis Group, **2008**.

184. Tamai, N.; Miyasaka, H. *Chem. Rev.* **2000**, *100*, 1875.

185. (*a*) Dixon, W.T.; Moghimi, M.; Murphy, D. *J. Chem. Soc. Faraday Trans. 2* **1974**, *70*, 1713; (*b*) Martin, K.; George, K.F. *J. Chem. Phys.* **1961**, *35*, 1312; (*c*) Dixon, W.T.; Foster, W.E.J.; Murphy, D. *J. Chem. Soc. Perkin 2* **1973**, 2124.

186. Tung, C.-H.; Yuan, Z.-Y.; Wu, L.-Z.; Weiss, R.G. *J. Org. Chem.* **1999**, *64*, 5156.

12

DELOCALIZATION AND MIGRATION OF EXCITATION ENERGY AND CHARGE IN SUPRAMOLECULAR SYSTEMS

Mamoru Fujitsuka and Tetsuro Majima

INTRODUCTION

Today a wide range of scientists engage in research on fabrication and characterization of nanosized materials, and their research strategies are quite diverse. For many scientists in the field of organic chemistry, covalent bonding of functional units, self-assembly, coordination, polymerization, and so on, are important approaches to fabricate nanosized materials. For fabricated nanosized materials, characterization of their functionality is an important subject. For nanosized materials aimed at molecular electro- and/or opto-functional devices, including artificial photosynthesis, energy conversion, and nonlinear optics, ultrafast phenomena in organized materials are quite important as well as bulk level functionality analysis; these ultrafast phenomena govern their functionalities in many cases. In particular, generation and migration of charge carrier and excitation energy are important elemental processes in these materials, although investigation of these processes in detail is quite a difficult task. The expectation is that phenomena possibly become faster than that observed in molecules in solution phase, due to strong interaction between chromophores in organized materials like those in solid phase. In addition,

Supramolecular Photochemistry: Controlling Photochemical Processes, First Edition. Edited by V. Ramamurthy and Yoshihisa Inoue.

inhomogeneity of materials makes analysis difficult. These factors in many cases can be negligible in conventional molecular chemistry in solution. To clarify these functionalities in nanosized materials, measurement methods with very high time resolution are needed.

Time resolution and stability of pulse lasers for various time-resolved measurements have been much improved compared with those in previous decades. The operation becomes easier; thus, even chemists who are not specialists of spectroscopic study can handle ultrafast laser systems. These highly stable ultrafast laser systems with subpicosecond pulse duration are potentially quite useful to obtain quantitative information on the above-mentioned processes in organized materials. High stability is an important factor because a subtle signal can be captured only by using such a highly stable laser. By using these highly stable lasers, various photoinduced processes in nanosized materials can be investigated.

In addition, the charge carrier in organized material should have different properties from that of a molecule in a solution, that is, as a radical cation or radical anion.[1] When the charge carrier is induced or generated in organized materials, it should be stabilized by various factors, such as polarization of surrounding chromophores or delocalization over other chromophores. The quantitative understanding of these factors is often a difficult subject because multiple and strong interactions due to high-density chromophores makes the problem quite complex. Furthermore, dynamic aspects such as migration of the charge carrier are also ultrafast processes and are affected by similar factors. To characterize the charged species in the organized materials, the factors governing such stabilization and dynamic aspects should be clarified.

In this chapter, we review recent works, including ours, that clarify ultrafast phenomena, in particular, migration and delocalization of excitation energy and charges in organized materials.

ENERGY MIGRATION IN SUPRAMOLECULAR SYSTEMS

Over the past few decades, fabrication of artificial photosynthesis systems has been a central subject for many research groups. Much effort has been put into research on the synthesis of charge separation systems, which realize a long-lived charge separated state with a high quantum yield. Nowadays, many donor–acceptor-type molecules such as dyad or triad have been reported to show a long-lived charge separated state with a high quantum yield, which is applicable to light-energy conversion systems.[2] Actually, some of such molecules have been successfully applied to light-energy conversion systems in the form of solar cells, although efficiency and durability need to be improved for actual application. To enhance the efficiency of such devices, efficient transport of energy absorbed by chromophores to charge generation sites will be essential for many systems, that is, energy-funneling systems.

In natural photosynthesis systems, the reaction center including the special pair is surrounded by light-harvesting complexes (LHs) composed of chromophores such as carotenoids and chlorophylls.[3–7] For example, photosynthetic purple bacteria have three kinds of LHs. The peripheral LHs, namely, LH2 and LH3, have absorption maxima at wavelengths shorter than that of LH1 and the reaction center, of which the latter absorbs the longest wavelengths and utilizes the collected light energy for the electron transfer. By combining these LHs, a quite wide range of photon energy can be utilized in the photosynthesis. In LHs, chlorophylls are arranged in wheel-like arrays, in which the transition dipoles of chlorophylls are nearly parallel to each other favorable to Förster-type energy transfer, although Dexter energy transfer is also possible through interactions mediated by lycopene and chlorophyll phytyl tails. Very fast energy transfer in subpicosecond regimes has been reported for these LHs.[8]

To achieve efficient light-energy harvesting in the artificial photosynthesis systems, well-organized arrays of chromophores have to be developed. In this section, the following three categories classified by fabrication methods to achieve artificial light-harvesting systems are summarized:

1. chromophore arrays organized by covalent linkage,
2. chromophore arrays organized by noncovalent linkage,
3. chromophore arrays by polymer systems.

Chromophore Arrays Organized by Covalent Linkage

Recent advanced organic synthesis enables various types of chromophore arrays, in which identical chromophores are connected by covalent bonds. Even if we are limited to porphyrin derivatives, which have often been employed to mimic natural LH, various kinds of arrays have been reported.

Energy Migration in Porphyrin Arrays. Osuka and coworkers synthesized various kinds of porphyrin arrays. In 1990s, they have reported porphyrin arrays in which porphyrins are connected by phenylene ring(s) and are aligned in a linear or face-to-face stacking form.[9,10] In 2000s, they have reported *meso–meso*-linked porphyrin arrays, which can be converted to triply *meso–meso*-, β-β-, and β-β-linked porphyrin tapes.[11] By using *meso–meso*-linked porphyrins as a building unit, they also succeeded in the construction of the wheel- and box-like porphyrin arrays.[12–17] These arrays give interesting information on the coupling of the transition dipole moments. Furthermore, by employing ultrafast laser spectroscopy, they measured the energy hopping rate between the porphyrins directly connected at *meso–meso* position to be ~0.2 ps in a linear porphyrin array.[12,13] A similar energy hopping rate has been reported for porphyrins connected at *meso–meso* position in a porphyrin wheel, although the energy hopping rate between porphyrins connected through phenyl rings was ~3 ps.[15,17]

Therien and coworkers studied various kinds of ethynyl-bridged porphyrin oligomers. They showed that depending on the bridging position, significant inter-ring conjugation can be obtained.[18]

Energy Migration in Dendrimers. Dendrimers are another class of molecules that possess light-harvesting functionalities. Yeow et al. reported on poly(propylene imine) dendrimers, whose branch terminals are connected with porphyrins.[19] They have observed fluorescence anisotropy decay due to energy migration among porphyrin units. Larsen et al. reported singlet–singlet annihilation due to singlet energy migration in the same dendrimers and found that the fast (18 ps) and slow (130 ps) annihilation rates are due to conformational distribution in the dendrimers.[20]

Aida and coworkers also reported on porphyrin dendrimers, in which free base porphyrin as the core is surrounded by zinc porphyrin (ZnP) arrays.[21,22] They observed efficient energy transfer from peripheral ZnPs to the central free base porphyrin.

Wasielewski and coworkers reported on the chlorophylls attached to 1,3,5-positions of benzene through ethynyl or phenylethynyl linkages.[23] They estimated energy transfer rates between chlorophylls based on the singlet–singlet annihilation rate to be 1.8 or 6.0 ps for the compounds with ethynyl or phenylethynyl linkages, respectively.

These examples show that by using recent advanced synthesis methods, many kinds of chromophore arrays have been realized. The energy migration rates in these arrays are of subpicosecond order when chromophores are arranged in the closest positions to each other. Especially in the case of the porphyrin arrays, rates tend to be faster because of their larger dipole interactions due to a strong transition dipole.

Chromophore Arrays Organized by Noncovalent Linkage

In natural LH, chlorophylls are placed at ideal positions for light-energy harvesting by noncovalent bonding with surrounding proteins. In particular, in plants, coordination of imidazoyl to magnesium centers of chlorophylls is an important motif to hold an ideal structure. Inspired by this, various researchers have constructed chromophore arrays based on noncovalent bonding formation.

Energy Migration in Porphyrin Arrays with Noncovalent Linkage. Kobuke and coworkers reported on porphyrin arrays arranged in a wheel-like configuration by using coordination to the zinc center of porphyrin of imidazoyl, which is attached to the *meso*-position of the porphyrin ring.[24–27] In their porphyrin arrays, neighboring porphyrins take a slipped cofacial orientation, in which the energy transfer occurred at ~200 fs. The energy transfer rate between porphyrins connected by 1,3-phenyl ring was ~4 ps.

Osuka and coworkers also reported on porphyrin arrays based on noncovalent bonding, in particular, porphyrin boxes and prisms.[28,29] For porphyrin boxes, the energy transfer rate from porphyrin to coordinated porphyrin was on the order of several tens of picoseconds, depending on the length of the phenyl ring linking porphyrin and the zinc center. For prisms, the energy transfer rate depends on the distance.

Wasielewski et al. prepared cyclic tetramer of Zn chlorophylls based on noncovalent bonding employing coordination of the pyridine ring at the *meso*-position to the zinc center.[30,31] The energy hopping rate was determined to be ~1 ps.

Energy Migration in Self-Assembled Nanoclusters. Self-assembling due to π–π interaction between chromophores is also useful to prepare chromophore arrays for light-energy harvesting. Wasielewski and coworkers reported various types of chromophore arrays based on π–π interaction between chromophores.[32–37] In particular, perylene-3,4:9,10-bis(dicarboximide) (PDI) was employed as a building block of such a self-assembly in their study. Its functionality as a LH has been examined by confirmation of single energy migration among PDI chromophores.

Energy Migration in Chromophore Arrays Constructed Using Biological Molecules as a Scaffold. Biological materials can also act as a scaffold for chromophore arrays. For example, the tobacco mosaic virus (TMV) has a hollow cylindrical structure with a high aspect ratio and is formed through the periodical self-assembly of the TMV coat protein (TMVCP) and RNA. Because of its well-characterized nanoscale structure, TMV can be utilized as a template for functionalization with inorganic and organic molecules both in the inner cavity and on the exterior surface. We have introduced a pyrene derivative to four specific amino acids in the cavity of the TMV rod structure (Fig. 12.1).[38] Rod-structure formation was examined by atomic force microscopy. Two pyrene-attached mutant (positions 99 and 100) assemblies showed a substantial increase in the length of the rod structures due to π–π interaction of pyrene. Furthermore, strong interaction between pyrenes in the cavity was confirmed by fluorescence spectroscopy via observation of strong excimer emission (Fig. 12.2).

In addition, TMV monomers conjugated with ZnP or free-base porphyrin (FbP) were synthesized and assembled.[39] In the ZnP/FbP mixed TMV assemblies, fluorescence due to ZnP was decreased, while that of FbP increased compared with the corresponding mixture of monomers, indicating the energy transfer from ZnP to FbP in the assembly. The quenching yield of ZnP emission gradually increased by increasing the content of FbP. Rod assemblies at pH 5.5 showed larger quenching yields than those in the disk assemblies at pH 7.0, indicating efficient light-energy harvesting functionality in the rod like assemblies. A similar light-harvesting functionality of the TMV

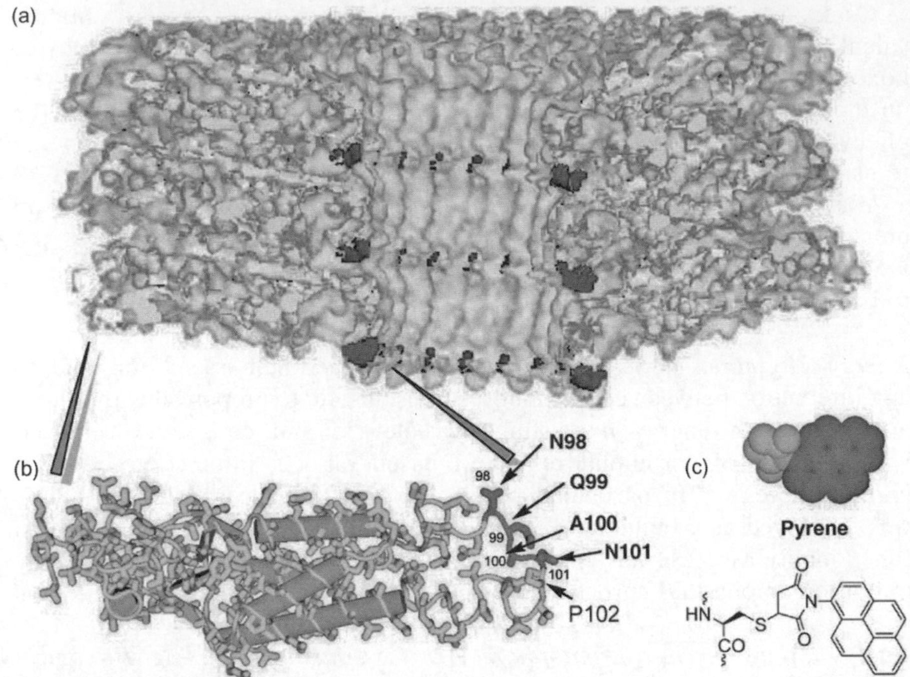

Figure 12.1. (a) Crystal structure of the TMV assembly (sectional view). Q99 amino acid is represented in dark gray. (b) TMVCP monomer and four amino acids to which pyrene derivatives were introduced. (c) Molecular structure of pyrene derivative.

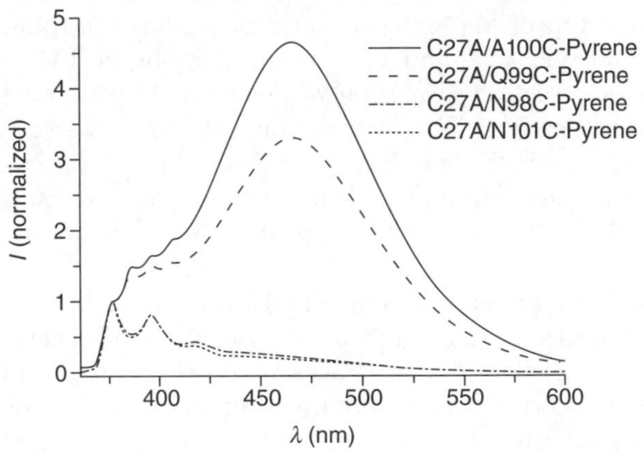

Figure 12.2. Fluorescence spectra of pyrene-attached TMVCP at pH 5.5.

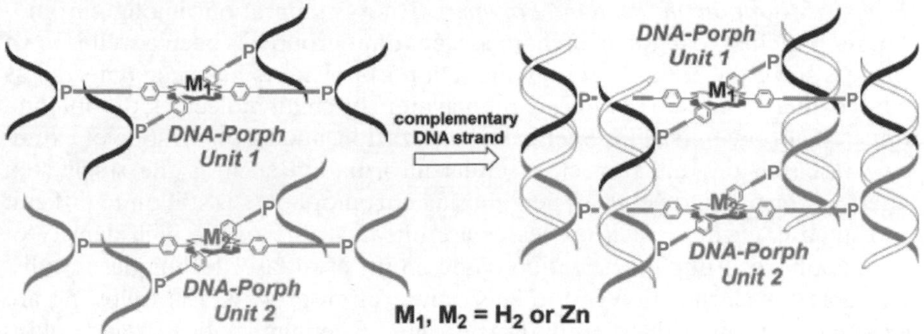

Figure 12.3. DNA–porphyrin conjugates constructed using DNA as a scaffold.

functionalized by various kinds of chromophores has been reported by Francis and coworkers.[40]

The helical chain of DNA can be also used as a scaffold of chromophore arrays. We have assembled four double helices using two DNA–porphyrin conjugates and their complementary strands (Fig. 12.3).[41] When ZnP and FbP were introduced, the energy transfer from ZnP to FbP was confirmed. This is an example of the formation and functionality of chromophore arrays constructed using DNA as a scaffold.[42–44]

These examples indicate that by using self-assembly based on intermolecular forces such as coordination, π–π interaction, and hydrogen bonding in biological materials, chromophore arrays can be constructed, and energy migration realized in such self-assembled chromophores is sufficiently fast when strong interactions between the chromophores are realized in the systems. Compared with the previous covalent-bonding systems, rather large chromophore arrays have been achieved. To remove undesired functionality due to inhomogeneity of a structure, a precise molecular design will be needed. For this reason, chromophore arrays constructed using biological materials as the scaffold may be a good strategy.

Chromophore Arrays by Polymer Systems

As indicated in the previous sections, various methods have been employed to fabricate chromophore arrays. For the construction of a relatively large nano-architecture with arrays of chromophores, polymerization is useful. Recently developed polymerization methods realize precise control of the polymerization degree and sequence of copolymers. For light-energy harvesting functionality, the following two kinds of polymers will be good candidates: (1) polymers obtained by polymerization of monomers possessing a chromophore as a pendant group and (2) polymers including largely expanded conjugated systems. In the following, investigations revealing the detailed energy migration process in such polymeric systems are reviewed.

Energy Migration in Pendant Polymers. Energy migration among chromophores attached to a polymer chain as a pendant group has been a well-known process since the 1960s.[45] Energy migration in polymers has been detected as a large quenching rate of fluorescent polymers by small molecules, the antenna effect of copolymers, fluorescence depolarization, and so on. In spite of extensive research on this subject, essential information such as the single-step energy transfer rate between neighboring chromophores is still quite difficult to estimate, because such processes are ultrafast due to the high density of chromophores. Thus, investigation based on ultrafast laser techniques is indispensable for clarification of the energy migration process. The following are examples revealing the energy transfer rate in pendant polymers and conjugated backbones by using ultrafast laser techniques.

Polypeptides can be regarded as polymers of which the polymerization degree and sequence can be controlled in detail by well-established synthetic procedures. If one modifies the constituting amino acid to have a chromophore useful to light-energy absorption, chromophore arrays for light-energy harvesting can be constructed. Figure 12.4 shows one example of such peptides. In the polypeptides, L-lysine is connected to tetraphenylporphyrin.[46,47] The energy-minimized molecular structure of octamer of porphyrin bearing L-lysine shows a helical backbone due to poly-L-lysine with porphyrins located around the backbone in a wheel-like configuration. The existence of the helical backbone was confirmed by induced circular dichroism in the Soret band region. The distance between the porphyrin units in these porphyrin peptides is longer even in the octamer; in the octamer center-to-center distance between porphyrin units ranges from 9.8 to 20.6 Å and the average distance is 14.9 Å, from which only a few hundreds cm^{-1} of exciton coupling are expected, in accordance with the observed bandwidth. This finding indicates that the interaction between porphyrin units is rather small. The distance between porphyrin units tends to increase for the shorter polypeptides.

The photoinduced processes in the porphyrin polypeptides were investigated using picosecond laser flash photolysis. The transient absorption spectrum of octamer at 20 ps after the laser flash (532 nm, full width at half maxima [FWHM] 30 ps) exhibited the characteristic features of porphyrin units in the singlet-excited state (Fig. 12.5). The singlet excited state of the octamer decayed according to a two-step decay in the picosecond region (Fig. 12.5a, inset). The slow decaying component can be attributed to the decay of the singlet-excited state ($\tau_f = 11.5$ ns). On the other hand, the decay rate constant of the fast decaying part was $(1.1 \pm 0.1) \times 10^{10}\,s^{-1}$. It is noteworthy that the fast decaying component almost vanished when the excitation laser fluence was low (Fig. 12.5a, inset). From these findings, the fast decaying component can be attributed to the deactivation of the singlet-excited state by the intramolecular singlet–singlet annihilation process, resulting from the singlet energy migration among the porphyrin units.[1] This assignment is supported by the fact that the fast decaying component was not observed in the kinetic trace of ΔOD at 445 nm of monomer (Fig. 12.5b). It is interesting that tetramer also

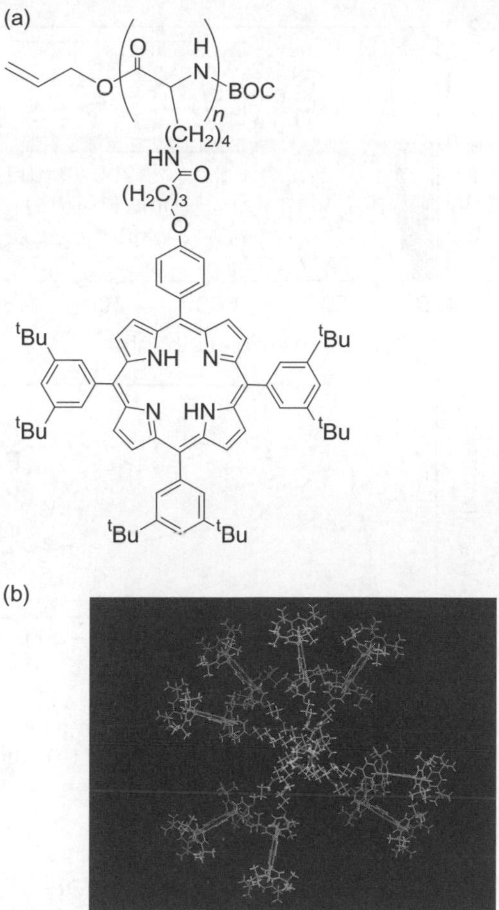

Figure 12.4. (a) Molecular structures of porphyrin peptides and (b) top view of energy minimized structure. *n*: 1, 2, 4, 8; ᵗBu, *tert*-butyl; BOC, *tert*-butoxycarbonyl.

showed the two-step decay due to the singlet–singlet annihilation (Fig. 12.5b), although the rate ($[7.3 \pm 0.4] \times 10^9\,s^{-1}$) was slower than that of the octamer. This finding indicates that the exciton–exciton annihilation becomes less efficient when the distances between the porphyrin units become large. Furthermore, the rate-determining step of the present deactivation can be attributed to the exciton migration, because the rate depends on the distance between the chromophores.

We found that the living polymerization of aryl isocyanides initiated by the Pd-Pt μ-ethynediyl complex generates polyisocyanides with well-defined polymer structure.[48] These polyisocyanides exhibit a stable 4_1 helical structure when bulky substitutes are introduced. Actually, by employing porphyrin as a pendant group, the helical porphyrin polymers have been synthesized

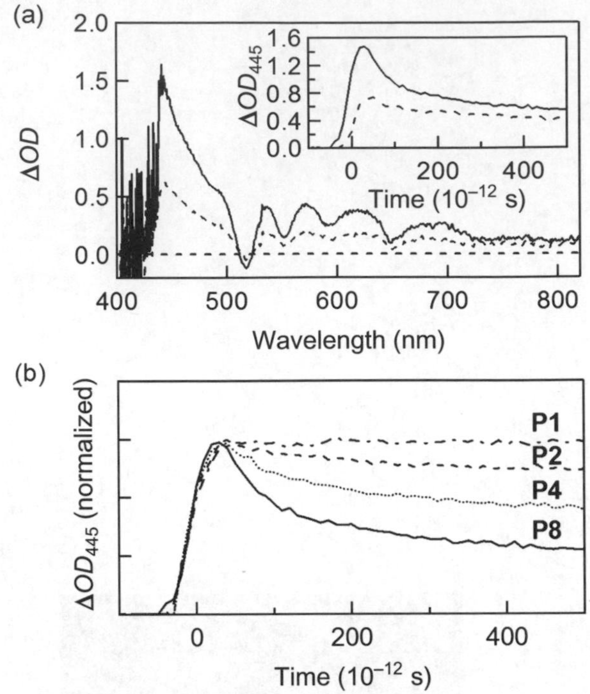

Figure 12.5. (a) Transient absorption spectra of porphyrin peptides obtained during laser flash photolysis using 532-nm picosecond laser pulse. (b) Kinetic traces of ΔOD at 445 nm during laser flash photolysis.

(Fig. 12.6). The B (Soret) band of the polymer showed splitting and substantial broadening with increasing the number of the repeating unit, indicating that the porphyrin units in the polymer take a face-to-face configuration.

Figure 12.7 shows transient absorption spectra of **HP200** in tetrahydrofuran (THF) during laser flash photolysis employing the picosecond laser (532 nm, FWHM 30 ps) as an excitation source. Immediately after laser excitation, the generation of the singlet-excited state was confirmed. The kinetic trace of ΔOD at 470 nm showed two-step decay due to singlet–singlet annihilation (Fig. 12.7, inset). The decay rate constant of the fast decaying part was $(3.9 \pm 0.4) \times 10^{10}\,s^{-1}$, almost the same as the present instrumental limit, indicating that singlet energy migration in the present polymers is quite fast due to closely stacked porphyrins in a face-to-face configuration.

The present living polymerization method allows synthesizing a block copolymer. In the block copolymer with a sequence of ZP_m-HP_n (m and n denote the number of repeating units), energy transfer from ZnP to FbP is expected. Time-resolved fluorescence and transient absorption spectroscopy confirmed the energy transfer. The energy transfer rates were almost independent of the polymer sequence, indicating that the singlet energy migration process is too

HPn: M = 2H, *n* = 1, 2, 20, 50, 100, 200
ZPn: M = Zn, *n* = 1, 2, 20, 50, 100

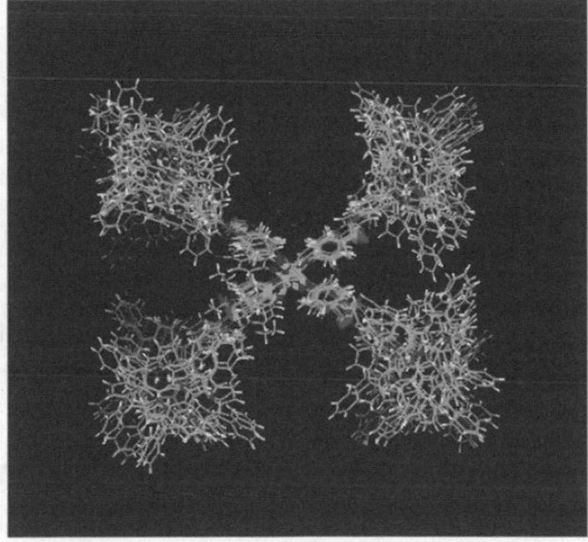

Figure 12.6. Molecular structure (top) and optimized structure of porphyrin polymer (bottom).

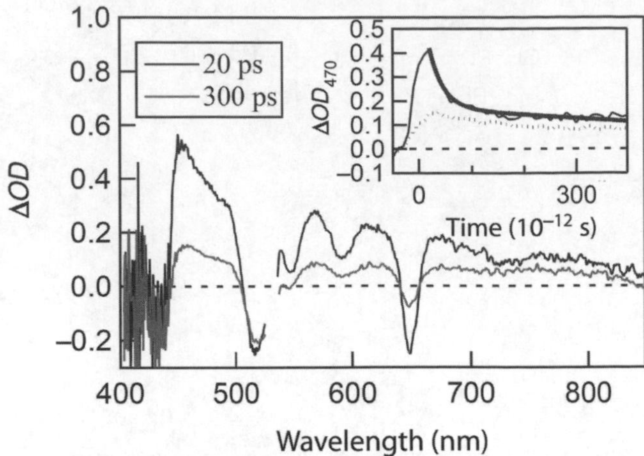

Figure 12.7. Transient absorption spectra observed at 20 and 300 ps after laser flash (532 nm, FWHM 30 ps) during laser flash photolysis of **HP200** in THF (1×10^{-4} M). Inset: kinetic trace of ΔOD at 470 nm (laser fluence: solid line, 20 mJ pulse^{-1}; dotted line, 5 mJ pulse^{-1}).

fast to affect the energy transfer rate at the interface between the ZnP and FbP units; that is, the energy transfer is the rate-determining step in the present block copolymers.

Energy Migration in Conjugated Polymers. Another good candidate for artificial light harvesting systems is a conjugated polymer because of its large absorption in UV-Vis region. Energy migration in conjugated polymers has also been reported for years. For example, time-resolved fluorescence spectra after the laser excitation clearly showed a longer wavelength shift due to energy relaxation, including energy migration from the segment with higher energy due to shorter conjugation length to the segment with lower energy with longer conjugation length.[49] But the obtained time constant for the spectral shift is not always equivalent to the single-step energy transfer from a segment to a neighboring segment. Furthermore, this spectral shift can also be explained on the basis of the molecular structural change, such as planarization. Thus, analysis of the energy transfer in a conjugated polymer has to be carried out using ultrafast laser techniques. The details of energy migration in a conjugated polymer became clearer in this decade.[50–53] Westenhoff et al. reported the energy migration in polythiophene in solution by means of ultrafast spectroscopy and Monte Carlo simulation. From the study, they revealed the effective conjugation length and disorder.[51]

We investigated the singlet energy migration along the polymer chain of block copolymer of oligothiophene and oligosilylene (DS*n*Ts, where *n* denotes the number of thiophene units; Fig. 12.8).[54] A well-defined conjugation length is one of the characteristics of the present block copolymers. The steady-state

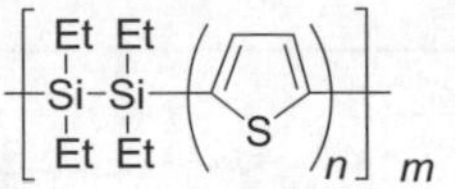

DS*n*T (n = 3, 4, 5)

Figure 12.8. Molecular structure of block copolymer of oligothiophene and oligosilylene.

absorption peak tends to shift to the longer wavelength side with an increasing number of thiophene units in the repeating unit, indicating the longer conjugation length of the longer oligothiophene. Furthermore, the peak position of the absorption band was on the longer wavelength side by 15–18 nm when compared with the corresponding oligothiophenes (nT, where n denotes the number of thiophene units). The peak shift from the corresponding oligothiophene can be attributed to the σ–π conjugation with the adjacent oligosilylene. A similar lower energy shift was observed for the fluorescence, S_n-S_1, and T_n-T_1 absorption bands.

To reveal the energy migration process along the polymer chain, the anisotropy of the fluorescence was estimated. By employing the fluorescence up-conversion method to the DSnTs in toluene, the parallel and perpendicular components of the fluorescence with respect to the pump pulse were obtained, as indicated in Figure 12.9. Applying the single exponential function to the anisotropy decay (r), the time constants for the anisotropy decay were estimated to be $(1.6\,\text{ps})^{-1}$, $(2.7\,\text{ps})^{-1}$, and $(4.5\,\text{ps})^{-1}$ for DS3T, DS4T, and DS5T, respectively. Since the pentamer of thiophene did not show any anisotropy decay in this time region, the anisotropy decay observed for DSnTs can be attributed to the singlet energy migration along the polymer chain. By using the Förster theory,[55] the time constants were calculated to be $(5.7\,\text{ps})^{-1}$, $(9.4\,\text{ps})^{-1}$, and $(19.2\,\text{ps})^{-1}$ for DS3T, DS4T, and DS5T, respectively. The calculated values are on the same order as the estimated ones. Furthermore, the ratio of the experimentally estimated rate constants (1.0:1.7:2.8) is almost the same as the calculated one (1.0:1.6:3.4). These results support the conclusion that energy migration along the present polymer is governed by the Förster theory. The effective σ–π conjugation in the present polymers is one of the reasons for the discrepancy between the theory and results. The enhanced conjugation makes the distance between the energy hopping slightly smaller, which enhances the hopping rate as observed in the present polymer systems.

Energy Migration in Branched Molecules. As model compounds of the block copolymers of oligothiophene and oligosilylene, the energy migration process in Si-linked oligothiophenes was also investigated (Fig. 12.10).[56] From the molecular orbital calculation, it was indicated that Si-linked oligothiophenes tend to take a configuration that reduces repulsion between oligothiophene entities and methyl groups on the silicon atom. From the anisotropy decay, excitation energy migration among the oligothiophenes was confirmed. The

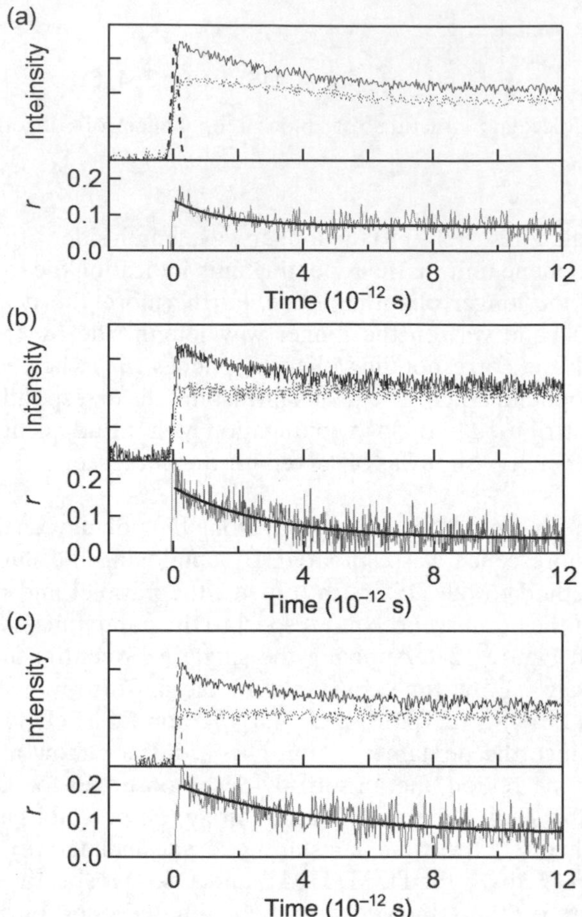

Figure 12.9. Fluorescence (upper panel) and anisotropy (*r*, lower panel) profiles of DS3T (a), DS4T (b), and DS5T (c) in toluene. Solid and dotted lines are parallel and perpendicular components of fluorescence with respect to the excitation pulse (430 or 390 nm), respectively. Broken lines in upper panels are the response function. Thick lines in the lower panels are fitted curves.

time constants for the anisotropy decay were estimated to be $(4.4\,\text{ps})^{-1}$, $(5.2\,\text{ps})^{-1}$, $(8.0\,\text{ps})^{-1}$, and $(9.4\,\text{ps})^{-1}$ for $3T_2$, $3T_3$, $5T_2$, and $5T_3$, respectively.

The incoherent energy hopping mechanism is usually employed to explain the energy migration process in various supramolecular systems such as porphyrin arrays. To obtain a better understanding, the Förster theory has been employed. Although the calculated values are 22–30 times smaller than the observed values, the calculated energy transfer rate constants are ordered as $3T_2 > 3T_3 > 5T_2 > 5T_3$, which is in the same order as the observed rate con-

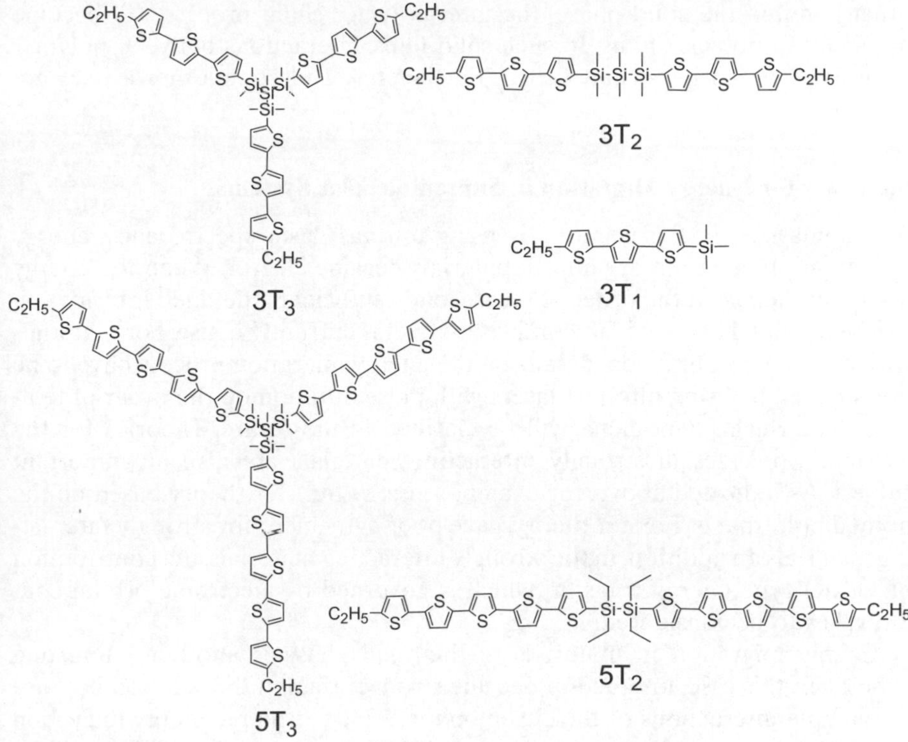

Figure 12.10. Molecular structures of Si-linked oligothiophenes (nT$_m$).

stants. Furthermore, the relative rate constants are similar to the observed ones. Therefore, the present energy migration process can be attributed to the incoherent hopping mechanism. The effect of σ–π conjugation on the present energy transfer was indicated.

Goodson and coworkers reported that the energy migration rate in the N-centered dendrimer is faster than those of the C- and P-centered dendrimers.[57–61] They attributed the fast energy migration rate to the coherent mechanism because of a strong interaction in the N-centered dendrimer, in which chromophores are connected by only one nitrogen atom. In the Si-linked oligothiophenes in the present study, the number of silicon atoms connecting oligothiophenes is three. The longer linker seems to diminish the interaction to yield incoherent hopping.

Here, we summarized energy migration processes in two kinds of polymers: (1) polymers obtained by polymerization of monomers possessing a chromophore as a pendant group and (2) polymers that include largely expanded conjugated systems. In both systems, fast energy migration processes were confirmed. At present, energy migration processes have been investigated for polymers in the solution phase, that is, the intra-polymer chain process. On the other hand, in the solid phase, the inter-polymer chain process will become dominant in polymer films. In such solid films, interactions between polymer chains will be an important factor. Further fast and complex processes are expected.

Summary for Energy Migration in Supramolecular Systems

As summarized in this section, by using ultrafast laser spectroscopy, energy migration processes in organized materials became clearer. When the energy migration rate is on the order of picoseconds, sufficiently detailed information can be obtained by subpicosecond laser systems currently in use. For materials with a stronger interaction, details on the energy migration process have to be investigated by using ultrafast lasers with pulse duration on the order of femtoseconds. Such phenomena will be clarified in the future. Theories for the dynamic processes in strongly interacting materials are also an important subject. As indicated above, for example, energy transfer theory based on the point dipole, that is, Förster theory, gave poor agreement for strongly interacting materials. In addition, in the strongly interacting materials, the contribution of Dexter-type energy transfer, which is governed by electronic orbital coupling, has to be considered.

Energy migration in materials in the solid phase is another interesting process. In this case, interaction became stronger than in the solution because of multiple interactions of the chromophores. Furthermore, energy migration to various directions has to be considered. Clarification of these issues will be useful to a wide range of researchers in the fields of both materials and basic science.

CHARGE DELOCALIZATION AND MIGRATION IN WELL-DEFINED MOLECULAR SYSTEMS

Conduction of charge in organic materials is one of the central issues in organic functional materials.[1] Based on information on the charge conduction in organic materials, adequate design of molecular electric devices will be possible. Conduction of carriers in organic materials, in which organic compounds are stacked, has been investigated extensively from the 1950s. Since interaction between organic molecules in organic crystals is weak, carrier mobility in these materials is rather small compared with metal and semiconductors. Thus, conduction in organic crystals has been treated on the basis of hopping and/or tunneling processes rather than the band conduction model used for semiconductors. In such systems, charge carriers have characteristics similar to charged state of constituting molecules, such as radical ions. Nevertheless, these charged species in organic materials are different from those in solution. For example, charge carriers in organic crystals are stabilized by various effects such as delocalization and polarization. In addition, it is expected that charge carriers induced from electrodes or generated by high-energy irradiation may be in a "hot" state; investigation of these states in a qualitative manner in crystal is quite difficult. Furthermore, the most basic and important parameter of charge conduction should be the single-step charge hopping or tunneling rate, which is also quite difficult to determine experimentally. To assess these issues, investigations on adequate model systems are required. In this section, we summarize the following topics based on our recent research results revealing delocalization and migration of charge in well-defined molecular systems:

1. charge delocalization in well-defined molecular systems,
2. excited-state behavior of dimer radical cation,
3. charge migration in DNA.

Charge Delocalization in Well-Defined Molecular Systems

As indicated above, charge carriers in organic crystals are stabilized by various effects such as delocalization and polarization. To evaluate these stabilization effects, the dimer model is useful, because dimer radical cations, that is, a complex of radical cation and a neutral molecule, can be regarded as the smallest units for the delocalized positive charge. Dimer radical cations are known to show characteristic absorption bands in the near-infrared (IR) region, which was found by Badger and Brocklehurst in the 1960s.[62,63] The absorption band is due to the transition between the states expressed as $\Phi_{\pm} = (1/\sqrt{2})(\varphi(M_A^{+})\varphi(M_B) \pm \varphi(M_A)\varphi(M_B^{+}))$, where $\varphi(M_A)$ and $\varphi(M_B)$ are the wave functions of neutral molecules in the dimer radical cation and the superscript + indicates the radical cation. Stabilization energy is equal to half the energy of the observed transition.

It is well known that stabilization of the dimer radical cation is affected by the conformation of its molecules, such as distance and orientation. However, structural analysis of the dimer radical cation by X-ray crystallography is rather limited.[64,65] Furthermore, in solution, it is quite difficult to determine the distance between the chromophores participating in intermolecular dimer radical cation formation. Although various dimer model compounds, in which two identical chromophores are linked by alkyl chain, have been investigated, the explanation of the relation between the structural parameters and observed stabilizing energy is still rather qualitative.[66–73] Therefore, a spectroscopic study of the dimer radical cations of molecules, in which two chromophores are connected by rigid bonds, will be beneficial to determine the factors that govern the properties of dimer radical cations. In addition, reports on the spectroscopic characterization of delocalization of negative charge are quite limited.[74,75]

We have selected cyclophanes, in which two benzene rings are connected by plural alkyl chains to hold benzene rings at a fixed distance and configuration. Theoretical investigation found that the benzene rings of cyclophanes

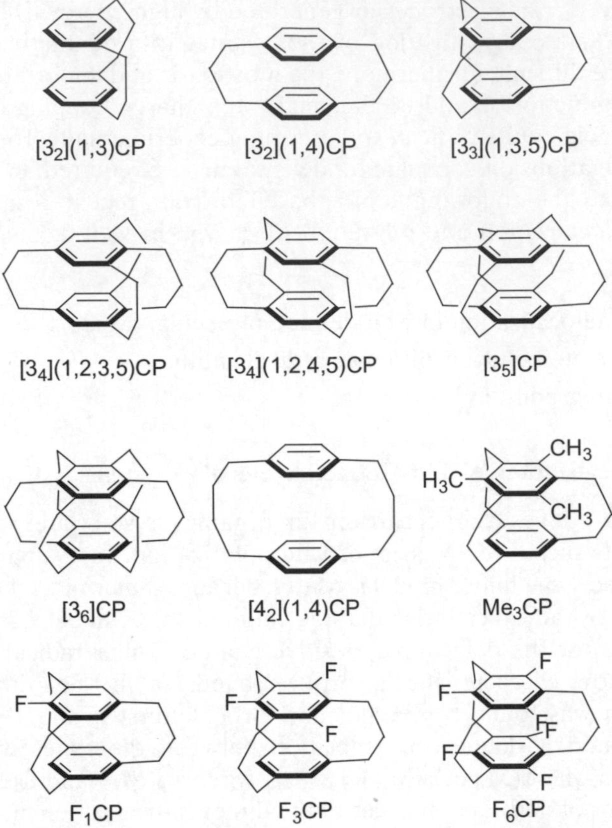

Figure 12.11. Molecular structures of cyclophanes.

have a completely overlapped configuration in neutral, oxidized, and reduced forms. In addition, we have employed pulse radiolysis and γ-ray irradiation to generate radical cations and radical anions of these species, because oxidants and reductants generated by these methods are strong enough to generate oxidized and reduced materials.

Positive Charge Delocalization in Cyclophanes. Figure 12.11 shows a series of [3_n]CPs investigated in the present study.[76–79] Figure 12.12 shows the transient absorption spectra of [3_3](1,3,5)CP in 1,2-dichloroethane during pulse radiolysis.[80,81] The spectrum of [3_3](1,3,5)CP at 50 ns after the pulse irradiation exhibited sharp and broad peaks at 510 and 730 nm, respectively. The absorption band at 510 nm was assigned to the local excitation (LE) band of radical cation of [3_3](1,3,5)CP. On the other hand, the absorption band of [3_3](1,3,5)CP at 730 nm can be assigned to the charge resonance (CR) band. Transient absorption spectra of dimer radical cation of cyclophanes listed in Figure 12.11 were confirmed similarly.

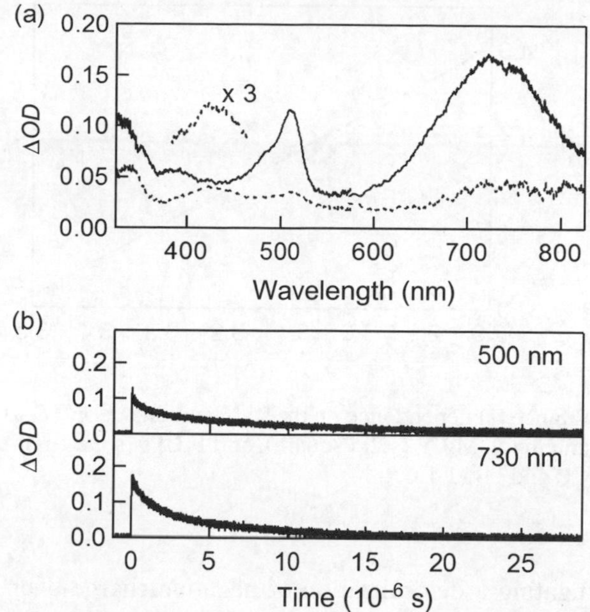

Figure 12.12. (a) Transient absorption spectra of $[3_3](1,3,5)$CP (1.0×10^{-2} M) in 1,2-dichloroethane at 50 ns (solid line) and 5 μs (broken line) after 8-ns pulse irradiation during pulse radiolysis. (b) Kinetic traces of ΔOD at 500 and 730 nm.

The peak position of the CR band tends to shift to the longer wavelength with an increase in distance between the two benzene rings. The interaction between two benzene rings of $[3_n]$CPs can be analyzed in terms of the exchange interaction. The interaction energy, that is, stabilization energy (E_{CR}), can be estimated from the relation $2E_{CR} = \nu_{CR}$, where ν_{CR} is the peak position of dimer radical cation. On the other hand, the E_{CR} value is a function of the distance between two chromophores (r) as $E_{CR} \propto \exp(-\beta r)$, where β is a constant. Thus, the stabilization energy estimated from the CR band was plotted against the transannular distance estimated by the density functional theory (Fig. 12.13). The linear relation in Figure 12.13 indicates that the exponential dependence of the stabilization energy on the transannular distance as expected for the exchange interaction. The β value was estimated to be 0.83 $Å^{-1}$. It was revealed that the electron donating or withdrawing nature of the substituents does not have a large effect on the stabilization energy.

Negative Charge Delocalization in Cyclophanes. In the 1960s, Ishitani and Nagakura reported the absorption spectrum of the radical anion of $[2_2]$paracyclophane at low temperatures, in which $[2_2]$paracyclophane showed a peak due to dimer radical anion at 760 nm.[74] Because of the absence of other absorption-spectrum measurements of the dimer radical anion, detailed

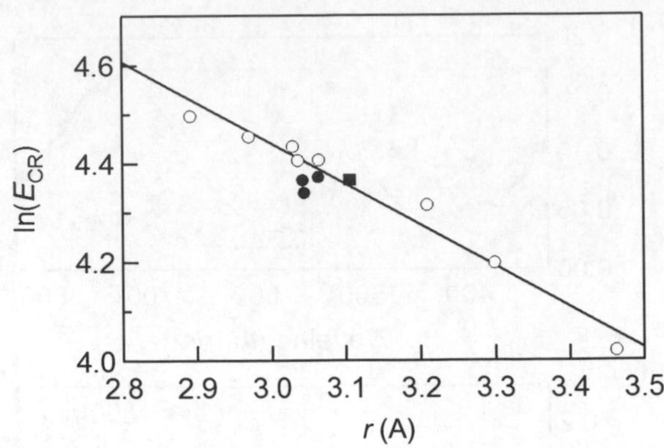

Figure 12.13. Distance (r) dependence on the stabilization energy (E_{CR}) for $[3_n]$CP and $[4_2](1,4)$CP (open circle), M_3CP (solid square), and F_nCPs (solid circle). The solid line is fitted for $[3_n]$CP and $[4_2](1,4)$CP.

analysis investigating a delocalization of negative charge over two chromophores on the basis of the molecular structures, such as transannular distance dependence, has not been reported in spite of its importance.

We investigated the intramolecular dimer radical anion of $[3_n]$CPs (Fig. 12.11) by measuring the absorption spectra after γ-ray irradiation.[82] It is well established that γ-ray irradiation to substrate in 2-methyltetrahydrofuran (MTHF) glassy matrix generates a radical anion of the substrate. The absorption spectrum of the radical anion of $[3_3](1,3,5)$CP was obtained, as shown in Figure 12.14a (the spectrum indicated by **3**). The absorption band at 1010 nm can be attributed to the intramolecular dimer radical anion of $[3_3](1,3,5)$CP, because the radical anion of benzene and trimethylbenzene did not show the absorption band in this region. Another absorption peak at 310 nm is the LE. The absorption band attributable to the dimer radical anion of cyclophanes was also observed with a series of $[3_n]$cyclophanes, as shown in Figure 12.14a, which also includes the absorption spectrum of the dimer radical anion of $[2_2]$ (1,4)CP generated by γ-ray radiolysis in MTHF glassy matrix. Furthermore, it became clear from theoretical calculations that the distance between two benzene rings became shorter with an increase in the number of propyl linkers. Thus, the peak position of dimer radical anion shifted to the shorter wavelength side with a decrease in the transannular distance.

The interaction can be analyzed in terms of the exchange interaction as explained in the section on dimer radical cation of cyclophanes, that is, $E_{CR} \propto \exp(-\beta r)$. In Figure 12.14b, the E_{CR} values of cyclophanes in this study were plotted against r estimated from the theoretical calculation. A linear relation between $\ln(E_{CR})$ and r was clearly confirmed, supporting the electronic delocalization over two benzene rings of $[3_n]$CPs. From the slope of Figure 12.14b, the β value was estimated to be 0.35 Å^{-1}. It is evident that the

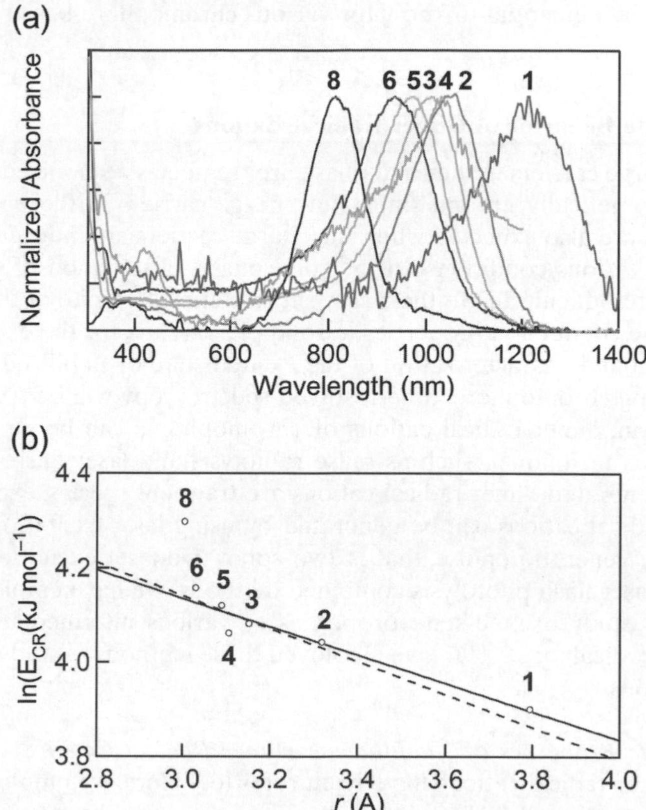

Figure 12.14. (a) Absorption spectra of cyclophanes in MTHF glassy matrix at 77 K after γ-ray radiolysis. (b) Distance (r) dependence of the stabilization energy (E_{CR}). The solid line was obtained using $[3_n]$cyclophane data. The dashed line was obtained by excluding the data of $[3_2](1,3)$CP (**1**: $[3_2](1,3)$CP; **2**: $[3_2](1,4)$CP; **3**: $[3_3](1,3,5)$CP; **4**: $[3_4](1,2,4,5)$CP; **5**: $[3_4](1,2,3,5)$CP; **6**: $[3_5]$CP; **8**: $[2_2](1,4)$CP).

stabilization energy of $[2_2](1,4)$CP is not on the lines of Figure 12.14b, indicating that other factors have to be considered for the stabilization of the dimer radical anion $[2_2](1,4)$CP. Possible factors are the absence of complete face-to-face structure and the large distortion of benzene rings. The estimated β value is smaller than that estimated for the intramolecular dimer radical cation of cyclophanes (0.83 Å^{-1}).[80,81] The present result indicates that the negative charge delocalizes in stacked chromophores in a manner similar to the positive charge, indicating that the negative charge is a potentially good carrier in organic assemblies, as is the positive charge.

These studies are the first examples that clarify the basic parameters for the delocalization of positive and negative charges in stacked chromophores. Although applicability of these parameters to other chromophores is unknown, these values are essential to know for various chromophore-based functional materials.

Excited-State Behavior of Dimer Radical Cations

Usually, charge carriers in the solid phase are treated as a stabilized form, that is, in an energetically ground state. But charge carriers in the energetically "hot" states are also expected when the charge carriers are induced or generated under various conditions. In the solid phase, observation of such "hot" states is quite difficult due to their very short lifetimes in spite of their contribution to the conduction properties in solid phase. Thus, for the experimental confirmation and characterization of the excited state of stabilized charge by a neighboring chromophore, time-resolved spectroscopy will be fruitful.

In solution, dimer radical cations of chromophores can be generated by using various techniques, such as pulse radiolysis and laser flash photolysis. Since the generated dimer radical cations are transient species, excited states of dimer radical cations can be generated by using laser excitation synchronized to the generation pulse, that is, two-color–two-laser excitation or pulse radiolysis–laser flash photolysis combined methods. We have employed these methods to study excited-state properties of various intermediates such as triplet and radical ions.[83] Here we employed these methods to study the dimer radical cations.

Excited-State Properties of Naphthalene Dimer Radical Cations. Up to now, various dimer radical cations have been reported. Here, we employed dimer radical cations of naphthalene ($Np_2^{\bullet+}$) to assess the excited-state properties of dimer radical cations, because the CR bands of the dimer radical cations appeared around 1000 nm, which can be easily excited by a fundamental pulse of an Nd:YAG laser.[84]

Dimer radical cations were generated by photoinduced electron transfer or pulse radiolysis. When $Np_2^{\bullet+}$ was excited at the CR band using the 1064-nm laser, bleaching and recovery of the transient absorption at 570 and 1000 nm, assigned to the LE and CR bands of $Np_2^{\bullet+}$, respectively, were observed together with the growth and decay of the transient absorption at 685 nm, assigned to $Np^{\bullet+}$ (Fig. 12.15). The immediate decay of the CR band and rise of the LE bands indicate that photodissociation occurred from the excited state of dimer radical cations. From the excitation laser intensity dependence, it was revealed that the dissociation of $Np_2^{\bullet+}$ proceeds via a one-photon process within the 5-ns laser flash to give $Np^{\bullet+}$ and Np in the quantum yield of 3.2×10^{-3} and in the chemical yield of 100%. The recovery time profiles of $Np_2^{\bullet+}$ at 570 and 1000 nm were equivalent to the decay time profile of $Np^{\bullet+}$ at 685 nm, suggesting that the association of $Np^{\bullet+}$ and Np occurs to regenerate $Np_2^{\bullet+}$ in 100% yield. Similar experimental results of the photodissociation and regeneration of $Np_2^{\bullet+}$ were observed during the pulse radiolysis–laser flash photolysis of Np in 1,2-dichloroethane.

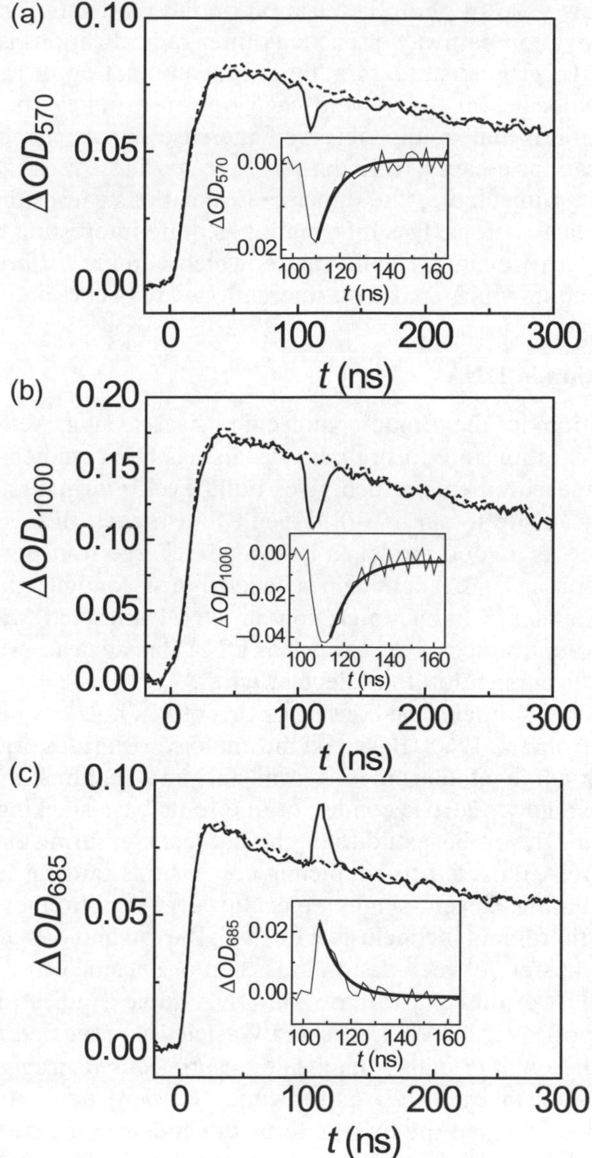

Figure 12.15. Time profiles of the transient absorption at 570 nm (a), 1000 nm (b), and 685 nm (c), assigned to $Np_2^{\bullet+}$, $Np_2^{\bullet+}$, and $Np^{\bullet+}$, respectively, during irradiation of one laser (dotted line) and two lasers with the 1064-nm laser power of 120 mJ flash^{-1} (solid line).

The quite low yield of photodissociation on laser excitation indicates that the main deactivation pathway of excited dimer radical cations is the internal conversion to the ground state. In addition, recombination of radical cations and neutral molecules in the solvent cage occurred quite efficiently within the pulse duration; that is, the observed increase in LE can be attributed to the radical cations escaped from the solvent cage. These results indicate that the excess energy injected to the dimer radical cations causes dissociation to neutral and radical cations. This information is quite interesting when considering the "hot" carrier in the solid phase, which is quite different from the ground state one in which attractive interactions are occurring.

Charge Migration in DNA

Charge migration in the organic molecules has been a well-investigated process. The migration rate in organic crystals has been studied by means of time-of-flight measurements, which gives bulk-level information. The charge transfer rate from site to site is still difficult to estimate. In this sense, DNA can be regarded as a good model to investigate charge transport, because in DNA, chromophores are aligned in a face-to-face configuration in a one-dimensional manner. Charge migration in DNA attracted wide attention because of the importance of topics such as DNA damage caused by oxidative materials and its application to molecular wires.

Whether DNA is an electric conductor or not has been an interesting scientific subject from the 1960s. Based on the analogy to various organic conductors, in which π-conjugated aromatics form stacked structures, DNA has been expected to be a good electric conductor due to its base-stacking structure in the duplex form. The earliest studies on this subject measuring electric current flow in bulk DNA films led to conflicting conclusions ranging from the conductive to insulating. Recent single molecular-level measurements of current flow in DNA still remain inconclusive. In 1993, Barton and coworkers showed that electron transfer between donor and acceptor became possible by attaching them to the 5′-termini of a 15-base pair DNA duplex, indicating long-range electron transport over 40 Å.[85] Lewis and Wasielewski revealed that the photoinduced oxidation of primary guanine donor proceeds via a single-step super exchange mechanism by using femtosecond transient absorption spectroscopy.[86] They also revealed the driving force dependence of oxidation rates of the nucleobase.[87] According to the oxidation potentials, G and A have lower oxidation potentials. Thus, charge hopping along G or A should be important in hole migration in DNA.

Charge Hopping on G. Giese investigated hole transfer from G^+ to GGG separated by A:T base pairs.[88] By changing the number of A:T pairs, they confirmed hole transfer through A:T base pairs by a single-step superexchange mechanism. The calculated β value in $k \propto \exp(-\beta r)$ was 0.7 Å^{-1}, which was similar to the reported value. Since the distance dependence of hole transfer was expressed by an exponential function, hole transfer was decelerated with the number of A:T pairs separating G^+ and GGG. At distances longer than 17 Å, hole transfer was not confirmed. But in DNA strands where G's are located between G^+ and a GGG sequence, long-distance charge transfer occurred by a multistep hopping mechanism. They have confirmed hole transfer over 54 Å. From kinetic analysis, Bixon et al. showed that hole transfer over 300 Å is also detectable.[89]

The measurements of hole transfer rate between G^+ and GG separated by a single A:T base pair were reported by Lewis et al. to be 5×10^7 and $4 \times 10^5\ s^{-1}$ for GAGG and GTGG sequences, respectively.[90]

Charge Hopping on A. Giese found that the hole transfer mechanism from G^+ to GGG separated by A:T base pairs largely depends on the number of A:T base pairs (n): When n is 1–3, hole transfer yield substantially decreased with an increase in n, while for $n > 4$, hole transfer yield decreased only very slightly. The observed distance dependence was explained on the basis of coherent super-exchange mechanism for the shorter distance and hopping mechanism for the longer distance.[91]

We investigated the charge separation process via the consecutive A-hopping mechanism using laser flash photolysis of DNA conjugated with naphthaldiimide (NDI) as an electron acceptor and phenothiazine (PTZ) as a donor (Fig. 12.16).[92] On laser excitation of NDI, charge separation and recombination processes between NDI and PTZ were observed. The yield of the charge separation via the consecutive A-hopping was slightly dependent on the number

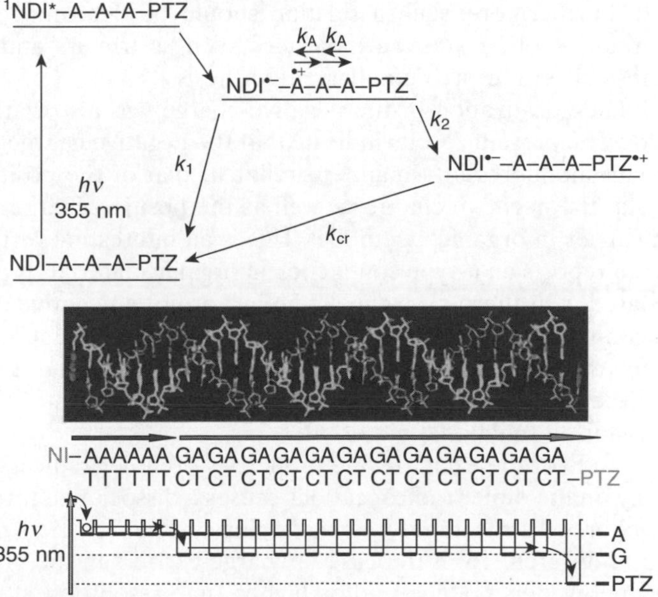

Figure 12.16. Kinetic scheme for charge separation and recombination in NDI- and PTZ-conjugated DNA.

of A bases between two chromophores, while charge recombination rate was strongly dependent on the distance. The charge-separated state persists over 300 μs when NDI was separated from PTZ by eight A bases. Furthermore, the rate constant of the A-hopping process was determined to be $2 \times 10^{10}\,s^{-1}$ from the analysis of the yield of the charge separation depending on the number of A-hopping steps. In addition, we found that such hole hopping promotes long-range hole transport over 100 Å.[93]

These studies indicate that G and A, which exhibit the lowest oxidation potentials in the four bases, mediate long-range hole transfer by a hopping mechanism. In particular, the hole hopping rate in consecutive A sequence is quite fast. As indicated above, the consecutive G sequence acts as hole trap, indicating that the conjugation over bases also has an important role in the electron transfer process in DNA. Polaron-like conduction in DNA is also considered as an important mechanism. Further study of the detailed mechanism of hole conduction in DNA would be a fascinating subject.

Summary for Charge Delocalization and Migration in Well-Defined Molecular Systems

From the relation between the stabilization energy of dimer radical cations and the distance between two chromophores estimated from theoretical calculations, factors governing the interaction between two chromophores can be determined. In the present case, conformation is limited to a face-to-face, completely overlapped structure. The relation between the distance and the stabilization energy in other conformations is another interesting issue to be investigated. Furthermore, such a relation should be determined for other chromophores and other stacked molecules such as trimers and tetramers, although these latter require complicated synthesis.

In addition, delocalization of the negative charge was also clarified in the present study. The present results indicate that the negative charge delocalizes in stacked chromophores in a manner similar to that of the positive charge, indicating that the negative charge, as well as the positive charge, is a potentially good carrier in organic assemblies. This is an interesting result, because the number of reports on n-type conduction in organic materials is quite scarce when compared with those on p-type. Whether or not a negative charge conduction is comparable to a positive charge conduction in stacked chromophores is an important question to be answered. In addition, the dynamic aspects of these stabilized carriers should also be determined for the detailed design of organized functional materials.

From the study of the excited-state properties of dimer radical cations, the excess energy on the dimer radical cations causes a dissociative force between the chromophores. This is an interesting finding on "hot" carriers in the solid, which is quite different from the case of charge carriers in the ground state. Excitation energy dependence and/or higher time-resolution study on this phenomenon will yield more detailed information.

From recent substantial studies on the charge transport in DNA, details of hole transfer have become clear. In particular, quite fast charge migration by A-hopping is relevant to molecular wire applications. Nevertheless, many problems remain. Charge carrier delocalization, structure dependence, and negative carrier transport are such examples. These issues will be cleared by investigations on well-designed DNA conjugates.

SUMMARY

In the present chapter, we reviewed works that clarify ultrafast energy migration process and delocalization of charges in the organized materials based on recent reports as well as our own work. Both processes are essential in functional organized materials, although their clarification in a quantitative manner is quite difficult. For the energy migration process in organized materials, by using ultrafast laser techniques, detailed information can be obtained. Clarification of the energy migration in strongly interacting materials, such as those in solid phase, is a subject for future investigation. As for charge carriers in the materials, quantitative analysis on stabilization by charge delocalization is realized only in cyclophanes. In addition, the dynamic aspects of charge carriers in organized materials are not completely understood. These investigations will make design and fabrication of advanced nanosized materials possible.

REFERENCES

1. Pope, M.; Swenberg, C.E. *Electronic Processes in Organic Crystals and Polymers*. New York: Oxford University Press, **1999**.
2. Fujitsuka, M.; Ito, O. In: Nalwa, H.S., editor. *Encyclopedia of Nanoscience and Nanotechnology*, Vol. 8. California, CA: American Scientific Publishers, **2004**, p. 593.
3. McDermott, G.; Prince, S.M.; Freer, A.A.; Hawthornthwaite-Lawless, A.M.; Papiz, M.Z.; Cogdell, R.J.; Isaacs, N.W. *Nature* **1995**, *374*, 517.
4. Koepke, J.; Hu, X.; Muenke, C.; Schulten, K.; Michel, H. *Structure* **1996**, *4*, 581.
5. McLuskey, K.; Prince, S.M.; Cogdell, R.J.; Isaacs, N.W. *Biochemistry* **2001**, *40*, 8783.
6. Roszak, A.W.; Howard, T.D.; Southall, J.; Gardiner, A.T.; Law, C.J.; Isaacs, N.W.; Cogdell, R.J. *Science* **2003**, *302*, 1969.
7. Liu, Z.; Yan, H.; Wang, K.; Kuang, T.; Zhang, J.; Gui, L.; An, X.; Chang, W. *Nature* **2004**, *428*, 287.
8. Bradforth, S.E.; Jimenez, R.; van Mourik, F.; van Grondelle, R.; Fleming, G.R. *J. Phys. Chem.* **1995**, *99*, 16179.
9. Nagata, T.; Osuka, A.; Maruyama, K. *J. Am. Chem. Soc.* **1990**, *112*, 3054.
10. Osuka, A.; Nakajima, S.; Maruyama, K. *J. Org. Chem.* **1992**, *57*, 7355.
11. Tsuda, A.; Osuka, A. *Science* **2001**, *293*, 79.
12. Cho, H.S.; Song, N.W.; Kim, Y.H.; Jeoung, S.C.; Hahn, S.; Kim, D.; Kim, S.K.; Yoshida, N.; Osuka, A. *J. Phys. Chem. A* **2000**, *104*, 3287.

13. Kim, D.; Osuka, A. *J. Phys. Chem. A* **2003**, *107*, 8791.
14. Cho, H.S.; Rhee, H.; Song, J.K.; Min, C.-K.; Takase, M.; Aratani, N.; Cho, S.; Osuka, A.; Joo, T.; Kim, D. *J. Am. Chem. Soc.* **2003**, *125*, 5849.
15. Peng, X.; Aratani, N.; Takagi, A.; Matsumoto, T.; Kawai, T.; Hwang, I.-W.; Ahn, T.K.; Kim, D.; Osuka, A. *J. Am. Chem. Soc.* **2004**, *126*, 4468.
16. Ha, J.-H.; Cho, H.S.; Song, J.K.; Kim, D.; Aratani, N.; Osuka, A. *Chem. Phys. Chem.* **2004**, *5*, 57.
17. Nakamura, Y.; Hwang, I.-W.; Aratani, N.; Ahn, T.K.; Ko, D.M.; Takagi, A.; Kawai, T.; Matsumoto, T.; Kim, D.; Osuka, A. *J. Am. Chem. Soc.* **2005**, *127*, 236.
18. Kumble, R.; Palese, S.; Lin, V.S.-Y.; Therien, M.J.; Hochstrasser, R.M. *J. Am. Chem. Soc.* **1998**, *120*, 11489.
19. Yeow, E.K.L.; Ghiggino, K.P.; Reek, J.N.H.; Crossley, M.J.; Bosman, A.W.; Schenning, A.P.H.J.; Meijer, E.W. *J. Phys. Chem. B* **2000**, *104*, 2596.
20. Larsen, J.; Bruggemann, B.; Polivka, T.; Sundstrom, V.; Akesson, E.; Sly, J.; Crossley, M.J. *J. Phys. Chem. B* **2005**, *109*, 10654.
21. Choi, M.-S.; Aida, T.; Yamazaki, T.; Yamazaki, I. *Angew. Chem. Int. Ed. Engl.* **2001**, *40*, 3194.
22. Choi, M.-S.; Aida, T.; Yamazaki, T.; Yamazaki, I. *Chem. Eur. J.* **2002**, *8*, 2668.
23. Kelley, R.F.; Tauber, M.J.; Wasielewski, M.R. *Angew. Chem. Int. Ed. Engl.* **2006**, *45*, 7979.
24. Ogawa, K.; Kobuke, Y. *Angew. Chem. Int. Ed. Engl.* **2000**, *39*, 4070.
25. Takahashi, R.; Kobuke, Y. *J. Am. Chem. Soc.* **2003**, *125*, 2372.
26. Ikeda, C.; Satake, A.; Kobuke, Y. *Org. Lett.* **2003**, *5*, 4935.
27. Hwang, I.-W.; Park, M.; Ahn, T.K.; Yoon, Z.S.; Ko, D.M.; Kim, D.; Ito, F.; Ishibashi, Y.; Khan, S.R.; Nagasawa, Y.; Miyasaka, H.; Ikeda, C.; Takahashi, R.; Ogawa, K.; Satake, A.; Kobuke, Y. *Chem. Eur. J.* **2005**, *11*, 3753.
28. Hwang, I.-W.; Cho, H.S.; Jeong, D.H.; Kim, D.; Tsuda, A.; Nakamura, T.; Osuka, A. *J. Phys. Chem. B* **2003**, *107*, 9977.
29. Hwang, I.-W.; Kamada, T.; Ahn, T.K.; Ko, D.M.; Nakamura, T.; Tsuda, A.; Osuka, A.; Kim, D. *J. Am. Chem. Soc.* **2004**, *126*, 16187.
30. Kelley, R.F.; Goldsmith, R.H.; Wasielewski, M.R. *J. Am. Chem. Soc.* **2007**, *129*, 6384.
31. Kelley, R.F.; Lee, S.J.; Wilson, T.M.; Nakamura, Y.; Tiede, D.M.; Osuka, A.; Hupp, J.T.; Wasielewski, M.R. *J. Am. Chem. Soc.* **2008**, *130*, 4277.
32. van der Boom, T.; Hayes, R.T.; Zhao, Y.; Bushard, P.J.; Weiss, E.A.; Wasielewski, M.R. *J. Am. Chem. Soc.* **2002**, *124*, 9582.
33. Ahrens, M.J.; Sinks, L.E.; Rybtchinski, B.; Liu, W.; Jones, B.A.; Giaimo, J.M.; Gusev, A.V.; Goshe, A.J.; Tiede, D.M.; Wasielewski, M.R. *J. Am. Chem. Soc.* **2004**, *126*, 8284.
34. Li, X.; Sinks, L.E.; Rybtchinski, B.; Wasielewski, M.R. *J. Am. Chem. Soc.* **2004**, *126*, 10810.
35. Rybtchinski, B.; Sinks, L.E.; Wasielewski, M.R. *J. Am. Chem. Soc.* **2004**, *126*, 12268.
36. Rybtchinski, B.; Sinks, L.E.; Wasielewski, M.R. *J. Phys. Chem. A* **2004**, *108*, 7497.
37. Kelley, R.F.; Shin, W.S.; Rybtchinski, B.; Sinks, L.E.; Wasielewski, M.R. *J. Am. Chem. Soc.* **2007**, *129*, 3173.
38. Endo, M.; Wang, H.; Fujitsuka, M.; Majima, T. *Chem. Eur. J.* **2006**, *12*, 3735.

39. Endo, M.; Fujitsuka, M.; Majima, T. *Chem. Eur. J.* **2007**, *13*, 8660.
40. Miller, R.A.; Presley, A.D.; Francis, M.B. *J. Am. Chem. Soc.* **2007**, *129*, 3104.
41. Endo, M.; Shiroyama, T.; Fujitsuka, M.; Majima, T. *J. Org. Chem.* **2005**, *70*, 7468.
42. Endo, M.; Fujitsuka, M.; Majima, T. *Tetrahedron* **2008**, *64*, 1839.
43. Endo, M.; Fujitsuka, M.; Majima, T. *J. Org. Chem.* **2008**, *73*, 1106.
44. Fendt, L.-A.; Bouarmaied, I.; Thoni, S.; Amiot, N.; Stulz, E. *J. Am. Chem. Soc.* **2007**, *129*, 15319.
45. Guillet, J. *Polymer Photophysics and Photochemistry*. Cambridge: Cambridge University Press, **1985**.
46. Fujitsuka, M.; Hara, M.; Tojo, S.; Okada, A.; Troiani, V.; Solladie, N.; Majima, T. *J. Phys. Chem. B* **2005**, *109*, 33.
47. Fujitsuka, M.; Cho, D.W.; Solladie, N.; Troiani, V.; Qiu, H.; Majima, T. *J. Photochem. Photobiol. A* **2007**, *188*, 346.
48. Fujitsuka, M.; Okada, A.; Tojo, S.; Takei, F.; Onitsuka, K.; Takahashi, S.; Majima, T. *J. Phys. Chem. B* **2004**, *108*, 11935.
49. Sato, T.; Fujitsuka, M.; Segawa, H.; Shimidzu, T.; Tanaka, K. *Synth. Met.* **1998**, *95*, 107.
50. Grage, M.M.L.; Zaushitsyn, Y.; Yartsev, A.; Chachisvilis, M.; Sundstrom, V.; Pullerits, T. *Phys. Rev. B* **2003**, *67*, 205207/1.
51. Westenhoff, S.; Daniel, C.; Friend Richard, H.; Silva, C.; Sundstrom, V.; Yartsev, A. *J. Chem. Phys.* **2005**, *122*, 094903/1.
52. Westenhoff, S.; Beenken, W.J.D.; Friend, R.H.; Greenham, N.C.; Yartsev, A.; Sundstrom, V. *Phys. Rev. Lett.* **2006**, *97*, 166804/1.
53. Zaushitsyn, Y.; Jespersen, K.G.; Valkunas, L.; Sundstrom, V.; Yartsev, A. *Phys. Rev. B* **2007**, *75*, 195201/1.
54. Fujitsuka, M.; Cho, D.W.; Ohshita, J.; Kunai, A.; Majima, T. *J. Phys. Chem. B* **2006**, *110*, 12446.
55. Lakowicz, J.R. *Principles of Fluorescence Spectroscopy*. New York: Kluwer Academic/Plenum Publishers, **1999**.
56. Fujitsuka, M.; Cho, D.W.; Ohshita, J.; Kunai, A.; Majima, T. *J. Phys. Chem. C* **2007**, *111*, 1993.
57. Varnavski, O.P.; Ostrowski, J.C.; Sukhomlinova, L.; Twieg, R.J.; Bazan, G.C.; Goodson, T. III *J. Am. Chem. Soc.* **2002**, *124*, 1736.
58. Wang, Y.; Ranasinghe, M.I.; Goodson, T. III *J. Am. Chem. Soc.* **2003**, *125*, 9562.
59. Ranasinghe, M.I.; Wang, Y.; Goodson, T. III *J. Am. Chem. Soc.* **2003**, *125*, 5258.
60. Wang, Y.; He, G.S.; Prasad, P.N.; Goodson, T. III *J. Am. Chem. Soc.* **2005**, *127*, 10128.
61. Goodson, T.G. III *Acc. Chem. Res.* **2005**, *38*, 99.
62. Badger, B.; Brocklehurst, B. *Nature* **1968**, *219*, 263.
63. Badger, B.; Brocklehurst, B. *Trans. Faraday Soc.* **1969**, *65*, 2582.
64. Rathore, R.; Kumar, A.S.; Lindeman, S.V.; Kochi, J.K. *J. Org. Chem.* **1998**, *63*, 5847.
65. Kochi, J.K.; Rathore, R.; Le Magueres, P. *J. Org. Chem.* **2000**, *65*, 6826.
66. Irie, S.; Horii, H.; Irie, M. *Macromolecules* **1980**, *13*, 1355.
67. Masuhara, H.; Tamai, N.; Mataga, N.; De Schryver, F.C.; Vandendriessche, J. *J. Am. Chem. Soc.* **1983**, *105*, 7256.

68. Masuhara, H.; Yamamoto, K.; Tamai, N.; Inoue, K.; Mataga, N. *J. Phys. Chem.* **1984**, *88*, 3971.
69. Tsuchida, A.; Tsujii, Y.; Ito, S.; Yamamoto, M.; Wada, Y. *J. Phys. Chem.* **1989**, *93*, 1244.
70. Tsujii, Y.; Tsuchida, A.; Ito, S.; Yamamoto, M. *Macromolecules* **1991**, *24*, 4061.
71. Ohkita, H.; Fushimi, T.; Atsumi, K.; Fujita, Y.; Ito, S.; Yamamoto, M. *Chem. Phys. Lett.* **2003**, *374*, 137.
72. Fushimi, T.; Fujita, Y.; Ohkita, H.; Ito, S. *J. Photochem. Photobiol. A* **2004**, *165*, 69.
73. Fushimi, T.; Fujita, Y.; Ohkita, H.; Ito, S. *Bull. Chem. Soc. Jpn.* **2004**, *77*, 1443.
74. Ishitani, A.; Nagakura, S. *Mol. Phys.* **1967**, *12*, 1.
75. Ganesan, V.; Rosokha, S.V.; Kochi, J.K. *J. Am. Chem. Soc.* **2003**, *125*, 2559.
76. Sakamoto, Y.; Miyoshi, N.; Shinmyozu, T. *Angew. Chem. Int. Ed. Engl.* **1996**, *35*, 549.
77. Sakamoto, Y.; Miyoshi, N.; Hirakida, M.; Kusumoto, S.; Kawase, H.; Rudzinski, J.M.; Shinmyozu, T. *J. Am. Chem. Soc.* **1996**, *118*, 12267.
78. Yasutake, M.; Koga, T.; Sakamoto, Y.; Komatsu, S.; Zhou, M.; Sako, K.; Tatemitsu, H.; Onaka, S.; Aso, Y.; Inoue, S.; Shinmyozu, T. *J. Am. Chem. Soc.* **2002**, *124*, 10136.
79. Nogita, R.; Matohara, K.; Yamaji, M.; Oda, T.; Sakamoto, Y.; Kumagai, T.; Lim, C.; Yasutake, M.; Shimo, T.; Jefford, C.W.; Shinmyozu, T. *J. Am. Chem. Soc.* **2004**, *126*, 13732.
80. Fujitsuka, M.; Samori, S.; Hara, M.; Tojo, S.; Yamashiro, S.; Shinmyozu, T.; Majima, T. *J. Phys. Chem. A* **2005**, *109*, 3531.
81. Fujitsuka, M.; Cho, D.W.; Tojo, S.; Yamashiro, S.; Shinmyozu, T.; Majima, T. *J. Phys. Chem. A* **2006**, *110*, 5735.
82. Fujitsuka, M.; Tojo, S.; Shinmyozu, T.; Majima, T. *Chem. Commun.* **2009**, 1553.
83. Fujitsuka, M.; Majima, T. In: Neckers, D.C.; Jenks, W.S.; Wolff, T., editors. *Advances in Photochemistry*, Vol. 29. Hoboken, NJ: Wiley, **2007**, p. 53.
84. Cai, X.; Tojo, S.; Fujitsuka, M.; Majima, T. *J. Phys. Chem. A* **2006**, *110*, 9319.
85. Murphy, C.J.; Arkin, M.R.; Jenkins, Y.; Ghatlia, N.D.; Bossmann, S.H.; Turro, N.J.; Barton, J.K. *Science* **1993**, *262*, 1025.
86. Lewis, F.D.; Wu, T.; Zhang, Y.; Letsinger, R.L.; Greenfield, S.R.; Wasielewski, M.R. *Science* **1997**, *277*, 673.
87. Lewis, F.D.; Kalgutkar, R.S.; Wu, Y.; Liu, X.; Liu, J.; Hayes, R.T.; Miller, S.E.; Wasielewski, M.R. *J. Am. Chem. Soc.* **2000**, *122*, 12346.
88. Giese, B. *Acc. Chem. Res.* **2000**, *33*, 631.
89. Bixon, M.; Giese, B.; Wessely, S.; Langenbacher, T.; Michel-Beyerle, M.E.; Jortner, J. *Proc. Natl. Acad. Sci. U.S.A.* **1999**, *96*, 11713.
90. Lewis, F.D.; Liu, X.; Liu, J.; Miller, S.E.; Hayes, R.T.; Wasielewski, M.R. *Nature* **2000**, *406*, 51.
91. Giese, B.; Amaudrut, J.; Kohler, A.-K.; Spormann, M.; Wessely, S. *Nature* **2001**, *412*, 318.
92. Takada, T.; Kawai, K.; Cai, X.; Sugimoto, A.; Fujitsuka, M.; Majima, T. *J. Am. Chem. Soc.* **2004**, *126*, 1125.
93. Takada, T.; Kawai, K.; Fujitsuka, M.; Majima, T. *Proc. Natl. Acad. Sci. U.S.A.* **2004**, *101*, 14002.

13

SUPRAMOLECULAR EFFECTS ON MECHANISMS OF PHOTOISOMERIZATION: HULA TWIST, BICYCLE PEDAL, AND ONE-BOND-FLIP

Robert S.H. Liu, Lan-Ying Yang, Yao-Peng Zhao, Akira Kawanabe, and Hideki Kandori

Photoisomerization in the form of interconversion of the *cis* and *trans* isomers around a double bond is the key triggering process for many photosensitive biopigments.[1] They include the visual pigment rhodopsin, the proton pumping bacteriorhodopsin (bR), the phototaxis of photoactive yellow protein (PYP), the seed-germinating trigger protein phytochrome, and many other related pigments. Much spectroscopic and chemical effort has gone into elucidation of the primary processes and products associated with these important pigments. These efforts along with photochemical and theoretical work with simple organic systems[2] have led to improved understanding of mechanisms of photoisomerization, often controlled by host molecules in dictating the exact mechanism participating in a given system. In this chapter, an attempt is made to review medium-controlled photoisomerization processes and to provide an overview of different types of mechanisms that take place in nature and under laboratory conditions.

Ever since the first picosecond time-resolved spectroscopic study on events revealing the rapid isomerization following photoexcitation of the visual

Supramolecular Photochemistry: *Controlling Photochemical Processes*, First Edition. Edited by V. Ramamurthy and Yoshihisa Inoue.

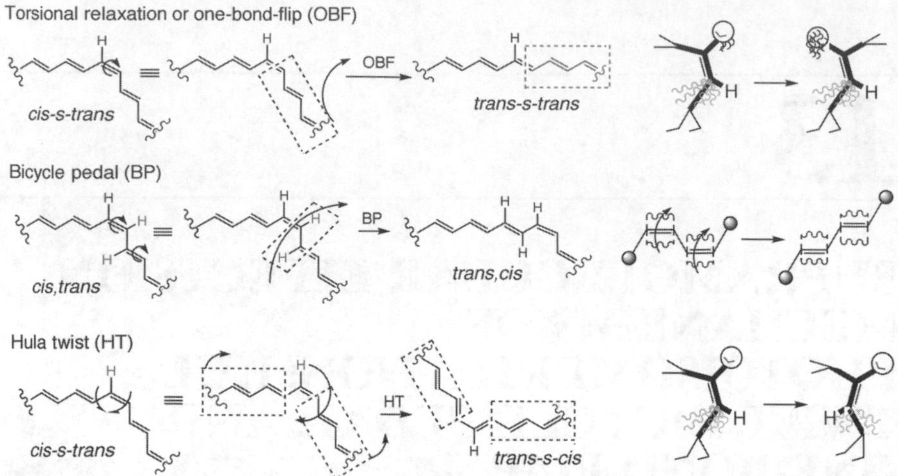

Figure 13.1. Three possible mechanisms of photoisomerization. Top: torsional relaxation where one half of the molecule flips around the formal double bond (or one-bond-flip [OBF]). Middle: the bicycle pedal (BP) process that results in isomerization at two formal double bonds. Bottom: the hula twist (HT) that results in configurational isomerization at one double bond and conformational isomerization at an adjacent single bond. The cartoon figures on the right emphasize the parts that change in space. See ftp site for color: ftp://ftp.wiley.com/public/sci_tech_med/supramolecular_photochemistry.

pigment rhodopsin,[3] doubt was raised as to how the seemingly volume-demanding torsional relaxation mechanism could take place at picosecond timescale within the well-defined binding cavity of rhodopsin. Improved short excitation pulses further narrowed the primary photochemical process of rhodopsin to 146 fs.[4] In 1978, Warshel postulated the interesting bicycle pedal (BP) process, in which two connecting C–H units of two alternating double bonds rotate concertedly.[5] The net result is simultaneous isomerization of two bonds with only a small part of the chromophore undergoing out-of-plane rotation (Fig. 13.1). Clearly, it is less volume demanding than the torsional relaxation process of which seemingly half of the retinyl chromophore will have to turn over around the 11,12 double bond (hence also known as one-bond-flip [OBF]). In order to account for the known 11-*cis* to all-*trans* photoisomerization (i.e., around one double bond only), Warshel further suggested an ensuing sequence of BP processes around single bonds starting from the primary photoproduct. Later, Warshel modified his model in which the double-bond rotation was accompanied by small twists of many nearby single and double bonds in the opposite direction.[6] Since, no bonds other than the 11,12 double bond underwent rotation for more than 90°, this modified BP process is more akin to a modified OBF process.

In 1985, the hula twist (HT) mechanism of photoisomerization was proposed in which an adjacent pair of single and double bonds rotate simultaneously.[7] To accomplish such rotations, only one C–H unit is required to translocate 180° across the plane of the polyene chromophore, while the remaining two halves of the molecule slide roughly in the plane of the polyene (Fig. 13.1). Hence, it is also a less volume-demanding process than OBF. The stereochemical consequence of such an isomerization is simultaneous isomerization of a double and a single bond, corresponding to formation of a conformationally unstable double-bond isomerized product.

Examples of BP isomerization (i.e., simultaneous isomerization of two double bonds) are rare (see below). But this is not the case for HT photoisomerization. In 1998, Fuss and coworkers in a study of photoisomerization of pre-vitamin D, **1**, in an organic glass at liquid nitrogen temperature showed that the primary photoproduct contained the structural feature of simultaneous isomerization of a pair of double and single bonds, the expected stereochemical consequence for HT.[8] This chapter made it clear to us when and where to look for new examples of HT and how to design new systems that will allow detection of the unusual structural transformation. In fact, a review of the literature revealed that many previously reported photoisomerization reactions that yielded unexpected products were in fact consistent with the involvement of the HT mechanism.[9,10] Since then, many unambiguous HT examples have appeared in the literature. Below we first describe ways to design test systems for establishing or eliminating the HT mechanism of isomerization. A review of examples of HT photoisomerization will be followed by a critical examination of other examples of photoisomerization carried out under perturbed conditions of organic glassy media or other confined media including polyene chromophores imbedded in protein-binding cavities.

R HO H $h\nu$ Organic glass at –194°C H R OH **1**

METHODS OF ESTABLISHING OR ELIMINATING THE HT MECHANISM OF PHOTOISOMERIZATION

Examples of BP, characterized by the two-bond isomerized product, are rare. Therefore, in the discussion that follows, emphasis will be placed on differentiating OBF and HT mechanisms of photoisomerization. The different stereochemical consequences from OBF and HT shown in Figure 13.1 should obviously be built into test systems for identifying the exact nature of the reaction mechanism. However, it is important first to highlight a few uniquely different structural features associated with the two mechanisms.[11]

The HT Marker

Many olefinic systems are not suitable test systems for distinguishing between the OBF and the HT mechanism of photoisomerization because they give identical products. For example, for *cis*-stilbene, the photoproducts from OBF and HT are *trans*-stilbene.

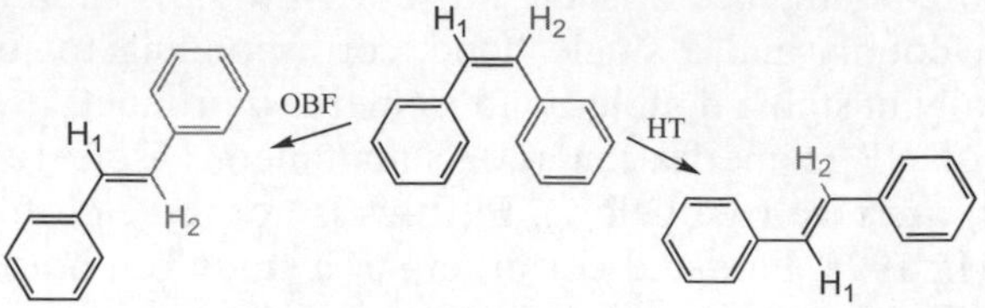

However, with the introduction of substituents to remove the twofold symmetry of the phenyl rings, photoproducts from the two processes are no longer the same. For example, in the case of *o,o′*-dimethylstilbene, **2**, OBF and HT should yield the following different products:

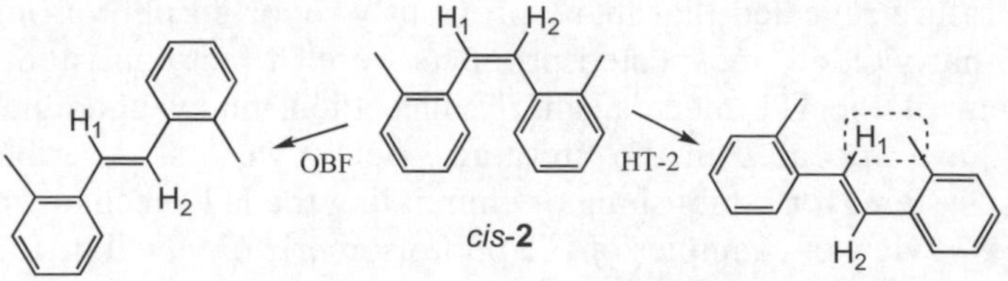

cis-**2**

The HT product is a high-energy conformer (methyl-H steric interaction) of the stable OBF product, with the former readily converting itself to the latter when permitted to do so (i.e., in solution). The corresponding *m*-substituted stilbene can, in principle, provide different products as well. However, the lack of steric crowding in the HT product will make it difficult to distinguish it from the OBF product. Nevertheless, it should be clear that *for demonstrating the formation of HT products, the introduction of the HT markers* (i.e., the *o*-substituents) *is essential*. Unfortunately, in the literature, there are purportedly HT studies where HT markers are nowhere to be found.[12]

For conjugated dienes or longer polyenes, an HT marker may not be necessary for detecting HT products. For example, for 1,3-butadiene, neither HT-1 nor OBF will cause detectable structural changes, while HT-2 will lead to the unstable s-*cis* conformer.

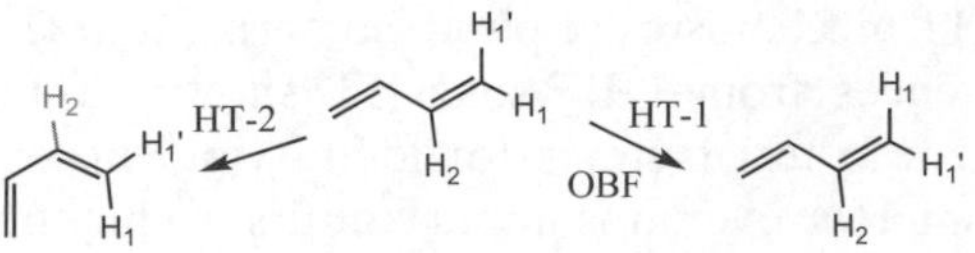

Regioselective HT

The last example, however, points out another important difference between OBF and HT mechanisms of photoisomerization. For OBF, it matters little whether an olefin is symmetrically or unsymmetrically substituted. One and only one photoproduct is available regardless which end of the double bond is flipped over. But this is not the case for HT photoisomerization. For example, for the unsymmetrical olefin *o*-methylstilbene, **3**, the stable product from HT-1 is clearly different from that of HT-2.

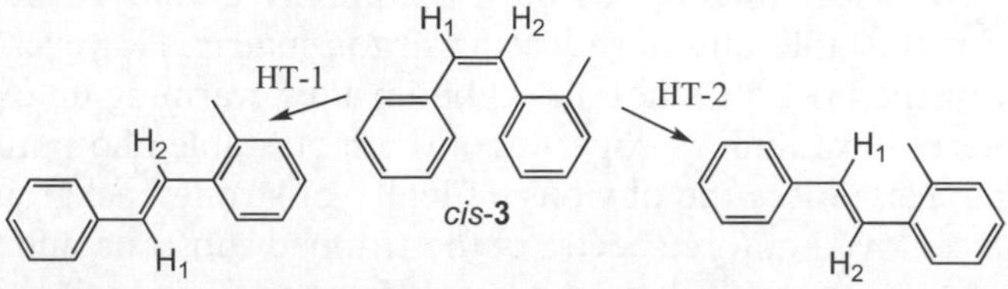

HT-1 HT-2 *cis*-**3**

Hence, *the possibility of different modes of HT photoisomerizations (i.e., regioselective HT) must be considered in the studies of unsymmetrically substituted olefins*.

Trapping the Metastable HT Product

An HT product is usually a high-energy conformer. The rate of rotational isomerization to the more stable conformer is facile taking place at temperatures below ~50 K[13] in fluid solution. Hence, it is necessary to use rigid media to trap the photoproduct for its detection. Yet, the cavity in the glass must be sufficiently large to permit movement of those atoms involved in an HT process. Thus, it should not be a surprise that HT in organic glasses was not uniformly detectable in all organic systems even when a vinyl-H is available for HT. A different parameter is involved in the argon matrix isolation studies. The reaction temperature (liquid helium temperatures) is lower than that needed for conformational equilibration. Hence, successful trapping of the unstable HT product need not be assisted by the restriction provided by the surrounding host molecules.

Fuss et al. first demonstrated the successful usage of an organic glass at liquid nitrogen temperature in trapping the HT products during irradiation of pre-vitamin D.[8] In reality, their work was preceded two decades ago by those of Alfimov et al.[14] and Castel and Fisher[15] who studied independently photochemical reactions of *cis*-diarylethylenes in organic glasses. The general conclusion then was that the method was a useful condition for the preparation of conformationally unstable *trans* products. However, its relation to the HT mechanism of photoisomerization was not appreciated then.

Low-Temperature Spectroscopic Methods

Because of high sensitivity and readily obtainable transparent solid media (argon matrix or organic glasses), UV-Vis spectroscopy has been the most frequently used method for detecting the metastable photoproduct formed in isomerization studies in spite of the fact that the method provides limited structural information (examples shown below). The broad line width encountered in solid-state nuclear magnetic resonance (NMR) makes the method not useful for obtaining meaningful structural information of conformational isomers.

Infrared (IR) spectroscopy has been commonly employed in studies of structures of metastable photoproducts in argon matrix. However, the ready evaporation of the host does not allow the ensuing warming up experiments for the purpose of examining properties of the metastable photoproducts. The use of organic glass faces the obvious difficulty of strong background signals that must be removed before spectra of the trapped substrate and the photoproduct can be examined. The method of difference Fourier-transform infrared (FT-IR) spectroscopy removes this difficulty and has been used successfully to study products of many photosensitive biopigments.[16] Recently, we have applied this method to a simple case of photoisomerization (Kawanabe, Kandori, and Liu, unpublished results). The results are summarized below.

The difference FT-IR spectra obtained during irradiation of the diester **4a** (10%) in EPA (ether:isopentane:ethanol = 5:5:2) glass at 77 K are shown in Figure 13.2. The spectra of the photoproduct (positive peaks) corresponded

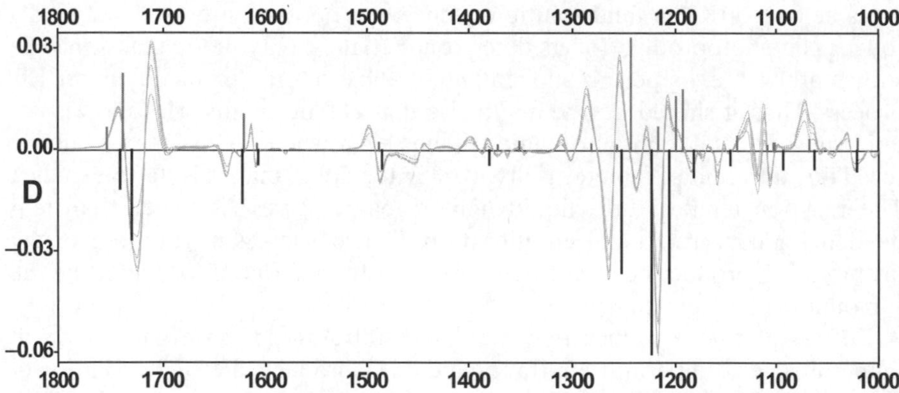

Figure 13.2. Difference FT-IR spectra obtained during the course of UV irradiation ($t = 0, 1, 2, 4, 8, 16$ min) of a 10% sample of the diester **4** in EPA glass at 77 K. The negative peaks correspond to the disappearance of starting material, and the positive peaks correspond to the appearance of photoproduct peaks. The strong peaks (>5% in intensity) of the simulated line spectra for the reactant and the high-energy conformer (the proposed HT product) are also shown. See ftp site for color: ftp://ftp.wiley.com/public/sci_tech_med/supramolecular_photochemistry.

to that of a new discreet species, clearly not due to a random collection of perturbed ground-state species caused by UV irradiation. In concert with the appearance of these peaks were diminished signals of the starting material (negative peaks). Upon mild warming, the IR spectrum of the photoproduct disappeared, reproducing that of the starting material. To assist in the interpretation of structures, the IR spectra of the reactant and the proposed photoproduct were simulated (shown as line spectra in Fig. 13.2).

Therefore, in this simple case, the IR data confirmed the two-site during low-temperature irradiation and the assignment of the high-energy conformer for the metastable photoproduct, in agreement with the HT mechanism of photoisomerization. The application of this method to more complicated cases involving simultaneous configurational and conformational transformations remains to be demonstrated.

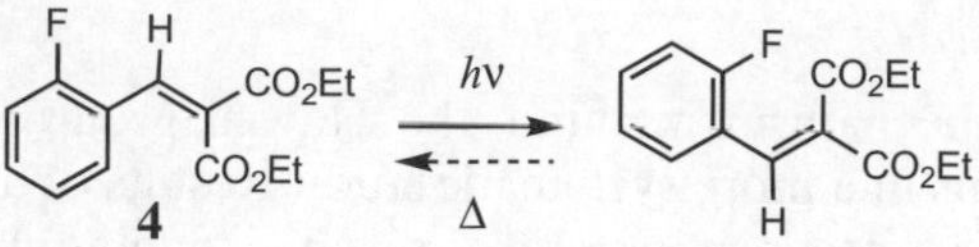

F H CO_2Et CO_2Et **4** $h\nu$ Δ F CO_2Et CO_2Et H

EXAMPLES OF HT PHOTOISOMERIZATION IN CONFINED MEDIA

The Styryl Photoswitches

Compound **4** is a simple example with a styrene chromophore. There are other related systems showing similar photoswitch characteristics: substituted nitriles, naphthyl derivatives, **5a,b**,[17] and other related compounds with larger aryl rings, **7–8** (Hirata, Zhao, and Liu, unpublished results). However, the observed spectral changes during low-temperature irradiation of these compounds were relatively small even though the extent of product formation was greater than 50% as revealed by the IR study (Fig. 13.2). Compound **6** was an exception. Its long wavelength absorption is a photochemically inactive charge-transfer band. The same band in the photoproduct is much lower in intensity. Hence, irradiation with light <370 nm led to a substantial change in UV absorption.

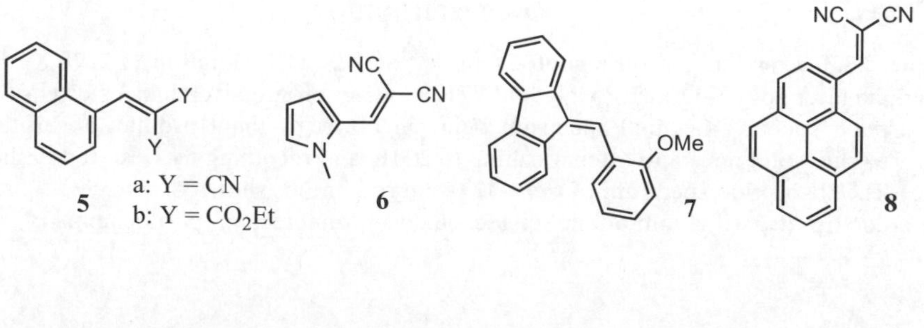

Y Y **5** a: Y = CN b: Y = CO_2Et NC CN N **6** OMe **7** NC CN **8**

1,2-Diarylethenes

Many examples of photoisomerization of 1,2-diarylethenes in organic glasses were reported by Alfimov and by Fisher and their coworkers.[14,15] The extensive list of compounds studied by these two research groups included, for example, naphthyl derivatives (**9a,b**), anthryl derivatives (**10**), and phenanthryl derivatives (**11**).

9a **9b** **10** **11**

With the HT mechanism in mind, the Hawaii group recently designed several new systems in a more systematic study. Results of a low-temperature photochemical–spectroscopic investigation on a naphthyl phenyl ethylene analog **12**[18] are shown in Figure 13.3 as a representative example. First, with the de-methyl analog, HT was shown not to take place at the site nearer the naphthalene. The methyl analog **12** then revealed that HT indeed took place at the site nearer the phenyl ring by forming, expectedly, an unstable photoproduct.

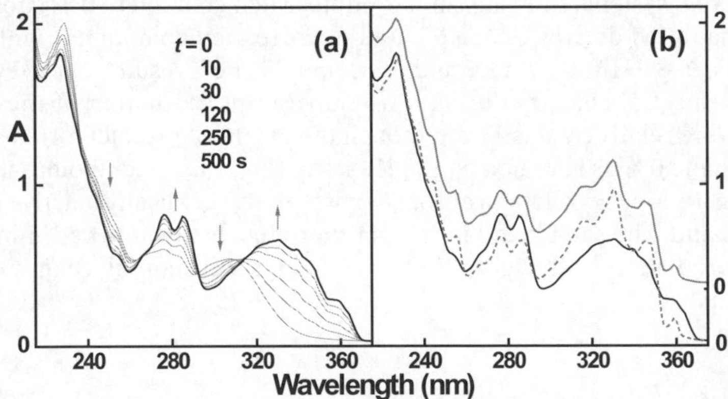

Figure 13.3. Low-temperature photochemistry of **12**. (a) Irradiation (100 W Xe-Hg lamp, no filter) of *cis*-**12** in EPA glass at 77 K showing progressive changes of UV-Vis absorption spectra, the final spectrum being in blue. (b) Final product absorption before (blue solid line) and after warming to 200 K and recooling to 77 K (red dashed line). The absorption spectrum of *trans*-**12** (solid red line) is shown on top. See ftp site for color: ftp://ftp.wiley.com/public/sci_tech_med/supramolecular_photochemistry.

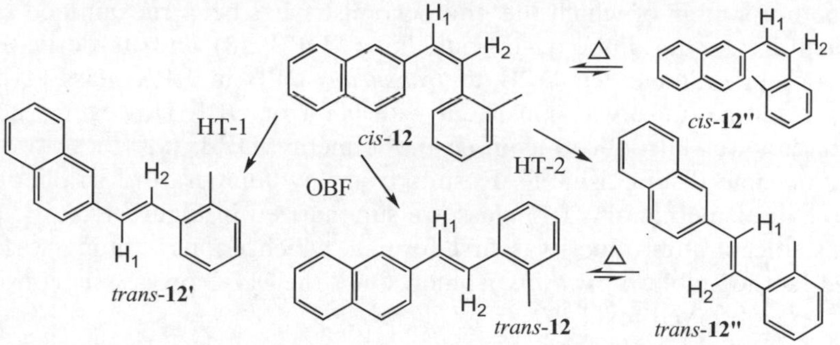

Other reported examples of 1,2-diarylethylenes included the following: substituted stilbene (**13 a,b**) and other miscellaneous compounds (e.g., **14**).[11,19,20]

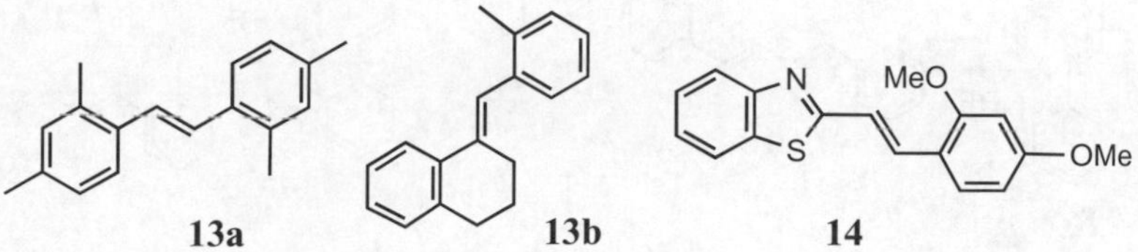

Conjugated Dienes and Other Simple Polyenes

Photochemical interconversion of conformers of 1,3-butadiene in argon matrix was first reported in 1975.[21] Subsequently, Squillacote et al. reported more detailed studies on substituted dienes (**15**).[22] Photochemical interconversion of all-*trans*-1,3,5,7-octatetraene (**16**) and its 2-S-*cis* conformer in *n*-octane crystal at liquid helium temperature was reported by Kohler and coworkers.[23]

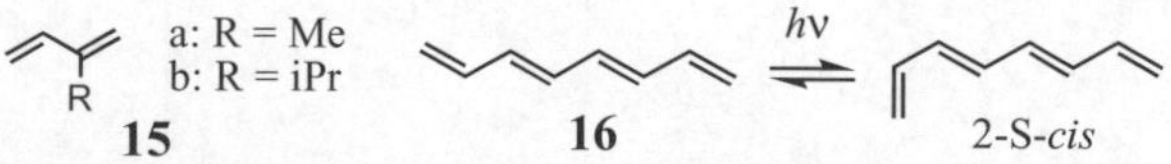

In 1985, when the HT concept was introduced,[7a] these examples along with those reported by Alfimov and by Fisher were recognized as early examples consistent with the volume-conserving mechanism of photoisomerization. Since then, other examples of conjugated polyenes have appeared in the literature. That of pre-vitamin D (**1**) reported by Fuss was already shown above.[8] 1,1-Diphenylbutadiene (**17a**) underwent HT photoisomerization in EPA glass[24]; however, higher substituted dienes (e.g., **17b**) was shown to be unreactive.

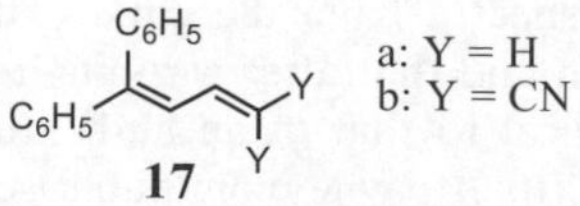

Another system of which the photochemistry has been reexamined at low temperatures is 1,4-diphenyl-1,3-butadiene (DPB, **18**) and its derivatives.[25] Photoisomerization of *cis*-DPB to *trans,trans*-DPB in EPA glass at liquid nitrogen temperature was consistent with HT-1 or OBF. However, when an HT marker was introduced as in *cis-o,o′*-dimethyl-DPB, **19**,[25] these two processes became distinguishable. Results from low-temperature irradiation in EM (ethanol : methanol = 1:1) glass are summarized in Figure 13.4.[26] Clearly, an unstable photoproduct was first formed, which upon warming was converted to the stable *trans,trans* isomer. Only the HT-1 process is consistent with these observed results.

cis-**18** HT/OBF → *t,t*-**18**

cis-**19** HT → *t,t*-**19′**; OBF → *t,t*-**19**

Similar low-temperature photochemical–spectroscopic studies of larger polyenes in organic glasses have met only limited success. Isomers (1-*cis*- or 3-*cis*) of 1,6-diphenyl-1,3,5-hexatriene (DPH, **20**) were found to be stable under UV irradiation in organic glasses.[26] We suspect the negative result was caused by the higher fluorescence yield of the DPH isomers,[27] and the space

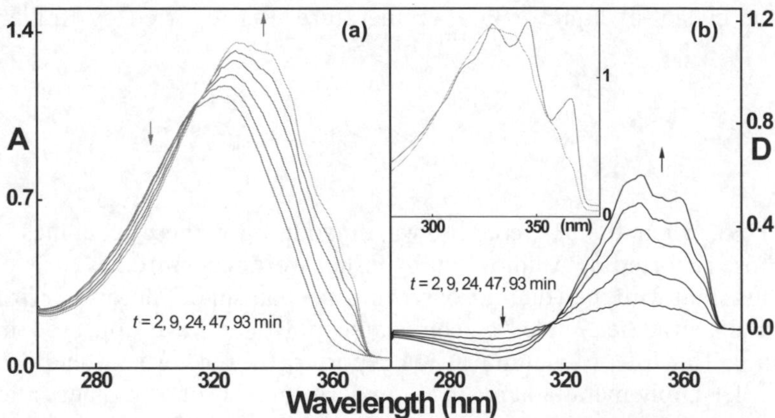

Figure 13.4. Photoisomerization of *cis*-**19** in EM glass at 77 K. (a) Left: changes in UV absorption during 93 min of UV irradiation reaching the final spectrum marked in green. (b) Right: difference spectra from the curves shown in (a). Insert: the final spectrum reproduced in green and that after warming to 200 K and recooled to 77 K in red, the latter being identical to that of an authentic sample of *trans,trans*-**19** in EM. See ftp site for color: ftp://ftp.wiley.com/public/sci_tech_med/supramolecular_photochemistry.

in the host cavity was insufficient for HT transformation of this molecule (i.e., the available "reaction volume"[28] is too small).

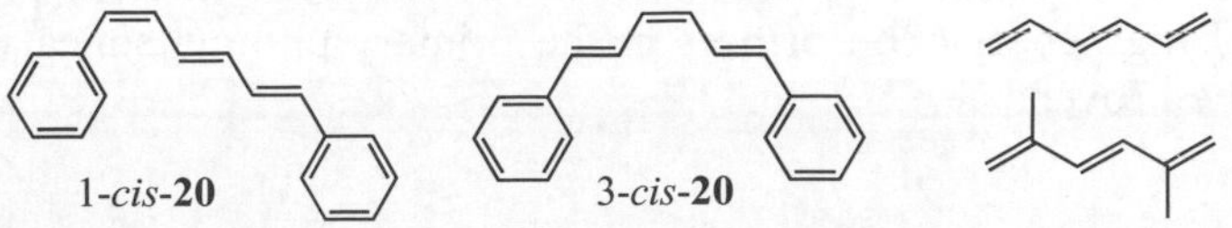

1-*cis*-**20** 3-*cis*-**20** **21** **22**

For the parent 1,3,5-hexatriene (**21a**), low-temperature irradiation of both the *trans* and the *cis* isomers during vapor deposition for argon matrix isolation conditions (at liquid helium temperature) analyzed by both UV and IR spectroscopic methods was reported.[29] The major product appeared to be the 2-s-*cis* conformer, which was identical to those reported for butadiene[21] and 1,3,5,7-octatetraene.[23] The observed single-bond isomerized product from the 3-*cis* isomer was consistent with an HT-2 process. The minor 3-*trans* product would be a product from competing OBF process (in the jet stream). Irradiation of hexatriene was also reported in argon matrix but limited to the relatively unreactive *trans* isomer. A similar study with 2,5-dimethyl-1,3,5-hexatriene (**22**) in argon matrix[30] had the additional complication of a less well-defined structure for the stable conformer of the *cis* isomer.

Other Systems

For larger polyenes, the study of HT photoisomerization is hampered by complex product mixtures (possibly more than one unstable isomers) that are difficult to unravel in the solid state. Limited success was found in low-temperature photochemistry of retinal.

The reported observation on the loss of the UV absorption of all-*trans*-retinal (**23**) when irradiated in a low-temperature glass[31] followed by regaining 70% of the original intensity upon warming to room temperature is consistent with formation of a highly twisted higher-energy conformer (by multiple HT photoisomerization). A recent reinvestigation of the photochemistry of 11-*cis*-retinal confirmed formation of a conformationally labile photoproduct, which was suspected to be either the 10-S-*cis* or 12-S-*cis* conformer of all-*trans*-retinal.[32]

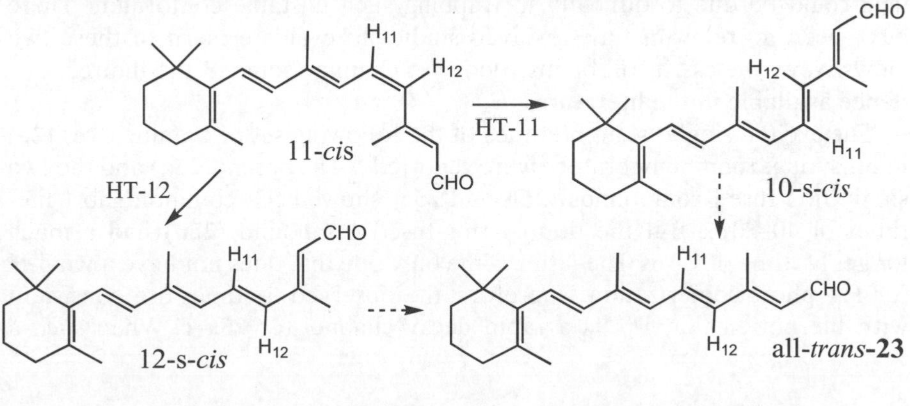

H_{11} H_{12} 11-*cis* CHO HT-11 CHO H_{12} H_{11} 10-s-*cis* HT-12 CHO H_{11} H_{12} 12-s-*cis* H_{11} CHO H_{12} all-*trans*-**23**

There are also other compounds in the literature where the unique structures appear to require HT to account for the photoisomerized product: the cyclic stilbene-cyclophane **24**[33] and a set of stilbene dendrimers.[34] However, formation of high-energy conformers in the primary photochemical process remains to be proven.

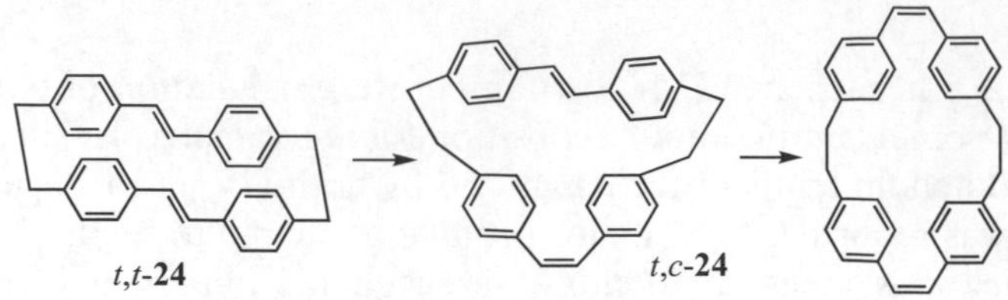

t,t-**24** → *t,c*-**24** → *c,c*-**24**

Theoretical Studies

Several theoretical papers confirming HT being a low-energy pathway of decay for an excited polyene are in the literature. Seltzer showed by modified neglect of differential overlap (MNDO) calculations[35] that both the OBF and HT photoisomerization of a model retinyl Schiff base did not require activation energy while isomerization by BP was activated. Higher-order calculations showed that HT was indeed a low-energy decay pathway for excited hexatriene[36] and stilbene.[37] And more recently, the decay pathways of excited butadiene and hexatriene[38] was shown to proceed via HT followed by pyramidalization of the HT carbon. The latter change nicely accounted for the known H-migration while not affecting the predicted stereochemical consequence of HT.

HT as a General Mechanism for Photoisomerization

As a working principle, Liu and Hammond[9] postulated that OBF was a preferred mechanism for photoisomerization in fluid media, but it could be suppressed by a viscosity-dependent barrier, making the volume-conserving HT process the dominant process in a restricted medium. Fuss et al., however, held a different view in that HT was a general mechanism for all photoisomerization reactions even in solution.[39] Failure to detect hindered cisoid photoproducts could be due to difficulty in trapping such unstable conformers. There have been no relevant time-resolved studies to evaluate merit of these two views. Nevertheless, it will be instructive to examine some of the indirect evidence available in the literature.

The excited singlet state lifetimes of the following set of cyanine dyes (**25**) in ethanol at room temperature were reported.[40] The parent (**25a**) and the two singly ring-fused compounds (**25b** and **25c**) showed the common short lifetimes of 40–80 ps. But the doubly ring-fused compound (**25d**) had a much longer lifetime of 1.4 ns. The latter is the only one that does not have a vinyl-H for HT photoisomerization. Therefore, the observed lifetimes are consistent with the notion that HT is a rapid decay channel for **25a–c**. When such a channel is blocked as in **25d**, another slower process (OBF) takes over. These solution data could be interpreted in agreement with Fuss' view. But it should be noted that this study was carried out in associated H-bonding solvent ethanol.

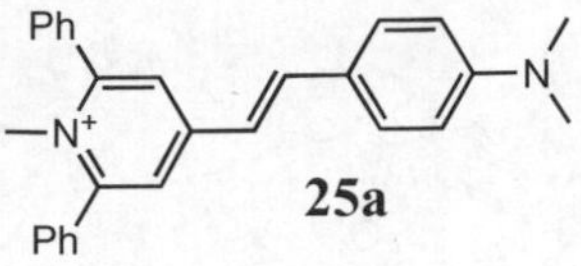

The detection of 2-s-*cis* conformer during UV irradiation and vapor deposition of hexatriene (**21a**, see above)[29] also suggested the formation of HT product in the vapor phase that was successfully trapped at the cold surface.

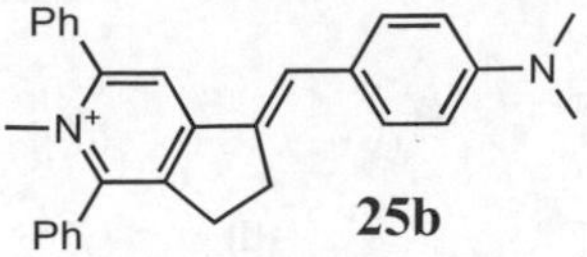

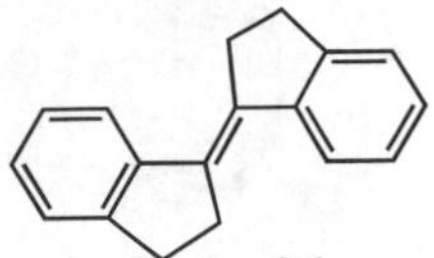

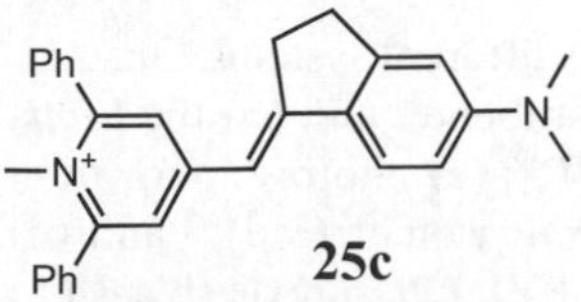

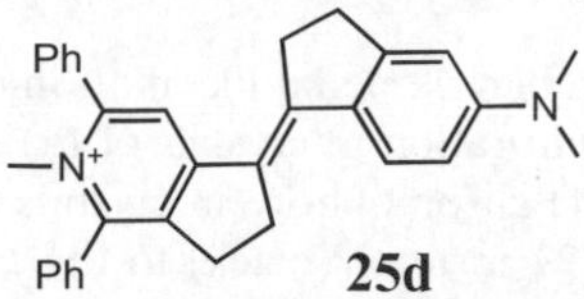

Calculated lifetimes of excited stilbenes as a function of medium viscosity (the Kramer equation) is known to deviate from the observed data in the high-viscosity range.[41] Elaborate corrections had been suggested in order to bring the calculated data in line with those observed. But the Kramer equation was based entirely on the OBF model of photoisomerization. It appeared to us that the introduction of an additional decay process (i.e., HT) in more viscous media could easily bring up the calculated data. In support of this suggestion, we note that for the ring-fused stilbene **26**, the calculated lifetimes did not deviate from those observed throughout the entire range of viscosity.[42] Clearly, for this case, HT photoisomerization is not possible. If this interpretation is correct, there appears room to consider a compromised view in that OBF is the dominant process in solvents of low viscosity, while HT gradually reveals itself in solvents of higher viscosity (or H-bonded solvent).

OTHER FORMS OF MECHANISMS OF PHOTOISOMERIZATION IN LESS RESTRICTED, CONFINED MEDIA

The BP and OBF modes of photoisomerization are also known when the photoirradiation was carried out in other confined media although definitive examples of simple organic systems are few. Thus, Matsumoto et al. showed that a *cis,cis*-muconate salt (**27**) photoisomerizes stereoselectively to the *trans,trans* isomer,[43] a process only consistent with the BP process (Fig. 13.5c). Similar observations were reported more recently for *cis,cis*-DPB (**28**) and a derivative[12b, 44] and for 1,5-di*cis*-1,6-diphenylhexatriene (**20**).[45]

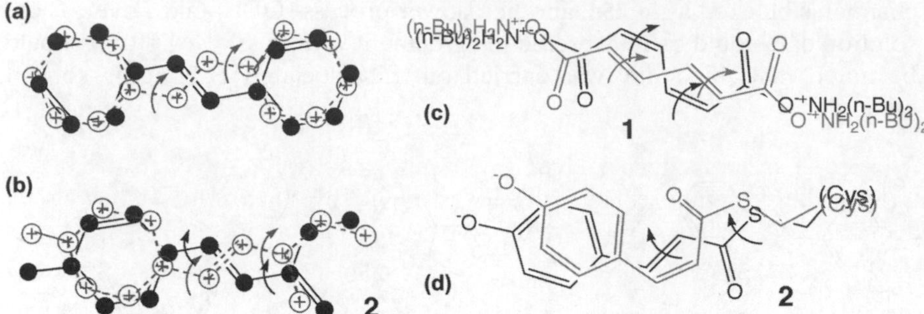

Figure 13.5. The common bicycle-pedal-like motion in four different systems. Ground-state conformational equilibration in crystals of (a) *trans*-stilbene[60] and (b) the *trans*-thiocinnamate (**29** in text) chromophore in crystals of PYP[49]; (c) photoisomerization of *cis,cis*-muconate salt (**27** in text) (black) to the *trans,trans* isomer (red),[43] and (d) photoisomerization of the thiocinnamate chromophore of PYP (**29** in text) (black) to the *cis* isomer (red).[49] See ftp site for color: ftp://ftp.wiley.com/public/sci_tech_med/supramolecular_photochemistry.

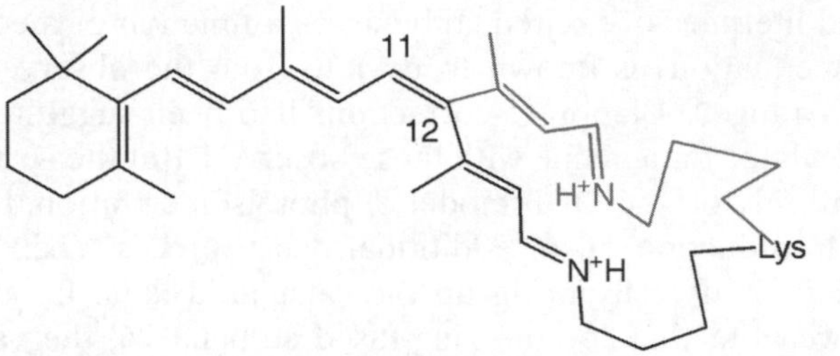

11 12 H^+N Lys N^+H

Scheme 13.1. A cartoon figure showing 11-*cis* (black) to all-*trans* (gray) in rhodopsin.

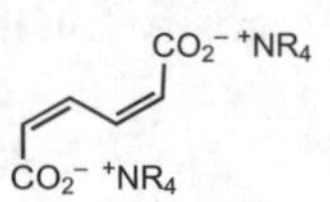

CO_2^- $^+NR_4$ CO_2^- $^+NR_4$

27 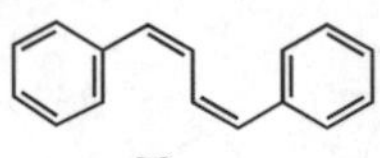 **28** 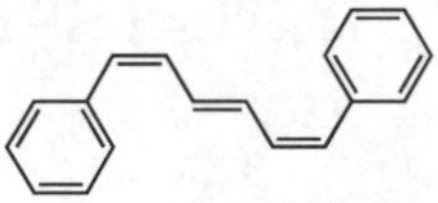 1,5-di*cis*-**20**

There have been several definitive studies where structures of the primary photoproducts of photosensitive biopigments were characterized by steady-state low-temperature X-ray crystallography or by nanosecond time-resolved crystallography revealing unusual mechanisms of photoisomerization. For the visual pigment rhodopsin, the primary photochemical reaction is the conversion of the 11-*cis*-retinyl chromophore (black) to a strained all-*trans*-retinyl structure.

The crystal structure of the first stable photoproduct, bathorhodopsin, has recently been reported.[46] When compared with the 11-*cis* chromophore in rhodopsin,[47] it became evident that the transformation actually corresponded to an OBF process. The key 11,12 double bond was found to have rotated by −112° while many nearby double and single bonds rotated by small angles in the opposite direction. The sum of the latter small changes totaled +112°, thus allowing the OBF process to take place without having to move the anchors of the attached 11-*cis*-retinyl chromophore nor turning over either half of the chromophore. The process was essentially that predicted by Warshel and Barboy.[6] It should be pointed out that such a modified OBF process is possible only because of the lengthy chromophore further tethered with the flexible butyl group from lysine.

The primary process in bR involves the conversion of the all-*trans*-retinyl (black) chromophore to a high-energy 13-*cis* isomer (gray). The crystal structure of the first stable photoproduct of the all-*trans*-retinyl chromophore of bR (the K intermediate) has also been reported.[48] Interestingly, its structure corresponds to simultaneous rotation of the 13,14 double bond and the 16,17 single bond (i.e., turning over three C–H units, Fig. 13.6). Such an "extended HT process" (a C–H unit extended by a double bond) allows the isomerization of the 13,14 bond while not moving the anchors of the pentaene Schiff base chromophore.

The structure of the primary photoproduct for the simple thiocinnamate chromophore (**29**) of PYP has been reported,[49] which showed the isomerization

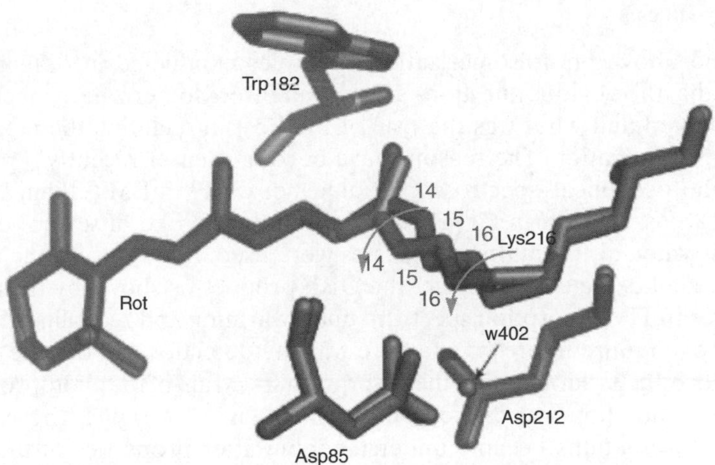

Figure 13.6. The all-*trans* and 13-*cis*-retinyl protonated Schiff base (PSB) as in the crystal structures of bR and its primary photoproduct K. The all-*trans*-retinyl chromophore from the crystal structure of bR (blue) superimposed with the 13-*cis* chromophore from the crystal structure of the K intermediate (maroon).[48] Most of the atoms remain unperturbed with the exception of the 14, 15, and 16 segments that turned over during the transition from bR to K, through rotation at the two bonds shown (green arrows). Three of the closest amino acid residues are also shown. See ftp site for color: ftp://ftp.wiley.com/public/sci_tech_med/supramolecular_photochemistry.

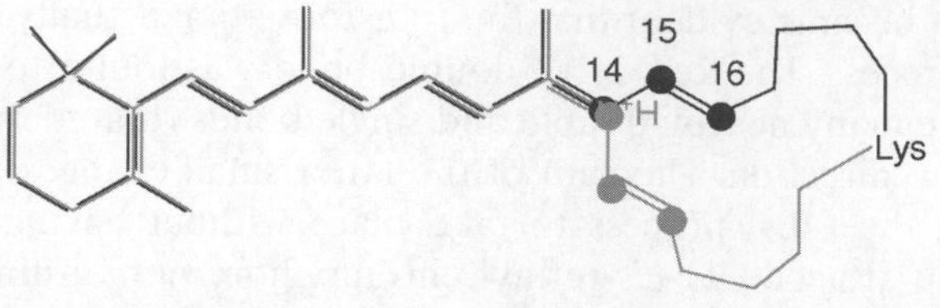

Scheme 13.2. A cartoon figure showing all-*trans* (black) to 13-cis (gray) conversion in bR.

of the *trans* double bond to *cis* and turning over of the adjacent carbonyl group, that is, simultaneous rotation of the adjacent single bond. Since only one double bond underwent *trans*/*cis* isomerization, the process should be considered an OBF process. However, it is also one that involved simultaneous rotation of the two connecting carbon units in a manner similar to a BP process. It can also be considered a modified BP process involving simultaneous rotation of an alternating pair of double and single bonds (Fig. 13.5d).

SUPRAMOLECULAR EFFECTS ON PHOTOISOMERIZATION REACTION MECHANISMS

Organic Glasses

Given the above photoisomerization examples conducted in different confined media, the obvious questions are why all three isomerization mechanisms were involved and what was the role of the host in dictating the mechanism of photoisomerization. The reasons have become clearer recently.

The photochemical–spectroscopic studies of *cis*-**19** in EM (ethanol:methanol = 1:1) glass (shown in Fig. 13.4) turned out to be one of several different outcomes when different organic glasses were used. In *iso*hexane, the primary photoproduct detected was in fact the OBF product, as shown by the absence of changes in UV absorption spectrum upon warming and recooling (Fig. 13.7, left). An even more interesting feature was the detection of both the HT and the OBF products ("doublet" in the absorption spectrum of the photoproduct[s]) when the irradiation was carried out in EPA (Fig. 13.7, right). These, at first, puzzling observations became understandable after properties of these low-temperature glasses under UV irradiation were taken into consideration.

For easy reference, the viscosities of the glasses used in this studies measured at low temperatures are listed in Table 13.1. Solvent viscosity decreases rapidly at temperatures above 77 K, particularly for the hydrocarbon solvents. Norman and Porter,[50] in discussing occasional failure to trap iodine atoms during photolysis of ethyl iodide in some of the glasses, stated that "The excess energy . . . in photolysis of ethyl iodide using radiation of 2537 Å, will amount to over 50 kcal mole^{-1}, and since a temperature rise of only 10°C is sufficient

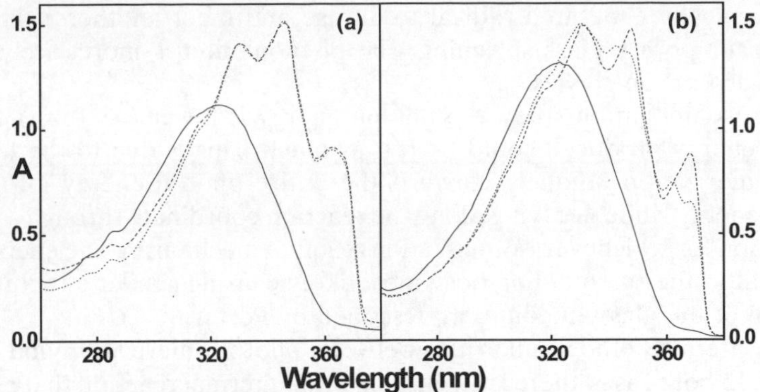

Figure 13.7. Photoisomerization of *cis*-**19** in other organic glasses. (a) Left: in *iso*hexane at 78 K. UV absorption spectra before (solid line) and after UV irradiation (dashed line) and after warming to 200 K and recooled to 78 K (dotted line), identical to that of *trans,trans*-**19**. (b) Right: in EPA glass at 78 K. UV absorption spectra before (solid line) and after 1520 s of UV irradiation (dashed line) and after warming to 200 K and recooling to 78 K (dotted line). Notice the change from the "doublet" features in the second case and their disappearance.

TABLE 13.1. Viscosity of Selected Organic Glasses as a Function of Temperature

Organic Glass	Temperature (K)	Viscosity (P)	Organic Glass	Temperature (K)	Viscosity (P)
3-Methylpentane	94.5[a]	1.8×10^4	EPA[e]	93[b]	9×10^5
	90.0[a]	1.1×10^7	EM[f]	115[b]	6.8×10^6
	85.5[a]	1.5×10^8	*n*-Propanol	125.0[a]	2.0×10^6
	80.5[a]	3.8×10^{10}		115.5[a]	1.1×10^7
	77.5[a]	1.9×10^{12}		112.5[a]	2.4×10^7
2-Methylpentane	101[b]	3×10^4		109.0[a]	2.0×10^8
(*iso*hexane)	97[b]	8×10^5		103.5[a]	1.5×10^{10}
	92[b]	6×10^6		100[c]	1×10^{13}
*iso*P-3MP[d]	77[c]	5.9×10^8			

[a]Ling, A.C.; Williard, J.E. *J. Phys. Chem*. **1968**, *72*, 1918.
[b]Greenspan, H.; Fischer, E. *J. Phys. Chem*. **1965**, *69*, 2466.
[c]Herkstroeter, W.G. In: Lamola, A., editor. *Creation and Detection of the Excited States*, Vol. 1. New York: Marcel Dekker, **1971**.
[d]Equal amounts of *iso*pentane and 3-methylpentane.
[e]Ether : isopentane : ethanol = 5:5:2.
[f]Equal amounts of ethanol and methanol.

to produce noticeable softening of the glasses . . . , each quantum absorbed by ethyl iodide should result in a viscosity-lowering of several hundred molecules of the solvent sufficient to enable the radicals to diffuse apart. Heat will rapidly be conducted away and the glass will again become rigid . . . " The message is clear. In spite of the use of so-called rigid organic glasses, it is possible for a photochemically generated radical to diffuse and meet another radical as a result of temporary glass softening. The phenomenon is more likely in the softer hydrocarbon glasses.

In the isomerization study, the photon energy is lower (>310 nm), but the thermal energy produced could be just as much. This is due to the fact that photoisomerization uniquely converts the entire photon energy into excess thermal energy while decaying along the reaction coordinate through a conical intersection for whichever isomerization reaction mechanism. The energy ends up as heat in the surrounding host, most likely causing similar structural perturbation of the glassy medium as described by Porter.

There are also other similarities between photoisomerization and radical trapping. In both cases, there is an ensuing rapid thermal reaction that removes the photochemically generated product: radical recombination in radical trapping and conformational equilibration in HT photoisomerization. The only difference is that the latter, being unimolecular, should be more sensitive to local glass softening than the bimolecular radical coupling. Conformational reorganization is a rapid process, for example, the activation energy of reaction of 2-S-*cis*-octatetraene to the relaxed all-*trans* form being 4.0 ± 0.2 kcal mol^{-1} (reacting rapidly at >50 K).[13] But successful trapping of high-energy conformers in rigid media had been amply described: for example, S-*cis*-1,3-butadiene in argon matrix[21] and 2-S-*cis*-1,3,5,7-octatetraene in *n*-octane crystal at <20 K[23] and high-energy conformers of aryl alkenes[13,14] and trienes[8] in organic glasses at liquid N_2 temperatures. The difference in the observed photochemistry of **19** in EM and *iso*hexane now becomes very understandable. EM is a harder alcoholic glass, not readily softened by the converted photon energy (Fig. 13.4), but *iso*hexane is a softer hydrocarbon glass, easily softened by the excess heat, thereby failing to trap the high-energy conformer (Fig. 13.7a), and EPA's viscosity is in between the above two, giving the result in the middle of the above two extremes (Fig. 13.7b).

To test the above explanation, we further showed that when irradiation of **19** in EM was carried out at 115 K (i.e., in its low-viscosity range), the only photoproduct detected was the stable *trans,trans* conformer, that is, identical to that in *iso*hexane glass at 77 K. Clearly at 115 K, EM behaved just like a low-viscosity hydrocarbon glass. We might mention that there was one report of isomerization of **19** in *iso*pentane glassy media giving an OBF type of product.[51] This observation while in agreement with that shown in Figure 13.7a is further complicated by the fact that *iso*pentane was reported to remain fluid at 77 K.[50]

Similar organic glass-dependent photoisomerization was also observed in stilbene derivatives. For example, in the case of *o,o′*-dimethylstilbene (*cis*-**2**), no HT product was detected when irradiation was carried out at 77 K in *iso*hexane (Zhao and Liu, unpublished results) (the change reported earlier nearer 250 nm[19] was most likely due to effect of glass cracking). Only when the harder *n*-propanol glass was used, an unstable photoproduct (as shown in changes of the long wavelength absorption band) was detected, which rearranged readily to the stable conformer upon warming (Zhao and Liu, unpublished results). Thus, it behaved quite similar to *cis*-**19** as shown in Figure 13.7. Interestingly, with a bulkier ring-fused analog of *o*-methylstilbene HT product was detected even in *iso*hexane glass.[19]

Low-temperature photoisomerization of DPB analogs containing methyl substituents on the connecting diene chain brought up the interesting possibility of HT around a C–Me unit.[52] However, the new knowledge of possible glass softening points out the need to examine further if the observed isomerization could be a result of OBF in a local melt.

In Crystals

HT is a medium-directed photoisomerization that takes place readily in organic glasses and in argon matrix at liquid helium temperature. In crystalline medium, there are no established examples of HT photoisomerization. It will be of interest to consider the structural differences among these hosts that show a preference, or otherwise, for the volume-conserving HT process.

Organic glasses involve random but close arrangement of the host molecules collapsing around the trapped substrate during glass formation. For argon matrix, the vapor deposition process probably does not lead to close interaction between the host and the guest molecules. However, the lower temperature (liquid helium temperature) encountered in argon matrix studies allows trapping unstable primary photoproduct without necessarily the benefit of additional help from the rigid host cavity as in organic glasses.

Organized hosts such as crystals present a different situation.[53] Close intermolecular contacts are often limited to a few points of each molecule, leaving substantial intermolecular empty space at other parts of the molecule. Reactions involving those parts in close molecular contact should be inhibited, while reactions at centers surrounded by empty space should be facilitated (perhaps even more so than that in solution). With this view in mind, the BP process of *cis,cis*-DPB (**18**) and analogs in crystals becomes not surprising. Below are the overlapped structures of *trans,trans*- and *cis,cis*-difluoro-DPB extracted from their crystal structures. The bulky phenyl group lies parallel to the same group of a second molecule. Hence, it will not be possible for any of them to turn over half of the molecule in OBF or to slide two halves across in HT. But in a BP motion, the bulky groups remain in place (or nearly so) while the middle butadiene fragment is free to move around because of the ample intermolecular empty space, resulting in turning over of the two C–H unit in a BP manner as an energetically downhill process (*cis,cis* to *trans,trans*).[44]

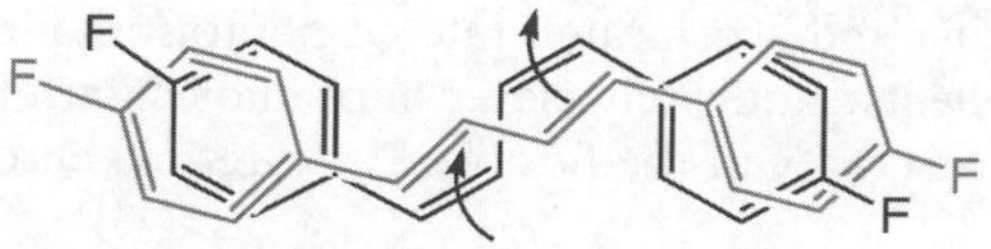

F F F F

Such a motion apparently can be extended to photoisomerization of crystals of 1,5-di*cis*-DPH (**20**)[45] where the direct conversion to the *trans,trans* isomer can be considered an extended BP motion involving the middle four C–H units. And the photoisomerization of a photochromic crystal (**30**)[54] and the chromophore of PYP crystals (Fig. 13.5) apparently followed a similar BP-like process.

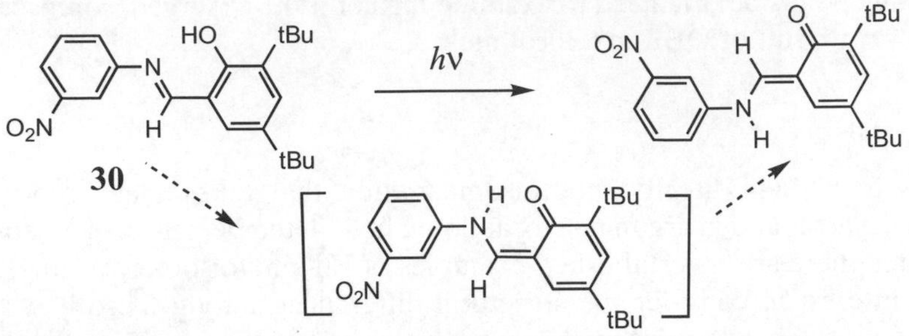

And the reported photoisomerization of crystalline *trans*-1,2-dibenzoylethylene (**31**) to the *cis* isomer[55] is better accommodated by a BP-like process (Fig. 13.5d) rather than the suggested HT process, which would require huge lateral displacement of a benzoyl group.

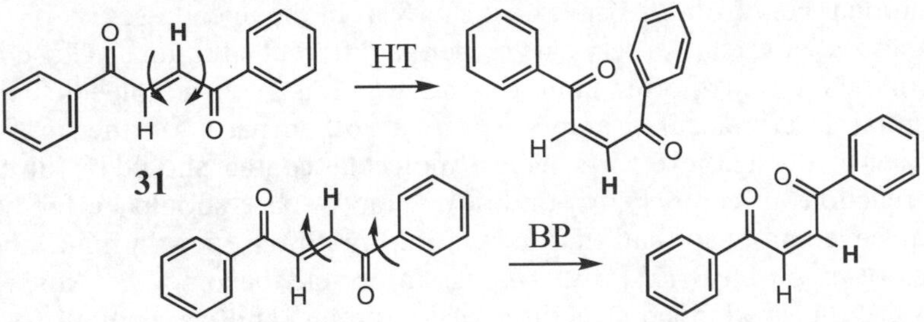

For rhodopsin, recent crystal structure[47] and fluorine-nuclear magnetic resonance (F-NMR) studies[56] have shown that the only parts of the 11-*cis*-retinyl chromophore in close contact with the opsin-binding cavity are the 11,12 vinyl-Hs, the methyl groups on the ring, and the side chain. In fact, one side of the polyene chain faces an empty space defined by the relatively rigid 3,4-transmembrane loop. It is the rotation of the C12-H unit[57] into this empty space[47] that is believed to lead to the observed OBF photoisomerization (Fig. 13.8), triggered by the initial bond-lengthening process[58] that was suggested for the stereospecific and accelerated rate of photoisomerization of rhodopsin.[56] The latter reflects exquisitely clever supramolecular effects of opsin on the structure and reactivity of the flexible 11-*cis*-retinyl chromophore.[56,59]

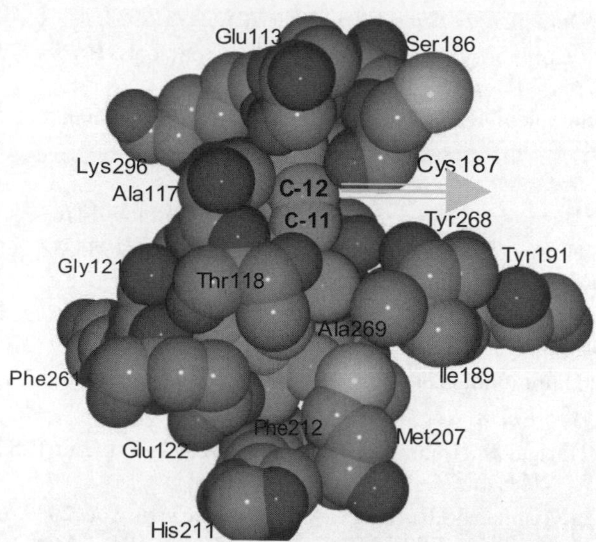

Figure 13.8. Partial crystal structure of rhodopsin[47] showing the amino acid residues surrounding the 11-*cis*-retinyl chromophore (green). Only atoms within 4.0 Å of the chromophore are shown. The chromophore leans against helix 3 (amino acids 113–122) on the left while facing an open space on the right. The yellow arrow indicates the preferred direction of isomerization. See ftp site for color: ftp://ftp.wiley.com/public/sci_tech_med/supramolecular_photochemistry.

SUMMARY

In this chapter, we have summarized the effects of organic hosts such as organic glasses and crystals as well as protein-binding cavities on photoisomerization processes. Clearly, these are genuine supramolecular effects that alter the photochemical properties of the trapped substrate.

ACKNOWLEDGMENTS

The work was partially supported by a grant from the National Science Foundation (CHE-0132250) and Japanese Ministry of Education, Culture, Sports, Science and Technology (19370067, 20044012). RSHL thanks Drs. W. Fuss and H. Tomioka for helpful discussions.

REFERENCES

1. See for example, van der Horst, M.A.; Hellingwerf, K. *J. Acc. Chem. Res.* **2004**, *37*, 13–20.

2. Dugave, C.; Demange, L. *Chem. Rev.* **2005**, *103*, 2475–2532.
3. Busch, G.D.; Applebury, M.L.; Lamola, A.A.; Rentzepis, P. *Proc. Natl. Acad. Sci. U.S.A.* **1972**, *69*, 2802–2806.
4. Wang, Q.; Schoenlein, R.W.; Peteanu, L.A.; Mathies, R.A.; Shank, C.V. *Science* **1994**, *266*, 422–424.
5. Warshel, A. *Nature (London)* **1976**, *260*, 679–683.
6. Warshel, A.; Barboy, N. *J. Am. Chem. Soc.* **1982**, *104*, 1469–1476.
7. (*a*) Liu, R.S.H.; Asato, A.E. *Proc. Natl. Acad. Sci. U.S.A.* **1985**, *82*, 259–263; (*b*) Liu, R.S.H.; Browne, D. *Acc. Chem. Res.* **1986**, *19*, 42–48.
8. Muller, A.M.; Lochbrunner, S.; Schmid, W.E.; Fuss, W. *Angew. Chem. Int. Ed. Engl.* **1998**, *37*, 505–507.
9. Liu, R.S.H.; Hammond, G.S. *Proc. Natl. Acad. Sci. U.S.A.* **2000**, *97*, 11153–11158.
10. Liu, R.S.H. *Acc. Chem. Res.* **2001**, *34*, 555–562.
11. Yang, L.-Y.; Harigai, M.; Imamoto, Y.; Kataoka, M.; Ho, T.-I.; Liu, R.S.H. *Photochem. Photobiol. Sci.* **2006**, *5*, 874–882.
12. (*a*) Saltiel, J.; Krishna, T.S.R.; Turek, A.M. *J. Am. Chem. Soc.* **2005**, *127*, 6938–6939; (*b*) Saltiel, J.; Krishma, T.S.R.; Turek, A.M.; Clark, R.J. *Chem. Commun.* **2006**, 1506–1508.
13. Ackerman, J.R.; Kohler, B. *J. Chem. Phys.* **1984**, *80*, 45–50.
14. Alfimov, M.V.; Razumov, V.F.; Rachinsky, A.G.; Listvan, V.N.; Scheck, Y.B. *Chem. Phys. Lett.* **1983**, *101*, 593–597 and references cited therein.
15. Castel, N.; Fisher, E. *J. Mol. Struct.* **1985**, *127*, 159–166 and references cited therein.
16. Kandori, H. *Biochim. Biophys. Acta* **2000**, *1460*, 177–191.
17. Krishnamoorthy, G.; Asato, A.E.; Liu, R.S.H. *Chem. Commun.* **2003**, 2170–2171.
18. Yang, L.-Y.; Liu, R.S.H. *Photochem. Photobiol.* **2007**, *84*, 1436–1440.
19. Imamoto, Y.; Kuroda, T.; Kataoka, M.; Shevyakov, S.; Krishnamoorthy, G.; Liu, R.S.H. *Angew. Chem. Int. Ed. Engl.* **2003**, *42*, 3630–3633.
20. Zhu, F.; Motoyoshiya, J.; Nishii, Y.; Aoyama, H.; Kakehi, A.; Shiro, M. *J. Photochem. J.* **2006**, *184*, 44–49.
21. Squillacote, M.E.; Sheridan, R.S.; Chapman, O.; Anet, F.A.L. *J. Am. Chem. Soc.* **1975**, *101*, 3657–3659.
22. (*a*) Squillacote, M.E.; Semple, T.C.; Mui, P.W. *J. Am. Chem. Soc.* **1985**, *107*, 6842–6846; (*b*) Squillacote, M.E.; Semple, T.C. *J. Am. Chem. Soc.* **1990**, *112*, 5546–5551.
23. Ackerman, J.R.; Forman, S.A.; Hossain, M.; Kohler, B. *J. Chem. Phys.* **1984**, *80*, 39–44.
24. Krishnamoorthy, G.; Schieffer, S.; Shevyakov, S.; Asato, A.E.; Wong, K.; Head, J.; Liu, R.S.H. *Res. Chem. Interm.* **2004**, *30*, 397–405.
25. Yang, L.-Y.; Liu, R.S.H.; Bowman, K.J.; Wendt, N.L.; Liu, J. *J. Am. Chem. Soc.* **2005**, *127*, 2404–2405.
26. Zhao, Y.-P.; Yang, L.-Y.; Liu, R.S.H. *Chem. Asian J.* **2009**, *4*, 754–759.
27. Saltiel, J.; Sears, D.J. Jr.; Ko, D.H.; Park, K.-M. Horspool, W.M.; Song, P., editors. *Handbook of Photochemistry & Photobiology*, Vol. 1. Boca Raton, FL: CRC Press, **1995**, pp. 3–15.
28. Ramamurthy, V.; Weiss, R.G.; Hammond, G.S. *Adv. Photochem.* **1993**, *18*, 67–234.

29. Furukawa, Y.; Takeuchi, H.; Harada, I.; Tasumi, M. *J. Mol. Struct.* **1983**, *100*, 341–350.

30. Brouwer, A.M.; Jacobs, H.J.C. *Recl. Trav. Chim. Pay-Bas.* **1995**, *114*, 449–456.

31. (*a*) Jurkowitz, L. *Nature* **1959**, *184*, 614–617; (*b*) Loeb, J.N.; Brown, P.K.; Wald, G. *Nature* **1959**, *184*, 617–620; (*c*) Wald, G. *Nature* **1959**, *184*, 620–624.

32. Yang, L.-Y.; Liu, R.S.H. *J. Ch. Chem. Soc.* **2006**, *53*, 1219–1224.

33. Ingjald, A.; Sandros, K.; Sundahl, M.; Wennerstrom, O. *J. Phys. Chem.* **1993**, *97*, 1920–1923.

34. Uda, M.; Mizutani, T.; Hayakawa, J.; Momotake, A.; Ikegami, M.; Nagahata, R.; Arai, T. *Photochem. Photobiol.* **2002**, *76*, 596–605.

35. Seltzer, S. *J. Am. Chem. Soc.* **1987**, *109*, 1627–1631.

36. Olivucci, M.; Bernardi, F.; Celani, P.; Rayazos, I.; Robb, M.A. *J. Am. Chem. Soc.* **1994**, *116*, 1077–1085.

37. (*a*) Fuss, W.; Lochbrunner, S.; Muller, A.M.; Schkarki, T.; Schmid, W.E.; Trushin, S.A. *Chem. Phys.* **1988**, *116*, 2034–2048; (*b*) Levine, B.G.; Ko, C.; Quenneville, J.; Martinez, T. *J. Mol. Phys.* **2006**, *104*, 1053–1060.

38. Wilsey, S.; Houk, K.N. *Photochem. Photobiol.* **2002**, *76*, 616–621; (*b*) Norton, J.E.; Houk, K.N. *Mol. Phys.* **2006**, *104*, 993–1008.

39. Ruiz, D.S.; Cembrau, A.; Garavelli, M.; Olivucci, M.; Fuss, W. *Photochem. Photobiol.* **2002**, *76*, 622–633.

40. Van der Meer, M.J.; Zhang, H.; Rettig, W.; Glasbeck, M. *Chem. Phys. Lett.* **2000**, *320*, 673–680.

41. See for example, Waldeck, D.H. *Chem. Rev.* **1990**, *91*, 415–436.

42. Doany, F.E.; Hochstrasser, R.M.; Greene, B.I. *SPIE* **1985**, *533*, 25–29.

43. Odani, T.; Matsumoto, A.; Sada, K.; Miyata, M. *Chem. Commun.* **2001**, 2004–2005.

44. Liu, R.S.H.; Yang, L.-Y.; Liu, J. *Photochem. Photobiol.* **2007**, *83*, 2–10.

45. Sonoda, Y.; Kawanishi, Y.; Tsuzuki, S.; Goto, M. *J. Org. Chem.* **2005**, *70*, 9755–9763.

46. Nakamichi, H.; Okada, T. *Angew. Chem. Int. Ed. Engl.* **2006**, *45*, 1–5.

47. Palczewski, K.; Kumasaka, T.; Hori, T.; Behnke, C.A.; Motoshima, H.; Fox, B.A.; Le Trong, I.; Teller, D.C.; Okada, T.; Stenkamp, R.E.; Yamamoto, M.; Miyano, M. *Science* **2000**, *289*, 739–745.

48. Edman, K.; Nollert, P.; Royant, A.; Belrahli, H.; Peboy-Peyoula, E.; Hadju, J.; Neutze, R.; Landau, E.M. *Nature* **1999**, *401*, 822–826.

49. Genick, U.K.; Borgstahl, G.E.O.; Ng, K.; Ren, Z.; Pradervand, C.; Burke, P.M.; Srajer, V.; Teng, T.Y.; Schildkamp, W.; McRee, D.E.; Moffat, K.; Getzoff, E.D. *Science* **1997**, *275*, 1471–1475.

50. Norman, I.; Porter, G. *Proc. Roy. Soc. London A* **1955**, *230*, 399–414.

51. Saltiel, J.; Bremwer, M.A.; Laohhasurayotin, S.; Krishna, T.S.R. *Angew. Chem. Int. Ed. Engl.* **2008**, *47*, 1237–1240.

52. Yang, L.-Y.; Liu, R.S.H.; Wendt, N.L.; Liu, J. *J. Am. Chem. Soc.* **2005**, *127*, 9378–9379.

53. Liu, R.S.H.; Hammond, G.S. *Acc. Chem. Res.* **2005**, *38*, 396–403.

54. Harada, J.; Uekusa, K.; Ohashi, Y. *J. Am. Chem. Soc.* **1999**, *121*, 5809–5810.

55. Kaupp, G.; Schmeyers, J. *J. Photochem. Photobiol.* **2000**, *59*, 15–19.

56. Liu, R.S.H.; Colmenares, L.U. *Proc. Natl. Acad. Sci. U.S.A.* **2003**, *100*, 14639–14644.

57. Mathies, R.A.; Lugtenburg, J. Stavenga, D.G.; deGrip, W.J.; Pugh, E.N. Jr., editors. *Handbook of Biological Physics*, Vol. 3. Amsterdam: Elsevier Science, **2000**, pp. 55–90.

58. Kukura, P.; McCamant, D.W.; Yoon, S.; Wandschneider, D.B.; Mathies, R.A. *Science* **2005**, *310*, 1006–1009.

59. Liu, R.S.H.; Hammond, G.S.; Mirzadegan, T. *Proc. Natl. Acad. Sci. U.S.A.* **2005**, *102*, 10783–10787.

60. Harada, J.; Ogawa, K. *J. Am. Chem. Soc.* **2004**, *126*, 3539–3544.

14

PROTEIN-CONTROLLED ULTRAFAST PHOTOISOMERIZATION IN RHODOPSIN AND BACTERIORHODOPSIN

Hideki Kandori

INTRODUCTION

Rhodopsin (Rh) and Bacteriorhodopsin (BR)

Photoreceptive proteins convert light into energy or signal; they contain a chromophore molecule inside a protein. It is well known that specific chromophore–protein interactions result in unique photophysical and photochemical processes, leading to functional optimization for light-energy and light-signal conversions. For example, our eye is an excellent light sensor, possessing such properties as high photosensitivity, low noise, wide dynamic range, and high spatial and time resolution. Since visual excitation is initiated by light absorption of visual pigments present in our eyes,[1–5] the excellent light-sensing ability of our vision is due to the properties of our light-sensor proteins. Humans have four photoreceptive proteins; one works for twilight vision, and the other three work for color vision. The former is present in the rod cells, and its photoreceptive protein is called Rh (λ_{max} ~ 500 nm). The latter is present in cone cells and are named according to the colors of absorption, such as "human blue" (λ_{max} ~ 425 nm), "human green" (λ_{max} ~ 530 nm), and "human red" (λ_{max} ~ 560 nm). Other animals have additional proteins, whose absorption

Supramolecular Photochemistry: Controlling Photochemical Processes, First Edition. Edited by V. Ramamurthy and Yoshihisa Inoue.

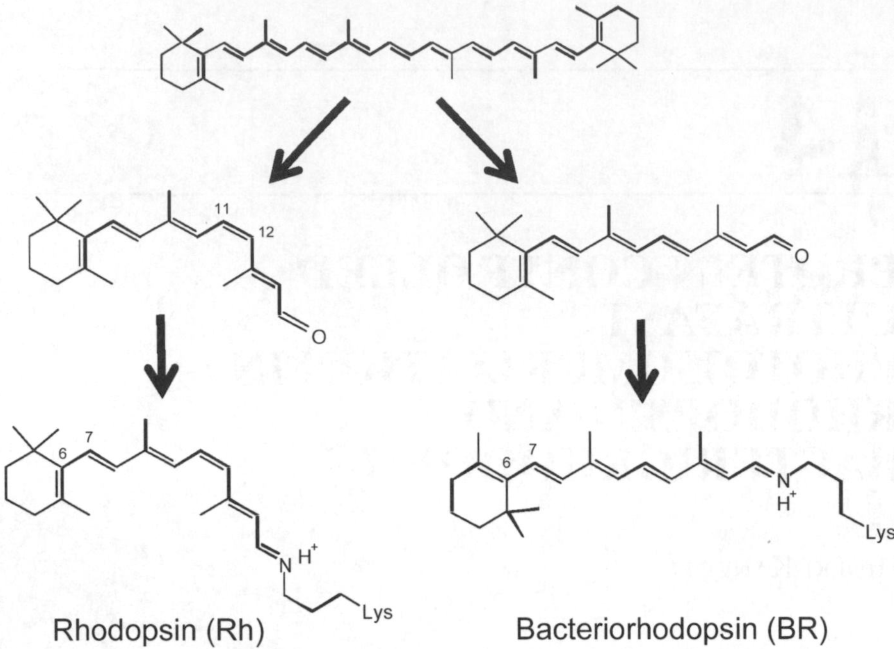

Figure 14.1. Chromophore molecules in rhodopsin (left) and bacteriorhodopsin (right). β-Carotene (top) is the source of the chromophore, and 11-*cis* and all-*trans* retinal are bound to protein to form rhodopsin and bacteriorhodopsin, respectively.

covers the ultraviolet (UV) region.[1] Although color pigments are not "rhodopsin," the word "rhodopsin" often represents visual pigments in general.

The chromophore molecule of Rh is 11-*cis* retinal. Figure 14.1 shows that the source of the visual chromophore is β-carotene. In our retina, thermally isomerized 11-*cis* retinal is supplied to a protein "opsin," and a retinal molecule forms a Schiff base linkage with a side chain of lysine residue in visual pigments. The Schiff base linkage is the only covalent bond between the chromophore and protein. In most visual pigments, the Schiff base is protonated (Fig. 14.1), and a negatively charged carboxylate (Glu113 for bovine Rh) stabilizes the protonated Schiff base by forming a salt bridge.

While all visual pigments use an 11-*cis* retinal as the chromophore molecule, some archaea and bacteria have retinal-binding proteins of the all-*trans* form. The most famous protein is BR, found in *Halobacterium salinarum*.[5–7] This archaeon contains four retinal-binding proteins: BR, halorhodopsin (HR), sensory rhodopsin I (SRI), and sensory rhodopsin II (SRII, also called phoborhodopsin). BR and HR are light-driven ion pumps, which act as an outward proton pump and an inward Cl^- pump, respectively.[5–7] SRI and SRII are photoreceptors of this halophilic archaeon, which are responsible for attractant and repellent responses in phototaxis, respectively.[8] Thus, the functions of these four proteins are different, but they have similar protein architecture. These proteins have been called archaeal Rhs because they were found only in the Archaea. However, recent genome projects have revealed the presence of similar proteins in Eubacteria and Eukaryotes. Thus, they are now called archaeal-type Rhs or microbial Rhs.

The chromophore molecule of archaeal-type Rhs is all-*trans* retinal. Despite the chromophore binding pocket being different from those of 11-*cis* retinal in visual Rhs, the all-*trans* retinal is also bound to a lysine residue via a protonated Schiff base linkage (Fig. 14.1). Positively charged chromophores are similarly stabilized by counterions, but in the case of archaeal-type Rhs, the counterions are more complex than visual Rhs. For example, BR has two negatively charged aspartates (Asp85 and Asp212) and a positively charged arginine (Arg82) in the Schiff base region. Thus, an electric quadrupole exists in the core region of BR. Figure 14.1 shows that the isomeric composition of visual and archaeal-type Rhs differs not only for the C11=C12 double bond, but also for the C6–C7 single bond.[5] In visual Rhs, the C6–C7 bond is in a *cis* form, and the polyene and β-ionone ring is not planar because of steric hindrance between C5-methyl group and C8-hydrogen. Consequently, the conjugation of π-electrons is not extended to the β-ionone ring. In contrast, the C6–C7 bond is in a *trans* form for archaeal-type Rhs, and the polyene and β-ionone ring is planar. Extended conjugation of π-electrons presumably contributes to the red-shifted absorption spectra in archaeal-type Rhs.

This chapter presents the current understanding of the photoreaction mechanism in Rh and BR. Ultrafast photoisomerization is particularly characteristic of Rh and BR, which takes place in a protein environment. It was argued 30 years ago that product formation in 6 ps is too fast for isomerization, but now isomerization in 200 fs for Rh and 500 fs (200–500 fs) for BR has been established. This was achieved by means of ultrafast spectroscopy with femtosecond time resolution, which allowed detection of the photophysics and photochemistry of the retinal chromophore in real time. In addition, atomic structures have been determined for Rh and BR in the last decade, which enabled us to understand the atomic details of ultrafast photoisomerization. The structures of the photoproduct states in Rh and BR have also been reported, whereas the reported structures for BR are not necessarily coincident among groups. On the basis of structure, theoretical investigation is also challenged to explain the mechanism of ultrafast photoisomerization. Here I introduce ultrafast photoisomerization in Rh and BR, where (1) the structural aspect, (2) real-time spectroscopy, and (3) supramolecular effects of protein are reviewed. Before starting each part, I show two supramolecular effects, color tuning and photochemistry, here in 17-1.

The Unique Color of the Retinal Molecule in Protein

One of the supramolecular effects of protein is color regulation. Our life would be far different if we had no color vision, and that is due to the protein control

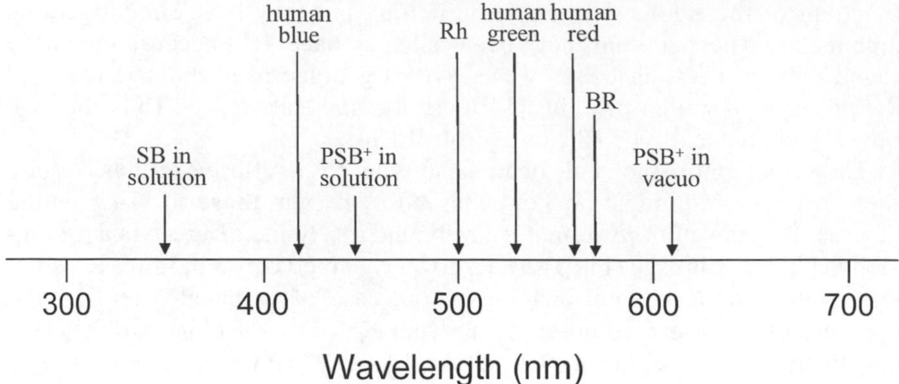

Figure 14.2. Absorption maxima of rhodopsin (Rh), human color pigments, and bacteriorhodopsin (BR), as well as the chromophore molecules in solution and vacuum. SB, Schiff base; PSB, protonated Schiff base.

of the energy gap between electronically ground and excited states of the retinal chromophore. For color regulation, protonation of the chromophore plays a crucial role. In fact, the absorption of the retinal Schiff base is in the UV region (λ_{max} ~ 350 nm) (Fig. 14.2), and the absorption is not very sensitive to the environment. In contrast, the protonated Schiff base of retinal exhibits a large variation in absorption that covers the visible region (400–700 nm). In other words, the "visible" region has been determined by the color tuning of our visual Rhs.

Wide color tuning in Rhs implies that the π–π transition of the retinal chromophore is highly sensitive to the protein environment. It should be noted, however, that such wide color tuning is not the case for the protonated Schiff base of retinal in solution. The chromophore in solution has been so far used as a standard model system of Rhs, where an acid such as HCl is normally added to stabilize the protonation state of the Schiff base.[5] The absorption spectra of the protonated Schiff base of retinal in solution have been limited (λ_{max}: 430–480 nm) under various counterions and solvents tested, which are also similar between 11-*cis* and all-*trans* forms. This indicates wide color tuning owing to the supramolecular effect of protein. The effect of protein on color tuning is normally measured as the difference in energy (cm^{-1}) between protein and solution (λ_{max}: 445 nm in methanol), which is called "opsin shift." For instance, the opsin shift of bovine Rh (λ_{max}: 500 nm) is calculated to be 2470 cm^{-1}.

What is the molecular mechanism of color tuning in protein? Among various interactions between the retinal chromophore and protein such as electrostatic effect of charged groups, dipolar amino acids, aromatic amino acids, hydrogen-bonding interactions, and steric contact effect, the effect of counterions near the Schiff base appears to be significant. The positive charge originally located at the Schiff base is more delocalized in the excited state, and more or less charge delocalization results in smaller or larger energy gap,

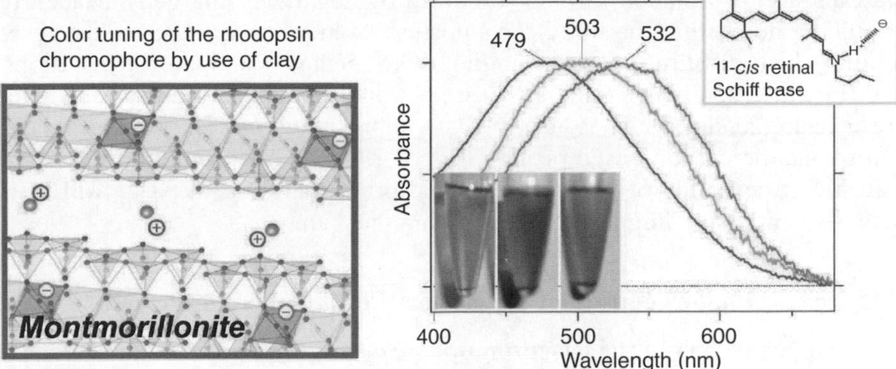

Figure 14.3. Artificial construction of color tuning in visual pigments by clay. When the protonated 11-*cis* retinal Schiff base is with montmorillonite in benzene, absorption spectra cover the entire visible range in clays from three different mining places. This figure is reproduced from Furutani et al.[11] See ftp site for color: ftp://ftp.wiley.com/public/sci_tech_med/supramolecular_photochemistry.

respectively. Accordingly, the presence of a counterion that localized the positive charge works for giving a large energy gap, that is, a spectral blue shift.

In solution, the counterion is located at the most stable position relative to the protonated Schiff base, where the Schiff base N–H group presumably forms a direct hydrogen bond. Under such strong interactions, the λ_{max} of the retinal chromophore is located at 430–480 nm. What is the opposite case, at an infinite distance? Recent successful measurement of the gas-phase absorption spectrum of the all-*trans* retinal protonated Schiff base reported the λ_{max} at 610 nm (Fig. 14.2).[9] This wavelength value might be unexpectedly short. The most red-shifted native protein is SRI (λ_{max}: 587 nm at neutral pH),[8] which does not exceed the value in vacuum. However, acid form and the O intermediate of BR have λ_{max} at 605 and 640 nm, respectively, the latter of which is even red shifted relative to that in vacuum. This suggests that the counterion effect is well screened in protein, which was also concluded from quantum chemical calculation.[10] Since the atomic structures of Rh and BR are known, highly accurate quantum chemical calculations are indeed challenging to reproduce and explain the color-tuning mechanism in protein.

There is no structural information for color pigments at present, and hence little is understood about the molecular mechanism of our color vision. On the other hand, to understand the color-tuning mechanism in protein, it is important to study the mechanism of the chromophore not in protein but in other systems. Although artificial construction of wide color tuning of the Rh chromophore in other materials had been unsuccessful for a long time, we recently found that the protonated Schiff base of 11-*cis* retinal exhibits various colors when adsorbed onto the clay montmorillonite.[11] In clays from three different mining regions, the absorption spectra cover the entire visible range (Fig. 14.3). The 11-*cis* chromophore is probably located between the layers in clay, and the protonated form is stabilized by negatively charged surfaces. It should be noted that the three clays possess an identical backbone structure, among which substitution of ions in the backbone layers and interlayer cations is different. Essentially, similar structure but fine structural modification yields color tuning in clay, which is also the case in proteins. Thus, silicate-based clay can mimic the carbon-based protein for color tuning of chromophores in our vision. Further study of both proteins and other materials like clay will lead to better understanding of the color-tuning mechanism.

The Unique Photochemistry of the Retinal Chromophore in Protein

The photochemistry of the Rh chromophore is unique in protein. For instance, the quantum yield of the photoreaction of Rh is known to be high, the molecular basis of which is responsible for the high sensitivity of our vision.[5] The quantum yield is essentially independent of temperature and excitation wavelength. Its fluorescence quantum yield was found to be very low ($\phi \sim 10^{-5}$), and the ultrafast reaction was inferred through a barrierless excited-state potential surface.[4] Product formation also takes place for Rhs at low temperatures such as liquid nitrogen (77 K) or helium (4 K) temperatures, where molecular motions are frozen.[12] In the early stage of investigation, *cis–trans* isomerization was questioned as the primary event in vision, because isomerization needs certain molecular motions of the chromophore. In fact, the first ultrafast spectroscopy of bovine Rh led to a conclusion that favors a reaction mechanism other than isomerization as the primary reaction of Rh, but ultrafast spectroscopy concluded that *cis–trans* photoisomerization is indeed a primary event.

Figure 14.4 shows photochemical reactions in Rh (a) and BR (b). In Rh, the 11-*cis* retinal is isomerized into the all-*trans* form. The selectivity is 100%, and the quantum yield is 0.67 for bovine Rh.[13] In archaeal-type Rhs, the

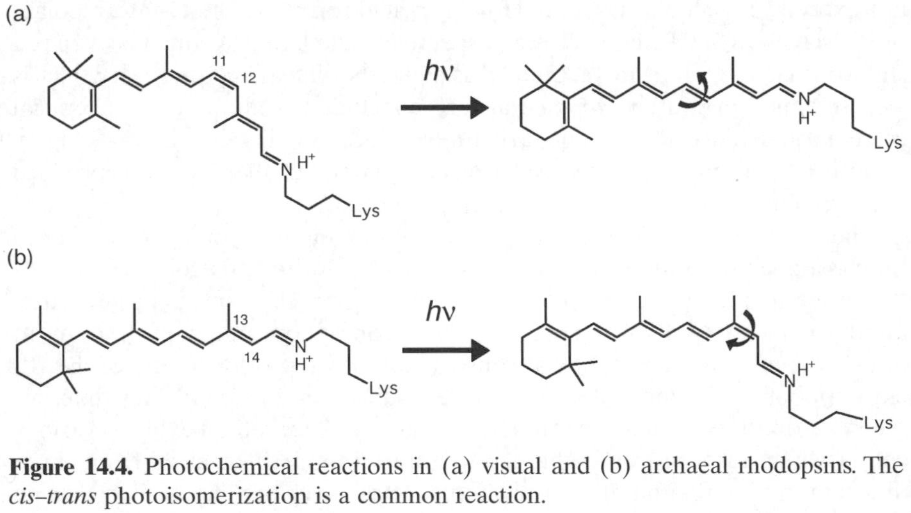

(a)

11 12 hν N H⁺ Lys

(b)

13 14 hν N H⁺ Lys

Figure 14.4. Photochemical reactions in (a) visual and (b) archaeal rhodopsins. The *cis–trans* photoisomerization is a common reaction.

all-*trans* retinal is isomerized into the 13-*cis* form. The selectivity is 100%, and the quantum yield is 0.64 for BR.[14] Squid and octopus possess a photoisomerase called retinochrome, which supplies an 11-*cis* retinal for their Rhs through the specific photoreaction. Retinochrome possesses all-*trans* retinal as the chromophore, and the all-*trans* retinal is isomerized into the 11-*cis* form with the selectivity of 100%.[15] Thus, photoproduct is different between archaeal-type Rhs and retinochrome, the all-*trans* form being converted into the 13-*cis* and 11-*cis* forms, respectively. This fact implies that the protein environment determines the reaction pathways of photoisomerization in their electronically excited states.

Previous high performance liquid chromatography (HPLC) analysis revealed that the protonated Schiff base of 11-*cis* retinal in solution is isomerized into the all-*trans* form predominantly, indicating that the reaction pathway in Rh is the nature of the chromophore itself.[16] On the other hand, the quantum yield was found to be 0.15 in methanol solution.[16] Therefore, the isomerization reaction is four to five times more efficient in protein than in solution. HPLC analysis also revealed that the protonated Schiff base of all-*trans* retinal in solution is isomerized into the 11-*cis* form predominantly (82% 11-*cis*, 12% 9-*cis*, and 6% 13-*cis* in methanol).[16] The 11-*cis* form as a photoproduct is the nature of retinochrome, not those of archaeal-type Rhs. This suggests that the protein environment of retinochrome serves as the intrinsic property of the photoisomerization of the retinal chromophore. In contrast, it seems that the protein environment of archaeal-type Rhs forces the reaction pathway of the isomerization to change into the 13-*cis* form. In this regard, it is interesting that the quantum yield of BR (0.64) is four to five times higher than that in solution (~0.15).[14,16] The altered excited-state reaction pathways in BR never reduce the efficiency. Rather, BR discovered that the reaction pathway from the all-*trans* to 13-*cis* form is efficient. Consequently, an efficient isomerization reaction is achieved as well as Rh.

In both Rh and BR, the first singlet-excited state (S1) is the Bu-like state. After the Franck–Condon excitation, stretching and torsional vibrations are coupled to relax in the S1 potential surface. The evolution along the torsional coordinate at C11=C12 in Rh and C13=C14 in BR leads to a conical intersection (CI) funnel, where the chromophore displays about 90° twisted double bond, and fully efficient decay takes place at the ground state.[17] Monitoring excited-state dynamics is challenging both experimentally and theoretically. Below, I show the investigation of Rh (17-2) and BR (17-3).

ULTRAFAST PHOTOISOMERIZATION IN RHS

The Structure of Rh: The Binding Pocket of 11-*cis* Retinal

The role of visual Rhs is to activate transducin, a heterotrimeric G protein, in the signal transduction cascade of vision.[1,2,18] Rh, a member of the G-protein-coupled receptor family, is composed of 7-transmembrane helices. At present,

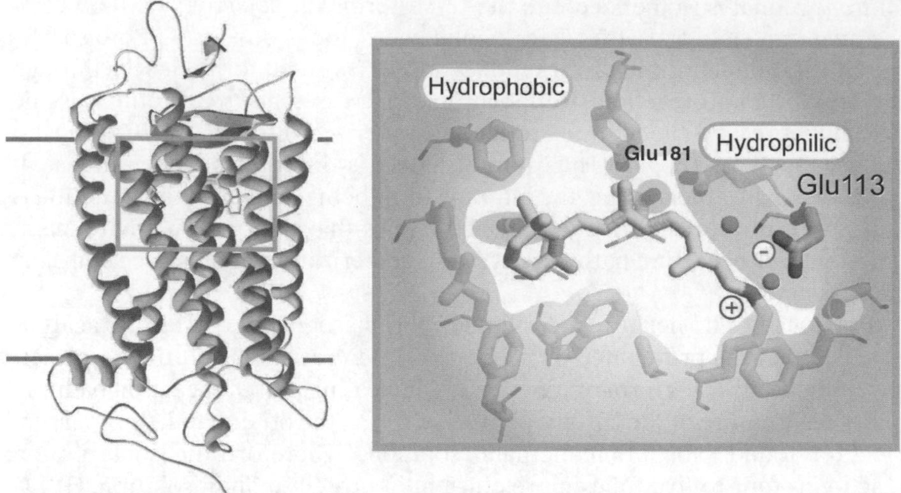

Figure 14.5. (Left) Protein structure of bovine rhodopsin.[19] The upper and lower regions correspond to the intradiscal and cytoplasmic sides, respectively. (Right) The highlighted structure of the chromophore (yellow, center stick drawing) and protein environment is shown. While hydrophobic amino acid residues surround β-ionone ring and polyene chain of the 11-*cis* retinal chromophore, the retinal Schiff base region is highly hydrophilic. The Schiff base is protonated, and negatively charged Glu113 (3.2 Å) stabilizes the protonation state of the chromophore. Protonation state of Glu181, 6.9 Å from the Schiff base nitrogen, is controversial (see text). See ftp site for color: ftp://ftp.wiley.com/public/sci_tech_med/supramolecular_photochemistry.

there are two atomic structures of Rhs, bovine Rh[19] and squid Rh,[20] both of which have essentially similar architecture. Figure 14.5 shows the structure of bovine Rh. An 11-*cis* retinal forms the Schiff base linkage with a lysine residue at the seventh helix (Lys296 for bovine Rh), where the Schiff base is protonated. The β-ionone ring of the retinal chromophore is coupled with hydrophobic region of opsin through hydrophobic interactions. Thus, the retinal chromophore is fixed by three kinds of chemical bonds—covalent bond, hydrogen bond, and hydrophobic interaction—in the retinal-binding pocket of Rh.

In bovine Rh, previous site-directed mutagenesis revealed a glutamate at position 113 (Glu113) to be the counterion (Fig. 14.5), which was determined by the fact that the substitution of Glu113 (E113Q) dramatically drops the pKa of the Schiff base.[21,22] Since the glutamate at this position is completely conserved among vertebrate photoreceptive pigments, the electrostatic interaction between the Schiff base (NH^+) and the glutamate (COO^-) is likely to be the common feature. The distance (3.2 Å) suggests the absence of an intervening water molecule, but surrounding polar amino acids and water molecules must stabilize the ion pair (Fig. 14.5).

Glu113 is not conserved in invertebrate Rh; for example, squid Rh possesses tyrosine (Tyr111) at the corresponding position. Although the counterion had been unknown for a long time, Terakita et al. revealed that Glu181 is a counterion of the Schiff base by site-directed mutagenesis.[23] Glu181 is highly conserved in the Rh family, and it was inferred that ancestral Rh has Glu181 as the counterion, but vertebrate Rh acquired both Glu113 counterion and efficient G-protein activation.[23] Recent structural determination of squid Rh further provides an insight about the counterion.[20] The Schiff base is located in a possible hydrogen-bonding distance with Asn87 (3.4 Å) and Tyr111 (3.1 Å), but Glu180 (corresponding to Glu181) is too distant to interact directly (5.0 Å). The molecular mechanism of color tuning (λ_{max}: 480 nm) and stabilization of the protonated Schiff base in squid Rh should be examined in the future. The structure of squid Rh also reported nine water molecules in an interhelical cavity,[20] which are more numerous than those in bovine Rh. Many of them probably change hydrogen bonds upon photoisomerization, because the Fourier-transform infrared (FT-IR) difference spectra between squid Rh and bathorhodopsin exhibited changes of eight water stretching vibrations.[24] An interhelical water cluster is probably involved in the photoisomerization process.

Glu181 is also one of the recent research topics in bovine Rh. Unlike invertebrate Rh, the E181Q mutant of bovine Rh has absorption similar to the wild type.[23,25] Nevertheless, the pKa of the Schiff base drops in the Meta-I state of E181Q, suggesting that Glu181 acts the Schiff base counterion in Meta-I.[25] In the "counterion switch model," it was proposed that a proton is transferred from Glu181 to Glu113 upon Meta-I formation. On the other hand, an FT-IR study of E181Q led to another proposal that Glu181 is deprotonated in the dark.[26] This idea was also supported by theoretical study.[27] FTIR spectroscopy is a strong tool to identify whether carboxylates are protonated, because vibrational frequencies at 1800–1700 cm^{-1} only originate from protonated carboxylic acids. It is noted, however, that the proposal is not from direct observation of deprotonation of Glu181 in the native Rh.[26] Therefore, it is still an open question if Glu181 is deprotonated or not in bovine Rh, and further experimental and theoretical contributions are necessary.

If Glu181 is indeed deprotonated, the "external point charge model," which attempted to explain color-tuning mechanism of Rh 30 years ago, may be relevant. Incorporation of a series of dihydroretinal was tested for bovine Rh, and from the unusual opsin shift for 11,12-dihydroretinal,[28] this model was proposed, where an additional negative charge is present at about 3 Å from the C12 and C14 atoms of the retinal chromophore.[29] Since site-directed mutagenesis of bovine Rh led to a conclusion that Glu113 is an exclusive counterion, the "external point charge model" appeared to be contradicted historically. According to the Rh structure,[19] however, the C12 atom is the nearest carbon in the chromophore from Glu181. The distance (4.5 Å) is larger than the predicted value (3 Å), but the proposal may be correct. This position is also important for red-sensitive pigments, because the corresponding histidine

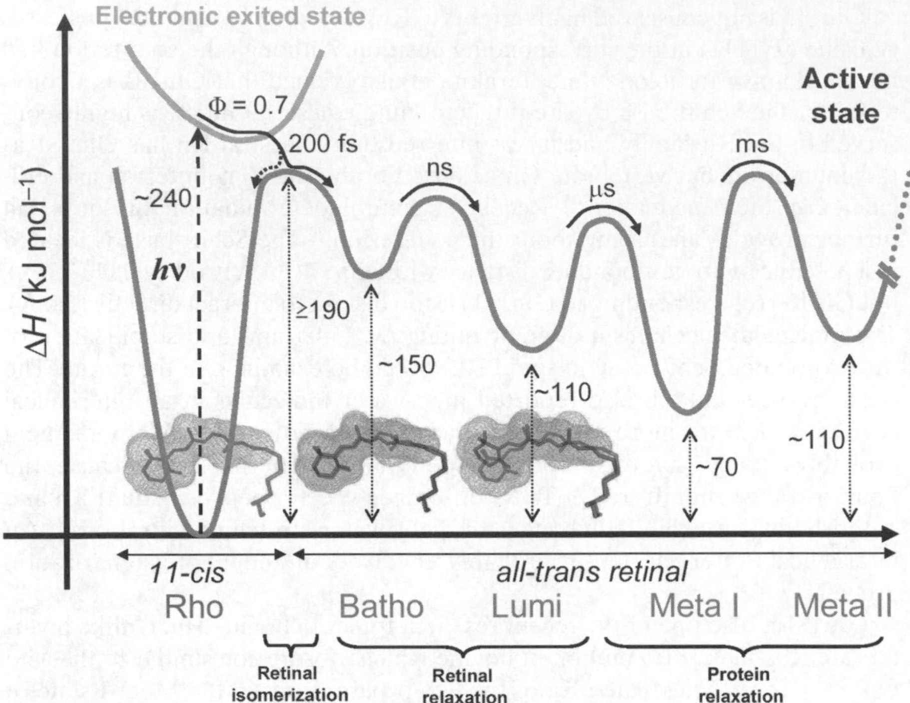

Figure 14.6. Energy diagram of the bleaching process of bovine Rh, adapted from Shichida and Imai.[1] The chromophore structures of bathorhodopsin[43] and lumirhodopsin[48] are also shown.

constitutes the binding site of Cl^- (His197 in human red and human green).[30] Thus, the role of Glu181 in both vertebrate and invertebrate Rh should be elucidated in more detail.

Ultrafast Spectroscopy of the Primary Process in Rh

Absorption of a photon by the chromophore induces the primary photoreaction, followed by conformational changes of protein, and eventually activation of the G protein (Fig. 14.6). This is called the "bleaching process" because Rh loses its color. To investigate the primary photoreaction processes in Rh, two spectroscopic approaches have been applied: low-temperature and time-resolved spectroscopy.[4,12] They can detect primary photointermediate states by reducing the thermal reaction rate at low temperature (low-temperature spectroscopy) or by directly probing dynamical processes at physiological temperatures (time-resolved spectroscopy). In 1963, low-temperature spectroscopy at 87 K detected a red-shifted product.[31] Since this photoproduct (now called bathorhodopsin) is converted to lumirhodopsin on warming and finally decomposed to all-*trans* retinal and opsin through several intermediates, it was proposed that bathorhodopsin has a "highly constrained and distorted" all-*trans* retinal as its chromophore and is on a higher potential energy level than Rh and subsequent intermediates.[32] According to the prediction of Yoshizawa and Wald, the process of Rh to bathorhodopsin should be a *cis–trans* isomerization of the chromophore.

The first picosecond experiment of bovine Rh in 1972 observed formation of bathorhodopsin within 6 ps after excitation at room temperature, and concluded that its formation would be too fast to be attributed to such a conformational change as the *cis–trans* isomerization of the retinal chromophore.[33] This could be reasonable because isomerization accompanies molecular motion that must take some time, whereas the recent view of photoisomerization in protein is entirely different. The experimental evidence of the 11-*cis* to all-*trans* isomerization as the primary reaction in Rh was given by our time-resolved study of the Rh analogs possessing 11-*cis*-locked ring retinals. In the case of five-membered Rh, only a long-lived excited state ($\tau = 85\,\text{ps}$) was formed without any ground-state photoproduct (Fig. 14.7d), giving direct evidence that the *cis–trans* isomerization is the primary event in vision.[34] Excitation of seven-membered Rh, on the other hand, yielded a ground-state photoproduct having a spectrum similar to photorhodopsin (Fig. 14.7c). These different results were interpreted in terms of the rotational flexibility along C11–C12 double bond.[34] This hypothesis was further supported by the results with an eight-membered Rh that possesses a more flexible ring. Upon excitation of the eight-membered Rh with a 21-ps pulse, two photoproducts, photorhodopsin-like and bathorhodopsin-like products, were observed (Fig. 14.7b).[35] Photorhodopsin is a precursor of bathorhodopsin found by picosecond transient absorption spectroscopy.[36] Thus, the picosecond absorption studies directly elucidated the correlation between the primary processes of Rh and the flexibility of C11–C12 double bond of the chromophore, and we eventually concluded that the respective potential surfaces were as shown in Figure 14.7. The structure of the intermediate states in Rh7 and Rh8 has been theoretically studied recently.[37]

In 1991, the *cis–trans* isomerization process in real time has been directly observed with the use of femtosecond pulses. Femtosecond transient absorption spectroscopy of bovine Rh showed that product formation was completed within 200 fs, suggesting that isomerization is an event in femtoseconds.[38] In addition, oscillatory features with a period of 550 fs ($60\,\text{cm}^{-1}$) were observed on the kinetics at probe wavelengths within the photoproduct absorption band of Rh, whose phase and amplitude demonstrate that they are the result of nonstationary vibrational motion in the ground state of photorhodopsin.[39] The observation of coherent vibrational motion in photorhodopsin supports the idea that the primary step in vision is a vibrationally coherent process and that the high quantum yield of the *cis–trans* isomerization in Rh is a consequence of the extreme speed of the excited-state torsional motion.[3]

Ultrafast photoisomerization in Rh was also probed by other experimental techniques. We performed fluorescence measurement of bovine Rh, which

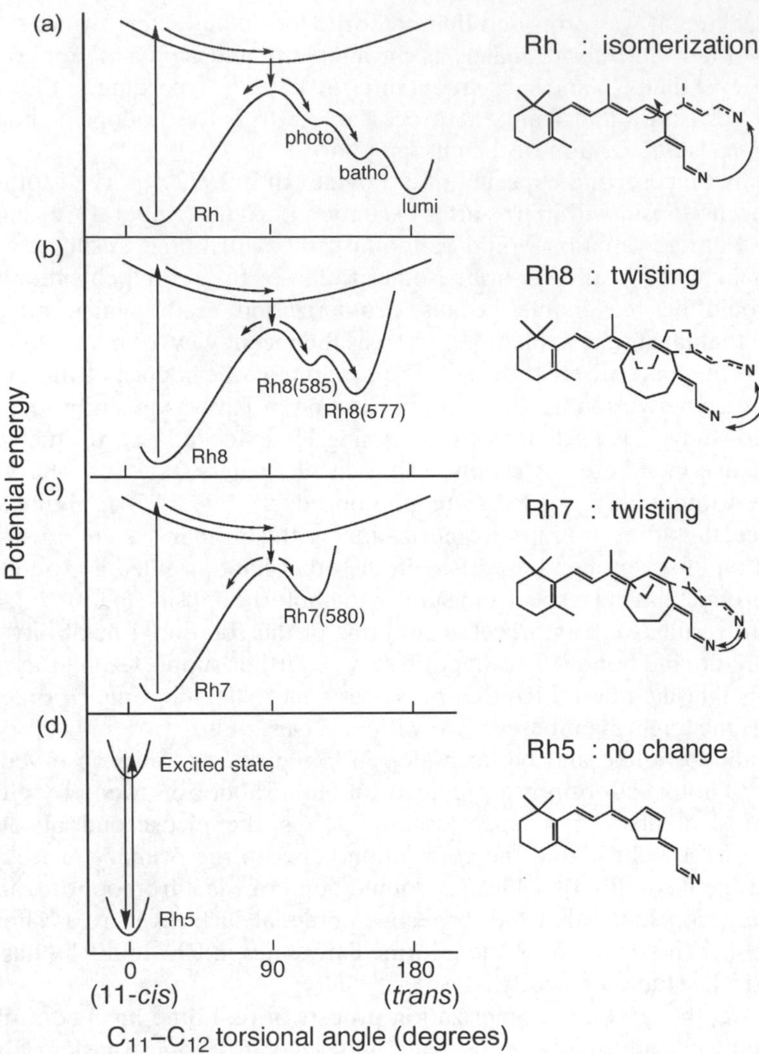

Figure 14.7. Schematic drawing of ground- and excited-state potential surfaces along the 11-ene torsional coordinates of the chromophore of rhodopsin (a), eight-membered rhodopsin (b), seven-membered rhodopsin (c), and five-membered rhodopsin (d). Modified from Mizukami et al.[35]

monitored decay components of 100–300 fs as well as small 1.0–2.5 ps components, and we concluded that the slow components (~30%) originate from the nonreactive excited state of Rh, while the fast components (~70%) come from the coherent isomerization observed in the femtosecond transient absorption spectroscopy.[40] In addition, femtosecond stimulated Raman spectroscopy reported time-resolved vibrational spectra from 200 fs to 1 ps, providing structural information of the retinal chromophore.[41] The highly twisted retinal molecule was successfully monitored in real time.

The Role of the Protein Environment: Energy Storage and Functional Expression

It is well known that Rh is an excellent molecular switch to convert a light signal into an electric response of the photoreceptor cell. As mentioned, the highly efficient photoisomerization of Rh (quantum yield: 0.67) is assured by the extremely fast *cis–trans* isomerization of the chromophore that is facilitated by the protein environment. Then, the question arises concerning the supramolecular effect on such photoisomerization. Figure 14.4a shows that the Schiff base side actually rotates, while there is no change at the side of β-ionone ring. This might be reasonable, because β-ionone ring is bulky and is difficult to move. However, molecular motion of the Schiff base in Figure 14.4a is also significant, and an entire rotation must not be the case for bathorhodopsin. The molecular shape has to be unchanged before and after photoisomerization from the 11-*cis* to all-*trans* form. The molecular mechanism is described in detail in the previous section.

Experimentally, important information was first obtained by resonance Raman spectroscopy,[3] which observed the C=N stretching vibration of the Schiff base for Rh and bathorhodopsin. The C=N stretch in H_2O is considerably up-shifted from the intrinsic stretching mode by coupling with the N–H in-plane bending vibration. The intrinsic C=N stretch can be measured in D_2O, where the coupling of the N–H bending vibration is removed. Since the frequency of the N–H in-plane bending increases if the hydrogen bond of the Schiff base is strengthened, the frequency difference of the C=N stretch in H_2O and D_2O has been regarded as the marker of the hydrogen-bonding strength of the Schiff base. That is, a large difference corresponds to strong hydrogen bond. The differences of the C=NH and C=ND frequencies were reported to be identical ($32\,cm^{-1}$) between bovine Rh and bathorhodopsin, implying no change of the Schiff base hydrogen bond upon retinal isomerization.[42]

About 30 years later, X-ray crystallography was applied to bathorhodopsin by illuminating the bovine Rh crystal at 105 K.[43] The difference in electron density between Rh and bathorhodopsin was small (see Fig. 14.6), as expected. Regarding the retinal chromophore, the displacement was very small at both sides—the β-ionone ring and the Schiff base—the latter of which is consistent with the Raman data[42] of no change in hydrogen-bonding strength of the Schiff base. In contrast, the dihedral angle around the C11=C12 bond changes from –40 to –155°, indicating that isomerization is certainly taking place. Thus, a greatly distorted polyene chain results in minimally changed overall chromophore shape, and this view is supported by femtosecond stimulated Raman spectroscopy[41] and quantum chemical calculation.[44]

Low-temperature photocalorimetric study revealed that about 60% of light energy ($\sim 150\,kJ\,mol^{-1}$) is stored in the structure of bathorhodopsin (Fig. 14.6).[45]

A highly twisted retinal chromophore contributes to the energy storage, and such structural deformation of the polyene chain has been experimentally monitored by enhanced hydrogen-out-of-plane (HOOP) vibrational modes at 1000–800 cm^{-1}.[46,47] It should be noted that electrostatic and hydrogen-bonding interactions are small, because the hydrogen-bonding strength of the Schiff base is unchanged before and after isomerization.[42] This is an interesting contrast to the case of BR (see 17-3). Relaxation of the distorted polyene chain leads to the formation of lumirhodopsin, which was monitored clearly by X-ray crystallography.[48] Formation of lumirhodopsin accompanies motion of the β-ionone ring toward the fourth helix (Fig. 14.6), which would be important for G-protein activation. Such motion was previously suggested by a photoaffinity labeling experiment.[49]

As is seen, the protein environment beautifully controls efficient *cis–trans* photoisomerization in Rh. The role of two methyl groups at the C9 and C13 positions is also noteworthy. Upon photoisomerization, the C9-methyl does not change its position, while the C13-methyl group actually moves (Fig. 14.6). The C13-methyl group has a steric conflict with the retinal C10-hydrogen in Rh, which causes the twists of the C11=C12 and C12–C13 bonds even in the dark; the important role of the C13-methyl for efficient isomerization has been reported experimentally[50] and theoretically.[51] In contrast, the C9-methyl group appears to function as a scaffold for the primary isomerization,[43] while its motion is important for the activation of G protein at late intermediates.[52]

The chromophore molecule in solution (particularly methanol) was useful to evaluate the "opsin shift" in color-tuning mechanism. Similarly, excited-state isomerization dynamics were compared for the chromophore of Rh between protein and solution by means of femtosecond fluorescence measurements, where five-membered 11-*cis*-locked retinal was also used. In methanol, the fluorescence lifetime of the five-membered chromophore was five times longer than that of the protonated Schiff base of 11-*cis* retinal.[53] This is in remarkable contrast to those in protein, because the fluorescence lifetime of five-membered Rh (85 ps) is two orders of magnitude longer than that of the native Rh.[34,40] These facts suggest protein moiety enhancing the isomerization rates of the chromophore, where functional optimization has been achieved during evolution. Finally, the recent structural determination of opsin should be noted. Unexpectedly, the surrounding amino acids are not much changed even in the absence of retinal,[54] which might be occupied by water molecules. The new structure further implicates the role of protein as the supramolecular effect.

ULTRAFAST PHOTOISOMERIZATION IN BR

Structure of BR: The Binding Pocket of All-*trans* Retinal

BR is a light-driven proton pump.[5–7] Figure 14.8 (left) illustrates the whole structure as well as proton pathways. In BR, seven helices constitute transmembrane portion of protein, and a retinal chromophore is bound to a lysine

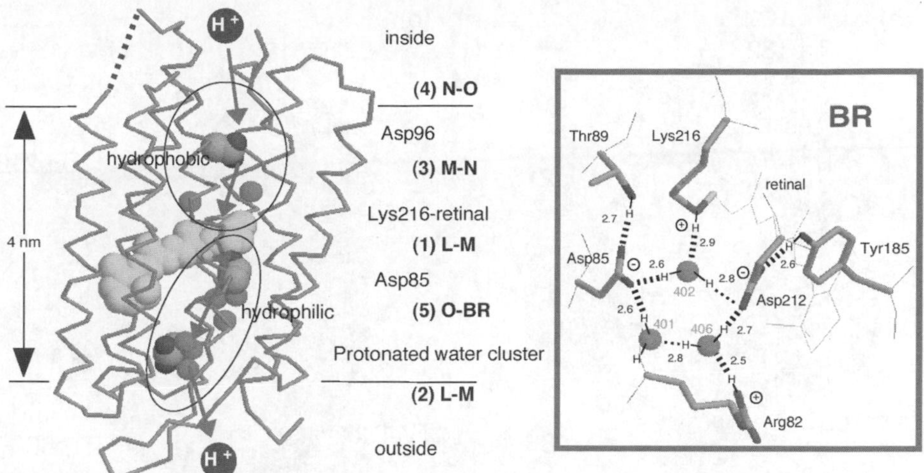

Figure 14.8. (Left) Protein structure of bacteriorhodopsin (BR).[55] All-*trans* retinal (large yellow structure at the center), internal water molecules (small gray or green spheres), and the amino acids constituting the proton pathway are shown. In BR, a proton is translocated by the sequential proton transfer reactions in the order shown in the figure. The upper and lower regions correspond to the cytoplasmic and extracellular sides, respectively. (Right) The structure of the Schiff base region in BR.[55] Green spheres represent oxygen atoms of water. Dotted lines are putative hydrogen bonds, whose distances are shown in angstroms. See ftp site for color: ftp://ftp.wiley.com/public/sci_tech_med/supramolecular_photochemistry.

residue of the seventh helix (Lys216) via a protonated Schiff base linkage.[55] The retinal chromophore is located at the center of the membrane, and the hydrophobicity is different between the cytoplasmic and extracellular domains. The cytoplasmic domain is highly hydrophobic, whereas the extracellular domain is composed of charged and polar amino acids that form a hydrogen-bonding network.[7] Figure 14.8 (left) shows the presence of seven to eight water molecules in the extracellular domain, but only two water molecules in the cytoplasmic domain.[55] Such an asymmetric hydrogen-bonded network could be the reason for the unidirectional proton transport in BR, where the proton transfer to the extracellular side occurs in 10^{-5} s, followed by reprotonation through a transiently formed proton pathway in the cytoplasmic domain on a slower timescale (10^{-4}–10^{-3} s).[6,7]

Figure 14.8 (right) shows that the Schiff base region has a quadrupolar structure with positive charges located at the protonated Schiff base and Arg82, and counterbalancing negative charges located at Asp85 and Asp212.[55] The quadrupole inside the protein is stabilized by three water molecules (water401, 402, and 406), where a water molecule (water402) bridges the Schiff base and counterions, unlike in the case of Rh. A notable structural feature is that Asp85 and Asp212 are located at similar distances from the Schiff base,

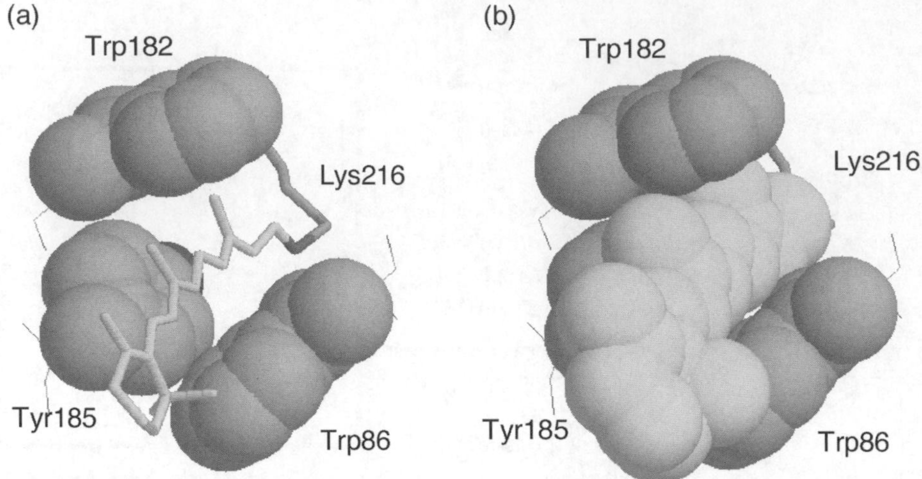

Figure 14.9. Crystallographic structure of bacteriorhodopsin (BR).[55] Trp86, Trp182, and Tyr185 are shown together with the retinal chromophore in the stick (a) or space-filling (b) drawing.

whereas the Schiff base proton is transferred only to Asp85 in microseconds. This fact suggests that the water molecules in the Schiff base region play important roles in the proton transfer reaction.[56] We succeeded in detecting vibrations of water molecules by means of low-temperature FT-IR spectroscopy,[57] and various mutant proteins were used to identify the vibrational bands. Consequently, it was revealed that water402 forms a very strong hydrogen bond with Asp85 (O-D stretch in D_2O: 2171 cm^{-1}), but interaction with Asp212 is weak (O-D stretch in D_2O: 2636 cm^{-1}).[58,59] As shown below, it has been found that such a strong hydrogen bond of water is a prerequisite for proton pump function.

Figure 14.9 illustrates another structural feature of BR. The linear polyene chain of the all-*trans* retinal is sandwiched vertically by two tryptophans, Trp86 and Trp182, while the phenol ring of Tyr185 is located parallel to the polyene chain of the retinal chromophore. Thr89 is located at the other side (not shown). The presence of three bulky groups, Trp86, Trp182, and Tyr185, presumably determines the isomerization pathway from all-*trans* to 13-*cis* form. As described above, the 11-*cis* form is the main photoproduct of the protonated Schiff base of all-*trans* retinal in solution.[5] Therefore, the reaction pathway is altered in the protein environment of BR, whereas the quantum yield of isomerization is four to five times higher in BR (0.64) than in solution (~0.15).[14,16]

The Primary Process in BR Studied by Ultrafast Spectroscopy

Unlike visual Rhs that bleach upon illumination, archaeal-type Rhs thermally return to the original state after functional processes. This is highly advantageous in the ultrafast spectroscopic studies. Therefore, various ultrafast spectroscopic techniques have been extensively applied in addition to low-temperature spectroscopy. In particular, BR has historically been regarded as the model system to test new spectroscopic methods. Like visual Rhs, light absorption of archaeal-type Rhs causes formation of red-shifted primary intermediates.[6,7] The primary K intermediate can be stabilized at 77 K. Time-resolved visible spectroscopy of BR revealed the presence of the precursor, called the J intermediate. The J intermediate is more red shifted (λ_{max} ~ 625 nm) than the K intermediate (λ_{max} ~ 590 nm). The excited state of BR possesses blue-shifted absorption, which decays in about 200 and 500 fs.[60] Since the J intermediate is formed in <500 fs, it is reasonable to consider photoisomerization taking place in femtoseconds.

To test when isomerization occurs, all-*trans*-locked five-membered retinal was incorporated into BR[61–63] as in the case of Rh, but the results were very different. In the experiments with picosecond time resolution, an intermediate was found with properties similar to those of the J intermediate.[61] Together with the ultrafast pump probe[62] and coherent anti-Stokes Raman[63] spectroscopic results, it was concluded that isomerization around C13=C14 is not a prerequisite for producing the J intermediate. More important, since the J intermediate is a ground-state species, isomerization does not take place in the excited state of BR according to their interpretation.[61–63]

However, other experimental data favor the common mechanism between visual and archaeal-type Rhs, namely, isomerization taking place in the excited state. Femtosecond visible-pump and infrared-probe spectroscopy detected the 13-*cis* characteristic vibrational band at 1190 cm^{-1} appearing with a time constant of ~0.5 ps, indicating that the all-*trans* to 13-*cis* isomerization takes place in femtoseconds.[64] This timescale is coincident with the formation of the J intermediate. Fourier-transform of the transient absorption data with <5-fs resolution also showed the appearance of the 13-*cis* form in <1 ps, supporting that the all-*trans* to 13-*cis* isomerization takes place in femtoseconds.[65] Previous anti-Stokes resonance Raman spectroscopy proposed that the J intermediate is a vibrational hot state of the K intermediate.[66] Thus, many experimental results are consistent with the isomerization model in the excited state.

Photoisomerization takes place in the excited state similarly for Rh and BR, but relaxation from the Franck–Condon state may be different. In Rh, the excited wave packet slides down the barrierless potential surface (Fig. 14.7). In contrast, several experiments of BR favor the three-state model that postulates a small potential barrier along the isomerization coordinate.[65,67–69] Actually, real-time spectroscopy of BR with <5-fs pulse showed that isomerization does not occur instantly, but involves transient formation of a tumbling state.[65] This clearly supports the three-state model and discounts the initially suggested two-state model. Essentially, I agree with this view, but I like to emphasize that the two- and three-state models are not sufficient, because both models adopt only one coordinate on isomerization. Fluorescence spectroscopy of bovine Rh[40] and transient absorption spectroscopy of HR[70,71] clearly detected a prolonged excited state after photoisomerization was complete, which can be explained by neither two- nor three-state models. Another coordinate uncoupled to isomerization has to be taken into account, and personally I think that BR also possesses such a state.

Unlike Rh, ultrafast spectroscopy has also been applied to various BR mutants, revealing that only the replacements of the charged residues reduced the photoisomerization rate, leading to less efficient photoisomerization.[72] This observation explains an important role of the electrostatic interaction of the counterion complex in the primary photoisomerization mechanism (Fig. 14.8, right). The prompt response of tryptophans (Fig. 14.9) after retinal photoexcitation was also reported by monitoring absorption of tryptophans.[73] Excited state is more long-lived in other archaeal-type Rhs,[70,71,74,75] and hence less efficient for photoisomerization. These observations suggest that BR possesses the optimized structure for the primary photoisomerization mechanism.

The Role of Protein Environment: Energy Storage and Functional Expression

It is noted that light-energy storage in archaeal-type Rhs is lower than in bovine Rh. Low-temperature photocalorimetric studies reported the energy stored in the primary intermediates to be ~150 and ~67 kJ mol^{-1} for bovine Rh[45] and BR,[76] respectively. This indicates that only 30% of light energy is stored in the structure of the primary K intermediate, which is about half of the case in Rh. In other words, about 70% of light energy is dissipated in the formation of the K intermediate. Nevertheless, such energy loss may not be serious from the functional point of view, because BR pumps a single proton by the use of one photon, and the free energy gain by pumping a proton is about 25 kJ mol^{-1}.[5] Interestingly, quantum yields of photoisomerization are not so different between bovine Rh (0.67) and BR (0.64). Therefore, such a difference in energy storage must be correlated with the structures of the primary intermediates.

X-ray crystallographic structures of the primary K intermediate have been reported by three groups.[77–79] As expected, protein structures are changed little before and after isomerization. One group concluded that the energy storage in the K state of BR is almost completely explained by the distortion at position C13.[78] However, it would be difficult to determine the bond angle accurately under the current resolution (>2.0 Å). In fact, three reported structures of the K intermediate of BR are considerably different among groups.[77–79] Vibrational spectroscopy reported that the enhanced HOOP vibrations in the K intermediate are mostly H-D exchangeable, implying that chromophore distortion is localized at the Schiff base moiety.[5] The stretching vibration of the Schiff base is significantly up-shifted (350 cm^{-1} for the N-D stretch and ~500 cm^{-1} for the N-H stretch), indicating the weakened hydrogen bond.[80] Thus, in the case of BR, retinal isomerization accompanies rotational motions of the Schiff base side, so that the hydrogen bond is significantly weakened; this is in contrast to visual Rhs.

Figure 14.8 shows that the hydrogen-bonding acceptor of the Schiff base is a water molecule, which bridges the ion pair. The motion of the Schiff base probably enforces rearrangement of the water-containing hydrogen-bonding network in the Schiff base region, and this was indeed probed by FT-IR spectroscopy. We observed the presence of strongly hydrogen-bonded water molecules in BR, and the frequencies are up-shifted on the formation of the K intermediate.[57,59] This observation strongly suggests that the hydrogen-bonding interaction is highly destabilized in the K state, which possibly contributes to the high energy state. In fact, quantum chemical/molecular mechanics normal mode calculations successfully reproduced the prominent spectral features for BR and the K intermediate.[81,82] This further suggests that the contribution of weakened hydrogen bonds to the energy storage is $46\,kJ\,mol^{-1}$, which is more than half of the total energy storage ($67\,kJ\,mol^{-1}$). Energy storage through the distorted retinal chromophore has been well accepted (this can be adopted for Rh), but energy storage in the hydrogen-bonding destabilization is a rather new concept.

The importance of hydrogen-bonding alteration in energy storage provided an unexpected finding, that is, positive correlation between the strong hydrogen bond of water and proton pump activity.[83] As mentioned, we measured FT-IR difference spectra of internal water molecules at 77 K, where measurement in D_2O is advantageous, because X-H and X-D stretching can be separated. Water O-D stretching vibrations appear at $2700–2100\,cm^{-1}$ dependent on their hydrogen-bonding strength, and we define the strong hydrogen bond at $<2400\,cm^{-1}$. The mutation study of BR showed that D85N and D212N only exhibit no water bands under strongly hydrogen-bonding conditions,[59] suggesting that such water molecules may be a prerequisite for the proton-pump function. Since then, we tested various Rhs including visual Rh, in view of whether the protein contains strongly hydrogen-bonded water or not in the unphotolyzed state (O-D stretch at $<2400\,cm^{-1}$).

Figure 14.10 summarizes the results, which clearly show a strong correlation between the presence of strongly hydrogen-bonded water molecule(s) and proton-pumping activity. For example, strongly hydrogen-bonded water molecules are observed for BR,[59] azide-bound HR,[84] SRII,[85] proteorhodopsin,[86] and *Leptosphaeria* Rh,[87] which all pump protons. Strongly hydrogen-bonded water molecules were not observed for HR,[88] *Anabaena* sensory Rh,[89] *Neurospora* Rh,[90] bovine Rh,[91] and squid Rh,[24] which have no proton-pumping activity. Cl^--pumping D85S BR has no strongly hydrogen-bonded water molecules,[92] but proton-pumping D212N(Cl^-) BR has strongly hydrogen-bonded water molecules.[93] These results were interpreted in terms of the proposal that the presence of a strong hydrogen bond of water is prerequisite for proton pumping in Rhs.[83] It is likely that destabilization of a water-containing hydrogen-bonding network plays an important role for light-energy storage in this case. Figure 14.10 shows two exceptions. The 13-*cis* form of BR[94] and SRII complex with the transducer protein[95] possess strongly hydrogen-bonded water, but no proton pump activity. However, the former has the 13-*cis*

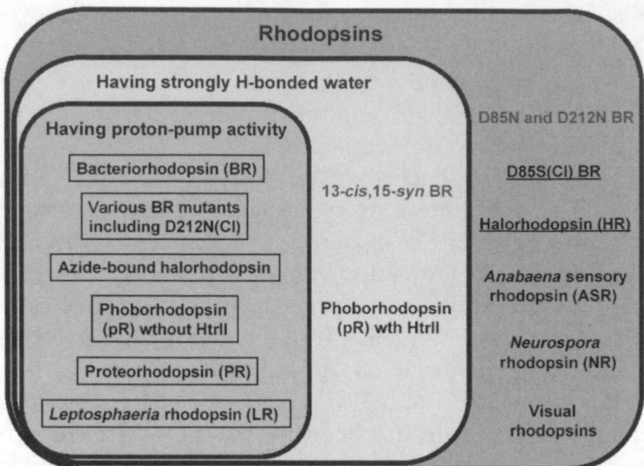

Figure 14.10. Various rhodopsins are classified according to (1) proton-pump activity and (2) whether they have strongly hydrogen-bonded water molecules (O-D stretch in D_2O at <2400 cm^{-1}). BR and various BR mutants including D212N(Cl), azide-bound HR, SRII without HtrII, PR, and LR are proton pumps, while D85S(Cl) BR and HR are chloride-ion pumps. On the other hand, SRII with HtrII, ASR, NR, and visual rhodopsins are light sensors, and 13-*cis*, 15-*syn* BR, D85N, and D212N BR are non-functional proteins.

chromophore, and it is known that only all-*trans* chromophore has proton pump activity.

Energy storage in SRII is particularly noted. SRII[85] and the SRII/HtrII complex[95] both possess strongly hydrogen-bonded water molecules, whereas the latter does not pump protons. Presumably, the specific interaction between SRII and HtrII ceases the proton-pump activity. Instead, the SRII/HtrII complex functions in light-signal transduction. Phylogenetic analysis suggests that Rh photosensors like SRI and SRII evolved from light-driven proton pumps.[96] Therefore, sensory Rhs can be thought of as pumps that have gained a mechanism to activate transducer proteins. We previously observed enhancement of the C14-D stretching vibration of the retinal chromophore at 2244 cm^{-1} upon formation of the K intermediate, and interpreted that a steric constraint occurs at the C14-D group in $SRII_K$ (Fig. 14.11).[97] By use of a mutant, the counterpart of the C14-D group was determined to be Thr204.[98] Although the K state of the wild-type BR does not possess the 2244-cm^{-1} band, the band newly appeared for the K state of a triple mutant of BR that functions as a light sensor (P200T/V210Y/A215T).[99] We found a positive correlation between the vibrational amplitude of the C14 atom at 77 K and the physiological phototaxis response. These observations strongly suggest that the steric constraint between the C14 group of retinal and Thr204 of the protein (Thr215 for a triple mutant of BR) is prerequisite for light-signal transduction by SRII.[98] Thus,

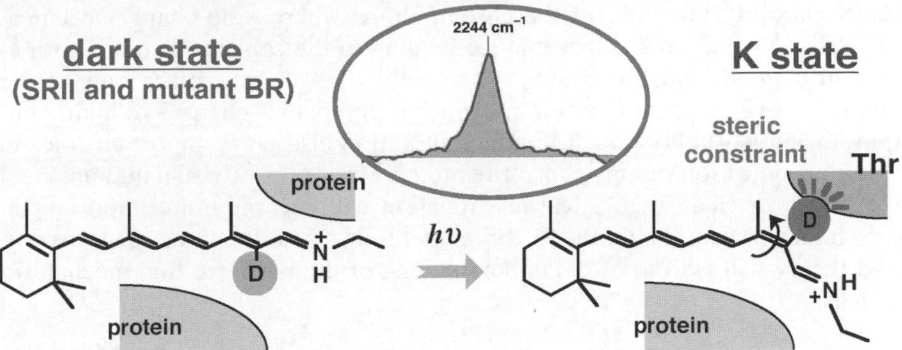

Figure 14.11. Schematic of the steric constraint formed in the K state of SRII and a mutant BR converted into a sensor. Among all seven monodeuterated all-*trans* retinal analogs (positions 7, 8, 10, 11, 12, 14, 15), only the C14-D stretching vibration at $2244\,cm^{-1}$ is significantly enhanced upon formation of the K state of SRII,[97] where the counterpart in the protein side is Thr204.[98] The corresponding amino acid in BR is Ala215. Although the wild-type BR shows no enhanced C14-D stretch, a similar band appeared in a triple mutant (P200T/V210Y/A215T) possessing Thr215.[98] The mutant BR has a sensor function,[99] and a positive correlation was observed between the C14-D stretching signal and phototaxis.[98]

photoisomerization-induced steric constraint between the C14-H of the retinal chromophore and Thr204 in SRII is one such mechanism. It is likely that a water-containing hydrogen-bonding network is important for the energy storage in proton pumps like BR, while steric constraint is important for the energy storage in sensor functions as seen for visual Rh, SRII, and a BR mutant as a light sensor.

SUMMARY AND PROSPECTS

This chapter has gathered together the current understanding of retinal photoisomerization in Rh and BR. Extensive studies by means of ultrafast spectroscopy of Rh and BR have shown that the primary process is isomerization: from 11-*cis* to all-*trans* form in Rh and from all-*trans* to 13-*cis* form in BR. Femtosecond spectroscopy of visual and archaeal Rhs eventually captured their excited states, and as the consequence, we have known that the unique photochemistry takes place in our eyes and in archaea. Such unique reactions are facilitated in the protein environment, and recent structural determination opened further understanding on the basis of structure.

In solution, photochemical properties are similar between the protonated Schiff bases of 11-*cis* and all-*trans* retinal. Nevertheless, we have learned the considerable difference in photochemistries of 11-*cis* (visual) and all-*trans* (archaeal) forms in protein (Rh). In visual Rhs, conformational distortion takes place at the center of the chromophore, whereas no changes occur at the Schiff base. Such changes may be leading to the coherent product formation in femtoseconds. In contrast, structural changes take place only at the Schiff base region of archaeal Rhs, which accompanies changes in hydrogen-bonding network. Hydrogen-bonding alteration also plays important roles in the function of ion pumps. Since the atomic structures of visual and archaeal Rhs are now clear, theoretical investigation will become more important in the future. A combination of three methods of diffraction, spectroscopy, and theory will lead to a real understanding of the isomerization mechanism in Rhs.

ACKNOWLEDGMENTS

I thank many collaborators, whose work appears in the references. For this review chapter, I thank Drs. Yuji Furutani and Victor A. Lorenz Fonfria for their help in preparing the figures.

REFERENCES

1. Shichida, Y.; Imai, H. *CMLS Cell. Mol. Life Sci.* **1998**, *54*, 1299.
2. Sakmar, T.P. *Prog. Nucleic Acid Res. Mol. Biol.* **1998**, *59*, 1.
3. Mathies, R.A.; Lugtenburg, J. In: Stavenga, D.G.; de Grip, W.J.; Pugh, E.N., editors. *Handbook of Biological Physics*, Vol. 3. Amsterdam: Elsevier, **2000**, pp. 1197–1209.
4. Kandori, H.; Shichida, Y.; Yoshizawa, T. *Biochemistry (Mosc.)* **2001**, *66*, 1197.
5. Kandori, H. In: Dugave, C., editor. *cis-trans Isomerization in Biochemistry*. Freiburg: Wiley-VCH, **2006**, pp. 53–75.
6. Haupts, U.; Tittor, J.; Oesterhelt, D. *Annu. Rev. Biophys. Biomol. Struct.* **1999**, *28*, 367.
7. Lanyi, J.K. *J. Phys. Chem. B* **2000**, *104*, 11441.
8. Spudich, J.L.; Yang, C.-S.; Jung, K.-H.; Spudich, E.N. *Annu. Rev. Cell Dev. Biol.* **2000**, *16*, 365.
9. Andersen, L.H.; Nielsen, I.B.; Kristensen, M.B.; El Ghazaly, M.O.A.; Haacke, S.; Nielsen, M.B.; Petersen, M.A. *J. Am. Chem. Soc.* **2005**, *127*, 12347.
10. Andruniów, T.; Ferré, N.; Olivucci, M. *Proc. Natl. Acad. Sci. U.S.A.* **2004**, *101*, 17908.
11. Furutani, Y.; Ido, K.; Sasaki, M.; Ogawa, M.; Kandori, H. *Angew. Chem. Int. Ed. Engl.* **2007**, *46*, 8010.
12. Yoshizawa, T.; Kandori, H. In: Osborne, N.; Chader, G., editors. *Progress in Retinal Research*. Oxford: Pergamon Press, **1992**, pp. 33–55.
13. Dartnall, H.J.A. *Vision Res.* **1967**, *8*, 339.
14. Tittor, J.; Oesterhelt, D. *FEBS Lett.* **1990**, *263*, 269.
15. Furutani, Y.; Terakita, A.; Shichida, Y.; Kandori, H. *Biochemistry* **2005**, *44*, 7988.

16. Koyama, Y.; Kubo, K.; Komori, M.; Yasuda, H.; Mukai, Y. *Photochem. Photobiol.* **1991**, *54*, 433.
17. Garavelli, M.; Celani, P.; Bernardi, F.; Robb, M.A.; Olivucci, M. *J. Am. Chem. Soc.* **1997**, *119*, 6891.
18. Hofmann, K.-P.; Helmreich, E.J.M. *Biochim. Biophys. Acta* **1996**, *1286*, 285.
19. Palczewski, K.; Kumasaka, T.; Hori, T.; Behnke, C.A.; Motoshima, H.; Fox, B.A.; Le Trong, I.; Teller, D.C.; Okada, T.; Stenkamp, R.E.; Yamamoto, M.; Miyano, M. *Science* **2000**, *289*, 739.
20. Murakami, M.; Kouyama, T. *Nature* **2008**, *453*, 363.
21. Zhukovsky, E.A.; Oprian, D.D. *Science* **1989**, *246*, 928.
22. Sakmar, T.P.; Franke, R.R.; Khorana, H.G. *Proc. Natl. Acad. Sci. U.S.A.* **1989**, *86*, 8309.
23. Terakita, A.; Koyanagi, M.; Tsukamoto, H.; Yamashita, T.; Miyata, T.; Shichida, Y. *Nat. Struct. Mol. Biol.* **2004**, *11*, 284.
24. Ota, T.; Furutani, Y.; Terakita, A.; Shichida, Y.; Kandori, H. *Biochemistry* **2006**, *45*, 2845.
25. Yan, E.C.Y.; Kazmi, M.A.; Ganim, Z.; Hou, J.-M.; Pan, D.; Chang, B.S.W.; Sakmar, T.P.; Mathies, R.A. *Proc. Natl. Acad. Sci. U.S.A.* **2003**, *100*, 9262.
26. Ludeke, S.; Beck, M.; Yan, E.C.; Sakmar, T.P.; Siebert, F.; Vogel, R. *J. Mol. Biol.* **2005**, *353*, 345.
27. Bravaya, K.; Bochenkova, A.; Granovsky, A.; Nemukhin, A. *J. Am. Chem. Soc.* **2007**, *129*, 13035.
28. Arnaboldi, M.; Motto, M.G.; Tsujimoto, K.; Balogh-Nair, V.; Nakanishi, K. *J. Am. Chem. Soc.* **1979**, *101*, 7082.
29. Honig, B.; Dimur, U.; Nakanishi, K.; Balogh-Nair, V.; Gawinowicz, M.A.; Arnaboldi, M.; Motto, M.G. *J. Am. Chem. Soc.* **1979**, *101*, 7084.
30. Wang, Z.; Asenjo, A.B.; Oprian, D. *Biochemistry* **1993**, *32*, 2125.
31. Yoshizawa, T.; Kito, Y. *Nature* **1958**, *182*, 1604.
32. Yoshizawa, T.; Wald, G. *Nature* **1963**, *197*, 1279.
33. Busch, G.E.; Applebury, M.L.; Lamola, A.A.; Rentzepis, P.M. *Proc. Natl. Acad. Sci. U.S.A.* **1972**, *69*, 2802.
34. Kandori, H.; Matuoka, S.; Shichida, Y.; Yoshizawa, T.; Ito, M.; Tsukida, K.; Balogh-Nair, V.; Nakanishi, K. *Biochemistry* **1989**, *28*, 6460.
35. Mizukami, T.; Kandori, H.; Shichida, Y.; Chen, A.-H.; Derguini, F.; Caldwell, C.G.; Bigge, C.; Nakanishi, K.; Yoshizawa, T. *Proc. Natl. Acad. Sci. U.S.A.* **1993**, *90*, 4072.
36. Shichida, Y.; Matuoka, S.; Yoshizawa, T. *Photobiochem. Photobiophys.* **1984**, *7*, 221.
37. De Vico, L.; Garavelli, M.; Bernardi, F.; Olivucci, M. *J. Am. Chem. Soc.* **2005**, *127*, 2433.
38. Schoenlein, R.W.; Peteanu, L.A.; Mathies, R.A.; Shank, C.V. *Science* **1991**, *254*, 412.
39. Wang, Q.; Schoenlein, R.W.; Peteanu, L.A.; Mathies, R.A.; Shank, C.V. *Science* **1994**, *266*, 422.
40. Kandori, H.; Furutani, Y.; Nishimura, S.; Shichida, Y.; Chosrowjan, H.; Shibata, Y.; Mataga, N. *Chem. Phys. Lett.* **2001**, *334*, 271.
41. Kukura, P.; McCamant, D.W.; Yoon, S.; Wandschneider, D.B.; Mathies, R.A. *Science* **2005**, *310*, 1006.

42. Eyring, G.; Mathies, R.A. *Proc. Natl. Acad. Sci. U.S.A.* **1979**, *76*, 33.
43. Nakamichi, H.; Okada, T. *Angew. Chem. Int. Ed. Engl.* **2006**, *45*, 1.
44. Frutos, L.M.; Andruniow, T.; Santoro, F.; Ferre, N.; Olivucci, M. *Proc. Natl. Acad. Sci. U.S.A.* **2007**, *104*, 7764.
45. Cooper, A. *Nature* **1979**, *282*, 531.
46. Eyring, G.; Curry, B.; Broek, A.; Lugtenburg, J.; Mathies, R.A. *Biochemistry* **1982**, *21*, 384.
47. Palings, I.; van den Berg, E.M.M.; Lugtenburg, J.; Mathies, R.A. *Biochemistry* **1989**, *28*, 1498.
48. Nakamichi, H.; Okada, T. *Proc. Natl. Acad. Sci. U.S.A.* **2006**, *103*, 12729.
49. Borhan, B.; Souto, M.L.; Imai, H.; Shichida, Y.; Nakanishi, K. *Science* **2000**, *288*, 2209.
50. Kochendoerfer, G.G.; Verdegem, P.J.E.; van der Hoef, I.; Lugtenburg, J.; Mathies, R.A. *Biochemistry* **1996**, *35*, 16230.
51. Sugihara, M.; Hufen, J.; Buss, V. *Biochemistry* **2006**, *45*, 801.
52. Ganter, U.M.; Schmid, E.D.; Perez-Sala, D.; Rando, R.R.; Siebert, F. *Biochemistry* **1989**, *28*, 5954.
53. Kandori, H.; Katsuta, Y.; Ito, M.; Sasabe, H. *J. Am. Chem. Soc.* **1995**, *117*, 2669.
54. Park, J.H.; Scheerer, P.; Hofmann, K.P.; Choe, H.-W.; Ernst, O.P. *Nature* **2008**, *454*, 183.
55. Luecke, H.; Schobert, B.; Richter, H.-T.; Cartailler, J.P.; Lanyi, J.K.J. *Mol. Biol.* **1999**, *291*, 899.
56. Kandori, H. *Biochim. Biophys. Acta* **2000**, *1460*, 177.
57. Kandori, H.; Shichida, Y. *J. Am. Chem. Soc.* **2000**, *122*, 11745.
58. Shibata, M.; Tanimoto, T.; Kandori, H. *J. Am. Chem. Soc.* **2003**, *125*, 13312.
59. Shibata, M.; Kandori, H. *Biochemistry* **2005**, *44*, 7406.
60. Dobler, J.; Zinth, W.; Kaiser, W. *Chem. Phys. Lett.* **1988**, *144*, 215.
61. Delaney, J.K.; Brack, T.L.; Atkinson, G.H.; Ottolenghi, M.; Steinberg, G.; Sheves, M. *Proc. Natl. Acad. Sci. U.S.A.* **1995**, *92*, 2101.
62. Zhong, Q.; Ruhman, S.; Ottolenghi, M.; Sheves, M.; Friedman, N. *J. Am. Chem. Soc.* **1996**, *118*, 12828.
63. Atkinson, G.H.; Ujj, L.; Zhou, Y. *J. Phys. Chem. A* **2000**, *104*, 4130.
64. Herbst, J.; Heyne, K.; Diller, R. *Science* **2002**, *297*, 822.
65. Kobayashi, T.; Saito, T.; Ohtani, H. *Nature* **2001**, *414*, 531.
66. Doig, S.J.; Reid, P.J.; Mathies, R.A. *J. Phys. Chem.* **1991**, *95*, 6372.
67. Gai, F.; Hasson, K.C.; McDonald, J.C.; Anfinrud, P.A. *Science* **1998**, *279*, 1886.
68. Du, M.; Fleming, G.R. *Biophys. Chem.* **1993**, *48*, 101.
69. Ruhman, S.; Hou, B.; Friedman, N.; Ottolenghi, M.; Sheves, M. *J. Am. Chem. Soc.* **2002**, *124*, 8854.
70. Kandori, H.; Yoshihara, K.; Tomioka, H.; Sasabe, H. *J. Phys. Chem.* **1992**, *96*, 6066.
71. Arlt, T.; Schmidt, S.; Zinth, W.; Haupts, U.; Oesterhelt, D. *Chem. Phys. Lett.* **1995**, *241*, 559.
72. Song, L.; El-Sayed, M.A.; Lanyi, J.K. *Science* **1993**, *261*, 891.

73. Schenkl, S.; van Mourik, F.; van der Zwan, G.; Haacke, S.; Chergui, M. *Science* **2005**, *309*, 917.
74. Kandori, H.; Tomioka, H.; Sasabe, H. *J. Phys. Chem. A* **2002**, *106*, 2091.
75. Lutz, I.; Seig, A.; Wegener, A.A.; Engelhard, M.; Boche, I.; Otsuka, M.; Oesterhelt, D.; Wachtveitl, J.; Zinth, W. *Proc. Natl. Acad. Sci. U.S.A.* **2001**, *98*, 962.
76. Birge, R.R.; Cooper, T.M. *Biophys. J.* **1983**, *42*, 61.
77. Edman, K.; Nollert, P.; Royant, A.; Belrhali, H.; Pebey-Peyroula, E.; Hajdu, J.; Neutze, R.; Landau, E.M. *Nature* **1999**, *401*, 822.
78. Schobert, B.; Cupp-Vickery, J.; Hornak, V.; Smith, S.O.; Lanyi, J.K. *J. Mol. Biol.* **2002**, *321*, 715.
79. Matsui, Y.; Sakai, K.; Murakami, M.; Shiro, Y.; Adachi, S.; Okumura, H.; Kouyama, T. *J. Mol. Biol.* **2002**, *324*, 469.
80. Kandori, H.; Belenky, M.; Herzfeld, J. *Biochemistry* **2002**, *41*, 6026.
81. Hayashi, S.; Tajkhorshid, E.; Schulten, K. *Biophys. J.* **2002**, *83*, 1281.
82. Hayashi, S.; Tajkhorshid, E.; Kandori, H.; Schulten, K. *J. Am. Chem. Soc.* **2004**, *126*, 10516.
83. Furutani, Y.; Shibata, M.; Kandori, H. *Photochem. Photobiol. Sci.* **2005**, *4*, 661.
84. Muneda, N.; Shibata, M.; Demura, M.; Kandori, H. *J. Am. Chem. Soc.* **2006**, *128*, 6294.
85. Kandori, H.; Furutani, Y.; Shimono, K.; Shichida, Y.; Kamo, N. *Biochemistry* **2001**, *40*, 15693.
86. Furutani, Y.; Ikeda, D.; Shibata, M.; Kandori, H. *Chem. Phys.* **2006**, *324*, 705.
87. Sumii, M.; Furutani, Y.; Waschuk, S.A.; Brown, L.S.; Kandori, H. *Biochemistry* **2005**, *44*, 15159.
88. Shibata, M.; Muneda, N.; Ihara, K.; Sasaki, T.; Demura, M.; Kandori, H. *Chem. Phys. Lett.* **2004**, *392*, 330.
89. Furutani, Y.; Kawanabe, A.; Jung, K.H.; Kandori, H. *Biochemistry* **2005**, *44*, 12287.
90. Furutani, Y.; Bezerra, A.G. Jr; Waschuk, S.; Sumii, M.; Brown, L.S.; Kandori, H. *Biochemistry* **2004**, *43*, 9636.
91. Furutani, Y.; Shichida, Y.; Kandori, H. *Biochemistry* **2003**, *42*, 9619.
92. Shibata, M.; Ihara, K.; Kandori, H. *Biochemistry* **2006**, *45*, 10633.
93. Shibata, M.; Yoshitsugu, M.; Mizuide, N.; Ihara, K.; Kandori, H. *Biochemistry* **2007**, *46*, 7525.
94. Mizuide, N.; Shibata, M.; Friedman, N.; Sheves, M.; Belenky, M.; Herzfeld, J.; Kandori, H. *Biochemistry* **2006**, *45*, 10674.
95. Furutani, Y.; Sudo, Y.; Kamo, N.; Kandori, H. *Biochemistry* **2003**, *42*, 4837.
96. Sharma, A.K.; Spudich, J.L.; Doolittle, W.F. *Trends Microbiol.* **2006**, *14*, 463.
97. Sudo, Y.; Furutani, Y.; Wada, A.; Ito, M.; Kamo, N.; Kandori, H. *J. Am. Chem. Soc.* **2005**, *127*, 16036.
98. Ito, M.; Sudo, Y.; Furutani, Y.; Okitsu, T.; Wada, A.; Homma, M.; Spudich, J.L.; Kandori, H. *Biochemistry* **2008**, *47*, 6208.
99. Sudo, Y.; Spudich, J.L. *Proc. Natl. Acad. Sci. U.S.A.* **2006**, *103*, 16129.

INDEX

Supramolecular Photochemistry: Controlling Photochemical Processes, First Edition. Edited by V. Ramamurthy and Yoshihisa Inoue.